ADVANCED LEVEL
STATISTICS
An Integrated Course

Second Edition

Also from the same publisher:

Crawshaw & Chambers A CONCISE COURSE IN A-LEVEL STATISTICS
Montagnon FOUNDATIONS OF STATISTICS
Greer STATISTICS FOR ENGINEERS
Thomas NOTES AND PROBLEMS IN STATISTICS
Yeats, White & Skipworth FINANCIAL TABLES
White, Yeats & Skipworth TABLES FOR STATISTICIANS
Bostock & Chandler MATHEMATICS — THE CORE COURSE FOR
    A-LEVEL
Bostock & Chandler FURTHER PURE MATHEMATICS
Bostock & Chandler MATHEMATICS — MECHANICS AND
    PROBABILITY
Bostock & Chandler FURTHER MECHANICS AND PROBABILITY
Bostock & Chandler APPLIED MATHEMATICS 1
Bostock & Chandler APPLIED MATHEMATICS 2
Bostock & Chandler PURE MATHEMATICS 1
Bostock & Chandler PURE MATHEMATICS 2
Hartley & Wynn-Evans A STRUCTURED INTRODUCTION TO
    NUMERICAL MATHEMATICS
Stroud FOURIER SERIES AND HARMONIC ANALYSIS
Stroud LAPLACE TRANSFORMS
Weltner, Grosjean, Schuster & Weber MATHEMATICS FOR ENGINEERS
    AND SCIENTISTS
Weltner, Grosjean, Schuster & Weber MATHEMATICS FOR ENGINEERS
    AND SCIENTISTS   STUDY GUIDE

ADVANCED LEVEL STATISTICS
An Integrated Course

Second Edition

A. FRANCIS
BSc, MSc, AFIMA

Senior Lecturer
Mathematics, Statistics and Data Processing
Clarendon College, Nottingham

Stanley Thornes (Publishers) Ltd

First published 1979 by Stanley Thornes (Publishers) Ltd.,
Old Station Drive, Leckhampton, CHELTENHAM GL53 0DN

Reprinted 1981
Reprinted 1983 twice
Reprinted 1986 with amendments
Second Edition 1988

British Library in Publication Data

Francis, A.
    Advanced level statistics.— 2nd ed.
    1. Statistical mathematics. For schools
    I. Title
    519.5

ISBN 0-85950-813-7

Typeset by Tech-Set, Gateshead, Tyne & Wear.
Printed and bound in Great Britain at The Bath Press, Avon.

PREFACE
TO SECOND EDITION

The following topics, mainly on the practical side, have been included:

Chapter 3: Logarithmic transformations of functions to linear form for least squares regression.

Chapter 4: An alternative practical approach to solving Bayes Theorem (conditional probability) problems.

Chapter 6: The introduction and use of Binomial and Poisson distribution function tables and special Poisson and Normal graph paper use.

Chapter 7: Acceptance sampling techniques.

Chapter 8: (Quality) Control Charts for the mean, range and number (or proportion) defective.

Chapter 10: Elements of Experimental Design.

Appendices 1 to 10: Statistical tables of:

Binomial, Poisson and Normal distribution functions;
percentage points for χ^2, T and F distributions;
Fisher Z-transformation;
range and control chart values;
negative exponential function;
random digits.

PREFACE
TO FIRST EDITION

My main reason for beginning this work was an inability to recommend a suitable text to students at or about the G.C.E. 'A' level to reflect the way I think Statistics should be presented.

Bearing in mind that the subject can still be considered as relatively new, the various examining boards seemed to have settled down to a broadly standard syllabus content.

My aim is to present an up to date text which covers the broad needs of all the examining boards and which hopefully will appeal to students and teachers alike. One of my reservations towards some of the present material on the market is the sometimes unclear presentation of significant sections of the basic content and its logical consequences in the applications. Important principles and results can easily be swamped by lengthy, verbose explanations and examples and consequently, basic important formulae and expressions can sometimes be difficult to find. Also, since a study of Statistics is necessary to not only student mathematicians but also (and increasingly) to students of the natural and social sciences and business studies, it is tempting to steer clear of the mathematics altogether or to use results completely 'out of the blue' without background or reference, each of which I feel is a mistake.

I have concentrated in this text on: a logical and ordered presentation of the complete material; well documented introductions to notations and terminology (in the form of *Definitions*) and fundamental results (in the form of *Statements*); fully worked examples to demonstrate each main point or different aspects of it; including necessary elementary mathematical results, together with simple examples, as an integral part of the text but in self-contained sections.

The need for well documented definitions of notation and terminology and statements of results, I feel, cannot be over emphasised. This is a recognised procedure in most science, mathematics and advanced statistics texts, but somehow is curiously lacking in many statistics books at an introductory level. My experience, both as a student and teacher of statistics, has shown a need for the learner to grasp initially the salient features of any section of the subject (formulae for mean and standard deviation, expressions for density functions of important distributions, forms for the various tests in particular situations, ... etc.), *true understanding* of their derivation and significance coming only after extensive practical exercise and/or further study. My own encounters with various levels of statistics have convinced me of this latter point. I am not advocating rote formula learning, rather the opportunity to identify significant features of the material quickly and easily.

The technique of highlighting all important features in the text is meant to be useful to: students wishing to prepare their own 'notes of notes' for an over perspective of the material; intensive revision students wishing to recall or relearn set techniques; teachers planning a course of study via the text; the casual user wishing to pinpoint a particular result quickly.

The Definitions in the text are meant as a formalised description of notation or terminology, very necessary in ordered, quantitative subjects. Statistics has many special symbols, phrases and expressions to describe standard situations and it is difficult for the learner to take all these in at a first meeting. So the need for a quick and easy back reference is important here.

Statements are made for all results that need to be recalled frequently. The results themselves are introduced in a natural way as an integral part of the text development and are then highlighted by 'boxing' (as are Definitions). Both are numbered consecutively by chapter. For example, Statement 2.11 (the 11th statement of Chapter 2) and Definition 7.2 (the 2nd definition of Chapter 7). Most statements are proved in the text, except where the proofs are either beyond the level of the course or will add nothing to the general development. No proofs are offered to statements in the few mathematical sections, since these particular results are meant simply to be used, not necessarily fully understood. A list of all Definitions and Statements follows this preface and is meant as a general reference or for final revision purposes prior to an examination.

As regards length and breadth of material covered, the book has been written mainly with an eye to the needs of 'A' Level G.C.E. students but it will undoubtedly prove to be a very valuable text for other courses. Many undergraduate courses at universities and polytechnics now cover basic Statistics in a fairly comprehensive manner as do the syllabuses of the professional bodies. Also medical and science research students without a statistical background, wishing to analyse and interpret statistical data, should find the text useful. It covers the present requirements of most boards for G.C.E. 'A' Level (particularly the AEB), together with a few natural extensions.

Both for general clarity and presentation of the material, the exercises have been placed in blocks at the end of each chapter, each block being referenced by the section to which it belongs. Also included is a set of miscellaneous exercises for each chapter, containing questions of an all round nature together with some that are more theoretical, intended in many cases as extensions of the text.

Each chapter is divided into sections, some of these further into subsections. Thus 6.4.2 refers to Chapter 6, section 4, subsection 2.

I would like to thank John White for his many invaluable comments and criticisms of the original typescript, Tony Stirk for the initial idea and ex-students Graham Lansdell and Ted Kaczmarek for some useful corrections. I am grateful to the following examination boards for permission to reproduce questions from past examination papers: The Institute of Statisticians (IOS); Associated Examining Board (AEB); Oxford and Cambridge Schools Examination Board (O & C); Southern Universities' Joint Board (SUJB); University of Cambridge Local Examinations Syndicate (Cambridge); Joint Matriculation Board (JMB); University of London (London).

ACKNOWLEDGEMENTS

The author and publisher would like to thank the Biometrika Trustees for permission to reproduce Tables 2, 3, 4, 5 and 6 of the Appendix.

CONTENTS

LIST OF DEFINITIONS

LIST OF STATEMENTS

Chapter 1   **Presentation of Data**

1.1 INTRODUCTION

The concept of 'number' is not new; it has been used by civilised man for thousands of years. However, the relative frequency with which the average person comes into contact with numbers today has increased dramatically. We cannot read a paper, watch television, or generally converse, without being confronted by figures — page numbers, opinion polls, death rates, unemployment, balance of payments, accident casualties, and so on. This development has occurred only over the past hundred years.

Our lives are largely controlled by the use of figures; we live by the time of day, tax codes, amounts of money, weights and measures, and so on. The need to be basically numerate is obvious. In these days of modern communications and information retrieval together with the development of powerful computers, where vast quantities of data are easily stored and are readily accessible, we now need, together with the skills of basic arithmetic, some techniques for sorting, analysing and presenting large quantities of information.

The subject of Statistics is aimed largely in this direction.

In the first instance, we attempt to obtain an overall picture from a set of unordered figures. Suppose we wished to obtain details concerning the wages of employees in a certain factory. It would be fairly easy to obtain, say, a list of 800 numbers representing the previous week's wages, but clearly it would take a 'superman' to glance at these and be able to satisfy himself as to the nature of this particular wage structure.

The first 'statistical' step would be to condense the figures into a form that the mind could comprehend fairly quickly and easily. This can be achieved quite simply by splitting the values of the data into ranges and quoting the total number of employees in each range. If the smallest wage was £20, and the largest £120, we might group together all the employees with wages in the ranges £20–£40, £40–£60, £60–£80, and so on. Of course this type of simplification sacrifices much information. For instance, although we might know that 32 employees draw wages within the range £80–£100, we do not know the distribution within this range. We therefore have to balance a certain amount of lost information against simplicity in order to obtain a clearer general picture of the situation.

We can simplify further by giving a diagrammatic representation of a set of data, especially if the need for the actual values of items is not as important as the relationship between them. For example, if we were investigating the type of accommodation the population of a particular district inhabited, we might obtain the following (hypothetical) result.

Type of accommodation			
House			
Detached	Semi-Detached or terrace	Flat or maisonette	Other
41	130	186	14

Instead of quoting the actual figures for each type of accommodation, we could display the data in the following form.

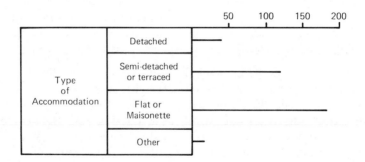

Here, we are sacrificing a certain amount of numerical accuracy to obtain a clearer relationship between the different types of accommodation. At a glance, we see that 'most people live in flats or maisonettes', or that 'there are about three times the number of people living in semi-detached or terraced houses than in detached houses', and so on. Quite often this might be the only sort of information we require.

So, at the outset, what we need to do is bring some sort of order into play. That is, put the data into such a form that we might make simple, general observations and prepare it for more detailed presentation and analysis later.

Of course we might not always be interested in an exhaustive analysis of a set of figures — a reasonable end in itself might be simply, a neat and ordered tabulation of data (see *Monthly Digest of Statistics* and *Annual Abstract of Statistics* published by H.M. Stationery Office for the Government Statistical Service).

But, whatever our reasons for collecting information, we need to take the first step of investigating various standard methods of sorting and presenting data.

1.2 TYPES OF DATA

For the purposes of this text, we divide data into two distinct types – quantitative and qualitative.

Quantitative data can be described as that type of data that may be represented by a numerical quantity. For example, the heights of a number of people, the lengths of a selected group of fish, the time taken to get to work (or school) in the morning, the number of books on each shelf in a library, and so on.

On the other hand, qualitative data measures attributes such as colour, rank, sex, etc. (i.e., quality rather than quantity).

Quantitative data can be further subdivided into two categories, discrete and continuous.

1.2.1 QUANTITATIVE DATA

Discrete Data   Consider the following set of numbers obtained by asking 20 students how many times they were late for morning lectures during a particular week:

$$1 \quad 0 \quad 1 \quad 2 \quad 3 \quad 0 \quad 1 \quad 1 \quad 0 \quad 1$$
$$2 \quad 0 \quad 0 \quad 2 \quad 2 \quad 5 \quad 1 \quad 0 \quad 0 \quad 1$$

This is an array of discrete, raw data. By *raw data*, we mean that as yet the data has not been processed or ordered in any way. By discrete, we mean that only definite values of the items we are observing are possible. Here, the only possible values are $0, 1, 2, 3, 4$ or 5.

However, we are not saying that only 'whole' (*integral*) numbers are allowed. Consider the following shoe sizes of a particular 10 male adults: $6, 7, 9, 7\frac{1}{2}, 9\frac{1}{2}, 5, 6, 8, 6\frac{1}{2}$ and 7. This is also an example of discrete data, where possible values could be any one of $5, 5\frac{1}{2}, 6, 6\frac{1}{2}, 7, 7\frac{1}{2}, \ldots, 11$, $11\frac{1}{2}, 12$, say. Other examples are the numbers of teachers in a particular set of schools, the number of books in a set of students' brief cases, and the number of electric light bulbs in each room of a house. The significant characteristic of data that is discrete is simply that the possible values are *known* quantities, each one of which may be identified exactly. All the previously given examples are what are known as *finite* discrete data, since the number of values the data can take is finite and we could physically list them if necessary.

An example of infinite discrete data would be, for example, the number of stars in a selected set of galaxies. Here, although the values that the data can assume is a known set, it is an infinite one: namely $0, 1, 2, 3, \ldots$

Continuous data   The following set of numbers represent the time taken, in minutes, to two decimal places (2D), for each of 20 people to solve a certain puzzle.

2.40 3.61 1.06 4.96 2.43 2.79 6.01 5.76 0.44 3.82
1.90 2.00 4.43 4.51 1.73 0.56 2.10 2.36 3.62 6.20

This is an array of continuous raw data. Consider the first value 2.40. Obviously no-one could measure the exact time that this particular person took to solve the puzzle — the best we can do is estimate the time by measuring it to a particular number of decimal places. Of course, as soon as we set a level of accuracy for the data (i.e., nearest whole number, 1D, 2D, etc.) then we can effectively treat the data as discrete, if desired, since the only recorded values for the data measured to 2D are 0.01, 0.02, 0.03, ... , 2.00, 2.01, ... , etc. A recorded value such as 2.016 would be impossible in this case. Quantities such as height, weight and temperature are all examples of continuous data.

The significant point about continuous data is that it cannot assume exact values, and the best we can do is to allocate a range within which a continuous item lies. The value of 2.40, measured to 2D, has thus been allocated the range '2.395 and up to (but excluding) 2.405'.

To summarise:

DEFINITION 1.1

We define:

(a) *discrete data* as data that can assume only fixed, known values, that may be identified;

(b) *continuous data* as data that (practically) have no single, exact values and can be identified only within a fixed, valid range.

EXAMPLE 1.1

State whether the following are examples of discrete or continuous data.

(a) Number of goals scored in each of 50 football matches.

(b) Time taken to score the first goal in each of 50 football matches.

(c) Number of spectators at each of 50 football matches.

(d) The heights of each of 500 blades of grass in millimetres.

SOLUTION

(a) The possible values for the data are 0, 1, 2, ... , i.e., *fixed* values in the range 0 to 500, say. Therefore the data is discrete.

(b) Since a game lasts for 90 minutes, 'waiting time to the first goal' could take *any* value between 0 and 90 minutes. Therefore the data is continuous.

(c) Suppose the maximum attendance at the largest ground is 50,000 then, theoretically, the attendance at any one match could be one of the *fixed* values $0, 1, 2, 3, \ldots, 49\,999, 50\,000$. Therefore the data is discrete.

(d)   Height (measured in any standard units) is a continuous quantity.

1.2.2 QUALITATIVE DATA

This is data that cannot be measured numerically.

The following is an array of types of 'other ranks' at an army unit.

Warrant Officers	Sergeants	Corporals	Privates
1	2	5	25

This is an example of 'strict' qualitative data since every member of this unit (and every other unit) can be classified into one of the given types without any ambiguity. Other examples are 'species of tree', 'sex', and 'certificate of cinema film'.

But consider the following data obtained by a particular person observing the colour of 6 people's eyes

light green, blue, lightish green, black, black/brown and blue.

These particular colours are what we might call 'subjective'. That is, they depend upon the definitions of colour of the observer. For example, what is the difference between 'light' and 'lightish' green? What if the observer is colour blind? Unless we have a standard colour chart upon which everyone concerned is in agreement, care must be taken when interpreting any results given by the data.

In this text, however, we concern ourselves mainly with the statistics of quantitative data.

1.3 TABULATION OF DATA

1.3.1 FREQUENCY DISTRIBUTIONS

Given a set of raw data we usually arrange it into what is known as a *frequency distribution* where we collect 'like' quantities and display them by writing down how many of each type there are, i.e., writing down their frequency.

For example, given the following discrete data of the number of chairs in each of 42 rooms of various houses,

$$8 \; 2 \; 4 \; 5 \; 2 \; 2 \; 3 \; 8 \; 5 \; 4 \; 6 \; 6 \; 3 \; 5 \; 7 \; 2 \; 7 \; 1 \; 3 \; 5 \; 6$$
$$6 \; 7 \; 2 \; 7 \; 8 \; 2 \; 3 \; 1 \; 4 \; 6 \; 5 \; 4 \; 2 \; 7 \; 7 \; 3 \; 6 \; 5 \; 6 \; 7 \; 4$$

we could count the number of 1's, 2's, 3's, etc., to obtain the following table.

No. of Chairs	1	2	3	4	5	6	7	8	Total
Frequency	2	7	5	5	6	7	7	3	42

Thus there are 2 rooms with the class '1 chair', 7 rooms with '2 chairs', and so on.

Some more examples of discrete frequency distributions follow.

The distribution of a certain 1000 families according to number of children.

No. of children in family	0	1	2	3	4	5	6	7	Total
No. of families	26	305	403	199	54	7	5	1	1000

The number of minor flying incidents reported for each of 1897 aircraft over a particular period of time.

No. of minor incidents	0	1	2	3	4	5	Total
No. of aircraft	1432	411	36	14	3	1	1897

The results of an experiment consisting of five stages, each of which can result in success or failure, conducted by 100 people.

No. of successes	0	1	2	3	4	5	Total
No. of people	2	3	14	38	33	10	100

When considering continuous data, the latter form of tabulation is inadequate since, as noted earlier, continuous quantities cannot be measured exactly. So we need to define the concept of a frequency distribution in a different way.

Consider the following continuous data, measured to 2D, 2.02, 2.11, 2.93, 2.31, 2.41, 2.13, 1.98. If we take the *range* 2.10 to 2.15 we see that it has a frequency of 2, since both 2.11 and 2.13 fall within it. Increasing the range so that it reads 2.10 to 2.40 increases the frequency by 1 since the value 2.31 is now added to the other 2.

In the continuous case then, we change the phrase 'items that are alike' to 'ranges (or classes) that are alike'.

Consider the following where measurements were taken in centimetres (cm) (to 2D) of the lengths of 18 blades of grass in an agricultural experiment to give:

$$
\begin{array}{cccccc}
8.24 & 10.86 & 8.31 & 9.82 & 8.63 & 9.41 \\
10.21 & 10.30 & 10.42 & 8.91 & 9.22 & 9.72 \\
10.03 & 10.17 & 9.41 & 9.24 & 8.86 & 10.07
\end{array}
$$

Taking ranges 8.00 to 8.50, 8.50 to 9.00, . . . , etc., we can form the following frequency distribution.

Length (in cm)	No. of blades of grass (frequency)
8.00 to 8.50	2
8.50 to 9.00	3
9.00 to 9.50	4
9.50 to 10.00	2
10.00 to 10.50	6
10.50 to 11.00	1 [Total 18]

Note that we consider the class '8.00 to 8.50' as meaning '8.00 and up to (*but excluding*) 8.50', the same being true for the other classes. (This notation will be used from here on unless otherwise specified.) This convention is used to ensure that each range of values is (a) unique and (b) contains all possible values between the two outer limits. That is, the (theoretical) exact value 9.50 would be included in the range '9.50 to 10.00' but not in the range '9.00 to 9.50'.

We can also group discrete data into classes, as shown above, if the total range of the data is large. For example, we could have tabulated the data of families by children as:

No. of children in family	0 to 1	2 to 3	4 to 5	6 to 7
No. of families	331	602	61	6

although we would not usually group data into only four classes.

Given a set of data to arrange into a frequency distribution there is no fixed rule as to how many classes we should split it up into but generally we would allocate not less than 6 or more than 12.

EXAMPLE 1.2

Represent the following data in the form of a frequency distribution.

5.1 7.7 2.4 0.3 4.5 9.3 3.0 5.8 0.3 5.8 6.4 9.3 1.5 6.3
0.9 4.4 2.1 6.3 9.1 0.9 4.7 5.5 6.2 8.7 5.0 5.4 3.9 6.5
5.3 6.5 6.2 2.1 5.5 3.6 5.6 8.6 8.5 6.5 5.0 5.5

SOLUTION

The numbers range from 0.3 to 9.3, so we can form ten natural classes as follows.

Class	Tally	Frequency
0.0 up to 1.0	1111	4
1.0 up to 2.0	1	1
2.0 up to 3.0	111	3
3.0 up to 4.0	111	3
4.0 up to 5.0	111	3
5.0 up to 6.0	ⅢⅠ ⅢⅠ 1	11
6.0 up to 7.0	ⅢⅠ 111	8
7.0 up to 8.0	1	1
8.0 up to 9.0	111	3
9.0 up to 10.0	111	3 [Total 40]

Normally a frequency distribution table would not include a 'tally' column but we add one here to illustrate a useful method of counting individual items using the 'five-bar gate' method.

1.3.2 INTERVALS, LIMITS AND BOUNDARIES

Consider a set of measurements, taken to 2D, and tabulated as a frequency distribution as follows.

Class	Frequency
0.00 to 1.99	1
2.00 to 3.99	11
4.00 to 5.99	33
6.00 to 7.99	16
8.00 to 9.99	3

The classes or groups 0.00 to 1.99, 2.00 to 3.99, ... , etc., are often called *intervals* and the values 1.99, 3.99, 5.99, 7.99 and 9.99 are the *upper class limits* of the five intervals.

Similarly, 0.00, 2.00, 4.00, 6.00 and 8.00 are the *lower class limits*.

Clearly, since the above data is inferred to be continuous, the classes do not contain all the theoretical values of the variable we are measuring. Consider the value 5.996 78. Although this will have been rounded to 6.00 (2D) and hence put in the interval 6.00 to 7.99, there is evidently no place for the exact value in the above distribution. Since this is not an ideal situation as regards continuous data, we introduce the concept of class boundaries.

The *lower class boundary* (l.c.b.) of a class (or interval) is defined as the lowest value that an item in the class can (theoretically) attain. Therefore, for the above distribution, the lower class boundaries of the classes are respectively 0.000, 1.995, 3.995, 5.995 and 7.995.

Similarly, the *upper class boundary* (u.c.b.) of a class is defined as the largest value that an item in the class can (theoretically) attain. In this case the upper class boundaries are respectively 1.995, 3.995, 5.995, 7.995 and 9.995.

Boundaries are defined so that the upper boundary of one class is identical to the lower boundary of the next highest class. Therefore a boundary value will always be included in two consecutive classes.

For grouped discrete distributions the setting up of boundaries is, to a certain extent, artificial. For instance, consider a distribution with classes 0 to 3, 4 to 7, 8 to 11, ... , etc., which can take only integral values. In this case the definition of a boundary falls down since the items in question cannot attain values between 3 and 4, 7 and 8, ... , etc., and in this sense these ranges can be thought of as a numerical 'no-man's land'. However, for statistical purposes, we need to set up boundaries and in this type of situation we simply apportion the area between the two classes equally. Hence the respective class boundaries are 3.5, 7.5, 11.5, ..., etc. Note that we take the l.c.b. of the class 0 to 3 as 0, since -0.5 would be meaningless in any context.

Finally we define the *length* (*size* or *width*) of a class as the difference between its two boundaries.

To summarise:

DEFINITION 1.2

For a continuous (or grouped discrete) frequency distribution divided into a number of classes, we define:

(a) *Lower and upper class limits* as the extreme values of each class quoted in the frequency table.

(b) *Lower and upper class boundaries*
    (i) *For continuous data* as the smallest and largest values respectively that an item in the class can (theoretically) attain,
    (ii) *For grouped discrete data* as the points midway between the upper and lower class limits of two consecutive classes.
Class limits and boundaries are often identical.

(c) The *length* (*size* or *width*) of a class as the numerical difference between the two class boundaries.

EXAMPLE 1.3

40 batches, each of 1000 electric light bulbs, produced by a certain process, were measured critically, those outside a certain diameter tolerance being labelled 'defective'. The results were as follows.

No. of defectives per batch	0 to 19	20 to 39	40 to 59	60 to 79	80 to 99
No. of batches	12	14	8	5	1

Calculate, for each class:

(a) the upper and lower class limits;

(b) the upper and lower class boundaries;

(c) the class widths.

SOLUTION

(a) The lower class limits are as given above, namely 0, 20, 40, 60 and 80. The upper class limits are 19, 39, 59, 79 and 99.

(b) The l.c.b.s are 0, 19.5 (halfway between 19 and 20), 39.5, 59.5 and 79.5. The u.c.b.s are, similarly, 19.5, 39.5, 59.5, 79.5 and 99.5.

(c) The common class width (except for the first class) is the difference between the boundaries, i.e., $99.5-79.5 = 79.5-59.5 = 59.5-39.5 = 39.5-19.5 = 20$. In putting the l.c.b. of the lst class as 0 we have an effective class width of $19.5-0 = 19.5$.

So far (except for lst class discrepancies), all the frequency distributions we have considered have had classes with the same width. This is not standard and although in general we would wish to have equal class widths, sometimes it is either impossible or impractical.

Consider the following frequency distribution of weekly wages in a factory.

Wage (in £)	No. of Employees
Under 30	5
30 and under 32	8
32 and under 34	17
34 and under 36	31
36 and under 38	22
38 and under 40	10
40 and over	12
	Total 105

The first and last classes are called *open* since they have no well-defined class limits. The reasons for presenting a distribution in this way should be clear. In both cases the frequencies are relatively small and splitting these classes up into intervals of length 2 until all the frequencies are accounted for would add nothing to the overall presentation.

The following distribution gives the number of fatal whooping cough cases in a particular area.

Age at death	No. of people
0 to 2 months	12
2 to 6 months	40
6 months to 2 years	8
2 to 10 years	2
Over 10 years	1
	Total 63

Notice here that every single class has a different length.

The problems associated with presenting data of this sort are obvious from the given distribution. Here, the majority of the observations (ages at death) fall at the beginning of the distribution although the final 11 are considered significant enough to be placed in three separate classes. That is, whooping cough is usually only fatal in babies but there are exceptions.

In these sort of cases, where we have unequal class intervals, care needs to be taken when presenting the data diagramatically as we discuss in Section 1.4.

1.3.3 RELATIVE FREQUENCY

There are many situations in Statistics where the actual frequencies of a distribution are not as important as the relative relations between classes. We have already noted that when employing a simple diagrammatic representation we usually forgo actual data values for relationships. However, in the case of frequency distributions, it is often necessary to express the frequencies as proportions or percentages of the total.

For example, given the data:

Class	10	11	12	13	Total
Frequency	5	7	9	3	24

each frequency expressed as a proportion of the total is calculated as $\frac{5}{24} = 0.208, \frac{7}{24} = 0.292, \frac{9}{24} = 0.375$ and $\frac{3}{24} = 0.125$ which is displayed in the form:

Class	10	11	12	13	Total
Proportion	0.208	0.292	0.375	0.125	1

To express the frequencies as percentages, we calculate thus:

$$\left(\frac{5}{24}\right) 100 = 20.8, \quad \left(\frac{7}{24}\right) 100 = 29.2,$$

$$\left(\frac{9}{24}\right) 100 = 37.5 \quad \text{and} \quad \left(\frac{3}{24}\right) 100 = 12.5$$

to give the table:

Class	10	11	12	13	Total
Percentage	20.8	29.2	37.5	12.5	100

1.4 GRAPHICAL REPRESENTATION

As mentioned earlier, it is often desirable to represent data pictorially in order to give an overall picture of the situation quickly. There are various standard methods of displaying data at our disposal.

1.4.1 BAR CHARTS AND HISTOGRAMS

Consider the following qualitative data of the types of tree predominant in selected areas of land.

Type of tree predominant	No. of areas
Oak	20
Ash	5
Chestnut	15
Larch	1
Total	31

This data could be represented pictorially as shown in Fig. 1.1 where both ways would be equally satisfactory.

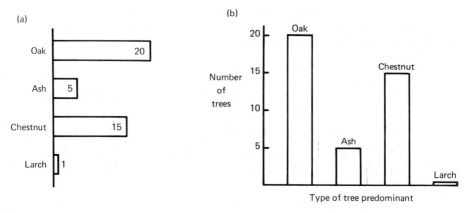

Fig. 1.1

This form of representation is commonly known as a *bar chart*, and is used for displaying qualitative data only. Notice, in particular, that the respective bars are drawn separated from one another emphasising the comparative isolation of one class from any other.

When representing quantitative data, however, the involvement of one class with another (bringing in the natural idea of common boundaries) is shown in a similar manner to pictorial representation except that the bars *are joined together*, the meeting points being the common class boundary. Representations of this sort are called *histograms*.

Consider the following discrete and continuous frequency distributions and their respective histograms (shown in Fig. 1.2).

(a)

Class	Frequency
10 to 13	2
14 to 17	10
18 to 21	4
22 to 25	3

(b)

Class	Frequency
10.00 to 10.99	20
11.00 to 11.99	30
12.00 to 12.99	56
13.00 to 13.99	14

(The above distribution is assumed to be rounded to 2D.)

Notice that, in both cases, we have plotted the end points of the classes as the common class boundaries (the figures in brackets) although we would generally show only class limits for presentation purposes.

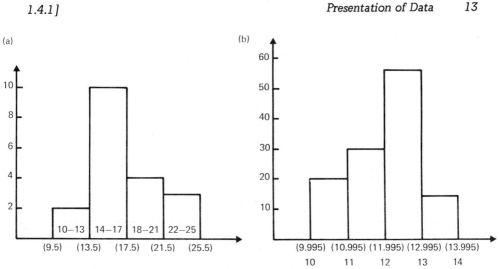

Fig. 1.2

It is quite natural, when drawing histograms, to take the height of each bar in direct proportion to the frequency of the class being represented, as we did in the two previous examples. However, when representing frequency distributions containing classes *with one or more unequal class intervals*, this method of directly relating the height of the bar to class frequency gives a decidedly false picture of the situation.

For example, consider the following data.

Class	0 to 2	2 to 4	4 to 6	6 to 8	8 to 12
Frequency	3	3	3	3	3

The histogram in Fig. 1.3(a) has been drawn using a height for each bar proportional to the common class frequency of 3, and so the bars are all the same height. But since the last class is twice the width of all the others, the histogram gives the impression that this class has twice the 'weight' of the others owing to its having double the area (12 square units as opposed to 6 square units), whereas this is not true.

In Fig. 1.3(b), we have taken account of this ambiguity by halving the height of the bar for the last class to $1\frac{1}{2}$ (as opposed to 3), thus resulting in the areas contained by all five bars being identical. This can be seen to give an intuitively correct picture of the situation.

So, if we choose a 'standard' class width (in this case a width of 2), a class having twice this width (the last class with a width of 4) will have a bar height of just $\frac{1}{2}$ of the corresponding frequency (the last class will have a bar height of $(\frac{1}{2}) 3 = 1\frac{1}{2}$).

To extend this rule, a class having a width 3 times the standard width will have a bar height of only $\frac{1}{3}$ of the corresponding frequency. In general, a class having a width n times the standard width will have a bar height of only $\frac{1}{n}$ of the corresponding frequency.

What we are doing, in effect, is ensuring that *the area of each histogram bar is proportional to the frequency of the class it is representing*. This is an important property of a histogram.

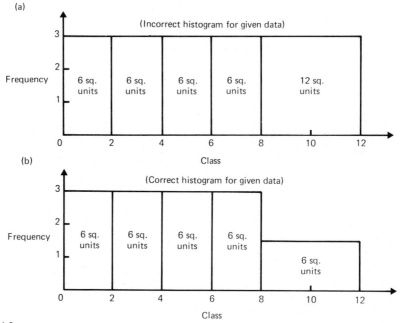

Fig. 1.3

DEFINITION 1.3

(a)   A *bar chart* is a diagrammatic representation used for qualitative data, consisting of separated, vertical bars of equal width whose *heights* are drawn proportional to the frequencies of the classes they represent.

(b)   A *histogram* is a diagrammatic representation used for quantitative data consisting of continuously joined vertical bars (meeting at class boundaries), the *areas contained by the bars* being drawn proportional to the frequencies of the classes they represent.

EXAMPLE 1.4

Illustrate the following data in the form of a histogram.

Wages (in £)	No. of employees
10 to 19.99	1
20 to 29.99	7
30 to 39.99	51
40 to 49.99	32
50 to 59.99	4
60 to 69.99	1

SOLUTION

The respective class boundaries are $9.995, 19.995, \ldots, 69.995$ with each class having the same width.

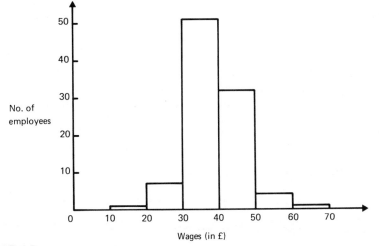

Wages (in £)

EXAMPLE 1.5

Calculate the heights of the bars of the histogram to represent the data given in the following frequency distribution.

Class	Frequency
118 to 121	2
122 to 128	6
129 to 138	14
139 to 148	31
149 to 158	63
159 to 168	28
169 to 175	11
176 to 178	3

SOLUTION

The respective class boundaries are $117.5, 121.5, 128.5, 138.5, 148.5, 158.5, 168.5, 175.5, 178.5$.

The first class has width $121.5 - 117.5 = 4$; the second class has width $128.5 - 121.5 = 7$; the next four classes have a common interval of 10 and the last two classes have widths 7 and 3 respectively.

We take the middle four class widths (10) as standard, giving the heights of these bars as $14, 31, 63$ and 28 corresponding respectively to the class frequencies. The first class has width 4, which is just $\frac{4}{10}$ of the standard. Hence, we multiply its frequency of 2 by $\frac{10}{4}$ giving a bar height of $\left(\frac{10}{4}\right)2 = 5$.

Similarly the second class frequency of 6 is multiplied by $\frac{10}{7}$ (having a class width of 7) to give the bar height $\left(\frac{10}{7}\right) 6 = 8.6$ (1D).

The heights of the bars for the last two classes are, respectively, $\left(\frac{10}{7}\right) 11$ and $\left(\frac{10}{3}\right) 3$, i.e. 15.7 and 10.

The complete set of heights are 5, 8.6, 14, 31, 63, 28, 15.7 and 10. Note that it is usual to take the standard class width as the one that occurs most frequently, since this entails the least number of calculations.

EXAMPLE 1.6

The following distribution relates to the age of students at a college.

Age (in years)	No. of students
16 to 16.9	80
17 to 17.9	50
18 to 18.9	20
19 to 19.9	12
20 to 23.9	8

Draw a histogram to represent this data.

SOLUTION

The width of the first 4 classes is 1 year while the width of the last class is 4 years (4 times the above standard width). Hence, if the first 4 bar heights are put at the respective frequencies 80, 50, 20 and 12, the height of the last bar needs to be $\frac{1}{4}$ of the frequency $8 = \frac{8}{4} = 2$.

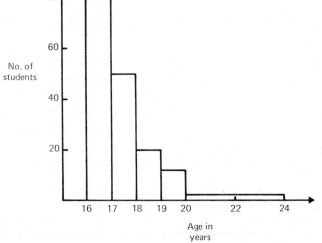

1.4.2 FREQUENCY POLYGONS AND CURVES

DEFINITION 1.4

(a)   The *mid-point* of a class is defined as that point lying mid-way
between the two class boundaries. It is calculated as $\dfrac{\text{l.c.b.} + \text{u.c.b.}}{2}$.

(b)   A *frequency polygon* for a frequency distribution having equal
class intervals is formed by plotting (as points) class frequencies
above the mid-points of the classes to which they relate and joining
these points *using straight lines*.

For example, given the distribution:

Class	10 to 15.9	16 to 21.9	22 to 27.9	28 to 33.9
Frequency	1	3	7	4
Class mid-point	12.95	18.95	24.95	30.95

we construct the frequency polygon shown in Fig. 1.4, where we have
also superimposed the corresponding histogram.

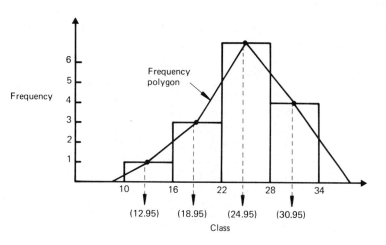

Fig. 1.4

Note that the plotted points for the frequency polygon are just the
centres of the top of the bars of the histogram.

We explained earlier that the area enclosed by a histogram bar is proportional to the frequency of the class it is representing and, so, the
total area enclosed by a histogram must be proportional to the total
frequency involved for the distribution.

It can be proved (using the geometry of the figure) that the total area enclosed beneath the polygon is equal to the area enclosed by the corre-sponding histogram.

Thus it is clear that the histogram and frequency polygon are naturally related, although we would choose one or the other (not both) to repre-sent a frequency distribution pictorially.

When dealing with distributions containing unequal class intervals we use the same technique for frequency polygons that we used for histograms. That is, we choose a standard width of class and adjust the heights of non-standard width classes accordingly.

One particular advantage of frequency polygons over histograms is the ability to compare several similar distributions on the same diagram.

If we have a continuous (or grouped discrete) frequency distribution with a large number of classes of relatively small size, the frequency polygon will be composed of correspondingly small line segments. Decreasing the size of the class, the number remaining large, will result in the polygon approaching a smooth curve. This curve is known as a *frequency curve*.

We can estimate a frequency curve for any distribution (without necessarily having a large number of classes with small widths) by joining the plotted points *with a smooth curve* rather than with straight line segments.

EXAMPLE 1.7

Draw a frequency polygon to display the following data of the weights of a sample of a certain 2000 items.

Weight	No. of items
6.21 up to 6.22	2
6.22 up to 6.23	5
6.23 up to 6.24	17
6.24 up to 6.25	31
6.25 up to 6.26	45
6.26 up to 6.27	53
6.27 up to 6.28	40
6.28 up to 6.29	6
6.29 up to 6.30	1

SOLUTION

We need first to determine the mid-points of each class using the class boundaries. These are: $\dfrac{6.21 + 6.22}{2} = 6.215$, $\dfrac{6.22 + 6.23}{2} = 6.225$ and similarly we obtain 6.235, 6.245, . . . , 6.295.

We now plot the corresponding class frequencies against the above mid-points (all class widths being equal), joining these points with straight lines to form the frequency polygon.

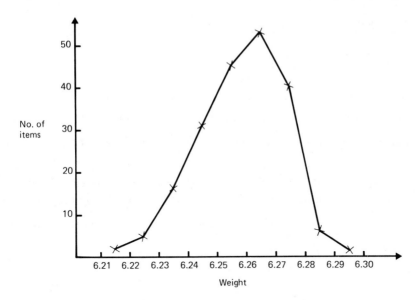

EXAMPLE 1.8

The following distribution gives the number of shares held by share-holders in a particular company.

No. of shares held	No. of people
1 to 49	75
50 to 99	145
100 to 149	132
150 to 199	64
200 to 249	38
250 to 299	25
300 to 449	14
450 to 599	11

Calculate:

(a)   the class boundaries,

(b)   the class mid-points,

(c)   the class widths,

and hence

(d)   draw an estimated frequency curve to represent this distribution.

SOLUTION

(a)   The class boundaries are 0.5, 49.5, 99.5, 149.5, 199.5, 249.5, 299.5, 449.5 and 599.5.

(b)   The class mid-points are respectively $\dfrac{0.5 + 49.5}{2} = 25.0;$

$$\frac{49.5 + 99.5}{2} = 74.5; \quad \frac{99.5 + 149.5}{2} = 124.5 \quad \text{and in a similar fashion}$$

we obtain 174.5, 224.5, 274.5, 374.5 and 524.5.

(c)   The class widths are calculated as u.c.b. -- l.c.b. for each class and are respectively 49, 50, 50, 50, 50, 50, 150 and 150.

(d)   If we take 50 as the standard width of the classes, the plotted frequency of the first class will be $\left(\frac{50}{49}\right) 75 = 76.5$. The next 5 frequencies are as given, namely 145, 132, 64, 38 and 25, the classes being of standard width. Since the last two classes have three times the width of the standard, the given frequencies are divided by three to give $\frac{14}{3} = 4.7$ and $\frac{11}{3} = 3.7$ respectively. The frequency curve is shown:

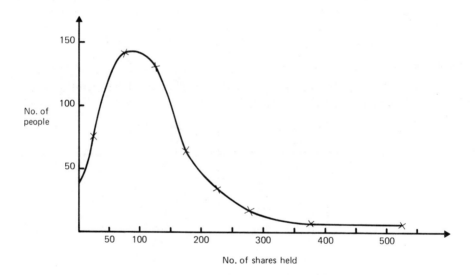

Frequency distributions can be classified according to the general shape of their frequency curves. A distribution having its largest frequencies concentrated at its centre with the frequencies of other classes sloping away evenly on either side is called *symmetric* as shown in Fig. 1.5(b). Frequency curves that are humped at one or the other end of the distribution are said to be *skewed* (to the right or left respectively, as shown in Fig. 1.5(a) or (c)).

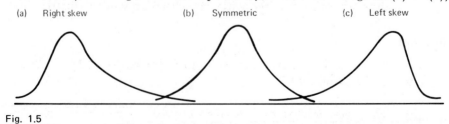

Fig. 1.5

Curves that are extremely skewed in one direction are sometimes referred to as *J-shaped*, while those that have their largest frequencies at both ends are called *U-shaped* (see Fig. 1.6).

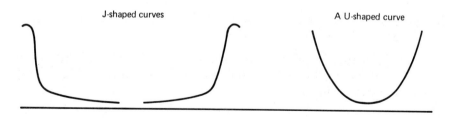

J-shaped curves   A U-shaped curve

Fig. 1.6

1.4.3 PIE DIAGRAMS

A popular method for displaying strictly qualitative data pictorially is by means of a *pie diagram*. This consists of drawing a circle and dividing it up into sectors of particular size corresponding to the frequencies of the classes of a given set of qualitative data (see Fig. 1.7).

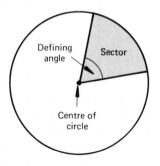

Fig. 1.7

Since a sector is completely defined by specifying an angle at the centre of a circle, the above procedure is equivalent to ensuring that the total angle at the centre of the circle $(360°)$ is split up according to the frequencies of the various classes. We do this by:

(a) finding the class frequency proportions of the total, and

(b) multiplying these proportions by $360°$.

As an example, consider the following data of the sizes (in student numbers) of the departments in a college.

		Department		
	Science	Business studies	General studies	Total
Student numbers	642	820	317	1779

The proportions in each department are: $\dfrac{642}{1779} = 0.3609$; $\dfrac{820}{1779} = 0.4609$ and $\dfrac{317}{1779} = 0.1782$.

Multiplying these three proportions by $360°$ gives $(0.3609)360° = 130°$ together with $166°$ and $64°$ respectively.

We can now construct a pie diagram consisting of a circle split into three sectors defined by angles $130°$, $166°$ and $64°$ as shown in Fig. 1.8(a), displaying the data as in Fig. 1.8(b).

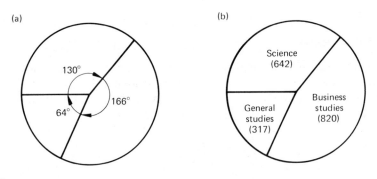

Fig. 1.8

Two sets of similar data can be conveniently compared using two pie diagrams where we ensure that the areas of the two circles are in the same proportion as the two respective total frequencies. That is, if one set of data had a total frequency that was double the other, we would construct the two pie diagrams in exactly the same way as shown previously except the area covered by the first circle would be just double that of the second circle.

In general, if r_1 and r_2 are the two radii and T_1 and T_2 are the two total frequencies in question, we would need that:

$$\frac{\pi r_1{}^2}{\pi r_2{}^2} = \frac{T_1}{T_2} \quad (\pi r^2 \text{ is the area of a circle, radius } r)$$

which simplifies to

$$r_2{}^2 = \frac{(r_1{}^2)T_2}{T_1} \quad \text{giving} \quad r_2 = r_1\sqrt{\frac{T_2}{T_1}} \qquad [1]$$

So, once we have decided on a convenient radius for the first circle (r_1), we can calculate the radius of the second circle (r_2) using expression [1], the compositions of both circles being calculated as discussed earlier.

To summarise:

DEFINITION 1.5

(a)   Given a set of qualitative data in the form of a frequency distribution:

Class	A_1 A_2 A_3 ... A_n	Total
Frequency	f_1 f_2 f_3 ... f_n	N

,

a *pie diagram* is constructed by:
 (i) drawing a circle of convenient radius;
 (ii) splitting the circle into n sectors (corresponding to each of the above classes) whose respective angles at the centre are calculated as:

$$\left(\frac{f_1}{N}\right) 360°, \left(\frac{f_2}{N}\right) 360°, \ldots, \left(\frac{f_n}{N}\right) 360°$$

(b)   To compare two qualitative distributions (with total frequencies T_1 and T_2 respectively) using two pie diagrams:
 (i) choose a convenient radius (r_1) for the first circle;
 (ii) the second circle radius (r_2) is calculated as:

$$r_2 = r_1 \sqrt{\frac{T_2}{T_1}}$$

(iii) the composition of both circles is decided as in (a).

EXAMPLE 1.9

The following data gives the distribution of seats in both houses of the Swedish parliament in 1968.

	Conservatives	Centre Party	Liberals	Social Democrats	Others
Upper House	25	21	25	79	1
Lower House	33	35	43	113	9

Display this data using proportional pie diagrams.

SOLUTION

The total for the Upper House is 151 $(T_1$, say) and for the Lower House is 233 (T_2). To find the relationship between the two radii of the circles we use:

$$r_2 = r_1 \sqrt{\frac{T_2}{T_1}} \quad \text{i.e.} \quad r_2 = r_1 \sqrt{\frac{233}{151}} = (1.24) r_1$$

Hence, whatever value we give to r_1, r_2 must be 1.24 times as large.

The sector angles for the five parties of the Upper House are calculated as: $\left(\dfrac{25}{151}\right)360° = 60°$ (to nearest degree), $\left(\dfrac{21}{151}\right)360° = 50°,$

$\left(\dfrac{25}{151}\right)360° = 60°,$ $\left(\dfrac{79}{151}\right)360° = 188°$ and $\left(\dfrac{1}{151}\right)360° = 2°$

respectively.

Similarly, the angles for the parties in the Lower House are:

$\left(\dfrac{33}{233}\right)360° = 51°,$ $\left(\dfrac{35}{233}\right)360° = 54°,$ $\left(\dfrac{43}{233}\right)360° = 66°,$

$\left(\dfrac{113}{233}\right)360° = 175°$ and $\left(\dfrac{9}{233}\right)360° = 14°.$

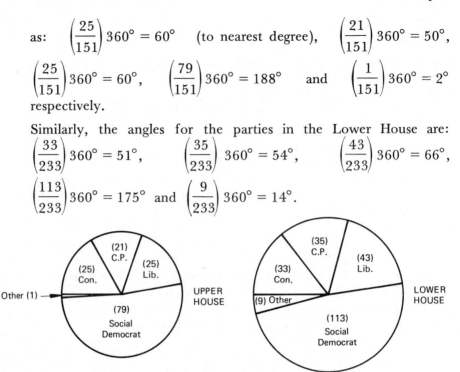

1.5 CUMULATIVE FREQUENCY

Here, we consider not individual frequencies of an item or class but the *total frequency up to* a particular item or upper class boundary.

For example, in the discrete distributions:

Class	10	11	12	13	14	15	16
Frequency	2	4	17	48	57	16	4

there are 2 items up to and including 10, 6 items up to and including 11, 23 items up to and including 12, and so on. In attempting to graph this cumulative information with what is called a *cumulative frequency curve,* we are alluding to continuous data (even though the data might be discrete). In this sense, we always consider cumulative data to be effective up to the upper class boundaries of respective classes. Hence, for the data given, we would form a *cumulative frequency distribution* as follows.

Upper class boundary	Frequency up to u.c.b. (cumulative)
9.5	0
10.5	2
11.5	6
12.5	23
13.5	71
14.5	128
15.5	144
16.5	148

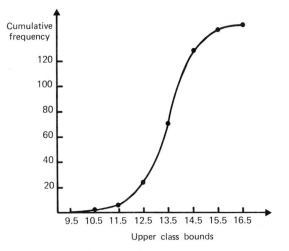

Fig. 1.9

The cumulative frequency curve is shown in Fig. 1.9. We can also form relative cumulative frequency distributions and curves (using proportions or percentages) if necessary.

EXAMPLE 1.10

Form a cumulative frequency distribution and curve to display the following information of service times at a certain post office.

Service time (in mins.)	No. of customers
2.5 to 2.9	3
3.0 to 3.4	21
3.5 to 3.9	11
4.0 to 4.4	6
4.5 to 4.9	4
5.0 to 5.4	1

SOLUTION

Identifying the upper class bounds as 2.95, 3.45, 3.95, 4.45, 4.95, and 5.45, the cumulative frequency distribution can be given as follows:

Upper class bound	Cumulative frequency
2.45	0
2.95	3
3.45	24
3.95	35
4.45	41
4.95	45
5.45	46

together with the cumulative frequency curve:

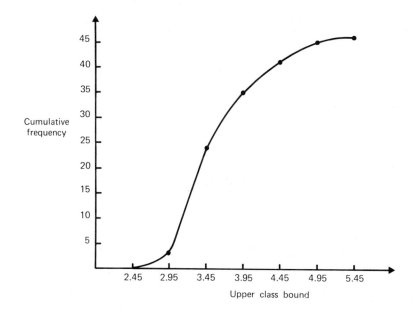

1.6 EXERCISES

SECTION 1.2

1. State, with reasons, whether the following types of data are continuous or discrete: (a) the temperature (in degrees centigrade) of a liquid; (b) the lifetime, in hours of writing, of a ballpoint pen; (c) the fare paid for a bus trip.

SECTION 1.3

2. Using the following data of students' marks in an examination:

    75 85 59 73 88 65 60 86 72 63 78 61 75 78 74 67
    60 68 74 66 82 78 94 73 71 83 79 96 67 62 75 81
    75 71 65 79 73 80 78 72 74 53 76 65 73 67 72 63

    (a) draw up a 'tally chart' using the classes 50–59, 60–69, ..., 90–99 (b) using the groupings in (a), draw a frequency distribution table for the given data.

3. The heights of a certain 60 men were measured (in inches to 2D) with the following results:

    66.2 67.4 68.8 68.2 69.4 67.3 64.1 60.8 61.2 67.3 67.0 67.2
    66.3 67.3 66.7 67.4 63.3 67.6 66.2 68.3 65.3 68.3 67.7 69.6
    66.3 67.6 72.7 70.2 68.5 64.2 66.4 72.8 61.3 67.9 67.6 67.9
    67.1 66.3 66.8 69.5 68.7 67.5 68.8 68.6 66.6 72.8 67.5 63.8
    73.5 65.3 67.4 60.2 67.6 71.4 70.1 63.6 64.3 65.3 67.9 67.3

    Draw up a frequency distribution table using the classes 60.0–62.9, 63.0–65.9, 66.0–68.9, ... , etc.

4.   The following are readings of the barometric pressure at different
        British towns on a particular day:

        29.877 29.996 30.535 29.954 29.674 30.082 29.558
        30.246 29.258 30.348 29.970 29.777 30.336 29.944
        30.060 30.480 30.040 30.640 30.033 30.166 30.442
        29.892 30.108 30.330 29.570 29.879 29.859 30.194
        29.384 30.318 30.186 29.649 30.297 29.466 29.694

        Illustrate the data by drawing up a *continuous* frequency distribution
        table consisting of five classes.

5.   Using the *Cambridge Elementary Statistical Tables* — Random Sampling
        Numbers (Table 8), count the frequencies of occurrence of the digits
        0 to 9 using the first two rows (a total of 80 observations). Draw up
        a frequency distribution table of the results.

6.   Identify the class limits, class boundaries and class widths for the
        following distribution classes:

(a) Age in (years)	Frequency	(b) Weight (in lbs)	Frequency
26 up to 31	4	100.0–109.5	102
31 up to 36	20	110.0–119.5	38
36 up to 41	38	120.0–129.5	27
41 up to 46	62	130.0–139.5	14
46 up to 56	51	140.0–149.5	2
56 up to 66	33		
66 up to 76	17		
76 up to 86	4		
86 up to 106	1		

7.   Construct a frequency distribution table for the following set of
        continuous data (measured to 3D) by splitting it into 5 *equal* classes
        beginning 0.375–0.381, 0.382–0.388, ... , etc. and calculate for
        each class: (a) upper and lower class limits;   (b) class boundaries;
        (c) the common class width:

        0.390 0.380 0.399 0.380 0.384 0.386 0.377 0.397
        0.379 0.387 0.404 0.376 0.390 0.404 0.396 0.384
        0.393 0.390 0.382 0.388 0.401 0.391 0.383 0.381

8.   Give the class widths and boundaries of each of the following distri-
        bution classes:

(a)	(b)	(c)
2– 4	109.5 up to 109.9	0.445–0.544
5– 7	110.0 up to 110.4	0.545–0.644
8–10	110.5 up to 110.9	0.645–0.744
11–13	111.0 up to 111.4	0.745–0.844
14–16	111.5 up to 111.9	0.845–0.944

9.   Using a calculating machine, construct a relative (percentage) frequency
        distribution table for each of the following distributions:

Age group	Male	Female
0– 4	1954	1864
5– 9	1701	1644
10–14	1659	1613
15–19	1743	1723
20–24	1828	1871
25–29	1813	1827
30–34	1916	1950
35–39	1945	1998
40–44	1844	1913
45–49	1574	1757
50–54	1353	1620
55–59	1237	1481
60–64	1070	1333
65–69	892	1128
70–74	651	840
75 and over	616	960

(Age distribution of the population of the United Kingdom in 1946 in thousands.)

Take the percentages correct to 2D.

Comment on the results, comparing the two distributions.

10.   The distance to 'take off' from a standing start of 400 aircraft was found to be distributed as follows:

Distance (yards)	No. of aircraft
700 up to 720	19
720 up to 740	20
740 up to 760	46
760 up to 780	93
780 up to 800	74
800 up to 820	58
820 up to 840	42
840 up to 860	33
860 up to 880	15

Draw up a (proportional) relative frequency distribution table (to 2D) for this data.

11.   Draw up a relative (proportion) frequency distribution table for the following data, using 3D accuracy:

Marks (out of 10) in a general knowledge test.

Mark	1–2	3–4	5–6	7–8	9–10
Frequency	9	46	79	70	23

12.   Convert the following frequency distribution table to: (a) a relative (proportion); and   (b) a relative (percentage) distribution table.

Number of racehorses trained by a selected set of stables.

No. of racehorses trained	0 to 4	5 to 9	10 to 14	15 to 19	20 to 24
No. of stables (frequency)	1	4	8	6	1

SECTION 1.4

13.   The areas of the recognized 'continents of the world' are given in the following data:

Continent	Area (millions of sq. miles)
Africa	11.7
Europe	1.9
Oceania	3.3
U.S.S.R.	7.9
N. America	9.4
S. America	6.9
Asia	10.4

Graph the data using a bar chart. Why would we not use a histogram to display this data?

14.   The following data shows the age distribution of a particular population:

Age (in years)	Frequency	Age (in years)	Frequency
15 up to 20	13	40 up to 45	78
20 up to 25	88	45 up to 50	92
25 up to 30	190	50 up to 55	74
30 up to 35	226	55 up to 60	23
35 up to 40	194	60 up to 65	22

Display this data by means of a histogram.

15.   The following table shows the marital status of males and females in a certain area at a particular time:

Marital status	Male (%)	Female (%)
Single	20.2	16.4
Married	68.1	67.0
Widowed	5.1	12.6
Divorced	6.6	4.0
Total	100.0	100.0

By considering the *totals* of each of the male and female, draw a bar chart, splitting each bar into 4 parts corresponding to the figures for 'single', 'married', 'widowed' and 'divorced' respectively. This is called a *component bar chart*. (Often the distinction is made between a *percentage* and *actual* component bar chart — the former having the comparing bars the same height, each representing 100%: the latter having the heights of the comparing bars proportional to the total frequencies.)

16.   In a shooting competition, the competitors have a total of 200 targets to aim for, scoring 0 or 1 depending on whether or not they hit the target. The results for 100 competitors were:

Score	No. of competitors	Score	No. of competitors
60 up to 79	4	120 up to 129	6
80 up to 89	5	130 up to 139	13
90 up to 99	8	140 up to 149	15
100 up to 109	6	150 up to 159	18
110 up to 119	7	160 up to 189	18

Display these results by means of a histogram.

17.   Draw a histogram for the following frequency distribution of a certain fatal disease in children:

Age at death	No. of children
0 to 2 months	12
2 months—6 months	40
6 months—2 years	8
2 to 10 years	2
10 years and over	1

State clearly the height of each bar, using the middle class width as standard.

18.   The weights of 50 industrial components of a certain type were measured and found to be distributed in the following way:

Component Weights (gm)	20—29	30—39	40—49	50—59	60—69	70—79
Frequency	4	7	20	9	5	3

Illustrate this data using a frequency polygon, showing clearly the class mid-points. (Assume the data is measured to the nearest gram.)

19.   The following frequency distribution table gives the income distribution of male employees with a certain company:

Yearly income (in £'s)	Number of men
500— 999	72
1000— 1999	117
2000— 2999	221
3000— 3999	92
4000— 4999	38
5000— 5999	11
6000— 9999	6
10000—20000	2

Construct a frequency polygon showing clearly the value (frequency) of the plotted points.

20.   Use a frequency polygon to display the following frequency distribution:

Weekly sales of umbreilas by a large department store in a year.

No. of umbrellas sold	No. of weeks	No. of umbrellas sold	No. of weeks
20—39	1	70— 79	12
40—49	5	80— 89	3
50—59	9	90—129	5
60—69	17		

21.   Use a pie diagram to represent the following data:

Working days lost (per 1000 inhabitants) due to labour disputes in 1965.

U.S.A. 119.5; France 20.0; U.K. 53.2; Argentina 26.4; Belgium 7.8; Denmark 48.4 (Total 275.3). (Data source: Enskilda Bank, Stockholm).

22.   The expenses for a year of a small business were as follows:

Expense	Amount
Staff wages	£15 000
Fuel	£ 1 400
Stocks	£23 000
Transport	£ 600
Held over as capital	£ 2 000
Total	£42 000

Represent this data by means of: (a) a bar chart; and (b) a pie chart.

23.   Telephones and T.V. licences (per 1000 inhabitants) in 1965.

	Telephones	T.V. licences
U.S.A.	480	362
Sweden	458	271
Canada	376	271
U.K.	195	248

Represent this data in the form of proportional pie diagrams.
(Data source: Enskilda Bank, Stockholm).

24.   Represent the following data in a convenient form:

Cloud cover over a particular town for a given period of time.

Cloud cover	Percentage of time
Very heavy	5
Heavy	18
Moderate	42
Light	14
None	21

SECTION 1.5

25.   Identify the boundaries and draw up cumulative frequency distribution tables for the following sets of data:

(a)
Class	f
10	2
11	8
12	26
13	41
14	11

(b)
Class	f
12–14	1
15–17	3
18–20	7
21–23	12
24–26	18
27–29	6

(c)
Class	f
0– 9.9	21
10–19.9	38
20–39.9	11
40–59.9	4

(d)
Class	f
− 0.5 to − 0.41	13
− 0.4 to − 0.31	26
− 0.3 to − 0.21	11
− 0.2 to − 0.11	10
− 0.1 to − 0.01	12
0 to 0.09	8
0.1 to 0.19	3
0.2 to 0.29	1

26.   Draw a cumulative frequency curve to illustrate the following data for U.K. merchant vessels of 500 gross tons and over in 1972:

Age (in years)	Under 5	5 up to 10	10 up to 15
Number	563	324	456

Age (in years)	15 up to 20	20 up to 25	25 and over
Number	271	125	59

(Data source: AAS)

MISCELLANEOUS

27.   (Practical.) Take out a fiction book from a library, select a page at random and, beginning from the top of the page, count how many letters in each of the first 300 consecutive words. (Use a tally chart.) Construct a frequency distribution table and draw up: (a) a histogram; and   (b) a frequency polygon, on a separate diagram.

28.   (Practical.) Repeat Exercise 27 using, say, a non-fiction text book (ignoring figures, dates, etc.), and plotting the frequency polygon on the same chart as the previous one. Comment on the differences.

29.   The figures given relate to pupils in public sector schools in January 1962:

Age (in yrs)	No. of boys	No. of girls	
16	63	55	
17	38	31	(Figures given to nearest thousand.)
18 and over	15	8	

Compile: (a) actual;   (b) percentage component bar charts to compare the two sets of figures. (See Exercise 15.) Discuss what further information might be needed in order to make the comparison more useful.

(Data source: AAS)

30.   The following data refers to personal incomes (before tax) for the period 1970/1971 as computed for tax purposes in Great Britain.

Income (£)	Number (in millions)	Income (£)	Number (in millions)
420– 500	0.9	1500– 2000	4.3
500– 600	1.2	2000– 3000	3.3
600– 700	1.3	3000– 5000	0.8
700– 800	1.2	5000–10000	0.3
800–1000	2.3	10000–20000	0.05
1000–1500	5.7		

(Data source: AAS)

Illustrate this data in an appropriate form.

31.   The following table of raw data is the result of taking a sample of 40 values   from a continuous population, the data being correctly measured to 2D:

18.34 17.63 19.02 19.13 18.71 19.63 18.73 18.38
18.22 18.84 19.02 17.91 18.54 18.84 19.04 18.91
18.73 18.51 17.71 18.81 19.21 18.81 19.20 18.78
18.62 18.81 18.74 17.84 18.24 17.98 18.38 18.09
18.43 18.02 17.91 18.53 19.14 18.68 18.81 18.47

Form a frequency distribution table with   *6 equal classes*   beginning '17.50 up to 17.90'.   Display the data using: (a) a histogram; (b) a *cumulative* frequency polygon.

32.   The figures below give the number of marriages in the United Kingdom in 1972 by age and sex:

Age (in yrs)	Male	Female
Under 21	77	176
21 to 24	185	154
25 to 29	104	64
30 to 34	34	24
35 to 44	33	26
45 and over	48	37
Totals	481	481

(The figures are given to the nearest thousand.)

(a) Display the figures for male and female separately using *histograms.*

(b) Compare the figures for male and female directly by using a *component bar chart* (one bar for each).

(c) Comment on the results.

(Data source: AAS)

33. The volume of a container (in cm^3) was estimated by 24 people as:

19.476 19.509 19.520 19.491 19.510 19.485 19.505 19.485
19.476 19.483 19.497 19.464 19.507 19.488 19.490 19.466
19.501 19.483 19.515 19.492 19.517 19.473 19.478 19.462

Using a common class width of $0.01\ cm^3$, draw a histogram to display this data.

34. Represent the following data in the most appropriate form.

Earnings of full time 'manual workers' aged 21 and over in Great Britain.

(The figures are given in millions as at April 1972.)

Gross weekly earnings	Under £15	Under £20	Under £30
Number of workers	0.1	0.5	3.1

Gross weekly earnings	Under £40	Under £50	Under £100
Number of workers	5.6	6.6	7.0

(Data source: AAS)

35. (a) Draw two pie diagrams, one for 1970 and one for 1971, to show how total expenditure on all foods per head per week in each year was shared by dairy products and other foods.

(b) Draw a bar chart to show how total dairy products expenditure was devoted to each class of product in 1970. (see Table A.)

Table A. Importance of dairy products in household food expenditure in 1970 and 1971.

	Pence per head per week	
	1970	1971
Liquid milk	19.50	23.79
Other milk	1.83	2.12
Cream	1.01	1.15
Cheese	4.67	5.71
Butter	6.57	8.23
Total dairy products	33.58	41.00
Total other foods	177.09	189.52
Total all foods	210.67	230.52

(Data source: U.K. Dairy Facts and Figures, 1972; Federation of U.K. Milk-Marketing Boards).

(IOS)

36.   A car owner records his total motoring costs over a year and from these calculates his average weekly expenditure as shown:

Items	Cost per week
Petrol and oil	£4.00
Tax and insurance	£1.50
Repairs and garaging	£2.00
Depreciation and interest charges	£2.50

Display this information on an accurate pie chart, marking the angles and naming each sector clearly.

During the following year the owner finds that petrol and oil increase by 62.5% but that depreciation and interest charges decrease by 20%, while the two remaining items remain unchanged. If a pie chart were constructed to show his motoring expenditure for this year, calculate the angle of each of the four sectors. (You need not make an accurate drawing of this chart.)

If the two pie charts were used together to compare the total expenditure of the motorist in these successive years, explain carefully why the radius of the second chart would have to be $\sqrt{(6/5)}$ times the first.

(London)

Chapter 2   **Statistical Measures**

2.1 INTRODUCTION AND NOTATION

2.1.1 INTRODUCTION

We look, first, at two useful types of measure which characterise any set of data or frequency distribution.

The first type, a measure of 'centralisation', attempts to locate a typical value about which the distribution clusters. This type of measure is called an *average* or *measure of location*.

The second type is a measure of how scattered or spread out a distribution is and is called a *measure of dispersion*.

Both of these measures are numerical quantities compatible with the data and are measured in the same units as the data itself.

Consider the two pairs of curves of frequency distributions shown in Figs. 2.1 and 2.2.

In Fig. 2.1, curve A spreads from 10 to 18 and curve B spreads from 15 to 22 (i.e., roughly the same spread), while curve A values tend to cluster round the value 14.5 and curve B round 18 (the two distributions cluster round completely different points). In Fig. 2.2, curve C spreads from 14 to 20 and curve D from 10 to 24 (different spreads), while both curves tend to be clustered round the same point, 17.

So, Fig. 2.1 illustrates two distributions with different measures of location but roughly the same spread and Fig. 2.2 shows two distributions with the same measure of location but different spreads.

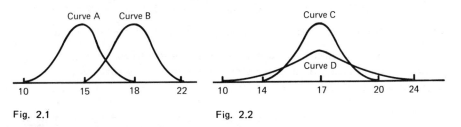

Fig. 2.1   Fig. 2.2

In this chapter we will be looking at five important averages which are the Arithmetic Mean, Median, Mode, Geometric Mean and Harmonic Mean. The first three are widely used in Statistics for all types of data, the latter two only in specific circumstances. The measures of dispersion we look at are the Range, Mean Deviation, Semi-Interquartile Range, Standard Deviation and Variance. The latter two are used most widely.

The two main types of measure mentioned here, however, are only particular examples of a more general 'description' of frequency distributions that we deal with in the final section of the chapter.

In the next subsection, we look at some important notation, a type of shorthand, without which many statistical techniques would be difficult to describe.

2.1.2 VARIABLE AND SUMMATION NOTATION

In statistics we are always considering quantities that vary. For example heights, weights, ages, times, and so on. In any particular context it is usual to denote a variable quantity by a letter (often x). So, for instance, if we were considering the lengths of some components, we could represent this quantity 'length' by the variable x. In particular, if we were not interested in the actual values of the lengths themselves (or did not know them) but wanted only to differentiate between the lengths of, say, six of them, we could label them $x_1, x_2, x_3, x_4, x_5, x_6$. Often it is found convenient to enclose them within brackets thus $(x_1, x_2, x_3, x_4, x_5, x_6)$; this is called a *set*. (We deal with the topic 'Set Theory' in a little detail in Chapter 4.) If we knew the actual lengths of the six components we could write the set as, say, $(2.85, 1.03, 1.92, 1.84, 2.06, 1.52)$.

Suppose we had a variable x that could take, say, ten values. If we were interested in their sum, without perhaps knowing the values of individual items, the best we could write would be $x_1 + x_2 + x_3 + \ldots + x_9 + x_{10}$. The notation to take care of sums such as this is now given.

DEFINITION 2.1

A convenient shorthand for the sum of n items $x_1 + x_2 + \ldots + x_n$

is $\sum_{i=1}^{n} x_i$ and translated as '*add up* all the x_i's beginning at x_1 (i.e.,

$i = 1$) and stopping at x_n $(i = n)$'. Unless stated otherwise it is

always assumed that i increases in steps of 1. $\sum$ (the Greek capital

S, pronounced 'sigma') is known as the *summation operator* (or *sigma operator*) and suffix i is often called a *dummy variable*.

For example, $\sum_{i=1}^{4} x_i$ represents the sum $x_1 + x_2 + x_3 + x_4$. It will be seen that this operator is useful in many contexts and will be used freely throughout the rest of the book. Some examples of its use now follow.

$$\sum_{i=1}^{3} y_i = y_1 + y_2 + y_3; \quad \sum_{i=1}^{5} (x_i^2) = x_1^2 + x_2^2 + x_3^2 + x_4^2 + x_5^2$$

(note here that no matter what letters or symbols the operator 'bracket' contains, only the variable described directly under the Σ sign is increased by 1 each time – both the 'x' and the '2' in this case remain constant);

$$\sum_{r=1}^{4} x_r^3 = x_1^3 + x_2^3 + x_3^3 + x_4^3$$

(here we are using the 'dummy' variable r instead of i);

$$\sum_{i=1}^{6} i = 1 + 2 + 3 + 4 + 5 + 6$$

(this is a useful notation for the sum of 'natural' numbers – positive integers – where, here, we are denoting the sum of the first 6 natural numbers). We would define the sum of the first n natural numbers by

$$\sum_{i=1}^{n} i = 1 + 2 + 3 + 4 + \ldots + n - 1 + n;$$

$$\sum_{i=1}^{4} i^2 = 1^2 + 2^2 + 3^2 + 4^2$$

(the sum of the squares of the first 4 natural numbers);

$$\sum_{i=3}^{7} x_i = x_3 + x_4 + x_5 + x_6 + x_7$$

(note that i does not have to begin at 1 but *unless otherwise specified always increases by* 1).

Using this notation we also have:

$$\sum_{i=1}^{4} a = a + a + a + a = 4a$$

(4 times, since i goes from 1 to 4). Note that i does not appear in the operand but we still need to add up as many times as the i dictates;

$$\sum_{i=1}^{5} (4 + x_i) = (4 + x_1) + (4 + x_2) + (4 + x_3) + (4 + x_4) + (4 + x_5)$$

$$\text{and } \sum_{i=1}^{4} (2x_i) = (2x_1) + (2x_2) + (2x_3) + (2x_4)$$

The principle then is to add up whatever is 'contained' by the operator bracket as many times as necessary.

The suffix i is often dropped *if it causes no ambiguity*. For instance, $\sum x$ or $\sum\limits_{i} x_i$ would mean 'add up all the x values (under consideration)'. Some important properties of the $\sum$ operator are now given.

STATEMENT 2.1

(a) $\sum\limits_{i=1}^{n} a = na.$ (b) $\sum\limits_{i} ax_i = a \sum\limits_{i} x_i.$

(c) $\sum\limits_{i} (x_i + y_i) = \sum\limits_{i} x_i + \sum\limits_{i} y_i.$

The proofs of these identities are fairly trivial as the following illustrations show.

$$\sum_{i=1}^{4} 2 = 2 + 2 + 2 + 2 = 4(2); \qquad \sum_{i=1}^{5} b = b + b + b + b + b = 5(b)$$

$$\sum_{i=1}^{4} 3x_i = 3x_1 + 3x_2 + 3x_3 + 3x_4 = 3(x_1 + x_2 + x_3 + x_4) = 3 \sum_{i=1}^{4} x_i$$

$$\sum_{i=1}^{3} (x_i + y_i) = (x_1 + y_1) + (x_2 + y_2) + (x_3 + y_3)$$

$$= (x_1 + x_2 + x_3) + (y_1 + y_2 + y_3) = \sum_{i=1}^{3} x_i + \sum_{i=1}^{3} y_i$$

EXAMPLE 2.1

Show that $\sum\limits_{i=1}^{n} (a + bx_i) = na + b \sum\limits_{i=1}^{n} x_i.$

SOLUTION

$$\sum_{i=1}^{n} (a + bx_i) = \sum_{i=1}^{n} a + \sum_{i=1}^{n} bx_i \quad \text{(by Statement 2.1(c))}$$

$$= na + b \sum_{i=1}^{n} x_i \quad \text{(by Statement 2.1(a) and (b))}$$

EXAMPLE 2.2

Find the value of

(a) $\sum\limits_{r=1}^{4} (3r^2 - 2).$

(b) $\sum\limits_{i=1}^{3} bx_i$ if $x_1 = 3, \quad x_2 = -2, \quad x_3 = 5 \quad$ and $\quad b = 4.$

SOLUTION

(a) $\sum_{r=1}^{4}(3r^2-2)=\sum_{r=1}^{4}3r^2-\sum_{r=1}^{4}2=3\sum_{r=1}^{4}r^2-4(2)$

$= 3(1^2+2^2+3^2+4^2)-8=3(1+4+9+16)-8$

$= 3(30)-8=82$

(b) $\sum_{i=1}^{3}bx_i=b\sum_{i=1}^{3}x_i=b(x_1+x_2+x_3)=4(3-2+5)=24$

EXAMPLE 2.3

Express the following sums in Σ notation as compactly as possible.
(a) $ax_1+ax_2+ax_3+ax_4+ax_5+ax_6$.
(b) $x_1y_1+x_2y_2+x_3y_3+x_4y_4$.
(c) $f_1x_1+f_2x_2+f_3x_3$.

SOLUTION

(a) $ax_1+ax_2+ax_3+ax_4+ax_5+ax_6=a(x_1+x_2+\ldots+x_6)$

$$=a\sum_{i=1}^{6}x_i$$

(b) $x_1y_1+x_2y_2+x_3y_3+x_4y_4=\sum_{i=1}^{4}x_iy_i$

(c) $f_1x_1+f_2x_2+f_3x_3=\sum_{i=1}^{3}f_ix_i$

2.2 THE ARITHMETIC MEAN

This is the first and most important of the averages we shall be examining and we begin with the basic definition, first for a set of numbers and secondly, for a frequency distribution.

2.2.1 FOR A SET

DEFINITION 2.2

The *arithmetic mean* (or just *mean*) of a set of numbers $(x_1, x_2, \ldots, x_n)$ is denoted by $\bar{x}$ (and read as 'x bar') and defined as:

$$\bar{x}=\frac{1}{n}(x_1+x_2+\ldots+x_n)=\frac{1}{n}\sum_{i=1}^{n}x_i$$

(i.e. $\bar{x}$ is the *sum* of the items divided by the *number* of items)

EXAMPLE 2.4

Find the arithmetic mean of the set $(-3, -1, 0, 2, 3, 4)$.

SOLUTION

The mean, $\bar{x} = (-3 - 1 + 0 + 2 + 3 + 4)/6 = 0.83$ (2D).

EXAMPLE 2.5

Find the mean of the numbers $1, 2, 256, 18, -72$ and 42.5.

SOLUTION

$$\bar{x} = (1 + 2 + 256 + 18 - 72 + 42.5)/6 = 41.25.$$

2.2.2 FOR A FREQUENCY DISTRIBUTION

STATEMENT 2.2

For a discrete frequency distribution taking values $(x_1, x_2, \ldots, x_n)$ with corresponding frequencies $(f_1, f_2, \ldots, f_n)$, the mean $\bar{x}$ is given by

$$\bar{x} = \sum_{i=1}^{n} f_i x_i \bigg/ \sum_{i=1}^{n} f_i$$

Proof Now, x_1 occurs exactly f_1 times, x_2 occurs f_2 times, $\ldots$, x_n occurs f_n times. So the total sum of all the items is

$$f_1 x_1 + f_2 x_2 + \ldots + f_n x_n = \sum_{i=1}^{n} f_i x_i$$

and the total number of items is clearly

$$f_1 + f_2 + \ldots + f_n = \sum_{i=1}^{n} f_i$$

But $\bar{x}$ is defined as the sum of all items divided by the number of items. Hence

$$\bar{x} = \sum_{i=1}^{n} f_i x_i \bigg/ \sum_{i=1}^{n} f_i$$

The arithmetic mean as calculated here is sometimes called a *weighted mean* since we can consider the frequencies $(f_1, f_2, \ldots, f_n)$ as a set of *weights* attached to the values $(x_1, x_2, \ldots, x_n)$. For example, if a student obtains 81 marks in an English exam and 73 in a History exam, where English is considered to be twice as important as (i.e., twice the weight of) History, we would calculate a *weighted average* mark as:

$$\frac{(2 \times 81) + (1 \times 73)}{3} = \frac{162 + 73}{3} = 78.33 \text{ (2D)}$$

that is, we treat the weights as frequencies and divide the total (weighted) mark by 3 (the total of the weights) to find the average.

EXAMPLE 2.7

Find the mean of the values (−3, −2, −1, 0, 1, 2, 3) which have corresponding frequencies (6, 5, 4, 3, 2, 1, 1).

SOLUTION

The method is to compile a frequency table as follows.

Value (x)	Frequency (f)	fx
−3	6	−18
−2	5	−10
−1	4	−4
0	3	0
1	2	2
2	1	2
3	1	3
	22	−25

$$\sum f = 22 \qquad \sum fx = -25$$

$$\bar{x} = \frac{\sum fx}{\left[\sum f\right]} = \frac{-25}{22} = -1.14 \text{ (2D)}$$

Note that we drop suffix i from the Σ operators, since their meaning is clear here.

EXAMPLE 2.8

Find the mean of the following set of 30 numbers by grouping them into a frequency distribution (4, 3, 6, 7, 5, 5, 3, 4, 9, 6, 5, 5, 6, 8, 3, 6, 6, 3, 5, 4, 7, 6, 4, 1, 9, 7, 8, 6, 4, 6).

SOLUTION

x	1	2	3	4	5	6	7	8	9
Tally	1		1111	~~1111~~	~~1111~~	~~1111~~ 111	111	11	11
(f)	1	0	4	5	5	8	3	2	2
(fx)	1	0	12	20	25	48	21	16	18

Totalling the last two rows gives $\Sigma f = 30$ and $\Sigma fx = 161$.

$$\therefore \qquad \bar{x} = \frac{161}{30} = 5.4 \ (1D)$$

EXAMPLE 2.9

A group of 10 numbers has a mean of 24 and a second group of 15 has a mean of 32. Find the mean of the combined group of 25.

SOLUTION

There are two ways of looking at the problem.

(a)   Since $\bar{x} = \dfrac{\Sigma x}{n}$, we have $\Sigma x = n(\bar{x})$, i.e., the number of items in a set, multiplied by their mean gives their total. Therefore the first group has a total of $10 \times 24 = 240$ and the second group has a total of $15 \times 32 = 480$. Hence the combined group must have a total of $240 + 480 = 720$ comprising 25 items.

$$\therefore \qquad \text{mean} = \frac{720}{25} = 28.8$$

(b)   We can treat this as a weighting problem. That is, 24 has a weight of 10; 32 has a weight of 15.

x	Weight (f)	fx
24	10	240
32	15	480
	25	720

$$\therefore \qquad \bar{x} = \frac{\Sigma fx}{\Sigma f} = \frac{720}{25} = 28.8 \quad \text{(as before)}$$

For a continuous frequency distribution (or a grouped discrete distribution) we clearly cannot use the previous method immediately since we do not have distinct x values but ranges of values of x. What we do here is simply take the mid-point of the class to represent the x value for the class and proceed in the usual way.

For example, we might have numbers in the range 6.50 to 6.99 with a corresponding frequency of 8. Here we take the x value at 6.745 (the mid-point) and the corresponding fx value for the class is $8(6.745) = 53.96$.

In this way we can use the method given in the last statement to find the mean.

EXAMPLE 2.10

Find the mean of the following distribution.

Class	5.00–5.49	5.50–5.99	6.00–6.49	6.50–6.99	7.00–7.49
Frequency	12	32	11	8	2

SOLUTION

The table for calculations is drawn up as follows.

Class	Mid-point (x)	Frequency (f)	fx
5.00–5.49	5.245	12	62.940
5.50–5.99	5.745	32	183.840
6.00–6.49	6.245	11	68.695
6.50–6.99	6.745	8	53.960
7.00–7.49	7.245	2	14.490
		65	383.925

We have $\Sigma f = 65$ and $\Sigma fx = 383.925$.

$$\therefore \qquad \bar{x} = \frac{383.925}{65} = 5.91 \; (2D)$$

EXAMPLE 2.11

Find the mean of the following grouped discrete distribution.

Class	0–4	5–7	8–10	11–12
Frequency	6	8	8	4

SOLUTION

Class	Mid-point (x)	Frequency (f)	fx
0–4	2	6	12
5–7	6	8	48
8–10	9	8	72
11–12	11.5	4	46
		26	178

$$\bar{x} = \frac{178}{26} = 6.85 \; (2D)$$

There are two points to note about the last example.

(a)   The fact that we have unequal class intervals makes no difference to the calculations for the mean.

(b)   The calculated mean (6.85) is not a typical member of the distribution, since it is not a discrete value as such (assuming the original data comprises whole numbers). However, when calculating statistical measures for 'discrete' distributions, we often give the answer in 'continuous' form unless otherwise specified or convenient.

EXAMPLE 2.12

500 people were asked how many coins they had in their pockets and the following results were obtained.

Number of coins	0	1	2	3	4	5	6—8	9—11	12 or more
Number of people	20	25	22	40	73	141	99	50	30

Find the mean number of coins

SOLUTION

We notice that the distribution is uneven in class interval and that the end class is open. We need to define an upper limit for this class that would be compatible with the data, and here, 20 would be acceptable (the choice should be based on common sense).

Number of coins	Mid-point (x)	f	fx
0	0	20	0
1	1	25	25
2	2	22	44
3	3	40	120
4	4	73	292
5	5	141	705
6—8	7	99	693
9—11	10	50	500
12 or more	16	30	480
(up to 20)		500	2859

$$\bar{x} = \frac{2859}{500}$$
$$= 5.7 \ (1D)$$

The arithmetic mean is the most commonly used of all statistical averages; it is easy to understand and fairly straightforward to calculate. Some particular characteristics are that it takes every item into account and it always exists no matter what set of data it is calculated for. It is readily suitable for arithmetic and algebraic manipulation.

Its main disadvantage is in the fact that it takes every item into account — even extreme values — and in certain circumstances could yield a 'locating' value at which few (or none) of the actual observations lie.

2.2.3 THE METHOD OF CODING

When dealing with large awkward values of a variable, the calculation of the mean by the methods so far employed can become tedious, simply because of unwieldy arithmetic, and for this reason *the method of coding* is introduced. The technique involves subtracting (or adding) a number from each of the original x values and, if possible and convenient, dividing (or multiplying) these new values by another number to obtain a set of X values which should be more manageable. We say that the x values have been *coded* (or *transformed*) into X values. We then find the mean of the X values ($\bar{X}$) and by using a suitable *decoding* formula, obtain $\bar{x}$.

Consider the set (3625.89, 4625.89, 5625.89, 6625.89) of x values. Subtracting 3625.89 from each value gives the set (0, 1000, 2000, 3000) and dividing this set by 1000 finally gives the X set (0, 1, 2, 3), considerably simplified as compared to the original x set. The coding we have used here can be expressed in the form $X = \dfrac{x - 3625.89}{1000}$. The mean of the X values, $\bar{X}$, is easily seen to be 1.5 and to decode (i.e., convert back to $\bar{x}$) we work in 'reverse' on the coding formula, first multiplying $\bar{X}$ by 1000 and secondly adding 3625.89.

Thus: $\bar{x} = (1.5)1000 + 3625.89 = 5125.89$

The method is exactly the same when applied to a frequency distribution. That is, the x values are coded into X values, the mean, $\bar{X}$, is found and then decoded to obtain $\bar{x}$.

In general we have:

STATEMENT 2.3

If (a) the set $(x_1, x_2, \ldots, x_n)$ is transformed to $(X_1, X_2, \ldots, X_n)$ or (b) the frequency distribution $\dfrac{x_1 x_2 \ldots x_n}{f_1 f_2 \ldots f_n}$ is transformed to $\dfrac{X_1 X_2 \ldots X_n}{f_1 f_2 \ldots f_n}$ by means of the *coding formula* $X = \dfrac{x - a}{b}$ and $\bar{X}$ is found, we can obtain $\bar{x}$ by means of the *decoding formula* $\bar{x} = a + b\bar{X}$.

Note a and b are chosen for convenience in order to make the X values as simple as possible.

Proof For case (a): We need to show that $\bar{x} = a + b\bar{X}$ where:

$$X = \frac{x - a}{b} \Rightarrow x = a + bX \qquad [1]$$

Now, $\bar{x} = \dfrac{\sum x}{n}$ (by definition)

$$= \frac{\sum (a + bX)}{n} \quad \text{(using [1])}$$

$$= \frac{\sum a}{n} + \frac{\sum bX}{n} = \frac{na}{n} + \frac{b\sum X}{n} \quad \left(\text{properties of } \sum\right)$$

$$= a + b\bar{X} \quad (\text{by definition})$$

The proof for case (b) is left as an exercise.

EXAMPLE 2.13

Find the mean of the set $(15, 18, 21, 24, 27, 30, 33, 36, 39, 42)$ using a method of coding.

SOLUTION

Subtracting a central value, 27 from each item gives the set $(-12, -9, -6. -3, 0, 3, 6, 9, 12, 15)$. Division by 3 (which is a common factor) further simplifies to give the set $(-4, -3, -2, -1, 0, 1, 2, 3, 4, 5)$.

Identifying the original set with the variable x and the final coded set with X, we have defined the coding $X = \dfrac{x - 27}{3}$, where here $a = 27$ and $b = 3$. Notice how this choice of coding dramatically simplifies the original values.

Now, $\bar{X} = \dfrac{-4 - 3 - 2 - 1 + 0 + 1 + 2 + 3 + 4 + 5}{10} = \dfrac{5}{10} = 0.5$

Hence, using Statement 2.3, we have $\bar{x} = 27 + 3(0.5) = 28.5$.

EXAMPLE 2.14

Calculate the mean of the set $(56, 72, 53, 64, 48, 70, 67, 72, 59, 63, 62)$ using a method of coding.

SOLUTION

The given set of x values can clearly be simplified by subtracting a central number 60, say, to give the set $(-4, 12, -7, 4, -12, 10, 7, 12, -1, 3, 2)$. Since these numbers are not ordered in any way (as in the last example) and have no common factor, it is pointless dividing by some quantity here (this would, in fact, only make the values 'more complicated' by introducing unwanted decimal places).

Identifying the new set with the variable X, we have effectively coded using $X = x - 60$ or *more specifically* $X = \dfrac{x - 60}{1}$ (i.e., $a = 60, b = 1$). $\bar{X}$ is easily calculated as 2.36 and hence, decoding, we have $\bar{x} = 60 + 1(2.36) = 62.36$ (2D).

EXAMPLE 2.15

Find the mean of the following frequency distribution:

x	1	4	7	10	13
f	21	30	25	16	8

using a method of coding.

(**Note**   We would not usually code for x values as small as this; we do so here as a simple example of the technique involved.)

SOLUTION

Here, we choose the assumed mean (a) as 7 since it lies centrally and the scaling factor (or b) as 3, the common difference between the x values.

The layout for calculations is given below.

x	$X = \dfrac{x-7}{3}$	f	fX
1	-2	21	-42
4	-1	30	-30
7	0	25	0
10	1	16	16
13	2	8	16
		100	-40

$\Sigma f = 100$

$\Sigma fX = -40$

Hence, $\bar{x} = a + b\bar{X} = a + b \left(\dfrac{\Sigma fX}{\Sigma f} \right)$ (by definition)

$$= 7 + 3 \left(\dfrac{-40}{100} \right) = 7 - \dfrac{120}{100}$$

i.e. $\bar{x} = 5.8$

The physical effect of this type of coding, $X = \dfrac{x-a}{b}$ (called a *linear transformation*) on a frequency curve for a distribution is described as follows.

(a)   Subtracting a quantity a from each of the x values effectively shifts the frequency curve through a distance a *without changing its shape*.

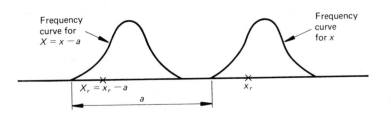

Fig. 2.3

Fig. 2.3 illustrates graphically what happens to a frequency curve in this situation. The x curve is moved to the left a distance a (the particular value x_r being shifted to the new value $X_r = x_r - a$). Note also that the mean of the x distribution, $\bar{x}$, will also be shifted by a distance a so that $\bar{X} = \bar{x} - a$.

(b)   Dividing each of the x values by b has the effect of squashing the curve and shifting it towards zero. For example, suppose the x frequency curve ranged from 5 to 10. If we divide every value by $b = 5$ we would have a new range of 1 to 2, a situation shown in Fig. 2.4 where the x curve has been squashed and shifted to the left. However, if the curve contains any negative values of x, this particular portion of the curve would have been shifted to the right (again, towards zero).

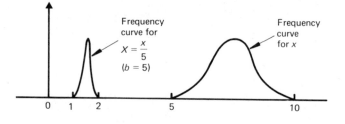

Fig. 2.4

(c)   Subtracting a and dividing by b $\left(\text{i.e., } X = \dfrac{x - a}{b} \right)$ will, of course, combine the two effects. That is, the curve will first be moved through distance a and then squashed by a factor b towards zero.

The problem of deciding which values of a and b to choose when coding, and indeed whether to code at all, is largely a matter of experience. Many of the examples used to demonstrate the method have been necessarily trivial in order to illustrate basic techniques and in practice coding might not have been considered. If the frequency distribution has a large number of classes (say 10 or more), it is often the case that the central mid-point does not have the largest frequency. This would certainly be true if we had a skewed distribution. If the central mid-point lay adjacent to the mid-point with the largest frequency, either could be chosen as a. Otherwise, a value for a somewhere between the two would be appropriate. The practical advantage of using a coding method can be clearly seen in the following example, where the given values of the involved variable are uncomfortably large.

The aim of coding is not only to reach an end result more quickly, but also to lessen the chance of errors. This is particularly important when using a modern calculator, which the majority of students have at their disposal. A calculator is only as accurate as its operator; the fewer operations performed, the more chance of an accurate result.

EXAMPLE 2.16

The following data relates to the heights (in cm) of a chosen 3810 army recruits:

Height	154–	156–	158–	160–	162–
No. of recruits	16	148	239	354	471

Height	164–	166–	168–	170–	172–
No. of recruits	542	538	425	414	256

Height	174–	176–	178–	180–	182–	184–190
No. of recruits	170	133	55	35	9	5

Use a method of coding to find the arithmetic mean of recruit heights.

SOLUTION

Note that the two central classes between them do not contain the largest frequency; this is attached to class 164–166. The mid-point of class 166–168 (i.e., 167) would probably be best for the choice of a. The scaling factor, b, is chosen as 2 (the common class width).

Mid-point (x)	$X = \dfrac{x - 167}{2}$	f	fX
155	− 6	16	− 96
157	− 5	148	− 740
159	− 4	239	− 956
161	− 3	354	− 1062
163	− 2	471	− 942
165	− 1	542	− 542
167	0	538	0
169	1	425	425
171	2	414	828
173	3	256	768
175	4	170	680
177	5	133	665
179	6	55	330
181	7	35	245
183	8	9	72
187	10	5	50
		3810	− 275

$$\therefore \qquad \bar{X} = -\frac{275}{3810}$$

Hence: $$\bar{x} = a + b\bar{X} = 167 + 2\left(-\frac{275}{3810}\right)$$

i.e.: $$\bar{x} = 166.86 \quad (2D)$$

2.3 THE MEDIAN

As already mentioned, the main disadvantage of the arithmetic mean (particularly for a small set of numbers) is that it takes extreme values into account to a degree.

For example, consider the following statement: 'The average wage of 9 staff in a certain small supermarket is £41.89'. On this evidence we would probably picture a symmetric frequency curve for the wages something like one of those shown in Fig. 2.5. However, the actual wages for the supermarket staff were £32, £29, £35, £38, £41, £35, £36, £28 and £103 (the latter being the wage of the manager). The mean of the wages is £41.89. Notice that 8 out of the 9 wages are below this value. In this sense the mean is not really representing or 'locating' the distribution in an acceptable manner. The Median is an average ideally suited to this type of situation, where there are extreme (unrepresentative) values present.

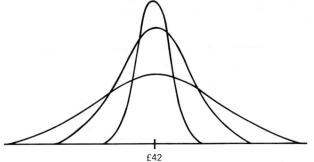

£42

Fig. 2.5

2.3.1 FOR A SET

DEFINITION 2.3

The *median* of a set of numbers $(x_1, x_2, \ldots, x_n)$ is defined as the *middle value* of the set *when arranged in size order*. If the set has an even number of items, the median is taken as the *mean of the middle two*.

For the supermarket wage example, the wages arranged in size order are £28, £29, £32, £35, £35, £36, £38, £41 and £103. Thus the median wage is £35 (the middle item) and, clearly, is more reasonable as an average than the mean of £41.89.

EXAMPLE 2.17

Find the median of the set (0.65, 0.68, 0.68, 0.66, 0.64, 0.65, 0.65, 0.67).

SOLUTION

Arranging the set in size order, we obtain 0.64, 0.65, 0.65, 0.65, 0.66, 0.67, 0.68 and 0.68 and since there are 8 items (an even number) we need to take the mean of the middle two, 0.65 and 0.66. Hence,

$$\text{median} = \frac{0.65 + 0.66}{2} = 0.655$$

EXAMPLE 2.18

Find the mean and median of the following set of numbers (2, 4, 3, 8, 17, 4, 5, 5, 8, 5, 3) and state why the median is a more 'reasonable' average than the mean in this case.

SOLUTION

There are 11 items and so the median will be the 6th in order of size (i.e., the middle item) and is easily seen to be 5. The total of the items is 64. Therefore

$$\bar{x} = \frac{64}{11} = 5.82 \ (2D)$$

The median is more appropriate here since there is an extreme value, 17, present.

An interesting interpretation of the median is that it can be said to describe 'the value of the average' rather than the 'average value'. For instance, if some data consisted of the heights of some people, we might describe the mean as the 'average height of a person' and the median as the 'height of the average person' — a thought-provoking distinction.

2.3.2 FOR A FREQUENCY DISTRIBUTION

STATEMENT 2.4

For a discrete frequency distribution taking the values $(x_1, x_2, \ldots, x_n)$ with corresponding frequencies $(f_1, f_2, \ldots, f_n)$, the median is the $\dfrac{N+1}{2}$ th value when the values are ranked, where $N = \Sigma f$.

Proof   Obvious from previous definition.

Here, we make no distinction as to whether there is an even or odd number of items. We note that sometimes the $\dfrac{N+1}{2}$ in Statement 2.4 is replaced by $\dfrac{N}{2}$ if N is fairly large, since the difference between the two will be negligible.

It is usually found convenient to include a column of cumulative frequencies (Cum f) when calculating the median for a discrete frequency distribution as the following examples show.

EXAMPLE 2.19

Find the median of the following discrete distribution.

x	0	1	2	3	4	5	6
f	5	5	10	20	30	20	10

SOLUTION

x	f	Cum f
0	5	5
1	5	10
2	10	20
3	20	40
4	30	70
5	20	90
6	10	100
	100	

We have $N = 100$ and we require the $\dfrac{N+1}{2}$ th $= \dfrac{101}{2}$ th $= 50.5$ th item, which falls at $x = 4$ using the Cum f column. Hence the median is 4.

EXAMPLE 2.20

Find the mean and median of the following distribution of the number of people living in each of 200 houses in a district.

No. of people	0	1	2	3	4	5	6
No. of houses	6	2	21	48	96	19	8

SOLUTION

No. of people (x)	No. of houses (f)	fx	Cum f
0	6	0	6
1	2	2	8
2	21	42	29
3	48	144	77
4	96	384	173
5	19	95	192
6	8	48	200
	200	715	

$$\bar{x} = \frac{715}{200} = 3.58 \ (2D)$$

The $\dfrac{N+1}{2}$ th item $(N = \Sigma f = 200) = 100.5$th item occurs at $x = 4$.

Therefore the median is 4.

When dealing with a continuous (or grouped discrete) distribution, we can only estimate a value for the median. There are two separate methods available and we will illustrate both of them with an example. Consider the following distribution.

x	f	Cum f
10–19.9	2	2
20–29.9	14	16
30–39.9	38	54
40–49.9	23	77
50–59.9	6	83
60–69.9	1	84
	84	

(Considering this as continuous data, we would take it as being measured to 1D.)

The first method is described as follows. Since $N = 84$, the median should be the $\dfrac{84+1}{2}$ th $= 42.5$th item which falls in the class 30–39.9 (called

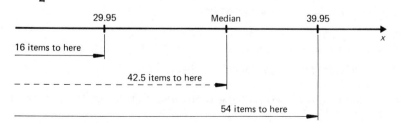

Fig. 2.6

the *median class*. From here we need to estimate just where in the class the median might be expected to lie — 30?, 35?, or 39.9? From the frequency distribution table given above we can see that there are 16 items up to 29.9 and 54 items up to 39.9. We require the (theoretical) 42.5th item.

We need to find the median m so that there are 42.5 items up to m (See Fig. 2.6). Since there are 16 items to 29.95 and 42.5 items to m, there must be $42.5 - 16 = 26.5$ items from 29.95 to m. (Similarly, there must be $54 - 42.5$ items $= 11.5$ items from m to 39.95.) Also, as there are 38 items total in the median class, m must lie a fraction $\dfrac{26.5}{38}$ of the way along from 29.95 to 39.95. Now, the actual distance into the class must be $\dfrac{26.5}{38} \times 10$ (since 10 is the class width), so that the median will lie at the point $29.95 + 10\left(\dfrac{26.5}{38}\right) = 36.92$. Note, finally, that all the numbers in the above expression are well-defined quantities. That is, 29.95 is the l.c.b. of the median class, 26.5 is $42.5 - 16$, i.e., $\dfrac{N+1}{2} - $ (Cum f up to l.c.b. of the median class), 38 is the median class frequency and 10 is the median class width.

This technique for estimating a median value is called the *method of interpolation*. We generalise with the following statement.

STATEMENT 2.5

Given a continuous (or grouped discrete) frequency distribution, having determined the median class, an estimate of the median is given by

$$m = L + \left[\frac{\dfrac{N}{2} - f_L}{f} \right] \cdot c$$

where $L = $ l.c.b. of median class
                  $N = $ total number of items (Σf)
                  $f_L = $ cumulative frequency up to point L
                  $f = $ median class frequency
                  $c = $ median class length

Proof   An extension of the previous discussion.

The second method is to operate graphically. This is similar to (but slightly more efficient than) the previous interpolation method. We need to draw a smooth cumulative frequency curve for the given data.

Upper class boundary (u.c.b.)	Cum f
9.95	0
19.95	2
29.95	16
39.95	54
49.95	77
59.95	83
69.95	84

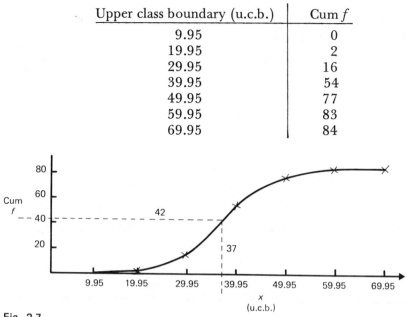

Fig. 2.7

We know that the median should be the 42nd item, so we simply see what value (using the curve) corresponds to a frequency of 42 and (see Fig. 2.7) we obtain 37 (approximately) which agrees quite well with the value 36.92 obtained by the interpolation method on the previous page.

Using this method, one should take care over the drawing of the curve; the more accurate the curve, the better the estimate.

It must be stressed that the plotted u.c.b.s are joined by a smooth curve and *not* by straight lines since this is the very reason why the graphical estimation method is more accurate than the method of interpolation which is (graphically) equivalent to joining the u.c.b. plotted points by straight lines.

We summarise the technique as follows.

STATEMENT 2.6

Given a continuous (or grouped discrete) frequency distribution, an estimate of the median is obtained by:

(a) drawing a smooth cumulative frequency curve for the data; and

(b) finding *the point on the x-axis* that corresponds to the value $\dfrac{N}{2}$ $(N = \Sigma f)$ on the f-axis via the curve.

This point is the *median estimate* and is, in general, a better estimate than that given by the previously stated interpolation method.

EXAMPLE 2.21

Estimate the median of the following distribution.

x	15–	20–	25–	30–	35–	40–	45–	50–	55–	60–	65–70
f	3	10	22	35	24	17	8	6	4	0	2

(AEB) 1964

SOLUTION

$$\Sigma f = 131 = N \qquad \therefore \frac{N}{2} = \frac{131}{2} = 65.5$$

The median class is the one that contains the 65.5th item, i.e., the 30–35 class.

x	f	Cum f
15–	3	3
20–	10	13
25–	22	35
30–	35	70
35–	24	.
40–	17	.
45–	8	.
50–	6	
55–	4	
60–	0	
65–70	2	

We have $L = 30$; $f_L = 35$; $f = 35$ and $c = 35 - 30 = 5$.

Hence,
$$m = 30 + \left(\frac{65.5 - 35}{35} \right) 5$$
$$= 34.4 \ (1D)$$

To end this discussion of the median average, we consider some of its characteristics.

As with the arithmetic mean, the median is easy to understand but, whereas in general, the mean will not represent any particular item, the median will in general if found exactly (not by interpolation or graphically).

The main advantages of the median over the mean are (a) its elimination of the effect of extreme items and (b) only the numerical values of the middle items need to be known specifically. This is a particularly practical property of the median when applied to 'life' data such as the life of electric light bulbs or electrical circuits, consumer goods, etc. Whereas we would have to wait for every single item to fail in order to find the mean length of life, we would only need to wait for the middle item (or items) to fail in order to determine the median.

A particular use of the median is its ability to average a set of strictly qualitative data (colour, rank, etc.), since these can always be ranked in some way enabling a middle item to be picked out — this cannot be done with any other of the common averages.

The main disadvantage of the median is the fact that it is generally unsuitable for arithmetic and algebraic manipulation.

2.4 OTHER AVERAGES

There are three other averages that are used occasionally in Statistics in particular circumstances. We shall deal with them fairly briefly here.

2.4.1 THE MODE

DEFINITION 2.4

The *mode* of a set of values is defined as that one which occurs with the greatest frequency.

Note that for a set that has no repeated values a mode *will not exist*.

For example, the mode of the set (2, 3, 3, 1, 3, 2, 4, 5, 8, 3, 2, 4, 4, 3) is 3 since it occurs most often.

For continuous (or grouped discrete) data a method not unlike that of interpolation is used. We demonstrate this by means of a practical example.

Consider the following distribution.

Class	f
20 and under 25	2
25 and under 30	14
30 and under 35	29
35 and under 40	43
40 and under 45	33
45 and under 50	9

We call the class '35 and under 40' the *modal class* since it has the largest frequency. Clearly, the modal value should be made to lie in this class. Furthermore, since the following class has a larger frequency than the one before the modal class, we would prefer that the mode should be larger than the modal class mid-point. In general, depending on whether the class following the modal class is larger or smaller than the class previous to the modal class, we would wish that the mode be greater or less (respectively)

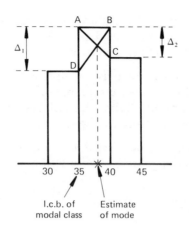

Fig. 2.8

than the modal class mid-point. For the example given, we show how to ensure this condition in Fig. 2.8.

We draw the three histogram bars for the three classes under consideration and by further superimposing the two lines shown (AC and BD), the mode is determined by the *x* value of their intersection.

The algebraic interpretation of this method of determining the mode is given in the following statement.

STATEMENT 2.7

For a continuous (or grouped discrete) frequency distribution, given the modal class, an estimate of the mode is given by:

$$L + \left[\frac{\Delta_1}{\Delta_1 + \Delta_2} \right] c$$

where: L = l.c.b. of modal class

Δ_1 = difference in frequencies between modal class and previous class

Δ_2 = difference in frequencies between modal class and following class

c = width of modal class   (see Fig. 2.8)

Note that the quantity $\frac{\Delta_1}{\Delta_1 + \Delta_2}$ is always strictly between 0 and 1 ensuring that the mode must lie in the pre-defined modal class.

No Proof

Using the given example we have $L = 35$, $\Delta_1 = 43 - 29 = 14$, $\Delta_2 = 43 - 33 = 10$ and $c = 5$. Thus:

$$\text{mode} = 35 + \left(\frac{14}{24}\right)5 = 37.9 \ (1D)$$

It should be noted that there could be more than one mode in a set of numbers or a discrete frequency distribution. For example, the set $(8, 6, 8, 5, 5, 7, 6, 8, 6, 9)$ has the two modes 6 and 8. We talk of distributions being unimodal, bimodal, etc.

EXAMPLE 2.22

Estimate the mode and median of the following distribution.

x	f	Cum f
9.3– 9.7	2	2
9.8–10.2	5	7
10.3–10.7	12	19
10.8–11.2	18	37
11.3–11.7	14	51
11.8–12.2	6	57
12.3–12.7	4	61
12.8–13.2	1	62

SOLUTION

For the mode, the modal class is 10.8 to 11.2 with an l.c.b. $L = 10.75$; $\Delta_1 = 18 - 12 = 6$; $\Delta_2 = 18 - 14 = 4$ and $c = 0.5$ (the modal class width).

$$\therefore \qquad \text{mode} = 10.75 + \left(\frac{6}{10}\right)(0.5) = 11.05$$

For the median, using the extra Cum f column in the above table and $\dfrac{N}{2} = \dfrac{62}{2} = 31$ gives the median class as 10.8 to 11.2 with with $L = 10.75$; $f_L = 19$; $f = 18$ and $c = 0.5$.

$$\therefore \qquad \text{median} = 10.75 + \left(\frac{(31 - 19)}{18}\right)(0.5) = 11.08 \ (2D)$$

The great advantage of the modal average is its simplicity both in concept and calculation. Also, as with the median, it tends to eliminate the effect of extreme items. However, it is generally unsuitable for any arithmetic and algebraic manipulation and is not used very widely.

2.4.2 THE GEOMETRIC MEAN

DEFINITION 2.5

The *geometric mean* of a set of n numbers $(x_1, x_2, \ldots, x_n)$ is defined as the *nth root of their product*,

i.e.   geometric mean $= \sqrt[n]{x_1 \cdot x_2 \cdot \ldots \cdot x_n}$

For example, the geometric mean of 2 and 4 is

$$\sqrt{(2)(4)} = \sqrt{8} = 2.83 \quad (2D)$$

The geometric mean of a set of numbers is conveniently computed using logarithms (base 10). To find the geometric mean of $(1, 2, 3, 4, 5)$ the technique would be:

geometric mean $= \sqrt[5]{(1) \cdot (2) \cdot (3) \cdot (4) \cdot (5)} = \sqrt[5]{120} \longrightarrow$

No.	Log.
120	2.0792
$\sqrt[5]{120}$	2.0792/5
	$= 0.4158$

i.e.   geometric mean $= 2.61 \quad (2D) \longleftarrow 2.605 \longleftarrow$

This average has a particular application to data that changes proportionally, as in population growth or quantities in the form of ratios. The arithmetic mean tends to give a larger value than is actually true when averaging ratios. Consider the following case.

In a particular village in year A there were 240 men to 300 women, while in year B there were 310 men to 400 women. Now the ratio of men to women in each year is $\dfrac{240}{300} = 0.8$ and $\dfrac{310}{400} = 0.775$ respectively.

The arithmetic mean of the two ratios is $\dfrac{0.8 + 0.775}{2} = 0.7875$. The geometric mean of the two ratios is $\sqrt{(0.8)(0.775)} = 0.7874$ to 4D. To see which of these two values is the more reasonable we can argue as follows. For the mean to be a 'representative average' we would like the *mean ratio* to be equal to the *ratio of the two separate means*. The mean for the men for the two years is $\dfrac{240 + 310}{2} = 275$ and for the women is $\dfrac{300 + 400}{2} = 350$. Hence the ratio of the two means is $\dfrac{275}{350} = 0.7857$ which is *not* equal to the mean ratio of 0.7875. In this sense the arithmetic mean can be considered as inconsistent here.

Now the geometric mean for the men for the two years is

$$\sqrt{(240)(310)} = 272.7636 \quad (4D)$$

and for the women is

$$\sqrt{(300)(400)} = 346.4102.$$

Therefore the ratio of the two geometric means is

$$\frac{272.7636}{346.4102} = 0.7874 \ (4D)$$

which is equal to the geometric mean of the ratios. Thus the geometric mean is the better of the two in this type of situation. Another advantage of the geometric mean over the arithmetic mean is that it pays less attention to extreme items (although the median and mode are still far better in this respect). However, the big disadvantage of this average is its difficulty in evaluation as compared with the ones so far considered.

Note that the geometric mean as defined here only exists for sets of data that are strictly positive (although it is possible, using a simple mathematical transformation to give a geometric mean for a set that has some (or all) zero or negative values).

EXAMPLE 2.23

Find the geometric mean of the set $(2, 3, 5, 7, 9)$.

SOLUTION

$$\text{geometric mean} = \sqrt[5]{(2) \cdot (3) \cdot (5) \cdot (7) \cdot (9)} \quad \text{(by definition)}$$

$$= \sqrt[5]{1890}$$

$$= 4.522 \ (3D)$$

No.	Log
1890	5)3.2765
4.522	0.6553

EXAMPLE 2.24

For three separate years the weekly amount of money spent on meat compared to the total weekly shopping bill for a particular family was

	Meat bill	Total bill
1970	£2.83	£18.41
1971	£3.24	£20.21
1972	£3.92	£24.05

Find the 'average' weekly proportion of the shopping bill spent on meat for the three years.

SOLUTION

We use the geometric mean here since we are averaging proportions.

For 1970 the proportion was $\dfrac{2.83}{18.41} = 0.154 \ (3D)$

For 1971 the proportion was $\dfrac{3.24}{20.21} = 0.160$ (3D)

For 1972 the proportion was $\dfrac{3.92}{24.05} = 0.163$ (3D)

The geometric mean of these proportions is:
$$\sqrt[3]{(0.154)(0.160)(0.163)} = 0.159 \ (3D)$$

2.4.3 THE HARMONIC MEAN

DEFINITION 2.6

The *harmonic mean* of a set of numbers $(x_1, x_2, \ldots, x_n)$ is given by
$$\text{harmonic mean} = \frac{1}{\dfrac{1}{n}\sum_i \dfrac{1}{x_i}} = \frac{n}{\sum_i \dfrac{1}{x_i}} \ ,$$

In words, the harmonic mean is the reciprocal of the arithmetic mean of the reciprocals of the numbers.

For example, the harmonic mean of $(2, 4, 6, 8)$ is
$$\frac{4}{\frac{1}{2} + \frac{1}{4} + \frac{1}{6} + \frac{1}{8}} = \frac{4}{(12 + 6 + 4 + 3)/24} = \frac{4}{25/24} = \frac{4 \times 24}{25} = 3.84$$
This measure of location is chiefly used for averaging averages.

EXAMPLE 2.24

A racing car covers five laps of a circuit in a race, each lap being covered at the following average speeds (in m.p.h.): 123.4, 132.8, 125.7, 126.9, 134.4. Find the average speed of the car for the whole race.

SOLUTION

$$\text{harmonic mean} = \frac{5}{\dfrac{1}{123.4} + \dfrac{1}{132.8} + \dfrac{1}{125.7} + \dfrac{1}{126.9} + \dfrac{1}{134.4}}$$

$$= \frac{5}{0.0081 + 0.0075 + 0.0080 + 0.0079 + 0.0074}$$

$$= \frac{5}{0.0389} = 128.5 \text{ m.p.h.}$$

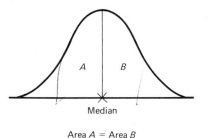

Area A = Area B

Fig. 2.9

2.4.4 QUANTILES

Since the median is defined as the 'middle' value of a set of numbers arranged in size order, when applied to a frequency distribution we can think of the median as splitting the area under a frequency curve into two equal portions, as in Fig. 2.9. Extending this idea, we can split a frequency curve into as many equal portions as we wish. Special names are given to those values that split a curve up into four, ten, and one hundred equal parts respectively, as follows.

DEFINITION 2.7

(a)   The three values that split a distribution into four equal portions are known as *quartiles*. In order of magnitude, they are usually represented by Q_1, Q_2, and Q_3 and called the first, second, and third quartiles respectively. Note that, by definition, the second quartile Q_2 is just the median, since it divides the area under the frequency curve into two equal portions (Fig. 2.10(a)).

(b)   The nine values that split a distribution into ten equal portions are known as *deciles* and are represented by $D_1, D_2, \ldots, D_8, D_9$. The fifth decile, D_5, is again the median (see Fig. 2.10(b)).

(c)   The ninety-nine values that split a distribution into one hundred equal portions are known as *percentiles* and are represented by P_1, $P_2, \ldots, P_{99}$, where, again, P_{50} is the median.

Collectively, all quantities that are defined as splitting a distribution into a number of equal portions (including the median, quartiles, deciles and percentiles) are called *quantiles*.

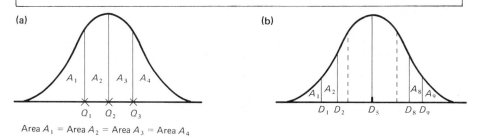

(a)

Area A_1 = Area A_2 = Area A_3 = Area A_4

(b)

Fig. 2.10

Clearly, for small sets of data, the value of calculating quantiles such as deciles or percentiles is questionable, but for frequency distributions with large numbers of items they are particularly useful. Their location in an ordered set or frequency distribution is calculated in a manner similar to that for the median. For instance, since quartiles split a set or distribution into *four* equal portions, the first and third quartiles, Q_1 and Q_3, will be the $\dfrac{1n}{4}$ th and $\dfrac{3n}{4}$ th items respectively in the size-ordered set or distribution. Similarly, D_7 will be the $\dfrac{7n}{10}$ th item and P_{23} the $\dfrac{23n}{100}$ th item.

In general, if a particular quantile splits a distribution into s equal parts, *the rth quantile* of the set will be the $\dfrac{rn}{s}$ *th item* of the size-ordered distribution.

Consider the following grouped distribution.

x	f	Cum f
70–72	5	5
73–75	18	23
76–78	42	65
79–81	27	92
82–84	8	100
	100	

The 1st and 3rd quartiles, Q_1 and Q_3 will be given by the $\dfrac{100}{4}$ th and $\dfrac{3(100)}{4}$ th items respectively (since $n = \Sigma f = 100$), i.e., Q_1 is the 25th item and Q_3 the 75th item and since Q_1 occurs in the class 76 to 78, this is the *1st quartile class* and 79 to 81 the *3rd quartile class* (since it contains Q_3).

To obtain specific values of quantiles within a particular class, we use a method of interpolation (analogous to the method to find the median). We give the general expressions for calculating Q_1 and Q_3 by interpolation in the following statement.

STATEMENT 2.8

For a grouped frequency distribution, given the 1st and 3rd quartile classes, an estimate of Q_1 and Q_3 using the method of interpolation is given by:

$$Q_1 = L_1 + \left[\frac{\frac{n}{4} - f_{L_1}}{f_1} \right] c_1$$

(continued)

and

$$Q_3 = L_3 + \left[\frac{\frac{3n}{4} - f_{L_3}}{f_3} \right] c_3$$

where: L_1 and L_3 are the l.c.b.s of the 1st and 3rd quartile classes respectively;

n is the total number of items in the distribution;

f_{L_1} and f_{L_3} are the cumulative frequencies up to the respective lower bounds L_1 and L_3;

f_1 and f_3 are the frequencies of the 1st and 3rd quartile classes respectively; and

c_1 and c_3 are the widths of the 1st and 3rd quartile classes respectively.

Proof A simple extension of the proof of the interpolation method to find the median (see Statement 2.5).

The two expressions given in this last statement are not meant for rote learning, but merely to compare with the expression given for the median and to extend these for the calculation of any other quantile.

We will calculate Q_1 and Q_3 for the previously given data.

For Q_1: $L_1 = 75.5$; $f_{L_1} = 23$; $f_1 = 42$; and $c_1 = 78.5 - 75.5 = 3$.

$$\therefore \qquad Q_1 = 75.5 + \left(\frac{25 - 23}{42} \right) 3 = 75.64$$

For Q_3: $L_3 = 78.5$; $f_{L_3} = 65$; $f_3 = 27$; and $c_3 = 3$.

$$\therefore \qquad Q_3 = 78.5 + \left(\frac{75 - 65}{27} \right) 3 = 79.61$$

To extend this method for the calculation of percentiles, we calculate P_{27} (the 27th percentile).

Now, P_{27} is the $\dfrac{27(100)}{100}$ th item = 27th item.

Therefore the P_{27} class is 76 to 78, giving $L = 75.5$ (l.c.b.); $f_L = 23$ (Cum f to L); $f = 42$ (frequency of P_{27} class) and $c = 3$ (width of P_{27} class).

$$\therefore \qquad P_{27} = 75.5 + \left(\frac{27 - 23}{42} \right) 3 = 75.79$$

We can use a cumulative frequency curve to determine quantities in the same sort of way that we did to find the median in the text of Section 2.3.2. However in this situation, it is most useful to use a cumulative *percentage* frequency curve. The following example demonstrates the technique.

EXAMPLE 2.25

The following data gives the weight of 1200 duck eggs:

Weight (mid-point) in grams	No. of eggs
57	7
60	13
63	68
66	144
69	197
72	204
75	208
78	160
81	101
84	54
87	25
90	13
93	4
96	2

Find the median, quartiles and P_{37}, the 37th percentile point using:

(a)   an interpolation formula method;

(b)   a graphical method.

SOLUTION

Weight			Cum	Cum
Mid-point	u.c.b.	f	f	$\% f$
	55.5		0	0
57	58.5	7	7	0.6
60	61.5	13	20	1.7
63	64.5	68	88	7.3
66	67.5	144	232	19.3
69	70.5	197	429	35.7
72	73.5	204	633	52.7
75	76.5	208	841	70.1
78	79.5	160	1001	83.4
81	82.5	101	1102	91.8
84	85.5	54	1156	96.3
87	88.5	25	1181	98.4
90	91.5	13	1194	99.5
93	94.5	4	1198	99.8
96	97.5	2	1200	100.0

The calculations necessary for both (a) and (b) have been put in the above table.

(a)   Median is the $\dfrac{1200}{2}$ th item = 600th item.

This lies in class '72'.

Thus: $\qquad L = 70.5; \quad f_L = 429; \quad f = 204; \quad c = 3$

$\therefore \qquad\qquad \text{median} = 70.5 + 3\left(\dfrac{600 - 429}{204}\right)$

$$= 73.01 \ (2D)$$

Q_1 is the $\dfrac{1200}{4}$ th item $= 300$th item.

This lies in class '69'.

Thus: $\qquad L = 67.5; \quad f_L = 232; \quad f = 197; \quad c = 3$

$\therefore \qquad\qquad Q_1 = 67.5 + 3\left(\dfrac{300 - 232}{197}\right)$

$$= 68.53 \ (2D)$$

Q_3 is the $\dfrac{3(1200)}{4}$ th item $= 900$th item.

This lies in class '78'.

Thus: $\qquad L = 76.5; \quad f_L = 841; \quad f = 160; \quad c = 3$

$\therefore \qquad\qquad Q_3 = 76.5 + 3\left(\dfrac{900 - 841}{160}\right)$

$$= 77.61 \ (2D)$$

P_{37} is the $\dfrac{37(1200)}{100}$ th item $= 444$th item.

This lies in class '72'.

Thus: $\qquad L = 70.5; \quad f_L = 429; \quad f = 204; \quad c = 3$

$\therefore \qquad\qquad P_{37} = 70.5 + 3\left(\dfrac{444 - 429}{209}\right)$

$$= 70.72 \ (2D)$$

The accompanying curve shows the comparative ease with which the required quantities are estimated.

2.5 ELEMENTARY MEASURES OF DISPERSION

2.5.1 THE RANGE

This is the simplest of all measures of dispersion and can be calculated very quickly and easily.

DEFINITION 2.8

The *range* of a set of numbers is the difference between the smallest and largest numbers in the set. For a grouped frequency distribution, an estimate of the range would be the difference between the l.c.b. of the first class and the u.c.b. of the last class.

For example, the range of the set (2, 3, 8, 9, 7, 5, 3, 8, 9, 2, 4)   is $9 - 2 = 7$ (since 9 is the largest, and 2 the smallest, number).

The range is particularly useful in cases where we need to calculate the dispersion of many small sets of numbers and to use other techniques would be far too time-consuming. However the simplicity of this measure, in particular the fact that it uses only two extreme values (ignoring all others), precludes its use in any extensive analysis.

2.5.2 THE MEAN DEVIATION

This is a measure of dispersion which utilises the arithmetic mean.

DEFINITION 2.9

The *mean deviation from the mean* of a set of numbers $(x_1, x_2, \ldots, x_n)$ with arithmetic mean $\bar{x}$ is defined as:

$$\text{mean deviation} = \frac{1}{n} \sum_{i=1}^{n} |x_i - \bar{x}|$$

where $|x_i - \bar{x}|$ is the *positive difference between* x_i *and* $\bar{x}$ called the *modulus* of $x_i - \bar{x}$. In words, the mean deviation from the mean is the mean of the positive differences between each number and the arithmetic mean of the set.

(continued)

(Note that we define the modulus of some number a, $|a|$ as:

$$|a| = \begin{array}{ll} a; & \text{if } a \text{ is positive (or zero)} \\ -a; & \text{if } a \text{ is negative} \end{array}$$

For example, $|-2| = 2$; $|2| = 2$; $|-6.34| = 6.34$, etc.)

As an example of Definition 2.9, consider the set $(2, 3, 5, 3, 4, 1)$.

$$\bar{x} = \frac{2 + 3 + 5 + 3 + 4 + 1}{6} = 3.0$$

So that the mean deviation from the mean is

$$\frac{|2 - 3| + |3 - 3| + |5 - 3| + |3 - 3| + |4 - 3| + |1 - 3|}{6}$$

$$= \frac{|-1| + |0| + |2| + |0| + |1| + |-2|}{6}$$

$$= \frac{1 + 0 + 2 + 0 + 1 + 2}{6} = 1$$

However, the mean deviation can be described more generally as follows.

DEFINITION 2.10

The mean deviation from some constant c of a set of numbers $(x_1, x_2, \ldots, x_n)$ is defined as:

$$\frac{1}{n} \sum_i |x_i - c|$$

where c can be any numerical value.

If $c = \bar{x}$, we just obtain the mean deviation from the mean (as in the previous definition) and in general, as a measure of dispersion, the mean deviation is best measured from a measure of location. That is, when c is either the mean, median or mode.

(Note that it is usual to interpret 'mean deviation' as 'mean deviation *from the mean*' unless otherwise stated.)

EXAMPLE 2.26

Find the mean deviation from:

(a) the mean;

(b) the median for the set $(1, 1, 1, 2, 7, 10)$.

SOLUTION

$$\text{mean} = \bar{x} = \frac{22}{6} = 3.67 \ (2D)$$

$$\text{median} = 1.5 \ (\text{the mean of } 1 \text{ and } 2)$$

The mean deviation from the mean:

$$= \frac{1}{n} \sum_i |x_i - \bar{x}|$$

$$= \tfrac{1}{6}[|1 - 3.67| + |1 - 3.67| + |1 - 3.67| + |2 - 3.67|$$
$$+ |7 - 3.67| + |10 - 3.67|]$$

$$= \tfrac{1}{6}(19.34) = 3.22 \ (2D)$$

The mean deviation from the median:

$$= \frac{1}{n} \sum_i |x_i - m| \quad (\text{where } m = \text{median}).$$

$$= \tfrac{1}{6}[|1 - 1.5| + |1 - 1.5| + |1 - 1.5| + |2 - 1.5|$$
$$+ |7 - 1.5| + |10 - 1.5|]$$

$$= (\tfrac{1}{6})16 = 2.67 \ (2D)$$

The mean deviation is amended in an obvious way for a frequency distribution.

STATEMENT 2.9

For a frequency distribution:

$$\frac{x_1 x_2 \ldots x_n}{f_1 f_2 \ldots f_n}$$

the mean deviation from a constant c is expressed as:

$$\text{mean deviation} = \frac{1}{N} \sum_i f_i |x_i - c| \quad \text{where} \quad N = \sum_i f_i$$

Proof   Left as an exercise.

The above form is used for grouped data also, where $x_1, x_2, \ldots, x_n$ represent respective group mid-points.

This measure of dispersion is not used widely because of the inconvenience in the calculation and theoretical handling of the quantities $|x_i - \bar{x}|$. Also, when $\bar{x}$ is rounded (rather than an exact quantity), the error involved in adding the various $|x_i - \bar{x}|$ can prove unacceptable in practice.

EXAMPLE 2.27

Find the mean deviation for the frequency distribution:

10 up to 20	20 up to 30	30 up to 40	40 up to 50
2	12	24	8

SOLUTION

Class	Mid-points (x)	(f)	(fx)
10 up to 20	15	2	30
20 up to 30	25	12	300
30 up to 40	35	24	840
40 up to 50	45	8	360
		46	1530

Now,
$$\bar{x} = \frac{\sum fx}{\sum f} = \frac{1530}{46} = 33.3 \ (1D)$$

So, using the class mid-points, the mean deviation is given by

$$\text{mean deviation} = \frac{1}{N} \sum_i f_i |x_i - \bar{x}| \quad \text{(by definition)}$$

$$= \frac{1}{46} [2|15 - 33.3| + 12|25 - 33.3| + 24|35 - 33.3|$$
$$+ 8|45 - 33.3|]$$

$$= \frac{1}{46} [2(18.3) + 12(8.3) + 24(1.7) + 8(11.7)]$$

$$= \frac{1}{46} (36.6 + 99.6 + 40.8 + 93.6)$$

$$= \frac{1}{46} (270.6) = 5.88 \ (2D)$$

2.5.3 MEASURES ASSOCIATED WITH QUANTILES

DEFINITION 2.11

(a) The measure of dispersion based on quartiles is the *semi-inter-quartile range* (or *quartile deviation*) and is calculated as $\frac{1}{2}(Q_3 - Q_1)$.

(b) The measure of dispersion based on percentiles is the *10 to 90 percentile range* and is calculated as $P_{90} - P_{10}$.

There are no other quantile measures of dispersion in use, the above finding only occasional use in specific contexts.

EXAMPLE 2.28

For the following frequency distribution:

Waiting time (in mins.)	0–5	6–11	12–17	18–23	24–29	30 and over
No. of occasions	2	6	8	2	1	1

calculate:

(a)   the 10th percentile;
(b)   the 90th percentile;
(c)   the 10 to 90 percentile range.

SOLUTION

Class	f	Cum f
0–5	2	2
6–11	6	8
12–17	8	16
18–23	2	18
24–29	1	19
30+	1	20

The 10th percentile is the $[10(20)/100]$th item $= $ 2nd item which falls in the class 0 to 5 (the P_{10} class). Similarly the 90th percentile is the 18th item which falls in the class 18 to 23.

The l.c.b.s. of the P_{10} and P_{90} classes respectively are 0 and 17.5.

Hence $$P_{10} = 0 + \left(\frac{2-0}{2}\right)6$$

i.e. $$P_{10} = 6$$

Similarly $$P_{90} = 17.5 + \left(\frac{18-16}{2}\right)6 = 23.5$$

Therefore, the 10 to 90 percentile range is:

$$P_{90} - P_{10} = 23.5 - 6 = 17.5$$

2.6 THE STANDARD DEVIATION AND VARIANCE

The standard deviation is the measure of dispersion used most widely in Statistics and is based on the arithmetic mean. The form it takes depends on the type of data considered; a set, frequency distribution or either using a method of coding (as with the mean). However, as will be shown, there are special forms (based on the definition) of this measure used to simplify calculations.

2.6.1 FOR A SET

DEFINITION 2.12

The *standard deviation* of a set of numbers $(x_1, x_2, \ldots, x_n)$ with mean $\bar{x}$ is denoted by s and defined:

$$s = \sqrt{\frac{(x_1 - \bar{x})^2 + (x_2 - \bar{x})^2 + \ldots + (x_n - \bar{x})^2}{n}} = \sqrt{\frac{\sum_i (x_i - \bar{x})^2}{n}}$$

In words, s is the square root of the mean of the squares of deviations from the mean (and, hence, is sometimes called the *root mean square deviation*).

For example, the set $(3, 4, 6, 2)$ has $\bar{x} = \dfrac{15}{4} = 3.75$ and:

$$s = \sqrt{\frac{(3 - 3.75)^2 + (4 - 3.75)^2 + (6 - 3.75)^2 + (2 - 3.75)^2}{4}}$$

$$= \sqrt{\frac{0.5625 + 0.0625 + 5.0625 + 3.0625}{4}} = \sqrt{\frac{8.75}{4}} = 1.479$$

i.e., $s = 1.48$ (2D).

For practical purposes, the form $\sqrt{\dfrac{\Sigma(x_i - \bar{x})^2}{n}}$ is clumsy to use and usually we use another expression to calculate the standard deviation which is useful to derive since it involves some elementary identities already introduced.

Ignoring the square root sign, we have:

$$\frac{\Sigma(x_i - \bar{x})^2}{n} = \frac{1}{n}\Sigma(x_i - \bar{x})^2 = \frac{1}{n}\Sigma(x_i^2 - 2\bar{x}x_i + \bar{x}^2) \quad \text{(expanding}$$
$$\text{the square)}$$

$$= \frac{1}{n}\left(\Sigma x_i^2 - \Sigma 2\bar{x}x_i + \Sigma \bar{x}^2\right) \quad \left(\Sigma \text{ operator properties}\right)$$

$$= \frac{1}{n}\left(\Sigma x_i^2 - 2\bar{x}\Sigma x_i + n\bar{x}^2\right) \quad \text{(since } \bar{x} \text{ does not depend on}$$
$$i, \text{ it is a constant with respect}$$
$$\text{to } i)$$

$$= \frac{\sum x_i^2}{n} - 2\bar{x}\frac{\sum x_i}{n} + \bar{x}^2$$

$$= \frac{\sum x_i^2}{n} - 2\bar{x}^2 + \bar{x}^2 \quad \left(\text{since} \quad \frac{\sum x_i}{n} = \bar{x} \right)$$

$$= \frac{\sum x_i^2}{n} - \bar{x}^2$$

i.e.
$$\frac{\sum (x_i - \bar{x})^2}{n} = \frac{\sum x_i^2}{n} - \bar{x}^2$$

$\therefore$
$$\sqrt{\frac{\sum (x_i - \bar{x})^2}{n}} = \sqrt{\frac{\sum x_i^2}{n} - \bar{x}^2}$$

To summarise:

STATEMENT 2.10

The standard deviation of a set of numbers $(x_1, x_2, \ldots, x_n)$ can be expressed using the *computational formula*:

$$s = \sqrt{\frac{\sum x_i^2}{n} - \bar{x}^2}$$

Proof   In preceding text.

For the set considered previously, $(3, 4, 6, 2)$ we have:

(x)	(x^2)
3	9
4	16
6	36
2	4
	65

We know $\bar{x} = 3.75$ with $n = 4$.

So $$s = \sqrt{\frac{\sum x^2}{n} - \bar{x}^2}$$

$$= \sqrt{\frac{65}{4} - (3.75)^2}$$

$$= \sqrt{16.250 - 14.063} = 1.48 \ (2D) \quad \text{(as before)}$$

EXAMPLE 2.29

Find the standard deviation of the set $(-1.2, 0.3, 2.1, -0.4, -2.1, 1.4, 0.1, 1.2)$.

SOLUTION

$$\bar{x} = \frac{-1.2 + 0.3 + \ldots + 0.1 + 1.2}{8} = \frac{1.4}{8} = 0.175$$

$$\sum x^2 = (-1.2)^2 + (0.3)^2 + \ldots + (0.1)^2 + (1.2)^2 = 13.92$$

$$\therefore \quad s = \sqrt{\frac{13.92}{8} - (0.175)^2} = \sqrt{1.740 - 0.031} = 1.31 \text{ (2D)}$$

2.6.2 FOR A FREQUENCY DISTRIBUTION

The previous definition and statement are easily amended to include a frequency distribution as follows.

STATEMENT 2.11

For a discrete frequency distribution, the standard deviation is defined by:

$$s = \sqrt{\frac{\sum f_i(x_i - \bar{x})^2}{\sum f_i}}$$

and written (for computational purposes) as:

$$s = \sqrt{\frac{\sum f_i x_i^2}{\sum f_i} - \left[\frac{\sum f_i x_i}{\sum f_i}\right]^2}$$

Note that:

$$\frac{\sum f_i x_i}{\sum f_i}$$

is the mean $\bar{x}$ of a frequency distribution.

Proof   Left as an exercise.

The above expressions are also used for a continuous (or grouped discrete) distribution where, as usual, the class mid-points are taken as the x values.

EXAMPLE 2.30

Find the standard deviation of the distribution:

x	0	1	2	3	4	5
f	1	3	11	9	5	2

SOLUTION

x	f	fx	x^2	fx^2	
0	1	0	0	0	
1	3	3	1	3	
2	11	22	4	44	$\Sigma f = 31$
3	9	27	9	81	$\Sigma fx = 82$
4	5	20	16	80	
5	2	10	25	50	$\Sigma fx^2 = 258$
	31	82		258	

$$s = \sqrt{\frac{\Sigma fx^2}{\Sigma f} - \left(\frac{\Sigma fx}{\Sigma f}\right)^2} = \sqrt{\frac{258}{31} - \left(\frac{82}{31}\right)^2} = \sqrt{8.323 - 6.997}$$

$$\therefore \qquad s = 1.15 \quad (2D)$$

EXAMPLE 2.31

Calculate the mean and standard deviation for the following distribution.

Class	2.45 up to 2.55	2.55 up to 2.65	2.65 up to 2.75	2.75 up to 2.85	2.85 up to 2.95
Frequency	3	9	11	8	4

SOLUTION

Class	Mid-point (x)	f	fx	fx^2	
2.45–2.55	2.5	3	7.5	18.75	$\Sigma f = 35$
2.55–2.65	2.6	9	23.4	60.84	
2.65–2.75	2.7	11	29.7	80.19	$\Sigma fx = 94.6$
2.75–2.85	2.8	8	22.4	62.72	
2.85–2.95	2.9	4	11.6	33.64	$\Sigma fx^2 = 256.14$
		35	94.6	256.14	

$$\bar{x} = \frac{94.6}{35} = 2.703 \ (3D)$$

$$s = \sqrt{\frac{256.140}{35} - (2.703)^2} = \sqrt{7.318 - 7.306} = 0.11 \ (2D)$$

EXAMPLE 2.32

Find the mean and standard deviation of the distribution:

x	35	45	55	65	75	85	95
f	6	20	30	20	10	8	6

SOLUTION

Continuing the rows from the above data:

fx	210	900	1650	1300	750	680	570
fx^2	7350	40 500	90 750	84 500	56 250	57 800	54 150

So, $\sum f = 100;$ $\sum fx = 6060;$ $\sum fx^2 = 391\,300$

and $$\bar{x} = \frac{6060}{100} = 60.6$$

with $$s = \sqrt{\frac{391\,300}{100} - (60.6)^2}$$

$$= \sqrt{3913 - 3672.36} = 15.51 \ (2D)$$

(Notice that when the x values are large, the calculations become tedious.)

2.6.3 THE METHOD OF CODING

The reasons for coding (and the methods involved) to find the standard deviation are the same as those for the mean, the only difference being in the form of the decoding expression. Since the standard deviation involves squaring x values, the method of coding takes on a more important role here, to overcome, possibly, very large values of x^2 (see Example 2.32).

The method involved is formally stated:

STATEMENT 2.12

To find the standard deviation, s of:

(a) a set of values $(x_1, x_2, \ldots, x_n)$; or

(b) a frequency distribution $\dfrac{x_1 x_2 \ldots x_n}{f_1 f_2 \ldots f_n}$;

the *coding* $X = \dfrac{x-a}{b}$ can be used to obtain S (the standard deviation of the X values) and *decoded* using the expression $s = bS$ where:

(*continued*)

$$S = \sqrt{\frac{\sum X^2}{n} - \left(\frac{\sum X}{n}\right)^2} \qquad \text{(case (a))}$$

or

$$S = \sqrt{\frac{\sum fX^2}{\sum f} - \left(\frac{\sum fX}{\sum f}\right)^2} \qquad \text{(case (b))}$$

Proof For case (a), to show that $s = bS$ when $X = \dfrac{x - a}{b}$.

If $X = \dfrac{x - a}{b}$ then $x = a + bX$ [1]

Now $s = \sqrt{\dfrac{\sum (x_i - \bar{x})^2}{n}}$ (by definition)

$= \sqrt{\dfrac{\sum (a + bX - (a + b\bar{X}))^2}{\sum n}}$ (using [1] and Statement 2.3)

$= \sqrt{\dfrac{\sum (bX - b\bar{X})^2}{n}} = b\sqrt{\dfrac{\sum (X - \bar{X})^2}{n}}$

$= bS$ (by definition)

The proof for case (b) is left as an exercise.

EXAMPLE 2.33

Use the coding $X = \dfrac{x - 125}{5}$ to find the mean and standard deviation of the following distribution.

x	105	110	115	120	125	130	135	140	145
f	14	26	42	70	68	35	11	2	1

SOLUTION

Note that the values of x are very large here.

x	f	$X = \dfrac{x-125}{5}$	fX	fX^2
105	14	-4	-56	224
110	26	-3	-78	234
115	42	-2	-84	168
120	70	-1	-70	70
125	68	0	0	0
130	35	1	35	35
135	11	2	22	44
140	2	3	6	18
145	1	4	4	16
	269		-221	809

Now,
$$\bar{X} = \frac{\sum fX}{\sum f} = \frac{-221}{269} = -0.822 \ \ (3\text{D})$$

$\therefore$
$$\bar{x} = a + b\bar{X} = 125 + 5(-0.822) = 120.89 \ \ (2\text{D})$$

Also,
$$S = \sqrt{\frac{\sum fX^2}{\sum f} - \left(\frac{\sum fX}{\sum f}\right)^2} = \sqrt{\frac{809}{269} - (0.822)^2}$$

$$= \sqrt{3.007 - 0.676} = 1.527 \ \ (3\text{D})$$

$\therefore$
$$s = bS = 5(1.527) = 7.64 \ \ (2\text{D})$$

Notice that, by choosing $a = 125$ and $b = 5$ (the x central value and common class interval respectively), we ideally simplify the x values in the table.

In the previous statement (the coding method for the standard deviation) the decoding formula $s = bS$ does not involve the constant a. This is because (as mentioned earlier) the effect of subtracting a constant a is to move every single value in the set (or distribution) through a distance a which, although 're-locating' the set, preserves the relationship of all the values to each other. That is, the spread or scatter of the set will not be affected by the subtraction of a constant. But since the scaling factor b has the effect of 'squashing' the curve (and hence the dispersion) by a factor of b, the standard deviation of the coded variable must change accordingly.

2.6.4 THE VARIANCE

DEFINITION 2.13

The *variance* of a set, or distribution, of numbers is defined as *the square of the standard deviation* and is denoted (in an obvious way) by s^2.

Depending on the type of data to hand (individual items, discrete or continuous frequency distributions or data to be coded), there are specific formulae for calculating the variance, obtained simply by removing the square root sign from the corresponding expressions for the standard deviation. We list these below for reference purposes.

STATEMENT 2.13

Precise expressions for the variance in particular situations are given by:

(a)   *For a set*

$$s^2 = \frac{\sum (x - \bar{x})^2}{n} \quad \text{(definition)}$$

$$= \frac{\sum x^2}{n} - \left(\frac{\sum x}{n}\right)^2 \quad \text{(computational formula)}$$

(b)   *For a frequency distribution*

$$s^2 = \frac{\sum f(x - \bar{x})^2}{\sum f} \quad \text{(definition)}$$

$$= \frac{\sum fx^2}{\sum f} - \left(\frac{\sum fx}{\sum f}\right)^2 \quad \text{(computational formula)}$$

(c)   *Coding, using* $X = \dfrac{x - a}{b}$

$$s^2 = b^2 S^2 \quad \text{(for all data)}$$

Proofs   See corresponding cases for standard deviation. The variance is most useful where algebraic manipulation is necessary since it does not involve the square root. However, the standard deviation is expressed in the same units as the original variable, the variance involving the square of the units.

EXAMPLE 2.34

Calculate the variance for the following frequency distribution.

Class	20 and up to 40	40 and up to 60	60 and up to 80	80 and up to 100	100 and up to 120
Frequency	1	2	6	3	3

SOLUTION

We take x as the mid-point of the classes.

Mid-point (x)	f	$X = \dfrac{x-70}{20}$	fX	fX^2
30	1	-2	-2	4
50	2	-1	-2	2
70	6	0	0	0
90	3	1	3	3
110	3	2	6	12
	15		5	21

For the coding here, $a = 70$ and $b = 20$.

$$s^2 = b^2 S^2 = b^2\left[\frac{\sum fX^2}{\sum f} - \left(\frac{\sum fX}{\sum f}\right)^2\right] = 400\left[\frac{21}{15} - \left(\frac{5}{15}\right)^2\right]$$

$$= 400(1.289) = 515.6 \ (1D)$$

2.7 MOMENTS AND FURTHER MEASURES

2.7.1 MOMENTS OF A FREQUENCY DISTRIBUTION

This sub-section is intended mainly to bring together some of the ideas discussed so far in this chapter under the more general concept of moments. We first define two important types of moment.

DEFINITION 2.14

(a) *The rth moment (about zero) of a frequency distribution* $\dfrac{x_1 x_2 \ldots x_n}{f_1 f_2 \ldots f_n}$ *is defined as* $m'_r = \dfrac{\sum fx^r}{\sum f}$.

So, the 1st moment about zero, $m'_1 = \dfrac{\sum fx}{\sum f} = \bar{x}$ (the mean).

(continued)

(b)   *The rth moment (about the mean) of a frequency distribution*
$\dfrac{x_1 x_2 \ldots x_n}{f_1 f_2 \ldots f_n}$ is defined as:

$$m_r = \frac{\sum f(x - \bar{x})^r}{\sum f}$$

The 2nd moment about the mean:

$$= m_2 = \frac{\sum f(x - \bar{x})^2}{\sum f} = s^2 \quad \text{(the variance)}$$

From the above we see that the mean and variance are only two of a whole (infinite) range of moments of a frequency distribution, and in particular $m'_0 = m_0 = 1$ is easily seen together with

$$m_1 = \frac{\sum f(x - \bar{x})}{\sum f} = \frac{\sum fx}{\sum f} - \frac{\sum f\bar{x}}{\sum f} = \bar{x} - \bar{x} \cdot \frac{\sum f}{\sum f} = 0$$

With regard to the definitions given of the two types of moment, there are two extensions that should be mentioned.

(a)   We have defined the moments in terms of a discrete frequency distribution only (this being the most applicable case). However, in terms of a set of values $(x_1, x_2, \ldots, x_n)$, we would define the rth moments as $m'_r = \dfrac{\sum x^r}{n}$ and $m_r = \dfrac{\sum (x - \bar{x})^r}{n}$ these being obvious from the previous definitions. Alternatively when dealing with a continuous (or grouped discrete) distribution, we take the mid-points of classes to represent the x values as usual.

(b)   Moments about the mean can always be expressed in terms of moments about zero (and vice versa). Both types of moments are used in their own right however, since, in particular circumstances, one will be found to be more convenient than the other. As a demonstration, consider the variance (m_2) of a distribution. By definition,

$$m_2 = \frac{\sum f(x - \bar{x})^2}{\sum f} = \frac{\sum fx^2}{\sum f} - \left(\frac{\sum fx}{\sum f}\right)^2 \quad \text{(Statement 2.11)}$$

$$= m'_2 - m'^2_1 \quad \text{(by definition)}$$

That is, we can express the second moment about the mean in terms of the 1st and 2nd moments about zero.

The corresponding expressions for m_3 and m_4 (in terms of moments about zero) are given as follows.

STATEMENT 2.14

The 3rd and 4th moments about the mean, m_3 and m_4, can be expressed in terms of moments about zero as follows.

(a) $m_3 = m_3' - 3m_1'm_2' + 2(m_1')^3$

(b) $m_4 = m_4' - 4m_1'm_3' + 6(m_1')^2 m_2' - 3(m_1')^4$

These results of course are valid for a set or distribution.

Proof for (a)   Using the definition applicable to a frequency distribution we have:

$$m_3 = \frac{\sum f(x - \bar{x})^3}{\sum f}$$

Now $(x - \bar{x})^3 = (x - \bar{x})^2(x - \bar{x}) = (x^2 - 2\bar{x}x + \bar{x}^2)(x - \bar{x})$

$$= x^3 - 3\bar{x}x^2 + 3\bar{x}^2 x - \bar{x}^3$$

$\therefore$ $\sum f(x - \bar{x})^3 = \sum f(x^3 - 3\bar{x}x^2 + 3\bar{x}^2 x - \bar{x}^3)$

$$= \sum fx^3 - 3\bar{x} \sum fx^2 + 3\bar{x}^2 \sum fx - \bar{x}^3 \sum f$$

Hence $\dfrac{\sum f(x - \bar{x})^3}{\sum f} = \dfrac{\sum fx^3}{\sum f} - 3\bar{x}\dfrac{\sum fx^2}{\sum f} + 3\bar{x}^2 \dfrac{\sum fx}{\sum f} - \bar{x}^3 \dfrac{\sum f}{\sum f}$

$$= \frac{\sum fx^3}{\sum f} - 3\frac{\sum fx}{\sum f}\frac{\sum fx^2}{\sum f} + 3\left(\frac{\sum fx}{\sum f}\right)^3 - \left(\frac{\sum fx}{\sum f}\right)^3$$

$$\left(\bar{x} = \frac{\sum fx}{\sum f}\right)$$

$$= \frac{\sum fx^3}{\sum f} - 3 \frac{\sum fx}{\sum f} \frac{\sum fx^2}{\sum f} + 2 \left(\frac{\sum fx}{\sum f} \right)^3$$

$\Rightarrow$

$$m_3 = m'_3 - 3m'_1 m'_2 + 2(m'_1)^3$$

The proof for (b) is left as an exercise.

EXAMPLE 2.35

Find the 3rd moment about the mean, m_3, of the following distribution.

Class	0–4	5–7	8–10	11–12
Frequency	6	8	8	4

SOLUTION

We will use the form $m_3 = m'_3 - 3m'_1 m'_2 + 2(m'_1)^3$ where:

$$m'_3 = \frac{\sum fx^3}{\sum f}; \quad m'_2 = \frac{\sum fx^2}{\sum f} \quad \text{and} \quad m'_1 = \bar{x} = \frac{\sum fx}{\sum f}$$

Class	Mid-point (x)	f	fx	fx^2	fx^3
0–4	2	6	12	24	48
5–7	6	8	48	288	1 728
8–10	9	8	72	648	5 832
11–12	11.5	4	46	529	6 083.5
		26	178	1489	13 691.5

So, $m'_1 = \bar{x} = \dfrac{178}{26} = 6.85$ $\qquad m'_2 = \dfrac{1489}{26} = 57.27$

$$m'_3 = \frac{13\,691.5}{26} = 526.60$$

$\therefore$ $m_3 = 526.60 - 3(6.85)(57.27) + 2(6.85)^3 = -7.5$ (1D)

Since the arithmetic involved in calculating higher moments is (as can be seen in the last example) fairly heavy, in practice we would use a coding method. The usual technique is employed where we use the relationship $X = \dfrac{x - a}{b}$ and, sticking to the convention of using capital letters for coded variables, we can calculate:

$$M_1' = \frac{\Sigma fX}{\Sigma f}; \quad M_2' = \frac{\Sigma fX^2}{\Sigma f}, \ldots \text{ etc.}$$

the (coded) moments about zero. To calculate (coded) moments about the mean we would use the relationships introduced earlier of:

$$M_2 = M_2' - (M_1')^2; \quad M_3 = M_3' - 3M_1'M_2' + 2(M_1')^3, \ldots \text{ etc.}$$

Finally, the decoding formula to evaluate $m_2, m_3, \ldots$, etc. is fairly straightforward and is given in the following statement.

STATEMENT 2.15

If a coding of the form $X = \dfrac{x - a}{b}$ is used to calculate coded moments about the mean $(M_1, M_2, \ldots)$, then the true moments about the mean $(m_1, m_2, \ldots)$ can be obtained using the decoding expression $m_r = (b^r)M_r$ (for all positive, integral r), i.e. $m_2 = (b^2)M_2$; $m_3 = (b^3)M_3$, and so on. Note that $m_1 = 0$ always.

Proof   Now:

$$m_r = \frac{\Sigma f(x - \bar{x})^r}{\Sigma f} \quad \text{(by definition)}$$

$$= \frac{\Sigma f(a + bX - (a + b\bar{X}))^r}{\Sigma f}$$

$$= \frac{\Sigma f(bX - b\bar{X})^r}{\Sigma f}$$

$$= b^r \frac{\Sigma f(X - \bar{X})^r}{\Sigma f}$$

i.e. $m_r = b^r M_r$ (by definition) *completing the proof*

Note that, putting $r = 2$ gives $m_2 = b^2 M_2$

i.e. $s^2 = b^2 S^2$, a result already known for the decoding of the variance (statement 2.13(c)).

EXAMPLE 2.36

Use a suitable coding to find the first 3 moments about the mean for the following distribution of the weights of a certain 40 students.

Weight (in lbs)	118 to 126	127 to 135	136 to 144	145 to 153	154 to 162	163 to 171	172 to 180	Total
Frequency	3	5	9	12	5	4	2	40

SOLUTION

Taking the variable x as the class mid-point and using the coding $X = \dfrac{x - 149}{9}$ $(a = 149, \ b = 9)$ we have:

x	f	$X = \dfrac{x - 149}{9}$	fX	fX^2	fX^3
122	3	-3	-9	27	-81
131	5	-2	-10	20	-40
140	9	-1	-9	9	-9
149	12	0	0	0	0
158	5	1	5	5	5
167	4	2	8	16	32
176	2	3	6	18	54
	40		-9	95	-39

First we calculate the first 3 coded moments about zero.

$$M'_1 = \frac{\Sigma fX}{\Sigma f} = \frac{-9}{40} = -0.225; \qquad M'_2 = \frac{\Sigma fX^2}{\Sigma f} = \frac{95}{40} = 2.375$$

$$M'_3 = \frac{\Sigma fX^3}{\Sigma f} = \frac{-39}{40} = -0.975$$

The coded moments about the mean are calculated as follows.

$$M_1 = 0 \quad \text{(always)}; \quad M_2 = M'_2 - (M'_1)^2$$
$$= 2.375 - (-0.225)^2 = 2.324$$

$$M_3 = M'_3 - 3M'_2 M'_1 + 2(M'_1)^3$$
$$= (-0.975) - 3(2.375)(-0.225) + 2(-0.225)^3$$
$$= 0.605$$

Finally, $m_1 = 0$; $m_2 = b^2 M_2 = 9^2(2.324) = 188.24$

and $m_3 = b^3 M_3 = 9^3(0.605) = 441.05$

It has already been noted just how the mean and variance 'describe' certain characteristics of a frequency distribution (namely location and spread). The same is true for other moments — they can each be said to describe a particular characteristic of a distribution. Practically through, other moments are not used frequently since in general the higher the moment (i.e., the larger the value of r) (a) the more arduous it is to calculate and (b) the more sensitive it is to small changes in the frequency distribution.

2.7.2 SKEWNESS AND KURTOSIS

The idea of skewness and kurtosis was mentioned briefly in Chapter 1.

Skewness might be described as a measure of non-symmetry.

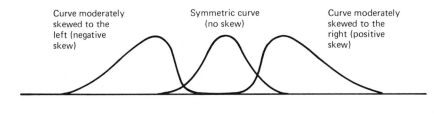

Curve moderately skewed to the left (negative skew)   Symmetric curve (no skew)   Curve moderately skewed to the right (positive skew)

Fig. 2.11

Representing skewness as a number we would wish ideally that a symmetric curve (distribution) have a measure of skewness of zero and it would seem reasonable to represent degrees of skewness to the left or right as negative or positive numbers respectively.

Now it can be shown empirically that only for a symmetric curve do the three main measures of location (mean, median and mode) coincide. For distributions that are moderately skewed (to the left or right) we have the approximate relation:

$$\text{mode} = \text{mean} - 3(\text{mean} - \text{median})\qquad [1]$$

which is shown pictorially in Fig. 2.12.

Note that the mean is the measure nearest the long tail of the distribution since, as we already know, it takes extreme values into account to a greater extent than the other two.

A simple measure of skewness then, would be mean — mode which is always negative for left skewed and positive for right skewed distributions.

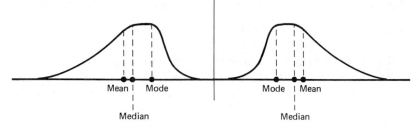

Fig. 2.12

However, dividing this quantity by the standard deviation makes this measure dimensionless (i.e., independent of the units in which the distribution is measured), a desirable property. So we have as a measure of skewness $\dfrac{\text{mean} - \text{mode}}{\text{standard deviation}}$ usually called *Pearson's measure of skewness.* Alternatively, we can write this as $\dfrac{3(\text{mean} - \text{median})}{\text{standard deviation}}$ using the relationship [1].

There are other measures of skewness, based mainly on moments about the mean, but we shall not consider them here.

Kurtosis is a property of a distribution describing how 'peaked' a distribution is. 'Platykurtic' is the name given to 'flat topped' distributions and 'leptokurtic' to more peaked distributions. The most common measure of kurtosis used is given by the quantity $\dfrac{m_4}{m_2{}^2}$.

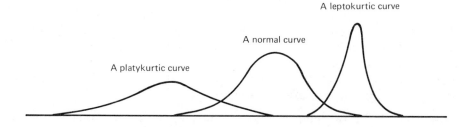

A platykurtic curve

A normal curve

A leptokurtic curve

Fig. 2.13

We take the 'standard' distribution as that which has $\dfrac{m_4}{m_2{}^2} = 3$, a property of a family of distributions referred to as *normal* (discussed extensively in Chapter 6). Since $m_2{}^2$ is positive and m_4 being essentially just a sum of squares and so also positive, the expression $\dfrac{m_4}{m_2{}^2}$ must also be positive. Hence, by considering whether $\dfrac{m_4}{m_2{}^2} - 3$ is either negative or positive, we have an idea how flat or peaked respectively the given distribution is compared to a 'normal' distribution.

To summarise:

DEFINITION 2.15

(a) *Pearson's measure of skewness* is given by $\dfrac{\text{mean} - \text{mode}}{\text{standard deviation}}$

or, alternatively by $\dfrac{3(\text{mean} - \text{median})}{\text{standard deviation}}$ for moderately skewed distributions. A negative, zero or positive value of these measures showing respectively left skew, symmetry (i.e., no skew), or right skew.

(b) *A measure of kurtosis* of a distribution is given by $\dfrac{m_4}{m_2^2} - 3$, a

negative or positive value showing how less or more peaked (respectively) the given distribution is compared to a 'normal' distribution.

2.8 EXERCISES

SECTION 2.1

1.   If the values that x can take are contained in the set $(1, 3, 2, 4, -3, 7)$, find the value of: (a) Σx; (b) Σx^2; (c) $\Sigma (x + 4)$.

2.   Calculate the values of the following:

      (a) $\displaystyle\sum_{i=1}^{3} (3 + 2i)$ (b) $\displaystyle\sum_{i=2}^{5} \left(\dfrac{i+1}{i-1}\right)$ (c) $\displaystyle\sum_{r=1}^{4} (3r - 5r^2)$

      (d) $\displaystyle\sum_{i=3}^{6} \left(\dfrac{1}{i}\right)$ (to 2D)

3.   If $x_1 = 4$, $x_2 = 4$, $x_3 = 3$, $x_4 = 2$ and $y_1 = 3$, $y_2 = 2$, $y_3 = -1$, $y_4 = 5$, find:

      (a) $\displaystyle\sum_{i=1}^{4} (x_i + y_i)$ (b) $\displaystyle\sum_{i=1}^{3} 3x_i$ (c) $\displaystyle\sum_{i=1}^{4} (x_i^2 + 2)$

      (d) $\displaystyle\sum_{i=1}^{4} (2y_i^2 - 4y_i + 3)$

4.   Write the following as compactly as possible using Sigma notation.

      (a) $x_1^2 + x_2^2 + x_3^2 + \ldots + x_{25}^2$.   (b) $3y_1^2 + 3y_2^2 + \ldots + 3y_n^2$.

(c) $2x_1y_1^2 + 2x_2y_2^2 + 2x_3y_3^2 + 2x_4y_4^2$.   (d) $\dfrac{f_1x_1 + f_2x_2 + \ldots f_nx_n}{f_1 + f_2 + \ldots + f_n}$

(e) $\dfrac{x_1y_1 + x_2y_2 + \ldots + x_ny_n}{n} - \dfrac{(x_1 + \ldots + x_n)(y_1 + \ldots + y_n)}{n^2}$

SECTION 2.2

5.   Find the mean of the sets:

(a) $(84, 92, 73, 67, 88, 74, 91, 74)$

(b) $(3.41, 2.86, -1.84, 2.31, -3.80, 2.14, 0.87)$

(c) $(5, 3, 6, 5, 4, 5, 2, 8, 6, 5, 4, 8, 3, 4, 5, 4, 8, 2, 5, 4)$

(d) $(0.53, 0.46, 0.50, 0.49, 0.52, 0.53, 0.44, 0.55)$

(e) $(4.6, 2.32, 0.8, 5.96, -1.4, 2.6, 8.33)$

6.   Find the mean of the first 20 positive integers (the set $1, 2, 3 \ldots 19, 20$).

7.   Find the mean of: (a) the first 10 odd numbers;   (b) the first 10 even numbers.

8.   The following list of numbers was the result of asking 30 people how many brothers and sisters they had:

2 1 1 2 0 1 0 3 2 0 1 1 5 1 2
2 2 1 1 3 0 2 4 0 0 2 1 0 1 2

Find the mean of this set. Assuming that each of the 30 people questioned were not related to each other, deduce the average number of children in the 30 families so defined.

9.   The following figures give the monthly averages of exports (in £ million) to 16 'developing countries' for the third quarter of 1974. Find the arithmetic mean.

2.7 3.5 3.0 1.2 11.7 3.0 1.0 5.2 6.4 8.4 1.4 12.8 3.8 1.5 4.6 2.6

10.   Find the mean of the following frequency distribution:

Class	2	3	4	5	6
Frequency	1	3	5	4	2

11.   Find the means of the following frequency distributions:

(a)
x	10	11	12	13
f	1	2	12	7

(b)
x	18.5	19.5	20.5
f	5	12	20

(c)
x	1	2	3	4	5	6
f	2	8	24	52	31	11

(d)
y	3	6	9	12
f	21	11	8	1

12.   A student's marks in the Aural, Oral and Practical parts of a French examination were 81, 86 and 68 respectively. If the Oral is considered to be twice as important as the Practical and the Aural half as

important again as the Oral, calculate an appropriate weighted average mark for the student.

(*Hint:* Calculate the ratio of importance of the three parts and use these as frequencies.)

13.   Of 100 people sitting an examination, 20 obtained 65 marks, 12 obtained 70, 8 obtained 75 and 2 obtained 80. The lowest mark was 25 which only two people obtained. If it is known that the remainder obtained a total of 2744 marks between them, what was the average mark for the 100?

14.   The mean of 13 numbers is 10 and the mean of 42 other numbers is 16. Find the mean of the combined set of 55.

15.   The wholesale prices of seven important minerals were indexed as follows: 110.91, 120.84, 73.00, 138.01, 95.12, 112.04 and 133.12. They were also weighted (according to industrial importance) respectively: 6, 5, 2, 9, 3, 4 and 1. Find the weighted average index.

16.   A man measured his waiting time (in minutes) for a number 185A 'bus on 20 occasions and obtained the distribution:

Waiting time	0–5	6–11	12–17	18–23	24–29	30 and over
No. of occasions	2	6	8	2	1	1

Find his mean waiting time using a method of coding.

17.   Find the means of the following frequency distributions:

(a) Class	Frequency	(b) Class	Frequency
10 to under 15	3	10–13	2
15 to under 20	7	14–17	21
20 to under 25	15	18–21	83
25 to under 30	20	22–25	14
30 to under 35	9	26–29	3
35 to under 40	4		

18.   The following data gives an estimate of the age of females in Northern Ireland (in thousands) as at 30 June 1973:

Age	Under 5	5–9	10–14	15–19	20–24	25–29
Number	72.1	76.0	72.1	63.6	53.1	51.3
Age	30–34	35–39	40–44	45–49	50–54	55–59
Number	44.7	41.6	42.5	42.9	43.2	38.2
Age	60–64	65–69	70–74	75–79	80–84	85 and over
Number	39.4	34.7	28.5	19.6	11.0	6.9

Find the mean age. (A reasonable upper limit for the end class would be 100.)

(Data source: MDS)

19.   For the following set of values of a variable x: 55, 60, 65, 70, 75 and 80, use the coding $X = \dfrac{x - 67.5}{5}$ to find the value of $\bar{X}$. Hence, by decoding, fing the value of $\bar{x}$.

20.   Find the mean of the set $x = (504, 498, 499, 504, 502, 496, 501, 501)$ by using the coding $X = x - 500$. (Notice that b, the scaling factor, is 1 here.)

21.   Using an arbitrary origin (assumed mean) of 67 and a scaling factor of 3, calculate the mean of the following distribution:

Class	60–62	63–65	66–68	69–71	72–74
Frequency	5	18	42	27	8

22.   Use an appropriate coding to obtain the mean of the distribution:

Class	3.5 up to 13.5	13.5 up to 23.5	23.5 up to 33.5	33.5 up to 43.5	43.5 up to 53.5
Frequency	8	12	20	15	5

23.   Find the mean of the following frequency distribution of the widths of 100 washers:

Width (in mm)	0–0.5	0.5–1.0	1.0–1.5	1.5–2.0	2.0–2.5	2.5–3.0	3.0–3.5
No. of washers	4	12	24	27	23	8	2

24.   A sample of 250 nuts manufactured by a certain company was inspected and the diameters of their heads were found to be as follows:

Diameter of head (in inches)	Frequency	Diameter of head (in inches)	Frequency
0.9747 to 0.9749	2	0.9765 to 0.9767	49
0.9750 to 0.9752	6	0.9768 to 0.9770	25
0.9753 to 0.9755	8	0.9771 to 0.9773	18
0.9756 to 0.9758	15	0.9774 to 0.9776	12
0.9759 to 0.9761	42	0.9777 to 0.9779	4
0.9762 to 0.9764	68	0.9780 to 0.9782	1

Find the mean diameter of the heads. (A coding method is suggested using an assumed mean of 0.9763 with scaling factor 0.0003.)

25.   Given the following continuous frequency distribution:

Weight (in gm)	No. of components	Weight (in gm)	No. of components
4.20 to 4.24	2	4.40 to 4.44	31
4.25 to 4.29	14	4.45 to 4.49	19
4.30 to 4.34	28	4.50 to 4.54	7
4.35 to 4.39	136	4.55 to 4.59	4

(a) state the common class width;

(b) find the mid-points of each class;

(c) denoting these mid-points by $x_1, x_2, \ldots, x_8$ respectively, transform the data using the coding $X_i = \dfrac{x_i - a}{b}$, where a is the *mid-point of the modal class* and b is the *common class width*; and

(d) hence find the mean weight of the components to 3D.

SECTION 2.3

26.   Find the median of the following sets of numbers:

(a) $(84, 91, 72, 68, 87, 78)$

(b) $(-3, -2, -1, 0, 1, 2, 3)$

(c) $(7.99, 8.01, 8.01, 8.02, 8.01, 8.02)$

(d) $(2.52, 3.96, 3.28, 9.20, 3.75)$.

27.   Determine the median of the following *discrete* distributions:

(a)

x	19	20	21	22	23
f	2	11	29	36	14

(b)

y	12.1	12.2	12.3	12.4
f	36	11	2	1

28.   For the following distribution:

Class	0–4	5–9	10–14	15–19	20–24	25–29
Frequency	1	14	23	21	15	6

(a) Draw a smooth cumulative frequency curve;   (b) using the curve, estimate a median value;   (c) compare this value with that obtained using the interpolation formula.

29.   Calculate a median value for the distribution of widths of washers given below:

Width (in mm)	0 up to 0.5	0.5 up to 1.0	1.0 up to 1.5	1.5 up to 2.0	2.0 up to 2.5	2.5 up to 3.0	3.0 up to 3.5
No. of washers	4	12	24	27	23	8	2

30.   The given data records the results of a general knowledge test given to 227 children:

Score	1–2	3–4	5–6	7–8	9–10
Frequency	9	46	79	70	23

Estimate the median score by interpolation.

31.   An observer visited an 'off-licence' on 20 occasions to note the number in the queue. The results were as follows:

No. of people in queue (x)	0	1	2	3	4 or more
No. of queues (f)	6	8	3	2	1

Find the mean and median queue length.

(*Notes:*   (a) An open-ended class does not affect the median in general;   (b) For the calculation of the mean an upper limit of 6 would seem reasonable for the last class.)

32.   Calculate the median for the data tabulated:

x	f	x	f	x	f
118–126	3	145–153	12	163–171	4
127–135	5	154–162	5	172–180	2
136–144	9				

33.   The following frequency distribution gives the heights of 1000 men. Calculate a median value using the interpolation formula.

Height (in inches)	No. of men	Height (in inches)	No. of men
Under 58	2	68 and under 70	263
58 and under 60	5	70 and under 72	121
60 and under 62	14	72 and under 74	36
62 and under 64	60	74 and under 76	7
64 and under 66	187	76 and over	1
66 and under 68	304		

SECTION 2.4

34.   Find the mode of the following set of numbers:

1.3 1.2 1.4 2.1 1.1 2.2 1.1 1.4 1.2 1.3 1.4 2.0 1.2 1.4
1.8 1.4 1.6 1.3 1.9 2.1 1.8 2.3 1.5 1.3 1.4 1.4 1.5

35.   Determine the mode of the set: $(-2.3, -2.0, -1.0, -2.0, 0.2, 0.2, -1.4, -3.2, -2.0, -1.0)$.

36.   Calculate a modal value for the following data, given: (a) the x values are discrete;   (b) the x values are continuous (but rounded to the nearest whole number).

x	2	3	4	5	6	7	8
f	2	1	2	3	1	2	2

37.   Find the mode of the discrete frequency distribution:

x	2.1	2.2	2.3	2.4	2.5	2.6	2.7	2.8	2.9	3.0	3.1	3.2	3.3	3.4
f	1	3	5	15	28	32	37	35	29	13	9	0	0	2

38.   (a) State the modal class; and   (b) calculate (using the interpolation method) the mode of the following distribution:

Marks in an examination.

Number of marks	Number of candidates	Number of marks	Number of candidates
20–24	3	50–54	73
25–29	8	55–59	60
30–34	31	60–64	32
35–39	49	65–69	11
40–44	92	70–74	8
45–49	95	75–79	1

39.   The distribution below gives the life (in hours) of 100 electric light bulbs. Find the mean, median and mode to 1D.

Life (in hours)	No. of bulbs	Life (in hours)	No. of bulbs
0 up to 250	0	1000 up to 1250	25
250 up to 500	3	1250 up to 1500	43
500 up to 750	7	1500 up to 1750	6
750 up to 1000	15	1750 up to 2000	1

40.   Find the geometric mean of the following sets of numbers:

(a) $(2, 3, 4, 5, 6)$ (b) $(1, 3, 5, 7, 9, 11)$ (c) $(1, 2, 4, 8, 16)$
(d) $(3, 9, 27, 81, 243)$

41.   The proportion of a certain government's total expenditure on defence for four successive periods was $0.37, 0.39, 0.42$ and 0.47. Find: (a) the arithmetic mean; and   (b) the geometric mean of these proportions (to 3D).

42.   Calculate the harmonic mean for the following sets:

    (a) $(1, 2)$ (b) $(1, 2, 3)$ (c) $(2, 4, 6, 9)$ (d) $(0.1, 0.2, 0.3)$

43.   The average weekly wages (in £) in a particular factory over five successive weeks were $35.26, 35.83, 34.22, 35.29, 36.04$. Find an appropriate average wage for the complete five week period.

44.   Calculate the three quartiles Q_1, Q_2 and Q_3 for the set $(2, 6, 1, 2, 4, 3, 5, 3, 1, 1, 8)$.

SECTION 2.5

45.   Find the mean deviation (from the mean) of the following sets of numbers:

    (a) $(1, 3, 9, 12)$ (b) $(3, 5, 7, 6, 12, 15, 10, 18)$
    (c) $(8, 8, 9, 8, 9, 9, 18, 3)$

46.   Calculate the mean deviation from: (a) the mean;   (b) the median for the set $(-1, 2, 1, -2, 4)$.

47.   Find the mean deviation from: (a) the mean;   (b) the median for the following distribution:

Class	10 up to 15	15 up to 20	20 up to 25	25 up to 30	30 up to 35	35 up to 40	40 up to 45
Frequency	3	7	16	12	9	5	2

48.   Calculate the mean deviation for the frequency distribution:

Class	2	3	4	5	6
Frequency	1	3	5	4	2

49.   For the set $(2, 2, 3, 4)$, find the mean deviation from: (a) the mean; (b) the median;   (c) the mode.

50.   Determine the mean and median for each of the sets $(1, 1.5, 3.2, 3.6, 1, 2.3, 3.2)$ and $(2.2, 1.7, 2, 3.1, 1.9, 2.4)$. For each of the sets, find the mean deviation from: (a) the mean;   (b) the median.

51.   The heights of a sample of milkmen were found to be as follows:

Height (in ins)	63 up to 65	65 up to 67	67 up to 69	69 up to 71	71 up to 73	73 up to 75	75 up to 77
No. of milkmen	8	21	48	81	32	19	2

Calculate the quantiles: (a) D_3 and D_7;   (b) Q_1 and Q_3;   (c) P_{10} and P_{90} and hence find: (d) the semi-interquartile range;   (e) the 10–90 percentile range.

52.   The weights of 96 children are tabulated below:

Weight (in lbs)	No. of children
100 up to 110	4
110 up to 120	11
120 up to 130	16
130 up to 140	15
140 up to 150	18
150 up to 160	20
160 up to 170	9
170 up to 180	3

Calculate the 10—90 percentile range and the semi-interquartile range.

53.   Find the semi-interquartile range for the distribution:

x	9.3—9.7	9.8—10.2	10.3—10.7	10.8—11.2	11.3—11.7
f	2	5	12	17	14

x	11.8—12.2	12.3—12.7	12.8—13.2
f	6	3	1

54.   Find the mean deviation 34.5 and the semi-interquartile range for the distribution:

x	10—19	20—29	30—39	40—49	50—59
f	3016	6894	9229	5714	3575

x	60—69	70—79	80—89	90 and over
f	1492	170	9	1

55.   In order to compare two distributions, it is often useful to have available measures of dispersion that are independent of the units involved in the two distributions. These are sometimes called 'coefficients of dispersion'. One of these, the *quartile coefficient* is given by $\dfrac{Q_3 - Q_1}{Q_3 + Q_1}$.

Compare the following distributions (w.r.t. dispersion) using this coefficient:

x	f
60—62	5
63—65	18
66—68	42
69—71	27
72—74	8

x	f
100—199	20
200—299	41
300—399	18
400—499	2

56.   Find the semi-interquartile range of the following distribution:

Class	Frequency
Under 8	2
8 up to 10	5
10 up to 12	14
12 up to 14	60
14 up to 16	187
16 up to 18	304

Class	Frequency
18 up to 20	263
20 up to 22	121
22 up to 24	36
24 up to 26	7
26 and over	1

57.   Find the three quartiles and the semi-interquartile range of the distribution:

x	f
12.5 to 13.0	1
13.0 to 13.5	2
13.5 to 14.0	6
14.0 to 14.5	23
14.5 to 15.0	49

x	f
15.0 to 15.5	16
15.5 to 16.0	2
16.0 to 16.5	0
16.5 to 17.0	1

SECTION 2.6

58.   Calculate the mean and standard deviation of the set $(5, 18, 10, 3, 15, 7, 12, 6)$.

59.   Find the standard deviation of the sets:

(a) $(5, 7, 1, 2, 6, 3)$ (b) $(0, 1, 0, 1, 0, 0, 0, 1, 2)$

(c) $(216, 203, 294, 246, 238, 280, 212, 197, 203, 262)$.

60.   Calculate the mean and standard deviation of the digits 0 to 9 inclusive.

61.   The following frequency distribution shows a sample from a certain population by age:

Age (in years)	Percentage of sample	Age (in years)	Percentage of sample
15 up to 20	1.3	40 up to 45	7.8
20 up to 25	8.8	45 up to 50	9.2
25 up to 30	19.0	50 up to 55	7.4
30 up to 35	22.6	55 up to 60	2.3
35 up to 40	19.4	60 up to 65	2.2

Find the mean and standard deviation.

62.   Find the mean and standard deviation of the distribution:

x	10	20	30	40	50
f	6	9	15	12	8

63.   Assuming that $1 + 2 + 3 + 4 + \ldots + n = \dfrac{n(n + 1)}{2}$

and $1^2 + 2^2 + 3^2 + 4^2 + \ldots + n^2 = \dfrac{n(n + 1)(2n + 1)}{6}$:

(a) find the mean and standard deviation of the frequency distribution:

x	1	2	3	4	$\ldots$	$n - 1$	n
f	a	a	a	a	$\ldots$	a	a

(b) explain why the value of a is irrelevant:

(c) verify the answer to (a) putting $n = 5$.

64.   Using an assumed mean of 67 and a scaling factor of 3 calculate the standard deviation of the following distribution:

Class	60–62	63–65	66–68	69–71	72–74
Frequency	5	18	42	27	8

(Note that 67 is the central mid-point and 3 is the common class interval.)

65.   Find the standard deviation of the following distribution of lengths
       (in cm):

Length	5.5 up to 7.5	7.5 up to 9.5	9.5 up to 11.5	11.5 up to 13.5	13.5 up to 15.5
Frequency	2	3	17	43	62

Length	15.5 up to 17.5	17.5 up to 19.5	19.5 up to 21.5	21.5 up to 23.5	23.5 up to 25.5
Frequency	81	70	61	18	3

66.   The weights of a sample of 40 students were found to be distributed
       as follows:

Weight (in lbs)	No. of students	Weight (in lbs)	No. of students
118 to 126	3	154 to 162	5
127 to 135	5	163 to 171	4
136 to 144	9	172 to 180	2
145 to 153	12		

Find, using a suitable coding method, the mean and standard deviation.

67.   Find the mean and variance of the distribution:

x	0	1	2	3	4	5	6	7
f	1	3	17	22	8	6	2	2

68.   The following frequency distribution gives the number of accidents for
       166 drivers over a certain period of time:

No. of accidents	0	1	2	3	4	5	6	7	8
No. of drivers	45	36	40	19	12	8	3	2	1

Find the mean and variance.

69.   (a) Calculate the mean and standard deviation of the diameter of ball
       bearings in Batch I from the following table:

Batch I	
Diameter (cm)	Number of ball bearings
0.160 up to 0.162	1
0.162 up to 0.164	3
0.164 up to 0.166	9
0.166 up to 0.168	20
0.168 up to 0.170	14
0.170 up to 0.172	2
0.172 up to 0.174	1
Total	50

(b) In Batch II, containing 100 ball bearings, the mean diameter was
0.165 cm and the standard deviation of diameter was 0.003 cm.
Calculate the mean and standard deviation of the diameter for Batches
I and II combined.   (IOS)

SECTION 2.7

70. Find the 1st four moments about the mean of the set (3, 4, 7, 7, 4, 5, 2, 4, 6).

71. Find the first three moments: (a) about zero; (b) about the mean of the following distribution:

Class	10 to under 15	15 to under 20	20 to under 25
Frequency	3	7	15
Class	25 to under 30	30 to under 35	35 to under 40
Frequency	20	9	4

72. For the distribution:

x	1	2	3	4
f	2	11	9	1

(a) find the 1st four moments about zero; and (b) if the frequency '9' is changed to '10', find the percentage change in these moments.

MISCELLANEOUS

73. For the following set of numbers (2, 3, 6, 2, 4, 5, 5, 8, 10, 7), find (a) the arithmetic mean $(\bar{x})$; (b) the geometric mean (gm); and (c) the harmonic mean (hm). Verify the relationship $\bar{x} > \text{gm} > \text{hm}$.

74. A tyre manufacturer conducts trials on a particular type of tyre. A sample of 100 tyres is put on test and the distances travelled by the tyres before reaching the legal limit of tyre wear are shown in the following table:

Distance in km	Number of tyres
5 000 to 25 000	8
25 000 to 35 000	14
35 000 to 45 000	24
45 000 to 55 000	26
55 000 to 65 000	16
65 000 to 85 000	12

(a) Plot these data as a histogram. (Note the range of the first and last class intervals.)
(b) Calculate the mean distance, showing all your working.
(c) Obtain the cumulative frequency distribution and estimate the median and interquartile range.

These data were collected by fitting the tyres to the front wheels of a fleet of hire cars of the same model. Suppose that the manufacturer had, instead, tested the tyres by running them on constant speed rollers in a simulated wear trial. Describe the effect you think this would have had on the distribution of distances travelled by the tyres.

Discuss briefly the relative merits of the two methods for examining tyre wear.

(JMB)

75. A firm which manufactures transistors supplies a number of radio repair shops. The numbers sent out during the year are given in the following table:

No. of transistors (x)	50	100	150	200	250	300
No. of shops (f)	10	58	70	40	13	9

Using a suitable scale factor to reduce the size of x, calculate the mean and the standard deviation of this distribution   (Cambridge)

76.   The following frequency distribution was obtained in an experiment:

x	f	
0 and up to 8.0	2	Find:
8.0 and up to 16.0	8	(a) the arithmetic mean;
16.0 and up to 24.0	18	(b) the median (by interpolation);
24.0 and up to 32.0	31	(c) the mode (by interpolation);
32.0 and up to 40.0	12	
40.0 and up to 48.0	4	

Verify also that the relationship   mode = mean — 3(mean — median) is approximately true.

77.   The life of electric light bulbs is tested by lighting 100 and observing at noon each day the numbers still burning or burnt out. The results of the observations are given in the following table:

Time in days since lighting	6	7	8	9	10	11	12	13	14	15
Number still burning	100	98	93	87	73	47	29	16	6	0

Draw a cumulative frequency curve and a histogram to illustrate the data. From your graphs find the median and the seventh decile of the distribution.

Calculate the approximate mean life and estimate the mode from Pearson's formula: (mean — mode) = 3(mean — median).   (O & C)

78.   (a) The following gives the heights of 40 boys measured in centimetres. Revise this table so that the intervals in centimetres are   145—160, 160—165, 165—170, 170—175, 175—185. Illustrate the new frequency table by means of a histogram. Label the axes and state the units used.

(b) Draw a cumulative frequency curve for the data given below and from the curve determine the median height and the interquartile range.

Height (cm)	145—150	150—155	155—160	160—165
Frequency	1	1	4	15
Height (cm)	165—170	170—175	175—180	180—185
Frequency	12	3	3	1

(London)

79.   For the following distribution:

x	60	61	62	63	64	65	66	67	68
f	2	0	15	29	25	12	10	4	3

find: (a) the mean and standard deviation;   (b) assuming the data is continuous (and measured to the nearest whole number), estimate the median.

80.   It is often stated that in frequency distributions there exists the approximate relation: $\dfrac{\text{mean deviation}}{\text{standard deviation}} = 0.8$. Test this statement

in the following distribution. The inner diameters were measured to the nearest 0.001 in.

Inner diameter of cylinders (0.001 in)	90	91	92	93	94	95	96	97	98	99	100
No. of cylinders	2	2	0	1	0	0	0	2	1	2	2

Inner diameter of cylinders (0.001 in)	101	102	103	104	105	106	107	108	109	110
No. of cylinders	2	1	3	2	5	2	1	0	0	2

81.   (a) Find the mean deviation for the following sets of numbers:

(i) 13, 15, 17, 19   (ii) 12, 20, 28, 36   (iii) 3, 5, 7, 9

(b) Associating the variable v with the set in (i), w with the set in (ii) and x with the set in (iii), find the relationship between w and x, v and x and, hence, w and v. Compare these three relationships using the results of (a).

(c) Deduce the following: If the set $x_1, x_2, \ldots, x_n$ has mean deviation of m, then the set $y_1, y_2, \ldots, y_n$ (where the coding $y_i = ax_i + b$ has been used) has mean deviation of am.

82.   The quantile that splits a frequency distribution into five equal parts is called a *quintile*.   (a) Write down expressions for the items (in terms of the total frequency n) that would represent the second and third quintiles of an *ordered* distribution; and   (b) find the values of these two quintiles (by developing appropriate formulae using an extension of Statement 2.8) for the following distribution:

x	0–2	2–4	4–6	6–8	8–10	10–12
f	1	8	19	17	6	1

83.   (a) Draw a cumulative frequency curve (an ogive) for the data given below and use this to estimate: (i) lowest decile;   (ii)   median; (iii) lower quartile;   (iv) upper quartile.

(b) Check your result for the median by an actual calculation.

(c) Use your cumulative frequency curve to estimate the proportion of married couples having a gross weekly income between £22 and £28.

Income of persons over retiring age in 1973. Gross weekly income of married couples

Income	Number of married couples (thousands)
£12 and under £14	144
£14 and under £16	330
£16 and under £18	329
£18 and under £20	247
£20 and under £25	412
£25 and under £30	206
£30 and over	391

(*Source:* Social Trends 1975)   (IOS)

84.   A measure of dispersion (independent of the units in which a distribution is measured) is the *coefficient of mean dispersion* given by $\dfrac{\text{mean deviation}}{\text{arithmetic mean}}$. Compare the following distributions using this measure:

(a)

Class	Frequency
35.0 up to 36.5	3
36.5 up to 38.0	7
38.0 up to 39.5	12
39.5 up to 41.0	26
41.0 up to 42.5	28
42.5 up to 44.0	13
44.0 up to 45.5	9
45.5 up to 47.0	2

(b)

Class	Frequency
59.5 up to 62.5	5
62.5 up to 65.5	18
65.5 up to 68.5	42
68.5 up to 71.5	27
71.5 up to 74.5	8

85.   In an attempt to devise an aptitude test for applicants for assembly work at a light engineering firm, the possibility of using a simple jigsaw puzzle was considered. As a first step, the time taken to complete the jigsaw by a group of the present employees was observed, with the following results:

Time to complete jigsaw (seconds)	Number of employees
10–29	8
30–39	12
40–44	11
45–49	13
50–59	19
60–79	16
80–109	9

(a) Draw a histogram of these data.

(b) Draw a cumulative frequency curve and estimate the median and semi-interquartile range.

(c) Calculate estimates of the arithmetic mean and the standard deviation.

(d) In addition to the employees recorded above, a further four employees attempted the jigsaw but gave up without completing it after 58, 89, 106 and 148 seconds respectively. Discuss, briefly, whether this additional information should be included in the calculation of the four numerical measures above. If it were included, indicate, without doing any further calculations, the effect it might have on the estimates above.   (AEB)

86.   By writing $\sqrt[n]{x_1 x_2 \ldots x_n}$ as $(x_1 x_2 \ldots x_n)^{1/n}$ and taking logarithms in the formula for the geometric mean (gm), show that the logarithm of the gm is the mean of the logarithms of the original values.

87.   If the values $(x_1, x_2, \ldots, x_n)$ occur with respective frequencies $(f_1, f_2, \ldots, f_n)$ with $\Sigma f_i = N$, derive a formula for the geometric mean of a frequency distribution.

88. Assuming that:

$$\sum_{i=1}^{n} i = \frac{n(n+1)}{2}; \quad \sum_{i=1}^{n} i^2 = \frac{n(n+1)(2n+1)}{6} \quad \text{and}$$

$$\sum_{i=1}^{n} i^3 = \frac{n^2(n+1)^2}{4}$$

show that the standard deviation of the frequency distribution:

x	1	2	3	4	...	$n-1$	n
f	1	2	3	4	...	$n-1$	n

is $\sqrt{\dfrac{(n-1)(n+2)}{18}}$

Verify this result when $n = 4$.

89. The variable x takes the values $a, a+d, a+2d, \ldots, a+(n-1)d$ (an arithmetic progression). Given that the sum of the first n natural numbers is $\dfrac{n(n+1)}{2}$ and the sum of the squares of the first n natural numbers is $\dfrac{n(n+1)(2n+1)}{6}$:   (a) find $\bar{x}$ by using the coding $X = x - a$;   (b) calculate the sum of the squares of the X values; and hence   (c) evaluate s_x^2, the variance of x.

90. Discuss the differences between discrete and continuous data with respect to presentation and the calculation of statistical measures of location such as the median and mode. Given the following frequency distribution:

Class	1	2	3	4	5	6	7
Frequency	2	5	12	23	18	7	1

calculate the median if: (a) the data is known to be discrete;   (b) the data is known to be continuous and measured to the nearest whole number.

91. Calculate the arithmetic mean and standard deviation of the following frequency distribution of ages (estimated) in the United Kingdom as at 30th June 1973.

Age (in yrs)	Frequency (in millions to 1D)	Age (in yrs)	Frequency (in millions to 1D)
Under 5	4.3	45–49	3.4
5–9	4.7	50–54	3.6
10–14	4.4	55–59	3.1
15–19	4.0	60–64	3.2
20–24	4.0	65–69	2.8
25–29	4.1	70–74	2.1
30–34	3.3	75–79	1.4
35–39	3.2	80–84	0.8
40–44	3.3	85 and over	0.5

(Data source: MDS)

92. By expanding $\Sigma f(x - \bar{x})$ (using the properties of the Σ operator), show that the first moment about the mean (m_1) for any frequency distribution is zero.

93. Show that $\Sigma f(x - \bar{x})^4$ can be written as:

$$\sum fx^4 - 4\bar{x} \sum fx^3 + 6\bar{x}^2 \sum fx^2 - 4\bar{x}^3 \sum fx + \bar{x}^4 \sum f$$

and hence prove that for any frequency distribution
$$m_4 = m_4' - 4m_1'm_3' + 6(m_1')^2 m_2' - 3(m_1')^4$$

94. A set $(x_1, x_2, \ldots, x_n)$ is transformed to the set $(X_1, X_2, \ldots, X_n)$ by using the coding $X = \dfrac{x - a}{b}$.

(a) Show that x^3 can be written as: $a^3 + 3a^2 bX + 3ab^2 X^2 + b^3 X^3$ and hence;

(b) $m_3' = a^3 + 3a^2 bM_1' + 3ab^2 M_2' + b^3 M_3'$ where $M_r' = \dfrac{\sum X^r}{n}$; and

(c) $m_3 = b^3 M_3$ where $M_3 = \dfrac{\sum (X - \bar{X})^3}{n}$.

95. Given the eleven numbers $2, 0, 1, 2, 1, 3, 0, 5, 8, 4, 7$, find the following statistics concerned with the location and spread of these numbers: (a) the mean;   (b) the median;   (c) the mean deviation from the mean;   (d) the standard deviation.

Comment briefly on the difference between your answers for the mean and the median.

When a twelfth number is added, the mean of the twelve numbers is 0.25 greater than the mean of the original eleven numbers. Calculate the twelfth number.   (London)

96. In the following table x is the number of grams of impurity in one-litre containers of a chemical solution.

x	0–25	26–50	51–75	76–100
f	20	73	85	114
x	101–125	126–150	151–175	176–200
f	106	54	36	12

By means of a cumulative frequency graph estimate the median impurity content and the tenth and ninetieth percentiles.   (Cambridge)

97. The following table gives an analysis by numbers of employees of the size of UK factories of less than 1000 employees manufacturing clothing and footwear.

Number of employees	11–19	20–24	25–99	100–199
Number of factories	1500	800	2300	700
Number of employees	200–499	500–999	Total	
Number of factories	400	100	5800	

Calculate as accurately as the data allow the mean and median of this distribution, showing your working. If 90% of the factories have less than N employees, estimate N.   (O & C)

98.   (a) Explain the term *weighted average*.

The following data relate to the basic weekly page paid in a factory to different types of workers.

Type of Worker	Number of Workers	Basic Weekly Wage
Skilled	110	£62.50
Semi-skilled	140	£56.50
Unskilled	170	£48.00

Find for the factory: (i) an unweighted weekly average wage;   (ii) a weighted weekly average wage.

(b) Find: (i) the mode;   (ii) the geometric mean, of the observations 8, 4, 2, 3, 4, 5, 6, 7.

                                                                        (IOS)

99.   Number of rooms per dwelling in the UK — 1961 and 1971.

	Percentages	
Number of rooms	1961	1971
1	1	2
2	5	4
3	12	9
4	27	23
5	35	30
6	13	23
7	3	5
8 or more	4	4

(*Source:* Office of Population Censuses and Surveys).

(a) For each distribution calculate the mean, standard deviation and coefficient of variation of the number of rooms per dwelling.

(b) Comment on your results.   (IOS)

Chapter 3

Regression and Correlation

3.1 INTRODUCTION TO BIVARIATE DATA

Up to now we have considered distributions (or populations) composed of only one variable. For example, populations of heights, weights, lengths, or ages, where every member of any one population is measured in terms of one and the same characteristic.

We now consider populations where each member is specifically described in terms of two characteristics.

Consider the following experimental data:

Experimental unit (individual)	Height (in ins.)	Weight (in lbs.)
A	71	166
B	87	162
C	70	163
D	68	148
E	76	179
F	75	168
G	67	151
H	71	176
J	74	154

For this set of data, each observation (experimental unit) has two measurements associated with it, in this particular case height and weight. We refer to this type of data as *bivariate* (two variables) as opposed to the type of data we have considered in Chapters 1 and 2 which is termed *univariate* (one variable).

The following table shows some other possible examples of sets of bivariate data.

Unit	Characteristics to be measured
Student	Height; age
Fish	Length; breadth
Tree	Height; root spread (average diameter)
Cat	Tail length; body length (excluding tail)
Flower	Stem length; stamen length (average)

Of course, we do not have to stop at data having only two measurable characteristics for each unit; we could have three or more. For example, if we were collecting data concerning housewives, we might sensibly consider the following influencing factors: husband's income, number of children, number of years married, value of house, and so on. This type of data is known as *multivariate* (each observation is measured in terms of many characteristics).

The aim in this chapter is first to sort and present bivariate data in a logical fashion and, secondly, to introduce appropriate statistical measures in order to characterise and compare bivariate distributions. A similar course was followed in Chapters 1 and 2 where we were dealing with data in a strictly univariate form.

3.1.1 SCATTER DIAGRAMS AND CORRELATION

The most common and convenient method of displaying a set of bivariate data is by means of a *scatter diagram*. This entails treating the bivariate pairs as sets of (x, y) coordinates and plotting them on a graph to obtain a set of points.

The scatter diagram for the previously quoted data of the heights and weights of nine individuals is shown in Fig. 3.1 where 'weight' is taken as the x-coordinate and 'height' as the y-coordinate.

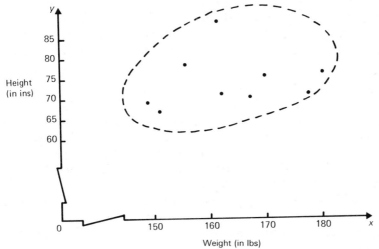

Fig. 3.1

In this particular diagram, notice that the observation pairs can be contained within an elliptical shaped boundary with smaller y (height) measurements tending to be paired off with smaller x (weight) measurements; larger y values with larger x values. Two variables related in this way are said to be *positively* (or *directly*) *correlated*.

On the other hand, consider the following data and its associated scatter diagram (Fig. 3.2):

Individual	1	2	3	4	5	6	7	8	9	10	11	12	13
Mathematics score	62	78	42	70	61	53	84	73	54	80	62	59	43
English score	58	41	63	39	47	58	62	63	85	59	71	68	78

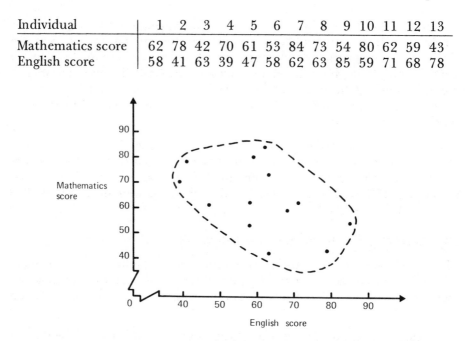

Fig. 3.2

In this diagram we can again enclose the observations within an approximate ellipse, although it is orientated differently to the previous one. Here, a small score at Mathematics tends to be associated with a large score at English and vice versa. This is known as *negative* (or *inverse*) *correlation*.

The concept of correlation, then, is concerned with a particular type of relationship between any two variables under discussion. In the two bivariate sets of data already considered the respective scatter diagrams helped to show fairly marked relationships between the two variables involved in each case. However, we can have a bivariate set of data in which one variable will bear little relationship to the other. For instance, consider measuring the variables 'height' and 'intelligence quotient' for a particular arbitrary set of people. Here, we would be very surprised indeed if these two variables showed any relationship whatsoever, since we know from experience that a person's height has nothing to do with his intelligence (and vice versa).

In this particular type of case (where the variables are not related in any way) we would obtain a scatter diagram of the sort shown in Fig. 3.3. Notice here that all the observations are contained within an approximate circle. High intelligence quotients are drawn from both tall and short people. Also, a tall person might have a high, medium or low intelligence quotient.

No correlation is said to exist between variables related in this way.

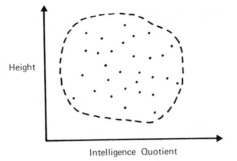

Fig. 3.3

EXAMPLE 3.1

Draw a scatter diagram for the following sample of bivariate data.

x	42	24	82	74	70	36	57	29	63	74	80	30
y	136	141	133	135	120	142	130	153	128	112	124	146

Using the diagram, state how the variables x and y are correlated.

SOLUTION

The scatter diagram for the given data is as follows.

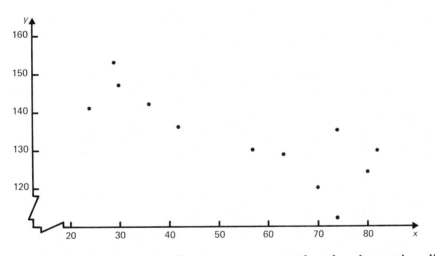

From the above scatter diagram, we can see that the observations lie roughly within an elliptical boundary inclined in such a way that high values of x are associated with low values of y and vice versa. We thus have that x and y are negatively correlated.

EXAMPLE 3.2

Display the following set of bivariate data in the form of scatter diagrams using different scales and comment on the results.

x	10	12	15	14	12	10	10	11	14
u	32	46	62	60	51	40	38	42	56

SOLUTION

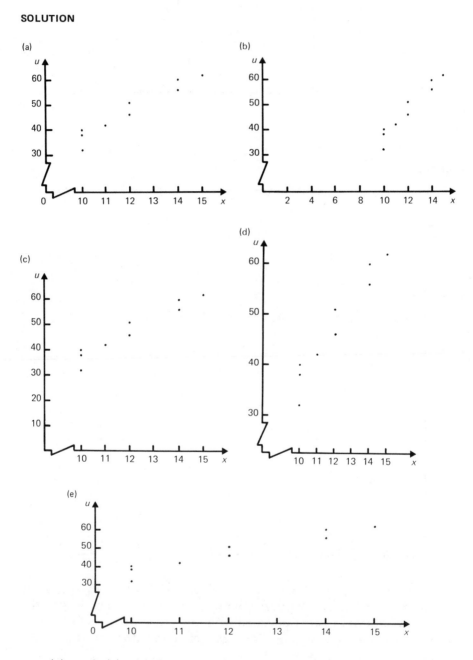

(a) and (c) simply move the cluster of points in space without changing their basic shape. However, (b), (d), and (e) change both the orientation and shape of the cluster and it is clear that, depending on how we choose the scales to plot the points, the picture appears to change (certainly to an untrained eye). The difference between (d) and (e) is quite startling. Of course, no matter which of the above diagrams is chosen to display the data, the correlation between the two variables is clearly shown to be positive.

However, one way of 'standardising the picture' shown by a scatter diagram is to ensure that the range of values covered by each variable occupy the same length on their respective axes. For the given data (the x-range is 5, the y-range is 30) we would ensure that 5 units on the x scale is represented by the same length as 30 units on the y scale, which has been done in (a).

Statistically, we measure the correlation between two variables as a number between -1 and $+1$ inclusive, where:

$$\begin{cases} +1 & \text{signifies 'perfect' positive correlation} \\ 0 & \text{signifies no correlation} \\ -1 & \text{signifies 'perfect' negative correlation} \end{cases}$$

An appeal to scatter diagrams will help to demonstrate how we might allocate numbers from the range -1 to $+1$ to particular sets of bivariate data.

Where no correlation existed, we saw that we could roughly enclose the points on a scatter diagram within a circle (see Fig. 3.4(a)). Suppose that the points fell within ellipses defined by 'squashing' the circle in the direction shown by the arrows in Figs. 3.4(b) and (c). Then, depending on the degree of 'squash', we could allocate an approximate number between 0 and 1 to describe the correlation. This could be done practically by dividing the 'width' of the ellipse by its 'length' and subtracting this quantity from 1. Clearly, the ideal state (of perfect positive correlation) would be obtained when the circle has been finally squashed into a straight line as in Fig. 3.4(d).

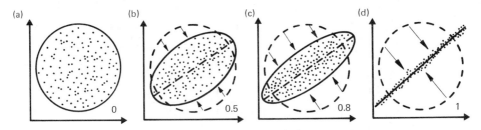

Fig. 3.4

Fig. 3.5 shows an analogous situation for allocating numbers between 0 and -1 to varying degrees of negative correlation.

Notice that in Fig. 3.5(d) the circle has been squashed into a straight line representing the 'perfect' state of negative correlation with a coefficient of -1.

We note at this point however that, although the use of an encompassing ellipse to ascertain a degree of correlation between two variables is useful for demonstrating a general concept, other more sophisticated (and objective) techniques exist as we will see in later sections.

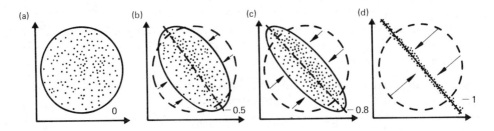

Fig. 3.5

3.1.2 BIVARIATE FREQUENCY DISTRIBUTIONS

We saw in Chapter 1 how to form frequency distributions for large sets of univariate data using a tally chart. Exactly the same type of considerations apply when we wish to present a large set of bivariate data. The analogue of a univariate tally chart, a bivariate tally chart, is shown as follows, the data consisting of 100 students' marks in both a Mathematics and French test.

		Mathematics mark				
		Below 50	50 up to 60	60 up to 70	70 up to 80	80 and above
	80 and above			1111	⊥⊥⊣1 11	111
	70 up to 80		1	⊥⊥⊣1 11	⊥⊥⊣1 ⊥⊥⊣1	⊥⊥⊣1 111
French mark	60 up to 70	11	1111	⊥⊥⊣1 ⊥⊥⊣1 ⊥⊥⊣1 1	⊥⊥⊣1 1	11
	50 up to 60	111	⊥⊥⊣1 111	⊥⊥⊣1 ⊥⊥⊣1	1	
	Below 50	1	⊥⊥⊣1	11		

A bivariate frequency distribution table can now be drawn up as follows.

	Below 50	50 up to 60	Mathematics 60 up to 70	70 up to 80	80 and above	Total
80 and above	0	0	4	7	3	14
70 up to 80	0	1	7	10	8	26
French 60 up to 70	2	4	16	6	2	30
50 up to 60	3	8	10	1	0	22
Below 50	1	5	2	0	0	8
Total	6	18	39	24	13	100

3.2 RELATED MATHEMATICS

3.2.1 THE EQUATION OF A STRAIGHT LINE

STATEMENT 3.1

The equation of a straight line (a *linear equation*) can always be written in the form $y = mx + c$ where m and c are constants.

Note m is called the *gradient* and c the *y-intercept* of the straight line.

Consider the equation $y = 2x + 3$ (here, $m = 2$ and $c = 3$).

Taking a few values of x, we can find the corresponding values of y defined by this equation.

For
$$x = -2, \ y = 2(-2) + 3 = -1$$
$$x = -1, \ y = 2(-1) + 3 = \ \ \ 1$$
$$x = \ \ \ 0, \ y = 2(0) \ \ + 3 = \ \ \ 3$$
$$x = \ \ \ 1, \ y = 2(1) \ \ + 3 = \ \ \ 5$$
and
$$x = \ \ \ 2, \ y = 2(2) \ \ + 3 = \ \ \ 7$$

Plotting these five (x, y) points on a graph, we see that by joining them together they form a straight line. This is the straight line defined by the equation $y = 2x + 3$ (see Fig. 3.6).

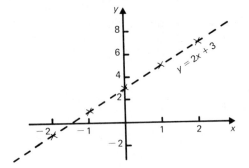

Fig. 3.6

If we had chosen different values of x, found the corresponding values of y and plotted these points, we would have obtained exactly the same straight line.

The equation of a straight line also has the following significance.

STATEMENT 3.2

Any point lying on a straight line satisfies the equation of that line. That is, if the point (h, k) lies on the line $y = mx + c$ then $k = mh + c$.

Consider the line:

$$y = 3x - 2 \tag{1}$$

and the two points $A(2, -3)$ and $B(1, 1)$.

For point A, $x = 2$ and $y = -3$. Substituting these values of x and y into [1] gives $-3 = 3(2) - 2$, i.e., $-3 = 4$ *which is untrue* (that is, the equation is not satisfied).

Therefore point A *does not* lie on the given line.

For point B, $x = 1$ and $y = 1$. Substitution into [1] gives $1 = 3(1) - 2$, i.e., $1 = 1$ *which is true* (that is, the equation is satisfied).

Therefore point B *does* lie on the given line.

Our choice of the x-values $-2, -1, 0, 1$ and 2 in plotting the line $y = 2x + 3$ (see Fig. 3.6) was purely arbitrary and it can be seen from the plotted points, together with the graph, that any two of the above x-values (together with their corresponding y-values) would equally well have defined the straight line. That is, it was unnecessary to select 5 points to graph the line, since only 2 are ever needed. Hence:

STATEMENT 3.3

Any two points plotted on a frame as (x, y) coordinates define a unique straight line.

EXAMPLE 3.3

Plot the graph of the straight line $y = 2 - x$.

SOLUTION

Notice that the equation is in the form $y = mx + c$, where, here, $m = -1$ and $c = 2$.

We choose two convenient x-values, -2 and 3.

$$x = -2 \quad \text{gives} \quad y = 2 - (-2) = 4$$
$$x = 3 \quad \text{gives} \quad y = 2 - 3 = -1$$

giving us the two defining points $(-2, 4)$ and $(3, -1)$ for the line.

Plotting these two points on an (x, y) frame and joining them with a straight line gives the required straight line graph.

Note that the straight line meets the y-axis at $y = 2$, confirming the c value (y intercept) of 2.

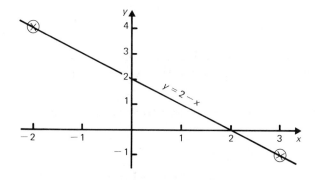

We now look at the problem of determining the equation of a straight line given only its graph. In effect, this means determining the values of m and c, so that we can write the line in the form $y = mx + c$.

Usually, finding the value of c is straightforward, since we simply need to read off the value of y where the line meets the y-axis. Determining m, the gradient, is a little more complicated.

DEFINITION 3.1

The gradient of the unique line joining two given points is defined as 'the distance travelled in the y direction' divided by 'the distance travelled in the x direction' in moving from one point to the other.

So that, given the graph of a straight line, we need only choose two points on it using these to determine the gradient of the line. An example will illustrate this clearly.

We will find the gradient of the line joining the two points A(2, 1) and B(4, 3). See Fig. 3.7.

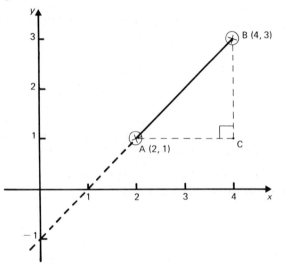

Fig. 3.7

The point C has been introduced (forming a right-angled triangle with A and B) in order to determine the x and y distances travelled more easily.

Travelling from A to B is equivalent to travelling along AC (x direction) and then along CB (y direction).

Here, AC = 2 and CB = 2.

$\therefore$ $$\text{gradient} = \frac{\text{distance travelled in } y \text{ direction}}{\text{distance travelled in } x \text{ direction}}$$

$$= \frac{\text{CB}}{\text{AC}} = \frac{2}{2} = 1$$

Now, if we wish to find the equation of the line in the form $y = mx + c$, we have $m = 1$. By extending BA (the dotted line in the figure) we obtain the y-intercept as -1, i.e., $c = -1$.

Hence, the equation of the line is $y = x - 1$.

Note that the same value of m would have been obtained if we had travelled from B to A, rather than from A to B. In this case we would travel along BC then CA, where $BC = -2$ and $CA = -2$ (since we would be travelling in the *negative* x and y directions).

$$\therefore \qquad \text{gradient} = \frac{BC}{CA} = \frac{-2}{-2} = 1$$

i.e., $m = 1$, as before.

EXAMPLE 3.4

Find the equation of the line (through points A and B) whose graph is given:

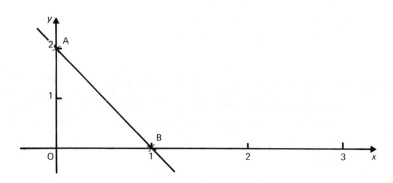

SOLUTION

We see immediately that the y-intercept is 2. That is, $c = 2$. We need to find m, in order to write the line in the form $y = mx + c$.

Now A has coordinates $(0, 2)$ and $B(1, 0)$. Using these two points to measure the gradient (and calling the origin of coordinates, O), we have:

$$\text{in moving from A to B, } y \text{ distance} = AO = -2$$

$$x \text{ distance} = OB = +1$$

$$\therefore \qquad \text{gradient} = m = \frac{AO}{OB} = \frac{-2}{1} = -2$$

Therefore the equation of the line is $y = -2x + 2$ or $y = 2 - 2x$

EXAMPLE 3.5

A variable straight line has equation $y = mx + 2$. Plot the lines for (a) $m = 0$ (b) $m = 1$ and (c) $m = 2$ on the same diagram.

SOLUTION

(a) For $m = 0$, the equation becomes $y = 2$.

(b) For $m = 1$, we have $y = x + 2$. When $x = 0$, $y = 2$ and when $x = 1$, $y = 3$.

(c)   For $m = 2$ we have $y = 2x + 2$. When $x = 0$, $y = 2$ and when $x = 1$, $y = 4$. The three lines are plotted in the adjoining diagram.

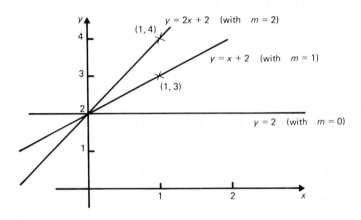

Notice that all the lines meet the y-axis at $y = 2$, the y-intercept point, which shows that different values of m (in the equation $y = mx + c$) simply 'swing' the line around about the point $(0, c)$ — in this case, $(0, 2)$.

Some manipulations of linear equations, necessary for the next section, are now mentioned.

STATEMENT 3.4

(a)   Any linear equation may be multiplied (or divided) throughout by any number without altering the graph of the straight line it represents.

(b)   Any two linear equations may be added or subtracted algebraically.

By way of demonstration of the above:

(a) $x + y = 1$ is the same as $2x + 2y = 2$
            (multiplying through by 2)

            $3x - y = 4$ is the same as $3y - 9x = -12$
            (multiplying through by -3)

            $4x - 8y = 16$ is the same as $x - 2y = 4$
            (dividing through by 4)

(b)   Consider the two lines

                    $x + y = 1$ [1]

and $3x - y = 4$ [2]

Adding these two equations gives:

$$(x + 3x) + (y - y) = 1 + 4$$

i.e. $$4x = 5$$

Also subtracting [2] from [1] gives:

$$-2x + 2y = -3$$

or subtracting [1] from [2] gives:

$$2x - 2y = 3$$

That is, if both $x + y = 1$ and $3x - y = 4$ are true relationships between x and y then so are $4x = 5$, $-2x + 2y = -3$ and $2x - 2y = 3$.

3.2.2 THE SOLUTION OF LINEAR, SIMULTANEOUS EQUATIONS

In this section we discuss methods to solve two linear (straight line) equations simultaneously. That is, given two straight line equations, we will find a value each for x and y that solves (satisfies) both equations.

Consider the following pair of linear equations:

$$\begin{cases} x + 2y = 12 & \text{[A]} \\ x - 3y = 2 & \text{[B]} \end{cases}$$

By subtracting equation [B] from equation [A] we obtain:

$$(x - x) + (2y - (-3y)) = 12 - 2$$

i.e., $5y = 10$. Therefore $y = 2$.

Here, we have *eliminated* x from the two equations, enabling us to solve for y.

We can now substitute this value of y $(y = 2)$ back into one of the original equations, [A] say, giving:

$$x + 2(2) = 12 \Rightarrow x = 12 - 4 = 8$$

Hence, the solution of the two equations is $x = 8, \ y = 2.$

(We can check this solution using equation [B],

i.e. $$8 - 3(2) = 8 - 6 = 2 \ \checkmark)$$

Consider the following pair of equations:

$$\begin{cases} x - 10y = 5 & \text{[D]} \\ 2x + 10y = 40 & \text{[E]} \end{cases}$$

Here, adding both equations gives $3x = 45$ (eliminating y), i.e., $x = 15$.

Substituting this value of x into [D] gives $15 - 10y = 5$, i.e., $y = 1$.

Therefore solution is $x = 15, \ y = 1.$

In general, a pair of linear equations will not be as 'convenient' as the above to solve for x and y.

For example, adding or subtracting the two equations:

$$\begin{cases} 2x + y = 5 & [1] \\ x + 3y = 5 & [2] \end{cases}$$

does not eliminate one of the variables immediately.

$$[1] + [2] \Rightarrow 3x + 4y = 10 \text{ and } [1] - [2] \Rightarrow x - 2y = 0$$

both of these equations containing both x and y.

We need to notice that the 'x term' in [1] is just twice the 'x term' in [2]. So that if we multiply [2] through by 2 we obtain $2x + 6y = 10$.

The two equations now stand as:

$$\begin{cases} 2x + y = 5 \\ 2x + 6y = 10 \end{cases}$$

Subtracting the second equation from the first eliminates x and we can then solve as usual.

Alternatively, noticing that the 'y term' in [2] is just 3 times the 'y term' in [1], we could multiply equation [1] through by 3 giving the two equations as:

$$\begin{cases} 6x + 3y = 15 \\ x + 3y = 5 \end{cases}$$

Upon subtraction the y term is eliminated and we can proceed to solve as usual.

Finally, we give a case where neither of the two previous methods can be employed to obtain a solution. Consider the equations:

$$\begin{cases} 7x - 3y = 31 & [3] \\ 9x - 5y = 41 & [4] \end{cases}$$

Here we notice that:

(a) the equations cannot be immediately added or subtracted (since we would still obtain an equation in both x and y),

(b) No one equation can be multiplied through by any number to obtain the same x or y terms.

However, multiplying [3] through by 5 and [4] through by 3 gives:

$$\begin{cases} 35x - 15y = 155 \\ 27x - 15y = 123 \end{cases}$$

(giving the y terms in both equations the same number). We can now subtract one equation from the other to eliminate the y term completely.

Alternatively, multiplying [3] through by 9 and [4] through by 7 gives:

$$\begin{cases} 63x - 27y = 279 \\ 63x - 35y = 287 \end{cases}$$

(giving the x terms in both equations the same number). Upon subtraction, the x term will be eliminated.

The particular technique used here is a matter of choice, but the first would generally be thought better since it involves smaller numbers.

We summarise the method of solving linear simultaneous equations as follows.

STATEMENT 3.5

Given two linear equations in the form:

$$\begin{cases} a_1 x + b_1 y = c_1 \\ a_2 x + b_2 y = c_2 \end{cases}$$

we solve them for x and y by:

(a)   *Eliminating one of the variables* by:
either   (i)   adding or subtracting the two equations immediately.
or   (ii) multiplying one of the equations through by a suitable number and then adding or subtracting the two equations;
or   (iii) multiplying each of the equations through by suitable numbers and then adding or subtracting the two equations.

(b)   Solving the equation obtained for the second variable.

(c)   Substituting back into one of the original equations to solve for the remaining variable.

(d)   Checking these two solutions by using the second of the original two equations.

This particular method of solving linear simultaneous equations is known as the *method of elimination*, and although other methods are available, this particular one is the simplest and most suitable for the purposes of this text.

We further demonstrate the technique with some worked examples.

EXAMPLE 3.6

Solve:

$$\begin{cases} y = 3x + 5 \\ 2y + 3x = 9 \end{cases} \quad \text{for } x \text{ and } y$$

SOLUTION

We first express the former equation in a more convenient form, giving:

$$\begin{cases} y - 3x = 5 & [1] \\ 2y + 3x = 9 & [2] \end{cases}$$

By adding [1] and [2] we obtain:

$$(y + 2y) + (-3x + 3x) = 5 + 9$$

i.e., $3y = 14$ (eliminating x). Therefore $y = \dfrac{14}{3}$.

Substituting this value of x into [1] gives:

$$\frac{14}{3} - 3x = 5$$

$$\therefore \quad 3x = \frac{14}{3} - 5 \Rightarrow x = \frac{1}{3}\left(\frac{14}{3} - 5\right) \Rightarrow x = -\frac{1}{9}$$

Thus

$$x = -\frac{1}{9}, \quad y = \frac{14}{3}$$

$$\left(\text{Check in [2]}: \; 2\left(\frac{14}{3}\right) + 3\left(-\frac{1}{9}\right) = \frac{28}{3} - \frac{1}{3} = \frac{27}{3} = 9 \; \checkmark. \right)$$

EXAMPLE 3.7

Solve the pair of equations

$$\begin{cases} 7x - 3y = 41 & [1] \\ 3x - y = 17 & [2] \end{cases}$$

for x and y.

SOLUTION

Notice that multiplying the second equation by 3 gives both equations the term $3y$.

We have:

$$7x - 3y = 41 \qquad [1]$$

$$9x - 3y = 51 \qquad [3]$$

$$[3] - [1] \Rightarrow 2x = 10 \Rightarrow x = 5$$

Substituting into [1] $\Rightarrow 7(5) - 3y = 41$:

$$\therefore \quad 3y = 35 - 41 \qquad \therefore \quad y = \frac{-6}{3} = -2$$

So that the solution is:

$$x = 5, \quad y = -2$$

(Check in [2]: $3(5) - (-2) = 15 + 2 = 17$ $\checkmark$.)

EXAMPLE 3.8

If $4x - 3y = 14$ and $3x - 4y = 0$,   find the values of x and y.

SOLUTION

$$4x - 3y = 14 \qquad [1]$$
$$3x - 4y = 0 \qquad [2]$$

Multiplying [1] by 3 and [2] by 4 gives a common term of $12x$ in both equations.

We have: $12x - 9y = 42$ [3]

and $12x - 16y = 0$ [4]

$$[3] - [4] \Rightarrow 7y = 42 \qquad \therefore \quad y = 6$$

Substituting in [1]:

$$4x - 3(6) = 14 \Rightarrow 4x = 14 + 18 = 32$$

i.e. $x = \dfrac{32}{4} = 8$

Therefore the solution is:

$$x = 8, \quad y = 6$$

(Check in [2]: $3(8) - 4(6) = 24 - 24 = 0$ $\checkmark$.)

We note finally the graphical significance of the solution of linear simultaneous equations. Each of the linear equations represents the graph of a straight line, and if these are plotted on the same diagram, the (x, y) coordinates of their point of intersection correspond to the x and y values found by solving the two equations simultaneously.

3.2.3 DIFFERENTIATION 1: THE DIFFERENTIAL OPERATOR

Differentiation can be described as a process that changes one function (of a particular variable) into another function (of the same variable) according to a particular, well-defined rule. Before stating this rule we give some simple examples of differentiation together with some notation.

(a)   The differential (with respect to x) of the function $6x^4$ is $24x^3$. (Note that $24 = 6 \times 4$ and $3 = 4 - 1$).

We write $\dfrac{d}{dx}(6x^4) = 24x^3$ where $\dfrac{d}{dx}$ is called the *differential operator*.

(b) $\dfrac{d}{dx}(7x^5) = 35x^4$ $(35 = 7 \times 5; \quad 4 = 5 - 1)$

(c) $\dfrac{\mathrm{d}}{\mathrm{d}x}(\frac{1}{2}x^8)$ $= 4x^7$ $(4 = \frac{1}{2} \times 8;\quad 7 = 8 - 1)$

(d) $\dfrac{\mathrm{d}}{\mathrm{d}x}(3x^{-4})$ $= -12x^{-5}$ $(-12 = 3 \times -4;\quad -5 = -4 - 1)$

DEFINITION 3.2

A function of x taking the form ax^b, where a and b are any two numbers (positive or negative, integral or fractional), is referred to as a *simple function of* x.

For example, $7x^2$, $-2x^{1/2}$, $0.76x^{-2.5}$ and $\frac{3}{8}x^{-5/8}$ are all simple functions of x.

Note   This definition is not general in mathematics; we introduce it here in order to present differentiation in the most concise way.

We are now in a position to define the process of differentiation.

DEFINITION 3.3

The *derivative* of the simple function ax^b with respect to x (w.r.t. x) is abx^{b-1}. We say that we have *differentiated* ax^b w.r.t. x to obtain abx^{b-1}.

We write $\dfrac{\mathrm{d}}{\mathrm{d}x}(ax^b) = abx^{b-1}$

Note a and b can be any numbers (positive or negative, integral or fractional).

The student should check the results of the previous examples (a) to (d) with the above definition.

In order to extend the range of functions that we are able to differentiate, we have the following statement.

STATEMENT 3.6

(a)   The derivative of a constant (a single number, independent of x) is zero, i.e., if c is a constant, then $\dfrac{\mathrm{d}}{\mathrm{d}x}(c) = 0$.

(b)   The derivative of a function that is expressible as the sum of any number of simple functions is just the sum of the separate derivatives of each function,   (*continued*)

---

i.e. $\dfrac{d}{dx}(a_1 x^{b_1} + a_2 x^{b_2} + \ldots + a_n x^{b_n})$

$$= \frac{d}{dx}(a_1 x^{b_1}) + \frac{d}{dx}(a_2 x^{b_2}) + \ldots + \frac{d}{dx}(a_n x^{b_n})$$

where $a_1, a_2, \ldots, a_n$ and $b_1, b_2, \ldots, b_n$ are any numbers.

For example,

$$\frac{d}{dx}(5) = \frac{d}{dx}(100) = \frac{d}{dx}(-8) = 0$$

and $\dfrac{d}{dx}(3x^6 + 5x^{-2}) = \dfrac{d}{dx}(3x^6) + \dfrac{d}{dx}(5x^{-2})$

$$= 18x^5 + (-10x^{-3}) = 18x^5 - 10x^{-3}$$

EXAMPLE 3.9

Differentiate: (a) $5x^4$ (b) $16x^{\frac{1}{2}}$ (c) $\frac{1}{4}x^{-\frac{4}{3}}$ w.r.t. x.

SOLUTION

(a) $\dfrac{d}{dx}(5x^4) = (4)(5)x^{4-1} = 20x^3$

(b) $\dfrac{d}{dx}(16x^{\frac{1}{2}}) = (16)(\frac{1}{2})x^{\frac{1}{2}-1} = 8x^{-\frac{1}{2}}$

(c) $\dfrac{d}{dx}(\frac{1}{4}x^{-\frac{4}{3}}) = (\frac{1}{4})(-4/3)x^{-\frac{4}{3}-1} = -\frac{1}{3}x^{-\frac{7}{3}}$

EXAMPLE 3.10

Evaluate:

(a) $\dfrac{d}{dx}(0.2x^{0.9})$ (b) $\dfrac{d}{dx}(2x^{-2})$ and (c) $\dfrac{d}{dy}(-3y^4)$

SOLUTION

(a) $\dfrac{d}{dx}(0.2x^{0.9}) = (0.2)(0.9)x^{0.9-1} = 0.18x^{-0.1}$

(b) $\dfrac{d}{dx}(2x^{-2}) = 2(-2)x^{-2-1} = -4x^{-3}$

(c) $\dfrac{d}{dy}(-3y^4) = -3(4)y^{4-1} = -12y^3$

Notice that the operator $\dfrac{d}{dy}$ means 'differentiate w.r.t. y' and operates on any function of y.

EXAMPLE 3.11

Differentiate the following functions w.r.t. the appropriate variable:

(a) $4x^3 + \frac{1}{2}x^{-2}$

(b) $4x^4 + 6x^3 - \frac{1}{2}x^2 + 3x - 10$

(c) $-0.2t^{-3.3} + 3.9t^{-0.2} - 4.2t^{0.4}$

SOLUTION

(a) $\dfrac{d}{dx}(4x^3 + \frac{1}{2}x^{-2}) = 4(3)x^{3-1} + (\frac{1}{2})(-2)x^{-2-1}$

$$= 12x^2 - x^{-3}$$

(b) $\dfrac{d}{dx}(4x^4 + 6x^3 - \frac{1}{2}x^2 + 3x - 10) = 16x^3 + 18x^2 - x + 3$

(c) $\dfrac{d}{dt}(-0.2t^{-3.3} + 3.9t^{-0.2} - 4.2t^{0.4})$

$$= 0.66t^{-4.3} - 0.78t^{-1.2} - 1.68t^{-0.6}$$

We have already seen two different notations for differentiation:

(a)   Stated in words, i.e., 'differentiate $3x^2$ w.r.t. x'.

(b)   The differential operator notation, i.e., $\dfrac{d}{dx}(3x^2)$.

There are two other forms used, usually in specific contexts:

(c)   If we label a function $f(x)$ or $g(y)$, for example, then we would write their derivatives as $f'(x)$ and $g'(y)$ respectively,

i.e. $f'(x) = \dfrac{d}{dx}(f(x))$ and $g'(y) = \dfrac{d}{dy}(g(y))$

(d)   We sometimes wish to label a function by a letter, for example, $y = 3x^2 + 4x$. This form is usually used for the purposes of drawing a graph of the function.

We label the derivative of this function as $\dfrac{dy}{dx}$

i.e. $\dfrac{dy}{dx} = \dfrac{d}{dx}(y)$ where y is a function of x

EXAMPLE 3.12

If $f(x) = x^3 - 10x$ and $h(y) = 3y^{-1} + 0.4y^{0.7}$, find:

(a) $f'(x)$ (b) $h'(y)$

SOLUTION

(a) $f'(x) = \dfrac{d}{dx}(f(x)) = \dfrac{d}{dx}(x^3 - 10x) = 3x^2 - 10$

(b) $\quad h'(y) = \dfrac{d}{dy}(h(y)) = \dfrac{d}{dy}(3y^{-1} + 0.4y^{0.7})$

$$= 3(-1)y^{-1-1} + (0.4)(0.7)y^{0.7-1}$$

$$= -3y^{-2} + 0.28y^{-0.3}$$

EXAMPLE 3.13

If $y = 3x^{-2} - 2x^3$, $z = 4x^3 - 3x^2 + 10x - 12$ and $S = 4t^3 - 3t^{-2} + 10$ find:

(a) $\dfrac{dy}{dx}$ (b) $\dfrac{dz}{dx}$ (c) $\dfrac{dS}{dt}$

SOLUTION

(a) $\dfrac{dy}{dx} = \dfrac{d}{dx}(y) = \dfrac{d}{dx}(3x^{-2} - 2x^3) = -6x^{-3} - 6x^2$

(b) $\dfrac{dz}{dx} = \dfrac{d}{dx}(z) = \dfrac{d}{dx}(4x^3 - 3x^2 + 10x - 12) = 12x^2 - 6x + 10$

(c) $\dfrac{dS}{dt} = \dfrac{d}{dt}(S) = \dfrac{d}{dt}(4t^3 - 3t^{-2} + 10) = 12t^2 + 6t^{-3}$

3.2.4 DIFFERENTIATION 2: PARTIAL DIFFERENTIATION

Sometimes, in Statistics, we need to consider functions that involve two different variables.

For example, $x^2 + 2xy - 10$, the two variables being x and y.

The function $t^3 - 4z^2 + 3t^2z^{-1} - z + 1$ involves the two variables t and z.

In these types of cases we can still differentiate using the same rules as laid down earlier except that we can only differentiate *w.r.t. one of the variables at a time* regarding the other as a *constant*.

For example, consider the function $z = x^2 + 2xy - 10$ (a function of the two variables x and y). We wish to differentiate it with respect to x.

The operator we use here is $\dfrac{\partial}{\partial x}$ and translated as 'differentiate partially w.r.t. x'.

We have: $\quad\dfrac{\partial z}{\partial x} = \dfrac{\partial}{\partial x}(x^2 + 2xy - 10)$

$$= \dfrac{\partial}{\partial x}(x^2) + \dfrac{\partial}{\partial x}(2xy) + \dfrac{\partial}{\partial x}(-10)$$

Now, $\quad \dfrac{\partial}{\partial x}(x^2) = 2x$

$$\dfrac{\partial}{\partial x}(2xy) = 2y \quad (2y \text{ is treated as a constant w.r.t. } x)$$

and $\qquad \dfrac{\partial}{\partial x}(-10) = 0 \qquad (-10$ is a constant$)$

i.e. $\qquad\qquad \dfrac{\partial z}{\partial x} = 2x + 2y$

Similarly, $\qquad \dfrac{\partial z}{\partial y} = \dfrac{\partial}{\partial y}(x^2) + \dfrac{\partial}{\partial y}(2xy) + \dfrac{\partial}{\partial y}(-10)$

Now, $\qquad \dfrac{\partial}{\partial y}(x^2) = 0 \qquad (\text{since } x^2 \text{ is a constant w.r.t. } y)$

$\qquad\qquad \dfrac{\partial}{\partial y}(2xy) = 2x \qquad (\text{since } 2x \text{ is a constant w.r.t. } y)$

and $\qquad \dfrac{\partial}{\partial y}(-10) = 0 \qquad (\text{since } -10 \text{ is a constant})$

i.e. $\qquad\qquad \dfrac{\partial z}{\partial y} = 2x$

All the usual rules of differentiation hold when we wish to differentiate partially, except for the fact that we can only ever differentiate w.r.t. one variable at a time, the other being treated as a constant.

EXAMPLE 3.14

Differentiate:

(a) $\quad 3x^2 - 2xy \qquad$ (b) $\quad 2y^2 - 4xy + 3x^2 - 10$

partially w.r.t. each variable.

SOLUTION

(a) $\quad \dfrac{\partial}{\partial x}(3x^2 - 2xy) = 6x - 2y \quad (\text{since } 2y \text{ is a constant here})$

$\qquad \dfrac{\partial}{\partial y}(3x^2 - 2xy) = -2x \quad (3x^2 \text{ and } 2x \text{ are constants})$

(b) $\quad \dfrac{\partial}{\partial x}(2y^2 - 4xy + 3x^2 - 10) = -4y + 6x$

$\qquad (2y^2, \ -10 \text{ and } 4y \text{ are constants})$

$\qquad \dfrac{\partial}{\partial y}(2y^2 - 4xy + 3x^2 - 10) = 4y - 4x$

$\qquad (3x^2, \ -10 \text{ and } 4y \text{ are constants})$

EXAMPLE 3.15

If $y = 4.2x^{-0.4} - 3.2t^{0.3}x^{0.2} + 9.3t^{3.4}$, find $\dfrac{\partial y}{\partial x}$ and $\dfrac{\partial y}{\partial t}$.

SOLUTION

$$\frac{\partial y}{\partial x} = (4.2)(-0.4)x^{-0.4-1} + (-3.2)(t^{0.3})(0.2)x^{0.2-1}$$

$$= -1.68x^{-1.4} - 0.64t^{0.3}x^{-0.8} \quad (t \text{ constant})$$

$$\frac{\partial y}{\partial t} = (-3.2)(0.3)x^{0.2}t^{0.3-1} + (9.3)(3.4)t^{3.4-1}$$

$$= 0.96x^{0.2}t^{-0.7} + 31.62t^{2.4} \quad (x \text{ constant})$$

3.2.5 DIFFERENTIATION 3: MAXIMA AND MINIMA

Consider a relationship of the form $y = f(x)$, where $f(x)$ is some given function of x. Given a particular set of values of x we can find a corresponding set of y values.

In this section we are concerned with finding maximum and/or minimum values of y as x varies over a particular range. (We note at this stage however, that the treatment of the topic in this section is necessarily elementary and we are concerned only with giving the student some insight into the general problem.)

Suppose that we wished to find the minimum value of the function $y = x^2 - 5x + 4$ for x ranging between 1 and 4. One particular method of doing this is to draw a graph of the curve corresponding to values of x in the given range.

Choosing values of x throughout the range at intervals of $\frac{1}{2}$ and calculating the corresponding values of y gives:

$$x = 1, \quad y = (1)^2 - 5(1) + 4 = 0;$$

$$x = 1.5, \quad y = (1.5)^2 - 5(1.5) + 4 = -1.25;$$

$$x = 2, \quad y = (2)^2 - 5(2) + 4 = -2;$$

similarly $x = 2.5$ gives $y = -2.25$; $x = 3$ gives $y = -2$; $x = 3.5$ gives $y = -1.25$; $x = 4$ gives $y = 0$.

Plotting these points on an (x, y) frame and joining them with a smooth curve, we obtain the graph shown in Fig. 3.8.

From the graph we can see that the minimum point is $(2.5, -2.25)$. That is, the minimum value of the function $y = x^2 - 5x + 4$ is $y = -2.25$.

This technique is adequate when finding maximum or minimum values of elementary functions that are 'well behaved', but in general the arithmetic proves arduous and results are often too approximate.

Before giving a more precise, mechanical method of obtaining certain maximum or minimum values of a relationship $y = f(x)$, we need to discuss briefly the difference between 'absolute' and 'local' maximum or minimum values.

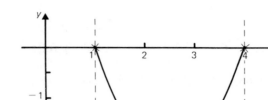

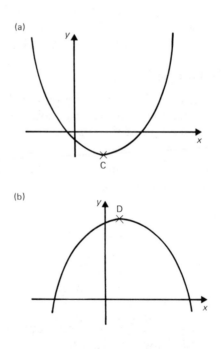

Fig. 3.8

Fig. 3.9

Consider the two sketches in Fig. 3.9.

(a) shows a point C known as a *local minimum*, since any other point 'close' to C (left or right) has a greater y value. (The minimum point found in the previous example was an example of a local minimum.)

Similarly, (b) shows a *local maximum*, point D. Here, any other point 'close' to D (left or right) has a smaller y value.

On the other hand, consider Fig. 3.10, showing a curve with both a local maximum and local minimum point, E and F respectively.

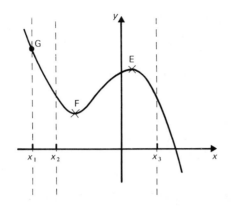

Fig. 3.10

Now, if we consider the values of y for the x range x_1 to x_3, it is clear that although E is a local maximum, the point G has a larger corresponding value of y. G is known as an *absolute maximum* point of the curve for the x range x_1 to x_3.

However, if we now consider the x range x_2 to x_3 only, we have E both as a local and absolute maximum point for this range of x values.

The method of differentiation enables us to find local maximum and minimum points of a function without involving needless numerical calculations.

STATEMENT 3.7

A procedure to find the local maximum and/or minimum points (if they exist) of a function $y = f(x)$ is:

(a)   find $\dfrac{dy}{dx}$;

(b)   solve the equation $\dfrac{dy}{dx} = 0$ (for x); and

(c)   substitute the value (or values) of the solution(s) of this equation into $y = f(x)$.

Note 1   This procedure will not find any absolute maximum or minimum values that are not also local maximum or minimum values of the function.

Note 2   For quadratic functions, functions of the form $y = ax^2 + bx + c$, where $a \neq 0$, any local maxima or minima found will automatically be absolute maxima or minima.

We will find the minimum value of the function $y = x^2 - 5x + 4$ considered earlier.

Here,

$$\frac{dy}{dx} = 2x - 5 \quad \text{and} \quad \frac{dy}{dx} = 0 \quad \text{gives} \quad 2x - 5 = 0 \Rightarrow x = \frac{5}{2}$$

Substituting this value of x into $y = x^2 - 5x + 4$ gives:

$$y = \left(\frac{5}{2}\right)^2 - 5\left(\frac{5}{2}\right) + 4 = \frac{25}{4} - \frac{25}{2} + 4 = \frac{41}{4} - \frac{50}{4} = \frac{-9}{4} = -2.25$$

i.e. $y_{\min} = -2.25$

EXAMPLE 3.16

Find the minimum value of the function $y = x^2 + 2x - 12$.

SOLUTION

$$\frac{dy}{dx} = 2x + 2.$$

$$\frac{dy}{dx} = 0 \text{ gives } 2x + 2 = 0, \quad \text{i.e.,} \quad x = -1.$$

When $x = -1$,

$$y = (-1)^2 + 2(-1) - 12 = -13$$

i.e. $y_{\min} = -13$

EXAMPLE 3.17

Find the greatest value of the function $y = -3x^2 + 15x - 13$ as x varies.

SOLUTION

We require the maximum value of y.

$$\frac{dy}{dx} = -6x + 15.$$

$$\frac{dy}{dx} = 0 \text{ gives } -6x + 15 = 0, \quad \text{i.e.,} \quad x = \frac{15}{6}.$$

Therefore $y_{max} = -3\left(\dfrac{15}{6}\right)^2 + 15\left(\dfrac{15}{6}\right) - 13$

i.e. $y_{max} = 5.75$

EXAMPLE 3.18

Find the least value of $S = 2.5t^2 - 16t$ as t varies.

SOLUTION

$\dfrac{dS}{dt} = 5t - 16.$

$\dfrac{dS}{dt} = 0$ gives $5t - 16 = 0,$ i.e., $t = \dfrac{16}{5}.$

Therefore $S_{min} = (2.5)\left(\dfrac{16}{5}\right)^2 - 16\left(\dfrac{16}{5}\right)$

$= 25.6 - 51.2 = -25.6$

Hence $S_{min} = -25.6$

The illustration and examples in this section have dealt only with maximum and minimum values of simple quadratic functions in order to keep the associated mathematical content to a minimum. Interested students wishing to study further are referred to any standard mathematical text dealing with the Differential Calculus.

3.3 REGRESSION LINES

3.3.1 FITTING REGRESSION LINES

We introduced the concept of correlation in the introductory section to this chapter by way of scatter diagrams and showed how a measure of correlation between two variables $(x$ and $y,$ say) depended on the approximate shape of a boundary encompassing the plotted (x, y) points. The closer to one of two particular straight lines the plotted points were, the closer to $+1$ or -1 we made the measure of correlation.

In this section we are concerned with finding a mathematical relationship between the two variables under consideration.

We begin with a simple case.
Consider the scatter diagram (Fig. 3.11) shown and notice that there is little scatter amongst the data pairs, showing a presence of strong positive correlation between the two variables.

A useful method for finding a mathematical relationship is to plot the point $(\bar{x}, \bar{y})$ on the diagram and then ensure that the line we fit passes through this point. This has been done in Fig. 3.12.

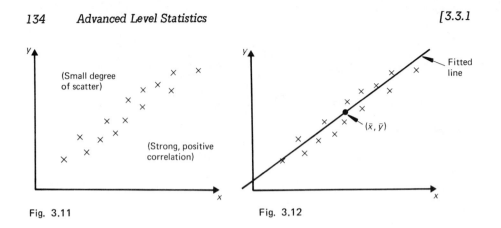

Fig. 3.11   Fig. 3.12

The equation of this line can then be found using the techniques discussed earlier in Section 3.2.1.

One of the uses of a fitted straight line, representing a relationship between two variables, is that it can be used to estimate a particular value of one of the variables given a corresponding value of the other.

For example, suppose that we had a set of marks in Geography (x) and History (y) for two separate examinations for a particular set of students and:

(a)   we plot the (x, y) points to form a scatter diagram;

(b)   we use this to ensure that the degree of correlation is high (i.e., close to $+1$ or -1; and

(c)   we estimate a straight line having the form $y = 0.7x + 5.1$, say, representing the relationship between the two variables.

If another student, who missed the History exam, obtained 48 marks in Geography, we could estimate the History mark using the above equation.

That is, estimated History mark,

$$y = (0.7)(48) + 5.1$$
$$= 38.7$$

Similarly, suppose that another student who obtained a History mark of 72 missed the Geography exam.

First, rearranging the given equation to put Geography (x) in terms of History (y), we have:

$$y = 0.7x + 5.1 \Rightarrow y - 5.1 = 0.7x$$

so that
$$x = \frac{1}{0.7}(y - 5.1)$$

Therefore, the estimated Geography mark,

$$x = \frac{1}{0.7}(72 - 5.1)$$

$$= 95.6 \quad (1D)$$

We note here that this particular method of obtaining a linear relationship:

(a)   is only approximate: and

(b)   only valid between variables showing a high degree of correlation.

To summarise:

STATEMENT 3.8

For a set of bivariate data $((x_1, y_1), (x_2, y_2), \ldots, (x_n, y_n))$ showing a high degree of correlation, we can *estimate* a linear mathematical relationship between the variables x and y by:

(a)   forming a scatter diagram for the (x, y) pairs;

(b)   calculating and plotting the point $(\bar{x}, \bar{y})$;

(c)   'fitting' a straight line, which passes through $(\bar{x}, \bar{y})$, 'by eye'.

If necessary, we can use the equation of the above line to estimate the value of one of the variables given a value of the other.

It should be clear that, even though the line is made to pass through the point $(\bar{x}, \bar{y})$, different observers will almost certainly obtain different equations for the line in question, this being the main disadvantage of this method.

EXAMPLE 3.19

(a)   Fit and find the equation of a straight line representing the relationship between the variables x and y, based on the following set of data.

x	1	3	4	6	8	9	11	14
y	1	2	4	4	5	7	8	9

(b)   Estimate a value for y, corresponding to $x = 10$.

SOLUTION

(a)   First, we plot the points, forming a scatter diagram.

Now

$$\bar{x} = \frac{\Sigma x}{n} = \frac{56}{8} = 7.0$$

and

$$\bar{y} = \frac{\Sigma y}{n} = \frac{40}{8} = 5.0$$

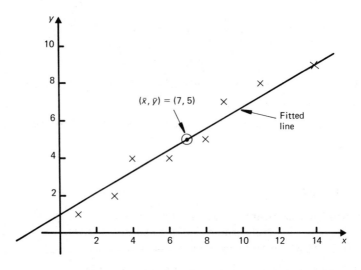

giving the mean point as (7.0, 5.0) which we have added to the scatter diagram.

Drawing in a line (through the mean point) to fit the data as closely as possible, we see that it crosses the y axis at $y = 1$.

So, putting the line as $y = mx + c$, we have $c = 1$,

i.e. $y = mx + 1$ [1]

Substituting the mean point $(7, 5)$ into [1] gives:

$$5 = m(7) + 1, \quad \text{i.e.,} \quad m = \frac{5 - 1}{7} = \frac{4}{7} = 0.57 \ (2D)$$

Thus the fitted line has equation $y = 0.57x + 1$.

(b)   When $x = 10$ we have:

$$y = (0.57)(10) + 1$$

$$= 6.7 \ (1D)$$

In general, lines fitted to data, describing the mathematical relationship between variables, are called *regression lines*.

We can now consider the following:

STATEMENT 3.9

For any bivariate set of data, connecting variables x and y, there are always two uniquely defined regression lines:

(a)   *the line of y on x* expressing y specifically in terms of x, which can be used for estimating y *given a value of x*;

(b)   *the line of x on y* expressing x specifically in terms of y, which can be used for estimating x *given a value of y*.

No Proof

We can demonstrate this statement as follows.

Assume that we have a large set of bivariate data whose points on a scatter diagram can be enclosed by an ellipse as in Fig. 3.13.

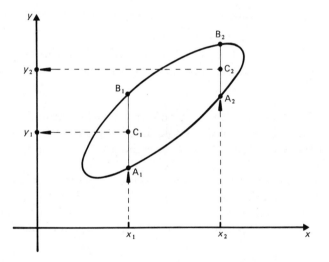

Fig. 3.13

We wish to form a regression line for estimating values of y for given values of x.

Consider a particular value of x, say x_1. The line $x = x_1$ meets the ellipse at points A_1 and B_1.

The chord $A_1 B_1$ represents the range of y values corresponding to the x value x_1. If we assume that the actual y values are distributed evenly along this line, then it is logical to take the *average y value* as a single, representative value of y corresponding to the given value of x, x_1, this value of y (y_1 say) lying at the mid-point of the chord, C_1.

Similarly, selecting another value of x, x_2 say, we would take as the representative y value (y_2) the y value that lies at point C_2 (the mid-point of the chord $A_2 B_2$).

If we carry on this procedure for all possible x values bounded by the ellipse, we will obtain a large (theoretically infinite) number of chords, whose mid-points will form a straight line (see Fig. 3.14).

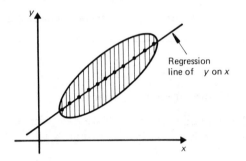

Regression
line of y on x

Fig. 3.14

Since this whole procedure was concerned with estimating y *given values of* x, the straight line obtained is the *regression line of* y *on x*.

However, this procedure was not concerned with estimating x given a particular value of y, so we would not be justified in rearranging the equation of the above line to use it for estimating x given y.

Fig. 3.15 shows the method of constructing the regression line of *x on y*. Here, we need to take specific values of y, draw the corresponding chords relating to relevant values of x (they will be horizontal here), choose the chord mid-points as representative x values and finally join these up.

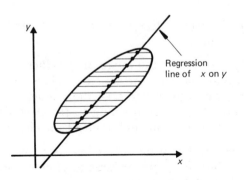

Regression
line of x on y

Fig. 3.15

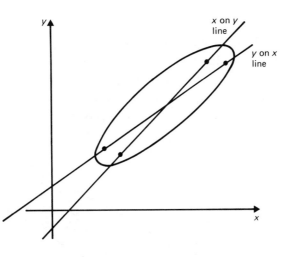

Fig. 3.16

Fig. 3.16 shows these two regression lines superimposed on the same diagram, where it can be seen that the two lines are markedly different.

This method of obtaining the two regression lines is only ideal of course if the ellipse contains a (theoretically) infinite number of data points, a situation that will never occur. This is because in choosing the mid-point of a horizontal or vertical chord as an average point we necessarily assume that we have points evenly spaced all along the chord, as previously mentioned. Given even a relatively large sample of points, together with a well-fitting ellipse, this assumption would be rather naive. In addition, the construction of an ellipse to fit a particular set of data is an arbitrary procedure anyway, and clearly no two people would agree on the exact shape of a particular ellipse. We need a more precise way of obtaining the two regression lines.

An important technique, and one that is most commonly used, is the 'method of least squares' which we deal with in the following sub-section.

However, this chord drawing technique, besides being a useful practical approach, is helpful in the understanding of why two regression lines are necessary.

EXAMPLE 3.20

Estimate the regression lines of y on x and x on y (using the 'chord drawing' technique) for the following set of bivariate data.

x	42	57	27	45	76	88	20	67	46	52	43	73	39	66	53
y	35	50	47	45	81	75	20	56	55	46	31	71	32	72	52

x	75	50	36	47	62	82	64	41	48	54	73	32	56	39	34
y	56	40	24	36	51	53	41	33	42	46	67	22	62	26	28

SOLUTION

First we plot these bivariate pairs on an (x, y) frame to give the associated scatter diagram and fit an ellipse to enclose the points as well as possible. (See the accompanying diagram.)

The next step is to construct the vertical and horizontal chords of the ellipse, which we have done for every five units of both x and y.

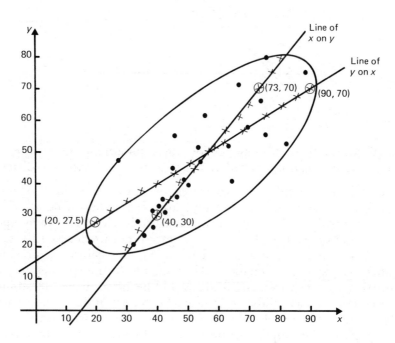

The crosses represent the mid-points of both vertical and horizontal chords. Joining up the two sets of crosses we obtain the two regression lines.

If we let $y = a_1 x + b_1$ represent the equation of the regression line of y on x, b_1 can be found immediately as the y intercept of the line.

From the figure, $b_1 = 15$.

To find the gradient, a_1, we need to select any two points on the line.

We choose $(20, 27.5)$ and $(90, 70)$. (These points are circled on the diagram.)

So that,

$$\text{gradient} = a_1 = \frac{70 - 27.5}{90 - 20} = \frac{42.5}{70} = 0.6 \ (1D)$$

Therefore, the regression line of y on x is $y = 0.6x + 15$.

Similarly, if we let the x on y regression line be $x = a_2 y + b_2$, we have $b_2 = 13$ (the x intercept).

Choosing the two points $(40, 30)$ and $(73, 70)$ (circled on the diagram) we calculate:

$$\text{gradient} = a_2 = \frac{73 - 40}{70 - 30} = \frac{33}{40} = 0.825 \quad (1D)$$

Therefore, the regression line of x on y is $x = 0.825y + 13$.

Note that since the ellipse encloses a relatively small number of points, the two regression lines calculated above must only be treated as approximate.

The connection between the two regression lines and the correlation coefficient for a set of bivariate data is shown explicitly in Section 3.4. However, we can see even at this stage how they relate to each other generally.

Consider the approximate elliptical boundary that can be drawn around the points plotted on a scatter diagram. We saw (in Section 3.1.1) that the 'longer and thinner' the ellipse, the stronger the measure of correlation between the two relevant variables. Also, from the previous discussion, it should be clear that the 'longer and thinner' the ellipse is, the closer the two regression lines will become.

Hence the more correlated the two variables are, the closer are the two regression lines.

Fig. 3.17 shows this relationship pictorially. For the cases of perfect positive or negative correlation, the two regression lines in each case will be coincident. In this light, the initial method of fitting a regression line 'by eye' to highly correlated data can only be an approximate method, but clearly the technique is reasonable for two variables that have a correlation coefficient close to $+1$ or -1.

3.3.2 THE LEAST SQUARES TECHNIQUE

The mathematical technique used to find the two regression lines is described as follows.

We are given the bivariate set of data $((x_1, y_1), (x_2, y_2), \ldots, (x_n, y_n))$.

Consider the two regression lines $y = a_1 x + b_1$ and $x = a_2 y + b_2$.

Let d_i be the vertical (y) distance from the point (x_i, y_i) to the line $y = a_1 x + b_1$ (see Fig. 3.18(a)) and let e_i be the horizontal (x) distance from the point (x_i, y_i) to the line $x = a_2 y + b_2$ (see Fig. 3.18(b)).

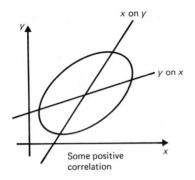

Some positive
correlation

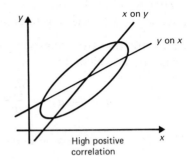

High positive
correlation

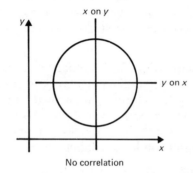

No correlation

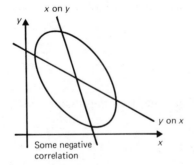

Some negative
correlation

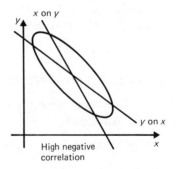

High negative
correlation

Fig. 3.17

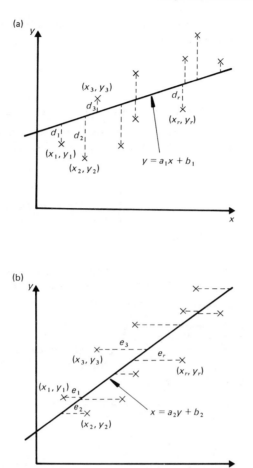

Fig. 3.18

Then we have:

DEFINITION 3.4

(a)   The *least squares regression line of* y *on* x is that line $y = a_1 x + b_1$ such that the quantity $\sum_{i=1}^{n} d_i^2$ is a minimum. See Fig. 3.18(a).

(b)   The *least squares regression line of* x *on* y is that line $x = a_2 y + b_2$ such that the quantity $\sum_{i=1}^{n} e_i^2$ is a minimum. See Fig. 3.18(b).

For practical purposes, and as a consequence of the above definition, we have the following statement.

STATEMENT 3.10

Given the bivariate set $((x_1, y_1), (x_2, y_2), \ldots, (x_n, y_n))$,

(a)   the constants a_1 and b_1 in the least squares regression line $y = a_1 x + b_1$ can be determined by solving the simultaneous equations:

$$\left. \begin{array}{l} \sum y = a_1 \sum x + nb_1 \\[2em] \sum xy = a_1 \sum x^2 + b_1 \sum x \end{array} \right\} \begin{array}{l} \text{called the } \textit{normal equations} \\ \text{for } y \text{ on } x \end{array}$$

(b)   the constants a_2 and b_2 in the least squares regression line $x = a_2 y + b_2$ can be determined by solving the simultaneous equations:

$$\left. \begin{array}{l} \sum x = a_2 \sum y + nb_2 \\[2em] \sum xy = a_2 \sum y^2 + b_2 \sum y \end{array} \right\} \begin{array}{l} \text{called the } \textit{normal equations} \\ \text{for } x \text{ on } y \end{array}$$

Proof   For (a).

We require the values of a_1 and b_1 that make $\sum d_i{}^2$ a minimum. (See Fig. 3.18(a).)

d_r is just the vertical distance between point (x_r, y_r) and line $y = a_1 x + b_1$,

i.e. $d_r = y_r - (a_1 x_r + b_1)$

So: $d_i = y_i - a_1 x_i - b_1$

$\therefore$ $\sum d_i{}^2 = \sum_i (y_i - a_1 x_i - b_1)^2 = S$ say   (for convenience)

To find the values of a_1 and b_1 that make S a minimum, we need to solve the two equations:

$$\frac{\partial S}{\partial a_1} = 0 \quad \text{and} \quad \frac{\partial S}{\partial b_1} = 0$$

Now:

$$\frac{\partial S}{\partial a_1} = \frac{\partial}{\partial a_1} \left\{ \sum_i (y_i + a_1 x_i - b_1)^2 \right\}$$

$$= \sum_i \left[\frac{\partial}{\partial a_1} (y_i - a_1 x_i - b_1)^2 \right]$$

$$= \sum_i \left[2(y_i - a_1 x_i - b_1) \cdot - x_i \right]$$

$$= \sum_i \left[-2 x_i y_i + 2 a_1 x_i^2 + 2 b_1 x_i \right]$$

$$= -2 \sum xy + 2 a_1 \sum x^2 + 2 b_1 \sum x \quad \text{(dropping subscripts)}$$

$\therefore \dfrac{\partial S}{\partial a_1} = 0$ gives:

$$-2 \sum xy + 2 a_1 \sum x^2 + 2 b_1 \sum x = 0$$

i.e. $\qquad \sum xy = a_1 \sum x^2 + b_1 \sum x \quad \text{(as required)}$

Also: $\qquad \dfrac{\partial S}{\partial b_1} = \dfrac{\partial}{\partial b_1} \left\{ \sum_i (y_i - a_1 x_i - b_1)^2 \right\}$

$$= \sum \left\{ \frac{\partial}{\partial b_1} (y_i - a_1 x_i - b_1)^2 \right\}$$

$$= \sum_i \left[2(y_i - a_1 x_i - b_1) \cdot - 1 \right]$$

$$= \sum_i \left[-2 y_i + 2 a_1 x_i + 2 b_1 \right]$$

$$= -2 \sum y + 2 a_1 \sum x + 2 b_1 n$$

$\therefore \dfrac{\partial S}{\partial b_1} = 0$ gives:

$$-2 \sum y + 2 a_1 \sum x + 2 n b_1 = 0$$

i.e. $\qquad \sum y = a_1 \sum x + n b_1 \quad \text{(as required)}$

For (b) a similar method is employed (using Fig. 3.18(b)).

EXAMPLE 3.21

Obtain the least squares regression line of y on x for the following bivariate data:

x	1	2	3	4
y	1	5	6	10

and estimate the value of y when $x = 2.5$.

SOLUTION

The normal equations for y on x are:

$$\Sigma y = a_1 \Sigma x + nb_1$$

$$\Sigma xy = a_1 \Sigma x^2 + b_1 \Sigma x$$

so that we need to calculate the quantities Σx, Σy, Σxy and Σx^2.

x	y	xy	x^2
1	1	1	1
2	5	10	4
3	6	18	9
4	10	40	16
10	22	69	30

$\Sigma x = 10 \quad \Sigma y = 22$

$\Sigma xy = 69 \quad \Sigma x^2 = 30$

$n = 4$

Substituting the calculated values into the normal equations gives:

$$\begin{cases} 22 = 10a_1 + 4b_1 & \qquad [1] \\ 69 = 30a_1 + 10b_1 & \qquad [2] \end{cases}$$

Solving [1] and [2] simultaneously, we obtain $a_1 = 2.8$ and $b_1 = -1.5$.

Hence, the least squares regression line of y on x is:

$$y = 2.8x - 1.5$$

When $x = 2.5$, we estimate a value of y by substituting this value into the above regression line,

i.e.
$$y_{est} = (2.8)(2.5) - 1.5$$
$$= 5.5$$

EXAMPLE 3.22

Obtain the two least squares regression lines for the given set of bivariate data.

x	10	12	8	13	12	11	11
y	27	41	27	32	34	26	38

SOLUTION

We set up a table as follows.

x	10	12	8	13	12	11	11	77
y	27	41	27	32	34	26	38	225
xy	270	492	216	416	408	286	418	2506
x^2	100	144	64	169	144	121	121	863
y^2	729	1681	729	1024	1156	676	1444	7439

So we have:

$$n = 7; \quad \sum x = 77; \quad \sum y = 225; \quad \sum xy = 2506;$$

$$\sum x^2 = 863; \quad \sum y^2 = 7439$$

The normal equations for y on x are:

$$\sum y = a_1 \sum x + nb_1$$

$$\sum xy = a_1 \sum x^2 + b_1 \sum x$$

and substitution gives

$$225 = 77a_1 + 7b_1 \qquad [1]$$
$$2506 = 863a_1 + 77b_1 \qquad [2]$$

Solving [1] and [2] simultaneously, we obtain $a_1 = 1.94$ (2D) and $b_1 = 10.83$ (2D).

Therefore the equation of the regression line of y on x is
$$y = 1.94x + 10.83.$$

The normal equations for x on y are:

$$\sum x = a_2 \sum y + nb_2$$

$$\sum xy = a_2 \sum y^2 + b_2 \sum y$$

and substitution gives:

$$77 = 225a_2 + 7b_2 \qquad [3]$$
$$2506 = 7439a_2 + 225b_2 \qquad [4]$$

Solving [3] and [4] simultaneously, we obtain $a_2 = 0.15$ (2D) and $b_2 = 6.18$ (2D).

Therefore the equation of the regression line of x on y is
$$x = 0.15y + 6.18.$$

A useful aid in remembering the form of the two normal equations is the following.

For the y on x line, $y = a_1 x + b_1$.

Step 1   *Sum both sides,*

giving
$$\Sigma y = \Sigma(a_1 x + b_1)$$

$$\Rightarrow \Sigma y = \Sigma a_1 x + \Sigma b_1$$

$$\Rightarrow \Sigma y = a_1 \Sigma x + nb_1$$

(the 1st normal equation)

Step 2   (a) *Multiply by* x *and* (b) *sum both sides,*

giving $xy = a_1 x^2 + b_1 x$ (from (a))

and $\Sigma xy = \Sigma(a_1 x^2 + b_1 x)$ (from (b))

$$\Rightarrow \Sigma xy = \Sigma a_1 x^2 + \Sigma b_1 x$$

$$\Rightarrow \Sigma xy = a_1 \Sigma x^2 + b_1 \Sigma x \quad \text{(the 2nd normal equation)}$$

Similarly, for the x on y line, $x = a_2 y + b_2$, we have the procedure:

Step 1   *Sum both sides.*

Step 2   (a) *Multiply by* y *and* (b) *sum both sides* which will give the corresponding normal equations for x on y.

The method of least squares as introduced in this section can be shown to be equivalent to the 'practical' technique, for obtaining the two least squares regression lines, as given at the end of Section 3.3.1.

Consider the regression line of y on x. Assuming that we have a large set of bivariate data to work with, let us consider further a particular, given value of x, x_r say.

Now, corresponding to the x value x_r there will be an associated set of y values, $y_1, y_2, \ldots, y_p$, say.

The 'practical' method took, as the single representative (estimate) of y, the average y value

$$\bar{y} = \frac{1}{p}\sum_{j=1}^{p} y_j = y_{\text{est}}$$

The 'least squares' method takes that value for y_{est} which minimises the expression:

$$\sum_{j=1}^{p} (y_{est} - y_j)^2$$

But it can be shown that the expression

$$\sum_{j=1}^{p} (y_{est} - y_j)^2$$

is a minimum when

$$y_{est} = \frac{1}{p} \sum_{j=1}^{p} y_j = \bar{y}$$

(The proof of this is left as a directed exercise.)

So the two methods are in fact equivalent.

At this stage, there are two points that need to be made concerning regression lines.

First, the use of a regression line for estimation purposes cannot be applied indiscriminately. It should only be used to estimate values of a variable within the range used in calculating the regression line. For example, consider the bivariate set of data shown in Fig. 3.19, where it can be seen that the x-values range between e and f. Given that the regression line $y = ax + b$ has been calculated using the given data pairs, we could use it to estimate values of y only for given x-values in the range e to f.

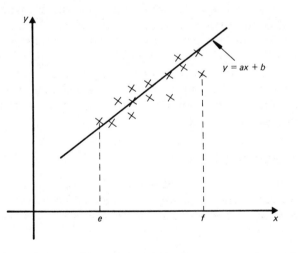

Fig. 3.19

However, in particular circumstances, regression lines are sometimes used for estimation purposes outside their valid ranges, this being acceptable as long as:

(a)   we do not 'stray too far' outside the valid range; and

(b)   there is no evidence to suppose that the given line is showing a false picture of the situation outside the range.

Any results obtained in this way (generally known as 'extrapolation') should be used cautiously.

Second, the given least squares technique for fitting regression lines assumes a linear (i.e., straight line) relationship between the two variables involved. Consider the scatter diagram shown in Fig. 3.20.

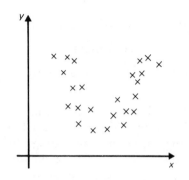

Fig. 3.20

Here, we notice immediately that although there *is* a definite relationship between x and y, it is clearly not of a linear form. We would not, therefore, be justified in fitting regression lines of the form $y = a_1 x + b_1$ or $x = a_2 y + b_2$. In this particular case, the relationship is 'parabolic' for y in terms of x and would be given by an equation of the form $y = ax^2 + bx + c$. The least squares technique is also used here to determine the constants a, b and c and follows a similar technique to that already given for a regression line, although we do not deal with this particular situation in this text.

3.3.3 THE METHOD OF CODING

In order to simplify the calculations involved in determining regression lines for bivariate data consisting of large, awkward numbers, we can employ a method of coding identical to that for calculating means and variances.

The technique is to code either (or both) of the variables, using the usual form $X = \dfrac{x - a}{b}$ and/or $Y = \dfrac{y - c}{d}$ where a, b, c and d are constants chosen for convenience. If now, for instance, we were interested in obtaining the least squares regression line for y on x, we would proceed to set up the normal equations in terms of X and Y as follows:

$$\sum Y = A_1 \sum X + n B_1$$

$$\sum XY = A_1 \sum X^2 + B_1 \sum X$$

(Note that the regression constants A_1 and B_1 will be, in general, different to those that would have been obtained — a_1 and b_1 — had we not coded.)

The normal equations are now solved for A_1 and B_1 and we thus form the regression line $Y = A_1 X + B_1$ $(Y$ on $X)$. Substitution of $Y = \dfrac{y-c}{d}$ and $X = \dfrac{x-a}{b}$, together with some rearrangement, will lead to the regression line of y on x, as required.

Consider the following data:

x	50	55	60	65	70
y	285	287	293	292	301

where we wish to find the least squares regression line of y on x. Since the values of x have clearly been controlled (and in particular are stepped in equal intervals and contain an odd number of values), we can choose a coding that simplifies these values dramatically; namely $X = \dfrac{x-60}{5}$. As regards the coding for y however, the best we can do is to subtract a single value from each of the observations. Using the median, 292, gives $Y = y - 292$.

The respective coded values become:

X	-2	-1	0	1	2
Y	-7	-5	1	0	9

The next step is to find the regression line of Y on X, $Y = A_1 X + B_1$, by solving the two normal equations:

$$\sum Y = A_1 \sum X + n B_1$$

$$\sum XY = A_1 \sum X^2 + B_1 \sum X \quad \text{for } A_1 \text{ and } B_1$$

Notice immediately (due to the choice of the x to X coding) that $\sum X = 0$, which when substituted into the above normal equations simplifies them to:

$$\sum Y = n B_1 \qquad\qquad [1]$$

$$\sum XY = A_1 \sum X^2 \qquad\qquad [2]$$

We calculate $\Sigma\,Y$, $\Sigma\,XY$ and $\Sigma\,X^2$ in the following table.

X	Y	XY	X^2
-2	-7	14	4
-1	-5	5	1
0	1	0	0
1	0	0	1
2	9	18	4

We thus have: $\quad\Sigma X\ =\ 0;\qquad \Sigma Y\ =\ -2;$

$$\Sigma XY\ =\ 37;\qquad \Sigma X^2\ =\ 10$$

Note that $n = 5$ here.

Equations [1] and [2] yield, on substitution $-2 = 5B_1$ and $37 = (A_1)10$ respectively. Hence:

$$B_1\ =\ -\frac{2}{5}\ =\ -0.4 \quad \text{and} \quad A_1\ =\ \frac{37}{10}\ =\ 3.7$$

Therefore the regression line of Y on X is:

$$Y\ =\ 3.7X - 0.4 \tag{3}$$

We can now decode, that is substitute $Y = y - 292$ and $X = \dfrac{x - 60}{5}$ into [3] giving

$$y - 292\ =\ (3.7)\left(\frac{x - 60}{5}\right) - 0.4$$

A little algebraic manipulation gives $y = 0.74x + 247.2$, which is the required *regression line of y on x*.

Had we not coded but worked with the original given data, we would have obtained the above regression line directly (by means of some arduous arithmetic!).

The regression line of x on y is calculated in a similar manner, but considering the line $X = A_2 Y + B_2$. We would solve the associated pair of normal equations for A_2 and B_2, then substitute back for X and Y in terms of x and y into $X = A_2 Y + B_2$. Some algebraic manipulation would then give the required form $x = a_2 y + b_2$, say.

We summarise as follows.

STATEMENT 3.11

To obtain the regression lines of y on x and/or x on y for the bivariate set:

x	x_1 x_2 x_3 $\ \ldots\ x_n$
y	y_1 y_2 y_3 $\ \ldots\ y_n$

(continued)

using the codings $X = \dfrac{x-a}{b}$ and $Y = \dfrac{y-c}{d}$, we proceed:

(a)   the Normal Equations of Statement 3.10 are adapted to read

$$\left.\begin{array}{l} \sum Y \;=\; A_1 \sum X + nB_1 \\[2ex] \sum XY \;=\; A_1 \sum X^2 + B_1 \sum X \end{array}\right\} \text{for } Y \text{ on } X$$

and/or

$$\left.\begin{array}{l} \sum X \;=\; A_2 \sum Y + nB_2 \\[2ex] \sum XY \;=\; A_2 \sum Y^2 + B_2 \sum Y \end{array}\right\} \text{for } X \text{ on } Y$$

and are solved for A_1, B_1 and/or A_2, B_2.

(b)   Substitution of $X = \dfrac{x-a}{b}$ and $Y = \dfrac{y-c}{d}$ into the regression
lines $Y = A_1 X + B_1$ and/or $X = A_2 Y + B_2$ (together with some
rearrangement) will give the desired regression lines.

EXAMPLE 3.23

Use a method of coding to calculate the regression lines of   (a) y on x
(b) x on y for the following data.

x	84	77	68	98	71	87	65	93	80	75
y	89	74	72	95	80	91	72	86	78	82

Estimate the value of x when y is 85.

SOLUTION

The best we can do with the given data is to subtract a central value
from the x and y values respectively.

In this case $X = x - 80$ and $Y = y - 82$ will suffice.

X	4	-3	-12	18	-9	7	-15	13	0	-5	-2
Y	7	-8	-10	13	-2	9	-10	4	-4	0	-1
XY	28	24	120	234	18	63	150	52	0	0	689
X^2	16	9	144	324	81	49	225	169	0	25	1042
Y^2	49	64	100	169	4	81	100	16	16	0	599

We have:

$$n = 10; \quad \sum X = -2; \quad \sum Y = -1;$$

$$\sum XY = 689; \quad \sum X^2 = 1042; \quad \sum Y^2 = 599$$

The Normal Equations for Y on X are:

$$\sum Y = A_1 \sum X + nB_1$$

$$\sum XY = A_1 \sum X^2 + B_1 \sum X$$

Substituting gives:	$-1 = -2A_1 + 10B_1$	[1]
and:	$689 = 1042A_1 - 2B_1$	[2]
[2] $\times -5$ gives:	$-3445 = -5210A_1 + 10B_1$	[3]
	$-1 = -2A_1 + 10B_1$	[1]
[1]$-$[3] gives:	$3444 = 5208A_1.$	

Hence:
$$A_1 = \frac{3444}{5208} = 0.661 \ (3D)$$

Substituting in [1] gives $-1 = -2(0.661) + 10B_1$

$$\therefore \qquad B_1 = \frac{2(0.661) - 1}{10} = 0.032 \ (3D)$$

Thus the Y on X line $(Y = A_1 X + B_1)$ is $Y = 0.661X + 0.032$.

Decoding, we have the line of y on x as:
$$y - 82 = (0.661)(x - 80) + 0.032$$
i.e. $\qquad y = 0.661x - 80(0.661) + 82 + 0.032$
$$\therefore \qquad \underline{y = 0.66x + 29.15 \ (2D)}$$

The X on Y Normal Equations are:

$$\sum X = A_2 \sum Y + nB_2$$

$$\sum XY = A_2 \sum Y^2 + B_2 \sum Y$$

Substituting gives:	$-2 = -A_2 + 10B_2$	[1]
and:	$689 = 599A_2 - B_2$	[2]
[2] $\times -10$ gives:	$-6890 = -5990A_2 + 10B_2$	[3]
	$-2 = -A_2 + 10B_2$	[1]
[1]$-$[3] gives:	$6888 = 5989A_2$	

Hence:
$$A_2 = \frac{6888}{5989} = 1.150 \ (3D)$$

Substituting in [1] gives $-2 = -1.150 + 10B_2$

$$\therefore \qquad B_2 = \frac{1.150 - 2}{10} = -0.085$$

Thus the line of X on Y $(X = A_2 Y + B_2)$ is $X = 1.150Y - 0.085$.

Decoding gives the line of x on y as:

$$x - 80 = (1.150)(y - 82) - 0.085$$

i.e. $\qquad x = 1.15y - 82(1.15) + 80 - 0.085$

$$\therefore \qquad \underline{x = 1.15y - 14.39 \quad (2D)}$$

To estimate the value of x given y we use the x on y line above.

Hence, when $y = 85$,

$$x = (1.15)85 - 14.39$$

i.e., $\qquad x = 83.4 \quad (1D)$

Returning to the data used earlier in the text, we noted that the x-values:

x	50	55	60	65	70
y	285	287	293	292	301

increase uniformly, and it is clear that these values have not been allowed to vary freely (that is, they are not random). They have been specially chosen.

In this sort of situation, x is called a *controlled variable*. One reason for controlling one of two variables in this way is to study how one behaves under set conditions (for the other). These 'controlled' situations occur quite often naturally in practice and are not only convenient but usually very necessary in many practical situations.

For example, in a medical study of the effectiveness of a drug for controlling high blood pressure, we might measure blood pressure against carefully chosen (i.e., controlled) doses of the drug administered to a particular subject. We would quite obviously not wish to administer 'randomly selected' doses in the test.

When a control variable is used $(x$, say), we are quite justified in calculating (and using for estimation purposes) a regression line of y on x. However, a regression line of x on y *must not be used* in an estimating capacity due to the fact that x has not been allowed to vary freely. For the above example of drug dosage (controlled x) versus blood pressure (y), we would use the regression line of y on x to estimate a blood pressure for a particular drug dosage. We would not be interested in (or justified in calculating) an estimated drug dosage corresponding to a particular level of blood pressure; it would be meaningless in this context.

Often a controlled variable occurs quite naturally as a time scale. For example, the scatter diagram and related regression line in Fig. 3.21 shows coal production (y) for a period of years (x). In this particular case, the regression line is that of y on x and we would use it practically to extrapolate for production in the future (assuming, of course, that there is reasonable evidence to suppose that the line is representative of the future relationship between the two variables).

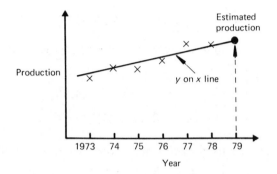

Fig. 3.21

EXAMPLE 3.24

Use a method of coding to find the least squares regression line of x on t for the following data.

t	1967	1968	1969	1970	1971	1972	1973	1974	1975
x	82	78	80	81	81	76	72	74	73

Estimate the value of x corresponding to $t = 1976$.

SOLUTION

Noticing that t is a controlled variable (with a given odd number of values), the obvious coding is $T = t - 1971$. A suitable coding for x is $X = x - 78$ (subtracting a central value).

We need to determine constants A and B to fit the X on T regression line $X = AT + B$, say. The corresponding Normal Equations take the form:

$$\sum X = A \sum T + nB \qquad [1]$$

$$\sum TX = A \sum T^2 + B \sum T \qquad [2]$$

T	-4	-3	-2	-1	0	1	2	3	4
X	4	0	2	3	3	-2	-6	-4	-5
TX	-16	0	-4	-3	0	-2	-12	-12	-20
T^2	16	9	4	1	0	1	4	9	16

Clearly, $n = 9$. Summing the above rows we obtain:

$$\Sigma T = 0; \quad \Sigma X = -5; \quad \Sigma TX = -69; \quad \Sigma T^2 = 60$$

Substituting into [1] and [2], we have:

$$-5 = 9B \quad \text{and} \quad -69 = 60A$$

(Note how simple the Normal Equations become with $\Sigma T = 0$.)

Thus:

$$B = -\frac{5}{9} = -0.556 \text{ (3D)} \quad \text{and} \quad A = -\frac{69}{60} = -1.15$$

So that the line of X on T is:

$$X = -1.15T - 0.556$$

Decoding, we have:

$$x - 78 = -1.15(t - 1971) - 0.556$$

i.e. $x = -1.15t + (1.15)(1971) + 78 - 0.556$

Thus: $x = -1.15t + 2344.1$ is the required line of x on t

When $t = 1976$, substitution into the above gives:

$$x = -(1.15)(1976) + 2344.1$$

i.e. $x = 71.7$ (1D).

3.4 PRODUCT MOMENT CORRELATION

3.4.1 THE COVARIANCE

In the bivariate situation, there is a measure analogous to the variance of a set of univariate data. This is defined as follows:

DEFINITION 3.5

Given the bivariate set $((x_1, y_1), (x_2, y_2), \ldots, (x_n, y_n))$, we define the *covariance* as:

$$s_{xy} = \frac{1}{n} \Sigma (x - \bar{x})(y - \bar{y})$$

For example, given the set $(10, 12), (8, 16), (18, 11)$ and $(12, 13)$, we will calculate the covariance, s_{xy}.

x	10	8	18	12	48
y	12	16	11	13	52

$\bar{x} = \dfrac{48}{4} = 12$

$\bar{y} = \dfrac{52}{4} = 13$

So, using the above definition, we have:

$$s_{xy} = \frac{1}{n} \sum (x - \bar{x})(y - \bar{y})$$

$$= \frac{1}{4} [(10 - 12)(12 - 13) + (8 - 12)(16 - 13)$$

$$+ (18 - 12)(11 - 13) + (12 - 12)(13 - 13)]$$

$$= \frac{1}{4} [(-2)(-1) + (-4)(3) + (6)(-2) + (0)(0)]$$

$$= \frac{1}{4} (2 - 12 - 12 + 0)$$

$$= \frac{1}{4} (-22)$$

i.e. $s_{xy} = -5.5$

In one sense, the covariance can be thought of as a 'generalisation' of the variance.

Consider: $s_x{}^2 = \sum (x - \bar{x})^2 = \sum (x - \bar{x})(x - \bar{x})$

$$s_y{}^2 = \sum (y - \bar{y})^2 = \sum (y - \bar{y})(y - \bar{y})$$

(suffices have been used simply to identify which variable is involved).

The covariance can be seen to be just 'half-way' between the above two variances. Alternatively, consider the following:

$$s_{xx} = \sum (x - \bar{x})(x - \bar{x}) = s_x{}^2$$

put $y = x$

s_{xy}

put $x = y$

$$s_{yy} = \sum (y - \bar{y})(y - \bar{y}) = s_y{}^2$$

One important characteristic of the covariance is that it can take both positive and negative values, as evidenced in the above example where the covariance was negative. Compare this fact with the case of the variance, which can take only positive values (trivially, it can take the value zero).

For computational purposes, and especially in the case where we have a large number of bivariate pairs, we can express s_{xy} in a more convenient and compact form.

STATEMENT 3.12

Given the set of bivariate data $((x_1, y_1), (x_2, y_2), \ldots, (x_n, y_n))$, the covariance, s_{xy}, is more easily computed using the form:

$$s_{xy} = \frac{1}{n} \sum xy - \bar{x}\bar{y} \quad \text{(the } \textit{computational formula}\text{)}$$

Proof

$$s_{xy} = \frac{1}{n} \sum (x - \bar{x})(y - \bar{y})$$

$$= \frac{1}{n} \sum (xy - \bar{x}y - \bar{y}x + \bar{x}\bar{y})$$

$$= \frac{1}{n} \left[\sum xy - \bar{x} \sum y - \bar{y} \sum x + n\bar{x}\bar{y} \right]$$

$$= \frac{1}{n} \sum xy - \bar{x} \frac{1}{n} \sum y - \bar{y} \frac{1}{n} \sum x + \bar{x}\bar{y}$$

$$= \frac{1}{n} \sum xy - \bar{x}\bar{y} - \bar{x}\bar{y} + \bar{x}\bar{y}$$

$$= \frac{1}{n} \sum xy - \bar{x}\bar{y} \quad \textit{as required}$$

The above computational formula is, of course, directly analogous to the computational formula for the variance of a set of values (Statement 2.13(a)). This analogy is stressed here (and earlier) simply as an aid to remembering the form that the covariance takes.

We will use the previous data to calculate the covariance with the aid of the above computational formula.

x	y	xy
10	12	120
8	16	128
18	11	198
12	13	156
48	52	602

As earlier, $\bar{x} = 12$ and $\bar{y} = 13$.

Also: $\sum xy = 602$

Hence: $s_{xy} = \frac{1}{n} \sum xy - \bar{x}\bar{y}$

$$= \tfrac{1}{4}(602) - (12)(13)$$

$$= 150.5 - 156$$

i.e. $s_{xy} = -5.5$ (the result obtained earlier)

We can also use a method of coding to calculate the covariance. The technique involved is identical to that used in all earlier sections. That is, we can code using the forms $X = \dfrac{x-a}{b}$ and $Y = \dfrac{y-c}{d}$. We then calculate the *coded covariance* as $s_{XY} = \dfrac{1}{n} \sum XY - \bar{X}\bar{Y}$. Finally, we decode using the form $s_{xy} = bds_{XY}$. (This will be formally stated and proved.) Note how similar this is to the decoding formula for the variance given in Statement 2.13(c), namely $s_x{}^2 = b^2 s_X{}^2$ or $s_y{}^2 = d^2 s_Y{}^2$ (depending on the variable and the coding form).

As a simple example of the method involved, consider the set:

x	5	10	15	20	25
y	30	32	28	35	35

We wish to calculate the covariance using the coding method.

The coding for x is obvious as $X = \dfrac{x-15}{5}$. In the case of y, the best we can do is $Y = y - 32$.

X	Y	XY
-2	-2	4
-1	0	0
0	-4	0
1	3	3
2	3	6
0	0	13

Here:

$$n = 5; \quad \bar{X} = \frac{1}{n}\sum X = \frac{0}{5} = 0; \quad \bar{Y} = \frac{1}{n}\sum Y = \frac{0}{5} = 0$$

Also $\sum XY = 13$

∴ $s_{XY} = \dfrac{1}{n}\sum XY - \bar{X}\bar{Y} = \dfrac{1}{5}(13) - (0)(0) = \dfrac{13}{5} = 2.6$

Note that here $\left(\text{using } X = \dfrac{x-a}{b}; \; Y = \dfrac{y-c}{d}\right)$ $b = 5$ and $d = 1$.

Thus, decoding, we have:

$$s_{xy} = bds_{XY} = (5)(1)(2.6) = 13$$

i.e., $s_{xy} = 13$

To summarise:

STATEMENT 3.13

Given the bivariate set $((x_1, y_1), (x_2, y_2), \ldots, (x_n, y_n))$ and using the codings $X = \dfrac{x-a}{b}$; $Y = \dfrac{y-c}{d}$, we calculate the covariance by:

(a) calculating $s_{XY} = \dfrac{1}{n} \sum XY - \bar{X}\bar{Y}$; and

(b) decoding, using $s_{xy} = bd\, s_{XY}$.

Proof of (b)   Now:

$$s_{xy} = \frac{\sum (x_i - \bar{x})(y_i - \bar{y})}{n}$$

$$= \frac{\sum [(a + bX - (a + b\bar{X}))(c + dY - (c + d\bar{Y}))]}{n}$$

$$= \frac{1}{n} \sum [(bX - b\bar{X})(dY - d\bar{Y})]$$

$$= bd\, \frac{1}{n} \sum (X - \bar{X})(Y - \bar{Y})$$

$$= bd\, s_{XY} \quad \text{(by definition)}$$

as required.

We end this sub-section with a brief return to least squares regression lines. By utilising the covariance, we can bypass the need to use the usual Normal Equations in calculating regression coefficients and constants as is stated in the following.

STATEMENT 3.14

For the bivariate set $((x_1, y_1), (x_2, y_2), \ldots, (x_n, y_n))$, the regression coefficients and constants for the least squares regression lines $y = a_1 x + b_1$ (y on x) and $x = a_2 y + b_2$ (x on y) can be calculated using:

(a) $a_1 = s_{xy}/s_x^2$ and $b_1 = \bar{y} - a_1\bar{x}$.

(b) $a_2 = s_{xy}/s_y^2$ and $b_2 = \bar{x} - a_2\bar{y}$.

Proof Left as a directed exercise.

EXAMPLE 3.25

Find the least squares regression line of y on x for the following data:

x	20	20	22	23	23	25	28	34
y	43	41	43	40	38	33	34	30

SOLUTION

The least squares line will be obtained using the form given in the previous statement. That is, $y = a_1 x + b_1$, where $a_1 = S_{xy}/S_x^2$ and $b_1 = (\Sigma y/n) - a_1(\Sigma x/n)$.

x	y	xy	x^2
20	43	860	400
20	41	820	400
22	43	946	484
23	40	920	529
23	38	874	529
25	33	825	625
28	34	952	784
34	30	1020	1156
195	302	7217	4907

Thus: $n = 8$; $\Sigma x = 195$; $\Sigma y = 302$; $\Sigma xy = 7217$; $\Sigma x^2 = 4907$

$$S_{xy} = \frac{\Sigma xy}{n} - \left(\frac{\Sigma x}{n}\right)\left(\frac{\Sigma y}{n}\right) = \frac{7217}{8} - \left(\frac{195}{8}\right)\left(\frac{302}{8}\right) = -18.0312$$

$$S_x^2 = \frac{\Sigma x^2}{n} - \left(\frac{\Sigma x}{n}\right)^2 = \frac{4907}{8} - \left(\frac{195}{8}\right)^2 = 19.2344$$

and therefore we have:

$$a_1 = \left(\frac{-18.0312}{19.2344}\right) = -0.9374$$

and

$$b_1 = \left(\frac{302}{8}\right) - (-0.9374)\left(\frac{195}{8}\right) = 60.5991$$

Thus the required y on x least squares regression line is

$$y = 60.60 - 0.94x$$

3.4.2 THE CORRELATION COEFFICIENT

DEFINITION 3.6

Given the bivariate set $(x_1, y_1), \ldots, (x_n, y_n)$, we define the *product moment correlation coefficient* as:

$$r = \frac{s_{xy}}{s_x s_y}$$

where s_{xy} is the covariance and s_x, s_y are the standard deviations of the x and y values respectively.

By virtue of the above definition, we can write:

$$r = \frac{\frac{1}{n}\sum(xy) - \bar{x}\bar{y}}{\sqrt{\left(\frac{1}{n}\sum x^2 - \bar{x}^2\right)}\sqrt{\left(\frac{1}{n}\sum y^2 - \bar{y}^2\right)}}$$

We already know that the covariance, s_{xy}, can take both positive and negative values. The reason it is divided by the product of the two standard deviations, is that the quantity so obtained (the correlation coefficient, r) lies between -1 and $+1$ (inclusive) and is independent of the units in which the data is measured.

Since we discussed the role and interpretation of the correlation coefficient early in the chapter we do not need to repeat the points here. However, there is an important relationship between the correlation coefficient and the coefficients of the two regression lines which is given in the following statement.

STATEMENT 3.15

If $y = a_1 x + b_1$ and $x = a_2 y + b_2$ are the two regression lines and r is the correlation coefficient for the bivariate set $(x_1, y_1), \ldots, (x_n, y_n)$, then:

(a) $r^2 = a_1 a_2$ and, further,

(b)
$$r = +\sqrt{a_1 a_2} \quad \text{for } a_1, a_2 \text{ positive}$$
$$r = -\sqrt{a_1 a_2} \quad \text{for } a_1, a_2 \text{ negative}$$

Proof   We have, from Statement 3.14, that the two regression coefficients a_1 and a_2 can be expressed in the form $a_1 = s_{xy}/s_x^2$ and $a_2 = s_{xy}/s_y^2$.

Hence:
$$a_1 a_2 = \frac{s_{xy} s_{xy}}{s_x^2 s_y^2} = \left(\frac{s_{xy}}{s_x s_y}\right)^2$$

$$= r^2 \quad \text{(from the previous definition)}$$

completing the proof.

(Note on (b)   The fact that $r^2 = a_1 a_2$ necessarily implies that $r = \pm\sqrt{a_1 a_2}$. We have not proved the exact form of (b) here due to the mathematics involved; however, there is no ambiguity in the statement since we always have either (a) both a_1 and a_2 positive or (b) both a_1 and a_2 negative.)

This relationship is a useful one in many practical circumstances, particularly when we need to calculate both the two regression lines together with the correlation coefficient for the same set of data.

EXAMPLE 3.26

Find the product moment correlation coefficient for the following bivariate set of data.

x	1	3	4	5	6
y	2	6	7	6	8

SOLUTION

We set up a table as follows.

x	y	xy	x^2	y^2
1	2	2	1	4
3	6	18	9	36
4	7	28	16	49
5	6	30	25	36
6	8	48	36	64
19	29	126	87	189

$$\sum x = 19 \quad \sum y = 29$$

$\therefore$
$$\bar{x} = 19/5 = 3.8$$

and
$$\bar{y} = 29/5 = 5.8$$

$$\sum xy = 126$$

$$\sum x^2 = 87 \quad \sum y^2 = 189$$

$$S_{xy} = \frac{1}{n} \sum xy - \bar{x}\bar{y} = 126/5 - (3.8)(5.8) = 3.16$$

$$S_x = \sqrt{\frac{1}{n} \sum x^2 - \bar{x}^2} = \sqrt{87/5 - (3.8)^2} = \sqrt{2.96} = 1.72 \text{ (2D)}$$

$$S_y = \sqrt{\frac{1}{n} \sum y^2 - \bar{y}^2} = \sqrt{189/5 - (5.8)^2} = \sqrt{4.16} = 2.04 \text{ (2D)}$$

$$\therefore \qquad r = \frac{S_{xy}}{S_x S_y} = \frac{3.16}{(1.72)(2.04)} = 0.9 \text{ (1D)}$$

EXAMPLE 3.27

For two sets of bivariate data, the regression lines for each are, respectively:

(a) $y = 1.94x + 10.83$ (y on x) and $x = 0.15y + 6.18$ (x on y)

(b) $y = -1.96x + 15$ (y on x) and $y = -2.22x + 15.91$ (x on y)

Find the product moment coefficient of correlation in each case.

SOLUTION

(a) r is given by $\sqrt{a_1 a_2}$. Here $a_1 = 1.94$ and $a_2 = 0.15$.

$\therefore \qquad r = +\sqrt{(1.94)(0.15)}$ since the regression

$\qquad \qquad = 0.54$ (2D)   coefficients are positive

(b) $a_1 = -1.96$ from the y on x line.

The line of x on y is given as $y = -2.22x + 15.91$.

In order to determine a_2, we need to transform the above equation as follows:

$$-2.22x = y - 15.91 \quad \text{giving} \quad x = \frac{15.91 - y}{2.22}$$

Finally we obtain:

$$x = -\frac{1}{2.22}y + \frac{15.91}{2.22}$$

Thus: $$a_2 = -\frac{1}{2.22} = -0.45$$

We now have:

$$r = -\sqrt{(-1.96)(-0.45)} = -0.94 \text{ (2D)}$$

(Note that r is negative since both regression coefficients are negative.)

3.4.3 THE METHOD OF CODING

For sets of bivariate data consisting of awkwardly large numbers we can use a method of coding in the usual way. As in the case of the regression lines and covariance, we can code for either one or both of the variables involved.

For instance, given the set:

x	25	30	35	40	45	50	55
y	82	73	81	85	79	87	95

it is obvious that x is a controlled variable and the best coding would clearly be $X = \dfrac{x - 40}{5}$. In the case of y, the best we could do is to subtract a 'central' value (the median, say). That is, $Y = y - 82$. Performing the arithmetic, we obtain:

X	-3	-2	-1	0	1	2	3
Y	0	-9	-1	3	-3	5	13

From here, we can proceed to calculate the correlation coefficient for the bivariate set of X and Y, r_{XY}, say. The interesting (and very useful) fact here is that we would have obtained exactly the same value for the correlation coefficient if we had used the original uncoded values. That is, $r_{XY} = r_{xy}$ (the latter denoting the coefficient obtained using the uncoded values).

The student should fully appreciate the practicalities of the situation here; namely, any coding can be used for one or both of the given variables (the exact form of the coding, of course, is chosen to simplify the values and hence the arithmetic as much as possible) and the resulting coefficient of correlation is the true one.

Returning to the given data, we calculate r as follows:

X	Y	XY	X^2	Y^2
-3	0	0	9	0
-2	-9	18	4	81
-1	-1	1	1	1
0	3	0	0	9
1	-3	-3	1	9
2	5	10	4	25
3	13	39	9	169

$\sum X = 0;\qquad \sum Y = 8$

$\sum XY = 65$

$\sum X^2 = 28;\qquad \sum Y^2 = 294$

$$s_{XY} = \frac{1}{n}\sum XY - \bar{X}\bar{Y} = 65/7 - (0)(8/7) = 9.286 \ (3D)$$

$$s_X = \sqrt{\frac{1}{n}\sum X^2 - \bar{X}^2} = \sqrt{28/7 - (0)^2} = 2$$

$$s_Y = \sqrt{\frac{1}{n}\sum Y^2 - \bar{Y}^2} = \sqrt{294/7 - (8/7)^2} = 6.379 \ (3D)$$

$$\therefore \quad r_{xy} = r_{XY} = \frac{s_{XY}}{s_X s_Y} = \frac{9.286}{(2)(6.739)} = 0.69 \ (2D)$$

To summarise:

STATEMENT 3.16

Given the bivariate set $(x_1, y_1), \ldots, (x_n, y_n)$ and using the codings $X = \dfrac{x-a}{b}$; $Y = \dfrac{y-c}{d}$, if we put $r_{XY} = \dfrac{s_{XY}}{s_X s_Y}$ then we have the relationship $r_{xy} = r_{XY}$ $\left(\text{where } r_{xy} = \dfrac{s_{xy}}{s_x s_y}\right)$.

That is, when using a method of coding to find the correlation coefficient, no decoding formula is necessary.

Proof   Now $r_{XY} = \dfrac{s_{XY}}{s_X s_Y}$.

But from Statement 3.13, $s_{XY} = \left(\dfrac{1}{bd}\right) s_{xy}$ and also from a simple extension of Statement 2.12, $s_X = \left(\dfrac{1}{b}\right) s_x$ and $s_Y = \left(\dfrac{1}{d}\right) s_y$.

Therefore:

$$r_{XY} = \frac{\dfrac{1}{bd} s_{xy}}{\dfrac{1}{b} s_x \, \dfrac{1}{d} s_y} = \frac{s_{xy}}{s_x s_y} = r_{xy} \quad \text{as required.}$$

We end the section with a note on the interpretation of a correlation coefficient. A high coefficient of correlation between two variables, while demonstrating a clear statistical relationship, might not necessarily mean there is a causal relationship present (one variable affecting the other directly).

For example, measuring the number of reported crimes against the consumption of milk in Britain over the past ten years might well result in a high coefficient of correlation. But no one would seriously suggest that after drinking a pint of milk we would be more likely to rob the nearest bank. It just so happens that society tends (for one reason or another) to produce more criminals (or perhaps we are made more aware of them) and at the same time the fact that the population grows yearly necessarily means that the general demand for milk is greater. Clearly, the two factors 'crime' and 'milk consumption' will almost certainly have no direct effect on each other.

An example of a relationship such as this is known as 'spurious correlation'.

3.5 BIVARIATE FREQUENCY DISTRIBUTIONS

The calculations for the regression lines and correlation coefficient in the case of bivariate frequency distributions are slightly more complicated by virtue of the fact that the data is in a more complex form and in general its volume will be greater. But we shall see that the formulae necessary for calculations are readily obtainable from the formulae already used in earlier sections.

As a model, we will take a general bivariate frequency distribution as follows:

	x_1	x_2	$\cdots$	x_n	
y_1	f_{11}	f_{12}	$\cdots$	f_{1n}	f_{y_1}
y_2	f_{21}	f_{22}	$\cdots$	f_{2n}	f_{y_2}
.	.	.		.	.
.	.	.		.	.
.	.	.		.	.
y_m	f_{m1}	f_{m2}	$\cdots$	f_{mn}	f_{y_m}
	f_{x_1}	f_{x_2}	$\cdots$	f_{x_n}	

where $f_{y_1}, f_{y_2}, \ldots, f_{y_m}$ are row totals and $f_{x_1}, f_{x_2}, \ldots, f_{x_n}$ are column totals.

Note f_{y_1} can be literally translated as 'the frequency of y_1' and so on.

The Normal Equations of Statement 3.10 are easily adapted (by writing Σfy for Σy; Σfx for $\Sigma x \ldots$ etc.) to give the Normal Equations for the case of a bivariate frequency distribution. They are written:

$$\left. \begin{aligned} \Sigma fy &= a_1 \Sigma fx + b_1 \Sigma f \\[2mm] \Sigma fxy &= a_1 \Sigma fx^2 + b_1 \Sigma fx \end{aligned} \right\} \quad (y \text{ on } x) \qquad [1]$$

$$\left. \begin{aligned} \Sigma fx &= a_2 \Sigma fy + b_2 \Sigma f \\[2mm] \Sigma fxy &= a_2 \Sigma fy^2 + b_2 \Sigma fy \end{aligned} \right\} \quad (x \text{ on } y) \qquad [2]$$

Referring to the model given earlier, Σfy would be calculated as:

$$f_{y_1} y_1 + f_{y_2} y_2 + \ldots + f_{y_m} y_m = \sum_{i=1}^{m} f_{y_i} y_i$$

Similarly, $\Sigma\, fx$ would be calculated as:

$$f_{x_1}x_1 + f_{x_2}x_2 + \ldots + f_{x_n}x_n = \sum_{i=1}^{n} f_{x_i}x_i$$

Σf is calculated as $\Sigma f_{y_i} = \Sigma f_{x_i}$ (i.e., the grand total of the table frequencies).

Also: $\qquad \Sigma fx^2 = \sum_{i=1}^{n} f_{x_i}x_i^2 \quad$ and $\quad \Sigma fy^2 = \sum_{i=1}^{m} f_{y_i}y_i^2$

$\Sigma\, fxy$ is the most complicated quantity to calculate.

(a)   We need to compute each xy value separately;
(b)   Multiply this value by its corresponding frequency; and
(c)   Sum all these products.

That is:

$$\Sigma\, fxy = f_{11}x_1y_1 + f_{12}x_2y_1 + f_{13}x_3y_1 + \ldots + f_{mn}x_ny_m$$

To demonstrate this procedure, we will calculate Σf, Σfx, Σfy, Σfx^2, Σfy^2 and Σfxy for the following bivariate frequency distribution.

x → y ↓	0	1	2	3	4
5	1	1	1	2	0
10	1	2	3	5	1
15	0	1	2	6	3
20	0	0	1	1	0

The table for calculations is set up as follows.

x → y ↓	0	1	2	3	4	f	fy	fy^2
5	⓪ 1 ⟋0	⑤ 1 ⟋5	⑩ 1 ⟋10	⑮ 2 ⟋30	⑳ 0 ⟋0	5	25	125
10	⓪ 1 ⟋0	⑩ 2 ⟋20	⑳ 3 ⟋60	㉚ 5 ⟋150	㊵ 1 ⟋40	12	120	1200
15	⓪ 0 ⟋0	⑮ 1 ⟋15	㉚ 2 ⟋60	㊸ 6 ⟋270	㊽ 3 ⟋180	12	180	2700
20	⓪ 0 ⟋0	⑳ 0 ⟋0	㊵ 1 ⟋40	㊿ 1 ⟋60	⑧⓪ 0 ⟋0	2	40	800
f	2	4	7	14	4	31	365	4825
fx	0	4	14	42	16	76	$\Sigma fxy = 940$	
fx^2	0	4	28	126	64	222		

Notice that we have split the boxes in the main body of the table into 3 sections. The middle figure is the frequency shown in the original table, the figure at top left is the product of the respective x and y values and the figure at bottom right is the product of the xy value and the frequency,

i.e.

The fxy values are then summed (the total in this case being 940), the total is usually shown at the bottom right of the completed table. A useful check that the frequencies of the table have been added correctly is that the sum of the 'row f' is the same as the sum of the 'column f' (in this case 31). So that, here, $\Sigma f = 31$, $\Sigma fx = 76$, $\Sigma fy = 365$, $\Sigma fx^2 = 222$, $\Sigma fy^2 = 4825$ and $\Sigma fxy = 940$.

We will now consider the calculation of the correlation coefficient. By adapting Statement 3.12 to take account of frequencies, we obtain a form for the covariance as:

$$s_{xy} = \frac{\Sigma fxy}{\Sigma f} - \left(\frac{\Sigma fx}{\Sigma f}\right)\left(\frac{\Sigma fy}{\Sigma f}\right) \qquad [3]$$

For the given data then, we have:

$$s_{xy} = \frac{940}{31} - \left(\frac{76}{31}\right)\left(\frac{365}{31}\right)$$

$$= 1.46 \ (2D)$$

Also: $s_x = \sqrt{\frac{\Sigma fx^2}{\Sigma f} - \left(\frac{\Sigma fx}{\Sigma f}\right)^2} = \sqrt{\frac{222}{31} - \left(\frac{76}{31}\right)^2} = 1.07 \ (2D)$

and $s_y = \sqrt{\frac{\Sigma fy^2}{\Sigma f} - \left(\frac{\Sigma fy}{\Sigma f}\right)^2} = \sqrt{\frac{4825}{31} - \left(\frac{365}{31}\right)^2} = 4.12 \ (2D)$

Thus, correlation coefficient:

$$r = \frac{s_{xy}}{s_x s_y} = \frac{1.46}{(1.07)(4.02)} = 0.34 \ (2D)$$

To find the two regression lines, we could use the Normal Equations of [1] and [2] which, after substitution, take the form:

$$\left.\begin{array}{l} 365 = 76a_1 + 31b_1 \\ 940 = 222a_1 + 76b_1 \end{array}\right\} \ (y \text{ on } x)$$

and
$$76 = 365a_2 + 31b_2 \atop 940 = 4825a_2 + 365b_2 \Bigg\} \ (x \text{ on } y)$$

The above can be solved to yield the 4 regression constants a_1, a_2, b_1 and b_2.

However, we will use the form of Statement 3.14 to obtain the regression constants.

That is:

$$a_1 = \frac{s_{xy}}{s_x{}^2}; \quad b_1 = \bar{y} - a_1\bar{x} \quad \text{and} \quad a_2 = \frac{s_{xy}}{s_y{}^2}; \quad b_2 = \bar{x} - a_2\bar{y}$$

Note that the above expressions can be used when we have bivariate data in the form of a frequency distribution.

We already have $s_{xy} = 1.46$ and we easily calculate $s_x{}^2 = 1.15$, $s_y{}^2 = 17.01$.

Also:
$$\bar{x} = \frac{\sum fx}{\sum f} = \frac{76}{31} = 2.45; \quad \bar{y} = \frac{\sum fy}{\sum f} = \frac{365}{31} = 11.77$$

Hence,

$$a_1 = \frac{1.46}{1.15} = 1.27 \quad \text{and} \quad b_1 = 11.77 - (1.27)(2.45) = 8.66$$

$$a_2 = \frac{1.46}{17.01} = 0.09 \quad \text{and} \quad b_2 = 2.45 - (0.09)(11.77) = 1.39$$

The two regression lines, then, are:

$$y = 1.27x + 8.66 \quad (y \text{ on } x)$$
$$x = 0.09y + 1.39 \quad (x \text{ on } y)$$

A method of coding can be used for bivariate frequency distributions, the form, use, and any decoding necessary being identical to the techniques used earlier.

Where a bivariate sample is given in the form of a grouped frequency distribution, the technique of using class mid-points as representative group values is employed.

The following formal example should demonstrate these points.

EXAMPLE 3.28

The following results were obtained by 73 students sitting both Mathematics and Computer Science examinations.

(a) Calculate the correlation coefficient for the Mathematics (x) and Computer Science (y) marks.

(b) A student was absent from the Computer Science exam, but obtained   37 marks   in Mathematics. Estimate a Computer Science mark, using the given data.

		Mathematics (x)					Total
		0–19	20–39	40–59	60–79	80–99	
	0–19			1	2		3
Computer	20–39		3	5	2	1	11
Science	40–59	1	1	4	6	4	16
(y)	60–79		2	8	14	6	30
	80–99			4	7	2	13
	Total	1	6	22	31	13	73

SOLUTION

Since the data is in grouped form and the values large, we will use a coding method (for the class mid-points). Since, for both x and y, the class width is 20 and the central value (mid-point of middle class) is 49.5, we can use the identical codings $X = \dfrac{x - 49.5}{20}$ and $Y = \dfrac{y - 49.5}{20}$.

(a)   We require $r = \dfrac{s_{XY}}{s_X s_Y}$ using the forms:

$$s_{XY} = \frac{\sum fXY}{\sum f} - \left(\frac{\sum fX}{\sum f}\right)\left(\frac{\sum fY}{\sum f}\right)$$

$$s_X^2 = \frac{\sum fX^2}{\sum f} - \left(\frac{\sum fX}{\sum f}\right)^2 \quad \text{and} \quad s_Y^2 = \frac{\sum fY^2}{\sum f} - \left(\frac{\sum fY}{\sum f}\right)^2$$

We now set up a table for calculations below.

Y \ X	-2	-1	0	1	2	f_Y	$f_Y Y$	$f_Y Y^2$
-2	4 / 0 / 0	2 / 0 / 0	0 / 1 / 0	−2 / 2 / −4	−4 / 0 / 0	3	−6	12
-1	2 / 0 / 0	1 / 3 / 3	0 / 5 / 0	−1 / 2 / −2	−2 / 1 / −2	11	−11	11
0	0 / 1 / 0	0 / 1 / 0	0 / 4 / 0	0 / 6 / 0	0 / 4 / 0	16	0	0
1	−2 / 0 / 0	−1 / 2 / −2	0 / 8 / 0	1 / 14 / 14	2 / 6 / 12	30	30	30
2	−4 / 0 / 0	−2 / 0 / 0	0 / 4 / 0	2 / 7 / 14	4 / 2 / 8	13	26	52
f_X	1	6	22	31	13	73	39	105
$f_X X$	−2	−6	0	31	26	49	$\Sigma fXY = 41$	
$f_X X^2$	4	6	0	31	52	93		

From the table we have:

$$\Sigma f = 73; \quad \Sigma fX = 49; \quad \Sigma fY = 39; \quad \Sigma fX^2 = 93;$$

$$\Sigma fY^2 = 105 \quad \text{and} \quad \Sigma fXY = 41$$

(the latter is the sum of the numbers in the bottom right hand corners in the table body, being shown in the extreme bottom right hand corner of the table).

Hence:

$$s_{XY} = \frac{41}{73} - \left(\frac{49}{73}\right)\left(\frac{39}{73}\right) = 0.203 \quad (3D)$$

$$s_X^2 = \frac{93}{73} - \left(\frac{49}{73}\right)^2 = 0.823 \quad (3D)$$

and: $$s_Y^2 = \frac{105}{73} - \left(\frac{39}{73}\right)^2 = 1.153 \quad (3D)$$

Also $s_X = 0.907$ and $s_Y = 1.074$.

Therefore:

$$r_{xy} = r_{XY} = \frac{s_{XY}}{s_X s_Y} = \frac{0.203}{(0.907)(1.074)} = 0.21 \quad (2D)$$

(b)   In order to estimate a value for y given an x value, we need to use the regression line of y on x. Since we are using coded values here, we first need to determine the line of Y on X, then decode. Since we have already calculated s_{XY} and s_X^2, we will use the form of Statement 3.14 (w.r.t. coded values).

That is, for $Y = A_1 X + B_1$, we have:

$$A_1 = \frac{s_{XY}}{s_X^2} \quad \text{and} \quad B_1 = \bar{Y} - A_1 \bar{X}$$

Here,

$$A_1 = \frac{0.203}{0.823} = 0.247 \quad (3D)$$

and $$B_1 = \frac{39}{73} - (0.247)\left(\frac{49}{73}\right) = 0.368 \quad (3D)$$

Hence the Y on X line takes the form $Y = 0.247X + 0.368$.

Decoding, we have:

$$\frac{y - 49.5}{20} = (0.247)\left(\frac{x - 49.5}{20}\right) + 0.368$$

$$\therefore \qquad y = 20 \left[\left(\frac{0.247}{20}\right)x - \frac{(0.247)(49.5)}{20} + 0.368\right] + 49.5$$

giving $$y = 0.25x + 44.63 \quad \text{(the line of } y \text{ on } x)$$

So, for a Mathematics (x) mark of 37, we estimate a Computer Science (y) mark using the above by:

$$y = (0.25)(37) + 44.63 = 53.9 \ (1D)$$

3.6 RANK CORRELATION

We now turn to a different method of calculating the correlation coefficient for a set of bivariate data. This method involves working with the 'rankings' of the variables, rather than their actual values.

Ranking a set of values simply entails allocating the rank '1' to the smallest value, '2' to the next largest . . . and so on.

For example,

x	12	93	41	2	15
x rank	2	5	4	1	3

In the case of variables that have some of their values repeated, the rank that is allocated is taken as the *average of the ranks* that would have been allocated (had the values not been repeated). That is, we require variables that have the same value to have the same rank.

Two simple examples should demonstrate this point.

(a)

x	3	17	12	12	8
rank	1	5	3.5	3.5	2

(b)

x	10	16	16	17	14	16
rank	1	4	4	6	2	4

In (a), the two values $x = 12$ should have been allocated ranks of 3 and 4 respectively. In this case we take the average of 3 and 4, namely 3.5, and give this rank to both variables. Notice that $x = 17$ still has rank 5.

In (b), ranks 3, 4 and 5 are available for $x = 16$. The average rank for all three is 4, which has been allocated to these values. Again, $x = 17$ must be given its correct rank of 6.

The method employed to calculate a coefficient of correlation using a ranking method is now described, using the following data.

x	2.3	2.7	2.3	2.5	2.8	2.2	2.3	2.3	2.9
y	18	26	20	20	27	21	19	22	31

The first step is to rank the x and y values separately, giving:

r_x	3.5	7	3.5	6	8	1	3.5	3.5	9
r_y	1	7	3.5	3.5	8	5	2	6	9

r_x and r_y signify the ranks of the x and y values respectively.

The next step is to find the sum of the squares of the differences $(\Sigma \, d^2)$ between each of the ranked pairs.

d	2.5	0	0	2.5	0	4	1.5	2.5	0
d^2	6.25	0	0	6.25	0	16	2.25	6.25	0

Hence $\Sigma d^2 = 37$ (notice that we can ignore '−' signs when calculating d values, since they would be eliminated anyway upon the calculation of d^2).

Finally, the coefficient of rank correlation is obtained using the form

$$r' = 1 - \left[\frac{6 \sum d^2}{n(n^2 - 1)} \right]$$

where n is the number of data pairs.

Here, then, we have:

$$r' = 1 - \frac{6(37)}{9(80)} = 1 - 0.308 = 0.692 \text{ (3D)}$$

To summarise:

DEFINITION 3.7

Given a set of bivariate data $(x_1, y_1), \ldots, (x_n, y_n)$, the *rank correlation coefficient* is calculated and defined by:

(a)   ranking the x and y values separately in order of size (ranks being averaged for identical values),

(b)   finding the squares of the differences of corresponding ranked pairs $d_1^2, d_2^2, \ldots, d_n^2$,

(c)   calculating the sum of the above squares, Σd^2, and

(d)   evaluating $r' = 1 - \dfrac{6 \sum d^2}{n(n^2 - 1)}$, where n is the number of ranked pairs.

This is sometimes referred to as *Spearman's* rank correlation coefficient.

The main advantage of the rank over the product moment correlation coefficient is its ease of calculation. However, it should be noted that the rank correlation coefficient is often used as an approximation to the product moment correlation coefficient, the latter being considered as the most satisfactory measure due to its unique relationship with the regression coefficients. When there is a high degree of correlation present, the two measures are usually very close numerically.

In its own right, the rank correlation coefficient is useful for measuring the correlation for bivariate data that shows a non-linear relationship. The product moment correlation coefficient can only be used for variables

displaying a definite linear relationship, and could well give a low correlation value for data of a slightly non-linear kind that does, however,
show a strong relationship.

Finally, in the case of bivariate data that has one or both of the variables
in qualitative form (non-numerical), it is always possible to rank the
values in some way and thus obtain a correlation coefficient. Indeed,
often data of this kind is originated in ranked form. For example, in the
case of beauty competitions or the form of racehorses. It should be clear
that in this type of situation a product moment correlation coefficient
cannot be calculated.

EXAMPLE 3.29

The following data gives United Kingdom debits in international
business transactions (current transfers in £ million) from personal
sector (x) and central government (y), for the quarters in the
period 1970 to 1972. Calculated Spearman's coefficient of rank
correlation.

x	56 57 55 58 51 56 56 58 57 57 57 57
y	52 40 37 43 57 45 47 51 68 49 43 48

SOLUTION

Rearranging the data in x-value order and allocating ranks, we have:

x	51 55 56 56 56 57 57 57 57	57 58 58
y	57 37 52 45 47 40 68 49 43	48 43 51
r_x	1 2 4 4 4 8 8 8 8	8 11.5 11.5
r_y	11 1 10 5 6 2 12 8 3.5	7 3.5 9
$\|d\|$	10 1 6 1 2 6 4 0 4.5	1 8 2.5
d^2	100 1 36 1 4 36 16 0 20.25	1 64 6.25

Here: $\sum d^2 = 285.5$

Hence: $r' = 1 - \dfrac{6\sum d^2}{n(n^2 - 1)}$

                       $= 1 - \dfrac{6(285.5)}{12(143)}$

i.e. $r' = 0.002$ (3D)

3.7 LOGARITHMIC TRANSFORMATIONS

Only *linear* regression and correlation techniques have been considered in
this chapter, although various other techniques (known as *curvilinear*)
exist. However, there are certain types of data which, although not
exhibiting a natural linear relationship, may be transformed to linear

form. This is done in order that linear regression and correlation analyses can be performed.

Two particular non-linear relationships which can be transformed to linear form, using a logarithmic transformation, are:

(a) $y = ab^x$

Taking the log of both sides gives:

$$\log y = \log a + x \log b$$

or $$Y = A + Bx$$

In other words, Y (i.e. $\log y$) and x have a linear relationship.

(b) $y = ax^b$

Taking the log of both sides gives:

$$\log y = \log a + b \log x$$

or $$Y = A + bX$$

In this particular case, Y (i.e. $\log y$) and X (i.e. $\log x$) have a linear relationship.

3.7.1 CONFIRMING THE RELATIONSHIP

In order to confirm relationships between variables x and y of the forms:

$$y = ab^x \quad \text{or} \quad y = ax^b$$

graphically, there are two equivalent methods that can be used, involving either ordinary (linear scale) graph paper or logarithmic scale graph paper.

(a)   To confirm the relationship $y = ab^x$ graphically.

(i)   Using ordinary graph paper.

In this case, the logarithms of all the y values are calculated and x is plotted against $\log y$. To verify the relationship, a straight line should be obtained.

(ii)   Using semi-log graph paper.

Semi-logarithmic graph paper has one of its two axes *logarithmically scaled*. Fig. 3.22 shows an example of this type of graph paper.

The four successive horizontal '1' lines on the printed right hand axes can be taken as: 1, 10, 100 and 1000 or 10, 100, 1000 and 10 000 ... etc, depending on the range of y values given. Some examples of scaling are shown on the graph.

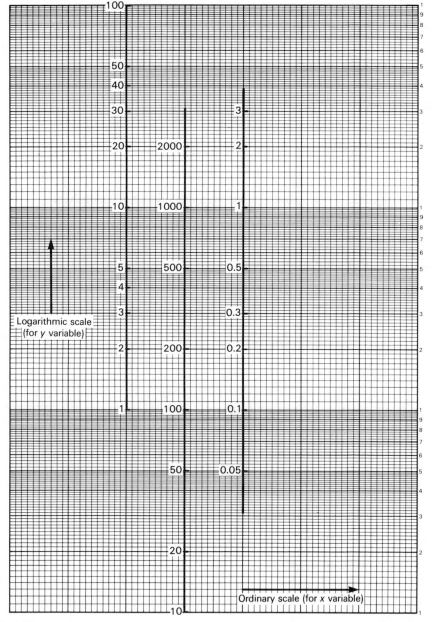

Fig. 3.22

The relationship is verified if, when y (on the vertical log scale) is plotted against x (on the ordinary scale), a *straight line* results.

The following values of x and y:

x	2	4.2	5	7.1	8
y	64	290	584	3210	7020

have been plotted in Fig. 3.23. Notice that the four '1' lines have been used as 10, 100, 1000 and 10 000 respectively in order to accommodate the range of y values given. It can be seen from the graph that there is a reasonable linear relationship (shown by the dashed line). Thus, the model $y = ab^x$ fits the given data reasonably well.

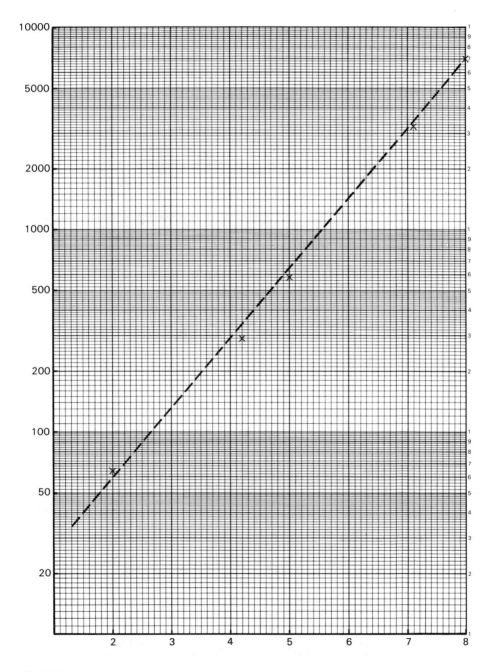

Fig. 3.23

(b)   To confirm the relationship $y = ax^b$ graphically.

(i)   Using ordinary graph paper.

In this case, the logarithms of both the y values and x values are calculated and $\log x$ is plotted against $\log y$. To verify the relationship, a straight line should be obtained.

(ii)   Using log–log graph paper.

Log–log graph paper has both its two axes *logarithmically scaled.* Fig. 3.24 shows an example of this type of graph paper. When x and y are related in the form $y = ax^b$, plotting x against y on this graph paper will result in a straight line.

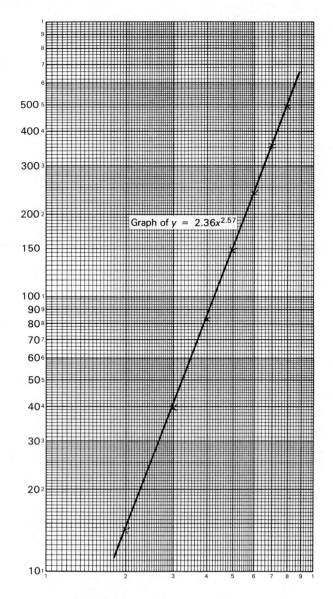

Fig. 3.24

As a demonstration, the table below shows values of y calculated for various values of x where $y = (2.36)x^{2.57}$ (i.e. $a = 2.36$ and $b = 2.57$).

x	2	3	4	5	6	7	8
y	14.0	39.7	83.2	147.7	235.9	350.6	494.1

These values of x and y are plotted in Fig. 3.24, resulting in a clear straight line.

3.7.2 IDENTIFYING THE RELATIONSHIP

The standard procedure for *estimating the values of a and b* for the above two types of relationship is given as follows:

(a) For $y = ab^x$ (via transformation $Y = A + Bx$):
   1. Obtain the values of $A(\log a)$ and $B(\log b)$, using least squares regression analysis on variables $Y(\log y)$ and x.
   2. Calculate $a = \text{antilog } A$ and $b = \text{antilog } B$.

(b) For $y = ax^b$ (via transformation $Y = A + bX$):
   1. Obtain the values of A and b, using least squares regression analysis on variables $Y(\log y)$ and $X(\log x)$.
   2. Calculate $a = \text{antilog } A$.

EXAMPLE 3.30

The number y of bacteria per unit volume present in a culture after x hours is given in the table below.

x	0	1	2	3	4	5	6
y	32	47	65	92	132	190	275

(a) Verify that $y = ab^x$ is a suitable model to describe the relationship between x and y by plotting values of $\log y$ against x on a scatter diagram.

(b) Use the method of least squares to obtain estimates of a and b.

(c) Estimate the number of bacteria present in the culture after 7 days.

SOLUTION

(a)

x	0	1	2	3	4	5	6
$Y = \log y$	1.51	1.67	1.81	1.96	2.12	2.28	2.44

The above values have been plotted below in Fig. 3.25.

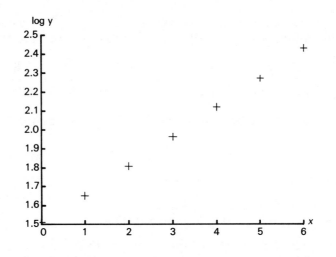

Fig. 3.25

The scatter diagram shows a clear linear relationship. Thus the model $y = ab^x$ is appropriate.

(b)

x	0	1	2	3	4	5	6
$Y = \log y$	1.51	1.67	1.81	1.96	2.12	2.28	2.44
xY	0	1.67	3.62	5.88	8.48	11.40	14.64
x^2	0	1	4	9	16	25	36

Thus $n = 7$; $\Sigma x = 21$; $\Sigma Y = 13.79$; $\Sigma x Y = 45.69$; $\Sigma x^2 = 91$

$$S_{xY} = \frac{\Sigma x Y}{n} - \left(\frac{\Sigma x}{n}\right)\left(\frac{\Sigma Y}{n}\right) = \frac{45.69}{7} - \left(\frac{21}{7}\right)\left(\frac{13.79}{7}\right) = 0.6171$$

$$S_x^2 = \frac{\Sigma x^2}{n} - \left(\frac{\Sigma x}{n}\right)^2 = \frac{91}{7} - \left(\frac{21}{7}\right)^2 \doteqdot 4$$

Thus: $B = S_{xY}/S_x^2 = 0.6171/4 = 0.1543$

and: $A = (13.79)/7 - (0.1543)(21/7) = 1.5071$

Hence: $b = \text{antilog } B = 10^{0.1543} = 1.43 \text{ (2D)}$

and: $a = \text{antilog } A = 10^{1.5071} = 32.14 \text{ (2D)}$

i.e. $y = (32.14)(1.43)^x$

(c)   When $x = 7$ (days), the number of bacteria in the culture is given by $y = (32.14)(1.43)^7 = 393$.

EXAMPLE 3.31

In an experiment, values of variables x and y were found to be:

x	10	10	12	13	13.5	14	16
y	6.75	6.58	6.33	6.15	6.10	5.96	5.76

If the two variables are known to be related in the form $y = ax^b$ estimate the values of a and b using linear least squares regression. Also estimate the value of y when $x = 11$.

SOLUTION

If $y = ax^b$, then we have $\log y = \log a + b \log x$ or $Y = A + bX$

x	10	10	12	13	13.5	14	16
y	6.75	6.58	6.33	6.15	6.10	5.96	5.76
$X = \log x$	1.00	1.00	1.08	1.11	1.13	1.15	1.20
$Y = \log y$	0.829	0.818	0.801	0.789	0.785	0.775	0.760
XY	0.829	0.818	0.865	0.876	0.887	0.891	0.912
X^2	1.0000	1.0000	1.1664	1.2321	1.2769	1.3225	1.4400

Thus:

$$n = 7; \quad \Sigma X = 7.67; \quad \Sigma Y = 5.557; \quad \Sigma XY = 6.078;$$
$$\Sigma X^2 = 8.4379$$

This gives: $S_{XY} = -0.0015$ and $S_X{}^2 = 0.0048$

and: $b = -0.31; \quad A = 1.1415$

and hence $a = $ antilog $1.1415 = 13.85$

Thus: $y = (13.85)x^{-0.31}$

When $x = 11$, $y = (13.85)(11)^{-0.31} = 6.59$ (2D)

3.8 EXERCISES

SECTION 3.1

1.   Display the following bivariate data in the form of a scatter diagram, commenting on the correlation:

x	2.9	3.5	1.8	1.3	2.4	0.7
y	9.4	11.6	6.3	5.1	8.6	4.0

2.   Construct a scatter diagram to illustrate the given set of bivariate data, stating how the two variables are correlated:

x	40	41	40	42	40	40	42	41	41	42
y	32	43	28	45	31	34	48	42	36	38

3.   The following data was obtained from a sample of 50 employees engaged by a particular company. This bivariate information consists of yearly income in £ (*x* variable) and age in years (*y* variable) for each of the 50:

(1765, 28) (1920, 31) (1730, 27) (1760, 47) (1300, 22)
(1490, 25) (1835, 30) (2950, 61) (1260, 20) (2000, 40)
(1620, 21) (1970, 46) (2300, 61) (2020, 41) (2500, 58)
(1720, 38) (2600, 60) (2200, 56) (1730, 38) (2040, 51)
(1900, 39) (2170, 40) (1800, 38) (1750, 41) (1800, 44)
(2600, 54) (1830, 31) (2240, 50) (2800, 63) (2200, 56)
(1820, 32) (2000, 53) (2600, 52) (2400, 58) (2180, 54)
(1625, 30) (2200, 48) (2950, 60) (1750, 56) (2620, 61)
(1325, 22) (2300, 52) (2050, 50) (1750, 37) (1410, 27)
(1250, 32) (1500, 30) (2350, 40) (2300, 47) (1900, 53)

Form a dot diagram and a bivariate frequency table grouping the *x* values according to 1250 up to 1500, 1500 up to 1750, . . . and the *y* values according to 20 up to 30, 30 up to 40, . . . Comment on the correlation.

4.   Display the following data in an appropriate form:

x	0	15	30	45	60	75	90	105	120
y	806	630	643	625	575	592	408	469	376

stating how the variables are correlated.

SECTION 3.2

5.   State the gradient and *y*-intercept of the following straight lines:

(a) $2y = 4x - 3$ (b) $y = \dfrac{x}{4} + 14$ (c) $x = 4y - 2$ (d) $\dfrac{x - 2}{4} = 4y$

6.   Plot the following straight lines: (a) $y = 3x + 2$;   (b) $y = \dfrac{x}{2} - 4$;   (c) $x = 4y - 2$;   (d) the line with gradient -2 and *y*-intercept 4.

7.   Obtain the gradients of the lines joining the following pairs of points: (a) (1, 2) and (2, 3);   (b) (−1, 2) and (−2, 3);   (c) (−1, −2) and (2, −3);   (d) (0, 4) and (−1, 0).

8.   Find the equation of the line that meets the *y*-axis at −2 and the *x*-axis at 3.

9.   Find the equation of the straight line passing through the points $(0, -1\frac{1}{6})$ and $(4, \frac{5}{6})$.

10.   A variable straight line has equation $y = 2x + c$. Plot those lines defined by: (a) $c = 0$;   (b) $c = 2$; and   (c) $c = 5$ on the same diagram.

11.   Solve the following pairs of simultaneous equations:
(a) $3x + 2y = 7; 5x + y = 7$ (b) $x + y + 8 = 0; x - y = 2$
(c) $3x + 2y = 2x - y - 56 = 0$

12.   Plot the points on a graph and find the equation of the line in each of the following:

(a)

x	1	3	5	7	9
y	3	9	15	21	27

(b)

x	-2	0	2	4	6
y	11	7	3	-1	-5

(c)

x	0	0.5	1	3	3.2	3.6
y	-5	-4	-3	1	1.4	2.2

(d)

x .	-4	-3	-2	-1	0
y	0	1.5	3	4.5	6

13.   Solve the following pairs of simultaneous equations for x and y:
(a) $x + y = 4; x + 3y = 8$ (b) $x - y = 1\frac{1}{2}; x + y = 3$
(c) $3x - y = 26; x - 5y = 4$ (d) $3x - 4y = 0; 4x - 3y = 14$
(e) $8x - 4y = 9x - 3y = 6$

14.   (a) Plot the graphs of $y = 5 - 5x$ and $y = 7x - 13$ on the same diagram and find the (x, y) coordinates of their point of intersection.
(b) Solve the pair of simultaneous equations $5x + y = 5; 7x - y = 13$.
(c) Comment on the results of (a) and (b)

15.   Differentiate the following functions with respect to x:
(a) $3x^2$ (b) $5x$ (c) $-2x^3$ (d) $3x^2 - 2x + 4$ (e) $2 - 8x + x^3$

16.   Find $\dfrac{dy}{dx}$ if:

(a) $y = 9x^3$;   (b) $y = 2 - 4x^4 + x^8$; and   (c) $y = 4x^{-1} + 6 - 3x$

17.   Evaluate $f'(x)$ for the following functions:
(a) $f(x) = 3x^3 - 10x^2$ (b) $f(x) = 2x^2 + 4 - 3x^{-3}$

18.   Differentiate the following functions partially w.r.t. x:

(a) $2x^2 - 4xy + y^2$ (b) $x^3 - y^3 + 2x^2y$ (c) $\dfrac{x}{y} - 1 + \dfrac{y}{x} - \dfrac{y^2}{x^2}$

19.   Find the maximum value obtained by the function $2 - 4x - 3x^2$ for all values of x.

20.   Find the local maximum and minimum values of the function $2x^3 - 4x^2 + 3$.

21.   Differentiate the function $3y^3 + y^2z - yz^2 + z^3$ w.r.t.:   (a) y;
(b) z.

SECTION 3.3

22.   30 students   sat two biology exams — a practical and a theoretical. Their marks were as follows:

Practical	30 75 58 34 52 70 50 81 62 30 57 60 45 46 48
Theoretical	22 59 78 50 41 71 32 60 49 40 55 60 22 40 60

Practical	22 62 74 38 80 43 60 38 64 62 28 70 42 54 85
Theoretical	25 39 79 35 82 50 50 45 60 68 50 83 65 70 74

(a) Draw a scatter diagram for this data.

(b) State how practical and theoretical marks are correlated (if at all).

(c) Fit an approximate ellipse around the points on the scatter diagram and using vertical and horizontal chords (at least ten of each) drawn at equal intervals, estimate the two regression lines of y on x and x on y.

(d) A student who missed the practical, obtained 61 in the theoretical exam. Estimate a practical mark using the appropriate regression line.

23.   Find the two least squares regression lines for the data of the previous exercise and compare these with the lines obtained above.

24.   Ten students sat two tests, their marks out of 10 being recorded as follows:

Test 1 (x)	8 6 5 7 10 8 10 8 6 7
Test 2 (y)	9 4 7 5 8 6 10 6 7 8

Find the two least squares regression lines.

25.   For the following set of bivariate data:

x	65 67 62 68 71 69 67 66 70 68 64 63
y	68 68 66 71 70 68 67 65 68 69 65 66

(a) construct a scatter diagram; and   (b) find the least squares regression lines of y on x and x on y.

26.   The figures in the table give the wine consumption in the United Kingdom in millions of gallons (y) for the years 1963 to 1972 (x).

Year (x)	1963 1964 1965 1966 1967
Consumption (y) (millions of gallons)	32.5 37.1 35.5 37.7 41.5
Year (x)	1968 1969 1970 1971 1972
Consumption (y) (millions of gallons)	46.4 44.8 45.8 53.9 62.0

Draw a scatter diagram to show these data.

Determine the least squares estimate of the regression line of y on x, showing all your working. Draw this line on your scatter diagram and use it to estimate the consumption for 1973.

Comment on the appropriateness of a *linear* regression model in this case, given also that the actual wine consumption in 1973 was 78.3 million gallons.   (JMB)

27.   Explain briefly how the principle of least squares is used to find a regression line based on a sample of size n. Illustrate on a rough sketch the distances whose squares are minimised, taking care to distinguish the dependent and independent variates.

The following table shows the yields (y) obtained in an agricultural experiment in tonnes per hectare after using x tonnes of fertiliser per hectare.

Fertiliser	(x)	0	0.2	0.4	0.6
Yield	(y)	1.26	1.47	1.87	2.00

(a) Find the means of x and y.

(b) Plot the measurements and their means on a graph of yield (y) against fertiliser (x).

(c) Calculate the regression line of y on x in the form $y = a + bx$.

(d) Draw the regression line you have calculated on the graph.

(e) Estimate the total yield in tonnes from a 2.2 hectare field if 0.5 tonne of fertiliser is used altogether.

(O & C)

28. To investigate a process which is carried out repeatedly in a chemical works, the amount, x, of a chemical added to a mixture is varied and the concentration, y, of the final product is noted. The results are as follows:

x (g)	10	10	15	15	20	20	25	25	30	30
y (%)	2.7	2.9	4.5	4.0	6.3	6.2	8.0	7.4	9.7	10.1

(a) Draw a scatter diagram of the data.

(b) Calculate the equation of the regression line of y on x and draw it on the scatter diagram.

(c) State, giving a reason, whether varying the amount of chemical added is an effective way of controlling the final concentration. What advice would you give to the works manager who requires the final concentration to be consistently in the range 3.5 to 4.5? (AEB) '86

29. For the following set of bivariate data:

x	10.03	10.04	10.06	10.03	10.07	10.07
y	252.3	252.4	251.7	252.5	251.3	251.8

x	10.04	10.05	10.08	10.09	10.06	10.05
y	252.8	252.0	251.7	251.1	252.0	252.5

Use the codings $X = (x - 10.05)100$; $Y = (y - 252.0)10$ to find the regression lines of Y on X and X on Y. Hence, by substitution, for Y and X in terms of y and x, find the regression lines of y on x and x on y.

30. Calculate the equation of the regression line of y on x for the following distribution:

x	25	30	35	40	45	50
y	78	70	65	58	48	42

Is it possible to calculate from the equation you have just found: (a) an estimate for the value of x when $y = 54$?   (b) an estimate for the value of y when $x = 37$?   In each case, if the answer is 'Yes', calculate the estimate. If the answer is 'No', say why not. (SUJB)

31. The following figures relate to the Index of Retail Prices in the United Kingdom:

Year (x)	1946	1947	1948	1949	1950	1951
Index (y)	153	168	181	186	192	201

Year (x)	1952	1953	1954	1955	1956
Index (y)	228	235	240	250	263

Using the codings $X = x - 1951$; $Y = y - 205$, calculate the least squares regression line of Y on X. Hence find the regression line of y on x and estimate the Price Index for 1957 and 1958. Compare with the actual Indices for 1957 and 1958 which were 273 and 281 respectively.

SECTION 3.4

32. Find the covariance of x and y (s_{xy}) for the sample:

x	2	4	5	3
y	3	6	11	8

33. Find the covariance of the following set of bivariate data:

x	10	12	10	9	11	14
y	12	11	15	18	14	8

34. If the set $((x_1, y_1), \ldots, (x_n, y_n))$ has Normal equations
$$\Sigma x = a_2 \Sigma y + nb_2 \quad \text{and} \quad \Sigma xy = a_2 \Sigma y^2 + b_2 \Sigma y$$
for the regression line of x on y, show that:

(a) $a_2 = \dfrac{s_{xy}}{s_y^2}$ (where s_y^2 is the variance of the y values); and

(b) $b_2 = \bar{x} - a_2 \bar{y}$.

(*Hint:* Solve the two equations simultaneously, multiplying the first by Σy and the second by n.)

35. Find the product moment coefficient of correlation for the following data:

x	67	82	44	46	68	24	75	58	66	32
y	10.4	8.0	9.0	9.3	8.6	12.3	10.5	9.9	10.5	14.5

using the codings $X = x - 60$ and $Y = 10(y - 10)$.

36.   (a)   Vehicles and Road Deaths (latest available figures for each country):

Countries	Vehicles per 100 population	Road deaths per 100 000 population
Great Britain	31	14
Belgium	32	30
Denmark	30	23
France	46	32
West Germany	30	26
Irish Republic	19	20
Italy	35	21
Netherlands	40	23
Canada	46	30
USA	57	35

(*Source:* Road Accident Statistics)

Calculate the product moment correlation coefficient between vehicle numbers and road deaths.

(b) Comment on your results and on the table above.

           (IOS)

37.   The marks of a set of candidates in two examination papers are tabulated below:

Paper 1	59 23 42 43 50 32 45 15 42 53 37 33 78 33 53
Paper 2	50 61 62 52 50 51 27 30 68 31 43 42 50 72 65
Paper 1	42 78 42 18 22 70 45 70 32 45 47 53 32 29
Paper 2	48 98 64 37 44 75 67 67 35 72 48 72 60 56

By using suitable codings, find the correlation coefficient.

38.   The following data gives the wheat yield (x) in millions of tons against the area planted (y) in millions of acres for England and Wales for the successive years 1964 to 1972:

x	3.7 4.1 3.4 3.8 3.4 3.3 4.2 4.7 4.7
y	2.2 2.5 2.2 2.3 2.4 2.1 2.5 2.7 2.8

Find the correlation coefficient.

SECTION 3.5

39.   344 students sat both an English and Mathematics examination the results of which are tabulated below:

	Maths mark (x)							
	0–35	35–45	45–50	50–55	55–60	60–70	70–100	Total
0– 35	0	3	5	4	4	2	0	18
35– 45	1	2	11	7	6	1	0	28
English 45– 50	1	5	14	22	16	3	1	62
mark 50– 55	4	10	24	32	21	12	1	104
(y) 55– 60	5	12	11	30	18	5	2	83
60– 70	2	3	8	11	9	2	1	36
70–100	1	1	4	5	2	0	0	13
Total	14	36	77	111	76	25	5	344

(a) Calculate a regression line of: (i) y on x; and   (ii) x on y based on the above results.

(b) Find the product moment coefficient of correlation.

40.   The following data relates to red blood cells stained with acid fuchsine. Observations are classified according to the deflection in mm (as observed using a galvanometer) versus the relative transparency.

Transparency (t)	\multicolumn{12}{c}{Area (a)}
	28	31	34	37	40	43	46	49	52	55	58	61
0.3155				1	1							
0.3455												
0.3755			1	3	1	1		1				
0.4055	2			3	3	3	3	1				
0.4355				5	12	7	2					
0.4655			3	5	8	8	8	4				
0.4955	1		2	6	10	17	10	4	4			
0.5255			2	4	13	29	21	12	3	2		
0.5555				4	8	14	21	15	6	3		
0.5855				4	5	13	16	11	12	3	1	
0.6155				1	3	3	6	5	4	2	2	3
0.6455					1	1	3	1	1			1
0.6755												
0.7055										1		

Use a method of coding $\left(\text{suggested is } T = \dfrac{t - 0.4955}{0.03} \text{ and } A = \dfrac{a - 43}{3} \right)$ to find the product moment correlation coefficient.
(Data source: Canadian Journal of Research)

SECTION 3.6

41.   Nine vegetable marrows were ranked by a panel of experts according to their size and succulence. The results are shown below.

Marrow	A B C D E F G H I
Size rank	8 3 1 9 4 2 7 5 6
Succulence rank	3 5 6 7 8 4 9 1 2

Find the rank correlation coefficient.   (AEB)'76

42.   (a) 'Ability in mathematics is the same thing as intelligence'. To test the validity of this statement 50 boys are given a test in mathematics and an intelligence test. When measured from suitable origins, their mathematics marks (x) and intelligence quotients (y) yield the following results:

$$\sum x = 25, \quad \sum y = 140, \quad \sum x^2 = 1713,$$
$$\sum y^2 = 6380, \quad \sum xy = 1990$$

Calculate the product moment correlation coefficient of x and y.

(b) Two embroiderers, A and B, are asked to place nine purple patches in order with the reddest first and the bluest last. The orders given are:

A	1 2 3 4 5 6 7 8 9
B	4 3 5 1 2 7 6 9 8

Calculate a coefficient of rank correlation for these data.   (Cambridge)

43.   In an agricultural experiment, a number of cultures were subjected to a particular treatment and their bacterial numbers (in millions per ml) were measured at a particular age (in days). Find the rank correlation coefficient.

Age	1	1	2	2	2	2	3
Bacterial No.	336	242	1058	1014	648	1048	1348
Age	7	7	7	14	14	14	16
Bacterial No.	2072	2925	2240	2825	2560	4900	3550

SECTION 3.7

44.   The variables x and y are supposed to conform to the relationship $y = ab^x$. Seven measurements were taken, giving the following results:

x	1.2	1.2	1.25	1.4	1.4	1.6	1.9	2.3
y	56.8	55.4	59.1	70.6	74.0	89.2	124.9	193.1

(a) By plotting $Y = \log y$ against x on a scatter diagram, show that the variables do conform to the above given relationship.

(b) Using the linear model $Y = A + Bx$ (where $A = \log a$ and $B = \log b$), find the values of A and B (and hence a and b) by the method of least squares.

(c) Estimate the value of y when $x = 1.5$.

45.   The following table gives experimental values of the pressure p of a given mass of gas corresponding to various values of the volume v, both measured in standard units. According to thermodynamic principles, a relationship having the form $p = av^{-b}$, where a and b are constants, should exist between the variables.

v	543	618	724	887	1186	1940
p	612	495	376	284	192	101

(a) Using the transformation $\log p = \log a - b \log v$, find the values of a and b by the least squares method.

(b) Estimate p when $v = 1000$.

46.   A process consists of mixing a number of volatile chemicals together and measuring the temperature (y, in degrees Celsius) after x hours of activity. Measurements of a particular mixture were obtained as follows:

x	0.84	0.91	1.12	1.41	1.55	1.68	1.84
y	5.7	6.5	9.6	16.2	21.0	26.6	35.6

It is believed that x and y conform to the law $y = ab^x$.

(a) By transforming y logarithmically, use the least squares technique to estimate values of a and b to 2D.

(b) Estimate the temperature of the mixture after exactly one hour.

(c) If the maximum temperature for the mixture in this process is $50°C$, what should the upper time limit be for which the chemicals are left active?

(d) Comment on the validity of the result of (c).

MISCELLANEOUS

47.   12 students were given a prognostic test at the beginning of a course
and their scores X_i in the test were compared with their scores Y_i
obtained in an examination at the end of the course $(i = 1, 2, \ldots, 12)$.
The results were as follows:

X_i	1 2 2 4 5 5 6 7 8 8 9 9
Y_i	3 4 5 5 4 8 6 6 6 7 8 10

Find the equation of the regression line of Y on X and determine the
correlation coefficient between X and Y.   (SUJB)

48.   The following data gives the numbers (in millions) of broadcast
receiving licences issued for black and white and colour televisions
for the period January to October 1974 inclusive:

Black/white (x)	12.05	11.92	11.77	11.68	11.56
Colour (y)	5.24	5.40	5.56	5.67	5.80
Black/white (x)	11.49	11.40	11.31	11.20	11.04
Colour (y)	5.89	5.99	6.11	6.25	6.42

(a) Calculate the correlation coefficient and the two regression lines
for the above data.

(b) Estimate the number of colour television licences that will be
issued when the number of black and white licences has fallen to
10 million.   (Data source: M.D.S.)

49.   Find the least squares regression lines of y on x and x on y given
the following data:

x	3 4 4 5 5 9 10 11 12 15 16 18
y	18 18 14 13 18 5 10 13 9 4 6 8

State without calculation how the variables are correlated.

50.   In an experiment to investigate the change in resistance of N-type
germanium with change in temperature, the following results were
observed:

Temperature (degrees Kelvin) (t)	Resistivity (ohms-metres) (x)
172	0.002 15
188	0.002 39
200	0.002 57
217	0.002 90
227.5	0.003 19
239	0.003 39
249	0.003 63
261	0.003 93
272	0.004 15
289	0.004 60
294.5	0.004 79
294.5	0.004 75
305	0.005 08
321	0.005 47
331.5	0.005 80
340.5	0.006 00

Using the codings $T = t - 261$
and $X = 100\,000(x - 0.003\,93)$,
calculate the product moment
coefficient of correlation.

51.   Given the following data:

x	-12	-8	-4	0	4	8	12
y	-10.95	-4.32	-2.51	-0.83	3.61	2.50	10.00

Use the least squares technique to find the regression line of y on x. Given no further information about the source of the above data, explain why the least squares regression line of x on y would almost certainly be invalid in an estimating capacity.

52.   (a) Find the product moment correlation for the following distribution:

x	18	33	26	25	28	22	32	24
y	53	40	45	44	40	51	37	50

(b) Two judges in a baby competition rank the 8 babies as follows:

Baby	A	B	C	D	E	F	G	H
Judge 1	2	6	1	5	7	4	3	8
Judge 2	3	1	5	2	7	8	4	6

Calculate a rank correlation coefficient between the two judges.

(SUJB)

53.   (a) From 10 pairs of values of x and y find the line of regression of y on x, given that:

$$\sum x = 35, \quad \sum y = 95, \quad \sum x^2 = 275, \quad \sum xy = 445$$

(b) The orders of merit of ten individuals at the start and finish of a course of training were:

Individual	A	B	C	D	E	F	G	H	I	J
Order at start	1	2	3	4	5	6	7	8	9	10
Order at finish	5	3	1	9	2	6	4	7	10	8

Find Spearman's coefficient of rank correlation between the two orders.

(O & C)

54.   The table below shows five pairs of values of y and t.

t	-2	-1	0	1	2
y	38	34	31	29	25

(a) Find the mean value of y.

(b) Find the line of regression of y on t in the form $y = a + bt$.

The values of x are exactly related to t by the equation $x = (t + 3)/2$.

(c) Deduce the line of regression of y on x.

(d) Estimate the value of x at which y becomes zero.

(O & C)

55.   (a) Describe circumstances in which you would use: (i) rank correlation coefficient;   (ii) product moment correlation coefficient.

(b) Calculate: (i) rank correlation coefficient; and   (ii) product
moment correlation coefficient between Mathematics mark and Statistics
mark for the following examination candidates.

	Mathematics Mark	Statistics Mark
Candidate 1	10	8
2	18	17
3	17	20
4	14	18
5	12	14
6	13	9
7	8	12
8	20	22

(IOS)

56.

Ounces of fertilizer per square yard (x)	Yield of crop (y)
0	160
1	168
2	176
3	179
4	183
5	186
6	189
7	186
8	184

(a) Calculate a least squares linear regression line of yield of crop on
ounces of fertilizer.

(b) Plot the data and the regression line on graph paper and comment
on your results.   (IOS)

57.   The heights h,   in cm, and weights W,   in kg, of   10 people are
measured. It is found that $\Sigma h = 1710$, $\Sigma W = 760$, $\Sigma h^2 = 293\,162$,
$\Sigma hW = 130\,628$ and $\Sigma W^2 = 59\,390$.   Calculate the correlation coefficient between the values of h and W.   What is the equation of the
regression line of W on h?   (O & C)

58.   The total monthly rainfall, correct to the nearest millimetre, and the
mean number of hours of sunshine per day, correct to the nearest
tenth of an hour, observed at a weather station in 1973 are given in
the following table.

Month	Rainfall (mm)	Sunshine (h)
January	39	1.1
February	35	2.7
March	20	4.5
April	63	5.1
May	76	5.5
June	65	7.6
July	88	5.2
August	54	5.7
September	77	4.8
October	51	2.9
November	44	2.8
December	60	1.8

(a) Plot the data given above as a scatter diagram.

(b) Rank the month of the year according to the amount of rainfall, giving rank 1 to the month with the highest rainfall.

(c) Repeat part (b) for the amount of sunshine, this time giving rank 1 to the month with the most sunshine.

(d) Calculate a coefficient of rank correlation for the given data, commenting briefly on your answer.

(e) Discuss how the diagram for part (a) and the answer to part (d) suggest the same conclusion.   (London)

59.   In an international competition for pianists two judges, A and B, placed the eight competitors in order of merit for (a) technical ability, (b) general musicianship, i.e., sensitivity, interpretation of the music, etc. The orders were as follows:

	Judge								
Technical ability	A	1	2	3	4	5	6	7	8
	B	2	3	1	7	4=	4=	6	8
General musicianship	A	1	2	3	4	5	6	7	8
	B	5	6	4	1	7	2	8	3

Calculate a coefficient of rank correlation for each of these sets of figures and comment on your results.   (Cambridge)

60.   Sugar Beet Crop and amount of Rainfall, 1970–1975.

Year	Crop of Sugar Beet (100 000 tons)	Rainfall (June, July and August) (cm)
1970	63	20
1971	77	26
1972	61	17
1973	73	22
1974	45	24
1975	52	14

(*Source:* Monthly Digest of Statistics).

Calculate: (a) the product-moment correlation coefficient,   (b) the rank correlation coefficient, between amount of rainfall and crop of sugar beet. Comment on your results.   (IOS)

61.   A research worker, Dr Lin Guistic, gave each of eight children a list of 100 words of varying difficulty, and asked them to define the meaning of each one. The table below gives for each child, their age in years and the number of correctly defined words.

Child	A	B	C	D	E	F	G	H
Age, x	2.5	3.1	4.3	5.0	5.9	7.1	8.1	9.4
Number of correct words, y	9	15	26	35	43	57	69	88

It is decided to fit a model of the form $\log_e y = a + b \log_e x$.

(a) Verify that this is a reasonable course to take by plotting the above data on log-log paper.

(b) Transform the data in the table, recording your values correct to one decimal place, to a form appropriate for fitting the above model.

(c) Calculate the equation of the regression line of $\log_e y$ on $\log_e x$ as given above.   (AEB) '83

62.   Given that $y = a_1 x + b_1$ [1] is the regression line of y on x for a set of bivariate data, and using the Normal equation $\Sigma y = a_1 \Sigma x + nb_1$ [2], show that:

(a) $\bar{y} = a_1 \bar{x} + b_1$ [3] (by using only equation [2]); and

(b) $y - \bar{y} = a_1 (x - \bar{x})$ (by using equations [1] and [3]).

Hence show that the regression line $y = a_1 x + b_1$ passes through the point $(\bar{x}, \bar{y})$.

Using a similar argument, show that the regression line $x = a_2 y + b_2$ (of x on y) also passes through $(\bar{x}, \bar{y})$ (i.e., both regression lines always pass through the mean point of the data pairs).

63.   For the bivariate set $((x_1, y_1), \ldots , (x_n, y_n))$, it is required that we form the regression line $y = a_1 x + b_1$ (y on x). The substitutions $Y = \dfrac{y - a}{b}$ and $X = \dfrac{x - c}{d}$ are made and the regression line of Y on X is found to be $Y = A_1 X + B_1$ [1]. Show, by substituting Y and X in terms of y and x into [1], that:

(a) $a_1 = \dfrac{bA_1}{d}$ and (b) $b_1 = a + bB_1 - ca_1$

Hence, if $x = a_2 y + b_2$ and $X = A_2 Y + B_2$, deduce that:

(c) $a_2 = \dfrac{dA_2}{b}$ and (d) $b_2 = c + dB_2 - aa_2$.

64.   In an experiment, ten bivariate observations are taken in terms of the two variables x and y. These values are then coded using $X = x - 103$ and $Y = y - 97$, to give the following totals: $\Sigma X = -3$; $\Sigma Y = 0$; $\Sigma XY = -444$; $\Sigma X^2 = 341$; $\Sigma Y^2 = 788$. Find: (a) the least squares regression lines of y on x and x on y; (b) the product moment correlation coefficient.

(*Hint:* See Exercise 63.)

65.   The set of bivariate data $((x_1, y_1), \ldots, (x_n, y_n))$ is coded according to the formulae $X = x - \bar{x}$ and $Y = y - \bar{y}$. By using the normal equations associated with the coded least squares regression line $Y = A_1 X + B_1$, show that:

(a) $B_1 = 0$ and (b) $A_1 = \dfrac{\sum XY}{\sum X^2}$

Hence, by decoding, show that the least squares regression line of y on x can be written as:

$$y = \frac{s_{xy}}{s_x^2} x + \left(\bar{y} - \frac{s_{xy}}{s_x^2} \bar{x} \right)$$

Similarly, show that the least squares regression line of x on y can be written as:

$$x = \frac{s_{xy}}{s_y^2} y + \left(\bar{x} - \frac{s_{xy}}{s_y^2} \bar{y} \right)$$

(The above constitutes the proof of Statement 3.14.)

66.   The regression lines of y on x and x on y are usually written in the form $y = a_1 x + b_1$ and $x = a_2 y + b_2$ respectively.

(a) Show that the line of x on y can be put into the form:

$$y = \frac{1}{a_2} x - \frac{b_2}{a_2}$$

Given that the correlation coefficient, r, is $+1$ and using the results $a_1 = \dfrac{s_{xy}}{s_x^2}$ and $a_2 = \dfrac{s_{xy}}{s_y^2}$ and the definition $r = \dfrac{s_{xy}}{s_x s_y}$, show that

(b) $s_{xy} = s_x s_y$ and (c) $a_1 = \dfrac{1}{a_2}$

Using also the results $b_1 = \bar{y} - a_1 \bar{x}$ and $b_2 = \bar{x} - a_2 \bar{y}$, show that

(d) $b_1 = -\dfrac{b_2}{a_2}$

Hence deduce that, in the case of perfect positive correlation, the two regression lines of y on x and x on y are identical.

Chapter 4   **Probability**

4.1 INTRODUCTION

Whether we are aware of it or not, chance plays an important part in our lives. Most of the daily decisions we make (and there are hundreds of them) are arrived at by means of our intuitive concept of chance.

For instance, when considering crossing a road, we automatically weigh up the traffic situation and, based on the results of this 'instant survey', we either cross immediately or wait. By waiting, in effect, we might well have come to the conclusion that 'the chance of our reaching the other side of the road in good health' is not very great! Even as we wait we are continually 'weighing up the chances' of crossing safely at that moment.

When leaving the house to go to work, we lock the doors because of the small 'chance' of someone walking in and helping himself to our furniture and valuables if we don't. However, we do not reinforce the doors with steel or put iron bars on the windows, since the 'chance' of anyone actually breaking in without being seen, say, by a neighbour, is so small that we are prepared to take the risk.

In this chapter we will be using words such as 'event', 'probability' (meaning chance) and 'experiment' — words which, in Statistics, have a particular meaning and will be defined accordingly.

Probability can be described as the 'nucleus' of Statistics (although perhaps this might not be evident to those studying the subject at a superficial level), and a great deal of the material covered in this chapter is the basis for much of the work in the rest of the book.

Probability can be viewed from three standpoints. There is *Classical Probability* and *Modern Probability Theory* which can be described as theoretical in that we do not need any practical data to achieve a working result. On the other hand, there is *Empirical* (or *Experimental*) *Probability*, a practical approach, using a given set of related data.

DEFINITION 4.1   (CLASSICAL PROBABILITY)

If an event, E, can happen in n different ways out of a total of N possible, equally likely, ways, then the *probability of event E occurring*, written as $\Pr(E)$, is given by:

$$\Pr(E) = \frac{n}{N}$$

The idea of an event here is left largely to intuition but, given this definition, we can go a long way to developing a reasonable working structure.

An objection to the above definition however is the use of the words 'equally likely'. This implies a knowledge of the concept of chance (i.e., probability), so that, in effect, we are defining probability in terms of probability! Modern probability theory does not solve this basic problem; it introduces probability as a set of numbers and considers various ways (given various situations) of assigning these numbers to well-defined statistical events, one of these ways embracing the above classical definition while putting the whole concept of probability inside a rigidly mathematical framework.

It is not our intention to study either of these approaches in isolation; rather to compromise and (in effect) use the classical definition within a more modern framework.

4.2 RELATED MATHEMATICS

4.2.1 ELEMENTARY SET THEORY

DEFINITION 4.2

A *set* is a list or collection of objects, these objects being called *elements* or *members*.

We usually denote sets by capital letters and elements by small letters or numerals.

For example, $A = (a, b, c, 1, 2)$ is translated as 'the set A has the elements a, b, c, 1 and 2'. It is usual for the elements of a set to be enclosed by brackets as shown above.

We can also describe the elements of a set in words as, for example, $M = $ (the letters of the alphabet) or $M = (a, b, c, d, \ldots, x, y, z)$.

Notes   (a)   All the elements of a set must be unique. That is, no element can be repeated.

(b)   The order in which elements of a set are written does not matter. So that, for instance, the set $(1, 2, 3)$ is just the same as set $(1, 3, 2)$ or $(3, 2, 1)$.

The symbol '$\in$' is used to mean 'is an element of'. For example, using set A above, $a \in A$, $1 \in A$ and $d \notin A$ (translated as 'd is not an element of A').

Sets can have particular relations with each other, and we describe two of these as follows:

DEFINITION 4.3

A set A is a *subset* of set B if all the elements of A are contained in B. We write $A \subset B$. (We can also say that B *contains* A.) Two sets A and B are *equal* if and only if both sets have identical elements.

For example, if $G = (1, 2, 3, c)$, $D = (1, 3)$, $E = (1, c)$ and $F = (1)$ then; $D \subset G$, $E \subset G$ and $F \subset G$. Also $F \subset E$ and $F \subset D$. (Note that any set is a subset of itself.)

There are two special sets that are important in set theory and have particular relations with any set (or sets) under consideration.

DEFINITION 4.4

(a)   The *empty set*, denoted by $\emptyset$ is a set that has no elements and is defined to be *a subset of every set*. $\emptyset$ is sometimes written as ( ).

(b)   The *universal set*, denoted by $\&$ is a set that contains any set (or sets) under consideration.

The universal set, in any particular context, is not a fixed set. For instance, if we were considering the set $A = (a, b, c)$ then a universal set could be $\& = (a, b, c, d)$ or $\& = (a, b, c, d, e, 7)$ or $\& = $ (the letters of the alphabet). If we were considering the sets $(1, 2, 3)$ and $(3, 5, 4)$, a universal set could be $\& = (1, 2, 3, 4, 5)$ or $\& = $ (positive integers) since both sets would be contained by each of the $\&$.

The need for universal and empty sets is not an obvious one but, mathematically, they are necessary in order to define a set theory algebra, and, although our interest is restricted to the particular use of sets in probability, we shall have occasion to refer to these special sets in certain contexts.

EXAMPLE 4.1

List all the possible subsets of the set $(2, 4, 6)$.

SOLUTION

There are eight subsets altogether, namely $(2), (4), (6), (2, 4), (2, 6)$, $(4, 6)$. The final two are $(2, 4, 6)$ (since any set is a subset of itself) and $\emptyset$ (since $\emptyset$ is a subset of every set).

DEFINITION 4.5

(a)   The *number* of a set A, written as $n(A)$, is defined as the number of elements in A.

(b)   The *complement* of a set A, written as A', is the set of all those elements that are not contained in A but are contained in the universal set. (Note that a complement cannot exist without some universal set for reference.)

For example, if $A =$ (vowels of the alphabet) $=$ (a, e, i, o, u) and $\mathscr{E} =$ (all letters of the alphabet), then $A' =$ (consonants), with $n(\mathscr{E}) = 26$, $n(A) = 5$ and $n(A') = 21$.

Sets can be combined in two different ways, described as follows.

DEFINITION 4.6

Let A_1 and A_2 be any two sets.

(a)   The *union* of A_1 and A_2, written as $A_1 \cup A_2$, is defined as that set which contains all the elements that belong to *either* A_1 or A_2 or both.

(b)   The *intersection* of A_1 and A_2, written as $A_1 \cap A_2$, is defined as that set which contains all the elements that belong to *both* A_1 *and* A_2. If $A_1 \cap A_2 = \emptyset$ (the empty set), A_1 and A_2 are said to be *disjoint*.

For example, if $A = (1, 2)$, $B = (2, 3)$ and $C = (1, 4, 6)$ then:

$A \cup B = (1, 2, 3)$ (since $1 \in A$, $2 \in A$ (also $2 \in B$) and $3 \in B$)

$A \cup C = (1, 2, 4, 6)$, $B \cup C = (1, 2, 3, 4, 6)$

$A \cap B = (2)$ (since 2 is the only element in both A and B)

$A \cap C = (1)$ and $B \cap C = \emptyset$ (since both B and C have no elements in common).

We can have both finite and infinite sets. For instance, if $A =$ (the positive integers), then A is an infinite set.

Sometimes it is found convenient to represent sets diagrammatically in the form of *Venn diagrams*.

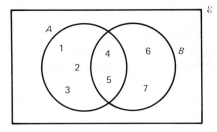

Fig. 4.1

Some examples follow.

(a)   If $A = (1, 2, 3, 4, 5)$ and $B = (4, 5, 6, 7)$, then we could represent this as a Venn diagram in the form shown in Fig. 4.1

Notes   (a)   We ensure that the areas representing both A and B enclose the elements 4 and 5 since $A \cap B = (4, 5)$.

(b)   We can take account of the fact that a universal set & exists, by enclosing both sets within a larger outer set.

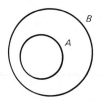

Here, we demonstrate the fact that $A \subset B$ (i.e., A is a subset of B).

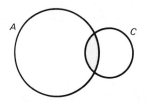

Here, we demonstrate the fact that A and C have a non-zero intersection (shaded area)

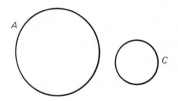

Here, $A \cap C = \emptyset$ (i.e., A and C are disjoint).

Fig. 4.2

EXAMPLE 4.2

Let $A = (1, 2, 3, 4, 5)$, $B = (2, 4, 6)$, $C = (1, 3, 5)$ and $\mathscr{E} = (1, 2, 3, 4, 5, 6, 7)$. List the elements of the sets:

(a) $A \cup B$ (b) $A \cap B$ (c) $A \cap C$ (d) $B \cup C$

(e) $B \cap C$ (f) $A \cup B \cup C$ (g) $(A \cap B)'$

SOLUTION

(a) $A \cup B$ is defined as the set that contains all the elements of A or B or both. Thus $A \cup B = (1, 2, 3, 4, 5, 6)$.

(b) $A \cap B$ is defined as the set that contains all the elements of both A and B, i.e., $A \cap B = (2, 4)$.

(c) $A \cap C = (1, 3, 5) = C$ (note that this is because $C \subset A$).

(d) $B \cup C = (1, 2, 3, 4, 5, 6)$.

(e) $B \cap C = \emptyset$ since the sets B and C have no elements in common.

(f) $A \cup B \cup C$ can be treated as $(A \cup B) \cup C$ (that is, evaluate the set $A \cup B$ first then union this with C) or $A \cup (B \cup C)$. For instance, $A \cup B = (1, 2, 3, 4, 5, 6)$.

$$\therefore \qquad (A \cup B) \cup C = (1, 2, 3, 4, 5, 6) \cup (1, 3, 5)$$
$$= (1, 2, 3, 4, 5, 6).$$

Otherwise $B \cup C = (1, 2, 3, 4, 5, 6)$.

$$\therefore \qquad A \cup (B \cup C) = (1, 2, 3, 4, 5) \cup (1, 2, 3, 4, 5, 6)$$
$$= (1, 2, 3, 4, 5, 6)$$

So that $A \cup B \cup C = (1, 2, 3, 4, 5, 6)$ whichever way we work it out.

(g) $A \cap B = (2, 4)$.

$$\therefore \qquad (A \cap B)' = (2, 4)' \quad \text{(w.r.t. } \mathscr{E})$$
$$= (1, 3, 5, 6, 7)$$

EXAMPLE 4.3

Correct each of the following statements, if $A = (a, b, c)$ and $M = (b, 3, \text{house})$:

(a) $b \subset A$ (b) $3 \in A$ (c) house $\in A \cap M$

(d) $(3) \in M$

SOLUTION

(a) The symbol '$\subset$' must connect two *sets* and b is an *element* of a set. We need to write either $b \in A$ or $(b) \subset A$.

(b) The elements of A are a, b and c only. Since 3 does not appear here, it is *not* an element of A. We may write $3 \notin A$ or $3 \in M$

(c)   Since $A = (a, b, c)$ and $M = (b, 3, \text{house})$, the elements of $A \cap M$ are those that are common to *both* A and M, i.e., $A \cap M = (b)$. Thus house $\notin A \cap M$, whereas house $\in M$.

(d)   '$\in$' means 'is an element of' and connects an element to a set. Here, both (3) and M are sets so we need to write $(3) \subset M$.

4.2.2 ARITHMETIC AND GEOMETRIC PROGRESSIONS

DEFINITION 4.7

(a)   An *ordered set* of numbers is a set whose elements (numbers) are arranged in a particular order.

(b)   A *sequence* (or *series*) is an ordered set of numbers, where each element is related to the previous element according to a certain specified rule.

We noted in the previous subsection that when dealing with sets it is not important what order the elements are written in.

When dealing with an ordered set, $(1, 2, 3)$ say, the positioning of the elements within the set should be unique and, for instance, the ordered set $(1, 2, 3)$ is not the same as the ordered set $(2, 3, 1)$.

Some examples of sequences are:

(a) $(1, 2, 3, 4, 5, 6, \ldots)$,

(b) $(2, 4, 16, 256, \ldots)$,

(c) $(16, 8, 4, 2, 1, \frac{1}{2}, \frac{1}{4}, \ldots)$,

(d) $(10, 7, 4, 1)$.

In (a), each element is obtained by adding 1 to the previous element, whereas in (b), each element is obtained by 'squaring' the previous one. In (c) we divide successive elements by 2 to obtain the following one and in (d) we subtract three from each element to obtain the next.

Note that the first three sequences are infinite, while the last is finite.

DEFINITION 4.8

A series of numbers is said to be in *Arithmetic Progression* if each term is formed by adding (or subtracting) a constant value to the preceding term.

It is usual to denote the first term by 'a' and the added constant by 'd'.

Hence, an arithmetic progression takes the form $(a, a + d, a + 2d, a + 3d, \ldots)$.

The previous example (a) of $(1, 2, 3, 4, 5, 6, \ldots)$ is an arithmetic progression with $a = 1$ and $d = 1$,

i.e., the sequence is $(1, 1 + 1, 1 + 2(1), 1 + 3(1), \ldots) = (1, 2, 3, 4, \ldots)$

Example (d) is also an arithmetic progression with $a = 10$ and $d = -3$.

The sequence here is:

$$(10, 10 - 1(3), 10 - 2(3), 10 - 3(3))$$
$$= (10, 7, 4, 1)$$

STATEMENT 4.1

(a)   *The nth term*, T_n, of an arithmetic progression is given by the expression:

$$T_n = a + (n - 1)d$$

(b)   The *sum of the first n terms*, S_n, of an arithmetic progression is given by the expression:

$$S_n = \frac{n}{2}(2a + (n - 1)d)$$

Consider the arithmetic progression $(1, 2, 3, 4, 5, \ldots)$ having $a = 1$ and $d = 1$.

Using (a) above, the 4th term, $T_4 = 1 + (4 - 1)1 = 4$ and the 12th term, $T_{12} = 1 + (11)1 = 12$.

Using (b) above, we have the sum of the first 5 terms as:

$$S_5 = \frac{5}{2}(2(1) + (5 - 1)1) = \frac{5}{2}(2 + 4) = 15$$

which can be verified, since $S_5 = 1 + 2 + 3 + 4 + 5 = 15$.

Also: $S_6 = \frac{6}{2}(2(1) + (6 - 1)1) = \frac{6}{2}(2 + 5) = 21$

and similarly

$$S_{20} = \frac{20}{2}(2(1) + (20 - 1)1) = 210$$

EXAMPLE 4.4

For the arithmetic progression $(-4, 1, 6, 11, \ldots)$, find:

(a)   the 11th term, T_{11},

(b) T_{31},

(c)   the sum of the first 4 terms, S_4, and

(d) S_{31}.

SOLUTION

We note immediately that $a = -4$ and $d = 5$.

(a)   Using the form $T_n = a + (n-1)d$ we have

$$T_{11} = -4 + 10(5) = 46.$$

(b)   Similarly $T_{31} = -4 + (30)5 = 146.$

(c)   Using the form $S_n = \dfrac{n}{2}(2a + (n-1)d)$ we have

$$S_4 = \dfrac{4}{2}(-8 + (3)5) = 14.$$

(this can be checked as $S_4 = -4 + 1 + 6 + 11 = 18 - 4 = 14.$)

(d)   Similarly $S_{31} = \dfrac{31}{2}(2(-4) + (30)5) = \dfrac{31}{2}(-8 + 150) = 2201.$

DEFINITION 4.9

A series of numbers is said to be in *Geometric Progression* if each term is formed by multiplying (or dividing) the preceding term by a constant.

We usually denote the first term by '*a*' and the multiplying constant by '*r*'.

Hence a geometric progression takes the form $(a, ar, ar^2, ar^3, \ldots)$.

The sequence $(3, 6, 12, 24, \ldots)$ is an example of a geometric progression with $a = 3$ and $r = 2$, i.e., it takes the form

$$(3, 3(2), 3(2^2), 3(2^3), \ldots) = (3, 3(2), 3(4), 3(8), \ldots)$$

$$= (3, 6, 12, 24, \ldots)$$

Also $(16, 8, 4, 2, 1, \frac{1}{2}, \frac{1}{4}, \ldots)$ is a geometric progression with $a = 16$ and $r = \frac{1}{2}$. It can be written as

$$(16, 16(\tfrac{1}{2}), 16(\tfrac{1}{2})^2, 16(\tfrac{1}{2})^3, \ldots) = (16, 8, 4, 2, \ldots)$$

Geometric progressions can, of course, be either finite or infinite.

As with arithmetic progressions, there are appropriate expressions for the nth term and the sum to n terms of a geometric progression.

STATEMENT 4.2

(a) The *nth term*, T_n, of a geometric progression is given by

$$T_n = ar^{n-1}.$$

(b) The *sum to n terms*, S_n, of a geometric progression is given by:

$$S_n = \frac{a(1-r^n)}{1-r}$$

For an infinite geometric progression, *the sum to infinity* (written S_∞) is given by:

$$S_\infty = \frac{a}{1-r} \qquad \text{(for } r \text{ only between } -1 \text{ and } +1\text{)}$$

$$= \infty \qquad \text{(otherwise)}$$

For example, in the sequence $(16, 8, 4, 2, 1, \frac{1}{2}, \ldots)$ we have a geometric progression with $a = 16$ and $r = \frac{1}{2}$.

Hence:
$$T_6 = ar^{6-1} = 16\left(\frac{1}{2}\right)^5 = \frac{16}{32} = \frac{1}{2}$$

Also: $S_4 = \dfrac{a(1-r^4)}{1-r} = \dfrac{16(1-\frac{1}{2}^4)}{1-\frac{1}{2}} = \dfrac{16(1-\frac{1}{16})}{\frac{1}{2}} = 32\left(\dfrac{15}{15}\right) = 30$

(which can easily be verified by addition).

Since r lies between -1 and $+1$, we can find the sum to infinity for this series as:

$$S_\infty = \frac{a}{1-r} = \frac{16}{1-\frac{1}{2}} = 32$$

Note that if r, the ratio of the series, is negative, the terms of the series will alternate in sign.

For example, in a geometric progression with $a = 3$ and $r = -2$, we obtain the sequence:

$$(3, -6, 12, -24, \ldots)$$

EXAMPLE 4.5

For the geometric progression $(9, 3, 1, \ldots)$, find:

(a) T_5 (b) T_6 (c) S_5 and (d) S_∞

SOLUTION

We are dividing each term by 3 each time, i.e., $r = \frac{1}{3}$ with $a = 9$.

(a) $T_5 = ar^4 = 9\left(\dfrac{1}{3}\right)^4 = \dfrac{9}{81} = \dfrac{1}{9}.$

(b) $T_6 = 9\left(\dfrac{1}{3}\right)^5 = \dfrac{9}{243} = \dfrac{1}{27}.$

(c) $S_5 = \dfrac{a(1 - r^5)}{1 - r} = \dfrac{9(1 - \frac{1}{3}^5)}{1 - \frac{1}{3}} = \dfrac{9(1 - \frac{1}{243})}{\frac{2}{3}}$

$= 9\left(\dfrac{3}{2}\right)\dfrac{242}{243} = \dfrac{242}{18} = 13\,\dfrac{4}{9}.$

(d) $S_\infty = \dfrac{a}{1 - r} = \dfrac{9}{1 - \frac{1}{3}} = 9\left(\dfrac{3}{2}\right) = 13\,\dfrac{1}{2}.$

EXAMPLE 4.6

Find T_4, S_4 and S_∞ (if applicable) for the following arithmetic or geometric progressions:

(a) $(9, -6, 4, \ldots)$ (b) $(-2, 1, 4, \ldots)$ (c) $(\frac{1}{4}, \frac{1}{2}, 1, \ldots)$

SOLUTION

(a)   Since the signs alternate, we suspect a geometric progression. Now $a = 9$, and to find r we can solve the equation $9r = -6$ (by definition of a geometric progression), i.e.,

$$r = \frac{-6}{9} = -\frac{2}{3}$$

(We can check this value of r using the second and third terms, since $-6 \cdot r$ must equal 4. Now $-6 \cdot r = -6 \cdot -\frac{2}{3} = \frac{12}{3} = 4$.)

Hence: $T_4 = 9\left(-\dfrac{2}{3}\right)^3 = 9 \cdot \dfrac{-8}{27} = -\dfrac{8}{3}$

$S_4 = \dfrac{9(1 - (-\frac{2}{3})^4)}{1 - (-\frac{2}{3})} = \dfrac{9(1 - \frac{16}{81})}{1 + \frac{2}{3}} = 9\left(\dfrac{65}{81}\right)\dfrac{3}{5} = 4\,\dfrac{1}{3}$

A sum to infinity exists since r is between -1 and $+1$:

$$S_\infty = \frac{9}{1 + \frac{2}{3}} = 9\left(\frac{3}{5}\right) = 5\,\frac{2}{5}$$

(b)   We have an arithmetic progression with $a = -2$ and $d = 3$.

∴ $T_4 = a + (4 - 1)d = -2 + 3(3) = 7$

$S_4 = \dfrac{4}{2}(2a + (4 - 1)d) = 2(-4 + 3(3)) = 10$

(c)   This is a geometric progression with $a = \frac{1}{4}$ and $r = 2$.

So, $T_4 = \dfrac{1}{4}(2)^3 = 2$

$S_4 = \dfrac{\frac{1}{4}(1 - 2^4)}{1 - 2} = \dfrac{\frac{1}{4}(1 - 16)}{-1} = \dfrac{15}{4} = 3\,\dfrac{3}{4}$

Since r is greater than 1, S_∞ is not a finite number.

4.2.3 INEQUALITIES

A *mathematical inequality* is a statement, similar to an equation, where 'equals' is replaced by such expressions as 'is less than' or 'is greater than', which we can conveniently call 'connecting expressions'.

There are four connecting expressions used, and these, together with their mathematical symbols are listed below:

$$'<': \text{ is less than}$$
$$'\leqslant': \text{ is less than or equal to}$$
$$'>': \text{ is greater than}$$
$$'\geqslant': \text{ is greater than or equal to}$$

$<$ and $>$ are usually termed *strict* inequalities.

For instance, the following statements are true:

$2 < 4$ (2 is less than 4)

$2 \leqslant 4$ (2 is less than or equal to 4)

$16 \ngeqslant 16.9999$ (16 is *not* greater than or equal to 16.9999)

$4 > 3.86$ (4 is greater than 3.86)

An inequality sign is said to be *reversed* if we change $<$ to $>$ (and vice versa) or $\leqslant$ to $\geqslant$ (and vice versa).

A basic result in the theory of inequalities is now stated.

STATEMENT 4.3

If $x > y$ and $y > z$ then $x > z$.

The above is also true if any one of the three other inequality signs is used in place of $>$.

This statement can easily be verified by the substitution of some simple numbers for x, y and z.

We can manipulate inequalities in much the same way as we manipulate equations.

For example, if $x + 2 = 3$ we have $x = 3 - 2$ giving $x = 1$ as a solution.

Similarly if $x + 2 < 3$, we have (subtracting 2 from both sides of the inequality) $x < 3 - 2$ giving $x < 1$ as the solution of the inequality.

However, there are pitfalls, and we state the set of rules applicable for *solving inequalities*.

STATEMENT 4.4

(a)   Any number may be added to or subtracted from both sides of an inequality.

(b)   Both sides of an inequality may be multiplied or divided by any *positive* number.

(c)   If both sides of an inequality are multiplied or divided by a negative number, *the inequality must be reversed.*

Some simple examples follow, using the above statement to solve three inequalities.

(a) $x + 3 > 5$ gives $x > 2$ (subtracting 3 from both sides).

(b) $2x < 6$ gives $x < 3$ (dividing both sides by 2).

(c) $3 - 2x \geqslant 27$ gives $-2x \geqslant 24$ (subtracting 3 from both sides)

   $\Rightarrow \quad x \leqslant -12$ (dividing both sides by -2, *reversing the inequality sign*).

Before looking at an important result involving inequalities, the student is reminded of the *modulus* of a number x, written as $|x|$ and defined as the absolute (positive) value of x,

i.e.: $|-4| = 4$; $|17| = 17$; $|-2.5| = 2.5$, etc.

STATEMENT 4.5

The solution of the inequality $x^2 < a^2$ is $-a < x < a$ which can be written as $|x| < a$.

Similarly: $x^2 \leqslant a^2 \quad \Rightarrow \quad |x| \leqslant a$

                    $x^2 > a^2 \quad \Rightarrow \quad |x| > a$

                    $x^2 \geqslant a^2 \quad \Rightarrow \quad |x| \geqslant a$

EXAMPLE 4.7

Solve the following inequalities:

(a) $x + 2 < 10$ (b) $\dfrac{x + 5}{2} \leqslant 4$ (c) $3 - 2x \geqslant 7$

SOLUTION

(a) $x + 2 < 10 \quad \Rightarrow \quad x < 10 - 2 \quad \Rightarrow \quad x < 8$

(b) $\dfrac{x + 5}{2} \leqslant 4 \quad \Rightarrow \quad x + 5 \leqslant 8 \quad \Rightarrow \quad x \leqslant 8 - 5 \quad \Rightarrow \quad x \leqslant 3$

(c) $3 - 2x \geqslant 7 \quad \Rightarrow \quad -2x \geqslant 7 - 3 \quad \Rightarrow \quad -2x \geqslant 4 \quad \Rightarrow \quad x \geqslant \dfrac{4}{-2}$

      $\Rightarrow \quad x \leqslant -2$

(Note that the sign has been reversed on division by a negative number.)

EXAMPLE 4.8

Solve the inequalities:

(a) $\dfrac{3}{x} - 2 > 4$ (b) $\dfrac{x^2 - 2}{3} < 5$

for x.

SOLUTION

(a) $\dfrac{3}{x} - 2 > 4 \quad \Rightarrow \quad \dfrac{3}{x} > 6$ (adding 2 to both sides).

Hence, multiplying both sides by x,

$$3 > 6x \quad \text{if} \quad x > 0 \qquad\qquad [1]$$

$$3 < 6x \quad \text{if} \quad x < 0 \qquad\qquad [2]$$

(Note that in [2] we have reversed the inequality sign since, here, x is assumed negative.)

So that [1] $x < \tfrac{1}{2}$ if $x > 0$ and [2] $x > \tfrac{1}{2}$ if $x < 0$.

Clearly [2] cannot be true, leaving [1] as the solution which can be written as $0 < x < \tfrac{1}{2}$.

(b) $\dfrac{x^2 - 1}{3} < 5 \quad \Rightarrow \quad x^2 - 1 < 15$ (multiplying both sides by 3).

Therefore $x^2 < 16$, and using the result of the previous statement we have $|x| < 4$.

EXAMPLE 4.9

Solve the inequality $-3 \leqslant 2x + 1 \leqslant 2$.

SOLUTION

It is convenient to work with the three parts of the inequality at the same time.

$-3 \leqslant 2x + 1 \leqslant 2$

$\Rightarrow -3 - 1 \leqslant 2x \leqslant 2 - 1$ (subtracting 1 from each part)

$\Rightarrow -4 \leqslant 2x \leqslant 1$

$\Rightarrow -2 \leqslant x \leqslant \tfrac{1}{2}$ (dividing each part by 2)

which is the desired solution.

4.3 BASIC PROBABILITY CONCEPTS

4.3.1 EXPERIMENTS AND EVENTS

The student should have an acquaintance with the idea of a mathematical set together with the basic set operations (see Section 4.2.1) for a complete understanding of many of the concepts in this (and subsequent) sections.

DEFINITION 4.10

(a)   An *experiment* can be defined (for convenience) as a procedure adopted in order to gain information about some process. It can usually be split up into an *action* and an *observation*.

(b)   An *event* of an experiment can be thought of as some particular situation that can arise during the experiment.

For instance, consider rolling a normal, six-sided die once. We can think of this as an experiment in that the physical rolling of the die is the action and looking at how many 'spots' are shown on the uppermost face as the observation. Possible events are 'the face shown uppermost is a 4', 'an even number is shown' or 'a number greater than 3 is shown'.

Suppose we were conducting an experiment to measure the number of cars passing a certain point in one minute. Here, our standing with paper, pencil, and stop-watch at the ready would be the action and the noting of how many cars actually passed in the minute interval would be the observation. Possible events are 'no cars pass', '36 cars pass', 'an odd number of cars pass', and so on.

Using set theory, however, we can develop the idea of experiments and events more rigorously and this is necessary for our discussion on probability.

DEFINITION 4.11

(a)   An *outcome* (or *elementary event*) of an experiment is a single way in which the experiment can result.

(b)   The set of all possible individual outcomes of an experiment is called the *outcome set*, usually denoted by S.

(c)   An *event* of an experiment is defined to be any subset of the outcome set, S.

The above terms and definitions are not 'universal', and many writers use different terms. However, the definition above together with some basic set theory serve as a good foundation for our particular discussions on probability.

We have that any experiment may be defined in terms of individual outcomes forming an outcome set S (serving as a universal set) and events,
defined as subsets of S.

In part (a) of the above definition we introduced two terms, 'outcome'
and 'elementary event', both meaning exactly the same. An outcome is an
intuitively reasonable word for an individual result of a single experiment.
On the other hand, defining an event as *any* subset of the outcome set
S, necessarily means that any single outcome will also be an event, if only
a trivial one — hence the use of 'elementary event'.

We will now put the previously mentioned 'die-rolling' and 'car-counting'
experiments into a set context.

For the die-rolling experiment, the individual outcomes are just the different faces that the die can show uppermost; that is 1, 2, 3, 4, 5 and 6.
Hence, $S = (1, 2, 3, 4, 5, 6)$.

The event 'a 4 is shown' will have the event set (4); the event 'an even
number is shown' will have the event set $(2, 4, 6)$ and so on.

For the car-counting experiment, the individual outcomes are 0, 1, 2, 3,
4, 5, ..., and although, theoretically, the number of cars passing any
point in a minute must have a limit, for all practical purposes we can
consider the outcome set as an infinite one. That is $S = (0, 1, 2, 3, 4, 5,
...)$.

The event 'no cars pass' has event set (0); the event 'an odd number of
cars pass' has event set $(1, 3, 5, 7, ...)$ which is also a practically infinite
set.

We can use Venn diagrams to illustrate these ideas. For example, Fig. 4.3
shows the outcome set S and some possible events for the die-rolling
experiment.

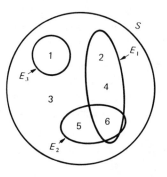

Fig. 4.3

E_1 is the event 'an even number', E_2 is the event 'a number greater than
4' and E_3 is the (elementary) event '1'.

DEFINITION 4.12

We say that an *event E has occurred* if the actual outcome of a performed experiment is contained in the event set for *E*.

Suppose we define the following events in the die-rolling experiment.
$E_1 = (2, 4, 6)$, $E_2 = (5, 6)$, $E_3 = (1)$ and $E_4 = (1, 2, 3, 4)$. If we actually
roll the die and observe the outcome '2', then both the events E_1 and E_4
will have occurred since $2 \in E_1$ and $2 \in E_4$. However, since $2 \notin E_2$ and
$2 \notin E_3$, the events E_2 and E_3 will not have occurred.

EXAMPLE 4.10

An experiment consists of drawing a card from a normal pack of 52
and reading it. Give details of the following sets.
(a)   The outcome set *S*.
(b)   The event $E_1 = $ 'a card lower than a 4 (Ace high)'.
(c)   The event $E_2 = $ 'an 8'.
(d)   The event $E_3 = $ 'a heart'.

SOLUTION

(a)   The outcome set $S = $ (the 52 cards in the pack). That is:
$$S = (2H, 3H, \ldots, KH, AH, 2C, 3C, \ldots, AC, \ldots AS)$$
using an obvious notation.
(b) E_1 is the set of all 2's and 3's. That is
$$E_1 = (2H, 3H, 2C, 3C, 2D, 3D, 2S, 3S)$$
(c) $E_2 = (8H, 8C, 8D, 8S)$.
(d) $E_3 = (2H, 3H, 4H, \ldots, QH, KH, AH)$.

Since we have defined an event of an experiment as any subset of the
outcome set *S*, and both $\emptyset$ and *S* are subsets of *S*, they are both valid
events of an experiment.

$\emptyset$ is often called the *impossible event* since, having no elements in its
event set, it can never occur. In a similar fashion *S* is often called the
certain event since, having all the outcomes of the experiment in its event
set, it will occur every time the experiment is performed.

Consider the 'die-rolling' event $E = $ '7'. We are quite happy to accept
that this event is impossible and yet we would feel it natural to write the
set for this event as (7), which of course is non-empty. But we define an
event strictly in terms of a particular experiment which is entirely specified by an outcome set *S*. Since $7 \notin S$ we conclude that the outcome 7
'does not exist.' Hence, under this *S*, the event *E* has an empty event
set, i.e. $E = \emptyset$.

DEFINITION 4.13

Let E be an event of some experiment having an outcome set S. The *complement of E*, written $\bar{E}$, is defined as that event that has all the outcomes in S that are not outcomes in E.

For example, in the die-rolling experiment with $S = (1, 2, 3, 4, 5, 6)$, let $E = (1, 2, 3)$. Then $\bar{E} = (4, 5, 6)$ (i.e., those outcomes that are in S but not in E). If $E_1 = (4)$, then $\bar{E}_1 = (1, 2, 3, 5, 6)$ being thought of as 'anything (in S) but 4' or 'not 4'.

In set theory we combine sets using the two operators ∪ and ∩. For the purposes of probability (dealing specifically with event sets) we change these two operators to 'or' and 'and' respectively. Consider the events E_1 and E_2 of some defined experiment. The event 'E_1 or E_2' (i.e., $E_1 \cup E_2$) occurs if any one of the elements in its event set is the actual outcome of the experiment. This is equivalent to saying that event 'E_1 or E_2' occurs if either one of 'E_1' or 'E_2' occurs or both. Similarly the event 'E_1 and E_2' will occur only if both the events 'E_1' and 'E_2' occur.

This is the reason why it is convenient and relevant to use 'or' and 'and' in this context.

So that if we define $E_1 = (1, 2, 3)$ and $E_2 = (2, 3, 4)$ we have:

$$\text{'}E_1 \text{ or } E_2\text{'} = (1, 2, 3, 4) \text{ and } \text{'}E_1 \text{ and } E_2\text{'} = (2, 3)$$

EXAMPLE 4.11

An experiment is defined as tossing 5 coins simultaneously and noting how many heads appear.

(a) Draw a Venn diagram for the outcome set S, showing the following events: E_1 = (at least 2 heads); E_2 = (at most 4 heads) E_3 = (an odd number of heads).

(b) If the experiment is actually performed and the outcome '2 heads' is observed, state which of the following events will have occurred: E_1, E_2, E_3, E_1 or E_2, E_1 and E_2, E_1 and E_3 and $\bar{E}_1$.

SOLUTION

When tossing five coins we can only obtain either 0, 1, 2, 3, 4, or 5 heads.

(a) Hence $S = (0, 1, 2, 3, 4, 5)$. The Venn diagram for the given events is shown in Fig. 4.4, with $E_1 = (2, 3, 4, 5)$, $E_2 = (0, 1, 2, 3, 4)$ and $E_3 = (1, 3, 5)$.

(b) 'E_1 or E_2' has event set $(0, 1, 2, 3, 4, 5) = S$; 'E_1 and E_2' has event set $(2, 3, 4)$ and 'E_1 and E_3' has event set $(3, 5)$ Also $\bar{E}_1 = (0, 1)$. Thus, if the actual outcome of the experiment is '2',

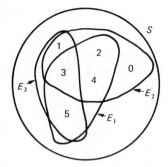

Fig. 4.4

E_1, E_2, E_1 or E_2, 'E_1 and E_2' will have occurred, since they all contain outcome '2' in their event sets. E_3, 'E_1 and E_3', and $\bar{E}_1$ will not have occurred in this case.

DEFINITION 4.14

Two events *of a single experiment* are said to be *mutually exclusive* if they cannot occur simultaneously as a result of the experiment. This is equivalent to saying that mutually exclusive events must have disjoint event sets.

It is clear that for two events to occur together their event sets must have at least one outcome in common. Conversely, if the two event sets are disjoint, there is no way in which both events can occur if the experiment is performed.

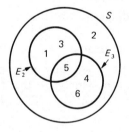

Fig. 4.5

Returning to the die-rolling experiment, let $E_1 = (2, 4, 6)$, $E_2 = (1, 3, 5)$, $E_3 = (4, 5, 6)$ and $E_4 = (1, 2, 3, 4)$. The only two events that are mutually exclusive are E_1 and E_2 since $E_1 \cap E_2 = \emptyset$. Fig. 4.5 shows the events E_2 and E_3. Here since $E_2 \cap E_3 = (5)$, E_2 and E_3 are not mutually exclusive. If the actual outcome of the experiment is '5' then clearly both E_2 and E_3 will occur. Of course, if the outcome was '3' instead, E_2 will occur while E_3 will not. So that if two events are *not* mutually exclusive they are not forced to occur together — but it *is* possible that they will.

Mutually exclusive events are such that no matter which outcome is realised, the two events *cannot* occur together.

We extend the idea of 'mutual exclusiveness' to three or more events in the following way:

DEFINITION 4.15

The n events $E_1, E_2, E_3, \ldots, E_n$ *originating from the same experiment* are *mutually exclusive* if and only if each and every pair of events is separately mutually exclusive.

Hence, the 3 events E_1, E_2 and E_3 will be mutually exclusive *only if* the three pairs of events E_1 and E_2, E_2 and E_3, E_1 and E_3 are all separately mutually exclusive.

EXAMPLE 4.12

Two normal six-sided dice are thrown as an experiment and the outcome set is defined as 'all possible pairs of numbers that can be obtained. So that $S = ((1,1), (1,2), (1,3), \ldots, (6,4), (6,5), (6,6))$. The following events are defined:

E_1 is 'the two dice show the same number'

E_2 is 'at least one of the numbers is a 5 or 6'

E_3 is 'the sum of the two numbers is less than 6'

(a)   Which of the pairs E_1 and E_2; E_1 and E_3, E_2 and E_3 are mutually exclusive.

(b)   If the outcome of the experiment is:

(i)   (2, 2)   (ii)   (5, 5)   (iii)   (5, 1)

state which of the events E_1, E_2 and E_3 will have occurred.

SOLUTION

(a) This is most clearly answered using a Venn diagram as shown:

We see that E_1 and E_2 have common outcomes. So also do E_1 and E_3. But the sets E_2 and E_3 are disjoint, i.e, E_2 and E_3 are mutually exclusive.

(b) (i) (2, 2) is an outcome of both E_1 and E_3. Therefore these two events occur if (2, 2) is the outcome of the experiment.

(ii) E_1 and E_2 occur if (5, 5) is the outcome of the experiment, since both these sets contain (5, 5) as an outcome.

(iii) (5, 1) is in E_2 only. Hence E_2 occurs if (5, 1) is an outcome of the experiment.

4.3.2 ELEMENTARY PROBABILITY

DEFINITION 4.16

Let an experiment have an outcome set S with E as any event. We define the *probability of E occurring*, written as $Pr(E)$, as a number satisfying the following conditions:

(a) $0 \leqslant Pr(E) \leqslant 1$.

(b) $Pr(S) = 1$.

(c) If E_1 and E_2 are two mutually exclusive events of S, then:

$$Pr(E_1 \text{ or } E_2) = Pr(E_1) + Pr(E_2)$$

(d) If $E_1, E_2, E_3, \ldots, E_n$ are n mutually exclusive events of S, then:

$$Pr(E_1 \text{ or } E_2 \text{ or } E_3 \text{ or } \ldots \text{ or } E_n) = Pr(E_1) + Pr(E_2) + \ldots + Pr(E_n)$$

The above sections are relevant to all types of outcome set, finite or infinite.

(e) If the outcome set S is *finite* with exactly n outcomes, $s_1, s_2, \ldots, s_n$ say, then $Pr(s_1) + Pr(s_2) + \ldots + Pr(s_n) = 1$,

i.e.
$$\sum_{i=1}^{n} Pr(s_i) = 1$$

(f) If the outcome set S is finite with *equally likely outcomes* (that is, the probability of occurrence of each of them is the same), then the probability of event E occurring is given by:

$$Pr(E) = \frac{n(E)}{n(S)}$$

Notes (i) Only when we have a finite, equally likely outcome set do we have a rule (contained in (f)) for assigning probabilities to events. In (a), (b), (c), (d) and (e) we are only asserting that the probabilities of events occurring exist as numbers and giving some rules for their scope.

(ii)   Since a probability of 1 is the largest possible by (a), we label the associated event a *certain event*, and by (b), this can only refer to the outcome set S.

Similarly (for an equally likely outcome set) we have that if E is a set containing no outcomes (i.e., $\emptyset$), then $n(E) = 0$ by definition, and (f) gives $\Pr(E) = \dfrac{0}{n(S)} = 0$, a probability associated with an event that cannot occur. Hence the description 'the impossible event' for $\emptyset$.

In this text we will often be dealing with finite, equally likely outcome sets, where all of (a), (b), (c), (d), (e) and (f) are valid, although there will be occasions when we cannot invoke (f) since the outcome set will not be composed of equally likely outcomes.

There is no practical definition available for the phrase 'equally likely' in the probability context and, as is conventional, we leave its meaning to intuition. However, consider the following situation. Suppose that we define an experiment as selecting a single letter of the alphabet. To set up this experiment physically we could write the 26 letters on 26 identical pieces of paper and place them in a box. Then, after thoroughly mixing the papers in the box, draw one (without looking). In this way we would ensure that no outcome (letter) would be more likely to be picked than any other, the outcomes being described here as equally likely.

In the context of this type of experiment, we often say that we are selecting an outcome at *random*, labelling the experiment a *random experiment*, with a *random event* used to describe any event associated with the corresponding equally likely outcome set.

EXAMPLE 4.13

An experiment is performed by tossing a normal coin and observing which side (head or tail) is shown uppermost.

(a)   Write down the outcome set S.

(b)   Calculate the probability that a 'head' occurs.

SOLUTION

(a)   The outcome set S is clearly (head, tail) or (H, T), each of the two outcomes being equally likely with $n(S) = 2$.

(b)   The event 'head' has event set (H).

$$\therefore \qquad \Pr(\text{head}) = \frac{n((\text{H}))}{n(S)} = \frac{1}{2} \quad (\text{using Definition 4.16(f)})$$

EXAMPLE 4.14

A bag contains four balls of the same size and texture but coloured respectively red, pink, green and brown. An experiment is performed by shaking the bag (in order to mix the balls up) and drawing a single ball.

(a)   Write down the outcome set S for the experiment,

(b)   Calculate the probability that the following are drawn:

(i)   a red ball   (ii)   a green ball   (iii)   a brown or pink ball

SOLUTION

(a) $S =$ (red, pink, green, brown), each of the outcomes being equally likely.

(b)   Denoting the event 'red' by R etc, we have:

(i) $\Pr(R) = \dfrac{n(R)}{n(S)} = \dfrac{1}{4}$ (ii) $\Pr(G) = \dfrac{n(G)}{n(S)} = \dfrac{1}{4}$

(iii)   Now we know that $\Pr(B) = \frac{1}{4} = \Pr(P)$. Also P and B are mutually exclusive events. Hence:

$$\Pr(B \text{ or } P) = \Pr(B) + \Pr(P) \quad \text{(Definition 4.16 (c))}$$

$$= \frac{1}{4} + \frac{1}{4} = \frac{1}{2}$$

EXAMPLE 4.15

A die is weighted so that a '4' is twice as likely to appear as a '1', '2', '3' or '5' and a '6' is twice as likely to áppear as a '4'. Find the probability of throwing an even number with a single roll of the die.

SOLUTION

$S =$ (1, 2, 3, 4, 5, 6) as usual, except that here the outcomes are not equally likely since the die has been weighted. However, if the die has not been altered in any other way besides that given, we must have $\Pr(1) = \Pr(2) = \Pr(3) = \Pr(5) = p$, say.

But $\Pr(4) = 2 \cdot \Pr(1) \quad \text{(given)}$

$$= 2p$$

Also $\Pr(6) = 2 \cdot \Pr(4) \quad \text{(given)}$

$$= 4p$$

Now, since we have a finite set S, we can use Definition 4.16(e), giving

$$\Pr(1) + \Pr(2) + \Pr(3) + \Pr(4) + \Pr(5) + \Pr(6) = 1$$

i.e. $p + p + p + 2p + p + 4p = 1 \Rightarrow 10p = 1 \Rightarrow p = \dfrac{1}{10}$

Hence: $\Pr(1) = \Pr(2) = \Pr(3) = \Pr(5) = \dfrac{1}{10}$

$\Pr(4) = \dfrac{2}{10} = \dfrac{1}{5}$ and $\Pr(6) = \dfrac{4}{10} = \dfrac{2}{5}$

$\therefore \Pr(\text{even number}) = \Pr(2 \text{ or } 4 \text{ or } 6)$

$\qquad\qquad\qquad\quad = \Pr(2) + \Pr(4) + \Pr(6)$ (mutually exclusive events)

$\qquad\qquad\qquad\quad = \dfrac{1}{10} + \dfrac{1}{5} + \dfrac{2}{5}$ (from above)

$\qquad\qquad\qquad\quad = \dfrac{7}{10}$

EXAMPLE 4.16

A die is rolled once as an experiment, with $S = (1, 2, 3, 4, 5, 6)$. Find the probabilities of the following events:

(a)   1 or 2.

(b)   An even number.

(c)   A number less than 4.

(d)   An even number or a number less than 4.

(e)   An even number and a number less than 4.

SOLUTION

Clearly here $\Pr(1) = \Pr(2) = \Pr(3) = \Pr(4) = \Pr(5) = \Pr(6) = \dfrac{1}{6}.$

(a) $\Pr(1 \text{ or } 2) = \Pr(1) + \Pr(2)$ (mutually exclusive events)

$\qquad\qquad\quad = \dfrac{1}{6} + \dfrac{1}{6} = \dfrac{1}{3}$

(b) $\Pr(\text{an even number}) = \Pr(2 \text{ or } 4 \text{ or } 6)$

$\qquad\qquad\qquad\qquad = \Pr(2) + \Pr(4) + \Pr(6)$ (mutually exclusive events)

$\qquad\qquad\qquad\qquad = \dfrac{1}{6} + \dfrac{1}{6} + \dfrac{1}{6} = \dfrac{1}{2}$

(c) $\Pr(\text{a number less than 4})$

$\qquad = \Pr(1 \text{ or } 2 \text{ or } 3)$

$\qquad = \Pr(1) + \Pr(2) + \Pr(3)$ (mutually exclusive events)

$\qquad = \dfrac{1}{6} + \dfrac{1}{6} + \dfrac{1}{6} = \dfrac{1}{2}$

(d)   Now the events (an even number) and (a number less than 4) are not mutually exclusive since:

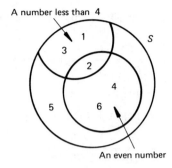

$$\text{(an even number)} \cap \text{(a number less than } 4)$$

$$= (2, 4, 6) \cap (1, 2, 3)$$

$$= (2)$$

which is not the empty set and so we cannot use Definition 4.16(c). Instead, we can write the event (an even number) or (a number less than 4) as:

$$(2, 4, 6) \cup (1, 2, 3) = (1, 2, 3, 4, 6) = E, \quad \text{say}$$

So that:

$$\text{Pr(an even number or a number less than } 4)$$

$$= \Pr(E) = \Pr((1, 2, 3, 4, 6))$$

$$= \frac{n((1, 2, 3, 4, 6))}{n(S)} \quad \text{(by definition with an equally likely}$$
$$\text{outcome set)}$$

$$= \frac{5}{6}$$

(e)   Now we already know that the event 'an even number' and 'a number less than 4' has event set (2).

Hence: $\text{Pr(an even number and a number less than } 4)$

$$= \Pr((2)) = \frac{n((2))}{n(S)} \quad \text{(by definition)}$$

$$= \frac{1}{6}$$

We can calculate the probability of the complement of any event occurring using the simple rule $\Pr(\bar{E}) = 1 - \Pr(E)$, where $\bar{E}$ is the complement of E w.r.t. S, the outcome set for the experiment.

This is easy to show as follows.

We have: $\Pr(S) = 1$ (by definition)

But $E \cup \bar{E} = S$ (see Fig. 4.6)

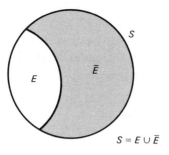

$$S = E \cup \bar{E}$$

Fig. 4.6

$\therefore$　　　　　　$\Pr(E \cup \bar{E}) = 1$　i.e.,　$\Pr(E \text{ or } \bar{E}) = 1$

giving　$\Pr(E) + \Pr(\bar{E}) = 1$,　since E and $\bar{E}$ are always mutually exclusive.

Hence　　　　　　　　$\Pr(\bar{E}) = 1 - \Pr(E)$

This rule is valid for any type of event including unions and intersections of events. That is, if E_1 and E_2 are two events of an experiment then $\Pr(\overline{E_1 \text{ or } E_2}) = 1 - \Pr(E_1 \text{ or } E_2)$ and $\Pr(\overline{E_1 \text{ and } E_2}) = 1 - \Pr(E_1 \text{ and } E_2)$.

For example, in the die-rolling experiment with $S = (1, 2, 3, 4, 5, 6)$, let $E_1 = (1, 2, 3)$ and $E_2 = (1, 5, 6)$. Then $\Pr(E_1) = \frac{1}{2}$, $\Pr(E_2) = \frac{1}{2}$, and

$$\Pr(E_1 \cup E_2) = \Pr((1,2,3,5,6)) = \frac{5}{6}, \quad \Pr(E_1 \cap E_2) = \Pr((1)) = \frac{1}{6}$$

So that:

$$\Pr(\overline{E_1}) = 1 - \Pr(E_1) = 1 - \frac{1}{2} = \frac{1}{2}$$

$$\Pr(\overline{E_2}) = 1 - \Pr(E_2) = 1 - \frac{1}{2} = \frac{1}{2}$$

$$\Pr(\overline{E_1 \cup E_2}) = \Pr(\overline{E_1 \text{ or } E_2}) = 1 - \Pr(E_1 \cup E_2) = 1 - \frac{5}{6} = \frac{1}{6}$$

Similarly

$$\Pr(\overline{E_1 \cap E_2}) = 1 - \Pr(E_1 \cap E_2) = 1 - \frac{1}{6} = \frac{5}{6}$$

To summarise:

STATEMENT 4.6

If E is any valid event of an experiment (having an outcome set S) then:

$$\Pr(\bar{E}) = 1 - \Pr(E)$$

where $\bar{E}$ is the complement of E w.r.t. S.

Proof:　in preceding text.

EXAMPLE 4.17

A box contains 5 red, 4 white and 6 blue balls which, except for their colour, are indistinguishable. Find, if a ball is drawn randomly from the bag, the probability that it is:

(a)   red   (b)   white   (c)   red or white   (d)   not white

and   (e)   neither red nor white.

SOLUTION

Putting the outcome set as $S = (\text{red, white, blue}) = (R, W, B)$ we do not have equally likely outcomes, since the balls are unevenly distributed. However, since the ratio of red to white balls is $5:4$, we must have:

$$\Pr(R):\Pr(W) = 5:4 \quad \text{giving} \quad \Pr(W) = \frac{4}{5} \cdot \Pr(R) \qquad [1]$$

Similarly

$$\Pr(B):\Pr(R) = 6:5 \quad \text{giving} \quad \Pr(B) = \frac{6}{5} \cdot \Pr(R) \qquad [2]$$

Now, since we are dealing with a finite outcome set, we have:

$$\Pr(R) + \Pr(W) + \Pr(B) = 1 \quad \text{(by Definition 4.16(e))}$$

$$\therefore \quad \Pr(R) + \frac{4}{5} \cdot \Pr(R) + \frac{6}{5} \cdot \Pr(R) = 1 \quad \text{(using [1] and [2])}$$

$$\Rightarrow \quad \Pr(R) \cdot \left(1 + \frac{4}{5} + \frac{6}{5}\right) = 1 \quad \Rightarrow \quad \Pr(R) = \frac{1}{\left(\frac{15}{5}\right)} = \frac{5}{15} = \frac{1}{3}$$

Hence:

(a) $\Pr(R) = \dfrac{1}{3}$

(b) $\Pr(W) = \dfrac{4}{5} \cdot \dfrac{1}{3} = \dfrac{4}{15}$ (using [1])

(c) $\Pr(R \text{ or } W) = \Pr(R) + \Pr(W)$ (mutually exclusive events)

$$= \frac{1}{3} + \frac{4}{15} = \frac{9}{15} = \frac{3}{5}$$

(d) $\Pr(\overline{W}) = 1 - \Pr(W)$ (Statement 4.6)

$$= 1 - \frac{4}{15} = \frac{11}{15}$$

(e) $\Pr(\text{neither red nor white}) = \Pr(\overline{R \text{ or } W})$

$$= 1 - \Pr(R \text{ or } W) \quad \text{(Statement 4.6)}$$

$$= 1 - \frac{3}{5} \quad \text{(from (c))}$$

$$= \frac{2}{5}$$

With reference to the above example, the solution to (a) was arrived at via a 'long-winded' approach. It should be clear that with 5 reds out of a total of 15 balls, $\Pr(R) = \dfrac{5}{15} = \dfrac{1}{3}$. However, the justification for this obvious truth still needs to comply with the rules for probability (as detailed in Definition 4.16).

One of the main pitfalls in solving problems involving probabilities is 'jumping in' and writing down a seemingly obvious statement. The only way to ensure it is correct (except in the more elementary situation and, possibly, the above is one) is to be aware of and follow the rules for forming probabilities. Returning to the above, another approach is to argue as follows. Theoretically, all 15 balls are distinct (no two reds, for example, will be physically identical). Hence, if a ball is to be drawn truly at random, the outcome set is $S = \{R_1, R_2, \ldots, R_5, W_1, W_2, \ldots, B_5, B_6\}$. Hence we have a finite, equally-likely outcome set S and we can use (f) of Definition 4.16.

Therefore:
$$\Pr(R) = \frac{n(R)}{n(S)}$$

Now, the event R has set $\{R_1, R_2, \ldots, R_5\}$.

Thus $n(R) = 5$, giving $\Pr(R) = \dfrac{5}{15} = \dfrac{1}{3}$.

We have seen that that the rule $\Pr(A \text{ or } B) = \Pr(A) + \Pr(B)$ only holds when A and B are mutually exclusive events of some experiment. A more general rule for calculating this type of probability is now given.

STATEMENT 4.7

If E_1 and E_2 are *any* two events of the same experiment, then

$$\Pr(E_1 \text{ or } E_2) = \Pr(E_1) + \Pr(E_2) - \Pr(E_1 \text{ and } E_2)$$

or: $$\Pr(E_1 \cup E_2) = \Pr(E_1) + \Pr(E_2) - \Pr(E_1 \cap E_2)$$

Proof   Left as an exercise.

Note that the above rule is valid whether or not E_1 and E_2 are mutually exclusive. However, if E_1 and E_2 are mutually exclusive, then $E_1 \cap E_2 = \emptyset$ giving $\Pr(E_1 \text{ and } E_2) = 0$, and the above rule becomes identical to that of Definition 4.16(c) for mutually exclusive events.

Returning to Example 4.16 we had the result: $\Pr(\text{even number and a number less than 4}) = \frac{1}{6}$.

So: $\Pr(\text{even number or number less than 4})$

$= \Pr(\text{even number}) + \Pr(\text{number less than 4})$

$- \Pr(\text{even number and a number less than 4})$

(using the above statement)

$$= \frac{1}{2} + \frac{1}{2} - \frac{1}{6} = \frac{5}{6} \quad \text{(agreeing with (d) of the example)}$$

EXAMPLE 4.18

A single card is drawn from a normal pack of 52. Calculate the probabilities of the following events occurring:

(a)　heart (i.e., the card drawn is a heart),

(b)　club,

(c)　seven (i.e., the card drawn is a seven of any suit),

(d)　heart or club, and

(e)　heart or seven.

SOLUTION

The equally likely outcome set for this experiment is $S = (52$ cards in the pack) with $n(\text{heart}) = n(\text{club}) = 13$ and $n(\text{seven}) = 4$.

(a)　$\Pr(\text{heart}) = \dfrac{n(\text{heart})}{n(S)}$　(by Definition)

$$= \frac{13}{52} = \frac{1}{4}$$

(b)　$\Pr(\text{club}) = \dfrac{n(\text{club})}{n(S)} = \dfrac{13}{52} = \dfrac{1}{4}$

(c)　$\Pr(\text{seven}) = \dfrac{n(\text{seven})}{n(S)} = \dfrac{4}{52} = \dfrac{1}{13}$

(d)　$\Pr(\text{heart or club}) = \Pr(\text{heart}) + \Pr(\text{club})$　(mutually exclusive events)

$$= \frac{1}{4} + \frac{1}{4} = \frac{1}{2}$$

(e)　$\Pr(\text{heart or seven}) = \Pr(\text{heart}) + \Pr(\text{seven}) - \Pr(\text{heart and seven})$
(by Statement 4.7)

$\Big($The event 'heart and seven' has just one element, namely the 'seven of hearts'.

∴　　　　　　　　$\Pr(\text{heart and seven}) = \dfrac{1}{52}.\Big)$

i.e.:　　$\Pr(\text{heart or seven}) = \dfrac{1}{4} + \dfrac{1}{13} - \dfrac{1}{52} = \dfrac{16}{52} = \dfrac{4}{13}$

Sometimes we wish to perform an experiment where the outcome set (together with a particular event set) is either unknown or undefinable. For instance, consider a machine in a factory which produces components within rigid tolerance limits. If a particular item is outside these limits it is labelled 'defective' and reprocessed. We can consider the whole process (including the tolerance check) as an experiment, where the outcome set can be thought of as comprising the two mutually exclusive and exhaustive (and infinite) event sets 'OK' and 'defective'. If we were interested in the event E, 'a defective item', and wished to find $\Pr(E)$, we clearly cannot use the form $\Pr(E) = \dfrac{n(E)}{n(S)}$ since numerator and denominator are both infinite numbers.

In this sort of case we use the following definition.

DEFINITION 4.17

The *empirical* or *experimental probability* of an event E occurring, written as $\Pr_{exp}(E)$, associated with some experiment, is obtained by:

(a)   performing the experiment a large number, N, of times,

(b)   observing how many times the event E occurs, n say,

(c)   calculating $\Pr_{exp}(E) = \dfrac{\text{No. of times event occurs}}{\text{No. of times experiment is performed}}$

$$= \frac{n}{N}$$

For example, in the previous situation of a machine producing 'OK' and 'defective' items, if we take a sample of 500 components and obtain, say, 20 defectives, we have that the experimental probability of obtaining a defective:

$$= \Pr_{exp}(E) = \frac{20}{500} = 0.04$$

Notes   (1)   The above expression for the empirical probability of an event E is, in general, taken as an estimate of the theoretical probability of E.

(2)   The greater the value of N (the number of experiments performed), the better the result is as an estimate of the theoretical probability.

EXAMPLE 4.19

A die was rolled 240 times and the number of 3s obtained after each of 24 rolls was: 4, 7, 7, 4, 2, 4, 3, 4, 1, 2. Draw a chart showing the cumulative proportion of 3s obtained, superimposing the corresponding theoretical probability.

SOLUTION

No. of rolls	24	48	72	96	120	144	168	192	216	240
Cumulative No. of 3s	4	11	18	22	24	28	31	35	36	38
Cumulative proportion	0.167	0.229	0.25	0.229	0.20	0.194	0.185	0.182	0.167	0.158

The theoretical probability of obtaining a 3 with a single roll of a die is $\frac{1}{6} = 0.17$ (2D).

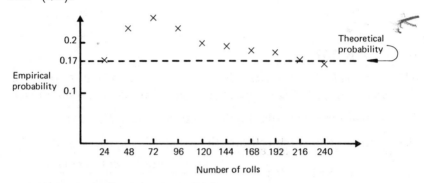

4.4 CONDITIONAL PROBABILITY

4.4.1 CONDITIONAL PROBABILITY AND INDEPENDENCE

DEFINITION 4.18

Let E_1 and E_2 be any two events, not necessarily from the same experiment. The *conditional probability* of E_1, given E_2 has occurred, is written as $\Pr(E_1/E_2)$ and defined:

$$\Pr(E_1/E_2) = \frac{\Pr(E_1 \text{ and } E_2)}{\Pr(E_2)}$$

This definition is applicable to all types of outcome sets both finite and infinite.

If we are dealing with an experiment having a finite, equally likely outcome set we can calculate conditional probabilities as numbers. Suppose that E_1 and E_2 are two events of this type of experiment.

$$\Pr(E_1/E_2) = \frac{\Pr(E_1 \text{ and } E_2)}{\Pr(E_2)} \quad \text{(by definition)}$$

But $\Pr(E_1 \text{ and } E_2) = \dfrac{n(E_1 \text{ and } E_2)}{n(S)}$ and $\Pr(E_2) = \dfrac{n(E_2)}{n(S)}$

Hence $\Pr(E_1/E_2) = \dfrac{n(E_1 \text{ and } E_2)}{n(S)} \dfrac{n(S)}{n(E_2)} = \dfrac{n(E_1 \text{ and } E_2)}{n(E_2)}$

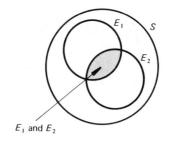

E_1 and E_2

Fig. 4.7

The significance of this result is as follows:

$$\Pr(E_1/E_2) = \frac{n(E_1 \text{ and } E_2)}{n(E_2)}$$

is just the probability of the event 'E_1 and E_2' occurring w.r.t. an outcome set E_2. See Fig. 4.7. In effect, we are ignoring everything except the event E_2 (which we know has occurred) and calculating the probability of the occurrence of that part of the event E_1 that is also contained in E_2.

Consider rolling a normal die as an experiment with the usual $S = (1, 2, 3, 4, 5, 6)$. Suppose that we were interested in obtaining a number less than 4 with a single roll. Putting $E_1 = (1, 2, 3)$, we require:

$$\Pr(E_1) = \frac{n(E_1)}{n(S)} = \frac{3}{6} = \frac{1}{2}$$

However, if we are told, after the die has been rolled, that an odd number has occurred, say $E_2 = (1, 3, 5)$, then to calculate $\Pr(E_1)$ now, given this extra information, we cannot consider the outcomes 2, 4 and 6 as possible ones. Only 1, 3 and 5 are possible outcomes here.

That is, the *effective* (or *reduced*) outcome set in this case is $E_2 = (1, 3, 5)$. Now, out of these three outcomes, only 1 and 3 will cause the event E_1 to occur, i.e., $E_1/E_2 = (1, 3)$ and we call E_1/E_2 a *conditional event* (see Fig. 4.8). So that, conditional on E_2, the probability of E_1 occurring is:

$$\frac{n((1, 3))}{n((1, 3, 5))} = \frac{2}{3} \quad \text{i.e.} \quad \Pr(E_1/E_2) = \frac{2}{3}$$

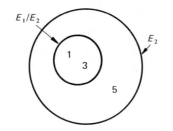

Fig. 4.8

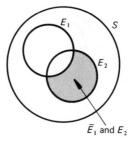

Fig. 4.9

To summarise:

STATEMENT 4.8

If S is a finite, equally likely outcome set of some experiment and E_1, E_2 are any two events of S, then:

(a) $\Pr(E_1/E_2) = \dfrac{n(E_1 \text{ and } E_2)}{n(E_2)}$.

(b) $\Pr(\bar{E}_1/E_2) = 1 - \Pr(E_1/E_2)$.

(c) If E_1 and E_2 are mutually exclusive, $\Pr(E_1/E_2) = 0$.

Proof   (a)   In preceding text.

(b)   See Fig. 4.9. The shaded region represents the set $\bar{E}_1$ and E_2.

$$
\begin{aligned}
\text{Now:}\qquad \Pr(\bar{E}_1/E_2) &= \frac{n(\bar{E}_1 \text{ and } E_2)}{n(E_2)} \quad \text{(using (a) above)} \\
&= \frac{n(E_2) - n(E_1 \text{ and } E_2)}{n(E_2)} \quad \text{(see Fig. 4.9)} \\
&= 1 - \frac{n(E_1 \text{ and } E_2)}{n(E_2)} \\
&= 1 - \Pr(E_1/E_2) \quad \text{(by (a) above)}
\end{aligned}
$$

(c)   Left as an exercise

EXAMPLE 4.20

A card is drawn from a normal pack of 52 as an experiment. Using an obvious notation for events, H, C, D, S and A, 2, 3, . . . , Q, K etc, find the values of the following:

(a) $\Pr(5)$ (b) $\Pr(H)$ (c) $\Pr(C)$ (d) $\Pr(\text{black suit})$
(e) $\Pr(5/H)$ (f) $\Pr(H/5)$ (g) $\Pr(C/\text{black suit})$

SOLUTION

The (equally likely) outcome set for this experiment is $S = $ (the pack of 52 cards) with $n(S) = 52$.

(a) $\Pr(5) = \dfrac{n(5)}{n(S)} = \dfrac{4}{52} = \dfrac{1}{13}.$

(b) $\Pr(H) = \dfrac{n(H)}{n(S)} = \dfrac{13}{52} = \dfrac{1}{4}.$

(c) $\Pr(C) = \dfrac{1}{4}$ similarly.

(d) $\Pr(\text{black suit}) = \Pr(C \text{ or } S)$

$$= \Pr(C) + \Pr(S) \quad \text{(mutually exclusive events)}$$

$$= \frac{1}{4} + \frac{1}{4} = \frac{1}{2}.$$

(e) $\Pr(5/H) = \dfrac{n(5 \text{ and } H)}{n(H)}$ (using previous statement)

$$= \frac{1}{13} \quad \text{(since the event '5 and H' has only the one outcome, 'the five of hearts').}$$

(f) $\Pr(H/5) = \dfrac{n(H \text{ and } 5)}{n(5)} = \dfrac{1}{4}.$

(g) $\Pr(C/\text{black suit}) = \dfrac{n(C \text{ and 'black suit'})}{n(\text{black suit})}$

$$= \frac{n(C)}{n(\text{black suit})} \quad \text{(since the event C and 'black suit' = C)}$$

$$= \frac{13}{26} = \frac{1}{2}$$

DEFINITION 4.19

(a)   Two experiments are *independent* if the result of one can in no way affect the possible result of the other.

(b)   Two events (E_1 and E_2, say) are *independent* if the probability that one of them occurs is in no way influenced by whether or not the other has occurred.

This enables us to write $\Pr(E_1) = \Pr(E_1/E_2) = \Pr(E_1/\bar{E}_2)$, and similarly $\Pr(E_2) = \Pr(E_2/E_1)$.

Two events originating from independent experiments will themselves be independent, while two events originating from the same experiment will not, in general, be independent.

To demonstrate (a) above, we could roll a die as one experiment and toss a coin as another. Here it is obvious that whatever the die shows could never affect what is shown on the coin and vice versa. In general, two *physically different* experiments (like the die roll and coin toss) will be independent.

We can also obtain independent experiments by repeating a single experiment a number of times *under exactly the same conditions*. This procedure is often referred to as conducting *independent trials* of an experiment.

For instance, consider a box containing 4 balls, coloured red, white, blue and black, say. If we describe an experiment as 'draw a ball, observe its colour and replace it', we can repeat this procedure as often as we like to obtain independent trials. However, if we alter the conditions of the experiment to 'draw a ball, observe its colour and discard (i.e., do not replace)', then we alter the physical conditions for the draw from trial to trial since, after one trial, there is one less ball to pick. Here the trials, and thus the separate experiments, are clearly not independent.

Experiments of these two types are generally referred to as 'sampling with (or without) replacement'.

Earlier in the chapter we had the result:

$$\Pr(E_1 \text{ or } E_2) = \Pr(E_1) + \Pr(E_2) - \Pr(E_1 \text{ and } E_2)$$

which simplified to

$$\Pr(E_1 \text{ or } E_2) = \Pr(E_1) + \Pr(E_2)$$

if E_1, E_2 were mutually exclusive.

We also have a general expression for $\Pr(E_1 \text{ and } E_2)$ which simplifies when E_1, E_2 are independent, as we now state.

STATEMENT 4.9

If E_1 and E_2 are *any two events*:

(a) $\Pr(E_1 \text{ and } E_2) = \Pr(E_2) \cdot \Pr(E_1/E_2)$
$$= \Pr(E_1) \cdot \Pr(E_2/E_1)$$

(b) $\Pr(E_1 \text{ and } E_2) = \Pr(E_1) \cdot \Pr(E_2)$, if and only if E_1, E_2 are independent.

Proof   (a)   Using Definition 4.18:

$$\Pr(E_1/E_2) = \frac{\Pr(E_1 \text{ and } E_2)}{\Pr(E_2)}$$

Hence: $\Pr(E_1 \text{ and } E_2) = \Pr(E_2) \cdot \Pr(E_1/E_2)$

Similarly: $\Pr(E_2 \text{ and } E_1) = \Pr(E_1) \cdot \Pr(E_2/E_1)$
$$= \Pr(E_1 \text{ and } E_2)$$

(b) $\Pr(E_1 \text{ and } E_2) = \Pr(E_1) \cdot \Pr(E_2) \Rightarrow \Pr(E_2) = \Pr(E_2/E_1)$
(since $\Pr(E_1 \text{ and } E_2) = \Pr(E_1) \cdot \Pr(E_2/E_1)$ using (a))
$$\Rightarrow E_1, E_2 \text{ independent}   \text{(by Definition 4.19)}$$

Conversely E_1, E_2 independent implies $\Pr(E_1) = \Pr(E_1/E_2)$ and substitution into (a) gives:

$$\Pr(E_1 \text{ and } E_2) = \Pr(E_2) \cdot \Pr(E_1)$$

For example, consider a bag containing 3 red, 2 blue and 4 white balls. We draw one ball and then a second, and we wish to find the probability of obtaining a red ball on the first draw followed by a blue on the second. *If the first ball is not replaced*, we have the situation of two *dependent* experiments. The first 'drawing a ball from 3 red, 2 blue and 4 white (without replacement)'; the second 'drawing a ball from 8 balls, the colours of which depend on the result of the first draw (experiment)'.

The important point here is that the two events R_1 (red on first draw) and B_2 (blue on second draw) are *not* independent. Hence, we use Statement 4.9(a)

So that:

$$\Pr(R_1 \text{ and } B_2) = \Pr(R_1) \cdot \Pr(B_2/R_1)$$
$$= \frac{3}{9} \cdot \frac{2}{8}$$

(since once a red ball has been drawn there are only 2 blues out of a total of 8 to choose from)

$$= \frac{1}{3} \cdot \frac{1}{4} = \frac{1}{12}$$

If the first ball is replaced, we have the situation of two independent experiments 'draw a ball from 3 red, 2 blue and 4 white', independent, since no matter which event occurs on the first draw (experiment), it cannot affect the result of the second draw in any way. But, independent experiments necessarily mean independent events. That is, the events R_1 and B_2 are independent.

Hence, using Statement 4.9(b), we have:

$$\Pr(R_1 \text{ and } B_2) = \Pr(R_1) \cdot \Pr(B_2)$$
$$= \frac{3}{9} \cdot \frac{2}{9} = \frac{6}{81} = \frac{2}{27}$$

We can extend the idea of event independence to three or more events.

Remember that to show three or more events to be mutually exclusive, it was only necessary to show that every possible pair was mutually exclusive. With independence it is a different matter.

STATEMENT 4.10

(a) $\Pr(E_1 \text{ and } E_2 \text{ and } E_3) = \Pr(E_1) \cdot \Pr(E_2/E_1) \cdot \Pr(E_3/E_1 \text{ and } E_2)$.

(b) $\Pr(E_1 \text{ and } E_2) = \Pr(E_1) \cdot \Pr(E_2)$, $\Pr(E_1 \text{ and } E_3) = \Pr(E_1) \cdot \Pr(E_3)$ and $\Pr(E_2 \text{ and } E_3) = \Pr(E_2) \cdot \Pr(E_3)$ *together with* $\Pr(E_1 \text{ and } E_2 \text{ and } E_3) = \Pr(E_1) \cdot \Pr(E_2) \cdot \Pr(E_3)$,

$\Leftrightarrow E_1, E_2, E_3$ are independent events.

No proof   (The proof of the above statement is not particularly difficult, it simply requires more knowledge of set theory than we have introduced in this text.)

Notes   (1)   The pattern of part (a) above should be easily recognised and we would extend as follows:

$\Pr(E_1 \text{ and } E_2 \text{ and } E_3 \text{ and } E_4)$

$= \Pr(E_1) \cdot \Pr(E_2/E_1) \cdot \Pr(E_3/E_1 \text{ and } E_2) \cdot \Pr(E_4/E_1 \text{ and } E_2 \text{ and } E_3)$

A little thought will show this result to be reasonable.

(2)   In part (b) we have that the independence of 3 events can be shown by demonstrating the independence of each pair and showing that

$$\Pr(E_1 \text{ and } E_2 \text{ and } E_3) = \Pr(E_1) \cdot \Pr(E_2) \cdot \Pr(E_3)$$

The extension of this to 4 events would then be: To show that any 4 events were independent we would need that:

(a)   every pair were independent,

(b)   every three were separately independent, and

(c) $\Pr(E_1 \text{ and } E_2 \text{ and } E_3 \text{ and } E_4) = \Pr(E_1) \cdot \Pr(E_2) \cdot \Pr(E_3) \cdot \Pr(E_4)$.

Conversely, if we knew that 4 particular events were independent then we would automatically have that any two or any three of the four are separately independent.

Consider a bag containing thirty balls of identical colour and shape numbered from 1 to 30. If we draw four balls (without replacement), the drawings can be considered as four non-independent experiments. To find Pr(all balls have even numbers) we first define the event E_1 as 'an even number with first ball', . . . , etc. We then find:

$\Pr(E_1 \text{ and } E_2 \text{ and } E_3 \text{ and } E_4)$

$= \Pr(E_1) \cdot \Pr(E_2/E_1) \cdot \Pr(E_3/E_1 \text{ and } E_2) \cdot \Pr(E_4/E_1 \text{ and } E_2 \text{ and } E_3)$

(by an extension of Statement 4.10 (a)).

Now, $\Pr(E_1) = \frac{15}{30}$ (since there are 15 even numbered balls out of the total of 30).

$\Pr(E_2/E_1) = \frac{14}{29}$ (since, given that an even numbered ball has been drawn, we are left with only 14 out of 29 to choose from.)

Similarly:

$$\Pr(E_3/E_1 \text{ and } E_2) = \frac{13}{28} \text{ and } \Pr(E_4/E_1 \text{ and } E_2 \text{ and } E_3) = \frac{12}{27}$$

$$\therefore \quad \Pr(E_1 \text{ and } E_2 \text{ and } E_3 \text{ and } E_4) = \frac{(15)(14)(13)(12)}{(30)(29)(28)(27)} = \frac{13}{261}$$

Alternatively, drawing the balls with replacement causes the events E_1, E_2, E_3 and E_4 to be independent.

Hence: $\Pr(E_1 \text{ and } E_2 \text{ and } E_3 \text{ and } E_4) = \Pr(E_1) \cdot \Pr(E_2) \cdot \Pr(E_3) \cdot \Pr(E_4)$

$$= \frac{(15)(15)(15)(15)}{(30)(30)(30)(30)} = \frac{1}{16}$$

The following examples are aimed at tying together the ideas introduced so far in this section. Although it is often possible to give answers directly (in one or two lines, say), the problems are put strictly into an 'experiment/ event/probability' context and solutions are explained in detail, particular steps being justified where necessary.

EXAMPLE 4.21

Three marksmen A, B and C fire at a target. The probabilities that they will hit the centre spot (independently) are $\frac{1}{4}, \frac{1}{2}$ and $\frac{2}{5}$ respectively. Find the probability that *exactly* two of them will hit the centre of the target, if all three fire at the same time.

SOLUTION

The probability we require is that exactly two hit the target. That is, A and B hit with C missing or A and C hit with B missing or B and C hit with A missing.

(*Note* that it is not correct to find the probability that A and B hit or A and C hit or B and C hit since, for example, $\Pr(A \text{ and } B \text{ hit})$ implies that we are not concerned with the result of Cs shot – which is not true. If C hits (as well as A and B), we would end up with three successes whereas we only require two.)

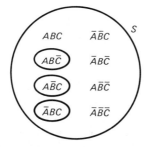

So, we need $\Pr((A \text{ and } B \text{ and } \bar{C})$ or $(A \text{ and } \bar{B} \text{ and } C)$ or $(\bar{A} \text{ and } B$ and $C))$ or, more compactly, $\Pr(AB\bar{C} \text{ or } A\bar{B}C \text{ or } \bar{A}BC)$. Here, we can put the situation into an experimental context by considering the experiment as 'the result of all three men firing', the outcome set being shown. In particular, the three ringed outcomes (elementary events) are mutually exclusive.

So we can write the required probability as:

$$\Pr(AB\bar{C}) + \Pr(A\bar{B}C) + \Pr(\bar{A}BC) \quad \text{(mutually exclusive events)}$$

Consider $\Pr(AB\bar{C})$. We can now take the situation as the three separate experiments A firing, B firing and C firing with outcome sets

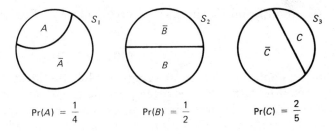

$$\Pr(A) = \frac{1}{4} \qquad \Pr(B) = \frac{1}{2} \qquad \Pr(C) = \frac{2}{5}$$

S_1, S_2 and S_3 as shown. We are given that these experiments are independent, so that the events A, B and C are also independent, with

$$\Pr(A) = \frac{1}{4}, \quad \Pr(B) = \frac{1}{2} \quad \text{and} \quad \Pr(C) = \frac{2}{5} \quad \text{(given)}$$

So that:

$$\Pr(AB\bar{C}) = \Pr(A) \cdot \Pr(B) \cdot \Pr(\bar{C}) \quad \text{(by Statement 4.9)}$$

$$= \frac{1}{4} \cdot \frac{1}{2} \cdot \left(1 - \frac{2}{5}\right) = \frac{3}{40}$$

Similarly:

$$\Pr(A\bar{B}C) = \Pr(A) \cdot \Pr(\bar{B}) \cdot \Pr(C) = \frac{1}{4}\left(1 - \frac{1}{2}\right)\frac{2}{5} = \frac{1}{20}$$

and

$$\Pr(\bar{A}BC) = \left(1 - \frac{1}{4}\right)\frac{1}{2}\left(\frac{2}{5}\right) = \frac{3}{20}$$

$\therefore \qquad \Pr(\text{exactly two hits}) = \frac{3}{40} + \frac{1}{20} + \frac{3}{20} = \frac{11}{40}$

EXAMPLE 4.22

Two cards are taken from a normal pack of 52 and examined simultaneously. Find the probability that:

(a)   they are both hearts,

(b)   just one card is a heart.

SOLUTION

Without loss of generality, we can consider that the cards are drawn one after the other, the first not being replaced. So that, we have a 'sampling without replacement' situation (i.e., two non-independent experiments)

(a)   Here we require Pr(heart with first draw and heart with second draw) = $\Pr(H_1 \text{ and } H_2)$, say.

Since H_1 and H_2 are not independent events, we use the expression of Statement 4.9(a), i.e.:

$$\Pr(H_1 \text{ and } H_2) = \Pr(H_1) \cdot \Pr(H_2/H_1)$$

Now:

$$\Pr(H_1) = \frac{n(H_1)}{n(S)} = \frac{13}{52} = \frac{1}{4}$$

Once a heart has been drawn, there are only 51 cards left in the outcome set and 12 cards in the event set 'heart', for the next draw.

Thus:
$$\Pr(H_2/H_1) = \frac{12}{51}$$

$\therefore$
$$\Pr(H_1 \text{ and } H_2) = \frac{1}{4}\left(\frac{12}{51}\right) = \frac{3}{51}$$

(Note that we cannot use the expression of Statement 4.8(a), since
H_1 and H_2 are not defined events within a single experiment. Here,
H_1 and H_2 originate from two separate non-independent experiments.)

(b)   In this case we require 'heart' on the first draw together with
'not heart' on the second draw *or* 'not heart' on the first draw
together with 'heart' on the second draw (these are two distinct
occurrences), i.e., we need to calculate $\Pr(H_1\bar{H}_2 \text{ or } \bar{H}_1 H_2)$ (using an
obvious notation). Here, we can think of an experiment 'the result
(w.r.t. 'hearts') of drawing two cards from a pack' with an outcome
set as shown. The two elementary events $H_1\bar{H}_2$ and $\bar{H}_1 H_2$ are clearly
mutually exclusive.

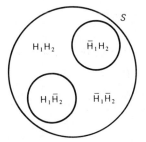

So that: $\Pr(H_1\bar{H}_2 \text{ or } \bar{H}_1 H_2) = \Pr(H_1\bar{H}_2) + \Pr(\bar{H}_1 H_2)$
$$\text{(mutually exclusive events)}$$
$$= \Pr(H_1)\Pr(\bar{H}_2/H_1) + \Pr(\bar{H}_1)\Pr(H_2/\bar{H}_1)$$
$$\text{(by Statement 4.9(a))}$$

Now: $\Pr(H_1) = \dfrac{1}{4}$ and $\Pr(\bar{H}_1) = 1 - \Pr(H_1) = 1 - \dfrac{1}{4} = \dfrac{3}{4}$

Also $\Pr(\bar{H}_2/H_1) = 1 - \Pr(H_2/H_1)$ (Statement 4.8(b))
$$= 1 - \frac{12}{51} = \frac{39}{51} = \frac{13}{17}$$

and $\Pr(H_2/\bar{H}_1) = \dfrac{13}{51}$ (since after the event 'not heart' there are still
13 hearts left out of a total of 51).

Hence: $\Pr(\text{just one card is a heart}) = \dfrac{1}{4}\left(\dfrac{13}{17}\right) + \dfrac{3}{4}\left(\dfrac{13}{51}\right) = \dfrac{13}{34}$

(Note that $\Pr(H_1\bar{H}_2) = \Pr(\bar{H}_1 H_2)$, which is not surprising, since the probability of obtaining only one heart out of two cards should not depend on which of the two cards we select first. So we needed only to calculate $\Pr(H_1\bar{H}_2)$, say, and then multiply by two to obtain the same result as above.)

EXAMPLE 4.23

The probabilities that the independent events $E_1, E_2, E_3, \ldots, E_n$ occur are respectively $p_1, p_2, p_3, \ldots, p_n$. Find the probability that *at least one* of these events occur.

SOLUTION

Putting the above problem in terms of the experiment 'the number of events that occur' we have an outcome set of $S = (0, 1, 2, 3, \ldots, n)$ since the experiment can only result in either $0, 1, 2, 3, \ldots, n$ of the given events occurring.

Putting $E = (1, 2, 3, \ldots, n)$ we require $\Pr$(at least one event occurs) $= \Pr(E)$. But $\Pr(E) = 1 - \Pr(\bar{E})$ (by Statement 4.6), where $\bar{E}$ is the event (0).

Now:

$$\begin{aligned} \Pr(\bar{E}) &= \Pr((0)) = \Pr(\text{none of the above events occur}) \\ &= \Pr(\bar{E}_1 \text{ and } \bar{E}_2 \text{ and } \bar{E}_3 \text{ and } \ldots \text{ and } \bar{E}_n) \\ &= \Pr(\bar{E}_1) \cdot \Pr(\bar{E}_2) \cdot \Pr(\bar{E}_3) \ldots \Pr(\bar{E}_n) \quad \text{(independent events)} \\ &= (1 - \Pr(E_1))(1 - \Pr(E_2))(1 - \Pr(E_3)) \ldots (1 - \Pr(E_n)) \\ &= (1 - p_1)(1 - p_2)(1 - p_3) \ldots (1 - p_n) \end{aligned}$$

$$\therefore \quad \Pr(E) = \Pr(\text{at least one event occurs})$$
$$= 1 - ((1 - p_1)(1 - p_2) \ldots (1 - p_n))$$

EXAMPLE 4.24

Two people A and B play a game by throwing a normal six sided die alternately until one of them throws a six to win. If A begins, find the probability that:

(a) A wins

(b) B wins the game.

SOLUTION

The problem can be thought of in terms of an experiment having an outcome set S (as shown overleaf) split into the two mutually exclusive and exhaustive events 'A wins' and 'B wins', both infinite sets, where A_r represents the outcome 'A wins on his rth roll' etc.

Now: $\Pr(A \text{ wins}) = \Pr(A_1 \text{ or } A_2 \text{ or } A_3 \text{ or } \ldots)$

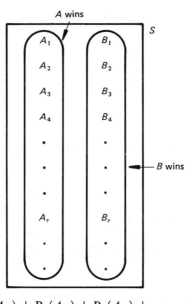

$$= \Pr(A_1) + \Pr(A_2) + \Pr(A_3) + \ldots \quad \text{(mutually exclusive events)}$$

$$\Pr(A_1) = \Pr(A \text{ wins with first roll}) = \frac{1}{6}, \quad \text{clearly}$$

If A and B represent the events 'A throws a six' and 'B throws a six' respectively, we have:

$$\Pr(A_2) = \Pr(\bar{A}\bar{B}A)$$

since both A and B need to 'not throw six' in order that A should win on his second throw,

i.e. $\quad \Pr(A_2) = \Pr(\bar{A}) \cdot \Pr(\bar{B}) \cdot \Pr(A) \quad \text{(independent events)}$

$$= \left(\frac{5}{6}\right)^2 \cdot \left(\frac{1}{6}\right)$$

Similarly,

$$\Pr(A_3) = \Pr(\bar{A}) \cdot \Pr(\bar{B}) \cdot \Pr(\bar{A}) \cdot \Pr(\bar{B}) \cdot \Pr(A)$$

$$= \left(\frac{5}{6}\right)^4 \cdot \left(\frac{1}{6}\right)$$

The pattern here is easy to see. $\Pr(A_4) = (\frac{5}{6})^6(\frac{1}{6})$, $\Pr(A_5) = (\frac{5}{6})^8(\frac{1}{6})$, and so on.

Hence: $\Pr(A \text{ wins}) = \frac{1}{6} + \left(\frac{5}{6}\right)^2\left(\frac{1}{6}\right) + \left(\frac{5}{6}\right)^4\left(\frac{1}{6}\right) + \left(\frac{5}{6}\right)^6\left(\frac{1}{6}\right) + \ldots$

The r.h.s. is just the sum to infinity of a geometric progression with $a = \frac{1}{6}$ and $r = (\frac{5}{6})^2$.

Using Statement 4.2(b):

$$\text{sum} = \frac{a}{1-r} = \frac{\frac{1}{6}}{1-(\frac{5}{6})^2} = \frac{6}{11}$$

i.e. $\Pr(A \text{ wins}) = \dfrac{6}{11}$

So that:

$$\Pr(B \text{ wins}) = \Pr(\overline{A \text{ wins}}) = 1 - \Pr(A \text{ wins}) = 1 - \dfrac{6}{11} = \dfrac{5}{11}$$

EXAMPLE 4.25

Two dice are thrown and it is known that at least one of them is greater than 3. Find the probability that at least one shows a 6.

SOLUTION

The outcome set S for the experiment is shown, each outcome being equally likely. We put E_1 = 'at least one die shows greater than 3' and E_2 = 'at least one die shows 6'.

11	21	31	41	51	61
12	22	32	42	52	62
13	23	33	43	53	63
14	24	34	44	54	64
15	25	35	45	55	65
16	26	36	46	56	66

By definition:

$$\Pr(E_2/E_1) = \frac{n(E_2 \text{ and } E_1)}{n(E_1)} = \frac{n(E_2)}{n(E_1)} \quad \text{since} \quad E_2 \subset E_1$$

$$= \frac{11}{27}$$

To end this section we look at a particular problem that might cause some difficulties unless one is 'pre-armed'.

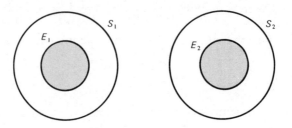

Fig. 4.10

Consider two experiments having respective outcome sets S_1 and S_2 with events E_1 and E_2 defined as shown in Fig. 4.10. (Note that we are *not* implying that the two experiments are independent, although they could be.)

Assuming that $Pr(E_1)$ and $Pr(E_2)$ are known, we wish to calculate $Pr(E_1$ or $E_2)$. It would be understandable (but incorrect) to assume here that $Pr(E_1$ or $E_2) = Pr(E_1) + Pr(E_2)$, the argument being 'the event sets E_1 and E_2 are disjoint therefore E_1, E_2 are mutually exclusive events'. However, a simple numerical example will show the falseness of this assumption.

Suppose

$$Pr(E_1) = \frac{1}{2} \text{ and } Pr(E_2) = \frac{3}{4}. \text{ Then } Pr(E_1) + Pr(E_2) = \frac{1}{2} + \frac{3}{4} = 1\frac{1}{4}$$

(an impossible value for a probability).

The argument is invalid because two events are defined to be mutually exclusive *only within the same experiment* (Definition 4.14).

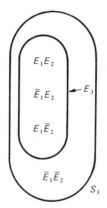

Fig. 4.11

To calculate $Pr(E_1$ or $E_2)$ in this context, we proceed as follows: Define a new experiment 'the result of the two (given) experiments with respect to events E_1 and E_2'. The outcome set for this experiment, S_3 say, is shown in Fig. 4.11, where we see that there are just four outcomes: both events occur (E_1E_2), E_1 occurs and E_2 does not $(E_1\bar{E}_2)$, E_2 occurs and E_1 does not $(\bar{E}_1E_2)$, both events do not occur $(\bar{E}_1\bar{E}_2)$. So, defining event E_3 as shown, we have:

$$Pr(E_1 \text{ or } E_2) = Pr(E_1E_2 \text{ or } \bar{E}_1E_2 \text{ or } E_1\bar{E}_2)$$

$$= Pr(E_3) = 1 - Pr(\bar{E}_3)$$

$$= 1 - Pr(\bar{E}_1\bar{E}_2)$$

Hence:

$$\Pr(E_1 \text{ or } E_2) = 1 - \Pr(\bar{E}_1) \cdot \Pr(\bar{E}_2) \qquad \text{if } E_1, E_2 \text{ are independent}$$
$$= 1 - \Pr(\bar{E}_1) \cdot \Pr(\bar{E}_2/\bar{E}_1) \quad \text{otherwise}$$

EXAMPLE 4.26

Three letters are to be chosen at random from all the letters of the word AEGEAN. Find the probability that:

(a)   the first letter is a consonant,

(b)   either the second or third letter is a vowel,

(c)   an E is not included.

SOLUTION

(a)   We can regard the choosing of the first letter as an experiment, where we require Pr(consonant). Clearly $n(S) = 6$, since this is the total number of letters we have to choose from. There are two consonants altogether and hence:

$$\Pr(\text{consonant}) = \frac{n(\text{consonant})}{n(S)} = \frac{2}{6} = \frac{1}{3}$$

(b)   Either the second or third letter being a vowel is equivalent to *both* of the second and third letters *not being consonants*.

That is: $\Pr(V_2 \text{ or } V_3)$ (V = vowel)

$$= 1 - \Pr(A_1 C_2 C_3) \qquad \text{(A = anything, C = consonant)}$$

(since $C_1 C_2 C_3$ is impossible)

$$= 1 - \Pr(V_1) \cdot \Pr(C_2/V_1) \cdot \Pr(C_3/V_1 C_2) \qquad \text{(Statement 4.9)}$$

$$= 1 - \left(\frac{4}{6}\right)\left(\frac{2}{5}\right)\left(\frac{1}{4}\right)$$

$$= 1 - \frac{1}{15}$$

$$= \frac{14}{15}$$

(c) $\Pr(\text{an E is not included}) = \Pr(\bar{E}_1 \text{ and } \bar{E}_2 \text{ and } \bar{E}_3)$

(where E_1 is the event 'E with first letter' etc)

$$= \Pr(\bar{E}_1) \cdot \Pr(\bar{E}_2/\bar{E}_1) \cdot \Pr(\bar{E}_3/\bar{E}_1 \text{ and } \bar{E}_2)$$

$$= \frac{4}{6}\left(\frac{3}{5}\right)\frac{2}{4} = \frac{1}{5}$$

4.4.2 BAYES THEOREM

DEFINITION 4.20

A set of events $E_1, E_2, \ldots, E_n$ of some experiment is said to be *exhaustive* if their union is the outcome set (S) of the experiment, i.e. if $E_1 \cup E_2 \cup E_3 \cup \ldots \cup E_n = S$.

For example, when rolling a die, $S = (1, 2, 3, 4, 5, 6)$.

If $E_1 = (2, 4, 5)$, $E_2 = (1, 2, 3, 4)$, $E_3 = (1, 3, 5, 6)$, $E_4 = (2, 4, 6)$ and $E_5 = (1, 3, 5)$ then (E_2, E_3) is an exhaustive set of events, since $E_2 \cup E_3 = S$. Also (E_1, E_3), (E_3, E_4), (E_1, E_2, E_3) and (E_4, E_5) are all sets of exhaustive events.

Notice also that events E_4 and E_5, besides being exhaustive, are also a mutually exclusive set.

Bayes Theorem (or formula) is used to solve a certain type of problem in probability, where an experimental outcome set is split into mutually exclusive and exhaustive events.

We will develop an experimental situation of the type where the theorem can be used, before stating and proving it.

Consider an experiment where we choose a single ball randomly from 13 of the same size and shape, seven being red and the rest white.

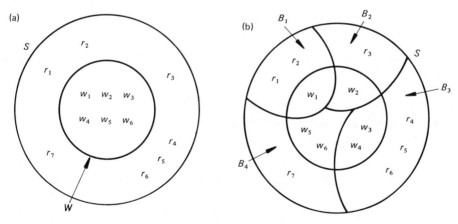

Fig. 4.12

In order to simplify and display this situation in Venn diagram form (Fig. 4.12(a)), we number the red balls from 1 to 7 and the white balls from 1 to 6. Identifying the event W as 'a white ball, we have:

$$\Pr(W) = \frac{n(W)}{n(S)} = \frac{6}{13}$$

We now split S into the four mutually exclusive and exhaustive events B_1, B_2, B_3 and B_4 as shown in Fig. 4.12(b). Physically, we can think of this as dividing the balls up into the four 'bags' B_1, B_2, B_3 and B_4. Here of course, before picking a ball we would need, randomly, to pick a bag. But, given a bag, we can easily calculate the probability of choosing say a white ball using conditional probability as follows:

$$\Pr(W/B_1) = \frac{n(W \text{ and } B_1)}{n(B_1)} \quad \text{(equally likely outcomes)}$$

$$= \frac{n((w_1))}{n((r_1, r_2, w_1))} = \frac{1}{3}$$

Similarly:

$$\Pr(W/B_2) = \frac{1}{2}, \quad \Pr(W/B_3) = \frac{2}{5} \quad \text{and} \quad \Pr(W/B_4) = \frac{2}{3}$$

Suppose now that we have already picked a white ball and wish to find the probability that it came from a particular bag, B_1 say. Then we calculate:

$$\Pr(B_1/W) = \frac{n(B_1 \text{ and } W)}{n(W)} = \frac{1}{6}$$

Similarly:

$$\Pr(B_2/W) = \frac{1}{6}, \quad \Pr(B_3/W) = \frac{2}{6} = \frac{1}{3} \quad \text{and} \quad \Pr(B_4/W) = \frac{1}{3}$$

These latter four probabilities were easy enough to determine since:

(a) we had an equally likely outcome set S, and

(b) each outcome was known.

However, if the only information we had about the experiment was the probabilities $\Pr(B_i)$ ($i = 1, 2, 3, 4$) and $\Pr(W/B_i)$ ($i = 1, 2, 3, 4$), we could not use the above approach.

In this type of situation we use Bayes theorem, which we now state and prove.

STATEMENT 4.11 (BAYES THEOREM)

If $E_1, E_2, \ldots, E_n$ form a set of mutually exclusive and exhaustive events for some experiment and A is any other event, the probability of event E_i occurring given that event A has occurred, is given by:

$$\Pr(E_i/A) = \frac{\Pr(E_i)\Pr(A/E_i)}{\Pr(E_1)\Pr(A/E_1) + \Pr(E_2)\Pr(A/E_2) + \ldots + \Pr(E_n)\Pr(A/E_n)}$$

$$= \frac{\Pr(E_i)\Pr(A/E_i)}{\sum_{j=1}^{n} [\Pr(E_j)\Pr(A/E_j)]}$$

See Fig. 4.13.

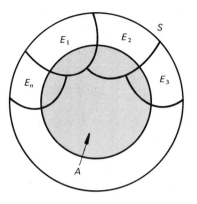

Fig. 4.13

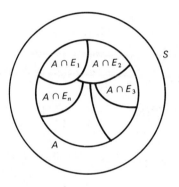

Fig. 4.14

Proof   Now $\Pr(E_i/A) = \dfrac{\Pr(E_i \cap A)}{\Pr(A)}$ and

(a) $\Pr(E_i \cap A) = \Pr(E_i)\Pr(A/E_i)$

(b)   We can split event A up into the mutually exclusive events:

$$A \cap E_1, A \cap E_2, A \cap E_3, \ldots, A \cap E_n \quad \text{(see Fig. 4.14)}$$

$\therefore \quad \Pr(A) = \Pr(A \cap E_1 \text{ or } A \cap E_2 \text{ or} \ldots \text{or } A \cap E_n) \quad \text{(exhaustive events)}$

$\qquad\qquad = \Pr(A \cap E_1) + \Pr(A \cap E_2) + \ldots + \Pr(A \cap E_n) \quad \text{(mutually}$
$\qquad\qquad\qquad\qquad\qquad\qquad\qquad\qquad\qquad\qquad\qquad\qquad \text{exclusive events)}$

$\qquad\qquad = \Pr(E_1)\Pr(A/E_1) + \Pr(E_2)\Pr(A/E_2) + \ldots + \Pr(E_n)\Pr(A/E_n)$

Hence result.

It is useful to think of Bayes theorem in terms of 'reversing conditional probabilities'. That is, if we know the conditional probabilities $\Pr(A/E_1)$, $\Pr(A/E_2)$, $\ldots$, etc., together with $\Pr(E_1)$, $\Pr(E_2)$, $\ldots$, etc., we can find any of the conditional probabilities $\Pr(E_1/A)$, $\Pr(E_2/A)$, $\ldots$, etc.

Returning to the previous example (red and white balls in bags), we had:

$$Pr(W/B_1) = \frac{1}{3}, \quad Pr(W/B_2) = \frac{1}{2}, \quad Pr(W/B_3) = \frac{2}{5} \quad \text{and} \quad Pr(W/B_4) = \frac{2}{3}$$

Also:

$$Pr(B_1) = \frac{n(B_1)}{n(S)} = \frac{3}{13}, \quad Pr(B_2) = \frac{2}{13}, \quad Pr(B_3) = \frac{5}{13} \quad \text{and} \quad Pr(B_4) = \frac{3}{13}$$

Assuming that the only information we have about the experiment is:

(a) these eight probabilities, and

(b) the events B_1 to B_4 are mutually exclusive and exhaustive

we can use the theorem to calculate $Pr(B_i/W)$ for $i = 1, 2, 3, 4$.

We have:
$$Pr(B_i/W) = \frac{Pr(B_i) \cdot Pr(W/B_i)}{\sum\limits_{j=1}^{4} Pr(B_j) \cdot Pr(W/B_j)}$$

Now:
$$Pr(B_1) \cdot Pr(W/B_1) = \frac{3}{13}\left(\frac{1}{3}\right) = \frac{1}{13}, \quad Pr(B_2) \cdot Pr(W/B_2) = \frac{2}{13}\left(\frac{1}{2}\right) = \frac{1}{13},$$

$$Pr(B_3) \cdot Pr(W/B_3) = \frac{5}{13}\left(\frac{2}{5}\right) = \frac{2}{13} \quad \text{and} \quad Pr(B_4) \cdot Pr(W/B_4) = \frac{3}{13}\left(\frac{2}{3}\right) = \frac{2}{13}$$

Hence:
$$\sum\limits_{j=1}^{4} Pr(B_j) \cdot Pr(W/B_j) = \frac{1}{13} + \frac{1}{13} + \frac{2}{13} + \frac{2}{13} = \frac{6}{13}$$

$$\therefore \quad Pr(B_1/W) = \frac{1}{13}\bigg/\frac{6}{13} = \frac{1}{6}, \quad Pr(B_2/W) = \frac{1}{13}\bigg/\frac{6}{13} = \frac{1}{6}$$

$$Pr(B_3/W) = \frac{2}{13}\bigg/\frac{6}{13} = \frac{1}{3} \quad \text{and} \quad Pr(B_4/W) = \frac{2}{13}\bigg/\frac{6}{13} = \frac{1}{3}$$

(as calculated earlier)

EXAMPLE 4.27

A town has three bus routes A, B and C. In the 'rush hour', A has twice as many buses on its route as both B and C. Over a period of time it has been found that, along a certain stretch of road, where the three routes converge, the buses on these routes run more than five minutes late $\frac{1}{2}$, $\frac{1}{5}$ and $\frac{1}{10}$ of the time respectively. If an inspector (standing near this particular stretch of road) finds that the first bus he sees is more than five minutes late, find the probability that it is a route B bus.

SOLUTION

Note here that we are given no details of any outcomes or event sets, just probabilities. We can still represent the experiment in the form of a Venn diagram as shown, where we have partitioned S into the three mutually exclusive and exhaustive events A, B and C, together with the event $L = $ 'late buses'.

We require the probability $Pr(B/L)$.

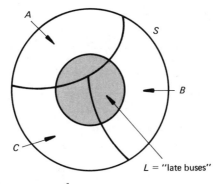

L = "late buses"

Using Bayes theorem we have:

$$\Pr(B/L) = \frac{\Pr(B) \cdot \Pr(L/B)}{\Pr(A) \cdot \Pr(L/A) + \Pr(B) \cdot \Pr(L/B) + \Pr(C) \cdot \Pr(L/C)}$$

Now, since route A has twice as many buses as both of the routes B and C, we can put $\Pr(A) = \dfrac{1}{2}$ and $\Pr(B) = \Pr(C) = \dfrac{1}{4}$.

Also we are given that:

$$\Pr(L/A) = \frac{1}{2}, \quad \Pr(L/B) = \frac{1}{5} \quad \text{and} \quad \Pr(L/C) = \frac{1}{10}$$

$\therefore \qquad \Pr(B/L) = \dfrac{(\frac{1}{4})(\frac{1}{5})}{(\frac{1}{2})(\frac{1}{2}) + (\frac{1}{4})(\frac{1}{5}) + (\frac{1}{4})(\frac{1}{10})} = \dfrac{\frac{1}{20}}{\frac{1}{4} + \frac{1}{20} + \frac{1}{40}} = \dfrac{2}{13}$

Conditional probabilities (and in particular Bayes theorem problems) can be solved using a relative frequency approach, by means of a two-way frequency table. Using this approach, probabilities are expressed as numbers of expected occurrences.

For example, suppose that a particular population was split up into half male and half female, where it is known that the probability of a male with driving experience is 0.75 and a female with driving experience is 0.6. Considering a theoretical population of 1000 gives a split of 500 men and 500 women. Of the 500 men, a proportion of 0.75 (= 375) would be expected to have driving experience, while a proportion of 0.6 of the 500 women (= 300) would be expected to have driving experience.

The following table can now be set up:

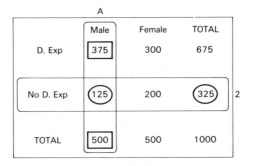

A	Male	Female	TOTAL
D. Exp	375	300	675
No D. Exp| 125	200	325	2
TOTAL	500	500	1000

The obvious probabilities obtainable from the table are, for example:

$$\Pr(\text{Male}) = \frac{n(\text{Male})}{n(\text{Total})} = \frac{500}{1000} = 0.5$$

$$\Pr(\text{D.Exp}) = \frac{n(\text{D.Exp})}{n(\text{Total})} = \frac{675}{1000} = 0.675$$

However, column A can be used to demonstrate the conditional probability:

$$\Pr(\text{D. Exp/Male}) = \frac{n(\text{D.Exp})}{n(\text{Male})} = \frac{375}{500} = 0.75$$

and row 2 to demonstrate the conditional probability:

$$\Pr(\text{Male/No D.Exp}) = \frac{n(\text{Male})}{n(\text{No D.Exp})} = \frac{125}{325} = 0.385 \text{ (3D)}$$

EXAMPLE 4.28

Rework Example 4.27, using a relative frequency approach.

SOLUTION

Consider a total of 1000 buses. Of these, 500, 250 and 250 would be A, B and C buses respectively. The numbers of late buses would be: for A, 50% of 500 (= 250); for B, 20% of 250 (= 50); for C, 10% of 250 (= 25). All other values in the defined 2×3 table are easily calculated and shown as follows:

	A	B	C	Total
Late	250	50	25	325
Not late	250	200	225	675
Total	500	250	250	1000

We require $\Pr(\text{B/Late}) = \dfrac{50}{325} = \dfrac{2}{13}$ (as previously calculated).

4.5 PERMUTATIONS AND COMBINATIONS

Suppose we wished to find the number of different ways of choosing three letters from A, B, C and D. First of all we could list all possible groups of these letters, where the same three do not appear in more than one group, i.e., ABC, ABD, ACD and BCD. Here, each group contains a unique set of letters. These are called *combinations* of three letters from a total of four (different letters), and we see that there are just four of them.

We write $^4C_3$ or $\binom{4}{3}$, which is translated as 'the number of combinations of three items from a total of four different items', and we see here that $^4C_3 = 4$.

If the order of letters was important then, for instance, the combination ABC could be split up into the separate groups ABC, ACB, BAC, BCA, CAB and CBA. Each of these six arrangements is called a *permutation* of three letters from the letters A, B and C. Note that, although the six arrangements above represent six *different* permutations, they still represent the *same single* combination of the letters A, B and C.

Therefore, any one combination gives rise to a number of different permutations.

In a similar fashion, we could split the combination ABD into the separate permutations ABD, ADB, BAD, BDA, DAB and DBA.

So, from these considerations, it is easy to see that each of the 4 combinations of three letters from the original four can be split into 6 permutations, giving rise to a total of $6 \times 4 = 24$ different permutations of three letters from a total of four (different letters).

We write $^4P_3$, which is translated as 'the number of permutations of three items from a total of four different items', and we see here that $^4P_3 = 24$.

If, for example, we wanted the number of combinations of two digits from 1, 2, 3 and 4, we could list as follows, 12, 13, 14, 23, 24 and 34, giving a total of 6, i.e., $^4C_2 = 6$.

To calculate the number of permutations of two digits from the given four, we have only to notice that each of the listed combinations of two digits can be made to give an extra permutation by simply reversing them. That is, 12 and 21, 13 and 31 and so on, giving a total of 12 different permutations, i.e., $^4P_2 = 12$.

To summarise:

DEFINITION 4.21

(a)   A *combination* of r items from a total of n different items is a selection of r items from the n, whereby two *distinct* combinations will have at least one of the r items selected not common to both groups. We define the *number* of combinations that are distinct as ^nC_r or $\binom{n}{r}$.

(continued)

(b)   Each combination of r items from the n has a number of different *permutations*, obtained by putting the elements of the combination in a different order. Two *distinct* permutations of a single combination will have at least two of their common elements in a different position. We define the *number* of distinct permutations (of r items from the n different items) as ^nP_r.

We shall shortly see how to evaluate the two numbers ^nC_r and ^nP_r, given values of n and r. First, however, we need to consider some other factors.

DEFINITION 4.22

We define the quantity n *factorial* (written $n!$) as:

$$n! = n \times (n-1) \times (n-2) \times \ldots \times 3 \times 2 \times 1$$

for n as a positive integer only.

For example, $3! = 3 \times 2 \times 1 = 6$; $4! = 4 \times 3 \times 2 \times 1 = 24$;

$$7! = 7 \times 6 \times 5 \times 4 \times 3 \times 2 \times 1 = 5040, \text{ etc.}$$

(Note that we restrict n to be a positive integer, and quantities such as $(-2)!$ and $(\frac{1}{2})!$ are not defined here.)

A useful relationship between factorial numbers is given as follows:

$$n! = n \times (n-1)!; \quad (n-1)! = (n-1) \times (n-2)!$$

giving, for instance, $n! = n \times (n-1) \times (n-2)!$

Also: $n! = n \times (n-1) \times (n-2) \times (n-3)!$,   etc.

This type of factorisation is often useful in simplifying expressions involving factorial numbers.

For example: $\dfrac{7!}{5!} = \dfrac{7 \times 6 \times 5!}{5!} = 7 \times 6 = 42$

EXAMPLE 4.29

Calculate: (a) $8!$ (b) $\dfrac{3!}{2!}$ (c) $\dfrac{6! \times 4!}{2! \times 5!}$

SOLUTION

(a) $8! = 8 \times 7 \times 6 \times 5 \times 4 \times 3 \times 2 \times 1 = 40\,320$

(b) $\dfrac{3!}{2!} = \dfrac{3 \times 2 \times 1}{2 \times 1} = 3$

(c) $\dfrac{6! \times 4!}{2! \times 5!} = \dfrac{6 \times 5! \times 4 \times 3 \times 2!}{2! \times 5!} = 6 \times 4 \times 3 = 72$

EXAMPLE 4.30

Write in terms of factorials only:

(a) 26×25

(b) 72

SOLUTION

(a) $26 \times 25 = \dfrac{26 \times 25 \times 24!}{24!} = \dfrac{26!}{24!}$

(b) $72 = 9 \times 8 = \dfrac{9 \times 8 \times 7!}{7!} = \dfrac{9!}{7!}$

STATEMENT 4.12

If one event can occur in a_1 different ways, a second event in a_2 different ways, ..., an rth event in a_r different ways, then the whole set of events can occur together in $a_1 \times a_2 \times a_3 \times \ldots \times a_r$ different ways.

No proof

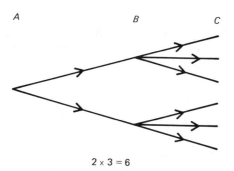

Fig. 4.15 $2 \times 3 = 6$

As a simple demonstration of the above statement, consider the following. There are two different routes from town A to town B and three routes from town B to town C. We could illustrate this situation with a *tree diagram* (Fig. 4.15). How many different ways can we travel from town A to town C, passing through town B? Clearly, with each of the two ways of getting from A to B, there are three ways of getting from B to C. Hence there are 6 (3×2) ways of getting from A to C via B.

STATEMENT 4.13

Any single combination of r items can produce $r!$ distinct permutations of the r items.

Proof   To make up the total number of different permutations, we can regard each single item as an event.

We can choose the first item (event) in r ways, clearly. The second item of the permutation can be chosen in only $r-1$ ways, since, of course, one has already been chosen. The third item can be chosen in $r-2$ ways, and so on. Finally, the last (or rth) item can be chosen in only one way, since this will be the only item left to choose.

Hence, by Statement 4.12, we have just

$$r \times (r-1) \times (r-2) \times \ldots \times 2 \times 1 = r!$$

different ways of choosing the r items.

That is, there are just $r!$ different permutations from a single combination of r items.

In the initial example of selecting three letters from a total of four (A, B, C and D), we examined the combination ABC and found that there were six different permutations of these letters. This could have been evaluated directly (i.e., without actually listing the permutations) by calculating $3! = 3 \times 2 \times 1 = 6$, since $r = 3$ here.

Note that the above statement is simply telling us that $^rP_r = r!$.

EXAMPLE 4.31

How many different three digit numbers can be made from the digits 0, 1, 2, 3, 4, 5 and 6, where we include arrangements such as 000 and 001 as numbers?

SOLUTION

There are seven digits to choose from altogether. We can regard the choice of each of the digits as an event, which can occur in any one of seven ways. Hence the choice of all three of the digits can occur in a total of $7 \times 7 \times 7 = 343$ ways. That is, there are 343 different three digit numbers possible.

(Note that we have not got permutations here, since we are allowing digits to be repeated.)

EXAMPLE 4.32

A penny is tossed five times and the result is recorded in the form HTHHT, for example.

(a)   In how many different ways can the result occur?

(b)   In how many different ways can the first three tosses show tails, and hence

(c)   Find the probability that the first three coins fall tails if a coin is tossed five times.

SOLUTION

(a)   Each toss of the coin can show either head or tails (i.e., 2 different ways of falling)

Hence, all five coins can fall in $2 \times 2 \times 2 \times 2 \times 2 = 32$ different ways.

(b)   The first three coins can fall tails together with the last two falling anything (head or tails) in just $1 \times 1 \times 1 \times 2 \times 2 = 4$ ways.

(c)   Pr(1st 3 coins out of 5 fall tails)

$$= \Pr(E), \quad \text{say}$$

$$= \frac{n(E)}{n(S)},$$

where S is the outcome set for the experiment 'tossing five coins'

$$= \frac{4}{32} \quad \text{(using parts (a) and (b) above)}$$

$$= \frac{1}{8}$$

We are now in a position to derive expressions to calculate the numbers $^{n}P_{r}$ and $^{n}C_{r}$.

STATEMENT 4.14

(a) $^{n}P_{r} = \dfrac{n!}{(n-r)!}$

(b) $^{n}C_{r} = \dfrac{n!}{r! \cdot (n-r)!}$

Proof   (a)   There are n different ways of choosing the first of the r items, $n-1$ different ways of choosing the second, $n-2$ different ways of choosing the third, . . . , and finally, there are just $n-r+1$ different ways of choosing the rth (last) item.

To get the value $n-r+1$, we need to notice the following pattern:

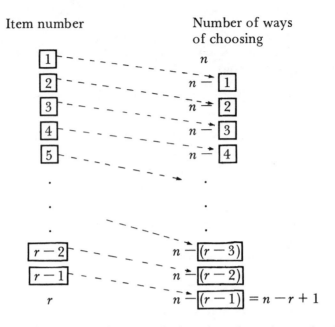

Hence, the total number of ways of choosing the r items is given by:

$$n(n-1)(n-2) \ldots (n-r+1) \quad \text{(using Statement 4.12)}$$

$$= \frac{n(n-1)(n-2) \ldots (n-r+1)(n-r)!}{(n-r)!}$$

$$= \frac{n!}{(n-r)!} \quad \text{(as required)}$$

(b) Now, every one of the ^nC_r combinations gives rise to $r!$ permutations (by Statement 4.13),

i.e. $\qquad ^nP_r = r! \, ^nC_r$

$\therefore \qquad ^nC_r = \frac{1}{r!} \, ^nP_r = \frac{n!}{r!(n-r)!} \quad \text{(as required)}$

For example, $^6P_2$ (the number of permutations of 2 from 6)

$$= \frac{6!}{(6-2)!} = \frac{6!}{4!} = \frac{6 \cdot 5 \cdot 4!}{4!} = 6.5 = 30$$

and $^8C_5$ (the number of combinations of 5 from 8)

$$= \frac{8!}{5! \cdot (8-5)!} = \frac{8!}{5! \cdot 3!} = \frac{8 \cdot 7 \cdot 6 \cdot 5!}{5! \cdot 3!} = \frac{8 \cdot 7 \cdot 6}{3 \cdot 2 \cdot 1} = 56$$

EXAMPLE 4.33

A soccer team is to be chosen from 14 players of which a particular 2 *must* be selected. In how many ways can the team be picked?

SOLUTION

It is combinations that we are interested in here, since the *order* that we pick any particular 11 in is, clearly, not important. (We are assuming here that any particular player can play in any position.)

Although we need 11 players in the team, since we are forced to pick a particular 2, the selection is effectively only 9 from 12.

∴ No. of ways of choosing team $= {}^{12}C_9$

$$= \frac{12!}{9! \cdot 3!} = \frac{12 \cdot 11 \cdot 10}{3 \cdot 2 \cdot 1} = 220.$$

EXAMPLE 4.34

A committee of 6 is to be chosen from 7 men and 4 women. Find the probability that:

(a)   exactly 5 men are chosen,

(b)   at least 3 women are chosen, and

(c)   no women are chosen (assuming that each of the 11 has an equal chance of being chosen).

SOLUTION

If we regard the setting up of the committee as an experiment, the outcome set S, will have just ${}^{11}C_6$ outcomes (namely, the number of different ways of choosing any 6 from the 11) $= \frac{11!}{6! \cdot 5!} = 462$ outcomes, all being equally likely.

(a)   The event 'exactly 5 men' is actually the event '5 men and 1 woman' $= E$, say.

Now, 5 men can be chosen in ${}^7C_5$ ways and 1 woman in ${}^4C_1$ ways.

(Note again here that it is combinations that we are interested in, since the order in which we arrange any selection is not important.)

So, we can choose 5 men and 1 woman together in exactly:

$${}^7C_5 \cdot {}^4C_1 = \frac{7!}{5! \cdot 2!} \cdot \frac{4!}{3! \cdot 1!} = 21 \cdot 4 = 84 \text{ ways}$$

∴ Pr(5 men and 1 woman chosen) $= \text{Pr}(E) = \frac{n(E)}{n(S)} = \frac{84}{462} = \frac{2}{11}$

(b)   Pr(at least 3 women are chosen)

$= \text{Pr}(3 \text{ women and } 3 \text{ men or } 4 \text{ women and } 2 \text{ men})$

$= \text{Pr}(3 \text{ women and } 3 \text{ men}) + \text{Pr}(4 \text{ women and } 2 \text{ men})$
   (mutually exclusive events)

$$= \frac{{}^4C_3 \cdot {}^7C_3}{462} + \frac{{}^4C_4 \cdot {}^7C_2}{462} = \frac{140}{462} + \frac{21}{462} = \frac{161}{462} = \frac{23}{66}$$

(c)   Pr(no women are chosen) = Pr(6 men are chosen)

$$= \frac{^7C_6}{462} = \frac{7}{462} = \frac{1}{66}$$

If we have a set of items that are not uniquely distinguishable (that is, some of them are exactly the same), we can make use of the following result.

STATEMENT 4.15

The number of permutations of *all of n items*, where n_1 of one type are alike, n_2 of a second type are alike, ..., n_r of an rth type are alike, is given by:

$$\frac{n!}{(n_1)! \cdot (n_2)! \cdot (n_3)! \dots (n_r)!} \quad \text{where } n_1 + n_2 + \dots + n_r \leqslant n$$

Proof   Left as an exercise.

For example, suppose we had 10 red, 3 white and 1 black, red and green billiard balls (a total of 16) to arrange in a row.

The number of different arrangements possible, using the above statement, is $\dfrac{16!}{10! \cdot 3! \cdot 1! \cdot 1! \cdot 1!}$, since there are 10 (identical) reds and 3 whites.

Finally we state a useful identity involving the combinatorial function, ^nC_r.

STATEMENT 4.16

$^nC_r = {}^nC_{n-r}$ for $0 \leqslant r \leqslant n$.

Proof $^nC_r = \dfrac{n!}{r!(n-r)!}$ (by definition)

$$= \frac{n!}{(n-r)!\, r!} = \frac{n!}{(n-r)!(n-(n-r))!}$$

$$= {}^nC_{n-r} \quad \text{(by definition)}$$

EXAMPLE 4.35

How many permutations can be made with all the letters from the word VEHICLE:

(a)   with no restrictions,

(b)   if the first letter must be C,

(c)   if the first and last letters must be L and E respectively?

SOLUTION

(a)   With no restrictions, we note first that there are two items alike (two Es). So, using Statement 4.15, there are

$$\frac{7!}{2!} = 7 \times 6 \times 5 \times 4 \times 3 = 2520$$

different permutations possible.

(b)   Since the first letter must be a C, we have only the six following letters to arrange as permutations. Again here, there are two Es.

$$\therefore \qquad \text{Number of permutations} = \frac{6!}{2!} = 6 \times 5 \times 4 \times 3 = 360$$

(c)   With the first letter as L and the last as E, we have only to permute the middle five letters, which are all different. So there are $^5P_5 = 5! = 5 \times 4 \times 3 \times 2 \times 1 = 120$ different permutations.

EXAMPLE 4.36

Six men compete in a track race. If only the 1st, 2nd and 3rd places count, how many different results are possible?

SOLUTION

We can think of the six as A, B, C, D, E and F respectively. We need to choose 3 from the six, but with *order counting* here, since, for example, ABC (i.e., A first, B second, and C third) would be a different result to BAC. That is, we are interested in permutations in this case.

Hence, the number of different possible results is:

$$^6P_3 = \frac{6!}{(6-3)!} = \frac{6!}{3!} = 6 \times 5 \times 4 = 120$$

4.6 EXERCISES

SECTION 4.2

1.   Write the following using set notation: (a) S is a subset of T;   (b) 2 is an element of S;   (c) The Universal set;   (d) 3 is not an element of the empty set;   (e) R contains $\emptyset$.

2.   If $A = (1, 2, 3, 4, 5)$, $B = (1, 2, 3, 4)$ and $C = (3, 4, 5, 6)$, find both the unions and intersections of every pair of sets.

3.   Construct Venn diagrams to illustrate the following two cases:
    (a) $A = (-1, 0, 1, 2, 3)$; $B = (1, 2, 3, 4, 5)$; $C = (3, 4)$
    (b) $M = (a, b, c)$; $N = (b, c, d)$; $P = (e, f, g, h)$; $Q = (a, b, e, f)$

4.   List all *sixteen* subsets of $A = (3, 4, 6, 7)$.

5.   Let $A = (2, 4, 6, 8)$; $B = (1, 3, 5, 9)$; $\& = (1, 2, 3, 4, 5, 6, 7, 8, 9)$.
        Evaluate the sets: A', B', $A \cup B$, $(A \cup B)'$ and $A \cap B$. If M is any set
        having $\&$ as its universal set, show, by using Venn diagrams, that:
        (a) $A \cup A' = \&$;   (b) $A \cap A' = \emptyset$.

6.   The sets P, Q, R, S and T are related according to the given diagram:

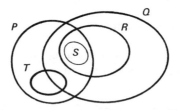

        The elements $1, 2, 3, \ldots, 9$ each belong to at least one of the above
        five sets. If $R = (5, 6, 8)$, $P = (1, 2, 3, 4, 5, 6)$ and the following
        relations are true: $4 \notin T$; $2 \notin Q$; $8 \notin S$; $2 \in T$; $T \cap Q \neq \emptyset$ and
        $n(T) = 2$, list the possible members of the sets Q, S and T.

7.   Write the following in set notation;   (a) a is a member of B;   (b) C is
        not a subset of B;   (c) 2 is not an element of either B or A;   (d) the
        set whose elements are 2 and 4 is a subset of the set whose elements
        are the positive even integers less than 12;   (e) the set whose elements
        are $1, 2$ and 7 contains B;   (f) 3 is not an element of the empty set.

8.   If $A = (-5, -4, -3, -2, -1, 0, 1, 2, 3, 4, 5)$, $B = (-4, -2, 0, 2, 4)$,
        $C = (-5, -3, -1, 1, 3, 5)$, $D = (-4, 4)$, $E = (-3, -2, -1, 0)$
        and $\emptyset$ is the empty set, state which of these sets (if any) could *replace*
        the set X in each of the following statements: (a) $X \cap B = C$;
        (b) $X \subset C$ and $X \not\subset A$;   (c) $X \subset B$ and $X \not\subset E$;   (d) $X \cap C \subset A$;
        (e) $A \cap X = B$;   (f) $B \cup X = A$. Is A a Universal set for B, C, D, E
        and $\emptyset$?

9.   Prove that, if $A \cap B = \emptyset$, then $n(A \cup B) = n(A) + n(B)$.

10.   Find: (a) the 12th term of $(3, 7, 11, \ldots)$;   (b) the 13th term
        of $(\frac{1}{2}, \frac{3}{4}, 1, \ldots)$;   (c) the last term and the sum of the set $(32, 29,$
        $26, \ldots,$ to 50 terms);   (d) the sum of $(1, 8, 15, \ldots,$ up to 40 terms).

11.   Find the 'common ratio' and the 6th term for each of the three series:
        (a) $(3, 9, 27, \ldots)$ (b) $(1, 3, 9, \ldots)$ (c) $(3, \frac{3}{2}, \frac{3}{4}, \ldots)$
        (d) $(1, \frac{1}{4}, \frac{1}{16}, \ldots)$

12.   Find the sum of the following g.p.'s: (a) $(1, 2, 2^2, 2^3, \ldots$ to 10
        terms);   (b) $(1, \frac{1}{2}, \frac{1}{4}, \frac{1}{8}, \ldots$ to 8 terms);   (c) $(3, \frac{3}{4}, \frac{3}{16}, \ldots$ to 5 terms);
        (d) $1, -\frac{1}{2}, \frac{1}{4}, -\frac{1}{8}, \ldots$ to 9 terms).

13.   The set of *natural numbers* is the set $(1, 2, 3, 4, 5, \ldots)$. By consider-
        ing the set as an a.p., find the sum of the first: (a) 5;   (b) 25;
        (c) 100;   (d) n natural numbers.

14.   Find the value of the following sums:
        (a) $101 + 111 + 121 + \ldots + 201$;
        (b) $3.0 + 4.2 + 5.4 + \ldots$ (to 20 terms);
        (c) $4 + 28 + 196 + \ldots + 9604$;
        (d) $-5 + 10 - 20 + 40 - \ldots + 2560$.

15.   Write the following using inequality symbols: (a) a is less than b; (b) 4 is greater than 3.836; (c) a is not greater than or equal to c; (d) 6 is not less than 2; (e) 4 is less than or equal to 4.

16.   Solve the following inequalities for x:

(a) $4x - 2 \leqslant 2x + 4$; (b) $x - 3 > 1 + 3x$;

(c) $2x - 3 \leqslant 5x - 9$; (d) $3x + 2 < 6x - 7$; (e) $\dfrac{3x}{2} \geqslant \dfrac{x}{3} + 1$.

17.   Solve the following for x: (a) $-1 \leqslant x + 3 < 2$; (b) $-6 < -2x \leqslant 4$; (c) $-9 \leqslant 3x \leqslant 12$.

18.   Solve the following inequalities:

(a) $-1 < 2x - 3 < 5$; (b) $-3 \leqslant 5 - 2x \leqslant 7$; (c) $4 \leqslant -2x \leqslant 10$; (d) $x^2 \leqslant 16$;   (e) $(y - 2)^2 < 25$.

19.   If $Z = \dfrac{X - m}{s}$ (where m and s are constants), show that $-1 \leqslant Z \leqslant 2$ is equivalent to $m - s \leqslant X \leqslant m + 2s$.

SECTION 4.3

20.   Two ordinary six-sided dice are thrown and the number of spots shown on each die is noted. By means of a Venn diagram, illustrate the outcome set S for the experiment together with the following events: $E_1 = $ 'at least one three'; $E_2 = $ 'total number of spots on the dice is greater than nine'. Are the two events mutually exclusive?

21.   An experiment is performed by opening the page of any book at random and noting the letter at the beginning of the fifth word.

(a) What is the outcome set for the experiment? If E_1 is the event 'a vowel'; E_2 is the event 'a consonant'; E_3 is the event 'a letter of the word NAIVE' and E_4 is the event 'one of the first ten letters of the alphabet', (b) which pairs of events are m.e.? (c) Find the complements of E_1, E_2 and E_4; (d) If the letter 'f' is the realised outcome of the experiment, which of the above four events will have occurred?

22.   Two teams A and B are to play a game of football and the result (not the score) noted. (a) Define an appropriate experiment and illustrate the outcome set; (b) What would the event be that corresponded to the statement 'team A does not lose'? (c) Name the outcomes in the complement of the above event.

23.   Three coins are tossed consecutively as an experiment. If HTH stands for the particular outcome "heads with first coin, tails on second and heads on third": (a) Write down the outcome set S for the experiment. Let E_1 be the event 'at least two heads'; E_2 be 'all coins show tails' and E_3 be 'at the most two tails'; (b) Illustrate these events, together with S, using a Venn diagram; (c) Which pairs of events are mutually exclusive?

24.   Show: (a) that an event E of some experiment and its complement $\overline{E}$ are always mutually exclusive; (b) that S (the certain event) is the complement of $\emptyset$ (the impossible event); and hence (c) that S and $\emptyset$ are mutually exclusive.

25.   An experiment is defined by "tossing a coin repeatedly until exactly three heads are obtained in total". Using an obvious notation, one outcome of S is HTTHH for example; another is TTTHHTH. Event E_1 is defined 'an odd number of tosses'; E_2 is 'an even number of tosses' and E_3 is 'less than nine tosses in all'.

(a) How many outcomes are there in S?   (b) Which pairs of E_1, E_2 and E_3 are mutually exclusive?   (c) Write down the elements of the event corresponding to 'the least number of tosses possible'.   (d) Which outcomes of S will end with a 'T'?

26.   Two dice are rolled as an experiment. A is the event 'sum of points is odd' and B is the event 'at least one 3 is shown'. Describe in words the events: (a) $A \cup B$;   (b) $A \cap B$;   (c) $(A \cap \bar{B}) \cup \bar{A}$.

27.   Two dice are thrown and the number of spots shown on each is noted. List the complete set of outcomes for the experiment and find: (a) Pr(double 3);   (b) Pr(total score of 9);   (c) Pr(total is an even number).

28.   A card is drawn from a normal pack of 52. Find the probability that the card is: (a) a heart;   (b) a spade;   (c) a four;   (d) the ace of clubs;   (e) not a 3;   (f) neither a heart nor a four.

29.   A sample of 50 married women was asked how many children they had in their family. The results are given in the table:

Number of children	0	1	2	3	4	5 or more
Number of families	6	14	13	9	5	3

Estimate the probability (using the above information) that if any married woman is asked the same question, she will answer: (a) none;   (b) between 1 and 3 inclusive;   (c) greater than 3;   (d) neither 3 nor 4;   (e) less than 2 or more than 4.

30.   Two players A and B have played 30 games of draughts in which A has won 20 and 2 were drawn. If they play one more game, estimate the probability that: (a) A wins;   (b) B does not lose.

31.   (a) Toss a coin 10 times and find and plot on a graph the empirical probability of the event 'head'.   (b) Toss the coin another 10 times and find the empirical probability of a 'head' (using the information from all 20 tosses), plotting this value.   (c) Repeat this procedure until the coin has been tossed a total of 100 times and there are 10 plotted points. (d) Superimpose a dotted line on the graph representing the theoretical probability of a 'head', commenting on the results. (See Example 4.19.)

32.   Let A and B be two events of an experiment with outcome set S such that the intersection of A and B is not empty. Display this information in the form of a Venn diagram and shade the area representing the event 'A or $\bar{B}$'.

33.   An experiment consists of tossing a coin and rolling a die simultaneously and noting both results.

(a) List the elements in the (equally likely) outcome set S.

(b) If E_1 is 'heads (on coin) and even number (on die)'; E_2 is '2, 3 or 5 on the die (anything on coin)' and E_3 is 'tails (coin) and odd number (die)', illustrate these events (together with S) on a Venn diagram.

(c) List the outcomes of the events: (i) E_1 or E_2 ;   (ii) E_2 and E_3 .

(d) Which pairs of events are mutually exclusive?

34.   An experiment is performed by drawing a ball from each of two bags, each bag containing four balls numbered from 1 to 4. Find the probability that the sum of the numbers on the two balls drawn is: (a) 6; (b) 3 using a Venn diagram.

35.   Three runners A , B and C compete in a race. If B and C have the same probability of winning and both are twice as likely to win as A , find the probability that: (a) either A or B wins;   (b) B does not win. (Assume the result cannot be a draw.)

36.   Two normal six-sided dice are tossed as an experiment and the following events are defined: E_1 is 'same number on both dice'; E_2 is 'at least one of the two dice shows a 5 or 6'; E_3 is 'the sum of the two dice is less than 6'.

(a) Find the probabilities of the following events: (i) E_1 ;   (ii) E_2 ; (iii) E_3 ;   (iv) E_1 and E_2 ;   (v) E_1 and E_3 .

(b) Using only the results of (a), find the probabilities of the following events: (i) E_1 or E_2 ;   (ii) E_2 or E_3 ;   (iii) E_1 or E_3 .

37.   In an experiment (with an equally likely outcome set S), A and B are two events. $\Pr(A \text{ or } B) = \frac{3}{4}$, $\Pr(B) = \frac{3}{8}$ and $n(A) = 4$. If $n(S) = 8$, find $\Pr(A \text{ and } B)$.

SECTION 4.4

38.   A fair octagonal (eight sided) die, with faces marked 1 to 8, is thrown as an experiment, the result being the number on the face on which the die rests. The following events are defined: $E_1 = (1, 2, 3, 4, 5)$; $E_2 = (2, 4, 6, 8)$; $E_3 = (1, 3, 5, 7)$. Find the following: (a) $\Pr(E_1/E_2)$; (b) $\Pr(E_1/E_3)$;   (c) $\Pr(\bar{E}_1/E_2)$;   (d) $\Pr(E_2/E_1)$;   (e) $\Pr(E_3/E_1 \text{ or } E_2)$; (f) $\Pr(E_2/E_3)$.

39.   If $\Pr(A) = \frac{1}{3}$, $\Pr(B) = \frac{1}{2}$ and $\Pr(A \text{ or } B) = \frac{3}{4}$, find: (a) $\Pr(A/B)$; (b) $\Pr(B/A)$;   (c) $\Pr(\bar{B}/\bar{A})$;   (d) $\Pr(\bar{A}/\bar{B})$.

(*Hint:* (a) Find $\Pr(A \text{ and } B)$ using Statement 4.7; (b) $\bar{A}$ and $\bar{B} = \overline{(A \text{ or } B)}$.)

40.   A normal coin and six sided die are thrown together. Find the probability that: (a) a 'head' and '4' are shown;   (b) a 'head' and 'even number' are shown;   (c) a 'tail' and 'not a 4' are shown.

41.   Two fair dice are rolled, the first an 'ordinary' one, the second having faces marked 1, 2, 3, 4, 4 and 6. Find the probability that: (a) die 1 shows 3 and die 2 shows 4;   (b) they both show 4;   (c) neither die shows 4;   (d) both dice show an even number;   (e) the total of the two dice is 10 or more.

42.   Three athletes compete in a mile race. The respective probabilities of each of them running it in less than 4 minutes are $\frac{1}{3}, \frac{1}{3}$ and $\frac{1}{12}$. If the running times of each are mutually independent, find the probability that the race is won in less than 4 minutes.

43.   Find the conditional probability that at least three heads are obtained on tossing four unbiased coins given that at least one head has been obtained.

44.   If E_1 and E_2 are mutually exclusive events of any experiment, show that $\Pr(E_1/E_2) = 0$. (Use Definition 4.18.) If $\Pr(E_1/E_2) = 1$, state the relationship between E_1 and E_2.

45.   If two dice are rolled, find the probability that their sum is even given that: (a) both numbers are the same;   (b) both numbers are different.

46.   A bowl contains eight chips, identical except for their colour. Three are red and the remaining five are blue. Two chips are to be drawn successively, at random and without replacement. Find the probability that the first chip drawn is red and the second blue.

47.   A cupboard contains eight pairs of shoes. If six shoes are taken at random and without replacement, find the probability that there is at least one matching pair among the six.

48.   An army cadet is given three routes for reaching a certain destination in a field exercise. It is calculated that, using these routes, he will arrive at the destination within a certain time with probabilities $\frac{3}{8}, \frac{2}{3}$ and $\frac{2}{5}$ respectively. If he chooses a route at random, and does in fact reach his destination within the given time, calculate the probability that he used route 1.

49.   60% of the people in a town are known to vote for party A, 50% of this group owning a car. If only 30% of non-party A voters own a car, find the probability that a car owner chosen at random will be a party A supporter.

50.   A high jump athlete has probability p of clearing a fixed bar at any attempt. If he jumps five times and clears only three times, calculate the conditional probability that he cleared on his first jump.

51.   Three bags each contain coloured discs as follows:

	Bag 1	Bag 2	Bag 3
White	1	2	3
Black	3	1	2
Red	2	3	1

A bag is chosen at random and two discs removed. If these discs are coloured red and white, which of the three bags are they most probably from?

52. E_1 and E_2 are two events with $\Pr(E_1) = \frac{1}{3}$ and $\Pr(E_2) = \frac{1}{4}$. If also $\Pr(E_1 \text{ or } E_2) = \frac{1}{2}$, find: (a) $\Pr(E_1/E_2)$;   (b) $\Pr(E_2/E_1)$;   (c) $\Pr(E_1 \text{ and } \bar{E}_2)$;   (d) $\Pr(E_1/\bar{E}_2)$.

53. E and F are two events with $\Pr(E) = \frac{1}{4}$, $\Pr(E \text{ or } F) = \frac{1}{3}$ and $\Pr(F) = p$. Find p, if: (a) E, F are mutually exclusive;   (b) E, F are independent;   (c) E is a subset of F.

54.   Bag A contains 5 red, 3 white and 8 blue discs while bag B contains 3 red and 5 white discs. A normal six sided die is thrown and if a 3 or 6 is obtained, a disc is chosen from bag B; otherwise a disc is chosen from bag A.

(a) If a red disc is chosen, what is the probability that it came from bag A.

(b) If a white disc is chosen, what is the probability that the die showed a 5?

55. A company owns a fleet of 20 cars, each having either manual or automatic transmission and either 2 or 4 doors. 13 cars are 2-door models and, of these, 12 have automatic transmission. There are only 4 cars with manual transmission. If a car is picked at random from the fleet, calculate the probability that it is: (a) automatic, (b) 4-door; (c) automatic or 2-door; (d) automatic and 2-door; (e) automatic/ 4-door; (f) 4-door/automatic.

56. Test drilling equipment has forecast whether or not a proposed new coal seam will provide economic quantities of coal *incorrectly* 20% of the time. The last 30 seams that were mined produced only 10 that were economic. The current test drilling has just given a favourable indication. What is the probability that this seam will be economic?

SECTION 4.5

57. Calculate: (a) $9!$; (b) $10!$; (c) $\dfrac{4!}{3!}$; (d) $\dfrac{6!\,7!}{8!\,4!}$; (e) $\dfrac{n!}{(n-1)!}$

(positive integer n).

58. Calculate: (a) $\dfrac{(n+1)!}{n!}$; (b) $\dfrac{(n+2)!}{n!}$; (c) $\dfrac{(n+r)!}{n!}$ (n and r positive integers; $n \geqslant r$).

59. How many different four letter words can be formed from the letters A, B, C, D, E and F if letters can be repeated?

60. A four digit number is to be chosen from the six digits 1, 2, 3, 4, 5 and 6 where any digit can be repeated. Find the probability that: (a) there are no 5's or 6's in the number; (b) the middle two digits are both even; and (c) the number is odd.

61. A singer is engaged at a concert to sing 5 songs. If he has a total repertoire of 10 suitable songs and he *must* include a particular 2 of these, in how many different ways can he present his programme?

62. How many different selections of 8 cards can be made from a pack of 52? How many of these will include the '6 of hearts'?

63. Calculate: (a) $^{10}C_4$; (b) $^8C_5$; (c) $^{15}C_4$; (d) $^{15}C_{11}$; (e) $^8P_3$.

64. Simplify: (a) $\dfrac{(n-r+1)!}{(n-r-1)!}$; (b) $\dfrac{(n-1)!}{(n-3)!}$.

65. A passenger compartment on a train has six seats with 3 facing forwards and 3 backwards. Three men and two women enter the compartment and seat themselves randomly.

(a) In how many different ways can they be seated? (b) In how many of these ways will the women be seated opposite each other? (c) In how many of these ways can two of the men be seated opposite each other?

66.   10 different items are to be placed in a row with the condition that a particular 3 of them must be together. In how many ways can this be done? In how many ways can 9 different items be placed in a row with the condition that a particular two do not come together?

67.   How many different 7-digit telephone numbers are possible, if all of the first three and all of the last four cannot be zeros?

68.   How many odd 4-digit numbers can be formed from the digits 0 to 9 if no digit can occur more than once?

69.   An Army company has 3 sergeants, 12 corporals and 113 privates. Find the probability that a random choice of five for guard duty will include no corporals.

70.   Find the number of ways that 10 different letters and 7 different numbers can be arranged in a row if the letters must be in alphabetical order and the numbers in strict numerical order.

71.   In a hand of whist, one player has 6 diamonds. Find the probability that his partner has exactly 4 diamonds.

72.   An experiment is performed by drawing two cards from a normal pack. In two trials of this experiment, find the probability that the first card is an ace and the second is a 10, J, Q or K each time.

73.   Two cards are drawn randomly from a pack. Given that a single 'black' card had been removed beforehand, find the probability that: (a) both cards are red;   (b) both cards are black;   (c) at least one of the cards is a heart;   (d) one card is a heart, the other a club.

74.   6 digits are drawn at random. Find the probability that they are all different.

MISCELLANEOUS

75.   A coin is fixed so that a head is twice as likely to appear as a tail. A die is also fixed so that the numbers 2 and 4 are each twice as likely to appear as an odd number and a 6 is twice as likely to appear as a 2 or 4. The coin and die are thrown together. Find the probability that the coin shows a tail and the die shows a number greater than 3.

76.   Two numbers are sampled randomly (with replacement) from the set (1, 2, 3, ... , 9). If it is known that the sum of the two numbers is even, find the probability that each of the numbers is odd.

77.   An experiment has probability p of succeeding. Three independent trials of the experiment are performed and it is found that if p is doubled, the probability of obtaining a total of two successes out of the three trials is also doubled. Find the value of p.

78.   A bag contains 150 red and 50 yellow discs. Find the probability that a disc picked at random will be red. If four discs are picked randomly, find the probability that there is: (a) 4 red discs;   (b) at least one yellow disc;   (c) 2 red and 2 yellow discs.

79.   An experiment is performed by tossing a coin continually until 3 heads are obtained in total. Using an obvious notation, two possible outcomes of the experiment are HTTHTH and HTHTTTH. Notice that:

(a) every sequence of H's and T's describing an outcome to the experiment must end with an H;   (b) every sequence has exactly 3 H's; and   (c) the outcome set S has an infinite number of elements. The following events are defined: E_1 is 'the coin is tossed exactly 5 times'; E_2 is 'H and T occur alternately' and E_3 is 'T's occur only as a single run of 2 or 3'.

(a) Illustrate these 3 events on a single Venn diagram; and hence (b) write down which pair of events are mutually exclusive.   (c) Explain how $\Pr(E_1)$ might be estimated (empirically).

80.   If E_1 and E_2 are any two events of an experiment, show that $\Pr(E_1 \text{ and } E_2) \leqslant \Pr(E_2)$, and hence prove that $0 \leqslant \Pr(E_1/E_2) \leqslant 1$.

81.   A bag contains eight identical balls (except for colour) of which 3 are red, the rest blue. 4 balls are to be drawn successively at random and without replacement. Find the probability that: (a) the colours alternate;   (b) the first blue ball appears on the 3rd draw.

82.   Successive drawings (with replacement) are made from an ordinary pack of 52 cards. How many draws are necessary so that the probability of at least one heart is at least 0.9.

83.   Using: (a) $\Pr(E_1 \text{ and } E_2) = \Pr(E_1) \Pr(E_2)$ if and only if E_1 and E_2 are independent;   (b) $\Pr(\bar{E}_1 \text{ and } \bar{E}_2) = \Pr(\overline{E_1 \text{ or } E_2})$,   show that if E_1 and E_2 are independent then E_1 and $\bar{E}_2$ are also independent.

84.   In a competition, an archer scores a point only if he hits the centre of a target, and it is found that (on the average) this happens only three times out of ten. How many archers does a team have to field in order to be at least 80% sure of scoring something?

85.   The ace to 10 of hearts are extracted from a normal pack of cards and three of these are chosen randomly. What is the probability that the smallest number shown on any of these three is 6? (The ace counts as 1.)

86.   If A and B are any two sets, then the new set $A - B$ is defined as "the set that contains all the elements of A *that are not in* B". (See the Venn diagram given.) State the relationship between A and B for the special cases: (a) $A - B = A$; and   (b) $A - B = \emptyset$.

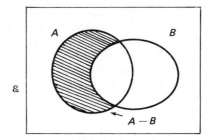

87.   (Continuation of Exercise 86.)   If A and B are any two events of an experiment;   (a)   show, by using a Venn diagram, that $A = (A - B) \cup (A \cap B)$ where events $A - B$ and $A \cap B$ are m.e.; (b)   Hence, using Definition 4.16(c), deduce that $\Pr(A - B) = \Pr(A) - \Pr(A \cap B)$.   (c) Show, using a Venn diagram, that $A \cup B = (A - B) \cup B$ where $A - B$ and B are m.e.   (iv) Hence, deduce that: $\Pr(A \cup B) = \Pr(A - B) + \Pr(B) = \Pr(A) + \Pr(B) - \Pr(A \cap B)$.

88.   *A* has 4 coins and *B* has 3. If they both toss all their coins, find the probability that they both throw the same number of heads.

89.   The probability that a certain beginner at golf will make a good shot is $\frac{2}{3}$ if he uses the correct club and $\frac{1}{4}$ if he uses an incorrect club. He chooses his club at random from a bag in which there are 5 clubs. He can reach the green in one shot if it is a good one or in two shots if the first one is bad but this is followed by a good shot. What is the probability that he will reach the green in two shots or less.

90.   Cards numbered from 1 to *N* are placed in a drum and *n* cards are drawn at random *with replacement*. Find the probability that: (a) all cards have numbers in the range 1 to *k* inclusive $(k \leqslant N)$;   (b) the largest number drawn is *k*.

91.   Two people *A* and *B* draw a single ball at a time (without replacement) from a bag with two white and three black balls. What are their respective chances of being the first one to draw white.

92.   If $p_x = \binom{n}{x} p^x (1-p)^{n-x}$ and $p_{x+1} = \binom{n}{x+1} p^{x+1} (1-p)^{n-x-1}$, show that $\dfrac{p_{x+1}}{p_x} = \dfrac{p}{1-p} \left(\dfrac{n-x}{x+1} \right)$.

93.   A bag contains *w* white balls (numbered from 1 to *w*) and *b* black balls (numbered from $w+1$ to $w+b$).

(a) How many different combinations are possible of drawing: (i) *r* white balls;   (ii) *r* black balls;   (iii) *r* balls of any colour.

(b) Show that $\displaystyle\sum_{s=0}^{r} \binom{w}{s}\binom{b}{r-s} = \binom{w+b}{r}$.

94.   Tom and Dick are playing with two unbiased dice, *X* and *Y*. The dice are thrown once and Tom's score (t) is found by adding the scores on the two dice. Dick's score (d) is found by subtracting the smaller score from the larger on the two dice. On graph paper draw a diagram to represent the possibility space with, for example, the point $(1, 6)$ to represent the event of there being a one on die *X* and a six on die *Y*.

Event *A* is '$9 \leqslant t \leqslant 11$', event *B* is '*d* is 0 or 3' and event *C* is '$t = 3$', On your diagram mark clearly the sets of points associated with each of these events. From your diagram find the probability that: (a) *C* occurs;   (b) both *A* and *B* occur;   (c) *C* occurs and *B* does not.

State a pair of the events *A*, *B* and *C* which are independent and also a pair which are mutually exclusive. Give a reason for each answer. Describe two events from this game which are exhaustive, but not necessarily *A*, *B* or *C*.   (London)

95.   Define the *conditional probability* of an event *A* given that an event *B* has occured.

A boy uses a home-made metal detector to look for valuable metallic objects on a beach. There is a fault in the machine which causes it to signal the presence of only 95% of the metallic objects over which it passes and to signal the presence of 6% of the non-metallic objects over which it passes. Of the objects over which the machine passes, 20% are metallic.

(a) Find the probability that a given object over which the machine passes is metallic and the machine gives a signal.

(b) Find the probability of a signal being received by the boy for any given object over which the machine passes.

(c) Find the probability that the boy has found a metal object when he receives a signal.

(d) Given that 10% of metallic objects on the beach are valuable, find the proportion of objects, discovered by a signal from the detector, that are valuable.   (JMB)

96.   (a) A number is chosen at random from the integers from 1 to 30 inclusive. Find the probability that the number chosen is: (i) a multiple of either 3 or 11;   (ii) a multiple of either 3 or 5 or both.

(b) A bag contains four white balls and five blue balls. If two balls are drawn at random one after the other, without replacement, calculate the probability that they are of the same colour.   (Cambridge)

97.   (a) What is the probability of at least one double six in three throws of two dice? What is the probability in n throws? Show that with 25 throws of 2 dice, one would feel that a wager on a double six occurring would be worthwhile.

(b) An electrical system has two components in it, X and Y. The system will not operate if either of the components fails. The probability of X failing is x, and the probability of Y failing is y. Find the probability that the system will not work. An extra Y component is now included in the system: this means that the system will fail to work if either X fails or both the Y fail. Find the probability now of the system failing.   (SUJB)

98.   (a) Two cards are withdrawn from a pack of playing cards, without replacement. Find the probability that the first is a 2 and just one of them is a heart.

(b) A coin whose diameter is 2.5 cm is rolled in a random way across a large chequered board whose squares are alternately black and white. When the coin stops rolling and falls on the board, the probability of its lying wholly within a white square is $\frac{1}{8}$. Calculate the length of the sides of the squares.   (London)

99.   In a gambling game between two players, the banker and his opponent, a fair die is thrown repeatedly. If a six is thrown in the first three throws the opponent wins that game, otherwise the banker does. If the opponent pays 10 p per game and receives 20 p each time he wins, show that his expected loss over 20 games is approximately $31\frac{1}{2}$ p.

The banker and his opponent agree to a modification of the game so that throwing continues for five throws; the opponent wins as before if a six is thrown in the first three throws but if a six appears at the fourth or fifth throw the game is a draw and the stake money is added to the prize money for the next game. In the event of successive draws the stake money is carried forward cumulatively. Determine the opponent's expected gain or loss from the first two games, giving your result to two significant figures.   (Cambridge)

100.   In a game, three cubical dice are thrown by a player who attempts to throw the same number on all three. What is the chance of the player: (a) throwing the same number on all three?   (b) throwing the same number on just two?

If the first throw results in just two dice showing the same number, then the third is thrown again. If no two dice show the same number, then all are thrown again. The player then comes to the end of his turn. What is the chance of the player succeeding in throwing three identical numbers in a complete turn?

What is the chance that all the numbers are different at the end of a turn?   (O & C)

101.   A random sample of 6 articles is taken from a large consignment and tested in two independent stages. The probability that an article will pass either stage is q. All six articles are first tested at stage 1; provided five or more pass, those which pass are retested at stage 2. The consignment is accepted if there is no more than one failure at each stage.

Find polynomials in q for: (a) the probability that stage 2 of the test will be required;   (b) the number of articles expected to undergo stage 2;   (c) the probability $P(q)$ of accepting the consignment.

Show that $dP/dq = 0$ when $q = 1$ and find $Pr(0.9)$ and $Pr(0.8)$. Sketch the graph of $Pr(q)$ for $0 \leqslant q \leqslant 1$.   (O & C)

102.   (a) The pages of a book are numbered from 1 to 200. If a page is chosen at random, what is the probability that its number will contain just two digits?

(b) The probability that a January night will be icy is $\frac{1}{4}$. On an icy night the probability that there will be a car accident at a certain dangerous corner is $\frac{1}{25}$. If it is not icy, the probability of an accident is $\frac{1}{100}$. What is the probability that: (i) 13 January will be icy and there will be an accident;   (ii) there will be an accident on 13 January?   (Cambridge)

103.   In a building programme the event that all the materials will be delivered at the correct time is M, and the event that the building programme is completed on time is F. Given that $Pr(M) = 0.8$ and $Pr(M \cap F) = 0.65$, explain in words the meaning of $Pr(F|M)$ and calculate its value. If $Pr(F) = 0.7$, find the probability that the building programme will be completed on time if all the materials are not delivered at the correct time.   (London)

104. A and B play a game as follows: A draws and keeps a card from a shuffled pack of six numbered 1 to 6. B then draws a card from the remaining five. The winner is the one with the card possessing the greater number. Determine whether there is any advantage in drawing first.

If the rules are altered so that A returns his card to the pack after noting its value and, also, if A and B draw the same card all the cards are shuffled and they draw again, find: (a) the probability that A wins on his first draw;   (b) the probability that A wins on his rth draw;   (c) the probability that A wins, if the game can continue for as long as is necessary.   (SUJB)

Chapter 5 Random Variables and Probability Distributions

5.1 INTRODUCTION

Everybody is familiar with the term 'variable'. It is simply a name given to any quantity that is likely to change. The concept was introduced quite naturally in Chapters 1, 2 and 4 when, for instance, we were considering the wages of employees in a certain factory. Here, the quantity 'wage of an employee' takes on different values as we go to different employees, and so is variable. The heights (or weights) of some members of a population can also be thought of as different values of the variable 'height (or weight) of a member'.

Variables need to be measurable quantities, otherwise we could not tell when and how they change, but need not necessarily be of a numerical nature. 'The state of the weather' at any particular time is a variable quantity (as we are all aware!), but we do not in general allocate numbers to measure its different possible states. We say 'cloudy', 'rainy', 'fine', 'sunny', 'stormy', and so on. Colour is also a variable quantity, and we measure it using the labels 'green', 'brown', 'light-blue', and so on, rather than $1, -5, 547$ or 7.69. We therefore have both quantitative and qualitative variables.

In this chapter we study variables that can be particularly associated with statistical experiments and investigate some characteristics of their behaviour.

We now define a particular type of variable, one that forms a very important part of the structure upon which many of the statistical ideas and techniques that follow depend.

DEFINITION 5.1

Consider an experiment, with an outcome set S split into the n mutually exclusive and exhaustive events $E_1, E_2, \ldots, E_n$. A variable, X say, which can assume exactly n numerical values each of which corresponds to one and only one of the given events is called a *random variable*.

Remember, that for a number of events to be mutually exclusive there must be no common intersections between any of them, and for them to be exhaustive we require that their combined union is just S, the outcome set. Note also that in the above definition, we require the associated random variable to take *only numerical values*.

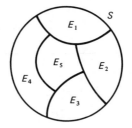

Fig. 5.1

Fig. 5.1 shows an outcome set S split into five mutually exclusive and exhaustive events. To define our random variable, we could allocate values to it according to the following rules.

$$X = 1 \quad \text{(corresponding to event } E_1\text{)}$$

$$X = 2 \quad \text{(corresponding to event } E_2\text{)}$$

. . . .

$$X = 5 \quad \text{(corresponding to event } E_5\text{)}$$

Alternatively, we could define our random variable (X) according to the following different rules.

Value of X	Corresponding event	Value of X	Corresponding event
5	E_1	-2	E_1
4	E_2	-1	E_2
3	E_3	0	E_3
2	E_4	1	E_4
1	E_5	2	E_5

Value of X	Corresponding event
1	E_1
2	E_2
3.294	E_3
17.621	E_4
-1.938	E_5

We see immediately that there is no unique way of defining a random variable given a particular experiment split up into a number of mutually exclusive and exhaustive events.

As another illustration of defining random variables, consider the experiment 'draw a card from a normal pack of 52'. We will define 3 different random variables on this experiment.

Random variable X_1 :	Random variable X_2 :	Random variable X_3
1 → (heart)	10 → (picture card or ace)	-1 → $(A, 2, 3, 4, 5)$
2 → (club)	20 → (a numbered card)	0 → $(6, 7, 8, 9)$
3 → (diamond)		1 → $(10, J, Q, K)$
4 → (spade)		

Note that in each of the three examples above, the set of corresponding events are mutually exclusive and exhaustive.

The reason we require a random variable to be defined in terms of mutually exclusive and exhaustive events is, given that a particular outcome of an experiment has occurred, the random variable will automatically have one and only one value associated with it, since the outcome of the experiment can lie in only one of the defined events.

EXAMPLE 5.1

Define 3 random variables for the experiment 'toss a normal coin once'.

SOLUTION

The outcome set for the experiment is S = (head, tail). We will label the 3 random variables X_1, X_2 and X_3.

X_1 could be defined as follows: $X_1 = 0$ (for a head); $X_1 = 1$ (for a tail).

X_2 could be defined: $X_2 = 1000$ (for a head); $X_2 = 2000$ (for a tail).

X_3 could be defined: $X_3 = 1.32$ (for S); $X_3 = 2.80$ (for $\emptyset$). Note that in the case of X_3, S and $\emptyset$ are mutually exclusive and exhaustive events, and the effect of this definition would be always to give X_3 the value 1.32 and never the value 2.80. This definition is a little extreme but nonetheless valid.

For the case where the outcome set of an experiment is itself defined in numerical terms, we have usually a 'natural allocation' of numbers to a random variable — namely the outcomes themselves.

For instance, in the experiment 'roll a single die once' we have S = (1, 2, 3, 4, 5, 6) clearly, and the natural allocation of values to a random variable X, say, would be: $X = 1$ (corresponding to outcome 1); $X = 2$ (corresponding to outcome 2); and so on. We are not forced to use this sort of correspondence of course, but generally we would.

EXAMPLE 5.2

Four cards are drawn from a normal pack of 52 as an experiment. The event E_i is defined as 'exactly i of the cards are hearts' and random variable X takes the value i corresponding to event E_i.

Show that X is a random variable and find the probability that X will not be less than 2 if the experiment is performed once.

SOLUTION

The outcome set S for the whole experiment is just all the permutations of 4 cards from a complete pack of 52. So that $n(S) = {}^{52}P_4$. Now, each one of these permutations contains either 0, 1, 2, 3 or 4 hearts, so that the set $E_0, E_1, \ldots, E_4$ are mutually exclusive and exhaustive for the experiment. Since X takes just one numerical value corresponding to a particular E_i, we have that X is a random variable by definition.

We require:

$\Pr(X \geqslant 2) = \Pr(X = 2 \text{ or } X = 3 \text{ or } X = 4)$

$\qquad = 1 - \Pr(X = 0 \text{ or } X = 1) \quad$ complementary events

$\qquad = 1 - (\Pr(\text{no hearts}) + \Pr(1 \text{ heart})) \quad$ mutually exclusive
$\qquad\qquad\qquad\qquad\qquad\qquad\qquad\qquad\qquad\qquad$ events

Now:

$\Pr(\text{no hearts}) = \dfrac{n(\text{no hearts})}{n(S)} = \dfrac{39 \times 38 \times 37 \times 36}{52 \times 51 \times 50 \times 49} = 0.3038 \ (4D)$

and:

$\Pr(1 \text{ heart}) = \dfrac{n(1 \text{ heart})}{n(S)} = \dfrac{4 \times 13 \times 39 \times 38 \times 37}{52 \times 51 \times 50 \times 49} = 0.4388 \ (4D)$

So,

$\qquad \Pr(\text{no hearts}) + \Pr(1 \text{ heart}) = 0.3038 + 0.4388 = 0.7426$

Hence:

$$\Pr(X \geqslant 2) = 1 - 0.7426 = 0.2574$$

$$= 0.257 \ (3D)$$

Finally, we considered event independence in some detail in the previous chapter and, since we have seen that random variables can be defined in terms of experiments and events, we now briefly 'tie in' these ideas. We will call two random variables independent, only if they are defined on independent experiments. For example consider the experiments 'rolling a normal six-sided die' and 'tossing a normal coin'. Clearly these two experiments are independent. If we associate random variable X with the die-rolling experiment and random variable Y with the coin-tossing experiment, then X and Y will automatically be independent random variables.

In Section 5.3, we define and discuss random variables in a more practical and convenient way.

5.2 RELATED MATHEMATICS

5.2.1 DIFFERENTIATION 4: PRODUCTS AND QUOTIENTS

We are concerned here with differentiating functions of the form $u \cdot v$ (a product) and $\dfrac{u}{v}$ (a quotient), where u and v are both *functions of the same variable*. For instance, consider differentiating a function of the form:

(a) $y = (2x^2 + 3)(4x - 10)^3$, or

(b) $z = \dfrac{2x^2}{(3 - x)^4}$.

The first would be extremely difficult and time consuming, the second would be impossible using any methods we have considered so far.

STATEMENT 5.1

Let u and v both be functions of a variable x.

(a)   If $y = u \cdot v$, then $\dfrac{dy}{dx} = u\left(\dfrac{dv}{dx}\right) + \left(\dfrac{du}{dx}\right)v$

(b)   If $y = \dfrac{u}{v}$, then $\dfrac{dy}{dx} = \dfrac{v\left(\dfrac{du}{dx}\right) - u\left(\dfrac{dv}{dx}\right)}{v^2}$

We demonstrate the use of (a) with a simple example $y = 3x(x + 2)$.

Using (a): $\dfrac{dy}{dx} = 3x \cdot 1 + 3(x + 2) = 6x + 6$

We can check this result as follows:

$$y = 3x(x + 2) = 3x^2 + 6x \quad \text{(sum of simple functions)}$$

Hence: $\dfrac{dy}{dx} = 6x + 6$

EXAMPLE 5.3

Differentiate $y = (3x + 1)(2x + 1)$.

SOLUTION

Here, $u = 3x + 1$ and $v = 2x + 1$

So that:

$$\frac{dy}{dx} = u\left(\frac{dv}{dx}\right) + \left(\frac{du}{dx}\right)v = (3x + 1)2 + 3(2x + 1) = 6x + 2 + 6x + 3$$

i.e.: $$\frac{dy}{dx} = 12x + 5$$

EXAMPLE 5.4

Find $\frac{dy}{dx}$ if $y = \frac{3}{2x - 1}$.

SOLUTION

We need to use the quotient rule here. Putting $u = 3$ and $v = 2x - 1$,

$$\frac{dy}{dx} = \frac{(2x - 1)0 - 3\cdot2}{(2x - 1)^2} = \frac{-6}{(2x - 1)^2}$$

EXAMPLE 5.5

Differentiate $\frac{x + 1}{x + 2}$.

SOLUTION

Put $y = \frac{x + 1}{x + 2}$ and $u = x + 1,\ v = x + 2$.

Then, using the quotient rule,

$$\frac{dy}{dx} = \frac{(x + 2)1 - (x + 1)1}{(x + 2)^2} = \frac{x + 2 - (x + 1)}{(x + 2)^2} = \frac{1}{(x + 2)^2}$$

EXAMPLE 5.6

If $y = (x^2 - x + 1)(x^2 + x - 1)$, find $\frac{dy}{dx}$.

SOLUTION

Let $u = x^2 - x + 1;\ v = x^2 + x - 1$.

Then:

$$\frac{dy}{dx} = (x^2 - x + 1)(2x + 1) + (2x - 1)(x^2 + x - 1)$$
$$= 4x^3 - 2x + 2 = 2(x^3 - x + 1)$$

5.2.2 DIFFERENTIATION 5: FUNCTION OF A FUNCTION

If y is a function of x, say, while x itself is a function of some other variable, t say, then y is called a *function of a function*.

For example, $y = 3x^2;\ x = \frac{2}{t}$. The problem is to find the derivative of y w.r.t. t. In this particular case we could substitute as follows:

$$y = 3\left(\frac{2}{t}\right)^2 = 3\left(\frac{4}{t^2}\right) = 12t^{-2}$$

Hence:

$$\frac{dy}{dt} = -24t^{-3} = -\frac{24}{t^3}$$

However, in general, this technique will prove arduous. Instead, we use the following.

STATEMENT 5.2

If $y = f(x)$ and $x = g(t)$, where f and g are any two functions, then:

$$\frac{dy}{dt} = \frac{dy}{dx}\frac{dx}{dt} = f'(x)g'(t)$$

For the example above, we had $y = 3x^2 = f(x)$ and $x = \frac{2}{t} = g(t)$.

But $\frac{dy}{dx} = 6x$ and $\frac{dx}{dt} = -2t^{-2}$.

Hence, using the statement, $\frac{dy}{dt} = 6x(-2t^{-2})$, that is, $\frac{dy}{dt} = \frac{-12x}{t^2}$.

But $x = \frac{2}{t}$.

$$\therefore \qquad \frac{dy}{dt} = -\frac{12}{t^2}\left(\frac{2}{t}\right) = -\frac{24}{t^3} \quad \text{(as found before)}$$

EXAMPLE 5.7

If $y = 3x^3 - 4x^2 + 10$ and $x = t^2 - 3$, find $\frac{dy}{dt}$.

SOLUTION

$$f(x) = 3x^3 - 4x^2 + 10 \quad \text{therefore} \quad f'(x) = 9x^2 - 8x$$
$$g(t) = t^2 - 3 \quad \text{therefore} \quad g'(t) = 2t$$

Hence:

$$\frac{dy}{dt} = f'(x)g'(t) = (9x^2 - 8x)2t = 2tx(9x - 8) \quad \text{factorising}$$
$$= 2t(t^2 - 3)(9t^2 - 27 - 8) \quad \text{substituting for } x$$
$$\therefore \qquad \frac{dy}{dt} = 2t(t^2 - 3)(9t^2 - 35)$$

EXAMPLE 5.8

Differentiate $(x^2 + 3x - 2)^5$ w.r.t. x.

SOLUTION

Put $y = (x^2 + 3x - 2)^5$. Now, we could multiply $x^2 + 3x - 2$ by itself 5 times, obtain an expression for y in terms of simple functions of x and then differentiate, but clearly this is complicated.

Putting, say $z = x^2 + 3x - 2$, we have $y = z^5$ (where $z = x^2 + 3x - 2$), i.e. y is now a function of z and z is a function of x.

$$\therefore \qquad \frac{dy}{dx} = \frac{dy}{dz}\left(\frac{dz}{dx}\right) \quad \text{(function of a function rule)}$$

$$= 5z^4(2x + 3)$$

$$= 5(x^2 + 3x - 2)^4(2x + 3)$$

EXAMPLE 5.9

Differentiate $\dfrac{1}{(4-x)^{\frac{3}{2}}}$.

SOLUTION

Put $y = \dfrac{1}{(4-x)^{\frac{3}{2}}} = (4-x)^{-\frac{3}{2}}$.

If we let $z = 4 - x$, then $y = z^{-\frac{3}{2}}$

Hence: $\qquad \dfrac{dy}{dx} = \dfrac{dy}{dz}\left(\dfrac{dz}{dx}\right) = -\dfrac{3}{2}(z^{-\frac{5}{2}})(-1)$

$$= \frac{3}{2}(4-x)^{-\frac{5}{2}}$$

Thus: $\qquad \dfrac{dy}{dx} = \dfrac{3}{2(4-x)^{\frac{5}{2}}}$

5.2.3 INTEGRATION 1: THE INTEGRAL OPERATOR

For the purposes of this sub-section, we can regard the process of 'integration' as *the reverse of differentiation*.

For example, we know that the derivative of x^4 w.r.t. x is $4x^3$. Therefore the *integral* of $4x^3$ w.r.t. x is $x^4 + C$ (where C is some constant). We see that we are really just going back to what we had before we differentiated. The reason for adding a constant is now shown.

Consider the following:

$$\frac{d}{dx}(x^4) = 4x^3; \quad \frac{d}{dx}(x^4 + 5) = 4x^3; \quad \frac{d}{dx}(x^4 - 100) = 4x^3$$

In fact $\dfrac{d}{dx}(x^4 + A) = 4x^3$, where A is any number.

So, in order to integrate $4x^3$ (w.r.t. x), we need to ask the question: 'what quantity, when differentiated, gives $4x^3$?' Clearly, the answer is not just x^4 but $x^4 + C$ (where C is some *arbitrary*, unknown constant).

We know that the integral of $4x^3$ (w.r.t. x) is $x^4 + C$.

We write:
$$\int 4x^3 \ dx \ = \ x^4 + C$$

(**Note** $\int$ (an elongated S) is supposed to represent the 'S' of 'SUM', since, in mathematics, an integral represents an infinite sum. But this need not concern us unnecessarily here).

The 'dx' can be thought of as representing 'w.r.t. x'. The above notation for the integral 'operator' is standard in mathematics.

In general, we have:

STATEMENT 5.3

$$\int c \cdot x^d \ dx \ = \ \frac{c}{(d+1)} x^{d+1} + E \ \text{where } E \text{ is an arbitrary constant.}$$

The above is true for all values of c and all values of d *except* -1.

Note the general rule from the above. We add 1 to the power of x (d became $d + 1$) and divide everything by this new power (divided by $d + 1$).

The proof of the above statement can be checked by differentiation.

EXAMPLE 5.10

Integrate $3x^2$ w.r.t. x.

SOLUTION

$$\int 3x^2 \ dx \ = \ \frac{3}{2+1} x^{2+1} + C$$
$$= \ x^3 + C$$

EXAMPLE 5.11

Find $\int \frac{3}{4} x^{\frac{2}{3}} \ dx$.

SOLUTION

$$\int \frac{3}{4} x^{\frac{2}{3}} \ dx \ = \ \frac{\frac{3}{4}}{(\frac{2}{3})+1} x^{(\frac{2}{3})+1} + C$$
$$= \ \frac{\frac{3}{4}}{\frac{5}{3}} x^{\frac{5}{3}} + C$$
$$= \ (9/20)x^{\frac{5}{3}} + C$$

EXAMPLE 5.12

Integrate　　(a)　$4.3x^{-0.4}$　　(b)　$\dfrac{3}{x^4}$　w.r.t. x

SOLUTION

(a)　　$\displaystyle\int 4.3x^{-0.4}\ dx = \dfrac{4.3}{-0.4+1}x^{-0.4+1} + C$

　　　　　　　　$= 7.17x^{0.6} + C$

(b)　　$\displaystyle\int \dfrac{3}{x^4}\ dx = \int 3x^{-4}\ dx = \dfrac{3}{-4+1}x^{-4+1} + C$

　　　　　　　$= -x^{-3} + C$

　　　　　　　$= -\dfrac{1}{x^3} + C$

STATEMENT 5.4

$$\int (ex^f + gx^h)\ dx = \int ex^f\ dx + \int gx^h\ dx$$

$$= \dfrac{e}{f+1}x^{f+1} + \dfrac{g}{h+1}x^{h+1} + C$$

i.e., the integral of a sum of simple functions is the sum of the separate integrals. Notice that, although we have integrated two simple functions, we only need to add one arbitrary constant.

We can now integrate polynomial expressions.

EXAMPLE 5.13

Integrate　(a)　$4x^3 - 3x^{-2}$　　(b)　$\dfrac{2}{3}x^{-\frac{1}{2}} - \dfrac{4}{5}x^{\frac{3}{4}} + 10$　w.r.t. x

SOLUTION

(a)　　$\displaystyle\int (4x^3 - 3x^{-2})\ dx = \dfrac{4}{4}x^4 - \dfrac{3}{-1}x^{-1} + C$

　　　　　　　　$= x^4 + 3x^{-1} + C$

(b)　　$\displaystyle\int \left(\dfrac{2}{3}x^{-\frac{1}{2}} - \dfrac{4}{5}x^{\frac{3}{4}} + 10\right)\ dx$

　　　$= \dfrac{\left(\frac{2}{3}\right)}{\left(\frac{1}{2}\right)}x^{\frac{1}{2}} - \dfrac{\left(\frac{4}{5}\right)}{\left(\frac{7}{4}\right)}x^{\frac{7}{4}} + 10x + C$

(Notice that the integral of a constant a, say, w.r.t. x is ax)

$$= \frac{2}{3}\left(\frac{2}{1}\right)x^{\frac{1}{2}} - \frac{4}{5}\left(\frac{4}{7}\right)x^{\frac{7}{4}} + 10x + C$$

$$= \frac{4}{3}x^{\frac{1}{2}} - \frac{16}{35}x^{\frac{7}{4}} + 10x + C$$

The integration we have considered so far is usually referred to as *indefinite integration* for reasons which will be made clear in the next sub-section.

5.2.4 INTEGRATION 2: DEFINITE INTEGRATION/AREA UNDER CURVE

In the case of indefinite integration, we found that we needed to add an arbitrary constant for reasons already explained. In this sub-section we look at integration in a more 'exact' manner where, for instance, the arbitrary constant met with previously will either be determinable or irrelevant.

For example, consider the equation $\frac{dy}{dx} = 2x + 3$. Integration (w.r.t. x) easily yields $y = x^2 + 3x + C$. Now, if we were given the extra information that $y = 12$ when $x = 2$, we can substitute these two values into this latter equation to give:

$$12 = 2^2 + 3(2) + C \quad \text{i.e.} \quad C = 2$$

Thus we have $y = x^2 + 3x + 2$. In this type of situation, C (the arbitrary constant) is *determinable*.

EXAMPLE 5.14

Find the value of y when $x = 1$, given the information:

(a) $\frac{dy}{dx} = 3x^2 - 2x + 8$, and

(b) $y = 2$ when $x = -1$.

SOLUTION

Integrating $\frac{dy}{dx}$ gives $y = x^3 - x^2 + 8x + C$.

Given that $y = 2$ when $x = -1$ enables the constant C to be eliminated.

Substitution gives:

$$2 = -1 - 1 - 8 + C, \quad \text{i.e.,} \quad C = 8$$

$$\therefore \qquad y = x^3 - x^2 + 8x + 8$$

Hence, when $x = 1$:

$$y = 1 - 1 + 8 + 8 = 16$$

At this stage we introduce a very useful and important piece of notation. The symbol $[f(x)]_a^b$ is defined as $f(b) - f(a)$. In words, it is the value of $f(x)$ at $x = b$ minus the value of $f(x)$ at $x = a$.

For example: $[2x^2]_1^2 = 2(2)^2 - 2(1)^2 = 8 - 2 = 6$

and:

$$[x^3 - 2x^2 + 4]_{-1}^1 = (1)^3 - 2(1)^2 + 4 - ((-1)^3 - 2(-1)^2 + 4)$$
$$= 1 - 2 + 4 - (-1 - 2 + 4) = 2$$

STATEMENT 5.5

A *definite integral* takes the form $\int_a^b f(x)\,dx$ where a and b are constants, and is evaluated as:

$$\int_a^b f(x)\,dx = [F(x)]_a^b$$

where $F(x)$ is the indefinite integral $\int f(x)\,dx$.

Note   We can write $F(x)$ without an arbitrary constant C, since the operation $[\ldots]$ cancels out constants automatically.

To demonstrate the note made in the above statement, consider the quantity:

$$[g(x) + a]_b^c = g(c) + a - (g(b) + a)$$
$$= g(c) - g(b) + a - a$$
$$= g(c) - g(b)   \text{with } a \text{ being eliminated}$$

EXAMPLE 5.15

Evaluate $\left[3x^3 - 4x^2 + \dfrac{3}{x}\right]_1^2$.

SOLUTION

$$\left[3x^3 - 4x^2 + \frac{3}{x}\right]_1^2 = 3(2)^3 - 4(2)^2 + \frac{3}{2} - \left(3(1)^3 - 4(1)^2 + \frac{3}{1}\right)$$
$$= 24 - 16 + 1.5 - (3 - 4 + 3)$$
$$= 7.5$$

EXAMPLE 5.16

Evaluate $\int_1^2 3x^2\,dx$.

SOLUTION

$$\int_1^2 3x^2 \, dx = \left[x^3 \right]_1^2 \quad \text{(we can leave out the arbitrary constant } C\text{)}$$

$$= (2)^3 - (1)^3 = 8 - 1 = 7$$

EXAMPLE 5.17

Find the value of $\int_{-1}^3 (3x^3 - 4x^2 + 3) \, dx$.

SOLUTION

$$\int_{-1}^3 (3x^3 - 4x^2 + 3) \, dx = \left[\frac{3}{4}x^4 - \frac{4}{3}x^3 + 3x \right]_{-1}^3$$

$$= \frac{3}{4}(3)^4 - \frac{4}{3}(3)^3 + 3(3)$$
$$- \left(\frac{3}{4}(-1)^4 - \frac{4}{3}(-1)^3 + 3(-1) \right)$$

$$= 34.67 \quad (2D)$$

One of the more common uses of the definite integral and one that we are particularly concerned with in Statistics is now given.

STATEMENT 5.6

The area bounded by the curve $y = f(x)$, the x-axis and the lines $x = a$, $x = b$ (the shaded area in the figure) is given by the definite integral:

$$\int_a^b f(x) \, dx$$

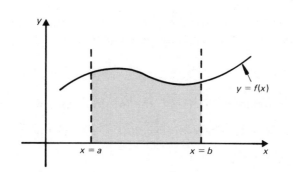

EXAMPLE 5.18

Find the area bounded by the curve $y = x^2 - 4x$, the x-axis and the lines $x = 5$ and $x = 6$.

SOLUTION

Here, $f(x) = x^2 - 4x$; $a = 5$ and $b = 6$.

Therefore the area required is:

$$\int_5^6 (x^2 - 4x)\, dx = \left[\frac{1}{3}x^3 - 2x^2 \right]_5^6 = \frac{(6)^3}{3} - 2(6)^2 - \left(\frac{(5)^3}{3} - 2(5)^2 \right)$$

$$= \frac{216}{3} - 72 - \left(\frac{125}{3} - 50 \right) = 0 - (-8.33)$$

$$= 8.33 \text{ (2D)}$$

EXAMPLE 5.19

Find the area enclosed between the curve $y = x + 4$, the x-axis and the lines $x = -2$ and $x = 2$.

SOLUTION

Here, $f(x) = x + 4$; $a = -2$ and $b = 2$.

So, area required is:

$$\int_{-2}^2 (x + 4)\, dx = \left[\frac{x^2}{2} + 4x \right]_{-2}^2$$

$$= \left(\frac{4}{2} + 8 \right) - \left(\frac{4}{2} - 8 \right)$$

$$= 16$$

EXAMPLE 5.20

If the area bounded by the curve $y = ax$, the x-axis and the lines $x = 0$ and $x = 2$ is 1, find the value of a.

SOLUTION

Using the previous statement, the information given can be translated as $\int_0^2 ax\, dx = 1$. Therefore $\left[\frac{a}{2}x^2 \right]_0^2 = 1$, giving $\frac{a}{2} \cdot 4 - 0 = 1$.

Therefore $a = \frac{1}{2}$.

5.2.5 THE EXPONENTIAL FUNCTION

DEFINITION 5.2

We define the general function $e^{f(x)}$ as a series of the form:

$$e^{f(x)} = 1 + f(x) + \frac{(f(x))^2}{2!} + \frac{(f(x))^3}{3!} + \ldots$$

$$= \sum_{r=0}^{\infty} \frac{(f(x))^r}{r!} \quad \text{where } f(x) \text{ is any function of a variable } x$$

In particular, putting $f(x) = x$ in the above, we have:

$$e^x = 1 + x + \frac{x^2}{2!} + \frac{x^3}{3!} + \frac{x^4}{4!} + \ldots$$

$$= \sum_{r=0}^{\infty} \frac{x^r}{r!} \qquad \qquad [1]$$

The constant e is a special mathematical number which occurs frequently in Statistics.

To obtain an approximate value for e (an exact value can never be determined) we simply substitute $x = 1$ into expression [1], to obtain:

$$e^1 = e = 1 + 1 + \frac{1}{2!} + \frac{1}{3!} + \frac{1}{4!} + \frac{1}{5!} + \ldots \qquad [2]$$

This series demonstrates the indeterminate nature of e, although we can approximate its value to as many decimal places as we wish by taking enough terms of the series above.

For example, taking the first 6 terms of [2] gives:

$$e = 1 + 1 + \frac{1}{2} + \frac{1}{6} + \frac{1}{24} + \frac{1}{120} + \frac{1}{720} + \ldots$$

$$= 2 + 0.5 + 0.167 + 0.042 + 0.008 + 0.001 + \ldots$$

i.e. $e = 2.72 \ (2D)$

The differentiation of the exponential function is contained in the following statement.

STATEMENT 5.7

(a)   In particular $\frac{d}{dx}(e^x) = e^x$.

(b)   In general $\frac{d}{dx}(e^{f(x)}) = f'(x)\, e^{f(x)}$ where $f(x)$ is any function of x.

Note   Given (a), we can prove (b) by using the differentiation of the function of a function technique.

Given (b), we can show (a) is true simply by substituting $f(x) = x$.

EXAMPLE 5.21

Differentiate   (a) e^{2x} (b) e^{-x} (c) $e^{1/x}$

SOLUTION

(a)   Here, putting $f(x) = 2x$ gives $f'(x) = 2$.

So, if $y = e^{2x}$, the previous statement gives:

$$\frac{dy}{dx} = f'(x)\, e^{f(x)}$$

$$= 2\, e^{2x}$$

(b) $f(x) = -x$ therefore $f'(x) = -1$.

Putting $y = e^{-x}$, we have $\dfrac{dy}{dx} = -e^{-x}$.

(c) $f(x) = \dfrac{1}{x} = x^{-1}$ therefore $f'(x) = -x^{-2} = -\dfrac{1}{x^2}$.

Hence: $\dfrac{dy}{dx} = -\dfrac{1}{x^2}\, e^{1/x}$

EXAMPLE 5.22

Differentiate $(1 + x^2)\, e^x$.

SOLUTION

Putting $y = (1 + x^2)\, e^x$, we can treat the right hand side as a product, with $u = 1 + x^2$ and $v = e^x$.

$$\therefore \qquad \frac{dy}{dx} = u\left(\frac{dv}{dx}\right) + \left(\frac{du}{dx}\right)v = (1 + x^2)\, e^x + 2x \cdot e^x$$

$$= e^x(1 + x^2 + 2x) = (1 + x)^2\, e^x$$

EXAMPLE 5.23

Find $\dfrac{dy}{dx}$ if $y = \dfrac{e^{2x}}{2 - 3x^2}$.

SOLUTION

Putting $u = e^{2x}$ and $v = 2 - 3x^2$, we will use the quotient rule for differentiation.

$$\frac{dy}{dx} = \frac{v\left(\frac{du}{dx}\right) - u\left(\frac{dv}{dx}\right)}{v^2} = \frac{(2 - 3x^2)2e^{2x} - e^{2x}(-6x)}{(2 - 3x^2)^2}$$

$$= \frac{2e^{2x}(2 - 3x^2) + 6xe^{2x}}{(2 - 3x^2)^2}$$

So that:

$$\frac{dy}{dx} = \frac{2e^{2x}}{(2 - 3x^2)^2} \cdot (2 + 3x - 3x^2)$$

5.3 RANDOM VARIABLES AND DISTRIBUTIONS

5.3.1 DISCRETE AND CONTINUOUS RANDOM VARIABLES

In Section 5.1, we defined a random variable in terms of mutually exclusive and exhaustive events of a statistical experiment by assigning numbers to the variable corresponding to different events.

An equivalent, and more convenient, definition is now given.

DEFINITION 5.3

Let X be a discrete variable that can assume only the values $x_1, x_2, \ldots, x_n$ with respective probabilities $p_1, p_2, \ldots, p_n$.

That is, $\Pr(X = x_i) = p_i$ for $i = 1, 2, \ldots, n$.

Then, if $\sum_{i=1}^{n} p_i = 1$, X is a *discrete random variable*.

We can tie the above definition in with Definition 5.1 as follows. We set up a hypothetical experiment having an outcome set S split up into the n mutually exclusive and exhaustive events $E_1, E_2, \ldots, E_n$. If we now associate value x_1 with event E_1, x_2 with $E_2, \ldots,$ and x_n with E_n, then X is a random variable (using the original definition).

But $\Pr(S) = 1$ (by definition).

Therefore $\Pr(E_1 \text{ or } E_2 \text{ or } \ldots \text{ or } E_n) = 1$ (exhaustive events).

Hence $\Pr(E_1) + \Pr(E_2) + \ldots + \Pr(E_n) = 1$ (mutually exclusive events).

Therefore $p_1 + p_2 + \ldots + p_n = 1$ (as required).

In the above definition, n does not have to be finite. That is, we can have a discrete random variable taking any one of an infinite number of values,

as long as the condition $\sum_{i=1}^{\infty} p_i = 1$ is satisfied.

Consider the following table of the probabilities associated with the values that a discrete random variable Y can take.

y	2	3	4	5	6
$\Pr(Y = y)$	$\dfrac{1}{2}$	$\dfrac{1}{4}$	$\dfrac{1}{8}$	$\dfrac{1}{16}$	$\dfrac{1}{16}$

The table translates as $\Pr(Y = 2) = \dfrac{1}{2}$, $\Pr(Y = 3) = \dfrac{1}{4}$, . . . , etc.

Since we have the sum of all the probabilities adding to 1

$$\left(\text{i.e.,} \qquad \frac{1}{2} + \frac{1}{4} + \frac{1}{8} + \frac{1}{16} + \frac{1}{16} = 1\right)$$

Y *is a discrete random variable.*

Now, we know already that it is impossible to give a continuous quantity an *exact* value. It is equally impossible to assign a probability to any single value that a continuous variable might take. Instead, we construct ranges of values for the variable and assign a probability to each of these ranges. For example, suppose that a continuous random variable X could only take values from 0 to 5 with the following given probabilities:

Range of X	0 up to 1	1 up to 2	2 up to 3	3 up to 4	4 up to 5
Probability	0.2	0.2	0.4	0.1	0.1

That is: $\Pr(0 \leqslant X < 1) = 0.2$; $\Pr(1 \leqslant X < 2) = 0.2$; . . . , etc.

Note here that:

$$\begin{aligned}
\Pr(0 \leqslant X < 5) &= \Pr(0 \leqslant X < 1 \text{ or } 1 \leqslant X < 2 \text{ or } \ldots \text{ or } 4 \leqslant X < 5) \\
&= \Pr(0 \leqslant X < 1) + \Pr(1 \leqslant X < 2) + \ldots + \Pr(4 \leqslant X < 5) \\
&= 0.2 + 0.2 + \ldots + 0.1 \\
&= 1
\end{aligned}$$

In this case we would call X a continuous random variable quite naturally and this would tie in with the discrete case defined previously.

Formally:

DEFINITION 5.4

Let X be a continuous variable that can assume values only in the ranges x_1 up to x_2, x_2 up to x_3, . . . , x_n up to x_{n+1} with respective probabilities $p_1, p_2, \ldots, p_n$. That is,

$$\Pr(x_1 \leqslant X < x_2) = p_1, \quad \Pr(x_2 \leqslant X < x_3) = p_2, \quad \ldots, \text{ etc.}$$

Then, if $\displaystyle\sum_{i=1}^{n} p_i = 1$, X is a *continuous random variable*.

In practice, however, the above definition is not very useful since the probabilities associated with particular ranges of values of a continuous random variable are not usually given in tabular form. The above definition has been included for completeness.

EXAMPLE 5.24

Let X be the discrete variable 'number of heads shown when two coins are tossed'. Show that X is a random variable.

SOLUTION

The values that X can assume are just the number of heads that can be obtained when 2 coins are tossed. That is, 0, 1 or 2.

We need to show that $\sum\limits_{\text{all } x} \Pr(X = x) = 1$ for $x = 0, 1, 2$.

Putting H for 'head' and T for 'tail', we have:

$$\begin{aligned}
\Pr(X = 0) = \Pr(\text{no heads}) &= \Pr(\text{T} \cdot \text{T}) \\
&= \Pr(\text{T}) \cdot \Pr(\text{T}) \ \text{(independence)} \\
&= \frac{1}{2}\left(\frac{1}{2}\right) = \frac{1}{4}
\end{aligned}$$

$$\begin{aligned}
\Pr(X = 1) = \Pr(1 \text{ head}) &= \Pr(\text{H} \cdot \text{T or T} \cdot \text{H}) \\
&= \Pr(\text{H} \cdot \text{T}) + \Pr(\text{T} \cdot \text{H}) \\
&\quad \text{(mutual exclusiveness)} \\
&= \Pr(\text{H}) \cdot \Pr(\text{T}) + \Pr(\text{T}) \cdot \Pr(\text{H}) \\
&\quad \text{(independence)} \\
&= 2\left(\frac{1}{2}\right)\left(\frac{1}{2}\right) = \frac{1}{2}
\end{aligned}$$

Similarly:

$$\begin{aligned}
\Pr(X = 2) = \Pr(2 \text{ heads}) &= \Pr(\text{H} \cdot \text{H}) = \Pr(\text{H}) \cdot \Pr(\text{H}) \\
&= \frac{1}{2}\left(\frac{1}{2}\right) = \frac{1}{4}
\end{aligned}$$

Thus:

$$\sum\limits_{\text{all } x} \Pr(X = x) = \sum\limits_{x=0}^{2} \Pr(X = x) = \frac{1}{4} + \frac{1}{2} + \frac{1}{4} = 1 \ \text{ as required}$$

Therefore X is a discrete random variable.

EXAMPLE 5.25

The variable X can assume the value r with probability $\dfrac{1}{4}\left(\dfrac{3}{4}\right)^{r}$ for $r = 0, 1, 2, \ldots$. Show that X is a discrete random variable.

SOLUTION

We can write the above information as:

$$\Pr(X = x) = \frac{1}{4}\left(\frac{3}{4}\right)^x \quad \text{for } x = 0, 1, 2, \ldots$$

Note that, here, the variable can assume an infinite number of values.

We need to show that $\displaystyle\sum_{x=0}^{\infty} \Pr(X = x) = 1$.

Now: $\displaystyle\sum_{x=0}^{\infty} \Pr(X = x) = \sum_{x=0}^{\infty} \frac{1}{4}\left(\frac{3}{4}\right)^x$

$$= \frac{1}{4} + \frac{1}{4}\left(\frac{3}{4}\right) + \frac{1}{4}\left(\frac{3}{4}\right)^2 + \frac{1}{4}\left(\frac{3}{4}\right)^3 + \ldots$$

But the right hand side is just an infinite geometric progression (see Section 4.2.2) with $a = \dfrac{1}{4}$ and $r = \dfrac{3}{4}$. Its sum to infinity is $\dfrac{a}{1-r}$.

$$\therefore \qquad \sum_{x=0}^{\infty} \Pr(X = x) = \frac{\frac{1}{4}}{1 - \frac{3}{4}} = \frac{\frac{1}{4}}{\frac{1}{4}} = 1$$

giving X as a discrete random variable.

At this stage, we look again briefly at the independence of two random variables. There is an exact statistical condition that needs to be satisfied before we can assert that two random variables are independent. In the discrete case, it is as follows. Let X and Y be discrete random variables. If $\Pr(X = x \text{ and } Y = y) = \Pr(X = x) \cdot \Pr(Y = y)$ for all valid x and y, then X and Y are independent.

However, for our purposes, we return to our previous description. That is, two random variables are independent if they are defined on two independent experiments.

Now, if two experiments are independent, any event of one is necessarily independent of any event of the other, and since the values that a random variable can take are associated with a particular set of mutually exclusive and exhaustive events of an experiment, we have that two random variables are independent if any value that one can take is in no way affected by any value the other might take. With respect to dependence of random variables, we need only be aware of two particular instances for the purposes of this text. First, if X and Y are two random variables *with one a function of the other* (i.e., $Y = 2X$, $X = Y^3$, $\ldots$, etc), then X and Y would be said to be dependent. Secondly, if two random variables are defined on the same experiment, they will, in general, be dependent.

As an example, consider the experiment of tossing n coins together. Define random variable X as 'number of heads' and random variable Y as 'number of tails'. Here, both random variables are defined on the same experiment and also, since from necessity $X + Y = n$, one is a function of the other. In this case, both the above-named instances of dependence are evident.

We summarise as follows.

DEFINITION 5.5

(a)   Two random variables are said to be *independent* if and only if they are each defined on independent experiments or, equivalently, if and only if any value that one might take is in no way affected by any of the values the other might take.

(b)   Two random variables are said to be *dependent* if one can be written as a function of the other.

Finally, at this stage we stress a particular convention that we have been using and will continue to use throughout the rest of the book and, without explanation, might cause some confusion. That is the use of large and small letters to denote variables and the values that they can take. A capital letter signifies the 'name of a variable'; a small letter denotes a 'value that the variable can assume'. Thus $\Pr(X = x)$ is translated as 'the probability that the variable called X takes a numerical value x'.

5.3.2 DENSITY AND DISTRIBUTION FUNCTIONS

DEFINITION 5.6

The *probability density function* (p.d.f.) of a discrete random variable X is a function that allocates probabilities to all the distinct values that the variable X can take.

The p.d.f. can take two forms.

(a)   A *tabulation* of probabilities corresponding to all the different values the variable X can take.

(b)   A *compact function* giving a general rule for assigning probabilities to all values of X.

We have already seen examples of both of the above forms of p.d.f.

In Example 5.24, we calculated the probabilities associated with the random variable X 'number of heads when 2 coins are tossed' and we obtained the following table.

x	0	1	2
$\Pr(X = x)$	$\dfrac{1}{4}$	$\dfrac{1}{2}$	$\dfrac{1}{4}$

By virtue of the fact that:

(a)   each possible value that X can take is included, and

(b)   each of these values has a probability associated with it (the total probability adding to 1), the above tabulation constitutes a p.d.f. for the random variable X in the form of (a) of Definition 5.6.

In Example 5.25, we had the rule $\Pr(X = x) = \dfrac{1}{4}\left(\dfrac{3}{4}\right)^x$ for assigning prob-

abilities to the values that random variable X can take. This, then, constitutes a p.d.f. (in the form of (b) of Definition 5.6) for X.

Both forms of p.d.f. mentioned in the definition are used practically, but it should be noted that if a p.d.f. is given in a compact form for a discrete random variable, the total probability must sum to unity. This is consistent with the fact that density functions are only associated with *random* variables.

In the case of continuous random variables, it is unusual to give a p.d.f. in a tabular form (as noted earlier). A compact function is usually more appropriate and convenient. In formal terms:

DEFINITION 5.7

The *probability density function* (p.d.f.) of a continuous random variable X is a function that allocates probabilities to all of the ranges of values that the random variable can take.

The p.d.f. can take two forms:

(a)   a tabulation of probabilities corresponding to an exhaustive and mutually exclusive set of ranges of values of X, and

(b)   a function of x ($f(x)$ say) which when *integrated* over a particular valid range of values of x gives the probability that the random variable X lies in that particular range.

To demonstrate (a), consider the following set of probabilities associated with the given range of values that a continuous random variable X can assume.

Range for X	0 up to 2	2 up to 3	3 up to 4	4 up to 5
Probability	0.2	0.4	0.1	0.3

Since (a) the four ranges given form a mutually exclusive and exhaustive set (for the total range 0 up to 5) and (b) each of the ranges has a probability associated with it (the total adding to 1 necessarily), the given tabulation forms a p.d.f. for the continuous random variable X.

To demonstrate (b), consider the function $f(x) = \dfrac{2}{5}(4 - x)$ for x in the range 1 to 2 only. (This does form a valid density function for a continuous random variable as we will show immediately after Statement 5.8.)

Suppose we require the probability that X lies in the range $\dfrac{5}{4}$ to $\dfrac{7}{4}$. Then, using the statement, we have:

$$\Pr\left(\frac{5}{4} \leqslant X \leqslant \frac{7}{4}\right) = \int_{\frac{5}{4}}^{\frac{7}{4}} \frac{2}{5}(4 - x)\,dx = \frac{2}{5}\left[4x - \frac{x^2}{2}\right]_{\frac{5}{4}}^{\frac{7}{4}}$$

$$= \frac{2}{5}\left[\left(\frac{28}{4} - \frac{49}{32}\right) - \left(\frac{20}{4} - \frac{25}{32}\right)\right] = \frac{2}{5}\left(\frac{40}{32}\right)$$

$$= \frac{1}{2}$$

Fig. 5.2 shows the physical interpretation of the result:

$$\Pr\left(\frac{5}{4} \leqslant X \leqslant \frac{7}{4}\right) = \int_{\frac{5}{4}}^{\frac{7}{4}} \frac{2}{5}(4 - x)\,dx$$

$$= \frac{1}{2}$$

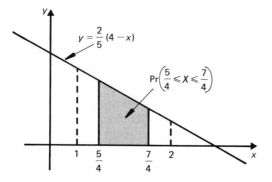

Fig. 5.2   (Not to scale)

That is, the area under the curve $y = \dfrac{2}{5}(4 - x)$ between the lines $x = \dfrac{5}{4}$, $x = \dfrac{7}{4}$ and the x-axis (the shaded area in the figure) represents the quantity $\Pr\left(\dfrac{5}{4} \leqslant X \leqslant \dfrac{7}{4}\right)$.

The connection between the area under a curve and probability is such an important aspect of the p.d.f. of a continuous random variable, that we re-iterate the point in general terms in the form of a statement.

STATEMENT 5.8

Let a continuous random variable X have a p.d.f. $f(x)$ valid over the range a to b only. Then, if $a \leqslant x_1 \leqslant x_2 \leqslant b$, $\Pr(x_1 \leqslant X \leqslant x_2)$ is represented by the area under the curve $y = f(x)$ between the lines $x = x_1$, $x = x_2$ and the x-axis (the shaded area in the figure),

i.e.: $$\Pr(x_1 \leqslant X \leqslant x_2) = \int_{x_1}^{x_2} f(x)\, dx$$

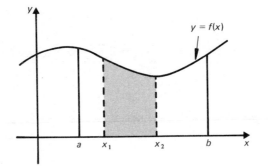

At this stage we note the significance of the 'total' probability associated with the p.d.f. of a random variable, discrete or continuous. We noted earlier the importance of the fact that the sum of the probabilities associated with a random variable must sum to unity. But since the p.d.f. of a random variable is responsible for allocating probabilities, we have:

STATEMENT 5.9

If the p.d.f. of a random variable X is given by

$$\begin{cases} \Pr(X = x); & \text{discrete } X \\ f(x); & \text{continuous } X \end{cases}$$

then we must have:

(continued)

$$\begin{cases} \sum\limits_{\text{all } x} p(X = x) = 1; & \text{discrete } X \\[2em] \int\limits_{\text{all } x} f(x)\, dx = 1; & \text{continuous } X \end{cases}$$

For example, earlier we considered the continuous probability function $f(x) = \dfrac{2}{5}(4 - x)$ for $1 \leqslant x \leqslant 2$. If X is to be a continuous random variable, we need $\int_1^2 \dfrac{2}{5}(4 - x)\, dx = 1$.

$$\begin{aligned} \text{Left hand side} &= \frac{2}{5}\left[4x - \frac{x^2}{2}\right]_1^2 = \frac{2}{5}\left[\left(8 - \frac{4}{2}\right) - \left(4 - \frac{1}{2}\right)\right] \\[1em] &= \frac{2}{5}\left(6 - \frac{7}{2}\right) \\[1em] &= \frac{2}{5}\left(\frac{5}{2}\right) = 1 = \text{right hand side} \end{aligned}$$

Hence in this particular case, X is a continuous random variable.

EXAMPLE 5.26

The continuous variable X has a probability function

$$f(x) = \frac{1}{4}(2x + 3)$$

(a)   Show that X is a continuous *random* variable (i.e., $f(x)$ is a valid p.d.f.) if X can take values in the range 0 to 1 only.

(b)   Find $\Pr\left(0 \leqslant X \leqslant \dfrac{1}{2}\right)$.

SOLUTION

(a)   We need to show that $\int_{\text{all } x} f(x)\, dx = 1$.

$$\begin{aligned} \text{l.h.s.} &= \int_0^1 \frac{1}{4}(2x + 3)\, dx = \frac{1}{4}\left[x^2 + 3x\right]_0^1 \\[1em] &= \frac{1}{4}[(1)^2 + 3(1)] = \frac{1}{4}(1 + 3) \\[1em] &= 1 = \text{r.h.s.} \end{aligned}$$

Therefore X is a continuous random variable.

(b)   Using Statement 5.8;

$$\Pr\left(0 \leqslant X \leqslant \frac{1}{2}\right) = \int_0^{\frac{1}{2}} \frac{1}{4}(2x + 3)\, dx = \frac{1}{4}\left[x^2 + 3x\right]_0^{\frac{1}{2}}$$

$$= \frac{1}{4}\left[\left(\frac{1}{2}\right)^2 + 3\left(\frac{1}{2}\right)\right] = \frac{1}{4}\cdot\frac{7}{4}$$

$$= \frac{7}{16}$$

EXAMPLE 5.27

A continuous random variable X has a p.d.f. as follows.

$$f(x) = c(x^2 - 2x + 3) \quad \text{for} \quad 0 \leqslant x \leqslant 2$$

$$= 0 \quad \text{otherwise}$$

Find   (a)   the value of c, and hence   (b) $\Pr(X \geqslant 1)$.

SOLUTION

(a)   Since we are given that X is a continuous random variable, we must have that $\int_{\text{all } x} f(x)\, dx = 1$.

$$\text{l.h.s.} = \int_0^2 c(x^2 - 2x + 3)\, dx = c\left[\frac{x^3}{3} - x^2 + 3x\right]_0^2$$

$$= c\left[\frac{(2)^3}{3} - (2)^2 + 3(2) - 0\right]$$

$$= c\left(\frac{8}{3} - 4 + 6\right) = \frac{14c}{3}$$

Hence, $\dfrac{14c}{3} = 1$. Therefore $c = \dfrac{3}{14}$.

(b)   In general mathematical terms, $\Pr(X \geqslant 1)$ can be translated as $\Pr(1 \leqslant X \leqslant \infty)$. But since $f(x) = 0$ outside the range 0 to 2, we can translate it practically as $\Pr(1 \leqslant X \leqslant 2)$.

Hence:

$$\Pr(X \geqslant 1) = \Pr(1 \leqslant X \leqslant 2) = \int_1^2 \frac{3}{14}(x^2 - 2x + 3)\, dx$$

$$= \frac{3}{14}\left[\frac{x^3}{3} - x^2 + 3x\right]_1^2$$

$$= \frac{3}{14}\left[\left(\frac{8}{3} - 4 + 6\right) - \left(\frac{1}{3} - 1 + 3\right)\right]$$

$$= \frac{3}{14}\cdot\frac{7}{3} = \frac{1}{2}$$

It is often necessary to consider cumulative probabilities associated with random variables. That is, we are interested in the probability that a random variable X, say, assumes a value less than or equal to some specified value t, say. A function that details this generally is called the *distribution function* of the variable w.r.t. t.

More specifically, if a continuous random variable X has a p.d.f. $f(x)$ for $a \leqslant x \leqslant b$, then the distribution function $\Pr(X \leqslant t)$ is defined as $\int_a^t f(x)\, dx$.

We usually denote the function by a capital letter. Thus:

$$F(t) = \Pr(X \leqslant t) = \int_a^t f(x)\, dx$$

Consider the continuous random variable X with p.d.f. $f(x) = \dfrac{2}{5}(4 - x)$ for $1 \leqslant x \leqslant 2$ discussed earlier. The distribution function is given by:

$$F(t) = \Pr(X \leqslant t) = \int_1^t \frac{2}{5}(4 - x)\, dx = \frac{2}{5}\left[4x - \frac{x^2}{2}\right]_1^t$$

$$= \frac{2}{5}\left[4t - \frac{t^2}{2} - \left(4 - \frac{1}{2}\right)\right]$$

i.e.: $F(t) = \dfrac{1}{5}(8t - t^2 - 7)$

So that, for instance, $F(1) = \dfrac{1}{5}(8(1) - (1)^2 - 7) = 0$, as expected since the variable cannot assume values less than 1.

Also $F(2) = \dfrac{1}{5}(8(2) - (2)^2 - 7) = 1$ as expected since the variable cannot assume values that are greater than 1.

To summarise:

DEFINITION 5.8

The *distribution function* of a random variable X (w.r.t. t) is written as $F(t)$ and defined as $\Pr(X \leqslant t)$. For:

(a) X discrete, with p.d.f. $\Pr(X = x)$; $x = x_0, x_1, x_2, \ldots$, we have:

$$F(t) = \Pr(X \leqslant t) = \sum_{x = x_0}^t \Pr(X = x)$$

(b) X continuous, with p.d.f. $f(x)$; $a \leqslant x \leqslant b$, we have:

(continued)

$$F(t) = \Pr(X \leqslant t) = \int_a^t f(x)\, dx$$

That is, the distribution function gives the (cumulative) probability that a random variable will assume a value less than or equal to some specified value.

EXAMPLE 5.28

Give the distribution function, $F(t)$, for the discrete random variable X with p.d.f. $\Pr(X = x) = \dfrac{1}{4}\left(\dfrac{3}{4}\right)^x$; $\quad x = 0, 1, 2, 3, \ldots$, assuming t is a positive integer.

Calculate $F(6)$ and hence find $\Pr(X \geqslant 7)$.

SOLUTION

By definition $F(t) = \displaystyle\sum_{x=0}^{t} \dfrac{1}{4}\left(\dfrac{3}{4}\right)^x$.

The right hand side is just the sum to $t+1$ terms of a geometric progression with $a = \dfrac{1}{4}$ and $r = \dfrac{3}{4}$. Using Statement 4.2(b), we have:

$$F(t) = S_{t+1} = \frac{\frac{1}{4}\left(1 - (\frac{3}{4})^{t+1}\right)}{1 - \frac{3}{4}} = 1 - \left(\frac{3}{4}\right)^{t+1}$$

In particular:

$$F(6) = 1 - \left(\frac{3}{4}\right)^7 = 0.867 \;\; (3\text{D})$$

Hence:

$$\Pr(X \geqslant 7) = 1 - \Pr(X \leqslant 6) = 1 - F(6) = 1 - 0.867$$
$$= 0.133 \;\; (3\text{D})$$

EXAMPLE 5.29

A continuous random variable X has a density function given by $f(x) = \dfrac{1}{4}$ for $0 \leqslant x \leqslant 4$ ($f(x) = 0$ elsewhere).

(a) Show that $f(x)$ is a valid p.d.f.

(b) Find the distribution function $F(t)$.

SOLUTION

(a) For $f(x)$ to be a valid p.d.f. we need $\displaystyle\int_{\text{all } x} f(x)\, dx = 1$.

l.h.s. $= \dfrac{1}{4}\displaystyle\int_0^4 1\, dx = \dfrac{1}{4}\big[x\big]_0^4 = \left(\dfrac{1}{4}\right)4 = 1 = $ r.h.s. as required

Hence, $f(x)$ as given, is a valid p.d.f.

(b) $F(t) = \Pr(X \leqslant t) = \int_0^t \frac{1}{4}\,dx = \frac{1}{4}[x]_0^t$

$$= \frac{1}{4}t \quad \text{(for } 0 \leqslant t \leqslant 4)$$

5.3.3 A PROBABILITY DISTRIBUTION

A probability distribution can be thought of as the exact theoretical analogue of a frequency distribution. The physical difference between the two is simply that one quotes the actual frequencies with which values of a variable occur, the other, theoretical probabilities.

Consider a statistical experiment, with an associated, well-defined random variable X. Physically performing the experiment a number of times and recording the frequency with which each value of X occurs will give us what we call a frequency distribution. On the other hand, without actually performing the experiment, we can often use the known experimental context to generate the probabilities associated with the various values that X can assume (this is not *always* possible). Hence, we obtain what is naturally called a *probability distribution*.

Note, of course, that if a variable, discrete or continuous, has a known probability density function, then a probability distribution is automatically defined.

As a practical illustration of the natural connection between these two types of distribution, consider the experiment 'roll a normal six sided die and count the number of spots on the uppermost face'. We can obtain a frequency distribution by simply repeating the experiment a number of times.

Let us associate the random variable X with the number of spots shown after a single roll. If we now roll the die 36 times, we might obtain the following result,

Number shown on die (x)	1	2	3	4	5	6	Total
Frequency (f)	4	6	8	7	5	6	36

which is our frequency distribution.

We now examine the theoretical situation of the same experiment. Here, the outcome set $S = (1, 2, 3, 4, 5, 6)$ is clearly 'equally likely' (assuming that the die is unbiased). The density function for X is easily seen to be

$\Pr(X = x) = \frac{1}{6}$; $x = 1, 2, 3, 4, 5, 6$ giving the table of probabilities,

Number shown on die (x)	1	2	3	4	5	6	Total
Probability (p)	$\frac{1}{6}$	$\frac{1}{6}$	$\frac{1}{6}$	$\frac{1}{6}$	$\frac{1}{6}$	$\frac{1}{6}$	1

constituting our probability distribution.

Comparing the two distributions directly, their apparent difference is in their totals. The sum of the frequencies must, of course, add to 36 — the number of repeats of the experiment. The sum of the probabilities, on the other hand, must sum to 1 because X is a random variable. However, the above two forms of distribution can be made directly comparable, as we now show.

Recall how, given a frequency distribution, we transform it to a relative (proportional) frequency distribution by dividing individual frequencies by the total frequency. Transforming the given frequency distribution in this manner we obtain:

Number shown on die (x)	1	2	3	4	5	6	Total
Proportion (p)	$\dfrac{4}{36}$	$\dfrac{6}{36}$	$\dfrac{8}{36}$	$\dfrac{7}{36}$	$\dfrac{5}{36}$	$\dfrac{6}{36}$	1

Notice that the total proportion adds to 1, enabling a direct comparison with the previous probability distribution. In this sense we can regard the proportions as probability estimates, useful in many situations.

On the other hand, consider the original probability distribution. If we wished to estimate the frequencies obtained *if the experiment were repeated* 36 *times*, we need only multiply each probability by 36 to obtain a set of what are called 'expected frequencies'. This transformation is now shown.

Number shown on die (x)	1	2	3	4	5	6	Total
Expected frequency (f)	6	6	6	6	6	6	36

Notice that the total (expected) frequency is 36. This now means that we can directly compare what we expect to happen theoretically (on the assumption that the die used is unbiased) with what actually happened when we physically repeated the experiment 36 times and obtained the previous frequency distribution.

So we have that frequency and probability distributions essentially differ only in the manner with which we operate on some statistical experiment. Theoretical considerations will give rise to a probability distribution; practical performances will result in a frequency distribution.

EXAMPLE 5.30

> A bag contains 6 white and 4 red counters, 3 of which are drawn at random and without replacement. If X is a random variable denoting the number of red counters drawn, construct a table showing the probability distribution of X. If this experiment is repeated 60 times, how many times would we expect to draw more than one red counter?

SOLUTION

X can take any one of the values $0, 1, 2$ or 3.

$$\Pr(X = 0) = \Pr(\text{no red counters})$$
$$= \Pr(W_1 \cdot W_2 \cdot W_3) \quad \text{using an obvious notation}$$
$$= \Pr(W_1) \cdot \Pr(W_2/W_1) \cdot \Pr(W_3/W_1 \cdot W_2)$$
non-independent events
$$= \frac{6}{10} \left(\frac{5}{9}\right) \frac{4}{8} = \frac{1}{6}$$

$$\Pr(X = 1) = \Pr(\text{1 red counter})$$
$$= \Pr(R_1 W_2 W_3 \text{ or } W_1 R_2 W_3 \text{ or } W_1 W_2 R_3)$$
$$= \Pr(R_1 W_2 W_3) + \Pr(W_1 R_2 W_3) + \Pr(W_1 W_2 R_3)$$
mutually exclusive events
$$= \Pr(R_1) \cdot \Pr(W_2/R_1) \cdot \Pr(W_3/W_2 R_1) + \ldots, \quad \text{etc.}$$
$$= \frac{4}{10} \left(\frac{6}{9}\right) \frac{5}{8} + \ldots, \quad \text{etc.}$$
$$= \frac{1}{6} + \frac{1}{6} + \frac{1}{6}$$

(Note that the probability of getting a single red in one position *must* be the same as the probability of getting a single red in any other position.)

$$= \frac{1}{2}$$

$$\Pr(X = 2) = \Pr(\text{2 red counters})$$
$$= \Pr(R_1 R_2 W_3 \text{ or } R_1 W_2 R_3 \text{ or } W_1 R_2 R_3)$$
$$= \Pr(R_1) \cdot \Pr(R_2/R_1) \cdot \Pr(W_3/R_2 R_1) + \ldots, \quad \text{etc.}$$
$$= \frac{4}{10} \left(\frac{3}{9}\right) \frac{6}{8} + \ldots, \quad \text{etc.}$$
$$= \frac{1}{10} + \frac{1}{10} + \frac{1}{10}$$
$$= \frac{3}{10}$$

$$\Pr(X = 3) = \Pr(\text{3 red counters})$$
$$= \Pr(R_1 R_2 R_3)$$
$$= \frac{4}{10} \left(\frac{3}{9}\right) \frac{2}{8}$$
$$= \frac{1}{30}$$

Hence, the probability distribution for X is given by the following table.

Number of red counters (x)	0	1	2	3	Total
Probability $\quad\quad\quad\quad\quad (p)$	$\dfrac{1}{6}$	$\dfrac{1}{2}$	$\dfrac{3}{10}$	$\dfrac{1}{30}$	1

Note that $\dfrac{1}{6} + \dfrac{1}{2} + \dfrac{3}{10} + \dfrac{1}{30} = 1$, as required for the total probability associated with a random variable.

Multiplying each probability by 60, we obtain an expected frequency distribution as follows.

Number of red counters (x)	0	1	2	3	Total
Expected frequency $\quad (f)$	10	30	18	2	60

From this table, we would expect more than 1 (i.e., 2 or 3) red counters 20 times out of a total of 60.

(Note that we obtain the same result by multiplying Pr(more than 1) by 60

$$= 60 \left(\frac{3}{10} + \frac{1}{30} \right) = 60 \left(\frac{10}{30} \right) = 20)$$

Not all experiments however will allow us to form a probability distribution. For example, consider experiments involving the following: the height of an adult male, the length of time taken to get to work and the number of people in a certain queue at particular times. These are all examples of situations where it is almost certainly impossible to form a probability distribution from purely theoretical considerations. The best we could do in such circumstances is to repeat trials of the experiment a large number of times and use the resulting frequency distribution (suitably transformed to a relative, proportional distribution) as an estimate of the true probability distribution.

Probability distributions can be represented pictorially in much the same way as frequency distributions, using histograms, frequency polygons and frequency curves. For a continuous probability distribution we have a natural frequency curve in the p.d.f., which can always be represented by a smooth curve.

EXAMPLE 5.31

For the continuous distribution defined by the p.d.f:

$$f(x) = \frac{1}{80}(3x^2 + 4) \text{ for } 0 \leqslant x \leqslant 4$$

(a)   Verify that the variable involved X, say, is a random variable.

(b)   Construct a probability distribution table for the ranges 0 to 1, 1 to 2, 2 to 3 and 3 to 4.

(c)   Display the results in the form of a histogram.

SOLUTION

(a)　For X to be a continuous random variable, we need:

$$\int_0^4 \frac{1}{80}(3x^2 + 4)\, dx = 1$$

left hand side $= \dfrac{1}{80}\left[x^3 + 4x\right]_0^4 = \dfrac{1}{80}\left[(4)^3 + 4(4) - 0\right]$

$$= \left(\frac{1}{80}\right) 80 = 1 \quad \text{as required}$$

Therefore X is a random variable.

(b)　$\Pr(0 \leqslant X < 1) = \int_0^1 \dfrac{1}{80}(3x^2 + 4)\, dx = \dfrac{1}{80}\left[x^3 + 4x\right]_0^1 = \dfrac{5}{80}$

$\Pr(1 \leqslant X < 2) = \int_1^2 \dfrac{1}{80}(3x^2 + 4)\, dx = \dfrac{1}{80}\left[x^3 + 4x\right]_1^2 = \dfrac{11}{80}$

$\Pr(2 \leqslant X < 3) = \int_2^3 \dfrac{1}{80}(3x^2 + 4)\, dx = \dfrac{1}{80}\left[x^3 + 4x\right]_2^3 = \dfrac{23}{80}$

$\Pr(3 \leqslant X < 4) = \int_3^4 \dfrac{1}{80}(3x^2 + 4)\, dx = \dfrac{1}{80}\left[x^3 + 4x\right]_3^4 = \dfrac{41}{80}$

The probability table can now be constructed:

Ranges of X	0 up to 1	1 up to 2	2 up to 3	3 up to 4	Total
Probability	$\dfrac{5}{80}$	$\dfrac{11}{80}$	$\dfrac{23}{80}$	$\dfrac{41}{80}$	1

(c)

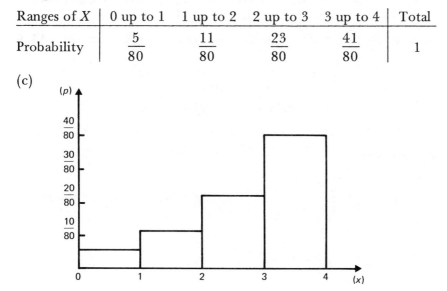

5.4 EXPECTATION

In the last section we noted the differences between frequency and probability distributions with regard to their origin and generation. We now go a step further in completing the picture between the two.

In Chapter 2, we examined certain statistical measures (mean, variance, etc.) employed to characterise frequency distributions. These measures have precise analogues in a probability distribution context.

The arithmetic mean and variance of a probability distribution can be calculated in a similar manner to that for a frequency distribution. The following section describes the techniques in detail.

5.4.1 THE EXPECTATION OF X

DEFINITION 5.9

Given a random variable X with a probability density function $\begin{cases} Pr(X = x); & \text{discrete } X \\ f(x); & \text{continuous } X \end{cases}$ we define the *expectation of* X, written as $E[X]$, by:

$$E[X] = \begin{cases} \sum_{\text{all } x} x \cdot Pr(X = x); & \text{for discrete } X \\[2mm] \int_{\text{all } x} x \cdot f(x) \cdot dx; & \text{for continuous } X \end{cases}$$

It is easy to demonstrate the fact that $E[X]$ is just the arithmetic mean of a discrete probability distribution as follows.

Suppose the probability distribution defined by the (discrete) p.d.f. is:

x	x_1 x_2 x_3 x_4 ... x_n	Total
p	p_1 p_2 p_3 p_4 ... p_n	1

i.e., $Pr(X = x_1) = p_1$; $Pr(X = x_2) = p_2$;..., etc.

Note that the total probability adds to 1 since X is a random variable.

By definition:

$$\text{mean} = \frac{\sum fx}{\sum f} = \frac{\sum px}{\sum p} \quad \text{(since } p \text{ replaces } f \text{ in the above)}$$

$$= \sum px \quad \text{(since } \sum p = 1\text{)}$$

$$= \sum_i x_i p_i$$

$$= \sum_{\text{all } x} x \, Pr(X = x) \quad \text{agreeing with the definition above}$$

For example, given the following p.d.f. of a discrete random variable X:

x	0	1	2	3	4
$\Pr(X = x)$	$\frac{1}{4}$	$\frac{1}{8}$	$\frac{1}{8}$	$\frac{1}{4}$	$\frac{1}{4}$

we would calculate the mean (expectation) of X as:

$$E[X] = \sum_{\text{all } x} x \, \Pr(X = x) = 0\left(\frac{1}{4}\right) + 1\left(\frac{1}{8}\right) + 2\left(\frac{1}{8}\right) + 3\left(\frac{1}{4}\right) + 4\left(\frac{1}{4}\right)$$

$$= 0 + \frac{1}{8} + \frac{1}{4} + \frac{3}{4} + 1$$

$$= \frac{17}{8} = 2.125$$

If a continuous random variable X has a p.d.f. $f(x) = \frac{3}{4}x(2 - x)$ for $0 \leqslant x \leqslant 2$, then:

$$E[X] = \int_{\text{all } x} xf(x) \, dx = \int_0^2 x \frac{3}{4}x(2 - x) \, dx = \frac{3}{4}\int_0^2 x^2(2 - x) \, dx$$

$$= \frac{3}{4}\int_0^2 (2x^2 - x^3) \, dx = \frac{3}{4}\left[\frac{2}{3}x^3 - \frac{1}{4}x^4\right]_0^2$$

$$= \frac{3}{4}\left(\frac{16}{3} - \frac{16}{4}\right) = \frac{3}{4}\left(\frac{16}{12}\right)$$

$$= 1$$

Note that $E[X]$ in both of the previous cases is a number (a constant) and not a function of X. This will always be the case.

It is often useful, when requiring the value of $E[X]$, to be able to recognise a symmetric distribution.

For instance, the distribution

x	1	2	3	4	5
$\Pr(X = x)$	$\frac{1}{16}$	$\frac{3}{16}$	$\frac{1}{2}$	$\frac{3}{16}$	$\frac{1}{16}$

is symmetric about the central value $x = 3$. Thus we have $E[X] = 3$.

The symmetry here was fairly easy to see, but for most continuous distributions it is not so obvious. Consider the continuous p.d.f. $f(x) = \frac{3}{4}x(2 - x)$ for $0 \leqslant x \leqslant 2$ again. With a fairly elementary knowledge of curve sketching we can obtain the density curve given in Fig. 5.3. Here, we can see that there is symmetry about the line $x = 1$. Hence $E[X] = 1$, as obtained previously.

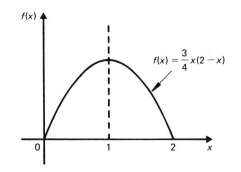

Fig. 5.3

EXAMPLE 5.32

Find the expectation of the random variable X whose density function is given as:

x	-1	0	1	2	3
$\Pr(X = x)$	0.3	0.1	0.1	0.3	0.2

SOLUTION

$$E[X] = \sum_{\text{all } x} x \, \Pr(X = x) \quad \text{by definition}$$

$$= (-1)(0.3) + 0(0.1) + 1(0.1) + 2(0.3) + 3(0.2)$$

$$= -0.3 + 0.1 + 0.6 + 0.6$$

i.e.: $E[X] = 1.0$

EXAMPLE 5.33

An experiment is defined by repeatedly tossing a coin until either one head or four tails in succession are obtained. Denoting H for head and T for tail,

(a)   write down the outcome set for the experiment.

(b)   Defining the random variable X as 'the number of tosses required', calculate $\Pr(X = x)$ for $x = 1, 2, 3$ and 4.

(c)   Hence find the expected number of tosses required to end the experiment.

SOLUTION

(a)   The outcome set for the experiment is:

$$S = (\text{H, TH, TTH, TTT}A)$$

where A represents 'anything', i.e, either H or T since TTTH and TTTT will both end the experiment.

(b)

$$\Pr(X = 1) = \Pr(H) = \frac{1}{2}; \quad \Pr(X = 2) = \Pr(TH) = \frac{1}{2}\left(\frac{1}{2}\right) = \frac{1}{4};$$

$$\Pr(X = 3) = \Pr(TTH) = \frac{1}{2}\left(\frac{1}{2}\right)\left(\frac{1}{2}\right) = \frac{1}{8};$$

$$\Pr(X = 4) = \Pr(TTTA) = \frac{1}{2}\left(\frac{1}{2}\right)\left(\frac{1}{2}\right)1$$

(since $\Pr(A) = \Pr(H \text{ or } T) = \Pr(H) + \Pr(T) = \frac{1}{2} + \frac{1}{2} = 1$), i.e.,
$\Pr(X = 4) = \frac{1}{8}$.

(c)   We require $E[X]$. The probability distribution for X can now be written:

x	1	2	3	4
$\Pr(X = x)$	$\frac{1}{2}$	$\frac{1}{4}$	$\frac{1}{8}$	$\frac{1}{8}$

Notice that the sum of the probabilities adds to 1 as required for a random variable.

Hence:

$$E[X] = \sum_{x=1}^{4} x\,\Pr(X = x) = 1\left(\frac{1}{2}\right) + 2\left(\frac{1}{4}\right) + 3\left(\frac{1}{8}\right) + 4\left(\frac{1}{8}\right)$$

$$= \frac{1}{2} + \frac{1}{2} + \frac{3}{8} + \frac{4}{8} = \frac{15}{8}$$

So:

$$E[X] = 1.875$$

EXAMPLE 5.34

If X is a continuous random variable with a p.d.f. $f(x) = kx$;
$0 \leqslant x \leqslant 5$,

(a) evaluate k,   (b) find $\Pr(2 \leqslant X \leqslant 4)$,   (c) calculate $E[X]$.

SOLUTION

(a)   Since X is a continuous random variable we need:

$$\int_{\text{all } x} f(x)\,dx = 1$$

$$\text{l.h.s.} = k\int_0^5 x\,dx = k\left[\frac{x^2}{2}\right]_0^5 = \frac{k}{2}(5^2 - 0) = \frac{25k}{2}$$

So that $\dfrac{25k}{2} = 1$. Hence $k = \dfrac{2}{25}$.

(b)   We can now write the p.d.f. as $f(x) = \dfrac{2x}{25}$ for $0 \leqslant x \leqslant 5$.

$$\Pr(2 \leqslant X \leqslant 4) = \int_2^4 f(x) \, dx \quad \text{(by definition)}$$

$$= \int_2^4 \frac{2x}{25} \, dx = \frac{2}{25} \left[\frac{x^2}{2} \right]_2^4 = \frac{1}{25}(4^2 - 2^2) = \frac{12}{25}$$

(c) $\quad \mathrm{E}[X] = \int_{\text{all } x} xf(x) \, dx \quad \text{(by definition)}$

$$= \frac{2}{25} \int_0^5 x^2 \, dx = \frac{2}{25} \left[\frac{x^3}{3} \right]_0^5 = \frac{2}{75}(5^3 - 0) = \frac{10}{3}$$

EXAMPLE 5.35

A person plays a game by throwing 3 normal six-sided dice. If he gets 3 sixes he wins £100; for 2 sixes he wins £10; for 1 six he wins £3. *Otherwise he loses £1.* Find his expected winnings per throw.

SOLUTION

Let the random variable X represent 'winnings per throw in pounds'. What we need to realise at the outset here is that X can assume four values, namely 100, 10, 3 or -1. Note that we are counting a loss of £1 as a *negative* gain.

The above four values of X correspond of course to the person obtaining 3, 2, 1 or 0 sixes as a result of the throw.

We now need to identify the distribution of X in order to calculate $\mathrm{E}[X]$.

$$\Pr(X = 100) = \Pr(666) = \frac{1}{6} \cdot \frac{1}{6} \cdot \frac{1}{6} = \frac{1}{216}$$

$$\Pr(X = 10) \;= \Pr(66\bar{6} \text{ or } 6\bar{6}6 \text{ or } \bar{6}66)$$

$$= \frac{1}{6} \cdot \frac{1}{6} \cdot \frac{5}{6} + \frac{1}{6} \cdot \frac{5}{6} \cdot \frac{1}{6} + \frac{5}{6} \cdot \frac{1}{6} \cdot \frac{1}{6} = 3 \left(\frac{1}{6} \right)^2 \left(\frac{5}{6} \right) = \frac{15}{216}$$

$$\Pr(X = 3) \;= \Pr(6\bar{6}\bar{6} \text{ or } \bar{6}6\bar{6} \text{ or } \bar{6}\bar{6}6)$$

$$= \frac{1}{6} \cdot \frac{5}{6} \cdot \frac{5}{6} + \frac{5}{6} \cdot \frac{1}{6} \cdot \frac{5}{6} + \frac{5}{6} \cdot \frac{5}{6} \cdot \frac{1}{6} = 3 \left(\frac{1}{6} \right) \left(\frac{5}{6} \right)^2 = \frac{75}{216}$$

$$\Pr(X = -1) = \Pr(\bar{6}\bar{6}\bar{6}) = \frac{5}{6} \cdot \frac{5}{6} \cdot \frac{5}{6} = \frac{125}{216}$$

In tabular form:

x	-1	3	10	100	Total
$\Pr(X = x)$	$\dfrac{125}{216}$	$\dfrac{75}{216}$	$\dfrac{15}{216}$	$\dfrac{1}{216}$	1

Therefore, expected winnings per throw:

$$E[X] = \sum_{\text{all } x} x \Pr(X = x)$$

$$= -1\left(\frac{125}{216}\right) + 3\left(\frac{75}{216}\right) + 10\left(\frac{15}{216}\right) + 100\left(\frac{1}{216}\right)$$

$$= \frac{-125 + 225 + 150 + 100}{216} = \frac{350}{216} = 1.62$$

i.e.,   expected winnings $= £1.62$.

5.4.2 THE EXPECTATION OF $g(X)$

The expectation operator E is not confined to acting on X alone. It can operate on any function of X. For example, we consider quantities such as $E[X^2]$, $E[3X]$, $E[4X^3 - 2X^2]$, etc.

The evaluation of the expectation of any function of X is contained in the following definition.

DEFINITION 5.10

Let $g(X)$ be any function of a random variable X having a p.d.f. of
$$\begin{cases} \Pr(X = x); & \text{discrete } X \\ f(x); & \text{continuous } X \end{cases}$$
Then the *expectation of* $g(X)$, written as $E[g(X)]$, is defined as:

$$E[g(X)] = \begin{cases} \displaystyle\sum_{\text{all } x} g(x) \Pr(X = x); & \text{discrete } X \\[2ex] \displaystyle\int_{\text{all } x} g(x) f(x) \, dx; & \text{continuous } X \end{cases}$$

Notice that the above forms are just generalisations of the two given in the previous definition. Putting $g(X) = X$ in the definition above gives $E[X]$ as in Definition 5.9.

EXAMPLE 5.36

For the discrete distribution:

x	0	1	2	3	4
$\Pr(X = x)$	$\dfrac{1}{4}$	$\dfrac{1}{8}$	$\dfrac{1}{8}$	$\dfrac{1}{4}$	$\dfrac{1}{4}$

find: (a) $E[X^2]$ (b) $E[3X]$.

SOLUTION

(a) $E[X^2] = \displaystyle\sum_{\text{all } x} x^2 \Pr(X = x)$

$$= 0^2 \left(\frac{1}{4}\right) + 1^2 \left(\frac{1}{8}\right) + 2^2 \left(\frac{1}{8}\right) + 3^2 \left(\frac{1}{4}\right) + 4^2 \left(\frac{1}{4}\right)$$

$$= \frac{1}{8} + \frac{4}{8} + \frac{9}{4} + \frac{16}{4} = \frac{55}{8}$$

i.e.: $E[X^2] = 6.875$

(b) $E[3X] = \sum_{\text{all } x} 3x \Pr(X = x)$

$$= (3)(0)\left(\frac{1}{4}\right) + (3)(1)\left(\frac{1}{8}\right) + (3)(2)\left(\frac{1}{8}\right)$$

$$+ (3)(3)\left(\frac{1}{4}\right) + (3)(4)\left(\frac{1}{4}\right)$$

$$= \frac{3}{8} + \frac{6}{8} + \frac{9}{4} + \frac{12}{4} = \frac{51}{8}$$

i.e.: $E[3X] = 6.375$

EXAMPLE 5.37

Find $E[X^2]$ for the continuous random variable X having the p.d.f. $f(x) = \frac{3}{4}x(2-x); \ 0 \leqslant x \leqslant 2$.

SOLUTION

$$E[X^2] = \int_{\text{all } x} x^2 \, f(x) \, dx \quad \text{by definition}$$

$$= \int_0^2 x^2 \, \frac{3}{4}x(2-x) \, dx = \frac{3}{4} \int_0^2 x^3 (2-x) \, dx$$

$$= \frac{3}{4} \int_0^2 (2x^3 - x^4) \, dx = \frac{3}{4} \left[\frac{1}{2}x^4 - \frac{1}{5}x^5\right]_0^2$$

$$= \frac{3}{4}\left(\frac{8}{5}\right) = \frac{6}{5}$$

i.e.: $E[X^2] = 1.2$

5.4.3 PROPERTIES OF OPERATOR E

STATEMENT 5.10

If X is a random variable (discrete or continuous) and a and b are any two constants, then:

(a) $E[a] = a$ (b) $E[aX] = a \, E[X]$

(c) $E[f_1(X) + f_2(X)] = E[f_1(X)] + E[f_2(X)]$ where $f_1(X)$ and $f_2(X)$ are any two functions of X.

Note In mathematics, these properties define E to be a *linear* operator.

Proofs (a) For discrete X:

$$E[a] = \sum_{\text{all } x} a \Pr(X = x) \quad \text{(by definition)}$$

$$= a \sum_{\text{all } x} \Pr(X = x) \quad \text{(since } a \text{ is a constant)}$$

$$= a \quad \left(\text{since } \sum_{\text{all } x} \Pr(X = x) = 1 \quad \text{for a random variable } X\right)$$

For continuous X:

$$E[a] = \int_{\text{all } x} a f(x) \, dx \quad \text{(by definition)}$$

$$= a \int_{\text{all } x} f(x) \, dx \quad \text{(since } a \text{ is a constant)}$$

$$= a \quad \left(\text{since } \int_{\text{all } x} f(x) \, dx = 1\right)$$

(b) For discrete X:

$$E[aX] = \sum_{\text{all } x} ax \Pr(X = x) \quad \text{(by definition)}$$

$$= a \sum_{\text{all } x} x \Pr(X = x) \quad \text{(since } a \text{ is a constant)}$$

$$= a E[X] \quad \text{(by definition)}$$

For continuous X:

$$E[aX] = \int_{\text{all } x} ax f(x) \, dx \quad \text{(by definition)}$$

$$= a \int_{\text{all } x} x f(x) \, dx \quad \text{(since } a \text{ is a constant)}$$

$$= a E[X] \quad \text{(by definition)}$$

(c) Left as an exercise.

In the above proofs, $\Pr(X = x)$ and $f(x)$ represent the p.d.f.s (discrete and continuous respectively) of a random variable X.

Notice in the above that the proofs are identical for the discrete and continuous cases, except for the method of summing, i.e., Σ for discrete X and $\int$ for continuous X.

Using part (c) of Statement 5.10, it is easy to see the extension to the expectation of the sum of any number of functions of a random variable.

That is:

$$E[f_1(X) + f_2(X) + \ldots + f_n(X)] = E[f_1(X)] + E[f_2(X)] + \ldots + E[f_n(X)]$$

So that now, in particular, we are able to calculate the expectation of any polynomial function of X.

Earlier, for the distribution

x	0	1	2	3	4
$\Pr(X = x)$	$\dfrac{1}{4}$	$\dfrac{1}{8}$	$\dfrac{1}{8}$	$\dfrac{1}{4}$	$\dfrac{1}{4}$

we calculated $E[X] = \dfrac{17}{8}$ and $E[X^2] = \dfrac{55}{8}$.

Hence:

$$E[X - 1] = E[X] - E[1] = \frac{17}{8} - 1 = \frac{9}{8}$$

and:

$$E[2X - 2] = E[2(X - 1)] = 2\,E[X - 1] = 2\left(\frac{9}{8}\right) = \frac{9}{4}$$

and:

$$\begin{aligned}
E[4X^2 - 3X + 2] &= E[4X^2] + E[-3X] + E[2] \\
&= 4\,E[X^2] - 3\,E[X] + 2 \\
&= 4\left(\frac{55}{8}\right) - 3\left(\frac{17}{8}\right) + 2 \\
&= \frac{185}{8}
\end{aligned}$$

Note   The expectation of any function of a random variable X, $E[g(X)]$ say, can be physically interpreted as the 'average' value of $g(X)$ as X varies and, as such, will always be evaluated as a number (and *not* a function of X).

Finally, here, we state without proof how the operator E can be used with a sum of functions of different random variables.

STATEMENT 5.11

If X and Y are any two random variables (independent or otherwise), and $g_1(X)$ and $g_2(Y)$ are any two functions of X and Y respectively, then:

$$E[g_1(X) + g_2(Y)] = E[g_1(X)] + E[g_2(Y)]$$

In particular, from the above we can assert $E[X + Y] = E[X] + E[Y]$ and $E[aX + bY] = a\,E[X] + b\,E[Y]$,   both of which we will have occasion to use in many contexts later in the book.

EXAMPLE 5.38

Two bags each contain 3 counters numbered from 1 to 3. A counter is drawn randomly from each bag. Let random variable X represent 'the sum of the two numbers on the counters'. Calculate:

(a) $E[X]$ (b) $E[X^2]$ (c) $E[2X^2 - 6X + 5]$.

SOLUTION

The equally-likely outcomes for the outcome set of the experiment together with the corresponding x-values are listed:

Outcome	(1, 1)	(1, 2)	(1, 3)	(2, 1)	(2, 2)	(2, 3)	(3, 1)	(3, 2)	(3, 3)
Total (X)	2	3	4	3	4	5	4	5	6

(The number pairs in the brackets represent the 1st and 2nd numbers drawn respectively.) Since each of the above outcomes has probability $\frac{1}{9}$ of occurring, we easily calculate the following distribution of X.

x	2	3	4	5	6
$Pr(X = x)$	$\frac{1}{9}$	$\frac{2}{9}$	$\frac{3}{9}$	$\frac{2}{9}$	$\frac{1}{9}$

(a)   From the symmetry of the distribution we have immediately $E[X] = 4$.

(b) $E[X^2] = 4\left(\frac{1}{9}\right) + 9\left(\frac{2}{9}\right) + 16\left(\frac{3}{9}\right) + 25\left(\frac{2}{9}\right) + 36\left(\frac{1}{9}\right)$

$= \frac{156}{9} = 17.33$ (2D)

(c) $E[2X^2 - 6X + 5] = 2\,E[X^2] - 6\,E[X] + 5$

$= 2\left(\frac{156}{9}\right) - 6(4) + 5$

$= 15.67$ (2D)

EXAMPLE 5.39

X, Y and Z are three random variables with $E[X] = 2$, $E[Y] = 3$ and $E[Z] = 5$. Calculate:

(a) $E[3X - 2Y]$ (b) $E[2Y - 3Z)$ (c)   the expectation of the average of Y and Z.

SOLUTION

(a) $E[3X - 2Y] = 3\,E[X] - 2\,E[Y] = 3 \cdot 2 - 2 \cdot 3 = 6 - 6 = 0$

(b) $E[2Y - 3Z] = 2\,E[Y] - 3\,E[Z] = 2 \cdot 3 - 3 \cdot 5 = 6 - 15 = -9$

(c)   This can be written as:

$$E\left[\frac{Y+Z}{2}\right] = \frac{1}{2} E[Y+Z]$$

$$= \frac{1}{2}(E[Y] + E[Z])$$

$$= \frac{1}{2}(3+5) = \frac{8}{2}$$

$$= 4$$

5.4.4 THE VARIANCE OF X

The variance of a probability distribution has the same significance and interpretation as the variance of a frequency distribution. It is defined solely in terms of the expectation of a particular function of random variable X and is given in the following definition.

DEFINITION 5.11

The *variance of a probability distribution* associated with a random variable X is written as $\text{Var}[X]$ and defined by:

$$\text{Var}[X] = E[(X-\mu)^2], \quad \text{where} \quad \mu = E[X]$$

As in the case of $E[X]$, we can show that the above definition (for the discrete case) is equivalent to the technique used in finding the variance of a frequency distribution.

For a frequency distribution, variance $= \dfrac{\sum f(x-\bar{x})^2}{\sum f}$

i.e.:

variance $= \dfrac{\sum \text{Pr}(x-\bar{x})^2}{\sum p}$ (changing f to p for a probability distribution)

$$= \sum p(x-\bar{x})^2 \quad (\text{since} \quad \sum p = 1 \quad \text{for random variable } X)$$

$$= \sum_i (x_i - \bar{x})^2 p_i$$

$$= \sum_{\text{all } x} (x-\bar{x})^2 \ \text{Pr}(X=x)$$

$$= E[(X-\mu)^2] \quad \text{by definition (for } X \text{ discrete)}$$

which of course agrees with the definition above.

As noted earlier, we can see that the variance of a random variable X is just the expectation of a special function of X, namely $(X - \mu)^2$.

Recall that, in the case of a frequency distribution, we hardly ever used the defining expression for s^2 in calculations. Instead we preferred to use the computational formula

$$s^2 = \frac{\sum fx^2}{\sum f} - \left(\frac{\sum fx}{\sum f}\right)^2$$

which is easier and more convenient. Exactly the same is true for a probability distribution. We have an analogous 'computational formula' for the variance, which we now state and prove.

STATEMENT 5.12

If X is a discrete or continuous random variable, we can calculate its variance using $\mathrm{Var}[X] = \mathrm{E}[X^2] - \mathrm{E}^2[X]$.

Note   that $\mathrm{E}^2[X]$ is a shorthand form for $[\mathrm{E}(X)]^2$.

Proof

$$
\begin{aligned}
\mathrm{Var}[X] &= \mathrm{E}[(X - \mu)^2] \quad \text{(by definition) (where } \mu = \mathrm{E}[X]\text{)} \\
&= \mathrm{E}[X^2 - 2\mu X + \mu^2] \quad \text{(expanding the square)} \\
&= \mathrm{E}[X^2] + \mathrm{E}[-2\mu X] + \mathrm{E}[\mu^2] \quad \text{(properties of E)} \\
&= \mathrm{E}[X^2] - 2\mu\,\mathrm{E}[X] + \mu^2 \quad (\mu = \mathrm{E}[X] \text{ is a constant w.r.t} \\
&\qquad\qquad\qquad\qquad\qquad\qquad\qquad \text{random variable } X) \\
&= \mathrm{E}[X^2] - 2\,\mathrm{E}^2[X] + \mathrm{E}^2[X] \\
&= \mathrm{E}[X^2] - \mathrm{E}^2[X] \quad \textit{completing the proof}
\end{aligned}
$$

Note that $\mathrm{E}[X^2]$ (probability-wise) corresponds to $\dfrac{\sum fx^2}{\sum f}$ (frequency-wise), just as $\mathrm{E}[X]$ corresponds to $\dfrac{\sum fx}{\sum f}$.

For the discrete distribution we have been considering, we obtained $\mathrm{E}[X] = \dfrac{17}{8}$ and $\mathrm{E}[X^2] = \dfrac{55}{8}$.

$$\therefore \qquad \mathrm{Var}[X] = \mathrm{E}[X^2] - \mathrm{E}^2[X] = \frac{55}{8} - \left(\frac{17}{8}\right)^2 = \frac{151}{64}$$

$$= 2.36 \ \ (2\mathrm{D})$$

For the continuous distribution $f(x) = \dfrac{3}{4}x(2-x)$, $0 \leqslant x \leqslant 2$, that we considered earlier, we obtained $E[X] = 1$ (in text) and $E[X^2] = 1.2$ (Example 5.37).

Hence:
$$Var[X] = E[X^2] - E^2[X] = 1.2 - 1^2$$
$$= 0.2$$

The variance operator has certain important properties (easily proved using the E operator) that we need to be aware of.

If a is any constant, we have:
$$Var[a] = E[a^2] - E^2[a] \quad \text{(by definition)}$$
$$= a^2 - a^2 \quad \text{(properties of E)}$$
$$= 0$$

This is an intuitively reasonable result since a constant cannot vary.

If X is any random variable and a is any constant, we have the following surprising result:
$$Var[aX] = E[a^2 X^2] - E^2[aX] \quad \text{(by definition)}$$
$$= a^2 E[X^2] - a^2 E^2[X] \quad \text{(properties of E)}$$
$$= a^2 [E(X^2) - E^2(X)]$$
$$= a^2 Var[X]$$

If we compare this with the corresponding result for the expectation of aX we can see its full significance, i.e.,
$$E[aX] = a E[X]$$
$$Var[aX] = a^2 Var[X]$$

Following on from this we have $Var[aX + b] = a^2 Var[X]$. This is a reasonable result if we remember that adding a constant to a variable simply moves its distribution in space without affecting the dispersion. We discussed this case for a frequency distribution in Section 2.3. So that we have the two distributions aX and $aX + b$ with the same variance; namely $Var[aX]$.

Finally, if X and Y are any two *independent* random variables, we have $Var[aX + bY] = a^2 Var[X] + b^2 Var[Y]$, which we state without proof.

As a consequence of the above, we have the following important identities:
$$Var[X + Y] = Var[X] + Var[Y]$$
$$Var[X - Y] = Var[X + (-Y)]$$
$$= Var[X] + Var[-Y]$$

$$= \text{Var}[X] + (-1)^2 \, \text{Var}[Y]$$

$$= \text{Var}[X] + \text{Var}[Y]$$

The results hold only for X, Y independent.

That is, the variance of *both a sum and a difference* of two random variables is always the *sum* of the two separate variances.

We summarise with the following statement.

STATEMENT 5.13

(a)   If X is a random variable and a and b are any two constants, then:

(i) $\text{Var}[a] = 0$ (ii) $\text{Var}[aX] = a^2 \, \text{Var}[X]$

(iii) $\text{Var}[aX + b] = a^2 \, \text{Var}[X]$

(b)   If X and Y are any two *independent* random variables and a and b are any two constants, then:

$$\text{Var}[aX + bY] = a^2 \, \text{Var}[X] + b^2 \, \text{Var}[Y]$$

Note   As a consequence of the above:

$$\text{Var}[X + Y] = \text{Var}[X] + \text{Var}[Y]$$

$$\text{Var}[X - Y] = \text{Var}[X] + \text{Var}[Y]$$

EXAMPLE 5.40

Let X be the discrete random variable 'number of heads when two coins are tossed'. Find:

(a) $\text{E}[X]$ (b) $\text{Var}[X]$ (c) $\text{Var}[3X - 2]$

SOLUTION

We calculated the distribution of X in Example 5.24 and obtained:

x	0	1	2
$\Pr(X = x)$	$\dfrac{1}{4}$	$\dfrac{1}{2}$	$\dfrac{1}{4}$

(a)   By the symmetry, $\text{E}[X] = 1$.

(b) $\text{Var}[X] = \text{E}[X^2] - \text{E}^2[X]$ (by definition)

$$\text{E}[X^2] = 0^2 \left(\frac{1}{4}\right) + 1^2 \left(\frac{1}{2}\right) + 2^2 \left(\frac{1}{4}\right) = \frac{3}{2}$$

$$\text{Var}[X] = \frac{3}{2} - 1^2 = \frac{1}{2}$$

(c) $\text{Var}[3X - 2] \;=\; 3^2 \,\text{Var}[X]$ (using the previous statement)

$$= \; 9\left(\frac{1}{2}\right) \;=\; 4.5$$

EXAMPLE 5.41

Independent continuous random variables X and Y have respective
p.d.f.s $f(x) = \dfrac{1}{4} x$ $(1 \leqslant x \leqslant 3)$ and $g(y) = \dfrac{1}{2}$ $(2 \leqslant y \leqslant 4)$.

(a)   Find the expectation and variance of X and Y.
(b)   Hence calculate the expectation and variance of $4Y - 3X$ to
2D.

SOLUTION

(a)
$$\text{E}[X] \;=\; \int_1^3 x\left(\frac{1}{4}\right) x \; dx \;=\; \frac{1}{4}\int_1^3 x^2 \; dx \;=\; \frac{1}{4}\left[\frac{1}{3}x^3\right]_1^3 \;=\; \frac{1}{4}\left(\frac{27}{3} - \frac{1}{3}\right)$$
$$= \; 2.17 \;\; (2\text{D})$$

$$\text{E}[X^2] \;=\; \int_1^3 x^2\left(\frac{1}{4}\right) x \; dx \;=\; \frac{1}{4}\int_1^3 x^3 \; dx \;=\; \frac{1}{4}\left[\frac{1}{4}x^4\right]_1^3$$
$$= \; \frac{1}{4}\left(\frac{81}{4} - \frac{1}{4}\right) \;=\; 5$$

$$\therefore \text{Var}[X] = \; \text{E}[X^2] - \text{E}^2[X] \;=\; 5 - \left(\frac{13}{6}\right)^2 \;=\; \frac{11}{36}$$

$$\text{E}[Y] \;=\; \int_2^4 y\left(\frac{1}{2}\right) dy \;=\; \frac{1}{2}\left[\frac{1}{2}y^2\right]_2^4 \;=\; \frac{1}{2}\left(\frac{16}{2} - \frac{4}{2}\right) \;=\; 3$$

$$\text{E}[Y^2] \;=\; \int_2^4 y^2\left(\frac{1}{2}\right) dy \;=\; \frac{1}{2}\left[\frac{1}{3}y^3\right]_2^4 \;=\; \frac{1}{2}\left(\frac{64}{3} - \frac{8}{3}\right) \;=\; \frac{28}{3}$$

$$\therefore \text{Var}[Y] = \; \text{E}[Y^2] - \text{E}^2[Y] \;=\; \frac{28}{3} - (3)^2 \;=\; \frac{1}{3}$$

(b)
$$\text{E}[4Y - 3X] \;=\; 4\,\text{E}[Y] - 3\,\text{E}[X]$$ (using Statement 5.11)
$$= \; 4(3) - 3\left(\frac{13}{6}\right) \;=\; 5.5$$

$$\text{Var}[4Y - 3X] \;=\; 16\,\text{Var}[Y] + 9\,\text{Var}[X]$$ (by previous statement)
$$= \; 16\left(\frac{1}{3}\right) + 9\left(\frac{11}{36}\right)$$
$$= \; 8.08 \;\; (2\text{D})$$

5.4.5 THE CALCULATION OF QUANTILES

In Sections 2.3 and 2.4.4 we defined and gave examples of quantiles such as the median, quartiles and percentiles for both discrete and continuous frequency distributions.

The definition of a quantile is exactly the same for probability distributions as for frequency distributions although the method of calculation differs slightly.

There are two distinct areas for consideration, discrete and continuous.

For discrete distributions, we can use just the same method as for frequency distributions. Consider the following discrete probability distribution, given with its (cumulative) distribution function:

x	0	1	2	3	4	5
$\Pr(X = x)$	0.134	0.362	0.204	0.146	0.103	0.051
$\Pr(X \leqslant x)$	0.134	0.496	0.700	0.846	0.949	1.000

Using an exact method, the median item is just one-half of the way along the distribution. This is the item that will contain the cumulative probability of $\dfrac{1}{2} = 0.5$. This is clearly 2, from the table, i.e., median = 2 (exact).

For a continuous probability distribution, given a well-defined p.d.f. $f(x)$ for $x_a \leqslant x \leqslant x_b$ say, quantiles may be determined by using their strict definition with respect to the splitting up of a distribution curve into a number of equal parts.

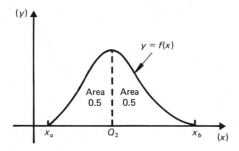

Fig. 5.4

For example, to calculate the median (Q_2), we would find that value that splits the distribution up into two halves (see Fig. 5.4). That is, the median (Q_2) is such that

$$\int_{x_a}^{Q_2} f(x) \, \mathrm{d}x = 0.5$$

or, in terms of the distribution function, $F(Q_2) = 0.5$. Similarly (see Fig. 5.5), the two quartiles would satisfy

$$\int_{x_a}^{Q_1} f(x)\, dx \;=\; F(Q_1) \;=\; 0.25$$

and

$$\int_{x_a}^{Q_3} f(x)\, dx \;=\; F(Q_3) \;=\; 0.75$$

where Q_1 and Q_3 represent the 1st and 3rd quartiles respectively.

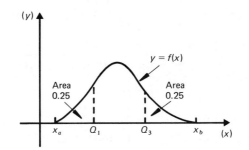

Fig. 5.5

Other quantiles, such as percentiles and deciles are calculated in a similar manner using these same techniques.

EXAMPLE 5.42

Using an interpolation method, calculate Q_1 and Q_3 (the 1st and 3rd quartiles) and hence the semi-interquartile range for the following probability distribution, assuming the values have been rounded to the nearest whole number.

x	0	1	2	3	4	5
$\Pr(X = x)$	0.2215	0.4114	0.2743	0.0816	0.0107	0.0005

SOLUTION

The distribution function for the distribution is given by:

x	0	1	2	3	4	5
$\Pr(X \leqslant x)$	0.2215	0.6329	0.9072	0.9888	0.9995	1.0000

Q_1 (0.25 of the way along the distribution) is in the class 0.5 to 1.5 giving:

$$Q_1 \;=\; 0.5 + \left(\frac{0.25 - 0.2215}{0.4114} \right) 1 \;=\; 0.569 \ \text{(3D)}$$

Similarly, we have:

$$Q_3 = 1.5 + \left(\frac{0.75 - 0.6329}{0.2743}\right)1 = 1.927 \ (3D)$$

Therefore the semi-interquartile range is:

$$\frac{Q_3 - Q_1}{2} = \frac{1.927 - 0.569}{2}$$

$$= 0.679 \ (3D)$$

EXAMPLE 5.43

Find an equation that the median satisfies for the continuous distribution given by the p.d.f. $f(x) = \frac{3}{13}(x^2 + 4)$ for $0 \leqslant x \leqslant 1$.

SOLUTION

Let m represent the median.

We have:

$$\int_0^m \frac{3}{13}(x^2 + 4) \, dx = 0.5 \quad \text{(by definition of the median)}$$

i.e.

$$\frac{3}{13}\left[\frac{1}{3}x^3 + 4x\right]_0^m = \frac{1}{2} \quad \text{giving} \quad \frac{3}{13}\left(\frac{1}{3}m^3 + 4m\right) = \frac{1}{2}$$

and after some algebraic rearrangement we easily obtain:

$$m(2m^2 + 24) = 13.$$

5.5 MOMENTS AND GENERATING FUNCTIONS

5.5.1 MOMENTS OF A PROBABILITY DISTRIBUTION

DEFINITION 5.12

Let X be a random variable. *The rth moment (about zero) of X is denoted by μ_r' and defined:* $\mu_r' = E[X^r]$.

Note the direct analogy between Definition 5.12 and Definition 2.14 for moments of a frequency distribution. That is, $E[X^r]$ is equivalent to the form $\frac{\Sigma f x^r}{\Sigma f}$ for a frequency distribution.

For instance, for the discrete distribution:

x	0	1	2	3	4
$\Pr(X = x)$	0.10	0.18	0.24	0.37	0.11

we have:

$$\mu_1' = \mathrm{E}[X] = \sum_{\text{all } x} x \Pr(X = x)$$

$$= 0(0.10) + 1(0.18) + \ldots + 4(0.11) = 2.21$$

Similarly:

$$\mu_2' = \mathrm{E}[X^2] = 0^2(0.10) + 1^2(0.18) + \ldots + 4^2(0.11) = 6.23$$

DEFINITION 5.13

Let X be a random variable. The rth moment (about the mean μ) of X is denoted by μ_r and defined $\mu_r = \mathrm{E}[(X - \mu)^r]$.

Note that the 2nd moment about the mean is just the variance, i.e.:

$$\mu_2 = \mathrm{E}[(X - \mu)^r] = \mathrm{Var}[X]$$

We can always express moments about the mean (which are usually awkward to calculate directly from the above definition) in terms of moments about zero (in general, far easier to calculate). For example, we have already noted the 'computational formula' for the variance, namely $\mathrm{Var}[X] = \mathrm{E}[X^2] - \mathrm{E}^2[X]$.   Or, in terms of moments, $\mu_2 = \mu_2' - (\mu_1')^2$.

Going a stage further, we have:

$$\mu_3 = \mathrm{E}[(X - \mu)^3] \quad \text{(by definition)}$$

$$= \mathrm{E}[X^3 - 3\mu X^2 + 3\mu^2 X - \mu^3] \quad \text{(expanding the cube)}$$

$$= \mathrm{E}[X^3] - 3\mu\, \mathrm{E}[X^2] + 3\mu^2\, \mathrm{E}[X] - \mu^3$$

(properties of E with μ as constant)

$$= \mathrm{E}[X^3] - 3\, \mathrm{E}[X]\, \mathrm{E}[X^2] + 3\, \mathrm{E}^3[X] - \mathrm{E}^3[X]$$

$$= \mathrm{E}[X^3] - 3\, \mathrm{E}[X]\, \mathrm{E}[X^2] + 2\, \mathrm{E}^3[X]$$

i.e. $\mu_3 = \mu_3' - 3\mu_1' \mu_2' + 2(\mu_1')^3$

Clearly we can extend this technique to express any general moment about the mean in terms of moments about the origin (zero).

In general, moments can be defined about any constant, c. That is, if X is a random variable, the rth moment (about c) of X would be given by the expression $\mathrm{E}[(X - c)^r]$. However, we are usually interested only in the two particular types of moments mentioned previously, i.e., when:

(a) $c = 0$ (giving moments about zero), and

(b) $c = \mu$ (giving moments about the mean).

We have seen that both the mean and variance of a distribution are just particular types of moments (be they probability or frequency). In general, any constant, descriptive measure of a distribution such as the mean and variance (indeed, any type of moment) is called a *parameter*.

5.5.2 THE MOMENT GENERATING FUNCTION

DEFINITION 5.14

The *moment generating function* of a probability distribution (defined by some random variable X) is written as $M(t)$ and defined by $M(t) = E[e^{tX}]$.

The reason we are interested particularly in this function is as follows:

$$E[e^{tX}] = E\left[1 + tX + \frac{t^2 X^2}{2!} + \frac{t^3 X^3}{3!} + \dots\right] \quad \text{(using the exponential expansion from Definition 5.2)}$$

$$= 1 + t\,E[X] + \frac{t^2}{2!}\,E[X^2] + \frac{t^3}{3!}\,E[X^3] + \dots \quad \text{(properties of E)}$$

That is, the coefficient of $\dfrac{t^r}{r!}$ is just $E[X^r]$, the r th moment about zero.

So that, if we can find an expression for $E[e^{tX}]$ in terms of a series for t , we can inspect the coefficients of $t, \dfrac{t^2}{2!}, \dfrac{t^3}{3!}, \dots$, etc., to obtain μ_1' , $\mu_2', \mu_3', \dots$, etc. From here, of course, we can now calculate $\mu_1, \mu_2,$ $\mu_3, \dots$, if desired, using the previously mentioned relationship between the two types of moment.

However, this technique is not always convenient. A far more practical way of obtaining a particular moment, given a moment generating function, is detailed in the following:

STATEMENT 5.14

If a random variable X has a moment generating function $M(t)$, then:

$$\mu_1' = E[X] = \left[M^{(1)}(t)\right]_{t=0}; \quad \mu_2' = E[X^2] = \left[M^{(2)}(t)\right]_{t=0}$$

and in general: $\mu_r' = E[X^r] = \left[M^{(r)}(t)\right]_{t=0}$.

Note $\left[M^{(r)}(t)\right]_{t=0}$ denotes the r th differential (w.r.t. t) of $M(t)$, evaluated at $t = 0$.

Proof

We prove first that $M^{(r)}(t) = E[X^r e^{tX}]$ [1]

The method of induction will be used.

Assume [1] is true for $r = k$,

i.e. $$M^{(k)}(t) = E[X^k e^{tX}]$$

But $$M^{(k+1)}(t) = \frac{\partial}{\partial t}\{M^{(k)}(t)\}$$

$$= \frac{\partial}{\partial t}\{E[X^k e^{tX}]\}$$

$$= E\left[\frac{\partial}{\partial t}\{X^k e^{tX}\}\right]$$

$$= E[X^{k+1} e^{tX}]$$

which satisfies [1] with $r = k + 1$.

That is, if [1] is true for $r = k$,   then it is true for $r = k + 1$.   [2]

Now $$M^{(1)}(t) = \frac{\partial}{\partial t}\{M(t)\} = \frac{\partial}{\partial t}\{E[e^{tX}]\}$$

$$= E\left[\frac{\partial}{\partial t}\{e^{tX}\}\right]$$

$$= E[X e^{tX}]$$

Therefore [1] is true for $r = 1$.

Hence (using [2]), it is true for $r = 2, 3, 4, \ldots$

We have then, $$M^{(r)}(t) = E[X^r e^{tX}]$$

Therefore: $$M^{(1)}(t) = E[X e^{tX}]$$

and $$[M^{(1)}(t)]_{t=0} = E[X e^{tX}]_{t=0}$$

$$= E[X] = \mu_1'$$

Similarly $$[M^{(2)}(t)]_{t=0} = E[X^2] = \mu_2'$$

and, in general, $$[M^{(r)}(t)]_{t=0} = E[X^r] = \mu_r'.$$

as required.

EXAMPLE 5.44

A continuous random variable X is defined by the p.d.f. $f(x) = a\,e^{-2x}$ for $x > 0$. Find:

(a)   the value of a (b) $M(t)$ and hence   (c) $E[X]$ and $\text{Var}[X]$

SOLUTION

(a)   Since X is a random variable, we must have $\int_{\text{all } x} f(x)\,dx = 1,$

i.e. $\int_0^\infty a\,e^{-2x}\,dx = 1.$

left hand side $= a\int_0^\infty e^{-2x}\,dx$

$$= a\left[-\frac{1}{2}e^{-2x}\right]_0^\infty = a\left(0 - \left(-\frac{1}{2}\cdot 1\right)\right) = \frac{a}{2}$$

Hence $\dfrac{a}{2} = 1$ giving $a = 2.$

So that the p.d.f. can now be written as $f(x) = 2\,e^{-2x}; \quad x > 0.$

(b) $M(t) = E[e^{tX}]$ by definition

$$= \int_0^\infty e^{tx}\,2\,e^{-2x}\,dx$$

$$= 2\int_0^\infty e^{-(2-t)x}\,dx$$

$$= 2\left[-\left(\frac{1}{2-t}\right)e^{-(2-t)x}\right]_0^\infty = \frac{2}{2-t}$$

(c)   Now $M(t) = \dfrac{2}{2-t}.$

$\therefore$

$$M^{(1)}(t) = \frac{(2-t)\dfrac{d}{dt}(2) - 2\dfrac{d}{dt}(2-t)}{(2-t)^2} \qquad \text{using the quotient rule}$$

$$= \frac{2}{(2-t)^2}$$

So:

$$E[X] = \left[M^{(1)}(t)\right]_{t=0} \quad \text{from previous statement}$$

$$= \frac{2}{(2)^2} = \frac{1}{2}$$

Also:

$$M^{(2)}(t) = \frac{d}{dt}(M^{(1)}(t)) = \frac{d}{dt}\left(\frac{2}{(2-t)^2}\right)$$

$$= \frac{(2-t)^2\dfrac{d}{dt}(2) - 2\dfrac{d}{dt}((2-t)^2)}{(2-t)^4}$$

$$= \frac{2\cdot 2\cdot(2-t)}{(2-t)^4} = \frac{4}{(2-t)^3}$$

$\therefore$

$$E[X^2] = \left[M^{(2)}(t)\right]_{t=0} = \frac{4}{2^3} = \frac{1}{2}$$

$\therefore$

$$\begin{aligned} \mathrm{Var}[X] &= E[X^2] - E^2[X] \\ &= \frac{1}{2} - \left(\frac{1}{2}\right)^2 \\ &= \frac{1}{4} \end{aligned}$$

We have already discussed (in Chapter 2) how the mean and the variance tend to 'locate' and 'confine' a distribution in particular ways. The mean fixes a point about which all other values fall according to some defined rule; the variance gives a measure of how these values spread out about the mean.

However, even though two distributions have the same mean and variance, this does not mean that they are identical. For instance, one distribution could be more skewed or peaked than the other. In fact, to prove that two distributions are identical, we need to show that each and every one of their corresponding moments are identical. But since the moments of any distribution are completely defined by a moment generating function, we have the following important result in the theory of distributions.

STATEMENT 5.15

Two random variables have identical distributions if and only if their moment generating functions are identical.

5.5.3 PROBABILITY GENERATING FUNCTIONS

A useful and important function connected with certain discrete distributions is defined in the following definition.

DEFINITION 5.15

The *probability generating function* of a discrete probability distribution defined by a random variable X, say, which can take only the values $0, 1, 2, 3, 4, \ldots$, is written as $G(t)$ and defined by $G(t) = E[t^X]$.

Note $G(t)$ is only defined for a particular type of discrete random variable, one that assumes only the values $0, 1, 2, 3, 4, \ldots$, etc.

The reason we are particularly interested in this function is as follows.

$$G(t) = E[t^X] = \sum_{\text{all } x} t^x \Pr(X = x)$$

$$= t^0 \Pr(X = 0) + t^1 \Pr(X = 1)$$
$$+ t^2 \Pr(X = 2) + \ldots + t^r \Pr(X = r) + \ldots$$

That is, the coefficient of t^r is $\Pr(X = r)$.

The probability generating function can also be used to obtain the mean and variance of the defined distribution in the following way:

STATEMENT 5.16

Let X be a discrete random variable which can assume only the values $0, 1, 2, 3, 4, \ldots$, etc., and let $G(t)$ be the probability generating function of the distribution defined.

Then:

$$\text{(a)} \quad [G^{(1)}(t)]_{t=1} = E[X]$$

and

$$\text{(b)} \quad [G^{(2)}(t)]_{t=1} = E[X^2] - E[X]$$

Proof   (a)   Putting $p_r = \Pr(X = r)$ for convenience, we have:

$$G(t) = p_0 + p_1 t + p_2 t^2 + p_3 t^3 + p_4 t^4 + \ldots \quad \text{(by definition)}$$

Differentiating both sides of the above w.r.t. t gives:

$$G^{(1)}(t) = \frac{d}{dt}(G(t)) = p_1 + 2p_2 t + 3p_3 t^2 + 4p_4 t^3 + \ldots$$

Substituting $t = 1$ into the above expression, we obtain:

$$[G^{(1)}(t)]_{t=1} = p_1 + 2p_2 + 3p_3 + 4p_4 + \ldots$$
$$= 1 \Pr(X = 1) + 2 \Pr(X = 2) + 3 \Pr(X = 3) + 4 \Pr(X = 4) + \ldots$$
$$= \sum_{\text{all } x} x \Pr(X = x)$$
$$= E[X] \quad \text{(by definition)}$$

The proof of (b) is left as an exercise. As a consequence of the above statement, it is easy to prove that

$$\text{Var}[X] = [G^{(2)}(t) + G^{(1)}(t) - (G^{(1)}(t))^2]_{t=1}$$

The proof of this is also left as an exercise.

EXAMPLE 5.45

A discrete probability distribution is defined by a random variable X having a p.d.f. $\Pr(X = x) = b(1-b)^x$; $x = 0, 1, 2, 3, \ldots$ where b is a constant.

(a)   Find the probability generating function (p.g.f.) for this distribution.

(b)   Use the above p.g.f. to obtain $E[X]$.

SOLUTION

(a)   The probability generating function for the distribution is given by:

$$G(t) = E(t^X) = \sum_{\text{all } x} t^x \Pr(X = x)$$

$$= \sum_{x=0}^{\infty} t^x b(1-b)^x$$

$$= b + bt(1-b) + bt^2(1-b)^2 + bt^3(1-b)^3 + \ldots$$

which is easily seen to be a geometric progression with a (the starting term) as b and r (the common ratio) as $t(1-b)$. The sum to infinity of this series is $\dfrac{a}{1-r} = \dfrac{b}{1-t(1-b)}$.

That is, $G(t) = \dfrac{b}{1-t(1-b)}$.

(b)   Now:

$$G^{(1)}(t) = \frac{d}{dt}(G(t)) = \frac{(1-t(1-b))\dfrac{d}{dt}(b) - b\dfrac{d}{dt}(1-t(1-b))}{(1-t(1-b))^2}$$

(using the quotient rule)

$$= \frac{b(1-b)}{(1-t(1-b))^2}$$

Hence:

$$E[X] = \left[G^{(1)}(t)\right]_{t=1} = \frac{b(1-b)}{(1-(1-b))^2} = \frac{b(1-b)}{b^2}$$

i.e.

$$E[X] = \frac{1-b}{b}$$

5.6 EXERCISES

SECTION 5.2

1.   Differentiate the following w.r.t. x, using either the product or quotient rule: (a) $(x+1)(x+2)$;   (b) $(2-x)(9-6x+x^2)$;   (c) $\dfrac{x}{x+1}$;
   (d) $\dfrac{x^3+5}{4x^3-3}$;   (e) $\dfrac{x^2}{(x-1)^2}$.

2. Differentiate $(3t - 2)^3$ w.r.t. t by putting $y = x^3$ and $x = 3t - 2$, then using the function of a function rule for $\dfrac{dy}{dt}$.

3. Find $\dfrac{dy}{dx}$ if: (a) $y = (2x + 4)^2$; (b) $y = (4x - 3)^4$; (c) $y = (3x^2 - 4)^3$;

   (d) $y = \dfrac{1}{(3x - 7)^3}$.

4. (a) Differentiate $(2x - 4)^3$ using the function of a function rule. Hence,
   (b) differentiate $(2x - 4)^3(4x^2 - 2)$ (using the product rule).

5. Find $\dfrac{dy}{dt}$ if $y = \dfrac{4t}{(2 - 2t)}$.

6. Differentiate $(3 + x^2)^3(2x + 4)$.

7. Integrate the following functions w.r.t. x:

   (a) $3x^3$ (b) $-5x^4$ (c) $\dfrac{3}{x^5}$ (d) $4x^{-4}$ (e) $(3.3)x^{-(4.3)}$

8. Evaluate: (a) $\displaystyle\int (2x^2 - 3x)\, dx$; (b) $\displaystyle\int \left(\dfrac{2}{x^3} + x - 4x^2\right) dx$.

9. Integrate $x^2(3 - 2x)$ w.r.t. x.

10. Evaluate: (a) $\displaystyle\int_0^1 2x^2\, dx$; (b) $\displaystyle\int_0^2 (x - 4)\, dx$; (c) $\displaystyle\int_1^2 (x^3 + x - 2)\, dx$;

    (d) $\displaystyle\int_{-a}^a 2x^3\, dx$.

11. Find the value of k if $\displaystyle\int_0^3 \left(\dfrac{x}{6} + k\right) dx = 1$.

12. Find the following areas (in square units):
    (a) Under the curve $y = x^2 - 7x + 6$ between the lines $x = 1$, $x = 6$.
    (b) Under the curve $y = 5x - x^2$ between the lines $x = 0$, $x = 5$.
    (c) Under the curve $y = x^3 - 3x^2 + 2x$ between the lines $x = 0$, $x = 1$.

13. Differentiate w.r.t. x: (a) e^{2x}; (b) e^{5x+2}; (c) $3e^{2x^2-4x}$; (d) e^{a-bx};

    (e) $\dfrac{1}{3e^{4x}}$.

14. Differentiate w.r.t. x: (a) $x\,e^x$;   (b) $x^2\,e^{-x}$;   (c) $x^3\,e^{-2x}$;

    (d) $(x^2 + 2)\,e^{2x^2}$; (e) $\dfrac{e^{2x}}{x}$; (f) $e^{3(x^2-2)^2}$.

SECTION 5.3

15. A bag contains five identical balls numbered 1, 2, 3, 4 and 5 respectively. An experiment is defined by drawing a single ball at random from the bag. Let X be the random variable "the number of the ball drawn". Show that X is a random variable.

16.   The probability that a variable Y takes the value y is given by $\dfrac{1}{3}\left(\dfrac{2}{3}\right)^{y}$ for $y = 0, 1, 2, 3, \ldots$ Show that Y is a discrete random variable.

17.   The probability density function (p.d.f.) of a discrete random variable X is given by $\Pr(X = x) = \dfrac{ax}{9}$ (for $x = 1, 2$ and 3). Find the value of a.

18.   A discrete random variable X has a p.d.f. of $\Pr(X = x) = \dfrac{A}{(x + 3)}$ for $x = 0, 1, 2,$ and 3.

       (a) Find the value of A;   (b) If E_1 is the event '$X = 1$', E_2 is the event '$X = 2$' and E_3 the event '$X = 3$', find: (i) $\Pr(E_1)$; (ii) $\Pr(E_2)$; (iii) $\Pr(E_3 \text{ or } E_1)$;   (iv) $\Pr(\overline{E_1})$.

19.   State which of the following functions could not represent the p.d.f. of a random variable, X say, where $x = 0, 1$ or 2. Give reasons for your answers.

       (a) $\dfrac{x^2 + 1}{8}$ (b) $\dfrac{2x - 1}{3}$ (c) $\dfrac{1}{2(1 + x)}$ (d) $\dfrac{15}{23(1 + 2x)}$

20.   A continuous variable X has a p.d.f. $f(x) = \dfrac{x}{4}$ (for $1 \leqslant x \leqslant 3$).

       (a) Show that X is a random variable.   (b) Find the probability that, if a single value of X is sampled, it will lie between 1 and 2.

21.   A continuous random variable Y has a p.d.f. of the form $f(y) = A(y^2 + 4)$ (for $0 \leqslant y \leqslant 1$).   (a) Calculate the value of A. (b) Find $\Pr(\tfrac{1}{2} \leqslant y \leqslant 1)$ to 2D.

22.   If the continuous random variable X has the p.d.f. $f(x) = \dfrac{x}{6} + C$ for $0 \leqslant x \leqslant 3$,   (a) evaluate C; and   (b) find $\Pr(1 \leqslant X \leqslant 2)$.

23.   A continuous variable X has the p.d.f. $f(x) = \tfrac{1}{2}$; $(2 \leqslant x \leqslant 4)$. (a) Show that X is a random variable.   (b) Find $\Pr(2 \leqslant X \leqslant 2.5)$.

24.   Draw (probability) histograms for the discrete probability distributions defined by the following p.d.f.s:

       (a) $\Pr(X = x) = \dfrac{1}{12} + \dfrac{x}{4}$ (for $x = 0, 1$ and 2).

       (b) $\Pr(Y = y) = \dfrac{1}{25}(x^2 - 4x + 5)$ (for $x = 0, 1, 2, 3, 4$ and 5).

25.   A pair of fair dice are thrown. If X is defined as "the minimum of the two numbers that appear", tabulate the p.d.f. of the random variable X.

26.   A continuous random variable X has a p.d.f. of the form $f(x) = kx$ for $2 \leqslant x \leqslant a$. Find the value of a (in terms of k).

27.   Continuous random variables X and Y have identical distributions, given by the p.d.f. $f(z) = \dfrac{z}{4}$; $1 \leqslant z \leqslant 3$. If X and Y are independent, find: (a) $\Pr(1.5 \leqslant X \leqslant 2)$;   (b) $\Pr(1.5 \leqslant Y \leqslant 2)$;   (c) $\Pr(1.5 \leqslant X \leqslant 2$ and $1.5 \leqslant Y \leqslant 2)$;   (d) $\Pr(at\ least\ one$ of X and Y lies between 1 and 2)

28.   Two dice are rolled and random variable X is defined as "the sum of the numbers shown". Tabulate: (a) the density function for X; (b) the distribution function for X.   (c) Denoting the distribution function by F, calculate and explain the meaning of $F(9)$.

SECTION 5.4

29.   Find $E[X]$ for the following distribution:

x	0	1	2	3	4	5
$Pr(X = x)$	0.01	0.08	0.23	0.36	0.21	0.11

30.   Use the symmetry of the following distributions to write down $E[X]$.

(a)

x	2	3	4	5	6
$Pr(X = x)$	0.20	0.25	0.10	0.25	0.20

(b)

x	0	1	2	3	4	5
$Pr(X = x)$	0.04	0.13	0.33	0.33	0.13	0.04

(c)

x	1	4	7	10	13	16
$Pr(X = x)$	a	b	c	c	b	a

31.   Calculate $E[X]$ for the continuous distribution defined by the p.d.f. $f(x) = 1 - \dfrac{x}{2}$ (for $0 \leqslant x \leqslant 2$).

32.   A discrete random variable X can take only the values 2 and 3. If $E[X] = \dfrac{11}{4}$, find the two probabilities $Pr(X = 2)$ and $Pr(X = 3)$.

33.   A discrete random variable X has a p.d.f. of the form

x	x_1	x_2	x_3	...	x_n
$Pr(X = x)$	p_1	p_2	p_3	...	p_n

where n is odd and   (i) $x_2 = x_1 + a$; $x_3 = x_1 + 2a$; $x_4 = x_1 + 3a$ ... etc. (ii) $p_1 = p_n$; $p_2 = p_{n-1}$; ... etc.
(a) Prove that $E[X] = x_{(n+1)/2}$ (*Hint:* $\Sigma p_i = 1$, since X is a r.v.)
(b) Deduce that the median value is also $x_{(n+1)/2}$.

34.   If X is a continuous random variable with a p.d.f. $f(x) = C$ (for $a \leqslant x \leqslant b$):   (a) Evaluate C; (b) Find $E[X]$.

35. A throws a fair six-sided die with faces marked $-10, 2, 3, 4, 5$ and a respectively and 'wins' the value shown on the die. (Note that he loses 10 if '-10' comes up). B throws a six-sided die with faces marked $-3, 2, 3, 4, 6$ and 8 respectively, (winning or losing depending on which face comes up), where the die is 'loaded' so that '8' comes up twice as often as any of the others. Find the value of a so that the expected gain of each of them $(A$ and $B)$ is the same.

36.   A cubical die is biased in such a way that the probability of scoring n ($n = 1$ to 6) is proportional to n. Determine the mean value and vari-

ance of the score obtained in a single throw. What would be the mean
and variance if the score showing were doubled?   (O & C)

37.   For the following discrete distribution:

x	1	3	5	8	9
$\Pr(X = x)$	0.1	0.4	0.3	0.15	0.05

find: (a) $E[X]$;   (b) $E[X-1]$;   (c) $E[3(X-1)]$.

38.   For the discrete distribution defined by the p.d.f.

x	0	1	2	3	4
$\Pr(X = x)$	$\frac{1}{16}$	$\frac{3}{16}$	$\frac{7}{16}$	$\frac{3}{16}$	$\frac{2}{16}$

find: (a) $E[X]$;   (b) $E[X-1]$;   (c) $E[X^2]$.

39.   Find: (a) $E[X]$;   (b) $E[X+1]$;   and   (c) $E[3X+3]$ for the continuous random variable X defined by the p.d.f. $f(x) = \dfrac{3x(2-x)}{4}$ $(0 \leqslant x \leqslant 2)$.

40.   If $f_1(x)$ and $f_2(x)$ are any two functions of a random variable X, prove that $E[f_1(X) + f_2(X)] = E[f_1(X)] + E[f_2(X)]$.

41.   For the discrete random variable defined by the p.d.f.

x	-3	2	4
$\Pr(X = x)$	0.4	0.3	0.3

find $E[f(X)]$ if $f(x) = 2(x-1)^2 + 3(x-1) - 5$.

42.   Find $E[2X^2 - 3X + 4]$ for the discrete probability distribution

x	2	3	4	5	6
$\Pr(X = x)$	0.20	0.25	0.10	0.25	0.20.

43.   A random variable has for its probability density function, $\phi(x)$, where:

$\phi(x) = 4(x-1)(2-x)(3-x)$ if $1 \leqslant x \leqslant 2$

$= 0$ otherwise

Find the mean and variance of X. What is the probability that a value of X taken at random will exceed 1.8?

44.   A random variable X has the p.d.f. $f(x) = \frac{1}{4}$ $(1 \leqslant x \leqslant 5)$. Calculate the median and the other two quartiles of this distribution.

45.   A discrete random variable X has a p.d.f. as follows:

x	0	1	2	a	6
$\Pr(X = x)$	0.1	0.1	0.3	0.2	0.3

Evaluate a in the two cases: (a) $E[X] = 3$;   (b) $E[X^2] = 18$.

46.   A retail ice cream salesman divides his working days into just 'sunny', 'medium' and 'cold'. He estimates that the probability of it being sunny is 0.2 and cold 0.3. If his takings on these three types of day are £50, £35 and £10 respectively on average and his average daily costs are £18 on any day, find his expected profit per day.

47.   An unbiased coin is tossed continually until a 'head' or five 'tails' occurs. Find the expected number of tosses.

48.   Two dice are rolled. Let X be the random variable "difference between the numbers shown on the two faces". Find the expectation and variance of X to 1D.

49.   (a) A gambler pays £1 to draw six cards, one from each of six ordinary packs. If at least four of the cards are spades, he receives £26. Otherwise he receives nothing. How much can he expect to lose in 100 tries?

        (b) The game is now changed, so that the gambler draws six cards from one pack, without replacement, and receives £30 if there are four or more spades among the six drawn. What is a fair price for him to pay?

                                       (AEB)'76

50.   Explain how a shift of origin and a change of scale was used, or could have been used, to simplify the calculation of mean and variance in one of your projects. Comment on the suggestion that, with the advent of computers, short cut calculations are out of date and no longer needed.

        The discrete random variable X has probability function $\Pr(X)$ defined as

$$\Pr(0) = \Pr(10) = \tfrac{1}{9} \qquad \Pr(1) = \Pr(6) = \tfrac{2}{9} \qquad \Pr(3) = \tfrac{1}{3} \qquad \Pr(X) = 0$$
elsewhere.

        Draw a sketch to illustrate this probability distribution and find $E[X]$ and $\mathrm{Var}[X]$.

        Find also the mean and variance of Y when $Y = 3X - 5$.   (London)

51.   Two dice are tossed and random variable X is defined as "the total shown on the two faces". Find: (a) $E[X]$;   (b) $\mathrm{Var}[X]$.

52.   For the discrete distribution:

x	0	1	2	3
$\Pr(X = x)$	0.209 72	0.367 00	0.275 25	0.114 69

x	4	5	6	7
$\Pr(X = x)$	0.028 67	0.004 30	0.000 36	0.000 01

        Calculate: (a) the 10th and 90th percentiles (P_{10} and P_{90}) using a method of interpolation; and   (b) the 10 to 90 percentile range.

53.   Find: (a) the 1st and 3rd quartiles; and   (b) the 10 and 90 percentile range of the continuous distribution defined by $f(x) = \dfrac{x}{4}$ $(1 \leqslant x \leqslant 3)$.

SECTION 5.5

54.   Find the probability generating function, $G(t)$, of the distribution defined by the p.d.f. $\Pr(X = x) = \begin{cases} p; & x = 1 \\ 1 - p; & x = 0 \end{cases}$ and use this to find $E[X]$.

MISCELLANEOUS

55. An unbiased coin is tossed exactly 4 times. Let the random variable X be 'the longest run of heads' (i.e., if HTHH occurs, then $X = 2$. If THTH occurs, then $X = 1$.) Tabulate the probability distribution of X and find: (a) $E[X]$;   (b) $Var[X]$.

(*Note*: $Pr(X = 0)$ is *not* zero.)

56. The continuous variable X has probability density (frequency) function $f(x)$, where

$$f(x) = 0 \qquad\qquad\quad x < 0$$
$$f(x) = kx^2(3 - x) \quad 0 \leqslant x \leqslant 3$$
$$f(x) = 0 \qquad\qquad\quad x > 3$$

Calculate: (a) the value of the constant k; (b) the mean (μ) and variance (σ^2) of X and verify that $\mu + 2\sigma = 3$; (c) the probability that X differs from its mean by more than 2σ.   (Cambridge)

57. A continuous probability distribution is defined by the p.d.f. $f(x) = 6x(1 - x)$ for $0 \leqslant x \leqslant 1$. If the random variable associated with this distribution is denoted by X: (a) Confirm that X is a random variable; (b) Calculate $E[X]$ and $Var[X]$; (c) If the median of X is a, find an equation that a satisfies.

58. Two bags each contain 3 counters numbered from 1 to 3. A counter is drawn randomly from each bag. Let X be the random variable 'the average of the two numbers' and Y be the random variable 'the largest of the two numbers'. Find: (a) $E[X]$; (b) $E[Y]$; and hence (c) the expectation of the mean of X and Y. See Example 5.38. Are X and Y independent random variables?

59. A gambler plays a game by rolling two dice and collecting a 'prize' of money only if two odd numbers result. In the case of two '5's, he collects £12; if only one '5', he collects £4; any other combination of odd numbers earns him £2. If there is a standing charge for each throw, how much should this charge be so that his *expected loss per throw* is £2.

60. (a) If a is a constant, show that $E[a] = a$ for both discrete and continuous distributions. (b) Show that $Var[X] = E[X^2] - E^2[X]$ for any random variable X for both discrete and continuous distributions.

(c) If X has p.d.f.

x	0	1	2
$Pr(X = x)$	$\frac{1}{2}$	$\frac{3}{8}$	$\frac{1}{8}$

find: (i) $E[X]$; (ii) $E[X^2]$; and hence (iii) $Var[X]$.

61. A random variable X has a p.d.f. given by:

$$f(x) = \begin{cases} 0.1; & 0 \leqslant x \leqslant 10 \\ 0; & \text{all other } x \end{cases}$$

(a) Calculate $E[X]$ and $Var[X]$. (b) A second (independent) random variable Y has an identical p.d.f. to that of X. Find $E[X + Y]$ and $Var[X + Y]$.

62. A random variable X has a p.d.f.

x	0	1
$Pr(X = x)$	q	p

Show that:

(a) $E[X] = p$; (b) $Var[X] = p(1 - p)$; (c) $E[X^r] = p$ for all integral r; (d) $E[(X - E[X])^3] = p(1 - p)(1 - 2p)$.

63.   A continuous random variable X has a p.d.f.:

$$f(x) = \begin{cases} x(x-1)(x-2) & 0 \leqslant x \leqslant 1 \\ a & 1 < x \leqslant 3 \\ 0 & \text{elsewhere} \quad \text{(where } a \text{ is a constant)} \end{cases}$$

(a) Determine a and calculate $E[X]$.

(b) Evaluate $\Pr(X \leqslant E[X])$.

(*Note*: For integration purposes, the two intervals $(0, 1)$ and $(1, 3)$ must be considered separately).

64.   Express $\text{Var}[Y]$ in terms of $E[Y]$ and $E[Y^2]$. Hence, putting $Y = aX - b$, prove that $\text{Var}[aX - b] = a^2 \cdot \text{Var}[x]$.

65.   A committee of five is to be chosen from 6 men and 3 women. Let X be the random variable 'number of women on the committee'.

(a) Show that $\Pr(X = r) = \binom{3}{r}\binom{6}{5-r} \Big/ \binom{9}{5}$ for $r = 0, 1, 2$ and 3 (see Example 4.33). A discrete random variable having this form of probability density function is said to have a *Hypergeometric distribution*.

(b) Deduce that $\displaystyle\sum_{r=1}^{3} \binom{3}{r}\binom{6}{5-r} = \binom{9}{5}$.

By generalising this situation, show that:

$$\sum_{r=0}^{n} \binom{W}{r}\binom{M}{n-r} = \binom{W+M}{n} \quad \text{where } n < W + M$$

66.   Using the series expansion of e^X, show that:

$$E[e^X] = \sum_{r=0}^{\infty} \left(\frac{1}{r!} E[X^r]\right)$$

67.   A variable X has a p.d.f. of $\Pr(X = x) = \dfrac{1}{e\,x!}$ for $x = 0, 1, 2, \ldots$.

Show that X is a random variable.

68.   Let X be a continuous random variable with a p.d.f. of $f(x)$. Show that $F(b) - F(a) = \Pr(a \leqslant X \leqslant b)$ where $F(t) = \Pr(X \leqslant t)$, the distribution function.

69.   Show that for any probability distribution:

$$\mu_4 = \mu_4' - 4\mu_3'\mu_1' + 6\mu_2'(\mu_1')^2 - 3(\mu_1')^4$$

(see Section 5.5.1) and write μ_5 in terms of moments about zero.

(*Hint*: Given that $(x-y)^3 = x^3 - 3x^2y + 3xy^2 - y^3$, $(x-y)^4$ can be expanded easily using $(x-y)(x-y)^3$.)

70.   (a) A random variable X can assume any one of the values $(0, 1, 2, 3, \ldots)$ with probability $\Pr(X = x) = \dfrac{c}{x!}$. Show that: (i) $c = \dfrac{1}{e}$; and   (ii) $E[X] = 1$.

(b) A random variable Y has an expectation and variance of m and v respectively. If a new random variable Z is now defined as

$Z = \dfrac{Y - m}{\sqrt{v}}$, show that $E[Z] = 0$ and $\mathrm{Var}[Z] = 1$.

(in this type of situation, Z is usually called a *standardised* random variable.)

71. A random variable X can take the values 0, 1, 2 and 3 only with probability function $\Pr(X = x) = \dfrac{c}{2 + x}$. Determine: (a) the value of c; (b) $\Pr(X = 2 \text{ or } X = 3)$; (c) $E[X]$.

72. A random variable X has the p.d.f.:

$$\Pr(X = x) = \begin{cases} c(4 - 2x) & 0 \leqslant x \leqslant 2 \\ 0 & \text{elsewhere} \end{cases}$$

(a) Find c; (b) Sketch the distribution; (c) Find the probability of the events: (i) $E_1 = \text{'} X \leqslant 1\text{'}$; and (b) $E_2 = \text{'} \tfrac{1}{2} \leqslant X \leqslant \tfrac{7}{8} \text{'}$.

73. The continuous variable X has probability density (frequency) function $f(x)$, where

$$\begin{aligned} f(x) &= 0 & x &< 2 \\ f(x) &= k(3 - x) & 2 &\leqslant x \leqslant 3 \\ f(x) &= 0 & x &> 3 \end{aligned}$$

Calculate: (a) the value of the constant k; (b) the median of X; (c) the probability that X exceeds 2.5; (d) the semi-interquartile range of X.   (Cambridge)

74. A random variable X takes values x such that $-1 \leqslant x \leqslant 1$ and its probability density function $f(x)$ is given by:

$$\begin{aligned} f(x) &= k(x^2 + 3x + 2) & \text{when } &-1 \leqslant x \leqslant 0 \\ f(x) &= k(x^2 - 3x + 2) & \text{when } &\ \ 0 \leqslant x \leqslant 1 \\ f(x) &= 0 & \text{otherwise} & \end{aligned}$$

where k is a positive constant.

(a) Sketch a graph of $f(x)$.

(b) Show by integration that $k = \tfrac{3}{5}$.

(c) Find the probability that X takes a value between $-\tfrac{1}{2}$ and $+\tfrac{1}{2}$.   (London)

75. The random variable X has a probability density function given by:

$$\Pr(x) = \begin{cases} kx(1 - x^2) & (0 \leqslant x \leqslant 1) \\ 0 & \text{elsewhere} \end{cases}$$

k being a constant. Find the value of k and find also the mean and variance of this distribution.

Find the median of the distribution.   (O & C)

76. A random variable X takes values x such that its probability density function is given by $f(x) = kx(10 - x)$, where k is a constant and $f(x) \geqslant 0$. Show that $k = 0.006$.

Calculate: (a) the probability that X takes a value between 0 and 3; (b) the expected value of X.   (London)

77. In a discrete probability distribution, the variable X takes the value $120/k$ with probability $k/45$, where k takes all positive integral values from 1 to n inclusive.

(a) Verify that $n = 9$.

(b) Calculate the probability that X takes the value 20.

(c) Find the probability that the value of X lies between 14 and 41.

(d) Calculate the expected value of X. (London)

78. A continuous variable x is distributed at random between the values 2 and 3 and has a probability density function of $\dfrac{6}{x^2}$. Find the median value of x. (SUJB)

79. A random variable X takes values X such that $-2 \leqslant X \leqslant 2$ and the probability density function $f(X)$ for X is given by:

$$f(X) = k(4 - X^2) \quad \text{when} \quad -2 \leqslant X \leqslant 2$$

$$f(X) = 0 \qquad\qquad \text{otherwise}$$

Sketch the graph of $f(X)$ and show that $k = 3/32$.

Explain why the expected value of X is 0 and calculate the standard deviation of this probability distribution. (London)

80. (a) Show that $\displaystyle\sum_{i=1}^{n} (x_i - m)^2$ can be written as:

$$\sum_{i=1}^{n} x_i^2 - 2m \sum_{i=1}^{n} x_i + nm^2$$

(b) Assuming that $\dfrac{\partial}{\partial a}\left(\displaystyle\sum_i f_i(a, b)\right) = \displaystyle\sum_i \left(\dfrac{\partial}{\partial a}(f_i(a, b))\right)$, where $f_i(a, b)$ is *any* function involving a and b, show that:

$$\dfrac{\partial}{\partial m}\left(\sum_{i=1}^{n} (x_i - m)^2\right) = -2 \sum_{i=1}^{n} x_i + 2nm.$$

(c) Hence show that the minimum value of the quantity $\displaystyle\sum_{i=1}^{n} (x_i - m)^2$ occurs when $m = \dfrac{1}{n} \displaystyle\sum_{i=1}^{n} x_i = \bar{x}$. That is, the sum of squares of deviations of a set of values from a constant is minimum when the constant is *the mean of the values.*

81. Show that X^r is obtained by successively differentiating the expression e^{tX} w.r.t. t, r times. (Remember that any expression involving X alone is a constant w.r.t. t.) Deduce that $M^{(r)}(t)|_{t=0} = \mathrm{E}[X^r] = \mu_r'$.

82. Let X be a discrete random variable assuming only the values $x = 0$, $1, 2, \ldots$ etc. with p_r denoting $\Pr(X = r)$.

Assuming that the probability generating function $G(t)$ can be expressed as $G(t) = p_0 + p_1 t + p_2 t^2 + \ldots + p_r t^r + \ldots$, show that $G''(t)|_{t=1} = \mathrm{E}[X^2] - \mathrm{E}[X]$.

By writing $\mathrm{Var}[X]$ as $\mathrm{E}[X^2] - \mathrm{E}^2[X]$ and $\mathrm{E}[X]$ as $G'(t)|_{t=1}$, show that $\mathrm{Var}[X] = (G''(t) + G'(t) - (G'(t))^2)|_{t=1}$.

Chapter 6

Some Special Probability Distributions

6.1 INTRODUCTION

In Statistics, there are certain special types of distributions that occur often enough in practical situations to deserve special attention theoretically. The most important of these are the binomial and Poisson (discrete) and the Normal (continuous) distributions. Some others are also studied briefly.

The attributes that characterise a distribution and give it a special importance are not always the same. For instance, the binomial distribution (Section 6.3) is derived from a particular type of experimental situation and, depending on the nature of the experiment, the actual form of the p.d.f. (and hence the probability curve) will change. However, a Normal distribution (Section 6.5) can be said to be characterised by the particular symmetric shape of its probability curve, the experimental situation being (relatively) irrelevant.

However, all these distributions have in common a flexibility in the exact form of their p.d.f.'s obtained by involving one, or more, special constants, called *parameters*. Upon taking particular values, the parameters fix a particular form to the distribution.

For example, consider the continuous p.d.f. $f(x) = kx$ for $0 \leqslant x \leqslant \sqrt{\dfrac{2}{k}}$.

For different values of k, we get different forms of the p.d.f. and probability curve. k, here, is a parameter.

When $k = 1$, we have $f(x) = x$ for $0 \leqslant x \leqslant \sqrt{2}$.

When $k = 2$, we have $f(x) = 2x$ for $0 \leqslant x \leqslant 1$ and so on.

6.2 RELATED MATHEMATICS

6.2.1 THE BINOMIAL THEOREM

This sub-section gives us a method of evaluating expressions such as $(x + y)^2$, $(x + y)^3$, $(x + y)^{10}$, etc., without needing to multiply the brackets together, and also some associated ideas for later use in the chapter.

STATEMENT 6.1   (THE BINOMIAL THEOREM)

If x and y are any quantities and n is a positive integer, then we can *expand* $(y + x)^n$ in the form:

$$(y + x)^n = \binom{n}{0} x^0 y^n + \binom{n}{1} x^1 y^{n-1} + \binom{n}{2} x^2 y^{n-2} + \ldots$$

$$\binom{n}{n-1} x^{n-1} y^1 + \binom{n}{n} x^n y^0$$

$$= \sum_{r=0}^{n} \binom{n}{r} x^r y^{n-r}$$

The right-hand side is known as the *binomial expansion of* $(x + y)^n$

Note $\binom{n}{r} = \dfrac{n!}{(n-r)!\, r!}$, the combinatorial function.

We can demonstrate this result easily for $n = 2$ and 3.

For $n = 2$, the theorem gives:

$$(y + x)^2 = \binom{2}{0} x^0 y^2 + \binom{2}{1} x^1 y^1 + \binom{2}{2} x^2 y^0$$

$$= y^2 + 2xy + x^2$$

which can be verified by multiplying out the brackets thus:

$$(y + x)^2 = (y + x)(y + x) = y^2 + 2xy + x^2$$

For $n = 3$, the theorem gives:

$$(y + x)^3 = \binom{3}{0} x^0 y^3 + \binom{3}{1} x^1 y^2 + \binom{3}{2} x^2 y^1 + \binom{3}{3} x^3 y^0$$

$$= y^3 + 3y^2 x + 3yx^2 + x^3$$

which is again verified by evaluating $(y + x)(y + x)(y + x)$.

Statistically, we are interested in the particular form of the theorem given by $y = 1 - p$ and $x = p$,

i.e.: $$(1 - p + p)^n = \sum_{r=0}^{n} \binom{n}{r} p^r (1 - p)^{n-r}$$

$$= 1 \quad (\text{since } (1 - p + p)^n = 1^n = 1)$$

There are numerous applications and identities concerned with this theorem but we limit ourselves to only a few of these that are directly applicable to the scope of this book.

Pascal's Triangle   The coefficients of the binomial expansion (i.e., the numbers pre-multiplying the x, y terms) form an interesting and useful pattern when looked at in isolation.

We will expand $(x + y)^n$ for a few values of n beginning at $n = 0$.

$$(x + y)^0 = \text{①}$$
$$(x + y)^1 = \text{①}x + \text{①}y$$
$$(x + y)^2 = \text{①}x^2 + \text{②}xy + \text{①}y^2$$
$$(x + y)^3 = \text{①}x^3 + \text{③}x^2y + \text{③}xy^2 + \text{①}y^3$$
$$(x + y)^4 = \text{①}x^4 + \text{④}x^3y + \text{⑥}x^2y^2 + \text{④}xy^3 + \text{①}y^4$$

Writing the above (ringed) coefficients in the form of a triangle gives the pattern shown in Fig. 6.1, and notice that any two adjacent values added together gives the value immediately below it in the following row. This particular characteristic of the number triangle enables us immediately to write down the next row of the triangle, and the one after that, and so on.

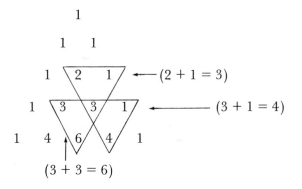

Fig. 6.1

Noting that each row always begins and ends with a 1, we have the next row as: $1, 5(= 1 + 4), 10(= 4 + 6), 10(= 6 + 4), 5(= 4 + 1)$ and 1. The next row is 1, 6, 15, 20, 15, 6 and 1. We can, of course extend this process indefinitely. Notice, in particular, that each row of coefficients forms a symmetric pattern, so that, for instance,

$$(x + y)^3 = x^3 + 3x^2y + 3xy^2 + y^3$$
$$= y^3 + 3y^2x + 3yx^2 + x^3$$
$$= (y + x)^3$$

as we would expect.

The ease with which we can generate binomial coefficients using Pascal's Triangle enables us to write down binomial expansions fairly rapidly. That is, the row 1, 5, 10, 10, 5 and 1 represents the coefficients in the expansion of $(x + y)^5$, i.e.: $(x + y)^5 = x^5 + 5x^4 y + 10x^3 y^2 + 10x^2 y^3 + 5xy^4 + y^5$

or, alternatively:

$$(y + x)^5 = y^5 + 5y^4 x + 10y^3 x^2 + 10y^2 x^3 + 5yx^4 + x^5$$

For a value of n as large as 20, say, Pascal's Triangle would grow rather large and in this case we could revert to the coefficients given in the theorem, namely $\binom{20}{0}$, $\binom{20}{1}$, ..., $\binom{20}{20}$.

By virtue of the fact that the numbers in Pascal's Triangle are just the binomial coefficients given by the combinatorial function $\binom{n}{r}$, we can write Pascal's Triangle as in Fig. 6.2.

Here we have the identities $\binom{2}{0} + \binom{2}{1} = \binom{3}{1}$; $\binom{2}{1} + \binom{2}{2} = \binom{3}{2}$ as before, and extending this we have in general $\binom{n}{r} + \binom{n}{r+1} = \binom{n+1}{r+1}$ a result quoted in a previous exercise.

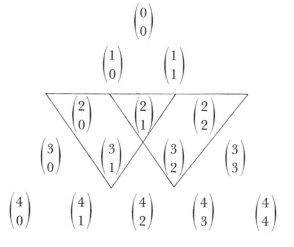

Fig. 6.2

Greatest term in a binomial expansion   When $(x + y)^n$ is expanded binomially as:

$$\binom{n}{0} y^n + \binom{n}{1} xy^{n-1} + \ldots + \binom{n}{n} x^n$$

upon substitution of numerical values for x, y and n, we obtain the expansion as a sum of numbers.

For example, with $x = 1$, $y = \dfrac{1}{2}$ and $n = 4$ we obtain:

$$\left(1 + \frac{1}{2}\right)^4 = \left(\frac{1}{2}\right)^4 + 4(1)\left(\frac{1}{2}\right)^3 + 6(1)^2\left(\frac{1}{2}\right)^2 + 4(1)^3\left(\frac{1}{2}\right) + (1)^4$$

$$= \frac{1}{16} + \frac{4}{8} + \frac{6}{4} + \frac{4}{2} + 1$$

$$= \frac{1}{16} + \frac{8}{16} + \frac{24}{16} + \frac{32}{16} + \frac{16}{16} \quad \text{(using a common denominator for comparison purposes)}$$

Here, we see that the 4th term is the largest, the rest dropping away in magnitude either side of it. A binomial expansion (for any values of x, y and n) always has this pyramid property, which we can use to find the largest term more quickly than having to generate every single term (which would be impractical for large values of n).

The technique is as follows. In the expansion of:

$$(y + x)^n = \binom{n}{0}x^0 y^n + \binom{n}{1}x^1 y^{n-1} + \ldots + \binom{n}{r}x^r y^{n-r} + \ldots + \binom{n}{n}x^n y^0$$

if we denote the first term by $T_1 = \binom{n}{0}x^0 y^n$; the second by $T_2 = \binom{n}{1}x^1 y^{n-1}$ then the $(r+1)$th term, $T_{r+1} = \binom{n}{r}x^r y^{n-r}$.

Consider the ratio:

$$\frac{T_{r+1}}{T_r} = \frac{\binom{n}{r}x^r y^{n-r}}{\binom{n}{r-1}x^{r-1} y^{n-r+1}}$$

$$= \frac{n!\,x^r y^{n-r}(r-1)!\,(n-r+1)!}{(n-r)!\,r!\,n!\,x^{r-1} y^{n-r+1}}$$

i.e.

$$\frac{T_{r+1}}{T_r} = \frac{x}{y}\,\frac{(n-r+1)}{r}$$

Now:

$$T_{r+1} > T_r \;\Rightarrow\; \frac{T_{r+1}}{T_r} > 1$$

$$\Rightarrow \frac{x(n-r+1)}{yr} > 1$$

i.e., $x(n-r+1) > yr$.

Therefore $x(n+1) - xr > yr$ giving $x(n+1) > (x+y)r$.

Hence $r < \dfrac{x(n+1)}{x+y}$. That is, T_{r+1}, the $(r+1)$th term, is greater than T_r, the rth term as long as $r < \dfrac{x(n+1)}{x+y}$.

For the example earlier with $x = 1$, $y = \dfrac{1}{2}$ and $n = 4$ we have:

$$T_{r+1} > T_r \quad \text{if} \quad r < \frac{1(4+1)}{1+\frac{1}{2}} \quad \text{i.e.} \quad r < \frac{10}{3} = 3\frac{1}{3}$$

Thus, since r must be integral, $r = 1, 2$ or 3 for $T_{r+1} > T_r$, i.e., $T_2 > T_1$, $T_3 > T_2$ and $T_4 > T_3$. So, for $r = 4$ and 5, $T_{r+1} \leqslant T_r$, i.e. $T_5 \leqslant T_4$.

So that T_4 must be the largest term $(r = 3)$ as found earlier.

We summarise:

STATEMENT 6.2

In the binomial expansion of $(y + x)^n$ with T_{r+1}, the $(r+1)$th term,

$$= \binom{n}{r} x^r y^{n-r}, \quad T_{r+1} > T_r \text{ if } r < \frac{x(n+1)}{x+y}$$

EXAMPLE 6.1

Find the largest term in the binomial expansion of $\left(\dfrac{1}{4} + \dfrac{3}{4}\right)^{20}$.

SOLUTION

Here $x = \dfrac{1}{4}$, $y = \dfrac{3}{4}$ and $n = 20$.

$$T_{r+1} = \binom{n}{r} x^r y^{n-r} = \binom{20}{r}\left(\frac{1}{4}\right)^r\left(\frac{3}{4}\right)^{20-r} = \binom{20}{r}\frac{3^{20-r}}{4^{20}}$$

Using the previous statement we have:

$$T_{r+1} > T_r \text{ if } r < \frac{x(n+1)}{x+y} \quad \text{i.e.} \quad r < \frac{\frac{1}{4}(21)}{1} \quad \text{giving} \quad r < 5\frac{1}{4}$$

Therefore $r = 5$ for the largest term.

Thus T_6, the largest term, is:

$$\binom{20}{5}\frac{3^{20-5}}{4^{20}} = 0.202 \quad \text{(3D)}$$

6.2.2 INTEGRATION 3: BY PARTS

Here we look at a technique that enables us to extend the range of functions we are able to integrate. For example, integrals of the form:

$$\int x\, e^{-x}\, dx, \int x^2\, e^{-x}\, dx, \int x^2\, e^{-x^2}\, dx$$

which have a particular use in Statistics cannot be evaluated using any methods so far considered. In this type of case, we use the method of Integration by Parts given in the following statement.

STATEMENT 6.3

The method of *Integration by Parts* takes the form:

$$\int u \frac{dv}{dx} \, dx = uv - \int v \frac{du}{dx} \, dx$$

where u and v are functions of x. If the integration is definite (from a to b say), the above is adapted to read:

$$\int_a^b u \frac{dv}{dx} \, dx = [uv]_a^b - \int_a^b v \frac{du}{dx} \, dx$$

The left-hand side of the above identity consists of the integrand $u \dfrac{dv}{dx}$, so that any integral we wish to evaluate needs its integrand in this form.

For example, to evaluate $\displaystyle\int x \, e^{-x} \, dx$, we could put $x = u$ and $e^{-x} = \dfrac{dv}{dx}$. The right-hand side of the identity consists of combinations of the functions u, v and $\dfrac{du}{dx}$. u we know already. v is obtained by integrating $\dfrac{dv}{dx}$ $\left(\text{i.e., } \displaystyle\int \dfrac{dv}{dx} \, dx = v\right)$ and $\dfrac{du}{dx}$ is obtained by differentiating u w.r.t. x (by definition).

In this particular case:

$$u = x, \quad v = \int \frac{dv}{dx} \, dx = \int e^{-x} \, dx = -e^{-x}$$

and

$$\frac{du}{dx} = \frac{d}{dx}(u) = \frac{d}{dx}(x) = 1$$

Hence, using the form given in the statement,

$$\int \underset{\substack{\| \| \\ u \frac{dv}{dx}}}{x \, e^{-x}} \, dx = \underset{\substack{\| \quad \| \\ u \quad v}}{x \, (-e^{-x})} - \int \underset{\substack{\| \quad \| \\ v \quad \frac{du}{dx}}}{(-e^{-x}) \, 1} \, dx$$

$$= -x \, e^{-x} + \int e^{-x} \, dx \quad \text{(notice that *this* integral can be determined)}$$

$$= -x \, e^{-x} - e^{-x} = -e^{-x}(1 + x)$$

If we wish to evaluate the above integral definitely (from 0 to ∞ say) we would have (using the above result):

$$\int_0^\infty x\, e^{-x}\, dx = \left[-e^{-x}(1+x)\right]_0^\infty = \left[-e^{-x}\right]_0^\infty + \left[-x\, e^{-x}\right]_0^\infty$$

$$= \left[0 - (-1)\right] + \left[0 - 0\right] \quad \text{(from Section 5.2.5)}$$

$$= 1$$

EXAMPLE 6.2

Evaluate $\int x^2\, e^{ax}\, dx$, using the method of integration by parts.

SOLUTION

Putting $x^2 = u$ and $e^{ax} = \dfrac{dv}{dx}$ gives $\dfrac{du}{dx} = 2x$ and

$$v = \int e^{ax}\, dx = \frac{1}{a} e^{ax}.$$

Now: $\displaystyle\int u\, \frac{dv}{dx}\, dx = uv - \int v\, \frac{du}{dx}\, dx$

i.e.: $\displaystyle\int x^2\, e^{ax}\, dx = x^2\, \frac{1}{a} e^{ax} - \int \frac{1}{a} e^{ax}\, 2x\, dx$

$$= \frac{x^2}{a} e^{ax} - \frac{2}{a} \boxed{\int x\, e^{ax}\, dx} \qquad [1]$$

We need to integrate by parts again on the right-hand integral (in order to eliminate the x). Here we put $x = u$ and $e^{ax} = \dfrac{dv}{dx}$ giving $\dfrac{du}{dx} = 1$ and $v = \dfrac{1}{a} e^{ax}$ again.

So that:

$$\int x\, e^{ax}\, dx = x\, \frac{1}{a} e^{ax} - \int \frac{1}{a} e^{ax} \cdot 1\, dx$$

$$= \frac{x}{a} e^{ax} - \frac{1}{a} \int e^{ax}\, dx = \frac{x}{a} e^{ax} - \frac{1}{a} \left[\frac{1}{a} e^{ax}\right] + C$$

$$= \frac{x}{a} e^{ax} - \frac{1}{a^2} e^{ax} + C$$

Hence, from [1]:

$$\int x^2\, e^{ax} = \frac{x^2}{a} e^{ax} - \frac{2}{a} \left[\frac{x}{a} e^{ax} - \frac{1}{a^2} e^{ax}\right] + D$$

$$= e^{ax} \left[\frac{x^2}{a} - \frac{2x}{a^2} + \frac{2}{a^3}\right] + D$$

(where D is an arbitrary constant).

6.3 THE BINOMIAL DISTRIBUTION

We begin by defining the particular form of p.d.f. that characterises this particular distribution. We then investigate some forms of the probability curve and (in Section 6.3.2) look at the experimental situation from which it can be derived.

6.3.1 DEFINITION AND NOTATION

DEFINITION 6.1

A discrete random variable X having a p.d.f. of the form:

$$\Pr(X = x) = \binom{n}{x} p^x (1-p)^{n-x}, \quad (x = 0, 1, \ldots, n-1, n)$$

is said to have a *Binomial Distribution*.

n (a positive integer) and p $(0 \leqslant p \leqslant 1)$ are parameters.

We demonstrate first that X with this distribution is a *random variable*.

That is, we need to show that $\sum_{\text{for all } x} \Pr(X = x) = 1$:

$$\sum_{\text{for all } x} \Pr(X = x) = \sum_{x=0}^{n} \binom{n}{x} p^x (1-p)^{n-x} \quad \text{(by definition)}$$

$$= (p + (1-p))^n \quad \text{(binomial expansion)}$$

$$= 1^n = 1, \quad \text{as required}$$

Hence, X *is* a random variable.

The exact form of the distribution will, of course, depend on the values of the two parameters n and p.

For instance, *when $n = 4$ and $p = \dfrac{1}{2}$*, we obtain the p.d.f.:

$$\Pr(X = x) = \binom{4}{x} \left(\frac{1}{2}\right)^x \left(\frac{1}{2}\right)^{4-x} \quad \text{since} \quad p = 1 - p = \frac{1}{2}$$

$$= \binom{4}{x} \left(\frac{1}{2}\right)^4 \quad \text{for} \quad x = 0, 1, 2, 3 \text{ and } 4 \quad \text{giving:}$$

$$\Pr(X = 0) = \binom{4}{0} \left(\frac{1}{2}\right)^4 = \frac{1}{16}; \quad \Pr(X = 1) = \binom{4}{1} \left(\frac{1}{2}\right)^4 = \frac{4}{16};$$

$$\Pr(X = 2) = \binom{4}{2} \left(\frac{1}{2}\right)^4 = \frac{6}{16}$$

$$\Pr(X = 3) \;=\; \binom{4}{3}\left(\frac{1}{2}\right)^4 \;=\; \frac{4}{16} \quad \text{and} \quad \Pr(X = 4) \;=\; \binom{4}{4}\left(\frac{1}{2}\right)^4 \;=\; \frac{1}{16}$$

We can now write the p.d.f. in the form of a probability table as follows:

x	0	1	2	3	4
$\Pr(X = x)$	$\dfrac{1}{16}$	$\dfrac{4}{16}$	$\dfrac{6}{16}$	$\dfrac{4}{16}$	$\dfrac{1}{16}$

and the histogram is shown in Fig. 6.3. Note that the distribution is symmetric, which will always occur when $p = \dfrac{1}{2}$ irrespective of the value of n. (The proof of this is left as an exercise.)

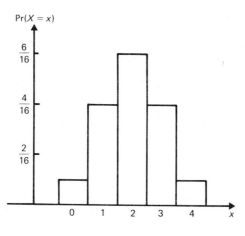

Fig. 6.3

When $n = 8$ and $p = \dfrac{1}{4}$, the p.d.f. takes the form:

$$\Pr(X = x) \;=\; \binom{8}{x}\left(\frac{1}{4}\right)^x\left(\frac{3}{4}\right)^{8-x} \quad \text{for} \quad x = 0, 1, 2, \ldots, 8$$

Hence:

$$\Pr(X = 0) \;=\; \binom{8}{0}\left(\frac{1}{4}\right)^0\left(\frac{3}{4}\right)^8 \;=\; \left(\frac{3}{4}\right)^8 \;=\; 0.10 \quad \text{(2D)}$$

$$\Pr(X = 1) \;=\; \binom{8}{1}\left(\frac{1}{4}\right)^1\left(\frac{3}{4}\right)^7 \;=\; 8\left(\frac{1}{4}\right)\left(\frac{3}{4}\right)^7 \;=\; 0.27 \quad \text{(2D)}$$

$$\Pr(X = 2) \;=\; \binom{8}{2}\left(\frac{1}{4}\right)^2\left(\frac{3}{4}\right)^6 \;=\; 28\left(\frac{3^6}{4^8}\right) \;=\; 0.31 \quad \text{(2D)}$$

and similarly we obtain:

$$\Pr(X = 3) \;=\; 0.21; \quad \Pr(X = 4) \;=\; 0.09; \quad \Pr(X = 5) \;=\; 0.02;$$

$$\Pr(X = 6) = 0.004 \quad (3D); \quad \Pr(X = 7) = 0.0004 \quad (4D)$$

and $\quad \Pr(X = 8) = 0.000\,02 \quad (5D)$

from which the histogram in Fig. 6.4 can be drawn up. Notice that the distribution is markedly skewed to the right. This will always be the case when $p < \dfrac{1}{2}$.

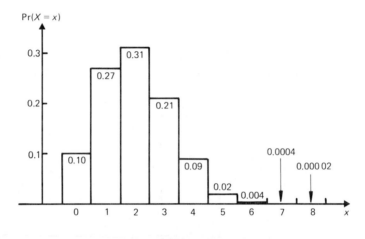

Fig. 6.4

Similarly, if $p > \dfrac{1}{2}$, the distribution is correspondingly skewed to the left. (The proof of this is left as a directed exercise.)

In general, for small n $(n \leqslant 10)$ the closer p is to 0 or 1 the more skewed (right or left) the distribution is. However, as n becomes larger the skew tends to be 'corrected' and the distribution becomes more symmetric.

That is, *for a fixed* n, a binomial distribution with, say, $p = 0.1$ will be more skewed (to the right) than one with, say, $p = 0.2$. On the other hand, *for a fixed* p, a binomial distribution with, say, $n = 100$ will be less skewed than one with, say, $n = 20$.

Finally, we introduce a convenient shorthand notation for a binomial distribution. Namely, if a random variable X has a p.d.f. of the form:

$$\Pr(X = x) = \binom{n}{x} p^x (1 - p)^{n-x} \quad \text{for} \quad x = 0, 1, 2, \ldots, n$$

then we write X is Bin(n, p) which is read as 'X has a binomial distribution with parameters n and p'.

6.3.2 DERIVATION

We can derive the p.d.f. of the binomial distribution in terms of an experimental situation.

Consider an experiment which has just two outcomes, 'success' S and 'failure' F, say. (Note that *any* experiment can be defined according to this rule simply by grouping all the experimental outcomes into two mutually exclusive and exhaustive sets, which then can be labelled S or F respectively).

Let $\Pr(S) = p$ giving $\Pr(F) = \Pr(\bar{S}) = 1 - p$. We now consider performing the experiment exactly n times (i.e., conduct n trials), where each trial is independent of all the others.

We call this set-up a *binomial situation*.

Let the variable X represent the number of 'successes' obtained in conducting the n trials. Clearly X can assume only one of the values $0, 1, 2, \ldots, n$ any others being impossible.

(We can argue at this stage that X is a random variable, since:

$$\Pr(X \text{ takes one of the values } 0, 1, 2, \ldots, n) = 1 \quad \text{(a certain event)}$$

$$\Rightarrow \Pr(X = 0 \text{ or } X = 1 \text{ or } X = 2 \text{ or } \ldots \text{ or } X = n) = 1$$

$$\Rightarrow \sum \Pr(X = x) = 1 \quad \text{(mutually exclusive events)}$$

which is the condition for a random variable.)

Now:

$$\Pr(X = 0) = \Pr(\underbrace{F \cdot F \cdot F \ldots F}_{n \text{ times}}) = \underbrace{\Pr(F) \cdot \Pr(F) \cdot \Pr(F) \cdot \ldots \cdot \Pr(F)}_{n \text{ times}}$$

(since the trials are independent).

i.e. $$\Pr(X = 0) = \underbrace{(1 - p)(1 - p)(1 - p)(1 - p) \ldots (1 - p)}_{n \text{ times}}$$

$$\Rightarrow \Pr(X = 0) = (1 - p)^n$$

To calculate $\Pr(X = 1)$ Now, we can obtain 1 success out of n trials in exactly $\binom{n}{1}$ different ways (the number of ways of arranging an S among $(n - 1)$ Fs). Each of these $\binom{n}{1}$ possible events are mutually exclusive (since one, and only one, can occur at a single trial) so that the probability that one of them occurs is just the sum of their separate probabilities of occurrence. Further, each of them has the same probability of occurrence since we require only one failure with $n - 1$ successes *irrespective of which trial the failure occurs on.*

That is:

$$\Pr(S \cdot F \cdot F \cdot \ldots \cdot F) = \Pr(F \cdot S \cdot F \cdot \ldots \cdot F) = \ldots = \Pr(F \cdot F \cdot F \cdot \ldots \cdot S)$$

But:

$$\Pr(S \cdot F \cdot F \cdot \ldots \cdot F) = \Pr(S) \cdot \Pr(F) \cdot \Pr(F) \cdot \ldots \cdot \Pr(F) \quad \text{(independent trials)}$$

$$= p(1-p) \cdot (1-p) \cdot \ldots \cdot (1-p)$$

$$= p(1-p)^{n-1}$$

$$= \text{the probability of a single success in } n \text{ trials}$$

Hence the sum of the $\binom{n}{1}$ probabilities is just $\binom{n}{1} p(1-p)^{n-1}$,

i.e.: $$\Pr(X = 1) = \binom{n}{1} p(1-p)^{n-1}$$

To calculate $\Pr(X = r)$ $X = r$ if exactly r Ss occur in the n trials. This can occur in $\binom{n}{r}$ different ways (the number of ways of arranging r things among n) and again each way has the same probability of occurrence, namely the probability of r Ss together with $(n-r)$ Fs $= p^r(1-p)^{n-r}$.

Hence:

$$\Pr(X = r) = \text{sum of probabilities of each of the different ways}$$

$$= \binom{n}{r} p^r (1-p)^{n-r}$$

So that if we perform n independent trials of an experiment which has a constant probability of success of p at each trial, then the probability of exactly x successes in the n trials is just $\binom{n}{x} p^x (1-p)^{n-x}$, the p.d.f. given for a binomial distribution.

DEFINITION 6.2

A *binomial situation* is described as an experiment, partitioned into two mutually exclusive and exhaustive event sets which can be labelled as 'success' and 'failure' (S and F) with $p = \Pr(S)$ known, repeated independently n times. Usually a random variable X is associated with the total number of 'successes' obtained with the n trials.

EXAMPLE 6.3

Six 'fair' coins are tossed. Find the probability of obtaining:

(a)   exactly 6 heads,

(b)   exactly 4 heads, and

(c)   at least 2 heads.

SOLUTION

We can regard the tossing of a single coin as an experiment and, since the tossing of one is independent of the tossing of any other, we can regard the tossing of the six coins as conducting six independent trials. If we label a 'head' as a 'success', we have $p = \dfrac{1}{2}$ $\left(\text{Pr(head)} = \text{Pr(success)} = \dfrac{1}{2}\right).$

So we have a binomial situation with $n = 6$ and $p = \dfrac{1}{2}.$

Defining X as 'number of heads' (= number of successes), X is $\text{Bin}\left(6, \dfrac{1}{2}\right)$ with:

$$\text{Pr}(X = x) = \binom{6}{x}\left(\frac{1}{2}\right)^x\left(\frac{1}{2}\right)^{6-x} = \binom{6}{x}\left(\frac{1}{2}\right)^6$$

(a) $\text{Pr(6 heads)} = \text{Pr}(X = 6) = \binom{6}{6}\left(\frac{1}{2}\right)^6 = \dfrac{1}{64}$

(b) $\text{Pr(4 heads)} = \text{Pr}(X = 4) = \binom{6}{4}\left(\frac{1}{2}\right)^6 = \dfrac{15}{64}$

(c) $\text{Pr(at least 2 heads)} = \text{Pr}(X \geqslant 2)$

$= 1 - \text{Pr}(X = 0 \text{ or } X = 1)$ (since $X = 0$ or $X = 1$ is the complement of the event $X \geqslant 2$)

$= 1 - (\text{Pr}(X = 0) + \text{Pr}(X = 1))$ (mutually exclusive events)

$= 1 - \left(\binom{6}{0}\left(\frac{1}{2}\right)^6 + \binom{6}{1}\left(\frac{1}{2}\right)^6\right) = 1 - \left(\frac{1}{2}\right)^6\left(\binom{6}{0} + \binom{6}{1}\right) = 1 - \dfrac{7}{64}$

$= \dfrac{57}{64}$

EXAMPLE 6.4

At a fairground stall there are a number of identical games and the probability that a single player wins a single game is 0.4. How many players need to play the game to ensure that the probability that at least one person wins is approximately 0.98.

SOLUTION

We have a binomial situation with n (the number of trials) as the number of people that need to play the game and p (the probability of success) as the probability of any person winning a single game, which is here 0.4. n is to be determined.

If X is the random variable 'number of people winning a game' then X is $\text{Bin}(n, 0.4)$ with:

$$\Pr(X = x) = \binom{n}{x}(0.4)^x (0.6)^{n-x} \quad \text{for } x = 0, 1, \ldots, n$$

We need that:

$$\Pr(\text{at least one person wins}) = \Pr(X \geqslant 1) \approx 0.98$$

i.e.:

$$1 - \Pr(X = 0) \approx 0.98 \quad (\text{since } X = 0 \text{ is the complement of } X \geqslant 1)$$

That is, we need $\Pr(X = 0) \approx 1 - 0.98 = 0.02$.

But:
$$\Pr(X = 0) = \binom{n}{0}(0.4)^0 (0.6)^n = (0.6)^n$$

Therefore we need to find n such that $(0.6)^n \approx 0.02$, and by trial and error we easily find $(0.6)^8 = 0.017 \approx 0.02$, i.e. $n = 8$.

Therefore eight people need to play the game.

EXAMPLE 6.5

A company manufactures screws which are sent out to customers in lots of 10 000. The manufacturer operates a sampling scheme whereby a random sample of 10 is taken from each lot ready for despatch and they are released only if the number of defective screws in the sample is less than 3. Otherwise, the whole lot of 10 000 is rejected and re-processed.

(a)   If 10% of all the screws produced are known to be defective, find the proportion of lots that will be rejected.

(b)   If the manufacturer replaces all his screw producing machines causing the proportion of defective screws to drop to only 4% and now releases lots only if the number of defective screws in the sample of 10 is less than 2, will the proportion of lots rejected decrease? If so, how many screws per lot will the manufacturer expect to save?

SOLUTION

(a)   This is a binomial situation with n (no. of trials = no. in sample) as 10 and p (probability of success = probability of defective screw) as 0.1. Let X be the random variable 'no. of defective screws in sample' giving X as $\text{Bin}(10, 0.1)$,

i.e.:
$$\Pr(X = x) = \binom{10}{x}(0.1)^x (0.9)^{10-x}$$

So that:

$$\Pr(\text{lot being rejected}) = \Pr(X \geqslant 3)$$

$$= 1 - (\Pr(X = 0) + \Pr(X = 1) + \Pr(X = 2))$$

$$= 1 - \left(\binom{10}{0}(0.1)^0 (0.9)^{10} + \binom{10}{1}(0.1)^1 (0.9)^9 \right.$$

$$\left. + \binom{10}{2}(0.1)^2 (0.9)^8 \right)$$

$$= 1 - (0.349 + 0.387 + 0.194)$$

$$= 0.07 \quad (2D)$$

Therefore a proportion 0.07 (7%) of lots will be rejected under this scheme.

(b) Here we have a binomial situation with $n = 10$ still, but now $p = 0.04$, i.e., X is Bin(10, 0.04) with:

$$\Pr(X = x) = \binom{10}{x}(0.04)^x (0.96)^{10-x}$$

So:

$$\Pr(\text{lot being rejected}) = \Pr(X \geqslant 2)$$

$$= 1 - (\Pr(X = 0) + \Pr(X = 1))$$

$$= 1 - \left(\binom{10}{0}(0.04)^0 (0.96)^{10} \right.$$

$$\left. + \binom{10}{1}(0.04)^1 (0.96)^9 \right)$$

$$= 1 - (0.665 + 0.277)$$

$$= 0.06 \quad (2D)$$

Hence a proportion 0.06 (6%) of lots will now be rejected which constitutes a saving of 1% per lot = 1% per 10 000 screws.

Therefore the manufacturer will expect to save $\dfrac{10\,000}{100} = 100$ screws per lot.

6.3.3 THE MEAN AND VARIANCE

We now find the mean and variance of a general binomial distribution in terms of its two parameters n and p.

Let X be Bin(n, p) giving:

$$\Pr(X = x) = \binom{n}{x} p^x (1-p)^{n-x} \quad \text{for } x = 0, 1, \ldots, n$$

Now:

$$\text{mean} = E[X] = \sum_{\text{all } x} x \cdot \Pr(X = x) \quad \text{(by definition)}$$

$$= \sum_{x=0}^{n} x \binom{n}{x} p^x (1-p)^{n-x} \qquad [1]$$

$$= \sum_{x=1}^{n} x \frac{n!\, p^x (1-p)^{n-x}}{(n-x)!\, x!} \qquad \text{(since } x=0 \text{ contributes nothing to the sum)}$$

$$= np \sum_{x=1}^{n} \frac{(n-1)!\, x}{(n-x)!\, x!} p^{x-1} (1-p)^{n-x} \qquad \text{(factorising } np \text{ out)}$$

(But:

(a) $\dfrac{x}{x!} = \dfrac{1}{(x-1)!}$,

(b) $(n-x)!$ can be written as $(n-1-(x-1))!$,

(c) $(1-p)^{n-x}$ can be written as $(1-p)^{n-1-(x-1)}$.)

Hence: $$E[X] = np \sum_{x=1}^{n} \frac{(n-1)!\, p^{x-1} (1-p)^{n-1-(x-1)}}{(n-1-(x-1))!\, (x-1)!}$$

Substituting $y = x-1$ and $N = n-1$ into the above gives:

$$E[X] = np \sum_{y=0}^{N} \frac{N!}{(N-y)!\, y!} p^y (1-p)^{N-y}$$

(notice that since x ranges from 1 to n, $y\ (=x-1)$ must range from 0 to $n-1$, i.e., 0 to N)

$$= np \left[\sum_{y=0}^{N} \binom{N}{y} p^y (1-p)^{N-y} \right]$$

$$= np \cdot 1$$

(since the bracketed expression is just the sum of the probabilities of a random variable Y which is $\text{Bin}(N, p)$)

$$\therefore \qquad E[X] = n \cdot p \qquad [2]$$

To find the variance we use $\text{Var}[X] = E[X^2] - E^2[X]$.

$$E[X^2] = \sum_{\text{all } x} x^2 \cdot \Pr(X = x) \qquad \text{(by definition)}$$

$$= \sum_{x=0}^{n} x^2 \binom{n}{x} p^x (1-p)^{n-x}$$

$$= \sum_{x=0}^{n} (x(x-1) + x) \binom{n}{x} p^x (1-p)^{n-x}$$

$$\text{(since } x(x-1) + x = x^2 \text{)}$$

i.e.:

$$E[X^2] = \sum_{x=0}^{n} x(x-1)\binom{n}{x} p^x (1-p)^{n-x} + \sum_{x=0}^{n} x\binom{n}{x} p^x (1-p)^{n-x}$$

$$= \sum_{x=0}^{n} \left[\frac{x(x-1)\,n!}{(n-x)!\,x!} p^x (1-p)^{n-x} \right] + np \quad \text{(from [1])}$$

Factorising $n(n-1)p^2$ out and writing $\dfrac{x(x-1)}{x!}$ as $\dfrac{1}{(x-2)!}$ and $n-x$ as $[(n-2)-(x-2)]$ we obtain:

$$E[X^2] = n(n-1)p^2 \left\{ \sum_{x=2}^{n} \frac{(n-2)!\,p^{x-2}(1-p)^{(n-2)-(x-2)}}{((n-2)-(x-2))!\,(x-2)!} \right\} + np$$

(x starts at 2 since $x = 0$ and 1 contribute nothing to the sum)

$$= n(n-1)p^2 \cdot 1 + np$$

(since the bracketed expression represents the sum of probabilities of a random variable which is $\text{Bin}(n-2, p)$)

i.e.: $E[X^2] = n^2 p^2 - np^2 + np$

$\therefore \qquad \text{Var}[X] = E[X^2] - E^2[X] = n^2 p^2 - np^2 + np - n^2 p^2$

$$(E[X] = np \quad \text{from [2]})$$

i.e.: $\qquad \text{Var}[X] = np - np^2$

$$= np(1-p)$$

We summarise the results:

STATEMENT 6.4

If random variable X has a binomial distribution with parameters n and p, i.e., X is $\text{Bin}(n, p)$ then:

(a) $E[X] = n \cdot p$

(b) $\text{Var}[X] = n \cdot p(1-p)$

Proof In previous text.

We can show these results using a more direct and sophisticated method – by means of the probability generating function.

The technique is:

(a) Obtain $G(t)$ for the binomial case.

(b) Calculate $E[X] = G'(1)$.

(c) Calculate $\text{Var}[X] = G''(1) + G'(1) - [G'(1)]^2$.

(Note that this procedure is valid here since X takes only the values $0, 1, 2, \ldots, n$ as required.)

(a)   Now:

$$G(t) = E[t^x] \quad \text{(by definition)}$$

$$= \sum_{x=0}^{n} t^x \binom{n}{x} p^x (1-p)^{n-x} \quad \text{(by definition)}$$

$$= \sum_{x=0}^{n} \binom{n}{x} (pt)^x (1-p)^{n-x}$$

$$= (1-p+pt)^n \quad \text{(using the binomial theorem)}$$

(b) $G'(t) = \dfrac{d}{dt}[(1-p+pt)^n]$

$$= n(1-p+pt)^{n-1}p \quad \text{(function of a function differentiation)}$$

Hence:

$$E[X] = G'(1) = np(1-p+p \cdot 1)^{n-1}$$

$$= np$$

(c)

$$G''(t) = \frac{d}{dt}[G'(t)] = \frac{d}{dt}[np(1-p+pt)^{n-1}]$$

$$= n(n-1)p(1-p+pt)^{n-2}p \quad \text{(func. of a func. differentiation)}$$

Therefore:

$$G''(1) = n(n-1)p^2$$

Hence:

$$\mathrm{Var}[X] = n(n-1)p^2 + np - n^2 p^2$$

$$= np(1-p)$$

EXAMPLE 6.6

A box contains 2 red balls out of a total of 7.

A ball is drawn randomly, its colour noted, the ball replaced and the box shaken. If 50 balls are drawn in this manner, find the expectation of the number of red balls drawn and their standard deviation.

SOLUTION

The drawing of a ball can be considered as an independent trial of an experiment. Calling the drawing of a red ball a success, we have a binomial situation with $n = 50$ (no. of trials) and

$$p = \mathrm{Pr}(\text{success}) = \frac{2}{7}.$$

Random variable X will represent 'no. of red balls drawn' with X as $\text{Bin}\left(50, \dfrac{2}{7}\right)$.

$$\therefore \qquad E[X] = np = 50\left(\frac{2}{7}\right) = \frac{100}{7} = 14.3 \ (1\text{D})$$

$$\text{Var}[X] = np(1-p) = 50\left(\frac{2}{7}\right)\left(\frac{5}{7}\right) = 10.20 \ (2\text{D})$$

$$\therefore \qquad \text{standard deviation of } X = \sqrt{\text{Var}[X]} = \sqrt{10.20} = 3.2 \ (1\text{D})$$

EXAMPLE 6.7

A binomially distributed random variable X is known to have $E[X] = 19.2$ and $\text{Var}[X] = 11.52$. Find the value of the two parameters n and p.

SOLUTION

Put X as $\text{Bin}(n, p)$ where n and p have to be determined.

Now: $\qquad\qquad E[X] = np = 19.2$ $\qquad\qquad$ [1]

and: $\qquad\qquad \text{Var}[X] = np(1-p) = 11.52$ $\qquad\qquad$ [2]

[1] gives $p = \dfrac{19.2}{n}$ and substituting this into [2] gives:

$$n\frac{19.2}{n}\left(1 - \frac{19.2}{n}\right) = 11.52$$

$$\Rightarrow \qquad 1 - \frac{19.2}{n} = \frac{11.52}{19.2}$$

$$\Rightarrow \qquad \frac{19.2}{n} = 1 - \frac{11.52}{19.2} = 0.4$$

$$\therefore \qquad n = \frac{19.2}{0.4} = 48$$

$$\therefore \qquad p = \frac{19.2}{n} = 0.4$$

6.3.4 RECURRENCE FORMULA

In general it is a fairly lengthy process calculating more than one or two probabilities of a binomial distribution using the p.d.f. form directly, involving a great deal of duplication. The recurrence formula method enables much arithmetic to be cut out.

STATEMENT 6.5

The relation between successive probabilities of a binomial distribution defined by X as $\text{Bin}(n, p)$ is given by:

$$\Pr(X = x + 1) = \frac{n - x}{x + 1} \frac{p}{1 - p} \Pr(X = x) \quad \text{for } x = 0, 1, 2, \ldots, n - 1$$

and known as the *binomial recurrence formula*.

Given $\Pr(X = 0) = p_0$, say, then we can calculate $\Pr(X = 1)$ by:

$$\Pr(X = 1) = \frac{n}{1} \frac{p}{1 - p} p_0 \quad \text{(putting } x = 0 \text{ in the recurrence formula)}$$

Then:

$$\Pr(X = 2) = \frac{n - 1}{2} \frac{p}{1 - p} \Pr(X = 1) \quad (x = 1 \text{ in recurrence formula})$$

$$= \frac{(n - 1)p}{2(1 - p)} \frac{np}{1 - p} \cdot p_0 \quad \text{and so on}$$

So that, initially, we need only calculate $\Pr(X = 0) = (1 - p)^n$ from the p.d.f. form, the other probabilities being obtained via the recurrence formula.

Proof   of the above statement:

If X is $\text{Bin}(n, p)$

then: $$\Pr(X = x) = \binom{n}{x} p^x (1 - p)^{n - x}$$

and also

$$\Pr(X = x + 1) = \binom{n}{x + 1} p^{x+1} (1 - p)^{n - x - 1}$$

So that the ratio:

$$\frac{\Pr(X = x + 1)}{\Pr(X = x)} = \frac{\left[\dfrac{n! \, p^{x+1} (1 - p)^{n - x - 1}}{(n - x - 1)! \, (x + 1)!} \right]}{\left[\dfrac{n! \, p^x (1 - p)^{n - x}}{(n - x)! \, x!} \right]}$$

$$= \frac{n! \, p^{x+1} (1 - p)^{n - x - 1} (n - x)! \, x!}{n! \, p^x (1 - p)^{n - x} (n - x - 1)! \, (x + 1)!}$$

$$= \frac{p}{1 - p} \frac{n - x}{x + 1}$$

Hence: $$\Pr(X = x + 1) = \frac{n - x}{x + 1} \frac{p}{1 - p} \Pr(X = x)$$

Although arithmetical computations are cut drastically by this method of calculating probabilities, great care must be taken with accuracy when using the recurrence formula.

For example, if an error is made in the initial calculation of $\Pr(X = 0)$, all other probabilities will be subject to the same degree of error. This is because $\Pr(X = 0)$, directly or indirectly, is used in all other calculations. This also applies to the initial value of p, except that any error involved here *will become successively larger* as further probabilities are calculated. Also, rounding to too few decimal places early in the calculations will clearly introduce a small error but, again, this will become successively larger.

A rule worth remembering is: when requiring answers to a certain number of decimal places, always work with *at least* one more place of decimals (preferably two or more) in all intermediate calculations. This procedure will help to minimise errors.

EXAMPLE 6.8

The random variable X has a binomial distribution with $n = 7$ and $p = 0.2$. Calculate the probability distribution for X using the recurrence formula.

SOLUTION

We use:

$$\Pr(X = x + 1) = \frac{n - x}{x + 1}\frac{p}{1 - p}\Pr(X = x)$$

and with $n = 7$ and $p = 0.2$ we have:

$$\Pr(X = x + 1) = \frac{7 - x}{x + 1}(0.25)\Pr(X = x) \qquad [1]$$

Since:

$$\Pr(X = x) = \binom{7}{x}(0.2)^x(0.8)^{7-x}$$

we have:

$$\Pr(X = 0) = (0.8)^7 = 0.209\ 715 \quad (6D)$$

Putting $x = 0$ in [1] gives:

$$\Pr(X = 1) = \frac{7}{1}(0.25)\Pr(X = 0) = 7(0.25)(0.209\ 715)$$

$$= 0.367\ 001$$

Similarly:

$$Pr(X = 2) \quad = \quad \frac{6}{2}(0.25)(0.367\,001) \quad = \quad 0.275\,251$$

$$Pr(X = 3) \quad = \quad \frac{5}{3}(0.25)(0.275\,251) \quad = \quad 0.114\,688$$

$$Pr(X = 4) \quad = \quad \frac{4}{4}(0.25)(0.114\,688) \quad = \quad 0.028\,672$$

$$Pr(X = 5) \quad = \quad \frac{3}{5}(0.25)(0.028\,672) \quad = \quad 0.004\,301$$

$$Pr(X = 6) \quad = \quad \frac{2}{6}(0.25)(0.004\,301) \quad = \quad 0.000\,358$$

and:

$$Pr(X = 7) \quad = \quad \frac{1}{7}(0.25)(0.000\,358) \quad = \quad 0.000\,013$$

In the form of a table, the probability distributions for X (to 5D) is:

x	0	1	2	3	4	5	6	7
$Pr(X = x)$	0.209 72	0.367 00	0.275 25	0.114 69	0.028 67	0.004 30	0.000 36	0.000 01

6.3.5 BINOMIAL TABLES

Definition 5.8 specified a distribution function for a discrete variable as:

$$F(t) \quad = \quad Pr(X \leqslant t)$$

For example, if X is Bin(4, 0.25), we have:

$$Pr(0) \quad = \quad (0.75)^4 \quad = \quad 0.3164$$

$$Pr(1) \quad = \quad 4(0.25)(0.75)^3 \quad = \quad 0.4219 \ldots \quad \text{etc.}$$

The following table shows the remaining probabilities, together with the values of the distribution function.

x	0	1	2	3	4
$Pr(x)$	0.3164	0.4219	0.2109	0.0469	0.0039
$F(x)$	0.3164	0.7383	0.9492	0.9961	1.0000

A probability distribution function can thus be obtained by accumulating successive probabilities.

Appendix Table 1 gives values of the distribution function for the Binomial distribution for n ranging from 2 to 13 and p ranging from 0.05 to 0.5. For example, reading from tables, $n = 4$ and $p = 0.25$ gives:

$$F(0) \quad = \quad 0.3164; \quad F(1) \quad = \quad 0.7383; \quad \text{etc.}$$

agreeing with the above calculations.

The tables can also be used for obtaining ordinary probabilities, since:

$$\Pr(0) \;=\; F(0); \quad \Pr(1) \;=\; F(1) - F(0); \quad \Pr(2) \;=\; F(2) - F(1); \quad \text{etc.}$$

For example, reading from tables, $n = 6$ and $p = 0.15$ gives (to 4D):

$F(0) \;=\; 0.3771$ giving $\Pr(0) \;=\; 0.3771$;

$F(1) \;=\; 0.7765$ giving $\Pr(1) \;=\; 0.7765 - 0.3771 \;=\; 0.3994$;

$F(2) \;=\; 0.9527$ giving $\Pr(2) \;=\; 0.9527 - 0.7765 \;=\; 0.1762$;

$F(3) \;=\; 0.9941$ giving $\Pr(3) \;=\; 0.9941 - 0.9527 \;=\; 0.0414$;

$F(4) \;=\; 0.9996$ giving $\Pr(4) \;=\; 0.9996 - 0.9941 \;=\; 0.0055$;

$F(5) \;=\; 1.0000$ giving $\Pr(5) \;=\; 1 - 0.9996 \;=\; 0.0004$;

$F(6) \;=\; 0$ (to 4D).

Notice that tables only go as far as $p = 0.5$. This is because, for example, a 35% chance of success is the same as a 65% chance of failure. In other words:

$$\Pr(X = r/\mathrm{Bin}(n, p)) \;=\; \Pr(X = n - r/\mathrm{Bin}(n, 1 - p)).$$

For example:

$$
\begin{aligned}
\Pr(X = 3/\mathrm{Bin}(7, 0.75)) &= \Pr(X = 4/\mathrm{Bin}(7, 0.25)) \\
&= F(4) - F(3) \\
&= 0.9871 - 0.9294 \quad \text{(from tables)} \\
&= 0.0577
\end{aligned}
$$

and

$$
\begin{aligned}
\Pr(X = 4/\mathrm{Bin}(12, 0.6)) &= \Pr(X = 8/\mathrm{Bin}(12, 0.4)) \\
&= F(8) - F(7) \\
&= 0.9847 - 0.9427 \quad \text{(from tables)} \\
&= 0.0420
\end{aligned}
$$

EXAMPLE 6.9

In a certain area, one person in 5 is unemployed. What is the probability that a random sample of 12 will contain at least 4 who are unemployed?

SOLUTION

This is a binomial situation with n, number in the sample, $= 12$ and p, probability of a person being unemployed, $= 0.2$.

If we associate the random variable X with the number of unemployed persons in the sample, we require:

$$\begin{aligned}
\Pr(X \geqslant 4) &= \Pr(X = 4 \text{ or } X = 5 \text{ or } \ldots \text{ or } X = 12) \\
&= 1 - \Pr(X = 0 \text{ or } X = 1 \text{ or } X = 2 \text{ or } X = 3) \\
&= 1 - \Pr(X \leqslant 3) \\
&= 1 - 0.7946 \quad \text{(from Appendix Table 1)} \\
&= 0.2054
\end{aligned}$$

6.3.6 FITTING A THEORETICAL DISTRIBUTION

At this stage we have a knowledge of the concepts of both frequency and probability distributions and their derivation. Also, given a specific binomial situation with known parameters n and p, we can generate a binomial probability distribution and, if necessary, transform this to a theoretical (expected) frequency distribution.

The problem here, is to *derive* (or *fit*) a theoretical distribution given only an observed frequency distribution with no prior information of the binomial situation that may have given rise to the observed data. In this sort of case we need to estimate n and p from the given frequency distribution.

We do so as follows: n is taken as the largest x value given, usually. For example, given no other information save the frequency distribution

x	0	1	2	3	4	5
f	2	5	8	10	4	1

we could only take n as 5. Now, for any binomial distribution (with parameters n and p) the mean is given by np. But we can calculate the mean of the given frequency distribution as $\bar{x}$, say.

By putting $\bar{x} = np$ we can calculate $p = \dfrac{\bar{x}}{n}$ since both $\bar{x}$ and n are now known.

For the distribution given we have $\Sigma fx = 72$, $\Sigma f = 30$.

Therefore $\bar{x} = \dfrac{72}{30} = 2.4$ and, since $n = 5$, $p = \dfrac{2.4}{5} = 0.48$.

We can now generate a binomial probability distribution defined by a random variable X as $\text{Bin}(5, 0.48)$ giving the probabilities $\Pr(X = 0) = 0.038$, $\Pr(X = 1) = 0.175$, $\Pr(X = 2) = 0.324$, $\Pr(X = 3) = 0.299$, $\Pr(X = 4) = 0.138$ and $\Pr(X = 5) = 0.025$.

Multiplying these probabilities by the sum of the given frequencies (30) gives the 'expected' distribution:

x	0	1	2	3	4	5	
f	1.1	5.3	9.7	9.0	4.1	0.8	(to 1D)

This distribution is said to be *fitted* and by comparing corresponding frequencies we say that the fit is good or not depending on how closely the frequencies match. In this case we would say that the fit is 'fairly' good.

The significance of comparing these two distributions in this manner can be described as follows: Given that (1) the theoretical distribution *is* binomial with parameters n and p and (2) the values of n and p were obtained from the given distribution, we can in a sense 'test' how close the given distribution is to a binomial by how well the two sets of frequencies agree.

Using the above results, since there is reasonably good agreement, we might conclude that the given distribution is binomial.

This latter deduction is reasonable as long as we bear in mind that the information on which we based the deduction (the comparison of frequencies) is purely subjective. That is, we have no criterion for good or bad agreement of the corresponding frequencies — what one person labels 'good', another might label 'bad'. (In Chapter 10 we develop a well-defined statistical test to deal with this type of situation.)

We finally note that, in order to estimate a value of p in order to generate a theoretical distribution, we are not bound to use the value of $\bar{x}$ from the given frequency distribution. This would occur if the specific binomial situation involved was known which would mean that both n and p would be known.

For example, if an experiment is performed by throwing three fair dice and observing how many times a '5' or '6' occurs with each throw, we might obtain the following frequency distribution:

No. of 5s or 6s	(x)	0	1	2	3	Total
	f	6	13	5	3	27

Here, the 3 dice have been thrown 27 times.

Since we know the situation, we have $n = 3$ (the number of dice thrown) and p = probability of 5 or 6 with a single throw, i.e., $p = \dfrac{1}{3}$, obviating the need to calculate $\bar{x}$ to estimate p.

We could now generate a binomial distribution and multiply the associated probabilities by 27 to obtain an expected frequency distribution for comparison.

To summarise:

DEFINITION 6.3

Given a frequency distribution, we *fit a binomial distribution* to it by:

(a) finding the value of parameters n and p either:
  (i)   from a known binomial situation, or
  (ii) using the frequency distribution, n, from the largest x value (usually) and p, by using the relation $\bar{x} = np$;

(b) generating a binomial probability distribution defined by X as Bin(n, p);

(c) using the total frequency (Σf) from the given distribution to multiply the probabilities, generating 'expected' frequencies.

The two corresponding sets of frequencies can then be compared.

EXAMPLE 6.10

A biased die is thrown 5 times as an experiment. The experiment is repeated 250 times and the number of 'even' numbers shown on the die in each experiment is recorded giving the results:

No. of 'evens' in experiment	0	1	2	3	4	5	Total
Observed frequency	11	41	83	73	36	6	250

Fit a binomial distribution to this data.

SOLUTION

In order to generate a theoretical binomial distribution we first need to find n and p. Although we have a binomial situation defined here, the value of p is unknown (since the die is biased and $p = \Pr(\text{even no.})$ is probably not $\dfrac{1}{2}$.

Since we are throwing the die 5 times, this constitutes the value of n. To find p, we need to use the relationship $\bar{x} = np$, i.e., $\bar{x} = 5p$ here.

x	f	fx
0	11	0
1	41	41
2	83	166
3	73	219
4	36	144
5	6	30
	250	600

Using the table we have:

$$\bar{x} = \frac{\sum fx}{\sum f} = \frac{600}{250} = 2.4$$

So that $5p = 2.4$, i.e. $p = \frac{2.4}{5} = 0.48$

Letting random variable X represent 'number of evens in experiment' gives X as $\text{Bin}(5, 0.48)$.

$\Pr(X = 0) = (0.52)^5 = 0.0380$ and using the recurrence formula $\left(\text{with } \frac{p}{1-p} = \frac{0.48}{0.52} = 0.9231\right)$:

$$\Pr(X = x + 1) = \frac{5 - x}{x + 1}(0.9231)\Pr(X = x) \text{ gives:}$$

$$\Pr(X = 1) = \frac{5}{1}(0.9231)(0.0380) = 0.1754$$

and similarly:

$$\Pr(X = 2) = 0.3240, \ \Pr(X = 3) = 0.2990,$$
$$\Pr(X = 4) = 0.1380 \text{ and } \Pr(X = 5) = 0.0255.$$

Multiplying these probabilities by 250 (and rounding to the nearest whole number) gives the expected frequency distribution:

No. of 'evens' in experiment	0	1	2	3	4	5	Total
Expected frequency	10	44	81	75	34	6	250

(Note that the above fit is quite good.)

6.4 THE POISSON DISTRIBUTION

This distribution has many things in common with the previous one. It is a discrete distribution taking only positive, integral values and in certain cases can be used as an approximation to a binomial, being easier and more convenient to calculate.

We begin by defining the form of its p.d.f. and finding its mean and variance.

6.4.1 DEFINITION AND NOTATION

DEFINITION 6.4

A discrete random variable X having a p.d.f. of the form:

$$\Pr(X = x) = e^{-\mu} \cdot \frac{\mu^x}{x!} \quad \text{for} \quad x = 0, 1, 2, 3, \ldots$$

where μ is a parameter, is said to have a *Poisson distribution*.

(Note that μ can take any positive value.)

We show first that a variable X with this p.d.f. is, in fact, a random variable, i.e., we need to show that $\sum\limits_{\text{all } x} \Pr(X = x) = 1$.

Now:
$$\sum_{\text{all } x} \Pr(X = x) = \sum_{x=0}^{\infty} e^{-\mu} \frac{\mu^x}{x!} \quad \text{(by definition)}$$

$$= e^{-\mu} \sum_{x=0}^{\infty} \frac{\mu^x}{x!}$$

$$= e^{-\mu} e^{\mu}$$

$\left(\text{since } \sum\limits_{x=0}^{\infty} \frac{\mu^x}{x!} = 1 + \mu + \frac{\mu^2}{2!} + \frac{\mu^3}{3!} + \ldots \text{ is just the series definition of } e^{\mu}.\right.$

See Section 5.2.5$\Big)$

$$= 1, \quad \text{as required}$$

i.e. X *is* a random variable.

As with the binomial, the exact shape of the distribution depends on the value of the parameter μ. However, since (theoretically) the distribution takes an infinite number of x values (however small the associated probabilities are), the distribution will always be skewed to the right.

Consider $\mu = 1$. The Poisson p.d.f. becomes:

$$\Pr(X = x) = e^{-1} \frac{1^x}{x!} = \frac{1}{e\,x!}.$$

Now:
$$\frac{1}{e} = \frac{1}{2.7183} \text{ (approx)} = 0.3679 \text{ (4D)}$$

Hence:
$$\Pr(X = 0) = (0.3679)\frac{1}{0!} = 0.3679$$

$$\Pr(X = 1) = (0.3679)\frac{1}{1!} = 0.3679$$

$$\Pr(X = 2) = (0.3679)\frac{1}{2!} = 0.1840$$

(Already the sum of these first three probabilities is nearly 0.92.)

$$\Pr(X = 3) = \frac{0.3679}{3!} = 0.0613$$

$$\Pr(X = 4) = \frac{0.3679}{4!} = 0.0153$$

$$\Pr(X = 5) = \frac{0.3679}{5!} = 0.0031$$

and the value of subsequent probabilities drop quite rapidly from here.

For $\mu = 3$, the Poisson p.d.f. becomes $\Pr(X = x) = e^{-3}\dfrac{3^x}{x!}$ and with $e^{-3} = 0.0498$ (4D) we have:

$$\Pr(X = 0) = 0.0498,\ \Pr(X = 1) = \frac{(0.0498)3}{1!} = 0.1494,$$

$$\Pr(X = 2) = \frac{(0.0498)9}{2!} = 0.2241,\ \Pr(X = 3) = \frac{(0.0498)27}{3!} = 0.2241$$

and also:

$$\Pr(X = 4) = 0.1680,\ \Pr(X = 5) = 0.1008,\ \Pr(X = 6) = 0.0504,$$

$$\Pr(X = 7) = 0.0216 \text{ and } \Pr(X = 8) = 0.0081.$$

The two frequency polygons for these distributions together with that for the distribution with $\mu = 6$ are shown in Fig. 6.5.

Notice that, with increasing μ, the distribution becomes less skew.

The shorthand notation used for the Poisson distribution is given as follows: If $\Pr(X = x) = e^{-\mu}\dfrac{\mu^x}{x!}$, then we write X is $Po(\mu)$, i.e., X has a Poisson distribution with parameter μ.

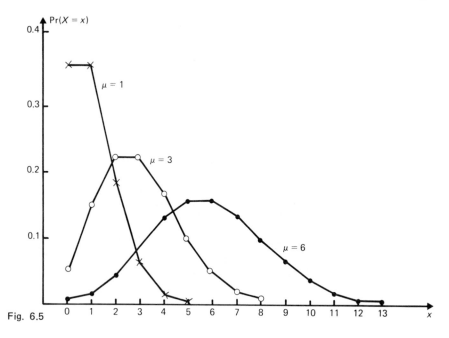

Fig. 6.5

EXAMPLE 6.11

If X is Po(1.6), find:
(a) $\Pr(X = 0)$,   (b) $\Pr(X = 1)$ (c) $\Pr(X = 2)$
(d) $\Pr(X \leqslant 2)$ and   (e) $\Pr(X > 2)$

SOLUTION

Since X is Po(1.6), $\Pr(X = x) = e^{-1.6} \dfrac{(1.6)^x}{x!}$ $(x = 0, 1, 2, \ldots)$.

(a) $\Pr(X = 0) = e^{-1.6} \dfrac{(1.6)^0}{0!} = e^{-1.6} = 0.2019$ (from expo-nential tables)

(b) $\Pr(X = 1) = e^{-1.6} \dfrac{(1.6)^1}{1!} = (1.6)(0.2019) = 0.3230$

(c) $\Pr(X = 2) = e^{-1.6} \dfrac{(1.6)^2}{2!} = (0.2019)\dfrac{(1.6)^2}{2} = 0.2584$

(d) $\Pr(X \leqslant 2) = \Pr(X = 0) + \Pr(X = 1) + \Pr(X = 2)$ (mutually exclusive events)

$$= 0.2019 + 0.3230 + 0.2584 = 0.7833$$

(e) $\Pr(X > 2) = 1 - \Pr(X \leqslant 2) = 1 - 0.7833 = 0.2167$

EXAMPLE 6.12

The random variable X is Po(0.8) and random variable Y is Po(0.8a).
If $\Pr(X = 2) = \Pr(Y = 0)$, find the value of a approximately.

SOLUTION

$$\Pr(X = 2) = e^{-0.8} \frac{(0.8)^2}{2!} = \frac{(0.4493)(0.8)^2}{2} = 0.1438$$

Also $\Pr(Y = 0) = e^{-0.8a}$. Therefore $e^{-0.8a} = 0.1438$.

From exponential tables, $e^{-1.94} \approx 0.1438$.

Hence $-0.8a = -1.94$, i.e., $a = 2.43$ (approximately).

6.4.2 THE MEAN AND VARIANCE

> **STATEMENT 6.6**
>
> If the random variable X has a Poisson distribution with parameter μ (i.e., X is $Po(\mu)$), then:
>
> (a) $E[X] = \mu$ and (b) $Var[X] = \mu$

Proof   If X is $Po(\mu)$ then:

$$\Pr(X = x) = e^{-\mu} \frac{\mu^x}{x!} \quad (x = 0, 1, 2, 3, \ldots)$$

$$\therefore \quad E[X] = \sum_{x=0}^{\infty} x\, e^{-\mu} \frac{\mu^x}{x!} \quad \text{(by definition)}$$

$$= \sum_{x=1}^{\infty} x\, e^{-\mu} \frac{\mu^x}{x!} \quad \text{(since } x = 0 \text{ contributes nothing to the sum)}$$

$$= e^{-\mu} \sum_{x=1}^{\infty} \frac{\mu^x}{(x-1)!} = \mu\, e^{-\mu} \sum_{x=1}^{\infty} \frac{\mu^{x-1}}{(x-1)!}$$

Making the substitution $y = x - 1$, we obtain:

$$E[X] = \mu\, e^{-\mu} \left[\sum_{y=0}^{\infty} \frac{\mu^y}{y!} \right]$$

$$= \mu\, e^{-\mu}\, e^{\mu} \quad \text{(since the bracketed term is } e^{\mu} \text{ by definition)}$$

$$= \mu \quad (e^{-\mu}\, e^{\mu} = 1)$$

i.e., $E[X] = \mu$ as required.

To find $Var[X]$ we use the identity $Var[X] = E[X^2] - E^2[X]$.

Now: $$E[X^2] = \sum_{x=0}^{\infty} x^2\, e^{-\mu} \frac{\mu^x}{x!}$$

$$= \sum_{x=0}^{\infty} (x(x-1) + x)\, e^{-\mu}\, \frac{\mu^x}{x!}$$

$$= \sum_{x=2}^{\infty} x(x-1)\, e^{-\mu}\, \frac{\mu^x}{x!} + \sum_{x=1}^{\infty} x\, e^{-\mu}\, \frac{\mu^x}{x!}$$

$$= e^{-\mu} \sum_{x=2}^{\infty} \frac{\mu^x}{(x-2)!} + \mathrm{E}[X]$$

$$= \mu^2\, e^{-\mu} \sum_{x=2}^{\infty} \frac{\mu^{x-2}}{(x-2)!} + \mu$$

$$= \mu^2\, e^{-\mu} \sum_{y=0}^{\infty} \frac{\mu^{y}}{y!} + \mu \quad \text{(with substitution } y = x - 2)$$

$$= \mu^2\, e^{-\mu}\, e^{\mu} + \mu$$

$$\therefore \qquad \mathrm{E}[X^2] = \mu^2 + \mu$$

Hence:
$$\mathrm{Var}[X] = \mu^2 + \mu - \mu^2$$
$$= \mu \quad \text{as required}$$

So that, the parameter (μ) involved in a Poisson distribution has the significance of being both the mean and variance of the distribution, this fact often characterising the Poisson among discrete, integral-valued distributions.

6.4.3 USES OF THE POISSON DISTRIBUTION

We have discussed and demonstrated how the binomial distribution is used practically. We now look at the two main practical uses of a Poisson distribution.

The number of events occurring in some given time (or space) interval follow approximately a Poisson distribution if:

(a)   the defined events occur at random, and

(b)   the event has a low probability of occurrence at any given instant.

Often, the events in question are known as *rare events*.

For convenience, we can call this set-up a *Poisson situation* or *process*. This result has been shown empirically for events such as telephone calls to a switchboard, blemishes in lengths of manufactured wire, car accidents and insurance claims. It can also be demonstrated mathematically.

For instance, given that calls coming to some telephone switchboard occur randomly with a mean of 5 per minute, then the probabilities of $0, 1, 2, \ldots$, etc. calls per minute is given (approximately) by a Poisson distribution with mean 5. So that, if we associate the random variable X with the number of calls per minute, we have:

$$Pr(X = x) = e^{-5} \frac{5^x}{x!} \quad (x = 0, 1, 2, 3, \dots)$$

If an insurance company is known to average 2 major claims each day then:

(a) assuming that major claims occur randomly (this may not be true in certain cases), and

(b) defining the random variable X as 'number of major claims per day', we have:

$$Pr(X = x) = e^{-2} \frac{2^x}{x!} \quad (x = 0, 1, 2, 3, \dots)$$

giving the probability that the company will receive $0, 1, 2, \dots$ major claims in any one day.

The second practical use of a Poisson distribution is as an approximation to a binomial distribution having the same mean.

If a binomial distribution has a large n (greater than 50) and a small p (less than $\frac{1}{10}$), it can be shown that the probabilities of $0, 1, 2, \dots$ successes given by a Poisson distribution with parameter $\mu = np$ approximates well to the true probabilities given by the defined binomial distribution. This is a reasonable result, since if the probability of success in a given binomial situation is small, it can be regarded as a relatively 'rare' event and, together with the fact that binomial trials are mutually independent, these events can be taken as occurring randomly, thus tying in with the previously defined Poisson situation.

The reasons we might use this approximation (if the conditions are satisfied) are:

(a) in general, Poisson probabilities are easier to calculate than their binomial counterparts, and

(b) given a binomial situation where *only* the mean, np, is known, it is possible to calculate probabilities using the Poisson method.

For example, given that the probability of a machine producing a defective item is 0.015, suppose that we wish to find the probability of discovering at least two defective items in a single lot of 100 produced by the machine. Here we have a binomial situation with $n = 100$ and $p = 0.015$, but since $n > 50$ and $p < \frac{1}{10}$, we can use a Poisson approximation with $\mu = np = 100(0.015) = 1.5$

Hence, putting the random variable X as 'no. of defects' we have X as $Po(1.5)$ approximately, and $Pr(\text{at least 2 defects}) = Pr(X \geqslant 2)$

$$= 1 - (Pr(X = 0) + Pr(X = 1)) = 1 - e^{-1.5} + e^{-1.5}(1.5))$$

$$= 1 - e^{-1.5}(1 + 1.5) \qquad\qquad = 1 - (2.5)(0.2231)$$

$$= 0.442 \ (3D)$$

We summarise with the following two statements.

STATEMENT 6.7

If an event with a low probability of occurrence at any instant (called a *rare event*) is known:

(a)   to occur randomly, and

(b)   to have a mean number of occurrences μ in a given time (or space) interval, then the number of occurrences *in this same interval* has a Poisson distribution (approximately) with parameter μ.

No Proof

Note   Care must be taken with 'unit' intervals. For instance, if we know that calls come to a switchboard at an average rate of 100 per hour, then the unit interval would be one hour. If we wanted to know the probabilities of $0, 1, 2, \ldots$ calls in any period of one hour, we could use a Poisson distribution with parameter 100. However, to find the probabilities of $0, 1, 2, \ldots$ calls in any half-hour period, we would need to use the *average rate per half-hour* as the Poisson parameter, which in this case is $100/2 = 50$. Similarly, the number of calls per quarter-hour and per minute would also have Poisson distributions with parameters $100/4$ and $100/60$ respectively.

STATEMENT 6.8

A binomial distribution with parameters n and p can be approximated by a Poisson distribution with parameter $np = \mu$ if n is large and p is small $\left(\text{in general } n > 50 \text{ and } p < \dfrac{1}{10}\right)$, the approximation being better for very large n and small p (i.e., $n \to \infty$ and $p \to 0$).

No Proof

EXAMPLE 6.13

Government statistics for a certain country show that the average number of major industrial accidents per year for large firms is 1.1 per 5000 employees. Find the probability that there will be at least one major accident in a year for a firm with:

(a)   5000 employees   (b)   10 000 employees.

State any assumptions made.

SOLUTION

We will associate the random variable X with the number of major accidents.

(a)   Here, we have a unit scale of 5000 employees and we are given that the average accident rate (per year) is 1.1 per 5000 employees. So that, approximately, X is Po(1.1).

$$\therefore \qquad \Pr(X \geqslant 1) = 1 - \Pr(X = 0)$$
$$= 1 - e^{-1.1} = 1 - 0.3329 = 0.6671$$

(b)   We now move to a new scale of '10 000 employees', and, since the accident rate is 1.1 per 5000, we 'expect' an accident rate of 2.2 per 10 000 employees. So that now, approximately, X is Po(2.2).

$$\therefore \qquad \Pr(X \geqslant 1) = 1 - \Pr(X = 0)$$
$$= 1 - e^{-2.2} = 1 - 0.1108 = 0.8892$$

The assumption made here is, of course, that we have a Poisson situation. That is, major industrial accidents of this particular kind occur randomly.

EXAMPLE 6.14

A machine producing thin metal strip is estimated to average one minor flaw per 24 metres. If flaws occur randomly, estimate the probability of:
(a)   2 flaws in a 3 metre length;
(b)   no flaws in a 12 metre length.

SOLUTION

Let the random variable X represent 'number of flaws'.

(a)   The unit length here is 3 metres. Therefore, since we expect 1 flaw in a 24 metre length, we should expect $\frac{1}{8}$ flaws in 3 metres, i.e., X is Po(0.125) (approximately).

$$\therefore \qquad \Pr(X = 2) = e^{-0.125} \frac{(0.125)^2}{2!} = 0.0069$$

(b)   The expected number of flaws in a 12 metre length will be $\frac{1}{2}$, so that X is Po(0.5) (approximately).

$$\therefore \qquad \Pr(X = 0) = e^{-0.5} = 0.6065$$

EXAMPLE 6.15

A large population is known to be 0.3% defective. If a sample of 350 is taken randomly, calculate the probabilities (to 3D) of obtaining 0, 1, 2, . . . , 5 defectives using a Poisson approximation.

SOLUTION

This is a binomial situation with $n = 350$ (each member of the population sampled is taken as a single independent trial of the experiment) and $p = 0.003$ (0.3%) where we are taking the occurrence of a defective as a 'success'. Since $n > 50$ and $p < \dfrac{1}{10}$, we are justified in using a Poisson approximation to the binomial.

The parameter for the Poisson, the mean, is taken as:

$$np = (350)(0.003) = 1.05.$$

So, if the random variable X represents 'no. of defects in sample' we have X as Po(1.05) with:

$$\Pr(X = x) = e^{-1.05} \frac{(1.05)^x}{x!} \quad (x = 0, 1, \dots)$$

$$\therefore \quad \Pr(X = 0) = e^{-1.05} = 0.3499 = 0.350 \text{ (3D)}$$

$$\Pr(X = 1) = e^{-1.05} \frac{(1.05)^1}{1!} = (0.3499)(1.05) = 0.367 \text{ (3D)}$$

$$\Pr(X = 2) = e^{-1.05} \frac{(1.05)^2}{2!} = (0.3499) \frac{(1.05)^2}{2} = 0.193 \text{ (3D)}$$

Similarly, $\Pr(X = 3) = 0.068$; $\Pr(X = 4) = 0.018$ and $\Pr(X = 5) = 0.004$.

Note that there will always be an error when using a Poisson distribution as an approximation to a binomial distribution, since the Poisson x values take an infinite range whereas binomial x values stop at $x = n$. In general however, the Poisson probabilities ignored for the approximation sum to a proportion so small that their omission has a negligible effect in most practical circumstances.

6.4.4 RECURRENCE FORMULA AND POISSON TABLES

Although the calculation of the probabilities for a Poisson distribution are considerably easier than those for a binomial distribution, we can still cut down on arithmetic by using an approximate recurrence formula, particularly when a number of successive probabilities need to be evaluated.

> **STATEMENT 6.9**
>
> The relation between successive probabilities of a Poisson distribution defined by X as $Po(\mu)$ is given by:
>
> $$Pr(X = x + 1) = \frac{\mu}{x + 1} Pr(X = x) \quad \text{for} \quad x = 0, 1, 2, 3, \ldots$$
>
> and known as the *Poisson recurrence formula*.

Proof If X is $Po(\mu)$, then:

$$Pr(X = x) = e^{-\mu} \frac{\mu^x}{x!}; \quad Pr(X = x + 1) = e^{-\mu} \frac{\mu^{x+1}}{(x + 1)!}$$

$$\therefore \quad \frac{Pr(X = x + 1)}{Pr(X = x)} = \frac{e^{-\mu} \mu^{x+1} x!}{(x + 1)! e^{-\mu} \mu^x} = \frac{\mu}{x + 1}$$

$$\therefore \quad Pr(X = x + 1) = \frac{\mu}{x + 1} Pr(X = x), \quad \text{as required}$$

This recurrence formula is used in exactly the same way as the binomial recurrence formula.

Also, for selected values of the mean m, Appendix Table 2 gives Poisson distribution function values.

For example if X is $Poi(3.2)$, then:

$$Pr(X \leqslant 4) = 0.7806 \quad \text{(from column } m = 3.2 \text{ and row 4)}$$

and $\quad Pr(X = 3) = F(3) - F(2)$

$$= 0.6025 - 0.3799$$

$$= 0.2226$$

EXAMPLE 6.16

Calculate the probabilities that $X = 0, 1, 2, 3, 4$ and 5 or more if X has a Poisson distribution with mean 1.8.

SOLUTION

For demonstration purposes, both the recurrence formula and tables will be used to calculate the probabilities required.

Recurrence formula method:

Since X is $Po(1.8)$, the recurrence formula takes the form:

$$Pr(X = x + 1) = \frac{1.8}{x + 1} Pr(X = x) \tag{1}$$

Now $\Pr(X = 0) = e^{-1.8} = 0.1653,$ and using [1] we have:

$$\Pr(X = 1) = \frac{1.8}{1} \Pr(X = 0) = (1.8)(0.1653) = 0.2975$$

$$\Pr(X = 2) = \frac{1.8}{2}(0.2975) = 0.2678$$

$$\Pr(X = 3) = \frac{1.8}{3}(0.2678) = 0.1607$$

$$\Pr(X = 4) = \frac{1.8}{4}(0.1607) = 0.0723$$

$$\Pr(X = 5 \text{ or more}) = 1 - (\Pr(X = 0) + \ldots + \Pr(X = 4))$$
$$= 1 - (0.1653 + \ldots + 0.0723) = 1 - 0.9636$$
$$= 0.0364$$

Poisson tables method:

Using Appendix Table 2, column 1.8, we have:

$$\Pr(X = 0) = F(0) = 0.1653$$
$$\Pr(X = 1) = F(1) - F(0) = 0.4628 - 0.1653 = 0.2975$$
$$\Pr(X = 2) = F(2) - F(1) = 0.7306 - 0.4628 = 0.2678$$
$$\Pr(X = 3) = F(3) - F(2) = 0.8913 - 0.7306 = 0.1607$$
$$\Pr(X = 4) = F(4) - F(3) = 0.9636 - 0.8913 = 0.0723$$

and:

$$\Pr(X = 5 \text{ or more}) = 1 - \Pr(X \leqslant 4)$$
$$= 1 - F(4)$$
$$= 1 - 0.9636$$
$$= 0.0364$$

which agree exactly with the probabilities obtained above.

6.4.5 FITTING A THEORETICAL DISTRIBUTION

We fit a theoretical distribution for the Poisson in a similar way to that for the binomial. Since a Poisson distribution has only one parameter, μ (the mean), given a frequency distribution, we simply calculate its mean and use this as the value of the parameter μ.

For example, given the frequency distribution:

x	0	1	2	3	4	Total
f	21	18	7	3	1	50

we will fit a theoretical Poisson.

$$\sum fx = 21(0) + 18(1) + 7(2) + 3(3) + 1(4) = 45$$

$$\therefore \qquad \bar{x} = \frac{45}{50} = 0.9$$

Putting $\mu = 0.9$, we can now generate a Poisson probability distribution.

Using Poisson tables, we have:

$$Pr(0) = F(0) = 0.4066$$
$$Pr(1) = F(1) - F(0) = 0.7725 - 0.4066 = 0.3659$$
$$Pr(2) = F(2) - F(1) = 0.9371 - 0.7725 = 0.1646$$
$$Pr(3) = F(3) - F(2) = 0.9865 - 0.9371 = 0.0494$$
and: $$Pr(4) = F(4) - F(3) = 0.9977 - 0.9865 = 0.0112$$

We ignore the remaining probabilities since the given distribution terminates at $x = 4$. (The sum of the above probabilities is 0.9977, so that the error involved in termination at $x = 4$ is only $1 - 0.9977 = 0.0023$ or 0.23%.)

Multiplying each of the above probabilities by the given total frequency of 50 gives the theoretical (expected) distribution:

x	0	1	2	3	4
f	20	18	8	2	1

the frequencies having been rounded to the nearest whole number. The fit here can be seen to be very good.

One of the significant characteristics of a Poisson distribution is, of course, that mean = variance, and for the previously given distribution:

$$\sum fx^2 = 21(0) + 18(1) + 7(4) + 3(9) + 1(16) = 89$$

giving: $$s^2 = \frac{89}{50} - (0.90)^2 = 0.97$$

comparing quite well with $\bar{x} = 0.90$.

Often, given a frequency distribution of the binomial/Poisson type the question of whether to fit a binomial or Poisson distribution can often be decided by how close the mean is to the variance. The closer these two are in value, the more likely that a Poisson distribution is applicable.

6.4.6 POISSON PROBABILITY PAPER

The form of this special graph paper is shown in Fig. 6.6.

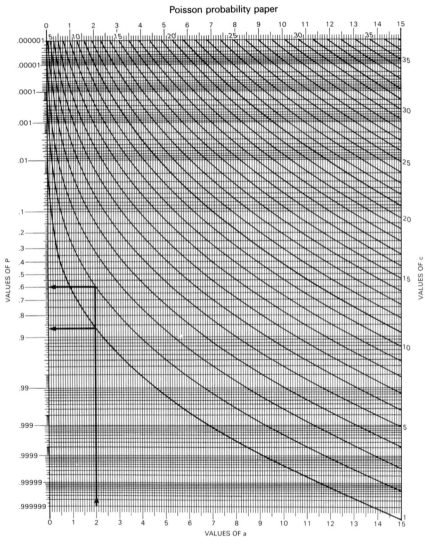

Poisson probability paper

Fig. 6.6

It is scaled on the vertical axis as *Poisson (reverse) probability* and on the horizontal axis as *Poisson mean (a)*. Also included is a set of curves representing '*c or more events*'.

There are two main uses of this special paper: (i) for calculating probabilities, and (ii) for testing a Poisson fit.

(i) Calculating probabilities
Given any Poisson mean, *a*, we can use the paper to estimate the probabilities of 0, 1, 2, 3, ... etc. events in the following way.

Where any c curve meets a vertical a value, a probability can be read off on the left vertical scale. This is the probability that c *or more events* will occur in a Poisson process with mean a.

For example, suppose X is Po(2).

Curve $c = 1$ meets $a = 2$ at corresponding probability 0.86 (read off the probability axis) and is shown on the graph in Fig. 6.6.

i.e. $\Pr(1 \text{ or more events}/a = 2) = 0.86$.

Thus $\Pr(0) = 1 - 0.86 = 0.14$

Curve $c = 2$ meets $a = 2$ at corresponding probability 0.60 and is also shown on the graph in Fig. 6.6.

Thus $\Pr(2 \text{ or more}) = 0.60$

Therefore $\Pr(2) = \Pr(1 \text{ or more}) - \Pr(2 \text{ or more})$

$$= 0.86 - 0.60$$

$$= 0.26$$

Other probabilities can be calculated using:

$$\Pr(3) = \Pr(2 \text{ or more}) - \Pr(3 \text{ or more})$$

$$\Pr(4) = \Pr(3 \text{ or more}) - \Pr(4 \text{ or more}) \dots \text{etc.}$$

(ii) Testing a Poisson fit

Given a frequency distribution, we can test whether a given frequency distribution is Poisson by rearranging the distribution to show c or more events against their (relative frequency) probability. If the distribution is Poisson, these plotted points should lie in a vertical line, the mean of the distribution.

For example, if we wished to test whether the following distribution was Poisson:

x	0	1	2	3	4	5	6	7 or more
f	1	2	5	8	9	7	4	4

we would proceed as follows:

x	0	1	2	3	4	5	6	7 or more
f	1	2	5	8	9	7	4	4
reverse F	40	39	37	32	24	15	8	4
p	1	0.975	0.925	0.8	0.6	0.375	0.2	0.1
c		1	2	3	4	5	6	7

Note that the 'reverse F' row is read as '40 items have value 0 or more; 39 items have value 1 or more, ... etc.' and the 'p' row is calculated as (reverse F) $\div$ (Σf).

The p probabilities need to be plotted using the appropriate c curve on the Poisson paper and this is shown in Fig. 6.7.

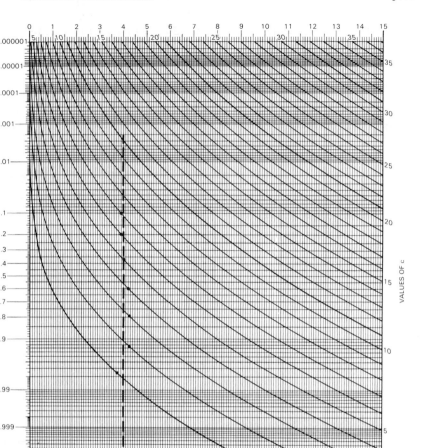

VALUES OF a

Fig. 6.7

Notice that the points do not form a perfectly straight vertical line so that the given distribution is not perfectly Poisson. However, the fit is not too unreasonable and an estimate of the respective Poisson mean is seen to be about 4.

6.5 THE NORMAL DISTRIBUTION

As the binomial and Poisson distributions were important examples of special distributions of the discrete kind, so the Normal distribution can be described as the single, most important continuous distribution in Statistics. The form of the distribution approximates very well to data of the 'natural phenomenon' type, such as weights, heights and ages; data that occurs naturally in all types of situations. Many of the frequency distributions given in the examples and exercises of Chapters 1 and 2 were of an approximate Normal type and much of the practical applications in later chapters assume that data to hand is of this type.

We begin with the equation of the p.d.f. which describes the distribution (Section 6.5.1) and examine some important characteristics. The two-parameter equation of the probability curve when used specifically for evaluating probabilities is so complex however that we need to use a special technique, peculiar to this distribution. We fix on a particular form of the curve (called the Standard Normal) and transform all other curves to this type for the purposes of problem solving.

6.5.1 DEFINITION, NOTATION AND CHARACTERISTICS

DEFINITION 6.5

A continuous random variable X having a p.d.f. of the form:

$$f(x) = \frac{1}{\sigma\sqrt{2\pi}} \exp\left(-\frac{(x-\mu)^2}{2\sigma^2}\right) \quad \text{for} \quad -\infty < x < \infty$$

and $-\infty < \mu < \infty$, $\sigma > 0$ is said to have a *Normal distribution*. Note that μ and σ (pronounced sigma) are *parameters*.

Since we are asserting that X with this distribution is a random variable, we must necessarily have that:

$$\int_{-\infty}^{\infty} \frac{1}{\sigma\sqrt{2\pi}} \exp\left(-\frac{(x-\mu)^2}{2\sigma^2}\right) dx = 1$$

(We will not attempt to prove this result here since the necessary mathematical techniques involved are beyond the scope of this text — however interested students are referred to Exercise 119.)

A convenient *shorthand notation* for a random variable distributed in this way is X is $N(\mu, \sigma^2)$ and read as 'X is distributed Normally with parameters μ and σ^2'.

The significance of these two parameters is shown in the following statement.

STATEMENT 6.10

If X is $N(\mu, \sigma^2)$, i.e., Normal with parameters μ and σ^2, then:

(a) $E[X] = \mu$ and   (b) $\text{Var}[X] = \sigma^2$.

Proof   See Section 6.7.4.

That is, the parameters involved in the p.d.f. of the Normal distribution (μ and σ^2) are simply the mean and variance respectively.

The proof of the above statement again involves mathematical techniques that are beyond the scope of the text.

The curve represented by a normal p.d.f. takes on a characteristic 'bell shape' for all values of μ and σ^2 and it can be shown that the distribution is symmetric about its mean, μ, as demonstrated in Fig. 6.8. That is, the portion of the curve to the left of $x = \mu$ is a mirror image of that portion to the right.

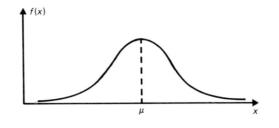

Fig. 6.8

Depending on the values of the parameters μ and σ^2, so the curve will alter in appearance. Fig. 6.9(a) shows the effect of altering μ with σ fixed and (b) the effect of altering σ with μ fixed.

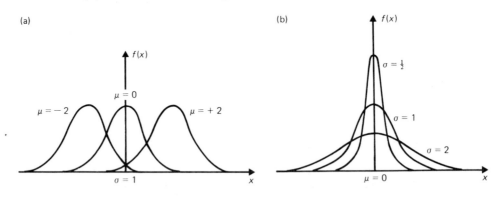

Fig. 6.9

To calculate probabilities associated with a Normal distribution we would need to integrate the relevant p.d.f. between the two limits that we were interested in.

For example, given that X is $N(100, 5)$, i.e., mean 100 and variance 5 we have:

$$f(x) = \frac{1}{\sqrt{5}\sqrt{2\pi}} \exp\left(-\frac{(x-100)^2}{(2)(5)}\right)$$

as the p.d.f. and, for instance, to calculate the probabilities $\Pr(X < 90)$, $\Pr(80 < X < 120)$ and $\Pr(X > 130)$ we would have to evaluate the integrals:

$$\int_{-\infty}^{90} f(x)\, dx, \quad \int_{80}^{120} f(x)\, dx \quad \text{and} \quad \int_{130}^{\infty} f(x)\, dx \quad \text{respectively}$$

This would be a far too long and complicated procedure to follow every time a probability is required, and the following sub-section develops an approach enabling the probabilities associated with any Normal distribution to be evaluated quickly and easily.

6.5.2 THE STANDARD NORMAL DISTRIBUTION

DEFINITION 6.6

A Normal distribution having $\mu = 0$ and $\sigma^2 = 1$ is called a *Standard Normal distribution*. The random variable associated with this distribution is usually denoted by Z, so that Z is $N(0, 1)$.

Putting $\mu = 0$ and $\sigma^2 = 1$ in the p.d.f. of the Normal distribution gives $\dfrac{1}{\sqrt{2\pi}} \exp\left(-\dfrac{x^2}{2}\right)$ usually denoted by $\phi(x)$ (ϕ is pronounced 'phi'), i.e., the p.d.f. of a standard Normal variable Z is $\phi(x) = \dfrac{1}{\sqrt{2\pi}} \exp\left(-\dfrac{x^2}{2}\right)$.

Recall from Section 5.3.2 that the distribution function of a continuous random variable X with p.d.f. $f(x)$ is written as $F(x)$ and defined as $\Pr(X < x)$. The distribution function of a standard Normal random variable Z is thus denoted by $\Phi(x)$ (where Φ is just 'capital' ϕ) and is given by:

$$\Phi(x) = \Pr(Z < x)$$

$$= \int_{-\infty}^{x} \frac{1}{\sqrt{2\pi}} \exp\left(-\frac{x^2}{2}\right) dx \quad \text{(by definition)}$$

Tables have been drawn up for $\Phi(x)$ for a wide range of values of x (see Appendix Table 3).

The area given by the tables, $\Phi(x)$, is shown in Fig. 6.10(a).

For example, the entry $\Phi(x) = 0.5398$ against $x = 0.10$ has the significance $\Pr(Z < 0.10) = 0.5398$ (see Fig. 6.10(b)).

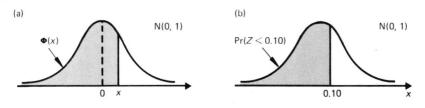

Fig. 6.10

Most Normal distribution function tables (and these in particular) only give values of $\Phi(x)$ for x in the range $x \geqslant 0$, and probabilities such as $\Pr(Z < -0.1)$ and $\Pr(Z > 0.3)$ have to be 'transformed' into probabilities of the type $\Pr(Z < x)$ where $x \geqslant 0$.

The results of the following statement enable us to calculate all types of probabilities connected with a standard Normal distribution using the symmetry property of the Normal curve.

STATEMENT 6.11

If Z is a standard Normal random variable (i.e., Z is $N(0, 1)$) and $\Phi(x) = \Pr(Z < x)$ then for $a \geqslant 0$:

(a) $\Pr(Z > a) = 1 - \Pr(Z < a) = 1 - \Phi(a)$;

(b) $\Phi(-a) = 1 - \Phi(a)$;

(c) $\Pr(Z > -a) = \Phi(a)$.

(d)   For any b and c (positive or negative)

$$\Pr(b < Z < c) = \Phi(c) - \Phi(b).$$

Proof   (a)   The area under the standard Normal curve we require is shaded (Area B) (see Fig. 6.11(a)). Since Z is a random variable, we have total area under curve as 1, i.e., $A + B = 1$.

Hence: $\Pr(Z > a) \;=\; \text{Area B} \;=\; 1 - \text{Area A}$

$$= 1 - \Phi(a) \quad \text{(by definition)}$$

(b)   The area we require is shaded (Area A) (see Fig. 6.11(b)). Now, since the curve is symmetric we must have Area A = Area B

i.e. $\Phi(-a) \;=\; \Pr(Z > a)$

$$= 1 - \Phi(a) \quad \text{using part (a)}$$

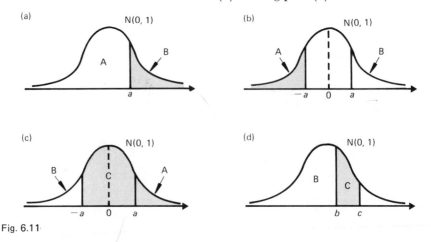

Fig. 6.11

(c)   The area we require is shaded   (Area C + Area A)   (see Fig. 6.11(c)).   Now,   Area A = Area B   (by symmetry).   Therefore Area (A + C) = Area (B + C).

∴ $\Pr(Z > -a) = \Pr(Z < a)$

                                        $= \Phi(a)$ (by definition)

(d)   The area required is shaded   (Area C) (see Fig. 6.11(d)).

So: $\Pr(b < Z < c)$ = Area C

                    = Area (B + C) − Area B

                    $= \Pr(Z < c) - \Pr(Z < b) = \Phi(c) - \Phi(b)$ (by definition)

In Fig. 6.11(d) we have assumed that both b and c are ≥ 0.

But, clearly, the same argument applies if either b is < 0 or both of b and c are < 0.

EXAMPLE 6.17

Find the values of:

(a) $\Phi(1.62)$ (b) $\Phi(-0.74)$ and   (c) $\Pr(Z > -1.0)$.

SOLUTION

(a) $\Phi(1.62) = 0.9474$ (directly from tables).

(b) $\Phi(-0.74) = 1 - \Phi(0.74)$ (Statement 6.11(b))

                    $= 1 - 0.7704$ (from tables)

                    $= 0.2296$

(c) $\Pr(Z > -1.0) = \Phi(1.0)$ (Statement 6.11(c))

                    $= 0.8413$ (from tables)

EXAMPLE 6.18

Find the values of:

(a) $\Pr(0.32 < Z < 1.24)$,

(b) $\Pr(|Z| < 0.4)$, and

(c) $\Pr(Z > 0.11 \text{ or } Z < -0.81)$ if Z is $N(0, 1)$.

SOLUTION

(a) $\Pr(0.32 < Z < 1.24) = \Phi(1.24) - \Phi(0.32)$ (Statement 6.11(d))

                    $= 0.8925 - 0.6255$

                    $= 0.2670$

(b) $\Pr(|Z| < 0.4) = \Pr(-0.4 < Z < 0.4)$ (by definition)

$\qquad = \Phi(0.4) - \Phi(-0.4)$

$\qquad = \Phi(0.4) - (1 - \Phi(0.4)) = 2\Phi(0.4) - 1$

$\qquad = 2(0.6554) - 1 = 1.3108 - 1$

$\qquad = 0.3108$

(c) $\Pr(Z > 0.11 \text{ or } Z < -0.81) = \Pr(Z > 0.11) + \Pr(Z < -0.81)$

$\qquad\qquad$ (mutually exclusive events)

$\qquad = 1 - \Phi(0.11) + 1 - \Phi(0.81)$

$\qquad = 2 - 0.5438 - 0.7910$

$\qquad = 0.6652$

We now require a method of transforming *any* Normal distribution into a standard normal form. This is shown in the following statement.

STATEMENT 6.12

If the random variable X has a Normal distribution with mean μ and variance σ^2, (i.e., X is $N(\mu, \sigma^2)$), then the random variable $\dfrac{X - \mu}{\sigma}$ has a *standard Normal distribution* $\left(\text{i.e., } \dfrac{X - \mu}{\sigma} \text{ is } N(0.1)\right)$.

No Proof although we demonstrate that $\mathrm{E}\left[\dfrac{X - \mu}{\sigma}\right] = 0$ and $\mathrm{Var}\left[\dfrac{X - \mu}{\sigma}\right] = 1$, as required for a standard Normal variable.

Now:

$$\mathrm{E}\left[\frac{X - \mu}{\sigma}\right] = \mathrm{E}\left[\frac{X}{\sigma}\right] - \mathrm{E}\left[\frac{\mu}{\sigma}\right] = \frac{1}{\sigma}\mathrm{E}[X] - \frac{\mu}{\sigma} \left(\text{since } \frac{\mu}{\sigma} \text{ is a constant}\right)$$

$$= \frac{\mu}{\sigma} - \frac{\mu}{\sigma} \quad (\text{since } \mathrm{E}[X] = \mu)$$

$$= 0$$

Also: $\quad \mathrm{Var}\left[\dfrac{X - \mu}{\sigma}\right] = \mathrm{Var}\left[\dfrac{X}{\sigma}\right] + \mathrm{Var}\left[\dfrac{\mu}{\sigma}\right]$

$\left(\text{since (a) } \dfrac{X}{\sigma} \text{ and } \dfrac{\mu}{\sigma} \text{ are independent and (b) variance of } \textit{difference} \text{ of two independent variables} = \textit{sum} \text{ of variances}\right)$

$$= \frac{1}{\sigma^2}\mathrm{Var}[X] + 0 \quad (\mathrm{Var}[\text{constant}] = 0)$$

$$= \frac{\sigma^2}{\sigma^2} = 1$$

Using the results of the last statement and a set of standard normal distribution function tables, we can calculate probabilities associated with any normal distribution. The following examples demonstrate a standard technique.

EXAMPLE 6.19

The random variable X is $N(100, 16)$. Find the values of the following:

(a) $\Pr(X < 90)$ (b) $\Pr(X > 108)$ and (c) $\Pr(95 < X < 105)$.

SOLUTION

(a) $\Pr(X < 90) = \Pr(X - 100 < 90 - 100)$

$\qquad\qquad\qquad$ (subtracting 100 from both sides)

$\qquad = \Pr(X - 100 < -10)$

$\qquad = \Pr\left(\dfrac{X - 100}{4} < -\dfrac{10}{4}\right)$ (dividing both sides by 4, the s.d.)

$\qquad = \Pr(Z < -2.5)$ $\left(\text{since } \dfrac{X - 100}{4} \text{ is } N(0, 1)\right)$

$\qquad = \Phi(-2.5)$ (by definition)

$\qquad = 1 - \Phi(2.5)$ (Statement 6.11(b))

$\qquad = 1 - 0.9938$ (from tables)

$\qquad = 0.0062$

(b) $\Pr(X > 108) = \Pr\left(\dfrac{X - 100}{4} > \dfrac{108 - 100}{4}\right) = \Pr(Z > 2.0)$

$\qquad\qquad\qquad\qquad\qquad\qquad\qquad\qquad$ (standardising)

$\qquad\qquad\qquad = 1 - \Phi(2.0)$ (Statement 6.11(a))

$\qquad\qquad\qquad = 1 - 0.9772$

$\qquad\qquad\qquad = 0.0228$

(c) $\Pr(95 < X < 105) = \Pr\left(\dfrac{95 - 100}{4} < \dfrac{X - 100}{4} < \dfrac{105 - 100}{4}\right)$

$\qquad = \Pr(-1.25 < Z < 1.25)$ (standardising)

$\qquad = \Phi(1.25) - \Phi(-1.25)$ (Statement 6.11(d))

$\qquad = \Phi(1.25) - (1 - \Phi(1.25))$ (Statement 6.11(b))

$\qquad = 2\Phi(1.25) - 1$

$\qquad = 2(0.8944) - 1$ (from tables)

$\qquad = 0.7888$

EXAMPLE 6.20

Over a long period of time a man has calculated that his morning walk to work is distributed approximately normally with mean 20 min and standard deviation 3.33 min. Find the probability that, if he leaves home 23 min before he is due at work one particular morning, he will be late.

SOLUTION

Let X be the random variable 'time taken to walk to work'. Then X is $N(20, (3.33)^2)$ approximately and we require the probability that X (the man's walk) is more than 23 min.

That is, we require:
$$\Pr(X > 23) = \Pr\left(\frac{X - 20}{3.33} > \frac{23 - 20}{3.33}\right) = \Pr(Z > 0.90)$$
$$= 1 - \Phi(0.90) = 1 - 0.8159$$
$$= 0.1841$$

EXAMPLE 6.21

It is known that the waist measurement of girls of a particular age is distributed normally with mean 66 cm and standard deviation 5 cm. If 6 girls of this age are picked randomly, find the probability that at least one will have a waist measurement of less than 64 cm.

SOLUTION

Let W be the random variable 'waist measurement of girls of this age'. Then W is $N(66, 25)$.

We first calculate the probability that a single girl picked at random will have a waist measurement less than 64 cm, i.e. we find:
$$\Pr(W < 64) = \Pr\left(Z < \frac{64 - 66}{5}\right)$$
$$= \Pr(Z < -0.4) = \Phi(-0.4) = 1 - \Phi(0.4)$$
$$= 1 - 0.6554 = 0.3446$$

Now, this probability will be the same for each of the 6 girls and we can regard the choice and waist measurement of the girls as 6 independent trials of a single binomial experiment, with probability of 'success', p, as 0.3446. (Here we are regarding a waist measurement of less than 64 cm as a 'success'.)

Hence, identifying the random variable X as 'no. of waist measurements less than 64 cm', we have X as $\text{Bin}(6, 0.3446)$.

$\therefore$

$$\Pr(X \geqslant 1) = 1 - \Pr(X = 0) = 1 - (0.6554)^6 = 1 - 0.0793 = 0.9207$$

EXAMPLE 6.22

A random variable X is $N(100, 16)$ (see Fig. 6.12). If $\Pr(X > a) = 0.5636$, find the value of a.

SOLUTION

Since the probability given is greater than 0.5, a must be less than the mean, 100.

Now, $\Pr(X > a) = 0.5636$.

$$\therefore \quad \Pr\left(\frac{X - 100}{4} > \frac{a - 100}{4}\right) = 0.5636 \quad \text{(standardising)}$$

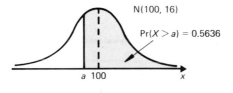

N(100, 16)

$\Pr(X > a) = 0.5636$

a 100 x

Fig. 6.12

$$\therefore \quad \Pr\left(Z > \frac{a - 100}{4}\right) = 0.5636 \quad \left(\text{since } \frac{X - 100}{4} \text{ is } N(0, 1)\right)$$

But $\dfrac{a - 100}{4}$ is a negative quantity, since $a < 100$, and can be expressed in the form $-\left(\dfrac{100 - a}{4}\right)$.

So, using Statement 6.11(c), we have:

$$\Pr\left(Z > -\left(\frac{100 - a}{4}\right)\right) = \Phi\left(\frac{100 - a}{4}\right)$$

Hence $\Phi\left(\dfrac{100 - a}{4}\right) = 0.5636$ and, from tables, we obtain

$\dfrac{100 - a}{4} = 0.16.$

$$\Rightarrow \qquad\qquad 100 - a = 4(0.16) = 0.64$$

$$\therefore \qquad\qquad a = 100 - 0.64 = 99.36$$

We end this sub-section with a useful property of any normal curve, utilising its symmetric nature.

STATEMENT 6.13

(a)   The central 95% of a standard Normal distribution lies between the limits ± 1.96 (see Fig. 6.13). Alternatively, the central 95% of any Normal distribution lies within 1.96 standard deviations of its mean.

(b)   The central $\begin{cases} 99\% \\ 99.8\% \end{cases}$ of a standard Normal distribution lies

between the limits $\begin{cases} \pm 2.58 \\ \pm 3.09 \end{cases}$

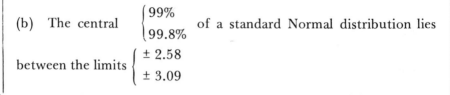

Fig. 6.13

Proof   (a)   Let the central 95% of a standard Normal distribution lie between $\pm a$.   Then we have $\Phi(a) = 0.95 + 0.025 = 0.975$ (see Fig. 6.14(a)). Therefore $a = 1.96$ (from tables).

Alternatively, if X is $N(\mu, \sigma^2)$, let the central 95% of the distribution lie between b and a (see Fig. 6.14(b)).

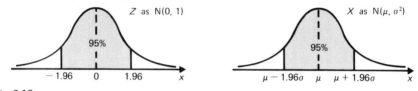

Fig. 6.14

We have: $\Pr(X < a) = 0.95 + 0.025$

$$= 0.975$$

Standardising, we have:

$$\Pr\left(Z < \frac{a - \mu}{\sigma}\right) = 0.975$$

$$\Rightarrow \quad \Phi\left(\frac{a - \mu}{\sigma}\right) = 0.975$$

$$\Rightarrow \quad \frac{a - \mu}{\sigma} = 1.96 \quad \text{(from tables)} \quad (a > \mu)$$

$$\Rightarrow \quad a = \mu + 1.96\sigma$$

Also $\Pr(X < b) = 0.025$ and standardising we have:

$$\Pr\left(Z < \frac{b - \mu}{\sigma}\right) = 0.025.$$

$\Rightarrow \qquad \Phi\left(\frac{b - \mu}{\sigma}\right) = 0.025$

$\Rightarrow \qquad 1 - \Phi\left(\frac{\mu - b}{\sigma}\right) = 0.025 \quad (\text{since } b < \mu)$

$\Rightarrow \qquad \Phi\left(\frac{\mu - b}{\sigma}\right) = 1 - 0.025 = 0.975$

$\Rightarrow \qquad \frac{\mu - b}{\sigma} = 1.96 \quad (\text{from tables})$

$\therefore \qquad b = \mu - 1.96\sigma$

The proof of (b) is left as an exercise.

EXAMPLE 6.23

A large number of students sit an examination. If the marks obtained are known to be normally distributed and:

(a)   14% of the students obtained less than 30 marks,

(b)   26% obtained more than 50 marks,

find the mean and variance of the mark distribution.

SOLUTION

We have a Normal distribution here with unknowns mean μ and variance σ^2.

Identifying the random variable X as 'no. of marks obtained in exam' we have X is $N(\mu, \sigma^2)$. The information given in (a) and (b) can be written in terms of probabilities as $\Pr(X < 30) = 0.14$ and $\Pr(X > 50) = 0.26$. The situation is shown in Fig. 6.15 where we note that $30 < \mu$ and $50 > \mu$ since both 0.14 and 0.26 are less than 0.5. Now, $\Pr(X < 30) = 0.14$.

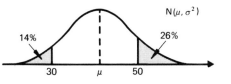

Fig. 6.15

$\Rightarrow \qquad \Pr\left(\frac{X - \mu}{\sigma} < \frac{30 - \mu}{\sigma}\right) = 0.14$

$\Rightarrow \qquad \Pr\left(Z < \frac{30 - \mu}{\sigma}\right) = 0.14$

i.e. $\Pr\left(Z < -\left(\dfrac{\mu - 30}{\sigma}\right)\right) = 0.14$

$\left(\text{where } \dfrac{\mu - 30}{\sigma} \text{ is positive since } 30 < \mu\right)$

$\Rightarrow \qquad \Phi\left(-\left(\dfrac{\mu - 30}{\sigma}\right)\right) = 0.14$

$\Rightarrow \qquad 1 - \Phi\left(\dfrac{\mu - 30}{\sigma}\right) = 0.14 \quad \text{(Statement 6.13(b))}$

$\Rightarrow \qquad \Phi\left(\dfrac{\mu - 30}{\sigma}\right) = 1 - 0.14 = 0.86$

Hence, from tables:

$$\frac{\mu - 30}{\sigma} = 1.08 \quad \text{(approximately)} \tag{1}$$

Similarly:

$$\Pr(X > 50) = 0.26 \quad \Rightarrow \quad \Pr\left(Z > \frac{50 - \mu}{\sigma}\right) = 0.26$$

$\Rightarrow \qquad 1 - \Phi\left(\dfrac{50 - \mu}{\sigma}\right) = 0.26 \quad \text{(Statement 6.13(a))}$

$\Rightarrow \qquad \Phi\left(\dfrac{50 - \mu}{\sigma}\right) = 1 - 0.26 = 0.7400$

But, from tables, $\Phi(0.64) = 0.7389$ and $\Phi(0.65) = 0.7422$.

Hence (by interpolation):

$$\frac{50 - \mu}{\sigma} = 0.643 \quad \text{(approximately)} \tag{2}$$

Now, [1] gives $\mu - 30 = 1.08\,\sigma$ and [2] gives $50 - \mu = 0.643\,\sigma$,

i.e. $\mu = 1.08\,\sigma + 30$ and $\mu = 50 - 0.643\,\sigma$

$\Rightarrow \quad 1.08\,\sigma + 30 = 50 - 0.643\,\sigma$

$\Rightarrow \qquad 1.723\,\sigma = 20$

Hence: $\sigma = \dfrac{20}{1.723} = 11.61 \quad \text{(2D)}$

Substitution back gives $\mu = (1.08)(11.61) + 30 = 42.54$ (2D).

Thus, mean $= 42.54$ (2D) and variance $= (11.61)^2 = 134.8$ (1D).

6.5.3 FITTING A THEORETICAL DISTRIBUTION

We now look at the problem of fitting a Normal distribution.

There are two possible situations to be considered:

(a)   *a given frequency distribution* – from which we would need to calculate $\bar{x}$ and s^2 in order to generate a theoretical distribution using $\mu = \bar{x}$ and $\sigma^2 = s^2$, and

(b)   *given only values of* μ *and* σ^2 – here the problem is determining numbers and sizes of classes in order to present a distribution adequately (this is distinctly non-trivial).

Although the techniques that follow might at first appear long and complex, the student needs only an understanding of the meaning and use of the standard Normal distribution function, the rest is a logical development (using basic techniques) up to, and after, its use.

Case (a)   We will go through the technique with a worked example. We wish to fit a Normal distribution to the following data:

Class	Frequency
Up to 15	3
15 up to 20	7
20 up to 25	15
25 up to 30	20
30 up to 35	9
35 and over	4
Total	58

The information we shall be using from this given distribution is as follows:

(a)   *the number and size of the classes* – this is in order that we can compare directly the above distribution with the one we fit theoretically;

(b)   *the total frequency* – since after generating a normal *probability* distribution we need to multiply each of the class probabilities by this total frequency in order to obtain a theoretical *frequency* distribution;

(c)   *the mean,* $\bar{x}$, *and variance,* s^2 – this is clearly necessary, since in order to fix a particular Normal distribution we need a knowledge of its two parameters μ and σ^2, which of course are estimated by $\bar{x}$ and s^2.

To construct the theoretical distribution we require the respective probabilities of a Normal variable (with parameters $\mu = \bar{x}$ and $\sigma^2 = s^2$) lying in the ranges given by the stated classes.

In general, we need to calculate the probabilities $p_1, p_2, \ldots, p_n, p_{n+1}$ (see Fig. 6.16(a)), where $u_1, u_2, \ldots, u_n$ are the respective upper bounds of each class. If we 'standardise' the class upper bounds $u_1, \ldots, u_n$ (by subtracting $\bar{x}$ and dividing by s), we obtain values $z_1, z_2, \ldots, z_n$, say (see Fig. 6.16(b)). Since these z-values comes from a $N(0, 1)$ distribution, we can now use standard Normal distribution function tables to find $\Pr(Z < z_1) = p_1$, $\Pr(Z < z_2) = p_1 + p_2$, $\Pr(Z < z_3) = p_1 + p_2 + p_3 \ldots$ etc. and by subtraction $p_2, p_3, \ldots, p_n$ can be determined.

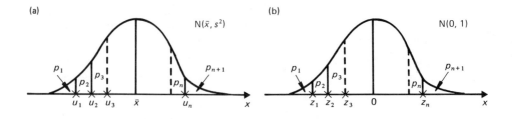

Fig. 6.16

Working with the frequency distribution given, we first calculate $\bar{x} = 25.7$ (1D) and $s^2 = 37.7$ (1D) with $s = 6.14$. (The lower bound of the first class was taken as 10, the upper bound of the last class as 40.)

Now, the upper bounds of the classes are respectively 15, 20, 25, 30, 35 and ∞ (since the last class is open-ended).

Standardising these bounds, we have respectively $\dfrac{15 - 25.7}{6.14} = -1.74$; $\dfrac{20 - 25.7}{6.14} = -0.93$; together with -0.11, 0.70, 1.51 and ∞, as the z-values.

Using the tables of the standard Normal distribution function we can now calculate $\Pr(Z < z) = \Phi(Z)$ for these values:

$$\Phi(-1.74) = 1 - \Phi(1.74) = 1 - 0.9591 = 0.0409$$

$$\Phi(-0.93) = 1 - \Phi(0.93) = 1 - 0.8238 = 0.1762$$

together with 0.4562, 0.7580, 0.9345 and 1 ($= \Phi(\infty)$).

Subtracting successive values of Φ, we obtain the required probabilities and hence (by multiplying by Σf) the required expected frequencies. The table summarising the procedure is shown below.

Class	Upper bounds	Standardised u.c.b.s (z)	$\Pr(Z < z)$ $\Phi(z)$	Probability (p)	Expected frequency $(p) \times \Sigma f$
Up to 15	15	-1.74	0.0409	0.0409	2.4
15 to 20	20	-0.93	0.1762	0.1353	7.9
20 to 25	25	-0.11	0.4562	0.2800	16.2
25 to 30	30	0.70	0.7580	0.3018	17.5
30 to 35	35	1.51	0.9345	0.1765	10.2
35 and over	∞	∞	1	0.0655	3.8

In this particular case (comparing the corresponding sets of frequencies) we would say the fit obtained is 'fairly good'.

The difference in the two distributions can be effectively shown by a comparison of their frequency curves (Fig. 6.17) where it is seen immediately that the given frequency distribution is markedly more peaked

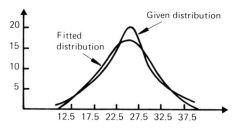

Fig. 6.17

than would be expected with a Normal distribution (the fitted distribution) having the same mean, variance and total frequency.

Note the use of upper class boundaries in fitting a Normal curve as shown in the second column of the previous table. If, for instance, we were originally given a frequency distribution with classes 20 to 29, 30 to 39, 40 to 49, ... , etc. (i.e., continuous data measured to the nearest whole number) we would enter 29.5, 39.5, 49.5, ... , etc. as the upper class bounds.

Case (b) Here, as already mentioned the problem is to determine the number and size of the classes for the presentation of the distribution, since the only information we have is the value of the two parameters.

For example, given $\mu = 120$ and $\sigma = 10$, would we construct classes 0 to 40, 40 to 80, 80 to 120, 120 to 160, ... , etc. or 100 to 105, 105 to 110, 110 to 115, ... , etc?

The danger of choosing too many classes is simply that, after fairly laborious calculations have been performed, we might find that extreme classes have no significant probability (or frequencies) falling within them, thus possibly needing to reconstruct class widths and begin calculations again. Too few classes is obviously unsatisfactory from a presentation and comparison point of view.

We can proceed to choose the class numbers and width logically (and sensibly) using Statement 6.13 which gives the result: the central 99.8% of a normal distribution lies within ± 3.09 s.d.s of the mean. We can translate this as 'most of a normal distribution lies within ± 3 s.d.s of the mean' giving (approximately) the range of a normal distribution as 6 s.d.s.

So that, given μ and σ^2, we can fix the 'practical' range within which all the classes lie as $\mu \pm 3\sigma$. As regards the number of classes (and hence their width), the choice is largely one of convenience although, in general, one would not choose less than 6 or more than 15. Too few classes do not give enough information about the distribution; too many is arduous to calculate and gives too much information to take in readily.

Consider generating a theoretical Normal distribution given only $\mu = 120$ and $\sigma = 10$. The bulk of the distribution lies between $120 \pm 3(10)$, i.e., $\mu \pm 3\sigma$, giving a range of 60 between 90 and 150. Choosing 10 classes, then, gives a common width of 6 to each class giving 90 to 96, 96 to 102, 102 to 108, . . . , 144 to 150.

However, we take account of the (theoretically) infinite range of the distribution by making the first and last classes open-ended so that we finally have the classes up to 96, 96 to 102, . . . , 138 to 144, 144 and over. The probability distribution is now generated in exactly the same way as in case (a) previously and, if a total frequency is also available, we can clearly generate an 'expected' frequency distribution on multiplication of the probabilities.

For reference purposes, we summarise these techniques.

DEFINITION 6.7

We *fit a Normal distribution* to:

(a)   a *given frequency distribution* by:
- (i)   calculating $\bar{x}$ and s^2 (as estimates of parameters μ and σ^2);
- (ii)   identifying the upper class bounds (u.c.b.s) for each class;
- (iii)   standardising the u.c.b.s, using $\bar{x}$ and s^2, to give the values $z_1, z_2, \ldots, z_n$;
- (iv)   finding $\Phi(z_1), \Phi(z_2), \ldots, \Phi(z_n)$;
- (v)   obtaining class probabilities by successive subtraction of the Φs;
- (vi)   obtaining 'expected' frequencies on multiplication of the probabilities by the total frequency of the given distribution.

(b)   A *given mean and variance* (μ and σ^2) by:
- (i)   determining the 'practical' range of the distribution as $\mu - 3\sigma$ to $\mu + 3\sigma$;
- (ii)   using this range to determine the number and size of classes (between 6 and 15 classes — usually about 10);
- (iii)   proceeding as in case (a) above.

EXAMPLE 6.24

Fit a Normal distribution given the following information:

(a)   Mean 132 and standard deviation 20.

(b)   Classes 80 to 100, 100 to 120, . . . , 160 to 180.

(c)   Total frequency 62.

SOLUTION

In tabular form:

Class	u.c.b.	Standardised u.c.b. (z)	Cumulative probability $\Phi(z)$	Probability (p)	Expected frequency $62 \times (p)$
80–100	100	−1.6	0.0548	0.0548	3
100–120	120	−0.6	0.2743	0.2195	14
120–140	140	0.4	0.6554	0.3811	24
140–160	160	1.4	0.9192	0.2638	16
160–180			1	0.0808	5

6.5.4 NORMAL APPROXIMATION TO THE BINOMIAL AND POISSON

We have seen already how, under certain conditions, the binomial distribution can be approximated by a Poisson distribution having the same mean. We can also use the Normal distribution as an approximating distribution as shown in the following statement.

STATEMENT 6.14

(a)   If X is distributed binomially with parameters n and p (i.e., X is $\text{Bin}(n, p)$), then, for large n and p not too small (or large), we can consider X as (approximately) distributed $N(np, np(1-p))$.

(b)   If X has a Poisson distribution with parameter μ (i.e., X is $\text{Po}(\mu)$) then, for large μ, we can consider X as (approximately) distributed $N(\mu, \mu)$.

That is, we can approximate both the binomial and Poisson distributions by a Normal distribution having the same mean and variance under the stated conditions.

No Proof

With regard to the above conditions, there are no set values for n, p and μ but, in general, for:

(a)   we would desire $n > 50$ and p close to $\dfrac{1}{2}$, and

(b) $\mu > 20$ for good approximations.

Before using the approximation practically, there is one important point to consider. We are approximating a discrete variable by a continuous one and we need to make a *continuity correction*. This simply entails transforming each discrete value into a corresponding continuous range. For example, the discrete values 3, 4 and 8 would be transformed to the corresponding ranges 2.5 to 3.5, 3.5 to 4.5 and 7.5 to 8.5.

To calculate the binomial probability $\Pr(X = 3)$ using the approximation would entail calculating $\Pr(2.5 < X < 3.5)$ using the appropriate Normal distribution.

As an illustration, we will take X as $\text{Bin}\left(12, \dfrac{1}{4}\right)$ and calculate $\Pr(X = 4)$
(a) exactly (b) using a Normal approximation.

(a) $\Pr(X = 4) = \dbinom{12}{4}\left(\dfrac{1}{4}\right)^4\left(\dfrac{3}{4}\right)^8 = 0.194$

(b)   Using the approximation, X is $N(np,\, np(1-p)) = N(3,\, 2.25)$ and $\Pr(X = 4)$ transforms to:

$$\Pr(3.5 < X < 4.5) = \Pr\left(\frac{3.5 - 3}{\sqrt{2.25}} < Z < \frac{4.5 - 3}{\sqrt{2.25}}\right) \quad \text{(standardising)}$$

$$= \Pr(0.33 < Z < 1.00)$$

$$= \Phi(1) - \Phi(0.33)$$

$$= 0.8413 - 0.6293 = 0.212$$

which represents a 9.3% error giving a fairly good approximation even though n is quite small.

It was pointed out previously that the approximation for the binomial was better for p close to $\dfrac{1}{2}$. We can demonstrate this by considering now X as $\text{Bin}\left(12, \dfrac{3}{8}\right)$ and calculating again $\Pr(X = 4)$ using both methods.

(a)   Exact: $\Pr(X = 4) = \dbinom{12}{4}\left(\dfrac{3}{8}\right)^4\left(\dfrac{5}{8}\right)^8 = 0.228$ (3D).

(b)   Approximation: We take X as $N(4.5, 2.8125)$. $\Pr(X = 4)$ transforms to:

$$\Pr(3.5 < X < 4.5) = \Pr\left(\frac{3.5 - 4.5}{\sqrt{2.8125}} < Z < \frac{4.5 - 4.5}{\sqrt{2.8125}}\right)$$

$$= \Phi(0) - \Phi(-0.60)$$

$$= 0.226 \quad \text{representing only a } 0.9\% \text{ error}$$

Here, the approximation is very good.

EXAMPLE 6.25

Certain emergency calls come to a telephone switchboard at an average rate of 2 per hour. Find the probability that there are:

(a)   2 or more calls in a 1 hour period,

(b)   4 or more calls in a 2 hour period, and

(c)   24 or more calls in a 12 hour period.

SOLUTION

Let X_i be the random variable 'no. of calls in a period of i hours'. Then X_i is $\text{Po}(2i)$, since we are regarding these calls as 'rare' events.

(a) $\quad \Pr(X_1 \geqslant 2) = 1 - (\Pr(X_1 = 0) + \Pr(X_1 = 1))$

$\qquad\qquad\quad = 1 - (e^{-2} + 2e^{-2}) = 1 - 3e^{-2}$

$\qquad\qquad\quad = 0.594 \quad (3D)$

(b) $\quad \Pr(X_2 \geqslant 4) = 1 - (\Pr(X_2 = 0) + \Pr(X_2 = 1)$

$\qquad\qquad\qquad\quad + \Pr(X_2 = 2) + \Pr(X_2 = 3))$

$\qquad\qquad\quad = 1 - \left(e^{-4} + 4e^{-4} + \dfrac{4^2}{2!}e^{-4} + \dfrac{4^3}{3!}e^{-4} \right)$

$\qquad\qquad\quad = 1 - e^{-4} \left(1 + 4 + 8 + \dfrac{64}{6} \right)$

$\qquad\qquad\quad = 0.567 \quad (3D)$

(c) Here, X_{12} is Po(24) and using a Normal approximation we have X_{12} as N(24, 24). We require $\Pr(X_{12} \geqslant 24)$ which transforms to $\Pr(X_{12} \geqslant 23.5)$ with a continuity correction.

Hence: $\quad \Pr(X_{12} \geqslant 23.5) = \Pr\left(Z > \dfrac{23.5 - 24}{\sqrt{24}} \right)$

$\qquad\qquad\qquad\qquad\quad = \Pr(Z > -0.102)$

$\qquad\qquad\qquad\qquad\quad = \Phi(0.102) = 0.541 \quad (3D)$

EXAMPLE 6.26

Bolts are manufactured by a machine and it is known that approximately 20% are outside certain tolerance limits. If a random sample of 200 is taken, find the probability that more than 50 bolts will be outside the limits.

SOLUTION

We have a binomial situation with $n = 200$ and $p = 0.2$. Identifying the random variable X as 'no. of bolts outside limits', X is Bin(200, 0.2) and we need to determine $\Pr(X > 50)$.

Using a Normal approximation, we put X as N(40, 32) and find $\Pr(X > 50.5)$. (Note that we have added a continuity correction here.)

$\Pr(X > 50.5) = \Pr\left(Z > \dfrac{50.5 - 40}{\sqrt{32}} \right) = \Pr(Z > 1.856) = 1 - \Phi(1.856)$

$\qquad\qquad\quad = 1 - 0.9683$

$\qquad\qquad\quad = 0.0317$

6.5.5 NORMAL PROBABILITY PAPER

This special graph paper is scaled as *Normal probability* (cumulative percentage) against *value of variable* and is shown in Fig. 6.18.

It is used for:

(a)   testing whether a given set or distribution can be considered as Normal.

(b)   estimating the mean and standard deviation of the given data.

The given data is transferred to the graph paper by plotting *cumulative frequency* (horizontal axis) against *upper class bound* (vertical axis) for each class and then fitting a STRAIGHT LINE through the points. The better the fit of the straight line to the points, so the more like a normal distribution the given data is. In other words, the plotted points corresponding to a perfect normal distribution would form an exact straight line.

The mean of the distribution is obtained by identifying the 50% point of the distribution $(Z_{50\%})$. The standard deviation can be estimated as $Z_{68\%} - Z_{50\%}$ since $Z_{68\%}$ lies at one standard deviation to the right of the mean.

The frequency distribution given in section 6.5.3 is shown below, together with necessary workings, and is then plotted on Normal probability paper in Fig. 6.18.

Upper class bound	f	Cum f	%
15	3	3	5.2
20	7	10	17.2
25	15	25	43.1
30	20	45	77.6
35	9	54	93.1

The straight line fit is good, thus confirming that the data is approximately normal.

Also shown on the graph is $Z_{50\%}$ and $Z_{68\%}$, enabling the mean and standard deviation to be estimated as 25.7 and 3.1 respectively (agreeing with that known from section 6.5.3).

When using Normal probability paper to plot a small set of values of a variable (less than 15 say), the following technique should be used:

(a)   arrange the n variable values in size order:

(b)   number them as $1, 2, 3, \ldots n$ which are equivalent to cum $f(F)$ values;

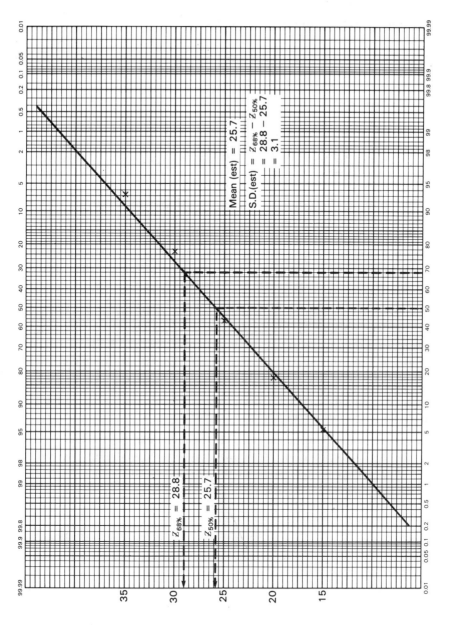

Fig. 6.18

(c)   divide each F value by $n+1$ (NOT n) and multiply by 100 to obtain $F\%$;

(d)   each original variable value is then plotted against its corresponding $F\%$ value.

The divisor of $n+1$ (instead of n) is a correction factor to take account of a small number of (apparently) discrete values being fitted with a Normal distribution.

6.6 SOME OTHER DISTRIBUTIONS

6.6.1 THE BERNOULLI DISTRIBUTION

DEFINITION 6.8

A discrete random variable X having a p.d.f. of the form:

$$\begin{cases} \Pr(X = 0) = 1 - p \\ \Pr(X = 1) = p \end{cases} \quad \text{for} \quad 0 \leqslant p \leqslant 1$$

is said to have a *Bernoulli distribution* with parameter p.

Notes   (a)   It is easily seen that X is a random variable, since the sum of probabilities of X taking its two values 0 and 1 is $(1 - p) + p = 1$ (as required).

(b)   Consider an experiment which can have only the outcomes 'success' or 'failure' where p is the probability of 'success'. Now associate the random variable X value 0 with a failure and 1 with a success and we can see that this distribution can be regarded as a special case of a binomial distribution with $n = 1$.

Indeed, the binomial distribution can be formulated solely in terms of *sums of independent Bernoulli random variables*.

For example, $n = 3$ in a binomial distribution can be represented by the sum of 3 (independent) Bernoulli random variables in the following way: Considering possible totals of the 3 random variables we have:

Total	Arrangements of r.v.s giving this total	Total no. of ways	Probability
0	000	1	$(1-p)(1-p)(1-p) = (1-p)^3$
1	001, 010, 100	3	$p(1-p)^2$
2	011, 101, 110	3	$p^2(1-p)$
3	111	1	p^3

If we now let the random variable Y_3 represent 'sum of the 3 random variables' we have (using the table):

$$\Pr(Y_3 = 0) = \Pr(000) = (1-p)^3$$

$$\Pr(Y_3 = 1) = \Pr(001 \text{ or } 010 \text{ or } 100)$$

$$= \Pr(001) + \Pr(010) + \Pr(100) \quad \text{(mutually exclusive events)}$$

$$= 3p(1-p)^2$$

and similarly:

$$\Pr(Y_3 = 2) = 3p^2 (1 - p)$$

and:

$$\Pr(Y_3 = 3) = p^3$$

which of course, is just the p.d.f. of a binomial random variable, i.e., Y_3 is $\text{Bin}(3, p)$.

An extension to the sum of n independent Bernoulli random variables (representing the n independent trials for a binomial distribution) can be shown in a similar way, giving the result Y_n is $\text{Bin}(n, p)$, where Y_n is the random variable 'sum of n independent Bernoulli random variables'.

We summarise:

STATEMENT 6.15

If $X_1, X_2, \ldots, X_n$ are n *independent Bernoulli random variables*

then, if $Y_n = \sum_{i=1}^{n} X_i$, Y_n *is* $\text{Bin}(n, p)$.

No Proof   (Demonstration for $n = 3$ in previous text.)

Since we have already noted that the Bernoulli distribution can be regarded as a special case of the binomial distribution with $n = 1$, it is easily seen that its mean and variance are p and $p(1 - p)$ respectively. However we state and prove this result using the expectation definition.

STATEMENT 6.16

If X has a Bernoulli distribution with parameter p, then:
(a) $E[X] = p$ and   (b) $\text{Var}[X] = p(1 - p)$.

Proof   (a)   By definition, $E[X] = \sum_{\text{all } x} x \cdot \Pr(X = x)$,

i.e. $$E[X] = 0 \cdot \Pr(X = 0) + 1 \cdot \Pr(X = 1)$$
$$= 0 \cdot (1 - p) + 1 \cdot p = p \quad \text{(as required)}$$

(b)   By definition, $\text{Var}[X] = E[X^2] - E^2[X]$

$$E[X^2] = \sum_{\text{all } x} x^2 \cdot \Pr(X = x)$$

$$= 0^2 \cdot \Pr(X = 0) + 1^2 \cdot \Pr(X = 1)$$

$$= 0 \cdot (1-p) + 1 \cdot p = p$$

Hence, $\mathrm{Var}[X] = p - p^2 = p(1-p)$ (as required).

6.6.2 THE GEOMETRIC DISTRIBUTION

DEFINITION 6.9

A discrete random variable X having a p.d.f. of the form:

$$\mathrm{Pr}(X = x) = (1-p)^x p \quad \text{with} \quad 0 \leqslant p \leqslant 1$$

is said to have a *geometric distribution* with parameter p.

This distribution can be derived as follows: Consider a binomial situation with p the probability of 'success'. We perform the associated experiment a number of times *until we obtain a success*.

Clearly, the experiment can be performed any (integral) number of times before the first success occurs. If we let the random variable X be defined as '*no. of failures* up to 1st success', then X can take values 0, 1, 2, 3, . . . , and:

$$\mathrm{Pr}(X = 0) = \mathrm{Pr}(\text{no failures to 1st success})$$

$$= \mathrm{Pr}(\text{success at 1st trial})$$

$$= p$$

$$\mathrm{Pr}(X = 1) = \mathrm{Pr}(\text{failure at trial 1 and success at trial 2})$$

$$= \mathrm{Pr}(\text{failure at trial 1}) \cdot \mathrm{Pr}(\text{success at trial 2})$$
$$\text{(independent events)}$$

$$= (1-p)p$$

Similarly $\mathrm{Pr}(X = 2) = (1-p)^2 p$ and $\mathrm{Pr}(X = 3) = (1-p)^3 p$.

So that, in general, $\mathrm{Pr}(X = x) = (1-p)^x p$ which is the required p.d.f. Usually the random variable X is described as 'waiting time (in trials) to first success'.

It is interesting and informative to note the characteristic difference between a geometric and a binomial random variable. They are both concerned with a simple binomial situation, the binomial distribution having the number of trials fixed and the *number of successes as the variable*, whereas the geometric distribution has the number of successes fixed (at 1) and the *number of trials as the variable*.

The shorthand notation used for this distribution is 'X is Geo(p)' and read as 'geometrically distributed with parameter p'.

For reference purposes, expressions for the mean and variance of this distribution are given in the following statement.

STATEMENT 6.17

If the random variable X has a geometric distribution with parameter p, i.e., X is $\text{Geo}(p)$, then:

(a) $E[X] = \dfrac{1-p}{p}$ and   (b) $\text{Var}[X] = \dfrac{1-p}{p^2}$

Proof   (See Section 6.7.3.)

It is noted finally that although we formulated this distribution in terms of 'waiting time to 1st success' in a given binomial situation, since separate trials are, by definition, independent, once a success has occurred, the waiting time to the next success is also geometrically distributed.

Therefore, we can, in general, describe a geometric random variable in terms of waiting time (w.r.t. number of trials) between successive binomial successes.

EXAMPLE 6.27

A person plays a game by rolling a die until he (or she) obtains a 'six', whereby £5 is received. Find the value that the game organiser should charge the thrower *per throw* in order that he may expect to gain 50 p *per game*.

SOLUTION

Since the charge per throw is unknown, let it be £a. If we call X the random variable 'number of non-winning throws', we have a 'geometric situation' with a success defined as 'throwing a six' with

$\text{Pr(success)} = p = \dfrac{1}{6}$, i.e., X is $\text{Geo}\left(\dfrac{1}{6}\right)$.

Hence: $E[X] = \dfrac{1-p}{p} = \dfrac{1-\frac{1}{6}}{\frac{1}{6}} = 5$

But, since the player also pays for a winning throw, the organiser should expect to collect £$6a$ per game (£$5a$ from losing throws, £a from the winning throw). So that, every game he will collect £$6a$ and pay out £5, and in order that he should make £0.5 profit, we need £$6a$ − £5 = £0.5, i.e., $6a = 5.5$. Therefore $a = \dfrac{5.5}{6} = 0.92$. Hence, the charge for each throw should be 92 p.

6.6.3 THE RECTANGULAR DISTRIBUTION

This distribution can be described as one of the simplest of continuous form, its probability curve consisting of a straight line parallel to the x axis. We first define the particular form of the p.d.f.

DEFINITION 6.10

A continuous random variable X having a p.d.f. of the form:

$$f(x) = \frac{1}{b-a} \quad \text{for} \quad a \leqslant x \leqslant b$$

where a and b are parameters, is said to have a *rectangular distribution*.

As soon as the limits for x are fixed (a to b in this case), the form $\frac{1}{b-a}$ for the p.d.f. is automatically fixed as is now shown.

Suppose $f(x) = c$ for $a \leqslant x \leqslant b$. Fig. 6.19 shows this situation.

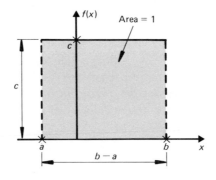

Fig. 6.19

Since X is a random variable, we must have that the area under the probability curve (the shaded area) is 1. That is, $c(b-a) = 1$.

Therefore $c = \dfrac{1}{b-a}$ as required.

The shorthand notation used for this distribution is X is $R(a, b)$, and read as 'random variable X has a rectangular distribution over the interval a to b'.

A characteristic feature of a rectangularly distributed random variable, due to its p.d.f. form, is the fact that the probability of the variable lying in a particular valid range is exactly the same as it lying in any other valid range *of the same length*.

For example, if X is $R(2, 5)$ then $\Pr(2 \leqslant X \leqslant 3) = \Pr(3.5 \leqslant X \leqslant 4.5)$ since both the ranges 2 to 3 and 3.5 to 4.5 have length 1 (see Fig. 6.20). This equality is obvious from the diagram, but we will show it by definition.

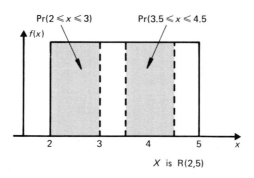

Fig. 6.20

X has the p.d.f. $f(x) = \dfrac{1}{b-a} = \dfrac{1}{5-2} = \dfrac{1}{3}$.

$\therefore \qquad\qquad \Pr(2 \leqslant X \leqslant 3) = \displaystyle\int_2^3 \dfrac{1}{3}\, dx$

$$= \dfrac{1}{3}\int_2^3 1\, dx = \dfrac{1}{3}\left[x\right]_2^3 = \dfrac{1}{3}(3-2) = \dfrac{1}{3}$$

and: $\qquad \Pr(3.5 \leqslant X \leqslant 4.5) = \displaystyle\int_{3.5}^{4.5} \dfrac{1}{3}\, dx$

$$= \dfrac{1}{3}\int_{3.5}^{4.5} 1\, dx = \dfrac{1}{3}\left[x\right]_{3.5}^{4.5} = \dfrac{1}{3}(4.5 - 3.5) = \dfrac{1}{3}$$

STATEMENT 6.18

If the random variable X has a rectangular distribution from a to b, i.e., X is $R(a, b)$, then:

(a) $E[X] = \dfrac{1}{2}(a+b)$ and (b) $\mathrm{Var}[X] = \dfrac{1}{12}(b-a)^2$

Proof (a) By definition:

$$E[X] = \int_{\text{all } x} x\, f(x)\, dx$$

$$= \int_a^b \dfrac{x}{b-a}\, dx \quad \left(f(x) = \dfrac{1}{b-a}\right)$$

$$= \dfrac{1}{b-a}\int_a^b x\, dx$$

$$= \frac{1}{b-a} \left[\frac{x^2}{2} \right]_a^b = \frac{1}{b-a} \left(\frac{b^2}{2} - \frac{a^2}{2} \right)$$

$$= \frac{1}{2(b-a)} (b^2 - a^2)$$

$$= \frac{(b-a)(b+a)}{2(b-a)} \quad \text{(difference of two squares)}$$

i.e. $E[X] = \dfrac{b+a}{2}$

(b)   Now $\text{Var}[X] = E[X^2] - E^2[X]$

$$E[X^2] = \int_a^b x^2 \frac{1}{b-a} \, dx = \frac{1}{b-a} \int_a^b x^2 \, dx$$

$$= \frac{1}{b-a} \left[\frac{x^3}{3} \right]_a^b = \frac{1}{b-a} \left(\frac{b^3}{3} - \frac{a^3}{3} \right)$$

$$= \frac{1}{3(b-a)} (b-a)(b^2 + ab + a^2) \quad \text{(difference of two cubes)}$$

$\therefore$ $E[X^2] = \dfrac{b^2 + ab + a^2}{3}$

Hence:

$$\text{Var}[X] = \frac{b^2 + ab + a^2}{3} - \left(\frac{a+b}{2} \right)^2 = \frac{b^2 + ab + a^2}{3} - \frac{a^2 + 2ab + b^2}{4}$$

$$= \frac{4(b^2 + ab + a^2) - 3(a^2 + 2ab + b^2)}{12}$$

$$= \frac{b^2 - 2ab + a^2}{12} = \frac{(b-a)^2}{12} \quad \text{(as required)}$$

This distribution has a discrete analogue. Consider rolling a normal six-sided die and associating the random variable X with 'no. of spots shown'. Then X can take values 1, 2, 3, 4, 5 and 6 where:

$$\Pr(X = 1) = \Pr(X = 2) = \ldots = \Pr(X = 6) = \frac{1}{6}$$

That is, the probability of the random variable taking any one (valid) value is the same as it taking any other.

This type of distribution is often described as *discrete rectangular*.

EXAMPLE 6.28

Find the probability that a random variable with a rectangular distribution over the interval 3 to 6.5 lies within one standard deviation of the mean.

SOLUTION

Here X is $R(3, 6.5)$ with $E[X] = \dfrac{3 + 6.5}{2} = 4.75$ and $Var[X] = \dfrac{(6.5 - 3)^2}{12} = 1.0208$.

Therefore standard deviation of $X = \sqrt{1.0208} = 1.01$ (2D).

We require the probability that X lies in the range 4.75 ± 1.01, i.e., we need:

$$\Pr(3.74 \leqslant X \leqslant 5.76) = \int_{3.74}^{5.76} \frac{1}{3.5} \, dx = \frac{1}{3.5}(5.76 - 3.74)$$

$$= 0.58 \quad (2D)$$

6.6.4 THE EXPONENTIAL DISTRIBUTION

We showed earlier that the geometric distribution could be regarded as the waiting time (in numbers of trials) between successive successes in a binomial situation. One of the uses of the following distribution is to describe the waiting time between successive events of a Poisson distribution.

DEFINITION 6.11

A continuous random variable X having a p.d.f. of the form:

$$f(x) = \mu \, e^{-\mu x}; \quad x > 0$$

is said to have an *exponential distribution* with parameter μ.

We write X is $\text{Exp}(\mu)$ in shorthand form.

STATEMENT 6.19

If a random variable has an exponential distribution with parameter μ, i.e., X is $\text{Exp}(\mu)$, then:

(a) $E[X] = \dfrac{1}{\mu}$ and (b) $Var[X] = \dfrac{1}{\mu^2}$

Proof (see Section 6.7.6.)

If this distribution is regarded as the waiting time between events in a Poisson process, then:

(a) the exponential parameter μ has just the same value as the respective Poisson parameter, and

(b) the units of time are identical in both distributions.

For example, suppose accidents on a particular stretch of road were known to occur in a Poisson fashion with mean 2 per week. Here, the time interval is '1 week' for the Poisson and the unit of time for the corresponding exponential distribution would be 'week'. Since the parameter (μ) here has value 2, the average waiting time between successive accidents should be $\frac{1}{2}$ week with variance $\frac{1}{2^2} = \frac{1}{4}$ week (using the above statement).

6.7 GENERATING FUNCTION TECHNIQUES

In this section we attempt briefly to relate particular generating function techniques to some of the special distributions studied in this chapter to find means and variances.

6.7.1 THE BINOMIAL DISTRIBUTION

We have already seen (Section 6.3.3) that the probability generating function (p.g.f.) for this distribution takes the form:

$$G(t) = E[t^X] = (1 - p + pt)^n$$

whereby the mean and variance can be determined.

The moment generating function (m.g.f.) for this distribution is:

$$M(t) = E[e^{tX}] = \sum_{x=0}^{n} e^{tx} \binom{n}{x} p^x (1-p)^{n-x}$$

$$= \sum_{x=0}^{n} \binom{n}{x} (p\, e^t)^x (1-p)^{n-x}$$

$$= (1 - p + p\, e^t)^n \quad \text{(by binomial theorem)}$$

We can find the mean and variance as follows:

$$M'(t) = n(1 - p + p\, e^t)^{n-1} p\, e^t$$

$$\therefore \quad [M'(t)]_{t=0} = E[X] = np$$

Also: $$M''(t) = np\, e^t (n-1)(1-p+p\, e^t)^{n-2} p\, e^t + np\, e^t(1-p+p\, e^t)^{n-1}$$

$$\therefore \quad [M''(t)]_{t=0} = E[X^2] = np(n-1)p + np = np^2(n-1) + np$$

Hence: $$\text{Var}[X] = E[X^2] - E^2[X]$$

$$= np^2(n-1) + np - n^2 p^2$$

$$= np - np^2 = np(1-p)$$

6.7.2 THE POISSON DISTRIBUTION

The m.g.f. for this distribution takes the form:

$$M(t) = E[e^{tX}] = \sum_{x=0}^{\infty} e^{tx} e^{-\mu} \frac{\mu^x}{x!} = e^{-\mu} \sum_{x=0}^{\infty} \frac{(\mu e^t)^x}{x!}$$

$$= e^{-\mu} e^{\mu e^t} \quad \text{(series definition)}$$

i.e. $\qquad M(t) = e^{\mu(e^t-1)}$

Using this to find the mean and variance, we have:

$$M'(t) = \mu e^t e^{\mu(e^t-1)}$$

$\therefore \qquad [M'(t)]_{t=0} = E[X] = \mu$

Also: $\qquad M''(t) = \mu e^t \mu e^t e^{\mu(e^t-1)} + \mu e^t e^{\mu(e^t-1)}$

$$= \mu e^t e^{\mu(e^t-1)}(\mu e^t + 1)$$

$\therefore \qquad [M''(t)]_{t=0} = E[X^2] = \mu(\mu+1)$

$\therefore \qquad \text{Var}[X] = E[X^2] - E^2[X] = \mu^2 + \mu - \mu^2$

$$= \mu$$

The p.g.f. for the Poisson is given by:

$$G(t) = E[t^X] = \sum_{x=0}^{\infty} t^x e^{-\mu} \frac{\mu^x}{x!}$$

$$= e^{-\mu} \sum_{x=0}^{\infty} \frac{(t\mu)^x}{x!}$$

$$= e^{-\mu} e^{\mu t} \quad \text{(series definition)}$$

$$= e^{\mu(t-1)}$$

6.7.3 THE GEOMETRIC DISTRIBUTION

The m.g.f. for this distribution is given by:

$$M(t) = E[e^{tX}] = \sum_{x=0}^{\infty} e^{tx} p(1-p)^x = p \sum_{x=0}^{\infty} ((1-p)e^t)^x$$

$$= \frac{p}{(1-(1-p)e^t)}$$

(the sum to infinity of a geometrical progression with $a=p$ and $r=(1-p)e^t$)

To find the mean and variance, we write $M(t)$ in the form:

$$M(t) = p(1-(1-p)e^t)^{-1}$$

$\therefore \qquad M'(t) = -p(1-(1-p)e^t)^{-2}[-(1-p)e^t]$

$$= p(1-p) \, e^t [1 - (1-p) \, e^t]^{-2}$$

Hence: $[M'(t)]_{t=0} = E[X] = p(1-p) \cdot (1 - (1-p))^{-2}$

$$= \frac{p(1-p)}{p^2} = \frac{1-p}{p}$$

Also: $M''(t) = p(1-p) \, e^t [1 - (1-p) \, e^t]^{-2}$

$$+ p(1-p) \, e^t \, \{-2[1 - (1-p) \, e^t]^{-3}\}\{-(1-p) \, e^t\}$$

$$= p(1-p) \, e^t [1 - (1-p) \, e^t]^{-2}$$

$$+ 2p(1-p)^2 \, e^{2t} [1 - (1-p) \, e^t]^{-3}$$

$\therefore \; [M''(t)]_{t=0} = E[X^2]$

$$= p(1-p)[1 - (1-p)]^{-2} + 2p(1-p)^2 [1 - (1-p)]^{-3}$$

$$= \frac{p(1-p)}{p^2} + \frac{2p(1-p)^2}{p^3}$$

$$= \frac{1-p}{p} + \frac{2(1-p)^2}{p^2}$$

$\therefore \qquad Var[X] = E[X^2] - E^2[X]$

$$= \frac{1-p}{p} + \frac{2(1-p)^2}{p^2} - \frac{(1-p)^2}{p^2}$$

$$= \frac{1-p}{p} + \frac{(1-p)^2}{p^2}$$

$$= \frac{p(1-p) + (1-p)^2}{p^2}$$

$$= \frac{1-p}{p^2}$$

(The preceding constitutes the proof of Statement 6.17.)

The p.g.f. for this distribution, using a similar technique to the above, is easily seen to be:

$$G(t) = E[t^X] = \sum_{x=0}^{\infty} t^x p(1-p)^x = \frac{p}{(1-(1-p)t)}$$

6.7.4 THE NORMAL DISTRIBUTION

The m.g.f. takes the form (writing e^{tX} as $\exp(tX)$ etc):

$$M(t) = E[\exp(tX)] = \frac{1}{\sigma\sqrt{2\pi}} \int_{-\infty}^{\infty} \exp(tx) \exp\left(-\frac{1}{2} \frac{(x-\mu)^2}{\sigma^2}\right) dx$$

$$= \frac{1}{\sigma\sqrt{2\pi}} \int_{-\infty}^{\infty} \exp\left(tx - \frac{1}{2} \frac{(x-\mu)^2}{\sigma^2}\right) dx$$

and this can be shown to be equal to:

$$\exp\left(\mu t + \frac{1}{2}\sigma^2 t^2\right)$$

(the proof of this is beyond the scope of the text).

So, with: $M(t) = \exp\left(\mu t + \frac{1}{2}\sigma^2 t^2\right)$,

$$M'(t) = (\mu + \sigma^2 t)\exp\left(\mu t + \frac{1}{2}\sigma^2 t^2\right)$$

∴ $\left[M'(t)\right]_{t=0} = E[X] = \mu$

Also: $M''(t) = (\mu + \sigma^2 t)(\mu + \sigma^2 t)\exp\left(\mu t + \frac{1}{2}\sigma^2 t^2\right)$

$$+ \sigma^2 \exp\left(\mu t + \frac{1}{2}\sigma^2 t^2\right)$$

∴ $\left[M''(t)\right]_{t=0} = E[X^2] = \mu^2 + \sigma^2$

Hence: $\text{Var}[X] = E[X^2] - E^2[X] = \mu^2 + \sigma^2 - \mu^2 = \sigma^2$

6.7.5 THE RECTANGULAR DISTRIBUTION

The form of the m.g.f. is:

$$M(t) = E[e^{tX}] = \frac{1}{b-a}\int_a^b e^{tx}\,dx$$

$$= \frac{1}{b-a}\left[\frac{1}{t}e^{tx}\right]_a^b$$

$$= \frac{1}{t(b-a)}(e^{bt} - e^{at})$$

However, this form is so complicated mathematically for calculating moments that it is rarely used.

6.7.6 THE EXPONENTIAL DISTRIBUTION

The m.g.f. takes the form:

$$M(t) = E[e^{tX}] = \int_0^\infty e^{tx}\,\mu\,e^{-\mu x}\,dx$$

$$= \int_0^\infty \mu\,e^{-(\mu-t)x}\,dx$$

which, using an integration by parts technique, can be shown to be equal to:

$$\left(1 - \frac{t}{\mu}\right)^{-1}$$

So that: $M'(t) = -1\left(1 - \dfrac{t}{\mu}\right)^{-2}\left(-\dfrac{1}{\mu}\right)$

∴ $\left[M'(t)\right]_{t=0} = \mathrm{E}[X] = \dfrac{1}{\mu}$

Also: $M''(t) = \dfrac{1}{\mu}\left(-2\right)\left(1 - \dfrac{t}{\mu}\right)^{-3}\left(-\dfrac{1}{\mu}\right) = \dfrac{2}{\mu^2}\left(1 - \dfrac{t}{\mu}\right)^{-3}$

∴ $\left[M''(t)\right]_{t=0} = \mathrm{E}[X^2] = \dfrac{2}{\mu^2}$

Hence: $\mathrm{Var}[X] = \mathrm{E}[X^2] - \mathrm{E}^2[X]$

$= \dfrac{2}{\mu^2} - \dfrac{1}{\mu^2}$

$= \dfrac{1}{\mu^2}$

6.8 EXERCISES

SECTION 6.2

1. Use the binomial theorem to expand the following:
    (a) $(2y + x)^6$; (b) $(2 + 3x)^{11}$; (c) $(3v + 4)^{12}$.

2. Calculate the coefficient of x^6 in the binomial expansion of:
    (a) $(a + bx)^6$; (b) $(3x + y)^{10}$; (c) $(x^2 + y)^9$.

3. Find the ratio of the term in x^5 to the term in x^6, in the expansion of $(2x + 3)^{20}$.

4. Find the greatest term in the following expansions:
    (a) $(2 + \frac{3}{4})^3$; (b) $(1 + \frac{1}{2})^{12}$; (c) $(\frac{1}{8} + \frac{7}{8})^{10}$; (d) $(\frac{1}{2} + \frac{1}{2})^{24}$.

5. Evaluate the following integrals:

    (a) $\displaystyle\int x\, e^{-ax}\, dx$ (b) $\displaystyle\int x^2\, e^x\, dx$

6. Given that $\displaystyle\int x\, e^{x^2}\, dx = \frac{1}{2} e^{x^2} + C$, evaluate: (a) $\displaystyle\int x^3\, e^{x^2}\, dx$;
    (b) $\displaystyle\int x^3\, e^{-x^2}\, dx$ using the method of integration by parts.

    $\left(\text{Hint: in (a) put } x^2 = u \text{ and } x\, e^{x^2} = \dfrac{dv}{dx}\right)$

7. Find the value of $\displaystyle\int_0^\infty x^2\, e^{-x}\, dx$.

SECTION 6.3

8. (a) If X is $\mathrm{Bin}(5, \frac{1}{3})$, find $\mathrm{Pr}(X = 2)$; (b) If X is $\mathrm{Bin}(6, \frac{1}{2})$, find $\mathrm{Pr}(X = 3)$; (c) If Y is $\mathrm{Bin}(4, \frac{1}{4})$, find $\mathrm{Pr}(Y = 3)$.

9. A random variable X is distributed binomially with parameters $n = 5$ and $p = \frac{1}{4}$. Find the probability (to 3D) that: (a) $X = 0$; (b) $X \geqslant 4$; (c) $X \not> 2$.

10. Write down the p.d.f. (simplifying where possible) of the r.v. X if: (a) X is $\mathrm{Bin}(10, \tfrac{1}{2})$; (b) X is $\mathrm{Bin}\left(n, \dfrac{1}{n}\right)$; (c) X is $\mathrm{Bin}((m+1)/2, 3/(1+y))$.

11. Three normal six sided dice are thrown. Find the probability of obtaining: (a) three odd numbers; (b) two odd numbers; (c) one odd number; (d) all even numbers.

12. Find the probability of obtaining $0, 1, 2, 3$ and 4 heads when four coins are tossed and draw a probability histogram for the binomial distribution so defined.

13. A large consignment of screws is known to be 1% defective. If a random sample of 20 screws is taken from the consignment, and taking this as a binomial situation with the r.v. X defined as "No. of *defective* screws in sample", evaluate: (a) n and p; (b) $\Pr(X = x)$ for $x = 0, 1, 2$ and 3 (to 4D); (c) $\Pr(X \geqslant 4)$.

14. 10 normal six sided dice are thrown. Find the probabilities of obtaining (to 3D); (a) no successes; (b) at least 8 failures; (c) at least twice as many successes as failures, if a success is '5' or '6'.

15. With n fair dice, the probability of throwing at least one '6' is $671/1296$. Find the value of n.

16. A company employs 8 representatives who work from the same building. If any one rep. spends just half of his day in the building, how many offices need be put at the reps' disposal so that each one will have an office at least 90% of the time.

17. A manufacturer sets up a 'double sampling' scheme as follows. A sample of 8 items is taken from a large lot ready for dispatch to customers. If there are no defectives, the lot is accepted (and sent out) and if there are 3 or more defectives the lot is rejected. If there is either 1 or 2 defectives in the sample, a second sample is taken from the same lot, and the lot is rejected only if there are 3 or more defectives in the *two samples combined*. (a) If the manufactured items are 12% defective, calculate the proportion of lots that will be accepted using the above double sampling scheme. (b) What proportion of lots will be accepted using only a single sampling scheme (with 3 or more defectives per sample causing a lot rejection) when the items are 12% defective?

18. 10% of the articles in a large bulk are defective. Calculate, by means of the binomial distribution, the probability that a sample of 50 items will include more than 2 defectives.

19. If the chance of a gibbon being male or female is even, what is the probability that: (a) a group of three gibbons are all female; (b) from a group of five gibbons, four are female.

20. (a) Find the probability of throwing at least two 'doubles' with four throws of a pair of dice. (b) Calculate the mean and variance of the number of heads expected if a coin is thrown ten times.

21. In 10 throws of a coin, what is the probability of obtaining 7 or more heads?

22.   A large batch of clay pots is moulded and fired. After firing, a random sample of 10 pots is inspected for flaws before glazing, decoration and final firing. If 20% of the pots in the batch have flaws, calculate correct to 2 significant figures, by using the appropriate binomial distribution, the probability that the random sample contains: (a) no pots with flaws; (b) exactly one pot with a flaw; (c) exactly two pots with flaws.

The batch is accepted without further checking if the random sample contains no more than two pots with flaws. Find the probability that the batch will be accepted without further checking.   (London)

23.   A large consignment of shelled peas is known to have (approximately) 1% discoloured. If a lot of 10 000 is dispatched for tinning, find the expected number of discoloured peas in the lot, and also the variance.

24.   Find the expected number of times a '6' will be shown together with the standard deviation, if a normal six sided die is thrown 1620 times.

25.   (a) If $ab = 1.25$ and $ab(1 - b) = 0.625$, find the values of a and b.

(b) A binomial distribution has mean 3.6 and variance 2.52. Find the values of the parameters n and p for the distribution.

(c) Explain why a binomial distribution could not have a mean of 1.25 and a variance of 0.625.

26.   A binomial distribution box contains a number of beads, some white, some coloured. Samples of 4 are taken from the box (and subsequently replaced), the number of coloured beads being noted on each occasion. The following results were obtained:

No. of coloured beads	0	1	2	3	4
No. of samples	64	85	40	9	2

Obtain the average number of coloured beads per sample, and, assuming the binomial distribution law holds, calculate the theoretical distribution for comparison with the original figures.

27.   Fit a binomial distribution to the following data:

x	0	1	2	3	4	5	6	Total
f	2510	3641	2440	823	214	27	3	9658

SECTION 6.4

28.   If X is Po(5/2), find: (a) $\Pr(X = 0)$; (b) $\Pr(X = 1)$; (c) $\Pr(X = 2)$; (d) $\Pr(X = 3)$; (e) $\Pr(X \leqslant 4)$; (f) $\Pr(X > 6)$.

29.   A random variable X has a Poisson distribution with parameter m, i.e., X is Po(m). If $\Pr(X = 2) = 0.2565$ (taking e as 2.7183), find the value of m.

30.   Find the first six terms of the Poisson distribution and plot histograms for: (a) $m = 4$; (b) $m = 2.1$ (m is the Poisson parameter).

31.   A sample of 200 items is taken from a large batch which is known to produce defectives with probability 0.005. Find the probabilities (to 3D) that there are 0, 1, 2, 3, 4 and 5 defectives in the sample using a Poisson approximation to the binomial.

32.   Minor accidents (which occur at random) on a certain stretch of road average 2 per day. Find the probability that: (a) more than two accidents occur in any day; (b) more than 4 accidents occur in any two day period.

33.   A large batch of items is known to have a proportion 0.03 defective. If a sample of 200 is taken, what is the probability that the sample will contain: (a) no defectives; (b) 4 defectives or less; (c) more than 5 defectives.

34.   A process for making electric light bulbs produces, on the average, 1 bulb in 100 defective when under control. Using the Poisson distribution to calculate the probability that a sample of 50 will contain: (a) 0; (b) 1; or (c) 2 defective bulbs. The bulbs are delivered in batches of 1000, from which 50 are taken at random. If at most 1 of the 50 is found to be defective, the whole batch is accepted. Calculate the probability of accepting a batch containing 50 defective bulbs.

35.   Use a Poisson approximation to calculate the probability of getting 10 successes in 1000 trials of an experiment for which the probability of a success at a trial is 0.01.

36.   Customers at a certain department store enter at an average of 60 per hour. (a) Find the probability that no-one enters the store during a particular five minute interval. (b) Find the length of the time interval for which the probability that no-one enters during it is 0.5.

37.   Certain mass produced articles, of which 0.5% are defective, are packed in cases of 100. What proportion of cases are free from defective articles and what proportion contain 2 or more defectives.

38.   Assuming that the probability of a coal miner being killed in a mine accident during a year is 1/1400, find the probability that in a mine employing 350 miners, there will be one fatal accident per year. State any assumptions made.

39.   If one in every 500 students has an accident during a college year, what is the probability that a college having 300 students has not more than 1 accident in a given college year.

40.   The table shows the results of a survey carried out by a class of children from an observation post which was set up by the side of a road near a town. The class recorded the number of vehicles moving towards the town which passed the post per minute during an afternoon.

Number of vehicles per minute	0	1	2	3	4	5	6
Frequency	3	6	13	12	3	2	1

(a) Find the time taken to collect these data.

(b) Calculate the mean number of vehicles per minute passing the post during this period.

(c) Construct a histogram to show these data.

(d) Using these data estimate the number of vehicles moving towards the town which you would expect to pass the post during the 5 minute interval immediately following the survey.

(e) Explain briefly why it would not be appropriate to use the results of the survey to estimate the number of vehicles passing the observation post and moving towards the town between 8 a.m. and 9 a.m. on a weekday.   (London)

41.   (a) I recorded the number of phone calls I received over a period of 150 days:

No. of calls	0	1	2	3	4
No. of days	51	54	36	6	3

(i) Find the average number of calls per day;   (ii) Calculate the frequencies of the comparable Poisson distribution.

(b) A firm selling electrical components packs them in boxes of 60. On average, 2% of the components are faulty. What is the chance of getting more than 2 defective components in a box? (Use the Poisson distribution.)   (SUJB)

42.   Failures of the braking system of a car occur at random (i.e., the number of failures is a Poisson variable) with, on average, one failure in 200 000 miles. Find the probability that: (a) the car completes 50 000 miles without a brake failure;   (b) there are more than 2 failures in 50 000 miles.

Two cars, A and B, with this type of braking system are bought and, while A is running its first 50 000 miles, B will run 100 000 miles. Find the probability that during this period there will be not more than one brake failure altogether.   (Cambridge)

43.   The number of accidents per day was recorded in a certain district for a period of 1500 days and the following results were obtained:

Number of accidents per day (x)	0	1	2	3	4	5 or more
Frequency (f)	342	483	388	176	111	0

Form a theoretical frequency distribution for comparison with the above data and having the same mean and total frequency. State any assumptions made and comment on the comparison of the two sets of frequencies.

44.   In the mass production of an article, 500 samples each of 30 articles are examined. The number of defective articles in the samples are shown in the following table:

Number of defectives in sample	0	1	2	3	4	Total
Frequency	309	142	40	8	1	500

(a) Find the mean number of defectives per sample; and   (b) show that the distribution is approximately the same as a Poisson distribution with this mean.   (c) Calculate the variances of both distributions.

45.   Fit a theoretical Poisson distribution to the following frequency distribution and compare the two, commenting on the result:

x	0	1	2	3	4	5	6	7
f	350	380	216	90	32	14	3	1

46.   Fit: (a) a Poisson;   (b) a binomial to the following distribution:

x	0	1	2	3	4	Total
f	531	354	99	15	1	1000

SECTION 6.5

47.   Find the values of: (a) $\Phi(0.21)$;   (b) $\Phi(2.15)$;   (c) $\Phi(-1.12)$;
(d) $\Pr(Z > 1.22)$;   (e) $\Pr(Z > -0.81)$;   (f) $\Pr(-1.12 < Z < 1.22)$;
(g) $\Pr(|Z| < 0.21)$ where Z is $N(0, 1)$ and $\Phi(x) = \Pr(Z < x)$.

48.   Show that, if Z is $N(0, 1)$, $\Pr(|Z| < a) = 2\Phi(a) - 1$ (a is a constant).

49.   If Z is $N(0, 1)$, use tables to find the values of b if:
(a) $\Phi(b) = 0.7967$;   (b) $\Pr(b \leqslant Z \leqslant 2.0) = 0.1$.

50.   If X is $N(2, 1/9)$, find: (a) $\Pr(X > 3)$;   (b) $\Pr(2 < X < 3)$.

51.   A random variable Y is $N(3, 16)$. Find the probability that a value of
Y taken at random will be negative. If 20 values are taken randomly,
how many would be expected to be negative?

52.   A machine in a factory produces components whose lengths are distributed normally, mean 102 mm, s.d. 1.5 mm.   (a) Find the probability
that if a component is selected at random and measured, its length
will be: (i) less than 100 mm;   (ii) greater than 104 mm.   (b) If
an output component is only accepted when its length lies in the range
100 mm to 104 mm,   find the proportion of components that are
accepted.

53.   A mass-produced cylindrical component is rejected if its diameter is
outside the 'tolerance range' 36.5 mm to 36.6 mm. If the diameters
are known to be distributed $N(36.56, 0.0004)$, how many, out of a
random sample of 1000 components, would be expected to be rejected?

54.   The lengths of 200 articles are known to be Normally distributed with
mean 155 cm and s.d. 20 cm. Calculate the number of articles with
lengths: (a) less than 100 cm;   (b) between 120 and 130 cm;
(c) between 150 and 175 cm;   (d) greater than 200 cm.

55.   It is known that for a Normally distributed variable X:
(i) $\Pr(X < 40) = 0.30$ and (ii) $\Pr(40 < X < 50) = 0.33$.
Find: (a) $\Pr(X > 50)$;   (b) $\Pr(30 < X < 40)$; and
(c) $\Pr(50 < X < 60)$.

56.   If Y is $N(5, 9)$, find the probability that a value of Y taken at
random will be negative.

57.   A large farm advertises potatoes in bags of 112 lb. If it is known that
the weights of the bags of potatoes are distributed Normally with mean
weight 114 lb and 10% of bags weigh 116 lb or more, calculate the
proportion of customers who will receive underweight bags.

58.   Last week $6\frac{2}{3}\%$ of the bars of soap made in a particular factory weighed
less than 90.50 grams and 4% weighed more than 100.25 grams.

(a) Find the mean and variance of the weights of the bars of soap
produced last week.

(b) What percentage of the bars of soap produced would you expect to weigh less than 88 grams?

(c) If the variance of the weights was reduced by one-third, what percentage of next week's production would you expect to weigh less than 88 grams, assuming that the mean weight was unchanged?

You may assume that the weights of the bars of soap are Normally distributed.   (IOS)

59.   An unbiased die is rolled and the result is recorded in the following way: Score 1 or 2 is denoted a success; Score 3, 4, 5 or 6 is denoted a failure.

An experiment consists of rolling the die four times.   (a) If the number of times a success is recorded is X, list the possible values of X. Calculate the probability $Pr(X)$ of each value of X occurring and list these in a table.   (b) Show that the expected value of X is $\frac{4}{3}$.

The die is rolled 100 times.   Using tables of the normal probability integral estimate the probability of obtaining between 30 and 40 successes inclusive.   (London)

60.   A cutting machine produces steel rods which must not be more than 100 cm in length. The mean length of a large batch of rods taken from the machine is found to be 99.80 cm and the standard deviation of these lengths is 0.15 cm.

(a) Assuming that the lengths of the rods are normally distributed, calculate, to one decimal place, the percentage of rods which are too long.

(b) The position of the cut can be adjusted without altering the standard deviation of the lengths. Calculate in cm, to 2 decimal places, how small the mean length should be if no more than 2% of the rods are to be rejected for being longer than 100 cm.

(c) If the mean length is maintained at 99.80 cm, calculate, to the nearest mm, by how much the standard deviation must be reduced if no more than 4% of the rods are to be rejected for being longer than 100 cm.   (London)

61.   In an artillery practice exercise, a gun is mounted on a railway truck on a straight railway line which passes over a bridge. From position A, the gun fires at the bridge which is y metres long. 35.2% of the shots fall beyond the bridge and 24.2% fall short of the bridge. From position B, which is nearer to the target than A, 12.3% of the shots fall beyond the bridge and 28.1% fall short of it. Given that the mean point of impact in the first series is 20 metres beyond the mean point of impact in the second series, and assuming a Normal distribution in each case, find: (a) the length, y, of the target; and   (b) the standard deviation for each of the distributions.

62.   As a result of tests on electric light bulbs, it was found that the lifetime of a particular make of bulb was distributed normally, with an average life of 2040 hours and a standard deviation of 60 hours. What proportion of bulbs can be expected to burn;   (a) for more than 2150 hours;   (b) for more than 1960 hours?

63.   An extinct species of animal was considered to have an average life of 25 years with a standard deviation of only two years. Of a family of 50 members of the species, born at the same time, how many would be

expected to have lived (assuming a Normal distribution): (a) between 20 and 26 years; (b) less than 22 years. (c) Calculate the age range (centrally located) within which 95% of the family would have been expected to die.

64.   Find between what limits the central: (a) 99%; (b) 99.8% of a standard Normal distribution lie.

65.   A Normal distribution has mean 5 and s.d. 3. (a) What is the probability that an item taken at random will have a negative value? (b) If five items are selected randomly, what is the probability that at least one is negative?

66.   Sacks of grain packed by an automatic machine loader have an average weight of 114 lb. It is found that 10% of the bags are over 116 lb. Find the standard deviation.

The machine is adjusted and the average weight per bag is now 113 lb. Assuming that the s.d. is unaltered, calculate the probability that a bag is (now) over 116 lb.

67.   A frequency distribution has mean 2.70, s.d. 0.22 and total frequency 209. Fit a Normal distribution to these data using the classes 2.05 up to 2.15, 2.15 up to 2.25, . . . , 3.25 up to 3.35.

68.   Use a Normal approximation to find the following probabilities: (a) 10 or more defectives in a sample of 150 from a population that is known to be 5% defective. (b) Between 17 and 21 defectives in a large sample, given that, on the average, samples have 15 defectives. (c) An item selected at random lying within one s.d. of the mean if a sample of 500 is taken from a 3% defective population.

69.   If the probability that an individual smokes cigarettes is $\frac{1}{2}$, and 1000 interviewers each choose a random sample of 100 individuals, how many interviewers would you expect to report: (a) 60 or more smokers; (b) more than 60 smokers; and (c) exactly 60 smokers? (Use a Normal approximation.)

70.   An article is being mass-produced by a machine and the specification demands that a certain dimension should not differ from 6.40 cm by more than 0.05 cm. 600 of the articles produced are measured to the nearest hundredth of a centimetre and the results are given as follows:

Centre of interval	6.36	6.37	6.38	6.39	6.40	6.41	6.42	6.43	6.44
Frequency	1	7	51	120	250	122	41	7	1

Find the corresponding expected frequencies for the Normal distribution with the same mean and standard deviation as the above distribution.

71.   Fit a Normal distribution to the following data of I.Q. scores for 500 students:

x	82–85	86–89	90–93	94–97	98–101	102–105	106–109
f	5	19	32	49	71	92	75

x	110–113	114–117	118–121	122–125	126–129	130–133
f	56	39	28	18	10	6

72.   Assuming the following data is continuous and measured to the nearest whole number, fit a Normal distribution having the same mean, variance and total frequency:

x	60	61	62	63	64	65	66	67	68
f	2	0	15	29	25	12	10	4	3

SECTION 6.6

73.   Show that, if X has a Bernoulli distribution with parameter p, $E[X^r] = p$ for all positive integral r.

74.   A rectangular r.v. X is distributed over the range 0.5 to 6.5, i.e., X is R(0.5, 6.5). Find $E[X]$, $Var[X]$ and $Pr(X < 4.5)$.

75.   If X is distributed rectangularly over the interval (a, b) and $a \leqslant c \leqslant d \leqslant b$, show that $Pr(c \leqslant X \leqslant d) = \dfrac{d - c}{b - a}$.

76.   A certain continuous value is recorded, to the nearest whole number, as 6. If it is known that the exact value is distributed rectangularly, find the probability that it is: (a) less than 5.7;   (b) between 5.7 and 6.2.

77.   A person rolls a die until he obtains a '6'. Calculate the average number of rolls he should expect to make if the experiment were repeated a large number of times.

78.   Minor accidents occur at random at the rate of three per day on a certain stretch of road. Find the probability that after a particular minor accident has occurred, at least a day will go by without another.

79.   Independent and identical trials of an experiment are performed until a 'success' occurs. If, on the average, it is found that 8 trials are necessary, estimate the probability that the experiment will be successful at any trial.

80.   A manufacturer of television sets finds from experience that his sets, on average, have one major breakdown every four years.   (a) If he guarantees his sets for one year against a major breakdown, what proportion of customers will be expected to use the guarantee for repairs?   (b) What guarantee would he need to give if he only wished 1 in 20 sets to be returned for major repairs under guarantee?

MISCELLANEOUS

81.   Colour blindness appears in 1% of the people in a certain population. How large must a random sample be (with replacement) if the probability of its containing a colour blind person is to be 0.95 or more?

82.   A book of 850 pages contains just 850 misprints in total. Estimate the probability that a page picked at random will contain at least 3 misprints.

83.   Random samples of electric light bulbs are taken from a moving belt. If 5% of all bulbs are defective, calculate the probability (to 3D) that there will not be more than one defective bulb in a sample of: (a) 5;   (b) 10.

84.   (a) Two variables X and Y can each take the values 0, 1, 2 or 3. If any of these 4 values can occur with equal probability for each variable, calculate the probability that the sum of the two variables' values adds to 3.   (b) If X is $\text{Bin}(3, \frac{1}{2})$ and Y is $\text{Bin}(4, \frac{2}{3})$, calculate $\Pr(X + Y = 3)$ to 3D.

85.   Find the mean and variance of the following frequency distribution and state which theoretical distribution it most likely is:

x	0	1	2	3	4	5	Total
f	20 978	2500	201	13	3	2	23 697

86.   The mean number of calls per hour to a particular telephone switchboard is 5. Assuming that calls come to the board randomly, estimate the probability that: (a) exactly 5 calls are received in a particular hour;   (b) more than half an hour elapses between successive calls.

87.   (a) Prove that if X is $\text{Bin}(n, \frac{1}{2})$, then the distribution of X is symmetric (i.e., show that $\Pr(X = 0) = \Pr(X = n) \ldots$ etc.).   (b) Hence write down $\text{E}[X]$ in this case.

88.   The numbers of bad oranges found in a shipment of 100 boxes, each containing 72 oranges, were as follows:

Number of bad oranges per box	0	1	2	3	4	5 or more
Number of boxes	45	36	14	4	1	0

(a) Calculate the arithmetic mean and variance of the number of bad oranges per box, correct to 2S.   (b) Name the most likely theoretical distribution corresponding to the above data with reasons.   (c) Write down numerical expressions for the first five terms of the theoretical distribution. Evaluate the theoretical probability that a box selected at random will contain exactly two bad oranges.   (d) What is the probability that one orange taken at random from the whole shipment will be bad.

89.   If X is $\text{Bin}(2, \frac{1}{2})$ and Y is $\text{Bin}(2, \frac{1}{4})$, find: (a) $\Pr(X + Y = 4)$; (b) $\Pr(X + Y = 3)$.

90.   If X is $\text{Bin}(3, \frac{2}{3})$ and Y is $\text{Po}(1)$, find: (a) $\Pr(X = 0)$; (b) $\Pr(X = 1)$; (c) $\Pr(Y = 0)$;   (d) $\Pr(Y = 1)$;   (e) $\Pr(X + Y < 2)$ and hence (f) $\Pr(X + Y \geqslant 2)$.

91.   An insurance company has 10 000 people insured with them against dying from a certain type of accident, from which it is known that 0.005% of the whole population die in any one year.   (a) Find the probability that less than 4 claims will be received by the company for this accident in any one year.   (b) If each of the insured pays a yearly premium of £x and the insurance company guarantees a payment of £y on death from this type of accident, find the value of x if the company should expect to make a profit of £5000 annually, given that their expected yearly payout (exclusive of premiums) is £9000.

92.   (a) Using the recurrence relation $\Pr(X = x + 1) = \dfrac{n - x}{x + 1}\left(\dfrac{p}{1 - p}\right).$

$\Pr(X = x)$ (for $x = 0, 1, 2, \ldots, n$) for the r.v. X as $\text{Bin}(n, p)$, deduce that $\Pr(X = x + 1) > \Pr(X = x)$ if $x + 1 < (n + 1)p$.

(b) If X is $\text{Bin}(12, \frac{1}{3})$, by using the above result show that $\Pr(X = x)$ has its greatest value when $x = 4$.

93.   The moment generating function of a random variable Y is given by $(\frac{3}{8} + \frac{5}{8} e^t)^4$. Give the distribution of Y and calculate $\Pr(Y \leqslant 2)$.

94.   On the average a Geiger counter reaches 30 counts per minute close to a certain radio-active material. Assuming a Poisson process to be applicable, find the probability that there will be exactly: (a) 5 counts in a 10 second period;   (b) x counts in a period of n seconds.

95.   Find the mean and variance of the following discrete frequency distribution:

x	0	1	2	3	4	5
f	1	7	23	21	11	1

Associating the random variable X_1 with the above frequency distribution and the random variable X_2 with the theoretical binomial distribution defined using the mean of the above distribution: (a) write down the recurrence relation between $\Pr(X_2 = r)$ and $\Pr(X_2 = r + 1)$. (b) Calculate $\Pr(X_2 = r)$ for $r = 0, 1, 2, 3, 4$ and 5.   (c) Find the percentage errors in using $\Pr(X_2 = r)$ for $\Pr(X_1 = r)$ for $r = 0, 1, \ldots, 5$ and comment on the results.

96.   Telephone calls coming in to a switchboard follow a Poisson distribution with mean 3 per minute. Find the probability that in a given minute there will be five or more calls. If the duration of every call is 3 minutes and at most 12 calls can be connected simultaneously, find an approximation to the probability that at a given instant the switchboard is fully loaded.

97.   Calculate Q_1 and Q_3 (the 1st and 3rd quartiles) for a standard Normal distribution. Deduce that, for a standard Normal distribution, $D_i = -D_{10-i}$; $i = 1, 2, \ldots, 4$ and $P_j = -P_{100-j}$; $j = 1, 2, \ldots, 49$ (where D_i and P_j are the ith decile and jth percentile respectively).

98.   Find the semi-interquartile range for the distribution of the r.v. X as $N(100, 16)$.

99.   (a) Find between what limits the central 95% of a standard Normal distribution lie.   (b) A soap powder manufacturer decides to reject only 5% of filled packets coming from a packing machine. If the weights of the packets of soap powder are distributed normally, mean 812 gm and variance 4 gm, and as many packets are rejected for underweight as overweight, what are the critical weight limits for rejection?   (c) If the manufacturer guarantees a minimum weight content of 809 gm per packet, what percentage of customers will have cause for complaint?

100.   A gun is firing at a target whose length parallel to the line of fire is 80 yards. 20% of the rounds fired fall short of the target and 35% fall beyond it. Estimate: (a) the distance of the mean point of impact from the centre of the target; and   (b) the length of the 50% (central) zone. Assume a Normal distribution for the calculations.

101.   (a) State under what circumstances the binomial distribution will arise. Explain how, and under what conditions a Normal distribution

can be used as an approximation to a binomial distribution. A machine produces articles of which on average 10% are defective. Use a suitable Normal approximation to calculate the probability that, in a random sample of 400 articles, more than 52 will prove to be defective.

(b) A telephone exchange receives calls at random at an average rate of 25 calls every 15 minutes. Use the Normal approximation to the Poisson distribution to calculate the probability that less than 30 calls, but not less than 15 calls, are received in a 15 minute period. (JMB)

102. A manufacturer of car batteries guarantees to replace them free if they fail within one year of purchase and to replace them at half-price if they fail in more than one but less than two years. Replacement batteries are not replaced if they fail. He knows that over the years the time to failure has had a Normal distribution with a mean of 3 years and a standard deviation of 0.8 years.

Calculate the probability: (a) that a battery fails in under one year; (b) that a battery fails in more than one but under two years.

The manufacturer sells a new battery for £14 of which £11 is the cost of production. Calculate his expected profit from the sale of a thousand batteries. (Cambridge)

103. In a large canteen a quarter of the customers buy a cup of coffee. (a) Find the probability that at least 3 out of the first 7 customers will buy a cup of coffee. (b) The probability that 500 customers will buy fewer than l cups of coffee is 0.99. Find l.

If overall one customer per thousand makes a complaint, find the probability of receiving no complaints from 500 customers, assuming complaints occur independently. (O & C)

104. The mass printed on a packet of cereals is given by the manufacturer as 250 g. In fact it is discovered that the packets coming from the factory have a mean mass of 255 g and a standard deviation of 2.5 g. On the assumption that the masses are normally distributed, estimate the percentage of packets weighing between 250.5 g and 260.5 g.

If the manufacturer decides to alter the mean mass so that 8% of the output is less than the intended mass of 250 g, what should the new mean be, assuming that the standard deviation remains unaltered? (Cambridge)

105. The number of road accidents per week in a certain town follows a Poisson distribution with parameter 10.5. Find the mean of this distribution from first principles. Calculate: (a) the probability that no accidents occur in a period of four weeks; (b) the probability that at least two accidents occur in a week; (c) the number of accidents most likely to occur in a week. (IOS)

106. (a) An unbiased coin is tossed 100 times. Write down an expression for the probability of obtaining 54 heads and estimate correct to 2 sig. fig. its value using the Normal approximation.

(b) Articles are turned off a production line in large batches and the probability of any article being faulty is p. An inspection scheme is used for each batch which requires a random sample of 10 to be tested. If no faulty article is found the batch is accepted; if two or more are faulty the batch is rejected. If one is faulty a further sample of 5 is tested and the batch accepted only if none of these is faulty. Show that the probability of accepting a batch is $q^{10}(1 + 10q^4 - 10q^5)$ where $q = 1 - p$ and find the expected number sampled per batch, giving the answer in terms of q. (SUJB)

107.

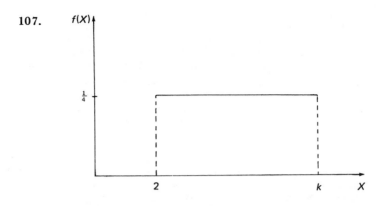

Describe the probability density function, $f(X)$, shown in the figure and state the type of random variable X which could be distributed in this way.   (a) Verify that $k = 6$;   (b) Calculate the probability of obtaining a value of X less than 3.2;   (c) Find the expected value of X;   (d) Show that the variance of this probability distribution is $\frac{4}{3}$.

(London)

108.   Define the binomial distribution, explaining clearly the symbols used.

A gun engages a target at a range at which the chance of a direct hit with any one round is $\frac{1}{3}$. How many rounds must be fired to give a 90 per cent probability of at least one hit? If, in fact, five rounds are fired and, failing a direct hit in these five rounds a further five rounds are fired, find the chance of two or more direct hits in all.

(O & C)

109.   The manager of an engineering factory believes that machine breakdowns occur at random at a constant average daily rate. The chief engineer claims that, although the weekly rate is constant, the daily average varies according to the day of the week. The following sets of data were extracted from records:

Breakdowns per day	0	1	2	3	4	5	6			
Frequency	112	78	29	12	7	10	2			

Breakdowns per week	0	1	2	3	4	5	6	7	8	9+
Frequency	1	6	13	27	38	37	26	22	13	17

Plot the two distributions on Poisson probability paper and comment on the manager's and the chief engineer's theories. (Use the same sheet of probability paper for both distributions.)

Assume the weekly number of breakdowns follows a Poisson distribution with mean 5.2. Using the probability paper, or otherwise, estimate for a particular week:

(a) the probability of 10 or more breakdowns,

(b) the number of breakdowns which would be exceeded with a probability of just less than 0.01.

(AEB) '85

110.   A manufacturer of fishing lines uses a machine to produce lengths of line whose breaking strain is normally distributed with standard deviation 1.50 newtons. Before buying a large batch of fishing lines, a retailer tests a random sample of 10 lengths of line from the batch, and accepts the batch provided the mean breaking strain of the sample is greater

than 39.5 newtons. What is the probability of accepting a batch with a mean breaking strain of 40.00 newtons?

The manufacturer installed a new machine to produce lines. The breaking strength of a random sample of 180 lengths of line produced by the new machine is given in the table below.

Breaking strain (newtons)	Frequency
38.1–38.5	6
38.6–39.0	11
39.1–39.5	26
39.6–40.0	46
40.1–40.5	43
40.6–41.0	31
41.1–41.5	13
41.6–42.0	4

Use arithmetic probability paper to verify that it is reasonable to assume that the sample comes from a normal population. Estimate from your graph the mean and standard deviation of the distribution.

Subsequent batches produced by the new machine are all found to have the same standard deviation, but not necessarily the same mean breaking strength. The retailer continues to test a sample of 10 lengths of line from each batch. Use your estimate of the standard deviation to calculate the minimum acceptable size of the sample mean to give a probability of at least 0.85 of accepting a batch with mean breaking strain of 40.00 newtons.   (AEB) '84

111.   (a) All schools in the county of Kentwall were involved in a 'Sponsored Pumpkin Grow' during the summer of 1980. Each pupil was given two pumpkin seeds and asked to come back at the end of the summer at which time the mass of their largest pumpkin was recorded in kilograms. The table below gives the results for two random samples of pupils, one from the Much Wapping School and one from Markum School.

Much Wapping   6.07, 8.32, 9.40, 6.68, 4.24, 8.45,
Primary School   10.00, 7.79, 5.79

Markum High   5.86, 2.78, 6.61, 4.57, 6.50, 7.72,
School   2.30, 8.45, 4.41, 5.86, 3.35

Using the *same* sheet of arithmetical probability paper

(i) plot the sample from Much Wapping Primary School,
(ii) plot the sample from Markum High School.

Compare the distributions.

(b) The number of misprints on 200 randomly selected pages from the 1981 editions of the Daily Planet, a quality newspaper, were recorded. The table below summarises these results.

Number of misprints per page x	0	1	2	3	4	5	6	7	8	8+
Frequency f	5	12	31	40	38	29	22	14	5	4

Use Poisson probability paper to verify that the Poisson distribution with mean 4 is a reasonable model for these data.

Determine graphically an estimate for the probability of more than 11 misprints on a page.   (AEB) '82

112. (a) If X is $N(a, b^2)$, we can transform this distribution to a standard
Normal by use of the coding $Z = \dfrac{X - a}{b}$ where Z is $N(0, 1)$.
Similarly, to transform back to $N(a, b^2)$ from $N(0, 1)$ we can re-arrange the above coding to give $X = a + bZ$. If X is $N(a_1, b_1{}^2)$ and is transformed to a new variable Y as $N(a_2, b_2{}^2)$, give an equation expressing Y in terms of X and interpret its meaning. (*Hint:* transform both X and Y to standard Normal distributions)

(b) The pass mark in an examination is 45. At a particular sitting, the marks of 1000 students were distributed Normally with mean 40 and 325 of the candidates passed. This mark distribution is transformed to one distributed with mean 43 and s.d. 10.   (i) How many more candidates will pass if the pass mark is still 45?   (ii) Find the transformed mark of a candidate obtaining 41 in the examination.

113. (a) Use the result of Statement 6.2 to show that if X is $\mathrm{Bin}(n, p)$ then for $p < \frac{1}{2}$, $T_{r+1} > T_r$ if $r < \frac{1}{2}(n + 1)$. Hence deduce that (for $p < \frac{1}{2}$) the distribution is skewed to the right.

(b) By writing $(x + y)^n = \sum\limits_{r=0}^{n} \binom{n}{r} y^r x^{n-r}$, use the above technique
to show that the distribution is skewed to the left for $p > \frac{1}{2}$.

114. A random variable T has an exponential distribution with mean $\dfrac{1}{a}$.
Show that the distribution function, $\mathrm{Pr}(0 \leqslant T \leqslant x)$, can be expressed as $1 - e^{-ax}$. Hence show that $\mathrm{Pr}(x_1 \leqslant T \leqslant x_2) = e^{-ax_1} - e^{-ax_2}$. (See Statement 6.19.)

115. The r.v. X is distributed rectangularly over the interval a to b, i.e., X is $R(a, b)$. Write down an expression for: (a) $\mathrm{Pr}(X \leqslant x_0)$;
(b) $\mathrm{Pr}(X \geqslant x_0)$ if $a \leqslant x_0 \leqslant b$.

116. If X is $\mathrm{Bin}(n, p)$:   (a) write down $E[X]$ and $\mathrm{Var}[X]$ in terms of n and p;   (b) Show that $E\left[\dfrac{X}{n}\right] = p$;   (c) By using the identity
$\mathrm{Var}[X] = E[(X - E[X])^2]$, show that $E\left[\left(\dfrac{X}{n} - p\right)^2\right] = \dfrac{p(1 - p)}{n}$.

117. If X is $\mathrm{Po}(m)$, *where m is an integer,* show that there exists an r such that $\mathrm{Pr}(X = r) = \mathrm{Pr}(X = r + 1)$ and find the value of r.
$\left(\textit{Hint: } \text{Use the recurrence formula } \mathrm{Pr}(X = x + 1) = \left(\dfrac{m}{x + 1}\right)\mathrm{Pr}(X = x)\right)$

118. By using the binomial theorem, show that m_n (the nth moment about the mean of any distribution can be expressed in the form:
$m_n = E\left[\sum\limits_{r=0}^{n} \binom{n}{r} X^{n-r} \mu^r\right]$ where X is the r.v. associated with the distribution concerned and $\mu = E[X]$. Using the properties of the expectation operator E, show that m_n can thus be expressed in terms of moments about the origin as: $m_n = \sum\limits_{r=0}^{n} \binom{n}{r}(m_1')^r m_{n-r}'$ (where m_r' is the rth moment about the origin).

119. (a) Using the substitution $y = \dfrac{x - \mu}{\sigma}$, show that the integral

$\displaystyle\int \exp\left(-\frac{1}{2}\left(\frac{x-\mu}{\sigma}\right)^2\right) dx$ can be written in the form $\displaystyle\int \exp\left(-\tfrac{1}{2}y^2\right)\sigma\,dy$.

(b) Given that $\displaystyle\int_{-\infty}^{\infty} \exp\left(-\tfrac{1}{2}y^2\right) dy = \sqrt{2\pi}$, show that if X is a normally distributed variable with parameters μ and σ, X is a *random* variable.

120. If X is $\mathrm{Po}(m_1)$ and Y is $\mathrm{Po}(m_2)$, where X and Y are independent, show that $\Pr(X + Y = r)$ can be written as:

$$e^{-m_1} e^{-m_2} \sum_{s=0}^{r} \frac{m_1^{\,s}}{s!}\left(\frac{m_2^{\,r-s}}{(r-s)!}\right).$$

By further considering the binomial expansion:

$$(m_1 + m_2)^r = \sum_{s=0}^{r} \binom{r}{s} m_1^{\,s} m_2^{\,r-s},$$

deduce that $X + Y$ is $\mathrm{Po}(m_1 + m_2)$.

121. If X is $\mathrm{Bin}(n_1, p)$ and Y is $\mathrm{Bin}(n_2, p)$ where X and Y are independent, use the method of Exercise 117 together with the identity:

$$\binom{n_1 + n_2}{r} = \sum_{s=0}^{r} \binom{n_1}{s}\binom{n_2}{r-s}$$ to prove that $X + Y$ is $\mathrm{Bin}(n_1 + n_2, p)$.

122. If X is $\mathrm{Bin}(n, p)$, show that the moment generating function $M(t) = E[e^{tX}]$ can be written as $M(t) = (1 - p + P e^t)^n$. By differentiating $M(t)$ partially w.r.t. t, prove that: (a) $M'(t) = np\,e^t \cdot (1-p+p\,e^t)^{n-1}$; (b) $M''(t) = np\,e^t(1-p+p\,e^t)^{n-2}(1-p+np\,e^t)$. Hence deduce that $E[X] = np$ and $\mathrm{Var}[X] = np(1-p)$.

123. (a) If the r.v. X is $\mathrm{Po}(m)$, show that $M(t) = E[e^{tX}]$ can be written

as $M(t) = \displaystyle\sum_{x=0}^{\infty} e^{-m}\frac{(m\,e^t)^x}{x!}$ and hence, that $M(t) = e^{m(e^t - 1)}$.

(b) Show that $M'(t) = m\,e^t\,e^{m(e^t-1)}$ and also that $M''(t) = m^2\,e^{2t}\,e^{m(e^t-1)} + m\,e^t\,e^{m(e^t-1)}$. (c) Hence prove that $E[X] = m$ and $\mathrm{Var}[X] = m$.

Chapter 7   **Random Samples and Their Statistics**

7.1 INTRODUCTION

In this chapter we will be concerned with the further development of random variables, particularly the sum (and difference) of normally distributed variables. We then extend this idea to an investigation of the behaviour of particular combinations of random variables sampled from (mainly) Normal distributions.

We have already discussed (in Chapter 4) the idea of a random selection of an item from a set or population. We require simply that each item has an equal chance of being selected. We can extend this idea quite naturally to include a 'random sample of n items' from the set or population. This procedure can be thought of as repeating the random selection of a single item n times. So that we necessarily need each item selection to be independent of another.

The above description of a random sample is adequate in practical circumstances where we have a manageable number of population items, but for sampling from some theoretical distribution, where we do not have (or cannot envisage) the population items physically on display, it is inadequate. In this case we employ the following definition.

DEFINITION 7.1

A *random sample of size* n (from some given distribution) is a set of n random variables X_1, X_2, X_3, $\ldots$, X_n, satisfying the following two conditions.

(a)   Each X_i $(i = 1, 2, 3, \ldots, n)$ has exactly the same distribution.

(b)   The set $X_1, X_2, X_3, \ldots, X_n$ are mutually independent.

For instance, given that X_1 and X_2 constitute a random sample of size 2 from N(50, 10), we necessarily have that:

(a) X_1 is N(50, 10) and X_2 is N(50, 10);

(b) X_1 and X_2 are independent random variables.

DEFINITION 7.2

Let $X_1, X_2, X_3, \ldots, X_n$ be a random sample of size n from some distribution. We define a *statistic* (of the sample) as any function of the set $X_1, X_2, X_3, \ldots, X_n$ involving no unknown quantities.

Note We require a statistic to be numerically identified once a sample has been drawn.

For example, suppose X_1, X_2 and X_3 form a random sample of 3 from $N(\mu, \sigma^2)$, where μ and σ^2 are both unknown parameters. We define quantities T_1, T_2 and T_3 as follows:

$$T_1 = 3X_1 - X_2; \quad T_2 = \frac{X_2 - \mu}{\sigma} \quad \text{and} \quad T_3 = \frac{X_1 + X_2 + X_3}{3}$$

T_1 and T_3 are both sample statistics, since they involve only sample variables and constants which, once the sample has been drawn, are known quantities. However, the parameters μ and σ^2 will still be unknown quantities after the sample has been drawn. Hence T_2 is not a sample statistic.

Sample statistics that will be of particular interest and importance to us are:

(a) the sample mean, $\displaystyle \bar{X} = \frac{X_1 + X_2 + X_3 + \ldots + X_n}{n} = \frac{1}{n} \sum_{i=1}^{n} X_i$

(b) the sample variance, $\displaystyle S^2 = \frac{1}{n} \sum_{i=1}^{n} (X - \bar{X})^2 = \frac{1}{n} \sum_{i=1}^{n} X_i^2 - \bar{X}^2$

Consider a random sample of size n namely $X_1, X_2, X_3, \ldots, X_n$ from some distribution. Let T be some statistic defined on the above sample. Performing the sample once will give us a numerical value for T, t_1 say. Repeating the sample a number of times will give a set of values for T, t_1, $t_2, t_3, \ldots$, etc. These values will of course, form a distribution and in this chapter we will also be concerned with investigating these distributions for particular statistics of interest.

EXAMPLE 7.1

Three normal six-sided dice are thrown. Discuss how we can regard this situation as a random sample of size 3 from some distribution.

SOLUTION

If we take as our variable X 'the number of spots displayed on the uppermost face of the die', then X is a variable taking the possible values $(1, 2, 3, 4, 5, 6)$. Since we are regarding each die as being 'fair', the probability set associated with the above values is $\left(\dfrac{1}{6}, \dfrac{1}{6}, \dfrac{1}{6}, \dfrac{1}{6}, \dfrac{1}{6}, \dfrac{1}{6} \right)$. Hence X is a random variable, since the probabilities sum to 1.

If we now associate random variable X_1 with die 1, X_2 with die 2 and X_3 with die 3, we have each of the random variables X_1, X_2 and X_3 distributed according to the p.d.f.

x	1	2	3	4	5	6
$\Pr(X = x)$	$\dfrac{1}{6}$	$\dfrac{1}{6}$	$\dfrac{1}{6}$	$\dfrac{1}{6}$	$\dfrac{1}{6}$	$\dfrac{1}{6}$

X_1, X_2 and X_3 are clearly independent.

Since:

(a) X_1, X_2, X_3 have the same distribution,

(b) X_1, X_2, X_3 are independent random variables,

by definition we have a random sample of size 3 from the given distribution.

7.2 SUMS AND DIFFERENCES OF RANDOM VARIABLES

In Chapter 5 (Sections 4.3 and 4.4) we stated some important results concerning the expectations and variances of sums and differences of random variables. Namely, if X and Y are any two random variables, then:

$$E[X + Y] = E[X] + E[Y]; \quad E[X - Y] = E[X] - E[Y]$$

Furthermore, if X and Y are independent, then:

$$\mathrm{Var}[X + Y] = \mathrm{Var}[X] + \mathrm{Var}[Y]; \quad \mathrm{Var}[X - Y] = \mathrm{Var}[X] + \mathrm{Var}[Y]$$

In this section we go further and say something about the distributions of sums and differences of Normally distributed random variables.

7.2.1 TWO RANDOM VARIABLES

STATEMENT 7.1

If X_1 is $N(\mu_1, \sigma_1^2)$ and X_2 is $N(\mu_2, \sigma_2^2)$ with X_1 and X_2 independent, then:

(a) $X_1 + X_2$ is $N(\mu_1 + \mu_2, \sigma_1^2 + \sigma_2^2)$

(b) $X_1 - X_2$ is $N(\mu_1 - \mu_2, \sigma_1^2 + \sigma_2^2)$

No Proof

The significant aspect of the above result is the fact that sums and difference of two Normal variables *are also Normal*.

A special case of the above is when X_1 and X_2 have the same Normal distribution. That is, when $\mu_1 = \mu_2 = \mu$, say, and $\sigma_1^2 = \sigma_2^2 = \sigma^2$, say. In this case we can regard the situation as X_1 and X_2 being a random sample of size 2 from $N(\mu, \sigma^2)$. Using the statement, we have the result:

$$X_1 + X_2 \text{ is } N(\mu + \mu, \sigma^2 + \sigma^2) = N(2\mu, 2\sigma^2)$$

and: $$X_1 - X_2 \text{ is } N(\mu - \mu, \sigma^2 + \sigma^2) = N(0, 2\sigma^2)$$

EXAMPLE 7.2

A man travels to work by walking part of the way and travelling the rest by bus. Over a period of time he calculates that his walking time and bus time (including waiting) are approximately distributed $N(10, 2)$ and $N(25, 7)$ minutes respectively, independent of each other. Find:

(a)   the distribution of total travelling time to work; and

(b)   the probability (if he leaves home 45 minutes before he is due at work) that he is late for work.

SOLUTION

Let X_1 be random variable 'walking time' and X_2 be random variable 'bus time' with X_1 as $N(10, 2)$ and X_2 as $N(25, 7)$.

(a)   Total time to work is 'walking time' + 'bus time' $= X_1 + X_2$ which is distributed $N(10 + 25, 2 + 7) = N(35, 9)$ (using the previous statement).

(b)   We require the probability that his total journey takes longer than 45 minutes,   i.e., we need:

$$Pr(X_1 + X_2 > 45) = Pr\left(Z > \frac{45 - 35}{3}\right) \text{ (standardising, using the result of (a))}$$

$$= Pr(Z > 3.33) = 1 - \Phi(3.33)$$

$$= 1 - 0.9995 \quad \text{(from tables)}$$

$$= 0.001 \quad \text{(3D)}$$

EXAMPLE 7.3

A random sample of 2 is taken from the distribution $N(100, 25)$. Find the probability that:

(a)   the sample sum is greater than 206,

(b)   the sample sum is less than 192 and hence,

(c)   the sample sum is between 192 and 206.

SOLUTION

Let the sample be represented by the random variables X_1 and X_2 with both distributed as $N(100, 25)$. Hence $X_1 + X_2$ is $N(100 + 100, 25 + 25) = N(200, 50)$.

(a)   We need:
$$\Pr(X_1 + X_2 > 206) = \Pr\left(Z > \frac{206 - 200}{\sqrt{50}}\right) = \Pr(Z > 0.85)$$
$$= 1 - \Phi(0.85) = 0.198$$

(b) $\Pr(X_1 + X_2 < 192) = \Pr\left(Z < \dfrac{192 - 200}{\sqrt{50}}\right) = \Pr(Z < -1.13)$
$$= 1 - \Phi(1.13) = 1 - 0.871 = 0.129$$

(c)   Clearly: $\Pr(192 < X_1 + X_2 < 206)$
$$= 1 - \Pr(X_1 + X_2 < 192) - \Pr(X_1 + X_2 > 206)$$
$$= 1 - 0.129 - 0.198$$
$$= 0.673$$

EXAMPLE 7.4

If men's heights are distributed   N(67, 3) inches   and women's N(64, 2) inches, find the probability that, if a married couple is randomly selected:

(a)   the man is more than 4 inches taller than the woman,

(b)   the woman is taller than the man.

SOLUTION

Let X_1 and X_2 be random variables representing men's and women's heights respectively, so that X_1 is N(67, 3) and X_2 is N(64, 2).

(a)   We require: $\Pr(X_1 > X_2 + 4) = \Pr(X_1 - X_2 > 4)$

But: $X_1 - X_2$ is   N(67 − 64, 3 + 2) = N(3, 5)   (difference of
two normal variables)

Hence:
$$\Pr(X_1 - X_2 > 4) = \Pr\left(Z > \frac{4 - 3}{\sqrt{5}}\right) \quad \text{(standardising)}$$
$$= \Pr(Z > 0.447) = 1 - \Phi(0.447) = 1 - 0.6725$$
$$= 0.33 \quad (2D)$$

(b)   We require:
$$\Pr(X_2 > X_1) = \Pr(X_1 - X_2 < 0)$$
$$= \Pr\left(Z < \frac{0 - 3}{\sqrt{5}}\right) \quad \text{(since } X_1 - X_2 \text{ is N(3, 5))}$$
$$= \Pr(Z < -1.342) = 1 - \Phi(1.342)$$
$$= 1 - 0.9102$$
$$= 0.09 \quad (2D)$$

(Note that we have assumed that the two distributions of heights are independent.)

7.2.2 AN EXTENSION

The extension of $E[X + Y] = E[X] + E[Y]$ for more than two variables is:

$$E[X_1 + X_2 + \ldots + X_n] = E[X_1] + E[X_2] + \ldots + E[X_n]$$

where $X_1, X_2, \ldots, X_n$ are any set of random variables. In the case of the variance, we require the set $X_1, X_2, \ldots, X_n$ to be mutually independent, and we then have the extension of $\text{Var}[X + Y] = \text{Var}[X] + \text{Var}[Y]$ as:

$$\text{Var}[X_1 + X_2 + \ldots + X_n] = \text{Var}[X_1] + \text{Var}[X_2] + \ldots + \text{Var}[X_n]$$

The proof of the above is trivial, being an extension of that for two variables.

We now extend Statement 7.1 to include the sum of any number of normal variables.

STATEMENT 7.2

Let $X_1, X_2, \ldots, X_n$ be independent random variables with respective distributions $N(\mu_1, \sigma_1^2)$, $N(\mu_2, \sigma_2^2)$, $\ldots$, $N(\mu_n, \sigma_n^2)$. That is, X_r is $N(\mu_r, \sigma_r^2)$. Then, the random variable:

$$X_1 + X_2 + X_3 + \ldots + X_n \quad \text{is}$$

$$N(\mu_1 + \mu_2 + \ldots + \mu_n, \sigma_1^2 + \sigma_2^2 + \ldots + \sigma_n^2)$$

More compactly, $\sum_{i=1}^{n} X_i$ is $N\left(\sum_{i=1}^{n} \mu_i, \sum_{i=1}^{n} \sigma_i^2\right)$.

Proof　We use repeated applications of Statement 7.1(a).

Let:

$$Y_2 = X_1 + X_2; \; Y_3 = X_1 + X_2 + X_3; \; \ldots; \; Y_n = X_1 + X_2 + \ldots + X_n$$

i.e. $Y_3 = Y_2 + X_3; \ldots; \; Y_n = Y_{n-1} + X_n$.

Now X_1 is $N(\mu_1, \sigma_1^2)$ and X_2 is $N(\mu_2, \sigma_2^2)$.

Therefore $Y_2 = X_1 + X_2$ is $N(\mu_1 + \mu_2, \sigma_1^2 + \sigma_2^2)$.

Also X_3 is $N(\mu_3, \sigma_3^2)$.

Hence $Y_3 = Y_2 + X_3$ is $N(\mu_1 + \mu_2 + \mu_3, \sigma_1^2 + \sigma_2^2 + \sigma_3^2)$, and so on.

Finally:　　　$Y_n = Y_{n-1} + X_n = X_1 + X_2 + \ldots + X_n$ is

$$N(\mu_1 + \mu_2 + \ldots + \mu_n, \sigma_1^2 + \sigma_2^2 + \ldots \sigma_n^2)$$

as required.

A special case of the previous result is when $X_1, X_2, \ldots, X_n$ have the same Normal distribution. That is when $\mu_1 = \mu_2 = \ldots = \mu_n = \mu$, say and $\sigma_1^2 = \sigma_2^2 = \ldots = \sigma_n^2 = \sigma^2$, say. In this case we can regard the situation as $X_1, X_2, \ldots, X_n$ being a random sample of size n from $N(\mu, \sigma^2)$.

Using the statement, we have the result:

$$X_1 + X_2 + \ldots + X_n \quad \text{is}$$

$$N(\mu + \mu + \ldots + \mu, \sigma^2 + \sigma^2 + \ldots + \sigma^2) = N(n\mu, n\sigma^2)$$

EXAMPLE 7.5

The 3 random variables X_1, X_2 and X_3 are distributed $N(50, 10)$, $N(100, 20)$ and $N(150, 30)$ respectively. Find the distributions of:

(a) $X_1 + X_2$ (b) $X_2 + X_3$

(c) $X_3 - X_1$ (d) $X_1 + X_2 + X_3$

SOLUTION

(a) $X_1 + X_2$ is $N(50 + 100, 10 + 20) = N(150, 30)$.

(b) $X_2 + X_3$ is $N(100 + 150, 20 + 30) = N(250, 50)$.

(c) $X_3 - X_1$ is $N(150 - 50, 30 + 10) = N(100, 40)$.

(d) $X_1 + X_2 + X_3$ is

$$N(50 + 100 + 150, 10 + 20 + 30) = N(300, 60).$$

EXAMPLE 7.6

A random sample of 6 items is taken from the distribution $N(50, 25)$. Find the probability that the sum of the sample is less than 280.

SOLUTION

We require: $\Pr(X_1 + X_2 + \ldots + X_6 < 280) = \Pr\left(\sum_{i=1}^{6} X_i < 280 \right)$

Since each X_i is distributed as $N(50, 25)$,

$$\sum_{i=1}^{6} X_i \text{ is } N(6(50), 6(25)) = N(300, 150).$$

Hence:

$$\Pr\left(\sum_{i=1}^{6} X_i < 280 \right) = \Pr\left(Z < \frac{280 - 300}{\sqrt{150}} \right) = \Pr\left(Z < \frac{-20}{\sqrt{150}} \right)$$

$$= \Phi(-1.633) = 1 - \Phi(1.633) = 1 - 0.949$$

$$= 0.051$$

7.2.3 MULTIPLES OF NORMAL RANDOM VARIABLES

In Chapter 5, we discussed and used the results $E[a \cdot X] = a \cdot E[X]$ and $Var[a \cdot X] = a^2 \cdot Var[X]$, where X is a random variable and a is a constant. For a normal variable, we have the following statement.

STATEMENT 7.3

If random variable X is $N(\mu, \sigma^2)$ and a is any positive constant, then the variable aX is $N(a\mu, a^2\sigma^2)$.

No Proof

If a is a positive integer $(a = 2$, say) and X is $N(\mu, \sigma^2)$, we have that $2X = X + X$ is $N(2\mu, 4\sigma^2)$, while, using an earlier result, we had that if X_1 and X_2 is a random sample of size 2 from $N(\mu, \sigma^2)$, $X_1 + X_2$ is $N(2\mu, 2\sigma^2)$. That is, $X + X$ has a larger variance than $X_1 + X_2$ although all the X values come from the same distribution. This apparent anomaly is explained by the fact that random variables X_1 and X_2 are independent (since they form a random sample) whereas X can never be independent of itself. Physically, $X_1 + X_2$ translates as 'sample *two items* independently and add their values'. On the other hand, $2X = X + X$ is interpreted as 'select *a single item* randomly and multiply its value by 2'. Thus the random variables $X + X$ and $X_1 + X_2$ are two different cases.

EXAMPLE 7.7

The independent random variables X_1 and X_2 are distributed $N(84, 8)$ and $N(85, 11)$ respectively. Find the distributions of:

(a) $X_1 + X_2$ (b) $3X_1$ (c) $X_2 - X_1$

(d) $4X_2 - 3X_1$ (e) $\dfrac{X_1 + X_2}{2}$

SOLUTION

(a) $X_1 + X_2$ is $N(84 + 85, 8 + 11) = N(169, 19)$.

(b) $3X_1$ is $N(3(84), 3^2(8))$ (using the previous statement)
$\qquad\qquad = N(252, 72)$.

(c) $X_2 - X_1$ is $N(85 - 84, 11 + 8) = N(1, 19)$.

(d) $4X_2$ is $N(4(85), 16(11)) = N(340, 176)$;
$\qquad 3X_1$ is $N(3(84), 9(8)) = N(252, 72)$

Hence $4X_2 - 3X_1$ is $N(340 - 252, 176 + 72) = N(88, 248)$.

(e) Now $X_1 + X_2$ has the distribution $N(169, 19)$ from (a).

Therefore $\tfrac{1}{2}(X_1 + X_2)$ is $N(\tfrac{1}{2}(169), (\tfrac{1}{2})^2(19)) = N(84.5, 4.75)$.

7.3 THE SAMPLE MEAN

Up to now we have been concentrating on distributions of sums (and differences) of random variables from mainly (infinite) Normal populations, since these are of most interest generally. In this section, we will be looking at the behaviour of $\bar{X}$, the sample mean, for random samples from:

(a)   any infinite population,

(b)   a Normal population (in particular), and

(c)   any finite population.

7.3.1 FROM ANY INFINITE POPULATION

STATEMENT 7.4

If $X_1, X_2, \ldots, X_n$ is a random sample of size n from any infinite population with mean μ and variance σ^2, then, with $\bar{X} = \dfrac{1}{n} \sum\limits_{i=1}^{n} X_i$, we have $E[\bar{X}] = \mu$ and $Var[\bar{X}] = \dfrac{\sigma^2}{n}$.

Proof $E[\bar{X}] = E\left[\dfrac{1}{n} \sum\limits_{i=1}^{n} X_i\right] = \dfrac{1}{n} E\left[\sum\limits_{i=1}^{n} X_i\right]$ (property of E)

$$= \dfrac{1}{n}\left[\sum\limits_{i=1}^{n} E[X_i]\right] \quad \text{(property of E)}$$

$$= \dfrac{1}{n} \sum\limits_{i=1}^{n} \mu = \dfrac{1}{n} n\mu = \mu$$

$Var[\bar{X}] = Var\left[\dfrac{1}{n} \sum\limits_{i=1}^{n} X_i\right]$

$$= \dfrac{1}{n^2}\left[Var\left(\sum\limits_{i=1}^{n} X_i\right)\right] \quad \text{(property of variance function)}$$

$$= \dfrac{1}{n^2}\left[\sum\limits_{i=1}^{n} Var[X_i]\right] \quad \text{(since } X_1, X_2, \ldots, X_n \text{ are independent)}$$

$$= \dfrac{1}{n^2}\left[\sum\limits_{i=1}^{n} \sigma^2\right] = \dfrac{1}{n^2}[n\sigma^2] = \dfrac{\sigma^2}{n}$$

completing the proof.

The above result also applies to a sample from a finite population, as long as the sampling is *with replacement*. Suppose the finite population was of size N with samples of size n being drawn. As long as the sample item is replaced after each observation, the choice is always from the fixed total N and the probabilities attached to values of the associated random variable will remain constant (and independent of N).

As an example, consider a finite population consisting of 10 counters in a bag; 2 numbered '1', 3 numbered '2' and 5 numbered '3'. We wish to take repeated samples of size 4 with replacement. Here, then, $N = 10$ and $n = 4$. Associating random variable X with the distribution, we have the following p.d.f.:

x	1	2	3
$\Pr(X = x)$	0.2	0.3	0.5

Notice that, since we replace each counter after a single selection, the above p.d.f. remains constant and independent of the size ($N = 10$ here) of the total population. That is, each of the sample variables X_1, X_2, X_3 and X_4 say, have the same distribution — that given for X above. Effectively then, we can consider the given finite population (with sample replacement) as an infinite one.

In this particular case, we have:

$$\mu = E[X] = 1(0.2) + 2(0.3) + 3(0.5) = 2.3$$

Also: $E[X^2] = 1^2(0.2) + 2^2(0.3) + 3^2(0.5) = 5.9$

Hence: $\sigma^2 = \text{Var}[X] = E[X^2] - E^2[X] = 5.9 - (2.3)^2 = 0.61$

We can now give the expectation and variance of the means of samples of size 4 from this population as:

$$E[\bar{X}] = \mu = 2.3 \text{ and } \text{Var}[\bar{X}] = \frac{\sigma^2}{n} = \frac{0.61}{4} = 0.1525$$

7.3.2 FROM A NORMAL POPULATION

STATEMENT 7.5

If $X_1, X_2, \ldots, X_n$ is a random sample of size n from $N(\mu, \sigma^2)$, then the statistic $\bar{X} = \dfrac{1}{n} \sum_{i=1}^{n} X_i$, the sample mean, has a Normal distribution with mean μ and variance $\dfrac{\sigma^2}{n}$, i.e. $\bar{X}$ is $N\left(\mu, \dfrac{\sigma^2}{n}\right)$.

Proof From Statement 7.2 we have $\sum_{i=1}^{n} X_i$ is $N(n\mu, n\sigma^2)$.

Hence: $\bar{X} = \dfrac{1}{n} \sum_{i=1}^{n} X_i$ is $N\left(\dfrac{1}{n} n\mu, \dfrac{1}{n^2} [n\sigma^2]\right)$ (from Statement 7.3)

$$= N\left(\mu, \frac{\sigma^2}{n}\right) \quad \text{as required}$$

Note that this is a particular case of the more general previous statement, where we are sampling specifically from a Normal distribution.

EXAMPLE 7.8

If a random sample of 20 is taken from a Normal distribution with mean 100 and variance 25, find the probability that the mean is greater than 101.

SOLUTION

Representing the distribution by X as N(100, 25), the sample mean, $\bar{X}$ is $N\left[100, \dfrac{25}{20}\right]$ using the result of the last statement.

$$\therefore \qquad \Pr(\bar{X} > 101) = \Pr\left(Z > \frac{101 - 100}{\sqrt{(25/20)}}\right) = \Pr(Z > 0.894)$$

$$= 1 - \Phi(0.894) = 1 - 0.814 = 0.186$$

EXAMPLE 7.9

An industrial process produces articles whose weights are distributed as N(100, 160) g. A control section examines samples of 10 from every large batch for output, and rejects the whole batch if the mean of the sample is greater than 105 g. Find what percentage of the whole machine output will be rejected.

SOLUTION

Let the random variable 'weight of an article' be represented by X as N(100, 160). For a sample of 10, $\bar{X}$ is $N\left(100, \dfrac{160}{10}\right) = N(100, 16)$.

$$\therefore \qquad \Pr(\bar{X} > 105) = \Pr\left(Z > \frac{105 - 100}{\sqrt{16}}\right) = \Pr(Z > 1.25)$$

$$= 1 - \Phi(1.25) = 1 - 0.8944 = 0.1056$$

So that proportion 0.1056 of sample means will be over 105 g.

Therefore proportion 0.1056 of batches will be rejected.

Hence 10.56% of the total output will be rejected.

7.3.3 FROM A FINITE POPULATION

We now consider taking a sample of n from a finite population size N, where the sampling is *without replacement*. The situation is different here for the following reason. Let us denote the sample variables, as usual, by $X_1, X_2, \ldots, X_n$ (the subscripts can denote the order in which the sample is drawn). The first variable X_1 has a specific distribution, determined by the composition of the N items. For example, there might be n_1 items associated with $X_1 = r_1$, n_2 items with $X_1 = r_2$, $\ldots$, n_s items with $X_1 = r_s$ (where $n_1 + n_2 + \ldots + n_s = N$). This will determine a probability distribution as mentioned. Once the first item has been drawn, and X_1 realises a value, x_1 say, there will only be $N-1$ items left in the population to choose from (since sampling is without replacement). That is, the composition of the population will have changed, resulting in a different distribution for X_2, and so on. Hence, each of the variables X_1, X_2, $\ldots$, X_n has a different distribution and, in particular, cannot form a 'random sample' (from Definition 7.1).

For this type of population, we have the following result for the sample mean:

STATEMENT 7.6

Let X_1, X_2, $\ldots$, X_n represent a sample of size n, without replacement, from a finite population of size N with mean μ and variance σ^2.

Then, if $\bar{X} = \dfrac{1}{n} \sum_{i=1}^{n} X_i$:

(a)　$E[\bar{X}] = \mu$　　　　(b)　$Var[\bar{X}] = \dfrac{\sigma^2}{n}\left(\dfrac{N-n}{N-1}\right)$

No Proof

It will be seen from (b) that, for very large N (i.e., $N \to \infty$), $\dfrac{N-n}{N-1} \to 1$ and hence $Var[\bar{X}] \to \dfrac{\sigma^2}{n}$, tying in with Statement 7.4 for an infinite population. That is, the larger N becomes, the closer $Var[\bar{X}]$ becomes to $\dfrac{\sigma^2}{n}$.

We now demonstrate the result of the above statement with a simple numerical case. Consider a population, size $N = 5$, consisting of the digits $1, 2, 3, 3, 6$ where we wish to take samples of size 2.

Firstly:

$$\mu = \frac{1 + 2 + 3 + 3 + 6}{5} = 3,$$

$$\sigma^2 = \frac{(1^2 + 2^2 + 3^2 + 3^2 + 6^2)}{5} - 3^2 = 2.8$$

We now calculate the expectation and variance of means of samples of size 2 using the above statement:

$$E[\bar{X}] = \mu = 3 \quad \text{and} \quad \text{Var}[\bar{X}] = \frac{\sigma^2}{n}\left(\frac{N-n}{N-1}\right)$$

$$= \frac{(2.8)(5-2)}{2(5-1)}$$

giving $\text{Var}[\bar{X}] = 1.05$.

These two results are now verified by considering the means of all possible samples of 2 from the given population. Denoting the values that the two sample variables $(X_1$ and $X_2)$ can take by x_1 and x_2, we list all possible combinations of two digits:

x_1	1	1	1	1	2	2	2	3	3	3	
x_2	2	3	3	6	3	3	6	3	6	6	Totals
$\bar{x}$	1.5	2	2	3.5	2.5	2.5	4	3	4.5	4.5	30
$\bar{x}^2$	2.25	4	4	12.25	6.25	6.25	16	9	20.25	20.25	100.5

Therefore the mean of $\bar{x} = E[\bar{X}] = \dfrac{30}{10} = 3$ and variance of

$\bar{x} = \text{Var}[\bar{X}] = \dfrac{100.5}{10} - 3^2 = 1.05$, both of these results agreeing with

those obtained using the statement (and calculated above).

EXAMPLE 7.10

Find the expectation and variance of the means of samples of 4 digits from the set $0, 1, 2, \ldots, 8, 9$.

SOLUTION

This is a finite population with $N = 10$ and sample size $n = 4$. We need to determine μ and σ^2, the mean and variance of the population.

Clearly, $\mu = 4.5$. The sum of squares of the population items is $0^2 + 1^2 + 2^2 + \ldots + 8^2 + 9^2 = 285$. Therefore

$$\sigma^2 = \frac{285}{10} - (4.5)^2 = 8.25.$$

Now, using the previous statement, $E[\bar{X}] = \mu = 4.5$:

$$\text{Var}[\bar{X}] = \frac{\sigma^2}{n}\left(\frac{N-n}{N-1}\right) = \frac{(8.25)(10-4)}{4(10-1)} = 1.375$$

7.3.4 THE CENTRAL LIMIT THEOREM

In Section 7.3.1 we found that the means of random samples of size n from any (infinite) population (mean μ, variance σ^2) had expectation and variance μ and $\dfrac{\sigma^2}{n}$ respectively.

We look now at one of the most important results in statistical theory with respect to the distribution of the above sample means.

STATEMENT 7.7 (THE CENTRAL LIMIT THEOREM)

If $X_1, X_2, \ldots, X_n$ is a random sample of size n from *any distribution* with mean μ and variance σ^2, then the sample mean,

$\bar{X} = \dfrac{1}{n} \sum_{i=1}^{n} X_i$, has an *approximate Normal distribution* with mean μ

and variance $\dfrac{\sigma^2}{n}$, i.e., $\bar{X}$ is $N\left(\mu, \dfrac{\sigma^2}{n}\right)$ (approximately).

Note that the larger the value of n, the better the approximation is.

No Proof

So that, whether we are taking random samples from, for instance, a binomial, Poisson, exponential or geometric distribution, the sample means (for reasonably large n) will have an approximately Normal distribution.

There is no set value for n (n_0 say) whereby we can assert that the approximation is good or bad if the sample size is larger or smaller than n_0. It will depend largely on the nature of the distribution being sampled. If, for instance, the original distribution is reasonably symmetric (without being too peaked or flat) we might only need a small sample size ($n = 10$ or 20) in order to obtain a good approximation. On the other hand, if the original distribution is excessively skew, we might need a sample size of 100 or more for a reasonable approximation. Of course, if the original distribution is normal, the result given by the Central Limit Theorem is exact and would tie in precisely with Statement 7.5.

The Normal approximation to the binomial (Statement 6.14(a)) is easily shown to be a particular application of the Central Limit Theorem as follows.

Recall that a random variable, X, with the p.d.f. $\Pr(X = 0) = 1 - p$ and $\Pr(X = 1) = p$ (for $0 \leqslant p \leqslant 1$) is said to have a Bernoulli distribution (see Section 6.6.1), and we had the result (Statement 6.15) that if $X_1, X_2, \ldots, X_n$ are n independent Bernoulli random variables, then

$\sum_{i=1}^{n} X_i$ is $\mathrm{Bin}(n, p)$.

We now let $X_1, X_2, \ldots, X_n$ be a random sample of size n from $\mathrm{Ber}(p)$ (i.e., a Bernoulli distribution with parameter p), which has mean $= \mu = p$ and variance $= \sigma^2 = p(1 - p)$. Using the Central Limit Theorem, we have:

$$\bar{X} \text{ is (approximately) } N\left(\mu, \frac{\sigma^2}{n}\right) = N\left(p, \frac{p(1-p)}{n}\right)$$

Hence, the binomial random variable $\sum_{i=1}^{n} X_i = n\bar{X}$ is (approximately):

$$N\left(np, n^2 \frac{p(1-p)}{n}\right)$$

$$= N(np, np(1-p)),$$

which is the Normal approximation. This argument then, using the Central Limit Theorem, constitutes a proof of Statement 6.14(a).

EXAMPLE 7.11

If $X_1, X_2, \ldots, X_{20}$ is a random sample from Po(3) (i.e., a Poisson distribution with parameter 3), find, using the Central Limit Theorem, the approximate probability that the sample mean will be greater than 4.

SOLUTION

If X is Po(3), then $E[X] = \mu = 3$ and $Var[X] = \sigma^2 = 3$ also. So, by the Central Limit Theorem, $\bar{X}$ is (approximately):

$$N\left(3, \frac{3}{20}\right) = N(3, 0.15)$$

$$\therefore Pr(\bar{X} > 4) = Pr\left(Z > \frac{4-3}{\sqrt{0.15}}\right) = Pr(Z > 2.58) = 1 - \Phi(2.58)$$

$$= 0.005 \quad (3D)$$

EXAMPLE 7.12

A rectangular distribution is defined over the interval 0 to 10. A random sample of size 30 is taken from this distribution. Find the approximate probability that the mean of the sample exceeds 5.5.

SOLUTION

Recall that a rectangular distribution defined over the interval a to b has mean $\frac{1}{2}(a+b)$ and variance $\frac{(b-a)^2}{12}$. Here, $a = 0$ and $b = 10$, giving mean $= \mu = 5.0$ and variance $= \sigma^2 = \frac{100}{12} = 8.33$ (2D). The sample size is $n = 30$. Using the Central Limit Theorem, $\bar{X}$, the sample mean, is (approximately) $N\left(5.0, \frac{8.33}{30}\right)$, i.e., $\bar{X}$ is N(5.0, 0.278).

Hence:

$$\Pr(\bar{X} > 5.5) = \Pr\left(Z > \frac{5.5 - 5.0}{\sqrt{0.278}}\right) = \Pr(Z > 0.95) \ (2D)$$

$$= 1 - \Phi(0.95) = 1 - 0.829$$

$$= 0.171 \quad (3D)$$

EXAMPLE 7.13

Ten dice are thrown as an experiment and the number of sixes obtained is recorded. If the ten dice are thrown 50 times, find the approximate probability that the average number of sixes obtained is less than 2.

SOLUTION

Throwing ten dice and observing the number of sixes is a well-defined binomial situation with $n = 10$ (number of dice) and $p = \dfrac{1}{6}$ (probability of a 'six'). The mean and variance of this binomial distribution are calculated as:

$$\mu = np = \frac{10}{6} = 1.667 \ (3D)$$

and

$$\sigma^2 = np(1 - p) = \frac{10}{6}\frac{5}{6} = 1.389 \ (3D)$$

We can regard the 50 throws as a random sample of 50, the variables involved $X_1, X_2, \ldots, X_{50}$, say, having independent and identical binomial distributions with mean 1.667 and variance 1.389.

Hence, using the Central Limit Theorem:

$$\bar{X} = \frac{1}{50} \sum_{i=1}^{50} X_i \text{ is (approximately) } N\left(1.667, \frac{1.389}{50}\right)$$

$$= N(1.667, 0.028) \quad (3D)$$

$$\therefore \qquad \Pr(\bar{X} < 2) = \Pr(\bar{X} < 1.5), \quad \text{using a continuity correction}$$

$$= \Pr\left(Z < \frac{1.5 - 1.667}{\sqrt{0.028}}\right)$$

$$= \Pr(Z < -1.00) \ (2D)$$

$$= 0.159 \ (3D)$$

7.4 THE χ^2 DISTRIBUTION

At this stage we introduce a new distribution, the χ^2 (chi squared) distribution, necessary in order to discuss how the variance of a random sample chosen from a normal distribution behaves.

We are not interested in this distribution for its own sake (as we are, for instance, with the normal, Poisson and binomial distributions), but for the various uses it has in Statistics.

The p.d.f. and the shape of the probability curve is briefly discussed. Some important characteristics of the distribution are then listed and, finally, (as with the Normal distribution) we interpret, and make use of, a special set of tables drawn up for the distribution.

7.4.1 PROBABILITY DENSITY FUNCTION

The distribution has a p.d.f. of the form:

$$f(x) = A_\nu \left(\frac{x}{2}\right)^{\frac{\nu}{2}-1} e^{-\frac{x}{2}}, \quad \text{for } x > 0$$

where A_ν is a special constant depending on ν and ν is an integral-valued parameter.

The graphs for $f(x)$ for some values of ν are given below.

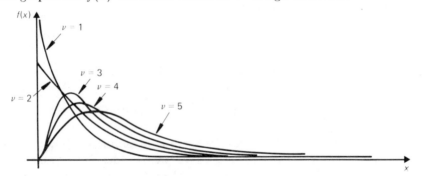

For $\nu \geqslant 3$, the density curve takes on a characteristic 'right-skew' shape, the skew easing for larger values of ν.

7.4.2 SOME SPECIAL CHARACTERISTICS

Some important results and identities for this distribution are now given.

(a)   The distribution is *continuous*.

(b)   There is *one* parameter, ν.

(c)   The shorthand notation used for the distribution is 'X is $\chi^2(\nu)$', translated as 'the random variable X has a χ^2 distribution with parameter ν'. The parameter ν is often referred to as 'degrees of freedom', for reasons we shall see in a later section.

STATEMENT 7.8

(a)   If X is $\chi^2(\nu)$, then $E[X] = \nu$ and $\text{Var}[X] = 2\nu$.

(b)   If Y is $N(0, 1)$, then Y^2 is $\chi^2(1)$.

(c)   If X_1 and X_2 are independent and distributed $\chi^2(\nu_1)$ and $\chi^2(\nu_2)$ respectively, then $X_1 + X_2$ is $\chi^2(\nu_1 + \nu_2)$.

No Proof

We see that the parameter involved with the distribution is just the mean, the variance always being just twice the mean. Part (b) above can be thought of as the definition of a χ^2 variable. That is, the square of a standard Normal variable is defined to be a χ^2 variable with parameter 1. Part (c) demonstrates the fact that the sum of two χ^2 variables is also a χ^2 variable having a parameter that is the sum of the original two.

An extension of (c) to more than two variables should now be obvious. That is, if $X_1, X_2, \ldots, X_n$ are independent χ^2 variables with respective parameters $\nu_1, \nu_2, \ldots, \nu_n$, then the variable $X_1 + X_2 + \ldots + X_n$ has a χ^2 distribution with parameter $\nu_1 + \nu_2 + \ldots + \nu_n$.

EXAMPLE 7.14

If random variable X_1 is $\chi^2(4)$ and X_2 is independently $\chi^2(6)$, give the distribution of the random variable $Y = X_1 + X_2$ stating its mean and variance.

SOLUTION

$Y = X_1 + X_2$ is $\chi^2(4 + 6) = \chi^2(10)$ by (c) of the above statement. Therefore $E[Y] = 10$ and $\mathrm{Var}[Y] = 2(10) = 20$ by (a) of the above statement.

EXAMPLE 7.15

If random variable X is $N(\mu, \sigma^2)$, show that the random variable $\dfrac{(X - \mu)^2}{\sigma^2}$ has mean 1 and variance 2.

SOLUTION

Since X is $N(\mu, \sigma^2)$, we have $\dfrac{X - \mu}{\sigma}$ is $N(0, 1)$, a standardised Normal variable. Therefore $\dfrac{(X - \mu)^2}{\sigma^2}$ has a $\chi^2(1)$ distribution, using (b) of the above statement. Hence:

$$E\left[\frac{(X - \mu)^2}{\sigma^2}\right] = 1 \quad \text{and} \quad \mathrm{Var}\left[\frac{(X - \mu)^2}{\sigma^2}\right] = 2(1) = 2$$

7.4.3 USE OF TABLES

The tables listed (Appendix Table 4) for a χ^2 distribution are slightly different from those we have been using for problems connected with a Normal distribution. These latter tables were distribution function tables for a standard Normal distribution. The tables for the χ^2 distribution (and for other distributions we shall deal with later) are called 'percentage point tables'.

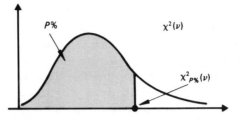

Fig. 7.1

A *percentage point* of a distribution is that value of the associated random variable that has a specified percentage of its distribution *lying to its left*. The $P\%$ point of a $\chi^2(\nu)$ distribution would be written as $\chi^2_{P\%}(\nu)$ and is illustrated in Fig. 7.1. That is, if X is $\chi^2(\nu)$ then the following is true:

$$\Pr(X < \chi^2_{P\%}(\nu)) = \frac{P}{100}$$

For example, $\chi^2_{23\%}(4)$ is that value of a $\chi^2(4)$ variable having 23% of the distribution lying to the left of it. $\chi^2_{50\%}(12)$ represents the median of a $\chi^2(12)$ distribution.

Returning to Table 4, notice that each row corresponds to a particular value of parameter ν and the columns give a selection of percentage values in the form of probabilities. For example, consider the value 23.69 given in column $P = 0.95 \, (= 95\%)$ and row $\nu = 14$. This is translated as 'the 95% point of a $\chi^2(14)$ distribution is 23.69', i.e., $\chi^2_{95\%}(14) = 23.69$.

The value 0.83 at the join of column $P = 0.025 \, (= 2.5\%)$ and row $\nu = 5$ has the significance $\chi^2_{2.5\%}(5) = 0.83$.

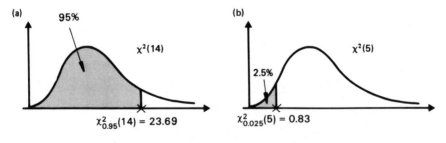

Fig. 7.2

Both these situations are shown pictorially in Fig. 7.2.

We shall be using these tables practically in our discussion of the sample variance distribution in the next section and for other purposes in later chapters.

Some simple illustrations of the use of the χ^2 percentage point tables follow.

EXAMPLE 7.16

If X and Y are independently distributed as $\chi^2(2)$ and $\chi^2(6)$ respectively, find the probability that:

(a) $X > 7.38$ (b) $X < 0.10$ (c) $Y > 16.81$

(d) $X + Y > 2.18$

SOLUTION

(a) X is $\chi^2(2)$. From Appendix Table 4, 97.5% of this distribution lies to the left of 7.38 (column $P = 0.975$, row $\nu = 2$)

$$\therefore \qquad \mathrm{Pr}(X > 7.38) \;=\; \frac{2.5}{100} \;=\; 0.025$$

(b) $\mathrm{Pr}(X < 0.10) = 0.05$ (from column $p = 0.05$, row $\nu = 2$)

(c) Y is $\chi^2(6)$. Therefore $\mathrm{Pr}(Y < 16.81) = 0.99$ (from column $P = 0.99$ and row $\nu = 6$)

Hence, $\mathrm{Pr}(Y > 16.81) = 1 - 0.99 = 0.01$.

(d) $X + Y$ is $\chi^2(2 + 6) = \chi^2(8)$ (sum of two χ^2 variables)

Hence:

$\mathrm{Pr}(X + Y > 2.18) = 1 - 0.025 = 0.975$ (from column 0.025, row $\nu = 8$).

7.5 SOME OTHER SAMPLE STATISTICS

7.5.1 THE SAMPLE VARIANCE

Given that the set $X_1, X_2, \ldots, X_n$ constitutes a random sample from some distribution, we have already seen that the sample variance is defined as the quantity $S^2 = \dfrac{1}{n} \sum_{i=1}^{n} (X_i - \bar{X})^2$. Note that S^2 is a statistic, since it depends only on the variables $X_1, X_2, \ldots, X_n$ and $\bar{X}$ (which itself depends on the variables $X_1, X_2, \ldots, X_n$), and hence is completely determined once the sample is taken.

We considered the expectation, variance and distribution of $\bar{X}$ in different situations of particular interest. However, due to the more complex nature of S^2, we consider its behaviour in only one situation as follows.

STATEMENT 7.9

Let $X_1, X_2, \ldots, X_n$ be a random sample of size n from $N(\mu, \sigma^2)$

with $S^2 = \dfrac{1}{n} \sum\limits_{i=1}^{n} (X_i - \bar{X})^2$, the sample variance.

Then, the quantity $\dfrac{nS^2}{\sigma^2}$ has a χ^2 distribution with parameter $n - 1$,

i.e., $\dfrac{nS^2}{\sigma^2}$ is $\chi^2(n-1)$.

Proof   Left as a directed exercise.

Note that we have not obtained a specific distribution for S^2, but $\dfrac{nS^2}{\sigma^2}$, and the technique employed to calculate probabilities for S^2 will be made clear in the following examples.

EXAMPLE 7.17

A random sample, size 6, is taken from a Normal distribution with variance 16. Find the probability that the sample variance is greater than 29.6.

SOLUTION

Denoting the sample variance by S^2, we have (from the previous statement) that $\dfrac{nS^2}{\sigma^2}$ (is distributed as $\chi^2(n-1)$.

In this case, $\dfrac{6}{16} S^2$ is $\chi^2(5)$ since, here, $n = 6$ and $\sigma^2 = 16$, i.e., $\dfrac{3}{8} S^2$ is $\chi^2(5)$.

Now we require: $\Pr(S^2 > 29.6) = \Pr\left(\dfrac{3}{8} S^2 > \dfrac{3}{8} (29.6)\right)$

$\left(\text{multiplying both sides of the inequality by } \dfrac{3}{8}\right)$

$\qquad = \Pr\left(\dfrac{3}{8} S^2 > 11.1\right)$

$\qquad = \Pr(\chi^2(5) > 11.1),$ since $\dfrac{3}{8} S^2$ is $\chi^2(5)$

So we need to find the probability that a $\chi^2(5)$ variable is greater than 11.1. From χ^2 percentage point tables, we have

$$\chi^2_{95\%}(5) = 11.07 = 11.1 \text{ (1D)},$$

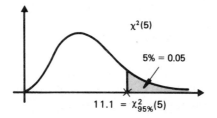

which is shown pictorially in the diagram above. Hence:

$$\Pr(\chi^2(5) > 11.1) = 5\% = 0.05 \text{ (approximately)}$$

EXAMPLE 7.18

Random samples of size 16 are repeatedly taken from a Normal population with variance σ^2 (unknown). Out of 1000 samples, the sample variance was found to exceed 35.91 only 25 times. Use this information to find an approximate value for σ^2.

SOLUTION

The information given can be put into a probability statement as

$$\Pr(S^2 > 35.91) = \frac{25}{1000} = 0.025$$

For a sample of size 16 from a Normal population with variance σ^2, we have $\dfrac{16S^2}{\sigma^2}$ is $\chi^2(15)$, using the previous statement. Adjusting the inequality in the above probability statement, we need to find

$$\Pr\left(\frac{16S^2}{\sigma^2} > \frac{16(35.91)}{\sigma^2}\right) = 0.025$$

i.e.,

$$\Pr\left(\chi^2(15) > \frac{16(35.91)}{\sigma^2}\right) = 0.025$$

This can be translated as $\dfrac{16(35.91)}{\sigma^2}$ is the 97.5% point of a $\chi^2(15)$ distribution, i.e.:

$$\chi^2_{97.5\%}(15) = \frac{16(35.91)}{\sigma^2} = 27.49 \quad \text{(from tables)}$$

Thus:

$$\sigma^2 = \frac{16(35.91)}{27.49} = 20.90 \quad \text{(2D)}$$

7.5.2 A SAMPLE PROPORTION

Here we look at a population of items in a slightly different way. Rather than measure each member numerically in terms of some defined random variable, we split the population into two mutually exclusive sections according to whether or not any item has a particular attribute. For instance, a population of bolts being produced by a machine can be classified as either 'good' or 'defective'; a population of people's heights can be split into '6 ft or less' and 'over 6 ft'.

The apparent connection between this type of situation and a binomial situation (where we split an experimental outcome set into 'success' or 'failure') is not coincidental, as we shall see.

Suppose the population in question has an exact proportion, π say, of its members having a particular attribute. We now draw a random sample of size n and observe the actual number y, say, of items with the attribute. (Thus the remaining $n-y$ items in the sample do not have the attribute.) Let us denote the sample proportion having the attribute by p.

Hence $p = \dfrac{y}{n}$ here.

Repeating this sampling procedure a number of times will give a distribution of values of p. It is this distribution that we consider here.

Let random variable Y denote the number of items in the sample with the attribute. (Remember the notation — Y denotes a variable name and y is a value it can take if the sample is actually performed.) Similarly for P and p.

We then have:

STATEMENT 7.10

Let an infinite population have a known proportion π of its members with some defined attribute. Let Y represent the random variable of the number of items in a random sample of size n with this attribute.
Then Y is $\text{Bin}(n, \pi)$.

Proof   We can regard the sample as n independent trials of an experiment, with the occurrence of the attribute counting as a 'success'.

Since the proportion of the population having the attribute is π, the probability of a success with any trial is π. Hence, by definition of a binomial situation, the number of sample items with the attribute, Y, is $\text{Bin}(n, \pi)$.

Using this result, we can now assert the following statement.

STATEMENT 7.11

Let a population have a known proportion π of its members with some attribute. The proportion P of a random sample of size n from the population having this attribute, is distributed approximately normally, with mean π and variance $\dfrac{\pi(1-\pi)}{n}$, i.e., P is

(approximately) $N\left(\pi, \dfrac{\pi(1-\pi)}{n}\right)$.

Note that the larger the value of n, the better the approximation.

Proof   We have, from the previous statement, that Y is $\text{Bin}(n, \pi)$. So, using the Normal approximation to the binomial, we have

$$Y \text{ is } N(n\pi, n\pi(1-\pi)) \text{ (approximately).}$$

But $P = \dfrac{Y}{n}$ by definition.

Hence: $P = \dfrac{1}{n} Y$ is $N\left(\dfrac{1}{n} n\pi, \dfrac{1}{n^2}[n\pi(1-\pi)]\right)$ (approximately)

i.e., P is (approximately) $N\left(\pi, \dfrac{\pi(1-\pi)}{n}\right)$, *completing the proof.*

Note that when we use the Normal approximation to the binomial, we need to introduce a continuity correction of $\pm \dfrac{1}{2}$. That is, if Y is a binomial variable, then $\Pr(Y \geqslant a)$, say, would be translated to $\Pr(Y \geqslant a - 0.5)$ and $\Pr(Y \leqslant a)$ would be translated to $\Pr(Y \leqslant a + 0.5)$. In exactly the same way, since (in the above proof) we take $P = \dfrac{1}{n} Y$, we need to use a continuity correction of $\pm \dfrac{1}{2}\dfrac{1}{n} = \pm \dfrac{1}{2n}$ whenever calculating probabilities involving proportion P.

EXAMPLE 7.19

1 in 8 of the residents of a particular city are known to own cars. Find the approximate probability that at least 15 of a random sample of 100 of the city population will own cars.

SOLUTION

We are given that proportion $\dfrac{1}{8} = 0.125$ of the city population own cars. So that $\pi = 0.125$.

For a random sample, size n, P is (approximately) $N\left(\pi, \dfrac{\pi(1-\pi)}{n}\right)$

$$= N\left(0.125, \frac{(0.125)(0.875)}{100}\right) \quad (n = 100)$$

So P is (approximately) $N(0.125, 0.001)$ (3D).

We require the probability that at least 15 out of 100 own cars.

i.e., we require: $Pr(P \geqslant 0.15) = Pr\left(Z > \dfrac{(0.15 - \frac{1}{200}) - 0.125}{\sqrt{0.001}}\right)$

$$= Pr(Z > 0.60)$$

$$= 1 - \Phi(0.60) = 0.2743$$

7.5.3 THE CORRELATION COEFFICIENT

The intention here is to look very briefly at particular statistics of some interest derived from a bivariate random sample. We consider a *bivariate random sample of size n* simply as a set of n independent pairs of random variables $((X_1, Y_1), (X_2, Y_2), \ldots, (X_n, Y_n))$.

We can now define the following statistics on the sample:

$$S_X{}^2 = \frac{1}{n} \sum X_i{}^2 - \bar{X}^2; \quad S_Y{}^2 = \frac{1}{n} \sum Y_i{}^2 - \bar{Y}^2;$$

$$S_{XY} = \frac{1}{n} \sum X_i Y_i - \bar{X} \cdot \bar{Y}$$

and hence

$$R = \frac{S_{XY}}{S_X \cdot S_Y}$$

Now, there are known distributions for statistic R, but in general these can only be used under certain conditions for both R itself and the sample size n.

The most useful distribution (from the point of view of this particular text), which unfortunately involves R being transformed in a particular way but nevertheless is valid for all values of R and large values of n, $(n \geqslant 50)$, is now described.

Given a correlation coefficient r, we can transform it according to the rule $r^* = Fi(r)$, which we shall call Fisher's transformation. Appendix Table 7 gives r^* for selected values of r.

For instance, row 0.40 and column .00 in Table 11 gives $r^* = 0.424$ for $r = 0.40$. That is, $Fi(0.40) = 0.424$. Similarly for the last row of the tables we have $Fi(0.992) = 2.759$.

Since a correlation coefficient r can range between the values -1 and $+1$, we need to be able to use the aforementioned tables for a negative value of r. To do this, we only have to be aware of the simple relationship $\mathrm{Fi}(-r) = -\mathrm{Fi}(r)$.

So, for instance: $\mathrm{Fi}(-0.93) = -\mathrm{Fi}(0.93)$

$$= -0.658 \quad \text{(from tables)}$$

The need for this particular transformation is made clear in the following statement.

STATEMENT 7.12

Let R be the correlation coefficient of a bivariate random sample of size n taken from a population having correlation coefficient ρ. Then, if $R^* = \mathrm{Fi}(R)$ and $\rho^* = \mathrm{Fi}(\rho)$ Fisher's transformation), we have:

$$R^* \text{ is } \mathrm{N}\left(\rho^*, \frac{1}{n-3}\right) \quad \text{approximately}$$

Notes   (a)   Fisher's transformation $r^* = \mathrm{Fi}(r)$ is given in Appendix Table 7.

(b) $\mathrm{Fi}(-r) = -\mathrm{Fi}(r)$.

No Proof

So transforming R to R^* gives an approximate Normal distribution for the new statistic. As an example of its use, consider a bivariate random sample of size 68 from a parent population having a coefficient 0.25. We wish to find $\Pr(R > 0.5)$.

We can only work with transformed quantities and we calculate first:

$$\rho^* = \mathrm{Fi}(\rho) = \mathrm{Fi}(0.25)$$

$$= 0.255 \quad \text{(from tables)}.$$

Now: $\Pr(R > 0.5) = \Pr(R^* > \mathrm{Fi}(0.5))$

(transforming both parts of the inequality)

$$= \Pr(R^* > 0.549) \quad \text{(from tables)}$$

But R^* is $\mathrm{N}\left(\rho^*, \dfrac{1}{n-3}\right) = \mathrm{N}(0.255, 0.015)$ since $n = 68$.

Hence:

$$\Pr(R^* > 0.549) = \Pr\left(Z > \frac{0.549 - 0.255}{\sqrt{0.015}}\right) \quad \text{(standardising)}$$

$$= \Pr(Z > 2.40) \quad \text{(2D)}$$

$$= 1 - \Phi(2.40)$$

$$= 1 - 0.9918$$

$$= 0.008 \quad \text{(3D)}$$

EXAMPLE 7.20

A bivariate random sample of size 50 is taken from a population whose correlation coefficient is -0.32. What is the probability that the sample coefficient will be positive? What is the minimum sample size required so that we expect a positive correlation coefficient only one time in a thousand?

SOLUTION

Here $\rho = -0.32$.

Therefore: $\rho^* = \text{Fi}(-0.32) = -\text{Fi}(0.32)$

$$= -0.332 \quad \text{(from tables)}$$

Also we have R^* is $N\left(\rho^*, \dfrac{1}{n-3}\right) = N(-0.332, 0.02)$.

Now we require:

$$\begin{aligned} \Pr(R > 0) &= \Pr(R^* > \text{Fi}(0)) \quad \text{(transforming)} \\ &= \Pr(R^* > 0) \\ &= \Pr\left(Z > \frac{0 - (-0.332)}{\sqrt{0.02}}\right) \\ &= \Pr(Z > 2.35) \\ &= 1 - \Phi(2.35) \\ &= 1 - 0.9906 \\ &= 0.009 \quad \text{(3D)} \end{aligned}$$

For the second part of the question, we need to find the value of n such that $\Pr(R > 0) = 0.001$ (i.e., 1 in a 1000).

We have R^* as $N\left(-0.332, \dfrac{1}{n-3}\right)$.

From the above statement $\Pr(R^* > 0) = 0.001$ (after transforming).

Hence: $\Pr\left(Z > \dfrac{0 - (-0.332)}{\sqrt{1/(n-3)}}\right) = 0.001$

giving: $\Pr(Z > (0.332)\sqrt{(n-3)}) = 0.001$

or: $1 - \Phi((0.332)\sqrt{(n-3)}) = 0.001$

Finally: $\Phi((0.332)\sqrt{(n-3)}) = 0.999$

From tables, we have therefore $(0.332)\sqrt{(n-3)} = 3.1$.

Rearranging and squaring

$$n - 3 = \left(\frac{3.1}{0.332}\right)^2$$

i.e. $$n = \left(\frac{3.1}{0.332}\right)^2 + 3 = 90.2 \quad (1D)$$

Hence, the minimum sample size is 91.

7.6 SIMULATED RANDOM SAMPLING

In this section we concentrate on a particular method of taking random samples from known distributions.

Often in statistical investigations, we need to take samples from populations that are (effectively) unknown to us. For example, sampling from the weights of British males between the ages of 20 and 40. It is clear that this particular distribution will never be exactly determinable.

The lengths of cod fish in the North Sea, the average diameter of stars in our galaxy and the number of live offspring per birth of a certain species of animal are other examples of populations that we have no complete knowledge of.

In order to take a random sample from this type of population, we must go to great lengths to ensure that the sample items are independent and representative of the total population.

On the other hand, there are distributions that are completely determined theoretically or are effectively known from past experience such as probability distributions with given p.d.f.s and large (sampling) frequency distributions known to be representative of some infinite population.

We will be concerned here with taking random samples from distributions such as these, using random number tables.

7.6.1 RANDOM NUMBER TABLES

These tables usually consist of consecutive listings of digits from the set 0 to 9 (inclusive), obtained under the condition that each digit has an equal chance of occurring at any stage. That is, each digit has probability $\frac{1}{10}$ of occurring. Representing the numerical value of any digit by random variable X, we have the distribution of digits described by the p.d.f. $\Pr(X = x) = \frac{1}{10}$; $x = 0, 1, 2, \ldots, 9$. Taking any row or column of the tables (at random!), we can regard the list of digits as a random sample $X_1, X_2, X_3, \ldots$, from the above distribution of X.

However, the tables can be used to represent any type of number, discrete or continuous.

Consider the discrete set $0, 1, 2, \ldots, 97, 98, 99$. Any member of this finite population can be represented by a 2-digit number. That is, 0 can be represented by 00, 1 by 01, 2 by 02, $\ldots$, and so on. Reading any list of digits from the tables in groups of two will represent a random sample from this population.

We can, of course, extend this process to include consecutive groups of n digits to be random selections from the population of all n digit numbers. (There would be 10^n members in this finite population.)

Referring to the Appendix specifically, Table 10 is a typical set of random number tables.

We will take the eighth row of the tables, given as 87024 74221 69721 ... 15430 and use the digits to take random samples from various populations.

(a)   We can take the above set as a sample of 50 numbers (single digits) from the range 0 to 9 inclusive, giving $(8, 7, 0, 2, 4, 7, \ldots, 4, 3, 0)$.

(b)   We can take the set as a random sample of 25 integers from the range 0 to 99, giving $(87, 02, 47, 42, \ldots, 54, 30)$.

(c)   Alternatively, we can group the digits in 'fives' and take a random sample from the integer range 0 to 99 999, giving (87 024, 74 221, 69 721, ...).

Examples of the use of the tables for samples from continuous ranges are given as follows.

(d)   Consider the continuous range 0 up to 10. We can take samples from this set to any desired degree of accuracy. If we required a sample to 3D accuracy, we would simply read the digits in groups of 'fours', inserting a decimal point between the first two digits. In this way, using the first row of the tables again, we obtain the sample $(8.702, 4.742, 2.169, \ldots)$.

(e)   From the range 0 up to 1, if we wished to take a sample to 5D accuracy, we would obtain $(0.870\,24, 0.74\,221, 0.69\,721, \ldots)$.

Note, in the case of sampling from a continuous population, the use of 'up to' in describing a continuous range. It is used in exactly the same way as in Chapters 1 and 2.

In (d), 0 up to 10 is translated as '0 and up to (but excluding) 10'. Thus the population is $(0.000, 0.001, 0.002, \ldots, 9.998, 9.999)$. Note that 10 is excluded from the population. Similarly in (e), the highest attainable value in the range up to 1 (5D) is 0.999 99.

EXAMPLE 7.21

Take a random sample of 20, from the discrete, integral range 0 to 999 using random number tables.

SOLUTION

We start at row 6 (Appendix Table 10), i.e., 78183 44396 ... etc.

Reading the digits in groups of three gives the following sample:

781 834 439 611 064 921 539 629 300 825
210 797 833 719 739 136 847 020 923 316

EXAMPLE 7.22

Using Appendix Table 10, take a random sample of 15 numbers (to 3D) from the continuous range 0 up to 100.

SOLUTION

We will start at column 16 reading down, i.e., 497369 ... , etc.

Since we require the sample values to have 3D accuracy, we need to consider groups of 5 digits, the decimal point being inserted between the 2nd and 3rd digits in each group. In this way, we obtain the following sample.

49.736 90.062 27.461 45.709 9.740 48.953 15.041 75.698
70.276 90.740 41.129 20.212 71.122 73.084 36.147

EXAMPLE 7.23

Take a random sample of 10 from the continuous range 0 up to 8 to 1D.

SOLUTION

We will use the last row and read backwards, i.e., 38894 37143 ... etc.

The technique here is to take our sample as usual (that is, for a range 0 up to 10), and simply discard any values outside the desired range.

The sample is 3.8, 8.9, 4.3, 7.1, 4.3, 9.3, 9.8, 9.9, 2.1, 9.4, 2.4, 8.6, 6.4, 2.4, 8.9, 2.9, 7.6.

The seven underlined values fall outside the range and are ignored.

7.6.2 SAMPLING FROM GIVEN DISTRIBUTIONS

Here we consider the problem of taking a random sample from a given distribution. We will illustrate the basic technique using the following frequency distribution.

x	0	1	2	3	4	5	6	Total
f	2	12	24	37	15	7	3	100

We wish to draw a random sample of 20 from this distribution.

The technique is not difficult as long as we bear in mind the basic principle involved in random sampling — *every member of the population must have an equal chance of being selected.*

In this case, we have 100 members to choose from and we need to choose 20 randomly. But we already know how to take a random sample of 20 from the integral range 0 to 99 inclusive by using random sampling numbers. What we do is to allocate an integer in this range to each of the 100 members of the frequency distribution. (Note that, for convenience, we would prefer to deal with numbers 1 to 100 rather than 0 to 99, and this is easily arranged by taking the sampling value 0 to represent 100.)

To allocate the 100 sampling numbers to the frequency distribution population is fairly straightforward. We examine each frequency of the given distribution and allocate the correct number of integers to this particular class.

For the given distribution, the first class ($x = 0$) has frequency 2. Thus, integers 01 and 02 are allocated here. The next class ($x = 1$) has frequency 12, so we allocate the next 12 integers, 03 to 14 inclusive, to this class. This is repeated for each class in the distribution. To help with the arithmetic of allocation, it is usually convenient to compile a cumulative frequency column. This is shown in the following table, together with the allocation of the sampling integers 1 to 100.

x	0	1	2	3	4	5	6
f	2	12	24	37	15	7	3
Cum f	2	14	38	75	90	97	100
Allocation	01 to 02	03 to 14	15 to 38	39 to 75	76 to 90	91 to 97	98 to 100

The following sample of 20 has been drawn from random number tables.

64 100 26 04 54 55 38 57 94 62
68   40 26 04 24 25 03 61 01 20

64 is in the range 39 to 75 which corresponds to $x = 3$

100 is in the range 98 to 100 which corresponds to $x = 6$

26 is in the range 15 to 38 which corresponds to $x = 2$

and so on.

Completing the sample and tabulating the results we finally obtain:

x	0	1	2	3	4	5	6
f	1	3	6	8	0	1	1

(The student might like to compare the mean and variance of this sample with the mean and variance of the distribution from which the sample was drawn.)

Variations on the above technique are used, the most important of which we need to mention. This is the case where the total frequency of the distribution from which we wish to sample does not add up to a convenient 10, 100, or 1000, etc. as the previous one did.

There are two techniques we can use. The first entails taking a sampling range that is convenient. That is, one whose upper limit is greater than the total frequency of the distribution being considered. We would then conduct the sample as usual, ignoring any sample values that fell outside the range of the frequency distribution. (This was done in Example 7,22.) For example, if the frequency distribution total was 285, we would use the sampling range 000 to 999. (In this case we would be discarding most of the sample items.)

Alternatively, we can transform the given distribution to a relative (proportional) frequency distribution giving a total proportion of 1, of course, convenient for the allocation of sampling numbers.

As an example, consider the following simple frequency distribution, tabulated, together with the necessary calculations.

x	f	Proportion (1)	Relative frequency (2)	Cumulative relative frequency (3)	Allocation of sampling numbers (4)
0	12	0.079	79	79	001 to 079
1	26	0.170	170	249	080 to 249
2	81	0.529	529	778	250 to 778
3	34	0.222	222	1000	779 to 1000
Total	153	1.000	1000		

The frequency proportions are shown (to 3D) in column (1). Column (2) gives the same figures with the decimal point removed for convenience. A cumulative column (3) then helps the final sampling number allocation in column (4).

Finally, it should be noted that the nature of the data of the given frequency distribution (that is, for instance, whether we have classes consisting of single, integral values or grouped, continuous values) will not alter the techniques mentioned here.

EXAMPLE 7.24

Draw a random sample of 30 from the following frequency distribution of the weights of 7749 English males.

Weights (in lbs)	90 up to 120	120 up to 150	150 up to 180	180 up to 210	210 up to 240	240 up to 270	270 up to 300
Frequency	188	2880	3672	846	142	20	1

SOLUTION

We notice first that the smallest frequency is 1 and, expressed as a percentage of 7749 (the total frequency) is 0.0001 (4D). So that when we transform the frequencies to proportions, we will need to use 4D accuracy since otherwise (i.e., 2D or 3D) we will cause the last proportion to be rounded to 0 with the effect of excluding the occurrence of the last class in the sample.

Converting frequencies to proportions, accumulating and finally allocating integers from the range 0000 to 9999 we obtain the following table.

Weights (in lbs)	90–	120–	150–	180–	210–	240–	270–
Proportions	0.0243	0.3716	0.4739	0.1092	0.0183	0.0026	0.0001
Cumulative	0243	3959	8698	9790	9973	9999	10 000
Allocation	0001 to 0243	0244 to 3959	3960 to 8698	8699 to 9790	9791 to 9973	9974 to 9999	9999 to 10 000

(The decimal point has been dropped from the cumulative row.)

We can now take a random sample of 30 from the integral range 0000 to 9999, which is listed including the classes to which each sample item corresponds.

6159 (150–) 3708 (120–) 0846 (120–) 5676 (150–) 2948 (120–)
3387 (120–) 7079 (150–) 0380 (120–) 9681 (180–) 7968 (150–)
6770 (150–) 1801 (120–) 6719 (150–) 2949 (120–) 5867 (150–)
0856 (120–) 2724 (120–) 2070 (120–) 4631 (150–) 0432 (120–)
2309 (120–) 0879 (120–) 1878 (120–) 0032 (90–) 8674 (150–)
7855 (150–) 5572 (150–) 5854 (150–) 7607 (150–) 5373 (150–)

(The sampling numbers are from random number tables.)

Tabulating this information, we have the sample given by:

Weight (in lbs)	90 up to 120	120 up to 150	150 up to 180	180 up to 210
Frequency	1	14	14	1

In the case of sampling from a probability distribution, we use the same technique already illustrated for frequency distributions.

Given a p.d.f. in the form of a probability distribution table, we need only form the (cumulative) distribution function and then allocate sampling numbers accordingly. In the accompanying table we are given a discrete p.d.f. for a random variable X and have added the distribution function $F(x)$.

x	3	4	6	10
$Pr(X = x)$	0.12	0.36	0.38	0.14
$F(x)$	0.12	0.48	0.86	1.00

Since the probabilities are given to 2D, the sampling numbers used are those from the range 00 to 99, and the allocations are easily seen to be

01 to 12, 13 to 48, 49 to 86 and 87 to 100 respectively. We could now sample in the usual way using random sampling numbers.

However, if we are given a p.d.f. in a discrete, compact form (i.e.,

$$\Pr(X = x) = \binom{10}{x} \left(\frac{1}{2}\right)^x \left(\frac{1}{2}\right)^{10-x} ; x = 0, 1, 2, \ldots, 10) \text{ or a continuous form}$$

$$\left(\text{i.e., } f(x) = \frac{1}{3} \text{ for } 4 \leqslant x \leqslant 7\right), \text{ we might be forced to decide on the}$$

number and length of classes and generate a probability distribution table for ourselves before we could begin the sampling procedure.

The following two examples demonstrate the approach.

EXAMPLE 7.25

Draw a random sample of 30 from a Poisson distribution with parameter 2.

SOLUTION

We have X as $Po(2)$ with $\Pr(X = x) = e^{-2}\left(\frac{2^x}{x!}\right)$ for $x = 0, 1, 2, 3,$ .... We clearly need to calculate the probabilities associated with each value of x. Immediately $\Pr(X = 0) = e^{-2} = 0.1353$ (4D).

Using the recurrence formula $\Pr(X = x + 1) = \dfrac{2}{x + 1}\Pr(X = x)$, we obtain:

$$\Pr(X = 1) = \frac{2}{1}(0.1353) = 0.2707; \; \Pr(X = 2) = \frac{2}{2}(0.2707) = 0.2707;$$

$$\Pr(X = 3) = \frac{2}{3}(0.2707) = 0.1804; \; \Pr(X = 4) = \frac{2}{4}(0.1804) = 0.0902;$$

$$\Pr(X = 5) = \frac{2}{5}(0.0902) = 0.0361; \; \Pr(X = 6) = \frac{2}{6}(0.0361) = 0.0120$$

Since the probabilities are now very small, we end with the final class:

$$\Pr(X = 7 \text{ or more}) = 1 - \sum_{x=0}^{6} \Pr(X = x) = 1 - 0.9954 = 0.0046$$

Since all the probabilities are to 4D, we can use the sampling range 0000 to 9999. The class probabilities, distribution function and allocations are tabulated below.

x	0	1	2	3	4	5	6	7+
$\Pr(X = x)$	0.1353	0.2707	0.2707	0.1804	0.0902	0.0361	0.0120	0.0046
Distribution function	0.1353	0.4060	0.6767	0.8571	0.9473	0.9834	0.9954	1.0000
Allocation	0001 to 1353	1354 to 4060	4061 to 6767	6768 to 8571	8572 to 9473	9474 to 9834	9835 to 9954	9955 to 10 000

Taking 30 sets of 4-digit numbers from random number tables (Appendix Table 10, starting at row 12) will give the following sample results:

x	0	1	2	3	4	5	6	7 or more
$f(X = x)$	3	10	8	4	5	0	0	0

EXAMPLE 7.26

Allocate 3-digit sampling numbers to classes of the distribution $N(100, 16)$ in order to accommodate random sampling using random digit tables.

SOLUTION

For a Normal distribution, we know from the theory that most of the distribution is covered by the range $\mu \pm 3\sigma$. Here, $\mu = 100$, $\sigma = 4$. Thus the range for our classes is $100 \pm 12 = 88$ to 112. If we use 8 classes altogether, each class width will be 3. The calculation of the probabilities and the allocation are shown below.

Class	u.c.b. (u)	Standardised u.c.b.s (z)	$\Pr(Z < z)$ $\Phi(z)$	Allocation
Up to 91	91	-2.25	0.012	001 to 012
91 to 94	94	-1.50	0.067	013 to 067
94 to 97	97	-0.75	0.227	068 to 227
97 to 100	100	0	0.500	228 to 500
100 to 103	103	0.75	0.773	501 to 773
103 to 106	106	1.50	0.933	774 to 933
106 to 109	109	2.25	0.988	934 to 988
109 and over	∞	∞	1.000	989 to 1000

7.7 ACCEPTANCE SAMPLING

7.7.1 INTRODUCTION

Acceptance sampling is a technique used with large consignments or batches of items. It enables a decision to be made regarding whether or not the consignment should be accepted or rejected as a complete lot, based on the evidence of the number of defectives found in either one or two small random samples taken from the consignment.

Subsection 7.7.2 describes the two most common ways of carrying out acceptance sampling. Namely, single and double sampling schemes.

Clearly, it is important that: (a) consignments that have a relatively large proportion of defectives *should not* be accepted; and (b) consignments that are relatively defective-free *should* be accepted. How well these two points are observed with a particular sampling scheme can be seen with an operating characteristic (curve). This plots the probability of accepting the consignment against the proportion of defectives in the consignment and is covered in the last subsection.

7.7.2 SINGLE AND DOUBLE SAMPLING PLANS

There are two common methods of acceptance sampling used.

The first is based on a single sample where, if the number of defectives found is less than or equal to a certain number, the batch is accepted; otherwise it is rejected.

For example, given a sample of 50 items, the rule might be: accept batch if 0 items are defective; reject batch otherwise. Thus the criterion for acceptance with this particular sample of 50 is at most 0 defectives.

Notice that just two parameters need to be specified here; n = number in sample and c = maximum number of defectives for accepting batch. For the example above, $n = 50$ and $c = 0$. This method is known as a *single sampling plan* (or *scheme*).

The second method involves using at most two samples and is known as a *double sampling plan* (or *scheme*). With this method, the results of drawing the first sample can be either outright acceptance or rejection of the batch or alternatively to take a second sample. Finally, if the second sample is taken the batch is either accepted or rejected according to the results of the two samples combined. The criterion for acceptance is based on the number of defectives found in the first and/or second samples.

For example:

Sample 1 = 12 items $(n_1 = 12)$:

Accept batch if 0 defectives $(c_{11} = 0)$.
Sample again if 1 or 2 defectives $(c_{12} = 2)$.
Reject batch otherwise.

Sample 2 = 12 items $(n_2 = 12)$:

Accept batch if 2 or less defectives from combined samples $(c_2 = 2)$.
Reject batch otherwise.

Notice that 5 parameters are necessary to specify a double sampling plan.

If the proportion of items that are defective in a batch is known, then, given a particular sampling plan, the probability of accepting or rejecting the batch can be calculated. For example, suppose that the proportion defective in a batch is known to be approximately 1 in 20 (i.e. $p = 0.05$). Then, for a single sampling plan with $n = 6$ and $c = 1$, we have:

$$\text{Pr(Accepting batch)} = \text{Pr(0 or 1 defectives)}$$

$$= 0.9672 \text{ (using Appendix Table 1)}$$

However, if the proportion of defectives in a batch is not known, we can still express the above probability in terms of p = proportion of defectives in batch as follows:

$$\text{Pr(Accepting batch)} = A(p) = \text{Pr (0 or 1 defectives/Bin}(6,p))$$
$$= (1-p)^6 + 6p(1-p)^5$$
$$= (5p+1)(1-p)^5$$

Note that substituting $p = 0.05$ in the above will yield the same value of 0.9672 which was obtained from tables.

To summarise:

DEFINITION 7.3

Acceptance sampling involves making a decision to accept or reject a large batch of items based on the results of drawing either one or two samples from the batch.

A *single sampling plan* is based on a sample size n and if the number of defectives found is less than or equal to value c, the batch is accepted; otherwise it is rejected.

A *double sampling plan* is based on an initial sample size of n_1. If the number of defectives found is c_{11} or less, the batch is accepted; if c_{12} or less, a second sample of size n_2 is taken; otherwise the batch is rejected. If a second sample is taken, the batch is only accepted if there are c_2 defectives or less in the two samples combined; otherwise the batch is rejected.

EXAMPLE 7.27

Large batches of electrical components are produced for use in computer manufacture. A double sampling plan of inspection is adopted as follows. Select 12 items from the batch and accept the batch if there are no defectives, reject the batch if there are 3 or more defectives, otherwise select another sample of 12 items. When the second sample is drawn count the number of defectives in the combined sample of 24 and accept the batch if the number of defectives is 2 or less, otherwise reject the batch. If the proportion of defectives in the batch is p find, in terms of p, the probability that it will be accepted and evaluate this probability when $p = 1/12$.

SOLUTION

This is a double sampling plan with $n_1 = n_2 = 12$, $c_{11} = 0$, $c_{12} = 2$ and $c_2 = 2$.

For Bin$(12, p)$, we have $\text{Pr}(0) = (1-p)^{12}$; $\text{Pr}(1) = 12p(1-p)^{11}$; $\text{Pr}(2) = 66p^2(1-p)^{10}$.

But accepting the batch means:

EITHER 0 defectives with sample 1 (i.e. 0_1)
OR 1 defective with sample 1 and 0 defectives with sample 2 (i.e. $1_1 \cdot 0_2$)
OR 1 defective with sample 1 and 1 defective with sample 2 (i.e. $1_1 \cdot 1_2$)
OR 2 defectives with sample 1 and 0 defectives with sample 2 (i.e. $2_1 \cdot 0_2$)

Thus $Pr(\text{Accepting batch}) = A(p)$

$$= Pr(0_1 \text{ or } 1_1 \cdot 0_2 \text{ or } 1_1 \cdot 1_2 \text{ or } 2_1 \cdot 0_2)$$

$$= Pr(0) + Pr(1) \cdot Pr(0) + Pr(1) \cdot Pr(1) + Pr(2) \cdot Pr(0)$$

$$= (1-p)^{12} + 12p(1-p)^{11} \cdot (1-p)^{12} + 12p(1-p)^{11} \cdot 12p(1-p)^{11}$$
$$+ 66p^2(1-p)^{10} \cdot (1-p)^{12}$$

$$= (1-p)^{12}[1 + 6p(2 + 33p)(1-p)^{10}] \quad \text{(after some factorisation)}$$

Substituting $p = 1/12$ into the above gives:

$Pr(\text{Accepting batch}) = A(1/12)$

$$= \left(\frac{11}{12}\right)^{12}\left[1 + \frac{1}{2}\left(2 + \frac{33}{12}\right)\left(\frac{11}{12}\right)^{10}\right]$$

$$= 0.702 \text{ (3D)}$$

Thus, $Pr(\text{Rejecting batch}) = 1 - 0.702 = 0.298$.

In other words, the risk of rejecting a batch which has only 1 in 12 defective is approximately 1 in 3.

Sometimes it is important to know the *expected number of items inspected* with a double sampling plan, particularly when comparing sampling schemes.

For example, if two schemes resulted in similar probabilities for rejecting batches with high defective rates and accepting batches with low defective rates, it would be advantageous to select that (double) sampling plan which on average requires less items to be sampled. This will obviously save time and money.

This expected number will vary with the proportion of the batch that is defective. Consider the double sampling plan: $n_1 = n_2 = 12$, $c_{11} = 0$, $c_{12} = 1$ and $c_2 = 2$, where the proportion defective in the batch is 0.1. Now, there are only two different values for the number of items inspected, 12 or 24.

$Pr(\text{Number inspected} = 24) = Pr(1 \text{ defective in first sample})$

$$= 12(0.1)(0.9)^{11} = 0.3766$$

$Pr(\text{Number inspected} = 12) = 1 - 0.3766 = 0.6234$.

Thus expected number inspected $= 24(0.3766) + 12(0.6234) = 16.52(2D)$.

EXAMPLE 7.28

The double sampling plan: $n_1 = n_2 = 7$, $c_{11} = 0$, $c_{12} = 2$ and $c_2 = 2$ is used for inspecting batches in a warehouse. If p is the proportion of defectives in the batch, calculate the expected number of items inspected for both cases $p = 0.2$ and 0.4:

(a)   if rejected batches are discarded;

(b)   if rejected batches have all their items inspected (and defective items replaced) where the batch size is 900.

Interpret the result in (b).

SOLUTION

Let variable I be 'number of items inspected'. We require $E[I]$, the expectation of I.

(a) A second sample is taken (and thus a total of 14 items inspected) only if 1 or 2 defectives are found in the first sample.

When $p = 0.2$,

$$\Pr(I = 14) = 7(0.2)(0.8)^6 + 21(0.2)^2(0.8)^5$$
$$= 0.3670 + 0.2753$$
$$= 0.6423$$
$$\Pr(I = 7) = 1 - 0.6423$$
$$= 0.3577$$

Therefore
$$E[I] = 7(0.3577) + 14(0.6423)$$
$$= 11.50 \text{ (2D)}$$

When $p = 0.4$,

$$\Pr(I = 14) = 7(0.4)(0.6)^6 + 21(0.4)^2(0.6)^5$$
$$= 0.1306 + 0.2613$$
$$= 0.3919$$
$$\Pr(I = 7) = 1 - 0.3919$$
$$= 0.6081$$

Therefore,
$$E[I] = 7(0.6081) + 14(0.3919)$$
$$= 9.74 \text{ (2D)}$$

(b) There are THREE distinct values for I in this case: $I = 7$ (on batch acceptance at the first sample), $I = 14$ (on batch acceptance at the second sample) and $I = 900$ (on batch rejection since, then, ALL items in the batch are inspected).

When $p = 0.2$,

$$\Pr(I = 7) = \Pr(\text{no defectives in first sample}) = \Pr(0_1)$$

$$= (0.8)^7 = 0.2097$$

$$\Pr(I = 14) = \Pr(1_1 \cdot 0_2 \text{ or } 1_1 \cdot 1_2 \text{ or } 2_1 \cdot 0_2)$$

$$= 7(0.2)(0.8)^6 \cdot (0.8)^7 + 7(0.2)(0.8)^6 \cdot 7(0.2)(0.8)^6$$
$$+ 21(0.2)^2(0.8)^5 \cdot (0.8)^7$$

$$= 0.0770 + 0.1347 + 0.0577 = 0.2694$$

Thus

$$\Pr(I = 900) = 1 - 0.2694 - 0.2097 = 0.5209$$

Therefore

$$E[I] = 7(0.2097) + 14(0.2694) + 900(0.5209) = 474$$

When $p = 0.4$,

$$\Pr(I = 7) = \Pr(0_1)$$

$$= (0.6)^7 = 0.0280$$

$$\Pr(I = 14) = \Pr(1_1 \cdot 0_2 \text{ or } 1_1 \cdot 1_2 \text{ or } 2_1 \cdot 0_2)$$

$$= 7(0.4)(0.6)^6 \cdot (0.6)^7 + 7(0.4)(0.6)^6 \cdot 7(0.4)(0.6)^6$$
$$+ 21(0.4)^2(0.6)^5 \cdot (0.6)^7$$

$$= 0.0037 + 0.0171 + 0.0073 = 0.0281$$

Thus

$$\Pr(I = 900) = 1 - 0.0281 - 0.0280 = 0.9439$$

Therefore

$$E[I] = 7(0.0280) + 14(0.0281) + 900(0.9439) = 850$$

That is, when the defective rate is as high as 0.4, most items will end up being inspected and even when the defective rate drops to 0.2, over half of all items will be inspected.

7.7.3 OPERATING CHARACTERISTIC

An *operating characteristic (curve)* is a graph which, for a particular sampling plan, shows the relationship between the proportion defective in a batch and the probability that a batch is accepted. In order to construct the operating characteristic, it is usually necessary to have a general expression for $\Pr(\text{Accepting batch})$ in terms of p, the proportion of the batch that is defective.

For example, in the single sampling plan given in the previous subsection, with $n = 6$ and $c = 1$, we had $\Pr(\text{Accepting batch}) = (5p + 1)(1 - p)^5$. Substituting particular values of p gives the values shown below and marked $\Pr(\text{Acc})$:

p	0	0.05	0.1	0.2	0.3	0.4
$\Pr(\text{Acc})$	1	0.967	0.886	0.655	0.420	0.233

p	0.5	0.6	0.7	0.8	0.9	1
$\Pr(\text{Acc})$	0.109	0.041	0.011	0.002	0	0

Plotting these figures on a graph gives the following:

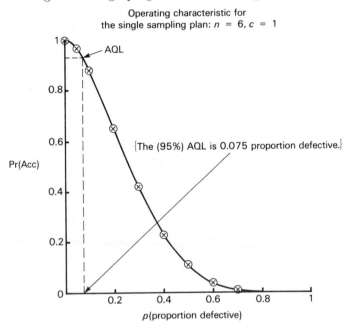

Operating characteristic for the single sampling plan: $n = 6, c = 1$

Notice that a special point has been identified on the graph, known as the *Acceptable Quality Level (AQL)* point. It is defined as that proportion of defectives which gives $\Pr(\text{Accepting batch}) = 0.95$. For the single sampling plan $(n = 6; c = 1)$ shown on the graph, the AQL is 0.075 or 7.5% defective.

The AQL of a batch (defined by a specific sampling plan) can be thought of practically as the maximum proportion of defectives that can be tolerated in an acceptable batch. Clearly, there is no natural universal criterion of what is acceptable, but it is generally agreed that an AQL set at a batch acceptance level of 0.95 is reasonable.

The operating characteristic should fall away from the AQL as steeply as possible in order for the customer (the recipient of the goods) to have a reasonable deal. The operating characteristic graphed above is not ideal since, for example, a batch with a defective rate of 0.15 (double the AQL) has only a 20% chance (read from the graph) of being rejected.

In Example 7.27, the double sampling plan defined by $n_1 = n_2 = 12$, $c_{11} = 0$, $c_{12} = 2$ and $c_2 = 2$ gave

$$\text{Pr(Accepting batch)} = (1-p)^{12}[1 + 6p(2 + 33p)(1-p)^{10}]$$

Substituting some values of p yields the following table values and associated operating characteristic.

p	0	0.05	0.1	0.2	0.3	0.4
Pr(Acc)	1	0.895	0.596	0.145	0.022	0.003
p	0.5	0.6	0.7	0.8	0.9	1
Pr(Acc)	0	0	0	0	0	0

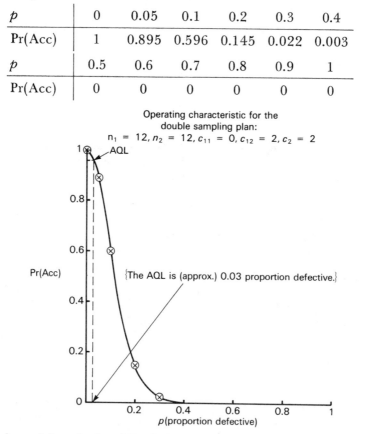

Operating characteristic for the double sampling plan:
$n_1 = 12, n_2 = 12, c_{11} = 0, c_{12} = 2, c_2 = 2$

{The AQL is (approx.) 0.03 proportion defective.}

Notice that, although the AQL is low at 0.03 (i.e. the maximum proportion of defectives that will be tolerated in an acceptable batch), the operating characteristic falls away from the AQL very rapidly. Note, however, the price paid for this is taking up to two samples, each much larger than the one in the previous single sampling scheme. Clearly, there must be a compromise between small convenient samples with a relatively high probability of accepting a defective batch and much larger, more costly sampling schemes with a relatively low probability of accepting a defective batch.

The consumer of the batch items normally sets not only the AQL but also the steepness of the operating characteristic. However, there are risks for the consumer, as was shown with the previous single sampling plan, where there was only a 20% chance of rejecting a batch that had a defective rate double that of the AQL. The normal way to reduce this risk (making the operating characteristic steeper) is to increase the sample size.

Sometimes the consumer will set a Lot Tolerance Proportion Defective (LTPD), which is the smallest batch defective rate to ensure almost certain rejection and is usually set so that the probability of rejection is 0.9. The LTPD can often work to the disadvantage of the supplier since batches with defective rates considerably lower than the LTPD might still have more chance of being rejected than accepted. For example, the operating characteristic for the single sampling plan: $n = 6$, $c = 1$ which was drawn earlier, yields the LTPD as 0.5 (since this gives $\Pr(\text{Acceptance}) = 0.1$). However, notice that if the batch defective rate is as low as 0.3 here, there is still a 60% chance that the batch will be rejected.

To summarise:

DEFINITION 7.4

An *operating characteristic* for a particular acceptance sampling plan is a graph of the probability of accepting the batch against the proportion of defectives in the batch.

An *Acceptable Quality Level (AQL)* of an operating characteristic is that proportion of defectives that will give $\Pr(\text{Accepting batch})$ = 0.95. Put another way, the AQL is the largest proportion of defectives that will be tolerated in an acceptable batch.

The *Lot Tolerance Proportion Defective (LTPD)* of an operating characteristic is that proportion of defectives that will give $\Pr(\text{Accepting batch}) = 0.1$. Put another way, the LTPD is the defective rate above which batches should normally be rejected.

EXAMPLE 7.29

Plot the operating characteristic for the single sampling plan having $n = 10$ and $c = 2$ and estimate the AQL and the LTPD from the graph.

SOLUTION

We have a sample size of 10 with batch acceptance if at most 2 defectives are found in the sample.

It is usually worthwhile to obtain a general expression for $\Pr(\text{Accepting batch})$ in terms of p and then substitute various values of p into the expression to obtain some points on the curve.

Thus, $\Pr(\text{Acc}) = \Pr(0, 1 \text{ or } 2 \text{ defectives in sample})$

$$= (1-p)^{10} + 10p(1-p)^9 + 45p^2(1-p)^8$$

$$= (1-p)^8[(1-p)^2 + 10p(1-p) + 45p^2]$$

$$= (1-p)^8(36p^2 + 8p + 1)$$

When $p = 0.1$,

$$\Pr(\text{Acc}) = (0.9)^8[36(0.1)^2 + 8(0.1) + 1] = 0.929.$$

Similarly, $p = 0.2$ gives $\Pr(\text{Acc}) = 0.678$

$p = 0.3$ gives $\Pr(\text{Acc}) = 0.383$, and

$p = 0.4$ gives $\Pr(\text{Acc}) = 0.167$

Plotting these points and joining them with a smooth curve gives the required operating characteristic for the single sampling plan; $n = 10$, $c = 2$.

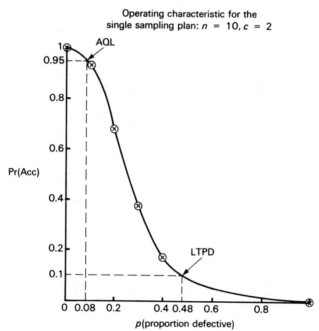

Operating characteristic for the single sampling plan: $n = 10$, $c = 2$

From the graph, the AQL can be seen to be approximately 0.08 and the LTPD is 0.48.

Extensive special tables are available which yield the parameters for either a single or double sampling scheme given (a) the size of the batch and (b) the desired AQL. A specimen row and column taken from a single sampling plan inspection table is shown below.

Batch size:	0.25%		0.4%		1.0%		2.5%	
	n	c	n	c	n	c	n	c
91–150					13	0		
151–280					50	1		
281–500					50	1		
501–1200					80	2		
1201–3200	200	0	125	1	125	3	50	3
3201–10 000					200	5		
10 001–35 000					315	7		
35 001–150 000					500	10		

Thus, for example, to have an AQL of 1.0% with a batch of 20 000 items, the sample size would need to be 315 with batch acceptance if no more than 7 items were defective (i.e. $n = 315$, $c = 7$). Similarly, to have an AQL of 2.5% with a batch of 3000 would mean a sample size of 50, accepting the batch if there were no more than 3 defectives.

7.8 EXERCISES

SECTION 7.1

1. State whether the tossing of two 'fair' coins can be considered as a random sample of size 2 from some distribution, giving justification for your answer.

   (*Note*: Remember that a random variable can take only *numerical values* and not values such as 'head' or 'tail'.)

2. Prove (using definition 7.1) that the tossing of an unbiased coin and the rolling of a die could be regarded as a random sample of size 2 from some distribution.

3. A specially constructed, seven-sided die can rest on any one of its faces with equal probability. State whether the rolling of this die and the tossing of an unbiased coin could be considered as a random sample of size 2 from some distribution.

SECTION 7.2

4. A random variable X_1 is N(100, 20) and X_2 is N(120, 30). Write down the distributions of: (a) r.v. $X_1 + X_2$; (b) r.v. $X_1 - X_2$; (c) r.v. $X_2 - X_1$.

5. The random variables X_1 and X_2 are distributed N(50, 10) and N(50, 5) respectively. Find: (a) the distribution of $X_1 + X_2$; (b) the distribution of $X_1 - X_2$; (c) the distribution of $X_2 - X_1$; and (d) $\Pr(X_2 - X_1 > 1)$.

6. A metal plate with a small centre hole is produced by two machines A and B. Machine A produces whole plates whose weights are distributed N(4.0, 0.003) g and machine B bores the holes removing a small amount of metal from each plate, the weights of which are distributed N(0.1, 0.001) g. If those final plates whose weights are greater than 3.92 g are rejected, find: (a) the distribution of the weights of the final bored plates; and (b) the percentage of final plates that will be rejected.

7. The time of arrival of a bus at a bus stop varies in a Normal distribution with a mean of 09.00 a.m. and a standard deviation of 2 minutes. Independently a second bus departs from the stop at a time which varies in a Normal distribution with a mean of 09.01 a.m. (i.e., one minute past 09.00 a.m.) and a standard deviation of 1 minute. Find the probability that: (a) the first bus arrives before the second bus leaves; (b) this happens on 5 given consecutive days. (O & C)

8.   In the assembly of a certain machine, circular bolts have to fit into circular holes. Continuous inspection has shown that the bolts are distributed Normally with a mean diameter of 1.509 cm and a standard deviation of 0.012 cm while the holes are distributed Normally with a mean diameter of 1.539 cm and a standard deviation of 0.018 cm. If bolts are selected at random, what proportion will not fit into the hole to which they are first applied? What is the probability that 100 successive bolts will all fit the holes to which they are first applied?

9.   Three lengths have Normal distributions with means 20.48, 35.97 and 62.34 cm and standard deviations of 0.21, 0.46 and 0.54 cm respectively. Give the distribution (with mean and standard deviation) of the sum of the lengths.

10.   The four random variables X_1, X_2, X_3 and X_4 are N(48, 5), N(43, 10), N(100, 30) and N(120, 10) respectively. Find the distributions of:
(a) $X_1 + X_2$;   (b) $X_1 + X_3$;   (c) $X_1 - X_2$;   (d) $X_1 + X_2 + X_3$;
(e) $X_2 + X_3 + X_4$;   (f) $X_1 + X_3 - X_2$;   (g) $X_2 + X_3 - (X_1 + X_4)$.

11. X and Y are independent Normal variables. State the means and variances of $X \pm Y$ in terms of those of X and Y. Cans of food have a mean weight of 274.5 g and a standard deviation of 11.6 g. The cans alone have a mean weight of 25.3 g and a standard deviation of 2.3 g. If the weights are Normally distributed in both cases, determine: (a) the mean and standard deviation of the weights of the contents of the cans; (b) the percentage of cans whose contents are underweight if the nett weight of the contents is quoted as 240 g; (c) the probability that the total weight of the contents of four cans exceeds 960 g.

12.   A wooden dog with synthetic hair is made by three machines A, B and C in the following way. Machine A outputs wooden cubes (with weight distribution N(8, 1.4) gm), Machine B takes the cubes and trims the wood into a 'dog' shape (taking off wood whose weight distribution is N(1.8, 0.12) gm) and finally Machine C takes the wooden dog shapes and sticks on the synthetic hair (whose weight distribution is N(0.2, 0.03) gm). Find the weight distribution of the completed wooden dogs.

13.   The independent random variables X_1, X_2, X_3 and X_4 are distributed as N(50, 10), N(60, 15), N(80, 12) and N(75, 10) respectively. Find the distributions of: (a) $X_2 + X_3$;   (b) $X_4 - X_1$;   (c) $X_1 + X_2 + X_4$;
(d) $3X_3$;   (e) $2X_1 - 3X_2 + 4X_4$.

14.   If X_1 and X_2 are independent r.v.s with X_1 as N(a, 10) and X_2 as N(100, b), find: (a) the distribution of $X_1 + X_2$;   (b) the distribution of $X_2 - X_1$;   (c) If $3X_1 - 2X_2$ is N(145, 122), find a and b.

15.   A large class of children are asked to draw (using compasses) a circle of radius 3 cm. The results are collected and it is found that the radii they have actually used are distributed Normally, mean 3 and variance 0.0025. If the perimeters of the circles are now measured, what might we expect their distribution to be? (Give your answer in terms of π.)

SECTION 7.3

16.   A random sample of size 10 is taken from an infinite population having mean 20 and variance 16 and a second random sample of size 15 is taken from an infinite population having mean 30 and variance 9.

Denoting the two sample means by $\bar{X}_1$ and $\bar{X}_2$ respectively, give the expectations and variances of the random variables: (a) $\bar{X}_1 + \bar{X}_2$; (b) $\bar{X}_2 - \bar{X}_1$;   (c) $4\bar{X}_1 - 2\bar{X}_2$.

17. A random sample of: (a) 10 is taken from $N(100, 20)$;   (b) 20 is taken from $N(50, 10)$;   (c) a is taken from $N(a^2, a)$.   Find the distribution of each of the above sample means.

18. The independent random variables X_1 and X_2 have distributions $N(160, 10)$ and $N(200, 50)$ respectively. A random sample of 5 is taken from the distribution $3X_1 + 2X_2$. Find the distribution of the sample mean.

19. A distribution is Normal with mean 68.0 inches and standard deviation 3.0 inches. A random sample of 25 items is taken. What is the probability that the mean of the sample is: (a) between 66.8 and 68.3 inches;   (b) less than 66.4 inches?

20. Electric light bulbs produced by a certain machine are known to have a mean lifetime of 900 hours with s.d. 80 hr. 1000 samples of size 100 are drawn from this assumed Normal distribution. How many of the samples would we expect to have a mean lifetime in excess of 910 hours?

21. A r.v. A is $N(1400, (200)^2)$ and r.v. B is $N(1200, (100)^2)$. If random samples of 120 items are taken from each distribution, find the probability that the sample from A will have a mean which is at least: (a) 160;   (b) 250 more than the mean from B.

22. Components made by a machine weigh 0.50 gm with an s.d. of 0.02 gm. If two samples are taken, both of 1000 components each, what is the probability that their means will differ by more than 0.002 gm? (Note that there are two cases to consider here. The first sample mean could be smaller *or* larger than the second.)

23. Samples of size 10 are taken (without replacement) from the following finite population:

x	0	1	2	3	4	Total
f	21	18	7	3	1	50

(a) Find the mean and variance of the theoretical distribution of sample means.   (b) Using a Normal approximation, find the probability that a sample mean exceeds 1.

24. It is known that the heights of 724 students are distributed approximately Normally with mean 65 inches, variance 12.5 inches. How many of 120 samples (taken without replacement) each of thirty students, would be expected to have means less than 63 inches?

25. A random sample of size 100 is taken from Bin(40, 0.5). Find the approximate probability that $\bar{X}$ is: (a) greater than 20.5;   (b) less than 19.3; and hence   (c) between 19.3 and 20.5.

26. A large number of samples of constant size are taken from a near-Normal population with mean 50 and variance 20 and it is found that 1% of the time the mean of the sample exceeded 55. Find the probable sample size, using the Central Limit Theorem.

SECTION 7.4

27.   If the random variables X_1, X_2 and X_3 are distributed as $\chi^2(1)$, $\chi^2(5)$ and $\chi^2(10)$ respectively, find the distributions of: (a) $X_1 + X_2$; (b) $X_1 + X_3$.

28.   The random variables X_1 and X_2 are distributed respectively as $N(\mu, \sigma_1^2)$ and $N(\mu, \sigma_2^2)$. Find the mean and variance of the variable $\dfrac{(X_1 - X_2)^2}{\sigma_1^2 + \sigma_2^2}$.

29.   Use tables of the χ^2 distribution function to:
       (a) find the value of P if: (i) $\chi^2_{P\%} = 19.02$ $(\nu = 9)$;
                                       (ii) $\chi^2_{P\%} = 24.43$ $(\nu = 40)$.
       (b) find the value of x if: (i) $\chi^2_{\frac{1}{2}\%} = x$ $(\nu = 29)$;
                                       (ii) $\chi^2_{99\%} = x$ $(\nu = 4)$.

30.   If X_1 and X_2 are independent random variables with X_1 as $\chi^2(3)$ and X_2 as $\chi^2(4)$, find the probability that: (a) $X_1 > 6.25$; (b) $X_1 < 0.115$; (c) $X_2 > 14.86$;   (d) $X_1 + X_2 > 14.07$.

31.   If r.v. X_1 is $N(\mu_1, \sigma_1^2)$ and r.v. X_2 is $N(\mu_2, \sigma_2^2)$, show that
$$E\left[\frac{(X_1 - \mu_1)^2}{\sigma_1^2} + \frac{(X_2 - \mu_2)^2}{\sigma_2^2}\right] = 2.$$

32.   If X is $N(100, 10)$ and Y is $N(20, 2)$, find the probability that:
       (a) $\dfrac{(X - 100)^2}{10} < 3.84$;   (b) $\dfrac{(Y - 20)^2}{2} > 2.71$;
       (c) $\dfrac{(X - 100)^2}{10} + \dfrac{(Y - 20)^2}{2} < 10.6$
       (assuming that X and Y are independent).

33.   If X_1 and X_2 are independent random variables and X_1 is $N(25, 20)$ and X_2 is $N(50, 45)$, find $\Pr\left(\dfrac{(X_1 - 25)^2}{20} + \dfrac{(X_2 - 50)^2}{45} < 4.61\right)$.

SECTION 7.5

34.   A random sample of 20 is taken from a Normal distribution having a variance of 28. Find the probability that: (a) $S^2 < 14.168$; (b) $S^2 > 42.196$; and hence   (c) $14.168 < S^2 < 42.196$.

35.   A random sample of 8 was taken from a Normal population and the following values were obtained: 8.24, 8.41, 9.02, 8.61, 8.82, 8.93, 8.98 and 8.51.   (a) Find the mean and variance of the sample.   (b) If the population variance is 0.271, find the probability of obtaining a sample variance as low as (or lower than) the sample variance obtained from the sample above.

36.   A random sample of 10 000 is taken from a population that is known to be 10% defective. Find the probability that less than 1000 are found defective.

37.   A certain candidate standing for election is known to have the support of proportion 0.46 of the electors. If a random sample of: (a) 200; (b) 1000 is taken from the electors, what is the probability that the candidate will have a majority?

38.   Find the probability that a sample of 12 drawn from a bivariate population with $\rho = 0.83$ will have a correlation coefficient as low as 0.5.

SECTION 7.6

39.   Use Appendix Table 10 to generate a random sample of size 15 from the discrete range 0 to 9999. (Begin at row 6.)

40.   Draw a random sample of 10 from the continuous range 0 up to (but excluding) 10 to 4D, using Appendix Table 10. (Begin at row 3.)

41.   Using Appendix Table 10, draw a random sample of 50 from the digits 0 to 9. Draw up a frequency distribution table for your results and find the mean and variance. Compare this with the *theoretical* mean and variance of the distribution.

42.   Take a sample of 30 from the following frequency distribution:

x	0	1	2	3	4	5	6	7
f	24	39	142	114	81	62	57	11

by drawing up a relative (proportion — to 4D) frequency distribution table and using the first 30 groups of 4 digits in Appendix Table 10 (beginning with 0174, 5206, ... etc.).

43.   Take another sample of 30 from the frequency distribution given in Example 7.24 (using a different set of 30 groups of 4 digits) and compare the two sample means.

44.   The following distribution gives the waiting time for a particular bus on 100 occasions.

Waiting time (in mins)	Frequency
0 up to 2.5	10
2.5 up to 5.0	21
5.0 up to 7.5	38
7.5 up to 10.0	18
10.0 up to 12.5	7
12.5 up to 15.0	4
15.0 up to 20.0	2

Take a random sample of 20 from the distribution and compare the distribution mean and variance with the sample mean and variance.

45.   Find the mean and variance of the simulated sample of 30 from the Poisson distribution obtained in Example 7.25. Compare this with the theoretical mean and variance. Take an extra reading of 30 4-digit numbers and add them to the sample already obtained to produce a new sample of 60 from the distribution. Find the mean and variance of this larger sample and compare it with the theoretical mean and variance.

SECTION 7.7

46.   A sampling scheme consists of sampling 6 items from a batch and only accepting the batch if there is at most one defective found in the sample. Calculate the probability that the batch will be accepted if the batch is known to have (a) 2%, (b) 5% defectives.

47.   A single sampling scheme has $n = 15$ and $c = 2$. Find an expression for the probability of accepting a batch in terms of p, the proportion of the batch that is defective.

48.   A double sampling scheme has parameters: $n_1 = 4$, $n_2 = 4$, $c_{11} = 0$, $c_{12} = c_2 = 1$. Find an expression for the probability of accepting a batch in terms of p, the proportion of the batch that is defective.

49.   A sampling scheme consists of selecting 7 items randomly from a large batch. The batch is accepted if no defectives are found and rejected if 2 or more defectives are found. Otherwise, another 7 items are sampled and the batch is accepted if there are no more than 2 defectives found out of the 14 items. Obtain an expression for the probability of accepting the batch in terms of p, the proportion of the batch that is defective.

50.   Two single sampling plans are being compared. Plan A has $n = 20$ and $c = 0$ (that is, a sample size of 20 with rejection of batch if one or more defectives are found) and plan B has $n = 20$ and $c = 2$. Calculate, for each plan, the probability of accepting a batch containing: (i) 2% defective; (ii) 10% defective.

51.   Two sampling plans are being compared. Plan 1 is a single sampling plan with sample size 8 and the batch being accepted if no more than 2 of the sample are defective. Plan 2 is a double sampling scheme with a first sample size of 4, batch acceptance with no defectives and rejection with 2 or more defectives. If 1 defective is found, a second sample is taken with the batch accepted only if the total number of defectives in the combined samples is no more than 2. Calculate, for both plans 1 and 2, the probability of batch acceptance if the proportion defective in the batch is (a) 0.2; (b) 0.6. Which plan is better and why?

52.   A double sampling scheme has $n_1 = 6$, $n_2 = 6$, $c_{11} = 1$, $c_{12} = 2$ and $c_2 = 2$. Assuming that rejected batches are discarded, calculate the expected number of items inspected for each of the following batch proportion defective: 0.05, 0.1, 0.3, 0.5 and 0.6.

53.   A single sampling scheme has $n = 5$ and $c = 1$. Draw the operating characteristic and use it to estimate the AQL.

54.   Graph the operating characteristics for the single sampling scheme: $n = 18$, $c = 2$ and the double sampling scheme: $n_1 = 9$, $n_2 = 9$, $c_{11} = 0$, $c_{12} = 1$, $c_2 = 2$ on the same chart. Which of the two plans is better?

55.   Large batches of similar components are delivered to a company. A sample of five articles is taken at random from each batch and tested to destruction. If at least four of the five articles are found to be good, the batch is accepted. Otherwise the batch is rejected. If the proportion defective in the batch is p, show that the probability of accepting the batch is:

$$A(p) = (1 - p)^4 (1 + 4p)$$

Use $A(p)$ to sketch the operating characteristic and hence estimate the AQL and LTPD for the batches with this sampling scheme.

MISCELLANEOUS

56. If: (a) X is N(0, 125); (b) $\bar{X}$ is the mean of a random sample of size 5 from the distribution given; and (c) $\Pr(\bar{X} < k) = 0.9$, find the value of k.

57. If X is N(52, 9) and Y is N(41, 16), where X and Y are independent, and random samples of size 25 are taken from both distributions, determine the distributions of the statistics: (a) $\bar{X} + \bar{Y}$; (b) $\bar{X} - \bar{Y}$.

58. A sample of 25 items is taken independently from both N(0, 16) and N(1, 9). Find the probability that the mean of the sample from the first distribution is greater than the mean of the sample from the second.

59. A certain factory manager regularly takes samples of the same size from a production line, and measures the standard deviation of the means of his samples. In his absence, his assistant carries on this process except for the fact that he takes samples of a different size to those of the manager. He finds that the standard deviation he obtains is half that of his superior's.

(a) Had he increased or decreased the sample size and by what factor?

(b) If the manager's original sample sizes had been 16 and his assistants had been 144, what would have been the effect on the standard deviation of the sample means?

60. The manager of a firm has calculated that his workers' time to finish a particular job is distributed N(3, 1) hours. In return for a pay increase and improved working conditions and benefits, the workers will finish the job in only two thirds of the usual time. If the manager considers that a time greater than 3.5 hours to complete the job is 'late', show that, under the new scheme, the men are about 25 times less likely to be late.

61. A firm produces plugs (having their diameters distributed as N(3.0, 0.1) mm) to go into circular holes (with diameters distributed N(4.0, a) mm). Find the value of a so that only 1% of the plugs are expected to be too large.

62. A steel rod is assembled in two parts, the length of each part being Normally distributed. It is known that the central 99% of the distribution of the first part of the rod lies between 1.688 and 1.712 and the central 99% of the distribution of the second part lies between 2.588 and 2.612, where all measurements are in metres. Find the limits between which the central 99% of the distribution of the assembled rods lie.

63. Fertilizer is packed, by a machine, into bags of nominal mass 12 kg. The mass of each bag may be regarded as a normally distributed random variable with mean 12.05 kg, standard deviation of 0.20 kg.

(a) What is the probability that the mass of a bag exceeds 12 kg?
(b) What mass is exceeded by exactly 95% of the bags?

A farmer buys 20 bags at a time.

(c) What is the probability that their mean mass will exceed 12 kg?
(d) What is the distribution of the total mass of the 20 bags and what is the probability that it lies between 239.5 kg and 240.5 kg?

The mass of the bags packed by a second machine may be regarded as a normally distributed random variable with mean 12.05 kg, standard deviation 0.05 kg. If the farmer's 20 bags are made up of n from the first machine and the rest from the second machine, what is the largest possible value of n which gives a probability of at least 0.95 of the total mass of the bags exceeding 240 kg?   (AEB) '85

64.   A random variable X is distributed N(50, 25). What must the size of a sample be if the mean, $\bar{X}$, has *at least* 0.9 probability of lying between 48 and 52?

65.   The diameters of 100 rods, measured in mm to the nearest mm, are classified in the following frequency table.

Diameters in mm	62–64	65–67	68–70	71–73	74–76	77–79
Class frequency	6	14	31	25	17	7

Taking the class marks to be 63, 66, . . . , etc., show that the mean and the standard deviation of this population are respectively 70.62 and 3.85 mm approx. Using the same class marks, allocate sampling numbers from 00 to 99 to these data. By reading consecutive pairs of digits from left to right along the following four rows of random numbers, select 10 samples each of 4 items.

50532   25496   95652   42457

73547   07136   40876   79971

54195   25708   27989   64728

10744   08396   56242   85184

Calculate the mean and standard deviation of the distribution of sample means so determined and compare them with the theoretical values expected for a large number of samples. Comment on the result.
                    (AEB)'73

66.   A baker makes digestive biscuits whose masses are normally distributed with mean 24.0 g and standard deviation 1.9 g. The biscuits are packed by hand into packets of 25. Assuming the biscuits included in each packet are a random sample from the population, what is the distribution of the total mass of biscuits in a packet and what is the probability that it lies between 598 g and 606 g?

Ten packets of biscuits are placed in a box. What is the probability that the total mass of biscuits in the box lies between 6010 g and 6060 g?

A new packer was including 26 biscuits in each packet. What is the probability that a packet selected at random from those containing 25 biscuits would contain a greater mass of biscuits than a packet selected at random from those containing 26 biscuits?   (AEB) '84

67.   The random variables X_1, X_2 and X_3 are distributed N(90, 16), N(80, 25) and N(a, b) respectively.

(a) Write down the distributions of the random variables: (i) $X_1 + X_2$; (ii) $X_1 - X_2$; (iii) $3X_2$; (iv) $4X_1 - 3X_2$.

(b) If the random variable $4X_3 - X_2$ is distributed N(120, 201): (i) find the values of a and b; (ii) find $\Pr(4X_3 - X_2 > 100)$.

68.   If X is N(0, 1) and Y is $\chi^2(1)$ find: (a) $\Pr(-1.96 < X < 1.96)$; (b) $\Pr(Y < 3.84)$. By writing $Y = X^2$, show that results (a) and (b) are equivalent.

69. The heights of a large number of shrubs of the same kind produced for sale by a horticultural nursery are Normally distributed with mean 1.14 m and standard deviation 0.25 m.

    Fifty samples, each consisting of 100 shrubs, are selected. In how many of these samples would you expect to find the mean sample height to be: (a) greater than 1.16 m;   (b) between 1.13 m and 1.18 m?
    (Cambridge)

70. (a) If X and Y are independent random variables with means μ_x, μ_y and variances σ_x^2, σ_y^2 respectively, show from first principles that the mean and variance of $aX + bY$ are $a\mu_x + b\mu_y$ and $a^2\sigma_x^2 + b^2\sigma_y^2$ respectively where a and b are any constants.

    (b) The diameters x of 110 steel rods were measured in centimetres and the results were summarised as follows: $\Sigma x = 36.5$, $\Sigma x^2 = 12.49$. Find the mean and standard deviation of these measurements.

    Assuming these measurements are a sample from a Normal distribution with this mean and this variance, find the probability that the mean diameter of a sample of size 110 is greater than 0.345 cm.   (O & C)

71. In a given adult human population, the heights of men have mean μ_1 and variance σ_1^2, while the heights of women have mean μ_2 and variance σ_2^2. If the proportion of men is p, calculate the mean and variance of height in the whole population.

    Assuming that a married couple is formed by choosing a man and a woman independently at random from the population, find the mean and variance of the difference in height between husband and wife.(IOS)

72. The random variable X is uniformly distributed on the sample space 1, 3, 5, 7. If $\bar{X}_2$ and $\bar{X}_3$ are the means of random samples size 2 and 3 respectively, tabulate the sampling distributions of $\bar{X}_2$ and $\bar{X}_3$. Display your results on a suitable diagram, and comment on them.   (London)

73. Batteries for a transistor radio have a mean life under normal usage of 160 hours, with a standard deviation of 30 hours. Assuming that battery life follows a normal distribution;   (a) calculate the percentage of batteries which have a life between   150 hours   and   180 hours;   (b) calculate the range, symmetrical about the mean, within which 75% of the battery lives lie;   (c) if a radio takes four of these batteries and requires all of them to be working, calculate the probability that the radio will run for at least 135 hours.   (O & C)

74. A factory produces shafts and gear wheels. A gear wheel is 'shrink fitted' to its shaft by heating the gear wheel to expand it, placing it on the shaft, and allowing it to cool. Shaft diameters (D cm) have a normal distribution with mean 1.0010 cm and standard deviation 0.0001 cm. The bores (B cm) of the gear wheels have a normal distribution with mean 1.0000 cm and standard deviation 0.0002 cm.

    Shafts and gear wheels are paired at random and the pair will be rejected if it is either too tight or insufficiently tight: before assembly the diameter of the shaft is greater than the bore of the gear wheel. Find the proportion of pairs which are rejected: (a) because they are too tight, i.e. $D - B > 0.0015$;   (b) because they are insufficiently tight, i.e., $D - B < 0.0006$.

    The manufacturing cost of shafts is £5 each and of gear wheels £30. Pairs which are too tight are modified at a further cost of £5 and pairs which are insufficiently tight are scrapped. Sufficient shafts and gears

are paired and modified where required to enable 100 acceptable pairs
to be available for sale. Calculate the expected cost of producing these
100 pairs.   (Cambridge)

75.   A sample from the continuous variable x (measured to the nearest
        whole number) has the following frequency distribution:

x	321	322	323	324	325	326
f	1	2	2	3	1	1

Calculate estimates for: (a) the mean of x;   (b) the variance of x;
(c) the median of x;   (d) the mean of x^2;   (e) the mean of $2x + 3$;
(f) the variance of $2x + 3$;   (g) the variance of $2x_1 + 3x_2$, where x_1
and x_2 are independent observations of x.   (Cambridge)

76.   A powder is manufactured in 10 kg batches. The iron contents of
        batches are Normally distributed about a mean of 200 ppm (= parts
        per million) with a standard deviation of 40 ppm. In a sample of 1000
        batches the number of them that contain less than x ppm of iron is
        eleven. Estimate x.

        The powder is made for an order which specifies that the iron content
        must be less than 250 ppm.   (a) Estimate the number of batches that
        contain too much iron;   (b) Find the probability that 4 batches
        chosen at random will *each* separately meet the specification;   (c) If 4
        randomly chosen batches are thoroughly mixed together, find the prob-
        ability that the 40 kg of mixture will meet the specification.   (O & C)

77.   Large batches of door hinges are produced at regular intervals by the
        Amalgamated Ironworks in Botherham. Let N be the number of
        hinges in each batch, and p the proportion of defectives. The company
        operates a single sample plan for its batch inspection by selecting 10
        hinges at random and accepting the batch if there are 2 or less defec-
        tives, otherwise rejecting the batch.
        (a) Show that the probability of accepting a batch is

        $$(1 - p)^8(1 + 8p + 36p^2)$$

        (b) Sketch the operating characteristic (OC) curve. (The values of this
        curve should be clearly indicated for two values of p other than 0
        or 1.)
        (c) When a batch is rejected, it is not discarded, but instead subjected
        to 100% inspection and all defectives are replaced. Thus if a batch is
        rejected, N hinges are inspected, whereas if the batch is accepted only
        10 hinges are inspected. If $N = 200$, determine the expected amount
        of inspection in terms of p, the proportion of defectives. Evaluate
        your expression for $p = 0.1, 0.2, 0.3$ and comment upon the answers.
                                                              (AEB) '83

78.   A manufacturer receives large batches of components daily and decides
        to institute an acceptance sampling scheme. Three possible plans are
        considered, each of which requires a sample of 30 components to be
        tested:

        Plan A. Accept the batch if no defectives are found, otherwise reject.
        Plan B. Accept the batch if not more than one defective is found,
                otherwise reject.

Plan C. Accept the batch if two or fewer defectives are found, otherwise reject.

(a) For each plan, calculate the probability of accepting a batch containing
(i) 2% defective
(ii) 8% defective

(b) Without further calculation, sketch on the same axes the operating characteristic of each plan.

(c) Which plan would be most appropriate in each of the following circumstances?
(i) There should be a high probability of accepting batches containing 2% defective.
(ii) There should be a high probability of rejecting batches containing 8% defective.
(iii) There should be a probability of at least 0.7 both of accepting a batch containing 2% defective and also of rejecting a batch containing 8% defective.

(d) Explain, briefly, to the manufacturer, how to carry out a double sampling scheme.   (AEB) '85

79.   Thatcher's Pottery produces large batches of coffee mugs decorated with the faces of famous politicians. They are considering adopting one of the following sampling plans for batch inspection.

Method A (single sample plan): Select 10 mugs from the batch at random and accept the batch if there are 2 or less defectives, otherwise reject the batch.

Method B (double sample plan): Select 5 mugs from the batch at random and accept the batch if there are no defectives, reject the batch if there are 2 or more defectives, otherwise select another 5 mugs at random. When the second sample is drawn, count the number of defectives in the combined sample of 10 and accept the batch if the number of defectives is 2 or less, otherwise reject the batch.
(i) If the proportion of defectives in a batch is p, find, in terms of p, for each method in turn, the probability that the batch will be accepted.
(ii) Evaluate *both* the above probabilities for $p = 0.2$ and $p = 0.5$.
(iii) Hence, or otherwise, decide which of these two plans is more appropriate, and why.   (AEB) '81

80.   By writing $(x_i - \mu)$ as $(x_i - \bar{x} + \bar{x} - \mu)$, where $\bar{x} = \dfrac{1}{n} \sum_{i=1}^{n} x_i$, show that

$$\sum_{i=1}^{n} (x_i - \mu)^2 = \sum_{i=1}^{n} (x_i - \bar{x})^2 + n(\bar{x} - \mu)^2. \quad \text{Note that } \sum_{i=1}^{n} (x_i - \bar{x}) = 0.$$

Hence show that s^2 can be written as $\dfrac{1}{n} \sum_{i=1}^{n} (x_i - \mu)^2 - (\bar{x} - \mu)^2$.

81.   Using Exercise 67, derive an expression for $\dfrac{n}{\sigma^2} s^2$. If $X_1, \ldots, X_n$ is a random sample from $N(\mu, \sigma^2)$, write down the distributions of

(a) $\dfrac{X_i - \mu}{\sigma}$ (b) $\dfrac{\bar{X} - \mu}{\sigma/\sqrt{n}}$ (c) $\dfrac{(X_i - \mu)^2}{\sigma^2}$ (d) $\dfrac{(\bar{X} - \mu)^2}{\sigma^2/n}$

Hence, show that the statistic $\dfrac{nS^2}{\sigma^2}$ is distributed as $\chi^2(n-1)$.

(Assume that $\dfrac{\Sigma(x - \mu)^2}{\sigma^2}$ and $\dfrac{n(\bar{x} - \mu)^2}{\sigma^2}$ are independently distributed statistics.)

Chapter 8

Estimation

8.1 INTRODUCTION

Statistical estimation is concerned with using the data obtained from a random sample to obtain information about unknown population parameters. The distributions of the sample statistics covered in the previous chapter are used throughout this chapter and two new distributions are introduced, the 'T' and the 'F'.

Where a sample is not detailed specifically, we will assume that it consists of the random variables $X_1, X_2, \ldots, X_n$, constituting a random sample of size n, unless otherwise specified.

If we draw a sample from some population and calculate $\bar{x}$, the sample mean, we quite naturally take this as a 'representative value' of the population mean, μ. Similarly, it would be quite natural to use the sample variance s^2 as being 'typical' of the population variance, σ^2. In this sense we are using the numbers $\bar{x}$ and s^2 as estimates of the true population parameters μ and σ^2, which are assumed unknown.

DEFINITION 8.1

Any statistic T, derived from a random sample and used to give information about some unknown population parameter θ (theta), is called an *estimator for* θ.

So that, the statistic $\bar{X}$ would be an estimator for the parameter μ, while S^2 would be an estimator for the parameter σ^2. (Note the distinction between the terms 'estimator for' — used for statistics — and 'estimate of' — used for a numerical value that a statistic can take.)

Of course, given a random sample $X_1, \ldots, X_n$, there is nothing stopping us taking, say, X_1 as an estimator for mean μ or, say, $X_1^2 + X_2^2$ as an estimator for σ^2, but we will show that certain estimators are better than others (in a well-defined way) for a particular parameter.

Estimation can be split into two general categories. The first (as we have discussed very briefly above) deals with statistics that 'locate' parameters, the second is concerned with constructing intervals within which a parameter is expected to lie with a given degree of confidence (probability). These two categories are referred to as *point* and *interval* estimation respectively.

8.2 POINT ESTIMATION

8.2.1 UNBIASEDNESS. THE MEAN AND VARIANCE

DEFINITION 8.2

An *unbiased estimator* for an unknown parameter is a statistic whose expectation is the parameter itself. That is if T is some statistic derived from a random sample taken from a population with unknown parameter θ, then T is an unbiased estimator for θ if and only if $E[T] = \theta$.

We can see from the above that we need some knowledge of the distribution of statistic T before being able to determine whether or not it is unbiased.

Consider a random sample $X_1, \ldots, X_n$ from a population with unknown mean μ. Then, by definition, $E[X_1] = E[X_2] = \ldots = E[X_n] = \mu$. So that, each of the variables $X_1, X_2, \ldots, X_n$ (considered as a 'trivial' statistic) is an unbiased estimator for μ. Consider also the statistic $\frac{1}{2}(X_1 + X_2)$.

Now:
$$E\left[\frac{1}{2}(X_1 + X_2)\right] = \frac{1}{2}E[X_1 + X_2]$$
$$= \frac{1}{2}\left(E[X_1] + E[X_2]\right)$$
$$= \frac{1}{2}(\mu + \mu)$$
$$= \mu$$

Hence $\frac{1}{2}(X_1 + X_2)$ is also an unbiased estimator for parameter μ.

It should be clear from the preceding examples that, given a sample, there are very many unbiased estimators for the mean, μ. The most important of these however is the sample mean, $\bar{X}$.

STATEMENT 8.1

Let $X_1, X_2, \ldots, X_n$ be a random sample of size n from a population with unknown mean μ. Then the sample mean, $\bar{X}$, is an unbiased estimator for μ, i.e., $E[\bar{X}] = \mu$.

Proof We already know that $E[\bar{X}] = \mu$, which constitutes the proof, but we will give the details again here for revision purposes.

$$E[\bar{X}] = E\left[\frac{1}{n}\sum_{i=1}^{n}X_i\right] \quad \text{(by definition)}$$

$$= \frac{1}{n}E\left[\sum_{i=1}^{n}X_i\right] \quad \text{(property of operator E)}$$

$$= \frac{1}{n}\sum_{i=1}^{n}E[X_i] \quad \text{(property of operator E)}$$

$$= \frac{1}{n}n\mu \quad \text{(by definition)}$$

$$= \mu \quad \textit{as required}$$

As regards the sample variance, S^2, we have the surprising result that $E[S^2] \neq \sigma^2$ (the unknown population variance). In other words, taking repeated samples from a population and finding the average of the respective sample variances, will not give us a reliable estimate of σ^2.

To pursue this point further, recall from Statement 7.9 that the quantity $\frac{nS^2}{\sigma^2}$ has a $\chi^2(n-1)$ distribution, if the population sampled is Normal.

So that:
$$E\left[\frac{nS^2}{\sigma^2}\right] = n-1 \quad \text{(mean of a } \chi^2 \text{ variable)}$$

giving:
$$\frac{1}{\sigma^2}E[nS^2] = n-1$$

Hence:
$$\frac{1}{n-1}E[nS^2] = \sigma^2$$

Thus:
$$E\left[\frac{nS^2}{n-1}\right] = \sigma^2$$

That is, the statistic $\frac{nS^2}{n-1}$ (and not S^2) is an unbiased estimator for σ^2.

STATEMENT 8.2

If $X_1, X_2, \ldots, X_n$ is a random sample from a population having unknown variance σ^2, then the statistic $\frac{nS^2}{n-1}$ is an unbiased estimator for σ^2, i.e., $E\left[\dfrac{nS^2}{n-1}\right] = \sigma^2$.

Notice in the above that the population from which the sample is drawn is not restricted to being Normal. (The proof of Statement 8.2 in the preceding text was confined to a sample from a Normal population.)

The following proof (although longer and more algebraic) is valid for a sample from *any* population.

Proof   (of Statement 8.2)

Now: $\dfrac{n}{n-1} S^2 = \dfrac{n}{n-1} \dfrac{1}{n} \Sigma (X_i - \bar{X})^2$ (by definition)

$= \dfrac{1}{n-1} \Sigma (X_i - \bar{X})^2$

So, we need to prove that:

$$E\left[\dfrac{1}{n-1} \Sigma (X_i - \bar{X})^2\right] = \sigma^2 \qquad [1]$$

We now write:

$$\Sigma (X_i - \bar{X})^2 \text{ as } \Sigma (X_i - \mu)^2 - n(\bar{X} - \mu)^2 \qquad [2]$$

(The proof of [2] was given as an exercise in the previous chapter.)

Note that μ is the (assumed unknown) mean of the parent population.

Since X_i $(i = 1, 2, \ldots, n)$ has expectation μ and variance σ^2, we have:

$$E[(X_i - \mu)^2] = \text{Var}[X_i] \quad \text{(by definition)}$$
$$= \sigma^2 \qquad [3]$$

and: $$E[(\bar{X} - \mu)^2] = \text{Var}[\bar{X}] \quad \text{(by definition)}$$
$$= \dfrac{\sigma^2}{n} \qquad [4]$$

We now proceed:

$$E\left[\dfrac{1}{n-1} \Sigma (X_i - \bar{X})^2\right] = \dfrac{1}{n-1} E\left[\Sigma (X_i - \bar{X})^2\right]$$

$$= \dfrac{1}{n-1} E\left[\Sigma (X_i - \mu)^2 - n(\bar{X} - \mu)^2\right] \quad \text{(from [2])}$$

$$= \dfrac{1}{n-1} \left\{E\left[\Sigma (X_i - \mu)^2\right] - n\, E[(\bar{X} - \mu)^2]\right\}$$

$$= \dfrac{1}{n-1} \left\{\Sigma E[(X_i - \mu)^2] - n\, E[(\bar{X} - \mu)^2]\right\}$$

$$= \dfrac{1}{n-1} \left(\Sigma \sigma^2 - n\, \dfrac{\sigma^2}{n}\right) \quad \text{(from [3] and [4])}$$

$$= \dfrac{1}{n-1} (n\sigma^2 - \sigma^2) = \dfrac{(n-1)\sigma^2}{n-1}$$

$$= \sigma^2$$

as required (from [1]) *completing the proof.*

EXAMPLE 8.1

The following random sample was obtained from a population with mean μ and variance σ^2: 12, 8, 11, 10, 8, 8, 13, 9, 11, 10. Obtain unbiased estimates of μ and σ^2.

SOLUTION

Unbiased estimates of μ and σ^2 are given by $\bar{x}$ and $\dfrac{nS^2}{n-1}$ respectively.

x	12	8	11	10	8	8	13	9	11	10	100
x^2	144	64	121	100	64	64	169	81	121	100	1028

$$\bar{x} = \frac{\sum x}{n} = \frac{100}{10} = 10$$

$$\frac{ns^2}{n-1} = \frac{1}{n-1}\left(\sum x^2 - n\bar{x}^2\right) = \frac{1}{9}(1028 - 1000) = \frac{28}{9} = 3.11 \quad (2D)$$

EXAMPLE 8.2

Show that the statistic $2\bar{X}$ is an unbiased estimator for θ for the rectangular distribution given by the p.d.f. $f(x) = 1/\theta$ $(0 \leqslant x \leqslant \theta)$, if a random sample of size n is drawn.

SOLUTION

Now: $\qquad E[2\bar{X}] = 2E[\bar{X}] = 2E[X]$

But: $\qquad E[X] = \int_0^\theta x\left(\frac{1}{\theta}\right) dx \quad \text{(by definition)}$

$$= \frac{1}{\theta}\left[\frac{x^2}{2}\right]_0^\theta$$

$$= \frac{1}{\theta}\left(\frac{\theta^2}{2}\right) = \frac{\theta}{2}$$

$\therefore \qquad E[2\bar{X}] = 2\left(\frac{\theta}{2}\right) = \theta$

Hence, $2\bar{X}$ is an unbiased estimator for θ.

8.2.2 EFFICIENCY AND CONSISTENCY

Consider a random sample of size n from some population and let T_1 and T_2 be two different unbiased estimators for some unknown parameter θ of the population. That is, T_1 and T_2 are both statistics derived from the sample such that $E[T_1] = E[T_2] = \theta$. Both T_1 and T_2 will have well-defined distributions and we will assume that their variances are not identical. Suppose, then, that $\text{Var}[T_1] < \text{Var}[T_2]$.

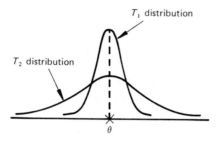

Fig. 8.1

Pictorially, their distributions might look something like those shown in Fig. 8.1.

We would say that T_1 is a *better* or *more efficient* estimator than T_2 since, due to its smaller variance, it is more likely (in the long run) to be closer to θ.

For example, consider a random sample of size n, $X_1, X_2, \ldots, X_n$, drawn from a population with mean μ and variance σ^2.

We saw earlier that the statistics X_4 and $\dfrac{1}{2}(X_1 + X_2)$ are both unbiased estimators for μ. That is:

$$E[X_4] = E\left[\frac{1}{2}(X_1 + X_2)\right] = \mu$$

Now:
$$\mathrm{Var}[X_4] = \sigma^2$$

Also:
$$\mathrm{Var}\left[\frac{1}{2}(X_1 + X_2)\right] = (\tfrac{1}{2})^2\,\mathrm{Var}[X_1 + X_2]$$

$$= \tfrac{1}{4}(\mathrm{Var}[X_1] + \mathrm{Var}[X_2])$$

$$= \tfrac{1}{4}(\sigma^2 + \sigma^2) = \frac{2\sigma^2}{4}$$

$$= \tfrac{1}{2}\sigma^2$$

So that, here, $\mathrm{Var}\left[\dfrac{1}{2}(X_1 + X_2)\right] < \mathrm{Var}[X_4]$.

Hence $\dfrac{1}{2}(X_1 + X_2)$ is more efficient than X_4 as an estimator for μ.

Of course, there will be many statistics derived from the same sample that will be unbiased estimators for some parameter. Out of all these we choose that one that has the smallest variance to call the best, or most efficient estimator.

DEFINITION 8.3

Given a random sample of size n from any population having an unknown parameter θ, there exists a set of all statistics that are unbiased estimators for θ. That unbiased estimator of the set that has the smallest variance is known as the *best* or *most efficient* estimator for θ.

The most efficient estimators for the mean and variance of any population are the (perhaps) obvious ones, as we now state.

STATEMENT 8.3

Given a random sample, size n, from a population with mean μ and variance σ^2:

(a)   The most efficient estimator for μ is $\bar{X}$.

(b)   The most efficient estimator for σ^2 is $\dfrac{n}{n-1} S^2$.

No Proof

Efficiency can be thought of as being concerned with the comparison of all unbiased estimators for some parameter θ *for a fixed sample size*, n.

Consistency, on the other hand, is concerned with fixing on a particular unbiased estimate for a parameter θ and considering its behaviour as the sample size n becomes large.

DEFINITION 8.4

Given a random sample, size n, from any population with (unknown) parameter θ, let T be an unbiased estimator for θ derived from the sample. If, as $n \to \infty$, $\text{Var}[T] \to 0$, then T is called a *consistent estimator* for θ.

Of course, given any statistic T that is an unbiased estimator for some parameter θ, we would expect it to vary about θ. If it is to be consistent, we simply require that its variance about θ becomes very small as the sample size increases. Put in another way, as the sample size increases we want the estimator to become more *reliable*.

In the limit (as n becomes infinitely large), the sample covers the whole population and here we would want the estimator to pinpoint θ exactly, which of course means $\text{Var}[T] = 0$.

Returning to the estimation of a mean μ from a random sample of size n, we saw earlier that $\frac{1}{2}(X_1 + X_2)$ is an unbiased estimator. We also calculated $\text{Var}\left[\frac{1}{2}(X_1 + X_2)\right] = \frac{\sigma^2}{2}$, which clearly (being independent of n) remains constant (strictly *non* zero) as the sample size, n, becomes infinitely large. Thus $\frac{1}{2}(X_1 + X_2)$ *is not* a consistent estimator for μ.

However, consider the statistic $\bar{X} = \frac{1}{n}\sum X_i$ (the most efficient unbiased estimator for μ).

We already know that $\text{Var}[\bar{X}] = \frac{\sigma^2}{n}$. But, as $n \to \infty$, $\frac{\sigma^2}{n} \to 0$. Therefore $\bar{X}$ *is* a consistent estimator for μ.

Also, in estimating the parameter σ^2 for the population, we stated that the quantity $\frac{n}{n-1}S^2$ is the most efficient unbiased estimator. $\frac{n}{n-1}S^2$ is also a consistent estimator, as we now state.

STATEMENT 8.4

Given a random sample size n from a population with mean μ and variance σ^2:

(a)   The statistic $\bar{X}$ is a consistent estimator for μ.

(b)   The statistic $\frac{n}{n-1}S^2$ is a consistent estimator for σ^2.

Proof   (a)   In preceding text.

(b)   Note that although this statement is true for *all* populations, we have the background theory to prove it only for a parent Normal population.

We need to show that $\text{Var}\left[\frac{n}{n-1}S^2\right] \to 0$ as $n \to \infty$.

Under the assumption of a parent Normal population, the quantity $\frac{nS^2}{\sigma^2}$ has a $\chi^2(n-1)$ distribution, from Statement 7.9.

Hence $\text{Var}\left[\frac{nS^2}{\sigma^2}\right] = 2(n-1)$.

Thus:
$$\frac{1}{\sigma^4}\text{Var}[nS^2] = 2(n-1)$$

giving:
$$\frac{1}{n-1}\text{Var}[nS^2] = 2\sigma^4$$

and:
$$\frac{1}{(n-1)^2} \text{Var}[nS^2] = \frac{2\sigma^4}{n-1}$$

Therefore $\text{Var}\left[\dfrac{nS^2}{n-1}\right] = \dfrac{2\sigma^4}{n-1}$, which tends to zero as n becomes large.

That is, $\text{Var}\left[\dfrac{nS^2}{n-1}\right] \to 0$ as $n \to \infty$ as required, *completing the proof.*

A statistic used as an estimator for some parameter θ is often denoted by $\hat{\theta}$ (theta 'hat') for convenience. For example $\hat{\mu} = \bar{X}$ and $\hat{\sigma}^2 = \dfrac{n}{n-1} S^2$.

EXAMPLE 8.3

A random sample $X_1, X_2, \ldots, X_n$ is drawn from a distribution with mean μ and variance σ^2 (both assumed unknown). Consider the statistic $T = \dfrac{1}{n+1} \sum X_i$. Show that $\text{Var}[T] < \text{Var}[\bar{X}]$ for all values of n. Also explain why this does not contradict the fact that $\bar{X}$ is the most efficient estimator for μ. Is T a consistent estimator for μ?

SOLUTION

Now:
$$\text{Var}[\bar{X}] = \frac{\sigma^2}{n}$$

$$\begin{aligned}
\text{Var}[T] &= \text{Var}\left[\frac{1}{n+1} \sum X_i\right] \\
&= \frac{1}{(n+1)^2} \text{Var}\left[\sum X_i\right] \\
&= \frac{1}{(n+1)^2} \sum \text{Var}[X_i] \\
&= \frac{1}{(n+1)^2} \sum \sigma^2 \\
&= \frac{n\sigma^2}{(n+1)^2}
\end{aligned}$$

So, we need to show that $\dfrac{n\sigma^2}{(n+1)^2} < \dfrac{\sigma^2}{n}$.

The above implies $\dfrac{n}{(n+1)^2} < \dfrac{1}{n}$ which means $n^2 < (n+1)^2$.

Hence, we need $n^2 < n^2 + 2n + 1$ or $2n + 1 > 0$.

But this must be true since $n \geqslant 1$ (n is the sample size).

So, $\text{Var}[T] < \text{Var}[\bar{X}]$ for all n.

Now, for an estimator to be efficient it must necessarily be unbiased. We already know that $\bar{X}$ is unbiased, i.e., $E[\bar{X}] = \mu$.

We also have:

$$E[T] = E\left[\frac{1}{n+1} \sum X_i\right] = \frac{1}{n+1} E\left[\sum X_i\right]$$

$$= \frac{1}{n+1} \sum E[X_i]$$

$$= \frac{1}{n+1} \sum \mu$$

$$= \frac{n\mu}{n+1} \neq \mu$$

i.e., T is a *biased* estimator for μ.

Thus the fact that $\bar{X}$ is the most efficient estimator for μ has not been contradicted. The 'biasedness' of T also precludes it from being a consistent estimator for μ.

8.2.3 SOME OTHER ESTIMATORS

When taking small samples from a Normal distribution, a very useful and practical technique for estimating the standard deviation is to use a statistic involving the range of the sample.

Briefly, consider a random sample $X_1, \ldots, X_n$ from a Normal population with (unknown) variance σ^2. Let $X_{(1)}$ denote the smallest and $X_{(n)}$ the largest random variable when $X_1, X_2, \ldots, X_n$ are arranged in order of size. The range of the sample is then defined by the statistic $R = X_{(n)} - X_{(1)}$, which has its own distribution (the form of which we are not concerned with — it is, mathematically, very complicated). It can be shown that $E[R] = b_n\sigma$ where b_n is a constant depending on n and σ is the standard deviation of the parent Normal population.

It is a simple matter to arrange the above to read $E\left[\left(\frac{1}{b_n}\right)R\right] = \sigma$. The quantity $\frac{1}{b_n}$ is usually written as k, for convenience. So that, we have $E[kR] = \sigma$. Hence the statistic kR is an unbiased estimator for σ.

Values of k are given in Appendix Table 8 for various values of n.

As a practical example, suppose a random sample of size 9 is taken from an assumed Normal distribution and the range, r, is found to be 21. Then, we have:

$$\hat{\sigma} = rk$$

$$= 21(0.3367) \quad \text{(Appendix Table 8)}$$

$$= 7.07 \quad \text{(2D)}$$

We would estimate the population standard deviation then as 7.07. The student will doubtless note the simplicity of this technique.

To summarise:

STATEMENT 8.5

Let $X_1, \ldots, X_n$ be a random sample of size n from a Normal distribution with unknown variance σ^2. Also, with $X_{(1)}$ and $X_{(n)}$ the smallest and largest respectively of the above random variables, put $R = X_{(n)} - X_{(1)}$ (the range). Then, an unbiased estimator for σ is given by kR, where k is a special constant depending on n (listed in Appendix Table 8).

That is, $E[kR] = \sigma$.

No Proof

It will be seen from the tables that k is only given for n up to 10. This is because the statistic kR is not very efficient for large values of n However, this fact, together with the relative simplicity of the estimation process, makes the technique particularly useful in industrial situations where small samples are taken at regular intervals from output for control purposes.

EXAMPLE 8.4

Estimate the standard deviation of the populations from which the following samples are drawn, using the range:

(a)   1.2, 1.8, 2.3, 2.1, 0.9, 1.3.

(b)   56, 59, 63, 55, 48, 52, 61, 64.

SOLUTION

(a)   Range of sample $= 2.3 - 0.9 = 1.4 = r.$

From Appendix Table 8, $k = 0.395$ $(n = 6)$.

Hence: $\hat{\sigma} = rk = (1.4)(0.395)$

$$= 0.55 \ (2D)$$

(b)   Range $= r = 64 - 48 = 16;$ $k = 0.351$ $(n = 8$ here$)$.

$\therefore$ $\hat{\sigma} = 16(0.351)$

$$= 5.62 \ (2D)$$

EXAMPLE 8.5

Estimate the variance of the population from which the following sample is drawn by:

(a) using the sample range, and

(b) using the best unbiased estimator:

$$14.2, 11.8, 12.3, 12.8, 11.2, 10.7, 13.4, 12.9$$

(c) Calculate the percentage error in using (a) rather than (b).

SOLUTION

Attributing variable x to the above sample values, we have:

x	14.2	11.8	12.3	12.8	11.2	10.7	13.4	12.9
x^2	201.64	139.24	151.29	163.84	125.44	114.49	179.56	166.41

giving $\Sigma x = 99.3$ and $\Sigma x^2 = 1241.91$. Range, $r = 14.2 - 10.7 = 3.5$.

(a) We have a sample size of 8, so from Appendix Table 8 $k = 0.351$.

Hence: $\hat{\sigma} = rk = (3.5)(0.351) = 1.229$ (3D)

$\therefore$ $\hat{\sigma}^2 = (1.229)^2 = 1.51$ (2D).

(b) $$s^2 = \frac{\Sigma x^2}{n} - \left(\frac{\Sigma x}{n}\right)^2$$

$$= \frac{1241.91}{8} - \left(\frac{99.3}{8}\right)^2$$

$$= 155.239 - 154.070 \quad (3D)$$

$$= 1.169 \quad (3D)$$

$\therefore$ best unbiased estimate $= \hat{\sigma}^2 = \dfrac{ns^2}{n-1} = \dfrac{8(1.169)}{7}$

$$= 1.34 \quad (2D)$$

(c) The absolute error involved in using (a) for (b) is

$$1.34 - 1.51 = -0.17.$$

Hence: the percentage error (excluding sign) $= \dfrac{0.17}{1.34}(100)$

$$= 13 \quad \text{(nearest whole number)}$$

Recall, from Statement 7.11, that if a population has a proportion π of its members with a certain attribute then, if a random sample of size n is considered, the proportion P of the sample with the attribute has an approximately Normal distribution. Namely, P is $N\left(\pi, \dfrac{\pi(1-\pi)}{n}\right)$ approximately.

Hence, since $E[P] = \pi$, P is an unbiased estimator for parameter π.

Further, P is a consistent estimator for π by virtue of the fact that $\mathrm{Var}[P] = \dfrac{\pi(1-\pi)}{n}$ tends to zero as n becomes infinitely large.

STATEMENT 8.6

If a population has an unknown proportion π of its members with some attribute and a random sample is drawn with P the proportion of the sample having the attribute, then:

(a) P is an unbiased estimator for π.

(b) P is a consistent estimator for π.

Proof   In preceding text.

8.2.4 POOLED ESTIMATES

Here we consider the estimation of certain parameters based on the results of more than one random sample, taken usually from the same population. We shall deal with just two standard situations of interest.

The first is where we take two random samples of sizes n_1 and n_2 from a population with mean μ and variance σ^2. Let us denote the respective sample means and variances by $\bar{X}_1, \bar{X}_2$ and S_1^2, S_2^2.

We know from previous considerations that $\bar{X}_1$ and $\bar{X}_2$ are both unbiased estimators for μ. Also both of $\dfrac{n_1 S_1^2}{n_1 - 1}$ and $\dfrac{n_2 S_2^2}{n_2 - 1}$ are unbiased estimators for σ^2.

We would like unbiased estimators for μ and σ^2 based on the two samples combined.

Now, for the combined sample, there is $n_1 + n_2$ observations. The sum of observations in the first sample is $n_1 \bar{X}_1$ and in the second, $n_2 \bar{X}_2$. Hence the grand sum of observations for the combined sample is $n_1 \bar{X}_1 + n_2 \bar{X}_2$.

Therefore:

$$\text{combined mean} = \frac{\text{sum of observations}}{\text{number of observations}} \quad \text{(by definition)}$$

$$= \frac{n_1 \bar{X}_1 + n_2 \bar{X}_2}{n_1 + n_2}$$

Since the above is a sample mean (for the combined sample), it is also an unbiased estimator for μ. We now state and prove this formally.

STATEMENT 8.7

If $\bar{X}_1$ and $\bar{X}_2$ are the respective means of random samples of size n_1 and n_2 from a population with mean μ, an unbiased estimator for μ is given by:

$$\hat{\mu} = \frac{n_1 \bar{X}_1 + n_2 \bar{X}_2}{n_1 + n_2}$$

Proof   We need to show that $E\left[\dfrac{n_1 \bar{X}_1 + n_2 \bar{X}_2}{n_1 + n_2}\right] = \mu.$

Now: $E\left[\dfrac{n_1 \bar{X}_1 + n_2 \bar{X}_2}{n_1 + n_2}\right] = \dfrac{1}{n_1 + n_2} E[n_1 \bar{X}_1 + n_2 \bar{X}_2]$

$$= \frac{1}{n_1 + n_2}\left\{ n_1 \, E[\bar{X}_1] + n_2 \, E[\bar{X}_2] \right\}$$

$$= \frac{1}{n_1 + n_2}\left(n_1 \mu + n_2 \mu \right)$$

$$= \frac{(n_1 + n_2)\mu}{n_1 + n_2}$$

$$= \mu$$

completing the proof.

For an unbiased estimator for the population variance based on the two samples, consider the following.

The two sample variances are S_1^2 and S_2^2 with respective sample sizes n_1 and n_2. We have, from Statement 8.2 that:

$$E\left[\frac{n_1 S_1^2}{n_1 - 1}\right] = E\left[\frac{n_2 S_2^2}{n_2 - 1}\right] = \sigma^2$$

Hence: $E[n_1 S_1^2] = \sigma^2(n_1 - 1)$ [1]

and: $E[n_2 S_2^2] = \sigma^2(n_2 - 1)$ [2]

Adding [1] and [2] gives:

$$E[n_1 S_1^2] + E[n_2 S_2^2] = \sigma^2(n_1 - 1) + \sigma^2(n_2 - 1)$$

so that: $E[n_1 S_1^2 + n_2 S_2^2] = \sigma^2(n_1 + n_2 - 2)$

Hence: $E\left[\dfrac{n_1 S_1^2 + n_2 S_2^2}{n_1 + n_2 - 2}\right] = \sigma^2$

Formally, we state:

STATEMENT 8.8

If S_1^2 and S_2^2 are the respective variances of random samples of size n_1 and n_2 respectively from a population with (unknown) variance σ^2, an unbiased estimator for σ^2 is given by

$$\hat\sigma^2 = \frac{n_1 S_1^2 + n_2 S_2^2}{n_1 + n_2 - 2}.$$

Proof In preceding text.

An extension of the above to three or more samples should be fairly clear. For example, if μ and σ^2 are the mean and variance of a population from which three random samples have been drawn, then (using an obvious notation):

$$\hat\mu = \frac{n_1 \bar X_1 + n_2 \bar X_2 + n_3 \bar X_3}{n_1 + n_2 + n_3}$$

and:

$$\hat\sigma^2 = \frac{n_1 S_1^2 + n_2 S_2^2 + n_3 S_3^2}{n_1 + n_2 + n_3 - 3}$$

are both unbiased estimators.

The proof in the general case (i.e., for r samples) is left as an exercise.

EXAMPLE 8.6

Two random samples of size 8 and 7 respectively are drawn from a population as follows:

10, 11, 10, 12, 9, 8, 10, 11   and   11, 10, 11, 8, 8, 11, 12

Calculate pooled, unbiased estimates of the population mean and variance.

SOLUTION

Sample 1		Sample 2	
x	x^2	x	x^2
10	100	11	121
11	121	10	100
10	100	11	121
12	144	8	64
9	81	8	64
8	64	11	121
10	100	12	144
11	121	71	735
81	831		

For sample 1: $n_1 = 8$; $\bar{x}_1 = 81/8 = 10.125$

$$s_1^2 = \frac{831}{8} - (10.125)^2 = 103.875 - 102.5156$$

$$= 1.3594$$

For sample 2: $n_2 = 7$; $\bar{x}_2 = 71/7 = 10.1429$

$$s_2^2 = \frac{735}{7} - \left(\frac{71}{7}\right)^2 = 105.0 - 102.8776$$

$$= 2.1224$$

Hence: $\hat{\mu} = \dfrac{n_1 \bar{x}_1 + n_2 \bar{x}_2}{n_1 + n_2} = \dfrac{81 + 71}{15}$

$$= 10.13 \quad \text{(2D)}$$

and: $\hat{\sigma}^2 = \dfrac{n_1 s_1^2 + n_2 s_2^2}{n_1 + n_2 - 2} = \dfrac{8(1.3594) + 7(2.1224)}{13}$

$$= \frac{10.8752 + 14.8568}{13}$$

$$= 1.98 \quad \text{(2D)}$$

EXAMPLE 8.7

Two variables X and Y are distributed $N(\mu, \sigma^2)$ and $N(3\mu, 4\sigma^2)$ respectively. Two random samples each of size n are drawn, one from each of these Normal populations, and have means and variances $\bar{x}, s^2$ and $\bar{y}, 3s^2$ respectively. Using the pooled data from these two samples, calculate an unbiased estimate of σ^2. Show that $a\bar{x} + b\bar{y}$ is an unbiased estimate of μ provided $a + 3b = 1$ and that, as b varies, the minimum value of the variance of this estimator is $\dfrac{4}{13}\left(\dfrac{\sigma^2}{n}\right)$.

                                                         (AEB)'73

SOLUTION

Now, we cannot use the method of Statement 8.8, since we are not combining samples in order to estimate a *single population* variance. Here, we already know that each sample comes from a separate distribution even though their respective variances are clearly related.

Consider the distribution of X. $\text{Var}[X] = \sigma^2$ and a sample of size n yields a variance of s^2. We can thus use:

$$\text{E}\left[\frac{nS^2}{n-1}\right] = \sigma^2 \qquad [1]$$

For the Y distribution, $\text{Var}[Y] = 4\sigma^2$ and the size n sample variance is $3s^2$. In this case we use:

$$\mathrm{E}\left[\frac{n3S^2}{n-1}\right] = 4\sigma^2 \tag{2}$$

In order to obtain a single expression only, involving σ^2, we can add [1] and [2]. Hence:

$$\mathrm{E}\left[\frac{nS^2}{n-1}\right] + \mathrm{E}\left[\frac{3nS^2}{n-1}\right] = \sigma^2 + 4\sigma^2$$

i.e. $\mathrm{E}\left[\dfrac{nS^2}{n-1} + \dfrac{3nS^2}{n-1}\right] = 5\sigma^2$ giving $\mathrm{E}\left[\dfrac{4nS^2}{n-1}\right] = 5\sigma^2$

Manipulating the constant 5, easily gives $\mathrm{E}\left[\dfrac{4nS^2}{5(n-1)}\right] = \sigma^2.$

Hence, using the information from both samples, $\dfrac{4ns^2}{5(n-1)}$ is an unbiased estimate of σ^2.

Now: $\mathrm{E}[a\bar{X} + b\bar{Y}] = a\,\mathrm{E}[\bar{X}] + b\,\mathrm{E}[\bar{Y}]$

$$= a\mu + 3b\mu$$

$$= (a + 3b)\mu$$

$$= \mu \quad \text{(only if } a + 3b = 1)$$

Hence $a\bar{x} + b\bar{y}$ is an unbiased estimate of μ if $a + 3b = 1$.

Also: $\mathrm{Var}[a\bar{X} + b\bar{Y}] = a^2\,\mathrm{Var}[\bar{X}] + b^2\,\mathrm{Var}[\bar{Y}]$

$$= a^2\frac{\sigma^2}{n} + b^2\frac{4\sigma^2}{n} \quad \begin{array}{l}\text{(using the distribution}\\ \text{of means theory)}\end{array}$$

$$= \frac{\sigma^2}{n}(a^2 + 4b^2)$$

$$= \frac{\sigma^2}{n}((1 - 3b)^2 + 4b^2) \quad \begin{array}{l}\text{(since } a + 3b = 1\\ \text{then } a = 1 - 3b)\end{array}$$

$$= \frac{\sigma^2}{n}(1 - 6b + 13b^2) \quad \begin{array}{l}\\ \text{(in terms of } b \text{ only)}\end{array} \tag{3}$$

We need to find the minimum of this variance with respect to b, which is done via the calculus as follows.

Now: $\dfrac{\partial}{\partial b}\,(\text{variance}) = \dfrac{\partial}{\partial b}\left(\dfrac{\sigma^2}{n}(1 - 6b + 13b^2)\right)$

$$= \frac{\sigma^2}{n}(-6 + 26b)$$

So, when $\dfrac{\partial}{\partial b}\,(\text{variance}) = 0$ we have $\dfrac{\sigma^2}{n}(26b - 6) = 0$;

i.e. $26b - 6 = 0 \quad \left(\text{since } \dfrac{\sigma^2}{n} \neq 0\right)$

Hence $b = \dfrac{3}{13}$ (solving the above equation) corresponds to the minimum value of variance.

Substituting this value into [3] gives:

$$\text{variance}_{\min} = \frac{\sigma^2}{n} \left(1 - 6 \left(\frac{3}{13} \right) + 13 \left(\frac{3}{13} \right)^2 \right) = \frac{\sigma^2}{n} \left(1 - \frac{18}{13} + \frac{9}{13} \right)$$

$$= \frac{4}{13} \left(\frac{\sigma^2}{n} \right)$$

The second standard situation of interest is where we need to pool samples taken from some population in order to estimate the proportion (π) of the population having a particular attribute.

Now, we know already that a sample proportion P is an unbiased estimator for a population proportion π. If two samples are pooled, the following result is easily obtained:

STATEMENT 8.9

Let two random samples of respective sizes n_1 and n_2 be taken from a population with proportion π having a certain attribute. Then, if P_1 and P_2 denote the respective sample proportions having the given attribute, an unbiased estimator for π is given by:

$$\hat{\pi} = \frac{n_1 P_1 + n_2 P_2}{n_1 + n_2}$$

Proof   Left as an exercise.

The above can be seen to be a simple weighted average of the sample proportions. An extension to three or more samples is obvious.

EXAMPLE 8.8

A random sample of 400 voters in district A gave 132 in favour of a particular political party. In district B, 56% out of the 500 sampled randomly supported the same party. Estimate the proportion of all voters in the two districts combined who support the above party. Use this proportion to calculate the approximate probability that this party would be certain to gain an overall majority (in districts A and B combined) if an election were held.

SOLUTION

For district A: $n_A = 400$ giving $p_A = \dfrac{132}{400} = 0.33$

For district B: $n_B = 500$ and $p_B = 0.56$ (given)

Hence: $\hat{\pi} = \dfrac{n_A p_A + n_B p_B}{n_A + n_B} = \dfrac{132 + 500(0.56)}{400 + 500}$

$= \dfrac{412}{900} = 0.458$ (3D)   (an estimate of the true proportion of party supporters)

To be certain of gaining an overall majority, the party must get more than 50% of all votes cast. Hence, we require $\Pr(P > 0.5)$.

Now, from Statement 7.11, P is (approximately) $N\left(\pi, \dfrac{\pi(1-\pi)}{n}\right)$.

So, writing $\hat{\pi}$ for π, we have:

$$P \text{ is } N\left(0.458, \frac{(0.458)(0.542)}{900}\right)$$

i.e. P is $N(0.458, 0.000\,275\,8)$

$\therefore \ \Pr(P > 0.5) = \Pr\left(Z > \dfrac{(0.5 + \frac{1}{1800}) - 0.458}{\sqrt{0.000\,275\,8}}\right)$ with continuity correction

$= \Pr(Z > 2.56)$

$= 1 - \Phi(2.56) = 1 - 0.9948$

$= 0.005$ (3D)

A natural question that may be asked in this sub-section is 'why would we bother to take two or more samples from a population'? 'Would it not be easier to take a single large sample'?

The answer to the second is obviously 'yes'. But sometimes, in order to take into account such important factors as efficiency, cost, and safety, it is necessary to take more than one sample.

We mention briefly two specific situations.

First, there is the changing population. People are apt to become tired; machines often become less efficient with age. Both of these factors are likely to change the nature of a specific population and 'control' measures can only be effectively taken as a result of periodic sampling.

Secondly, there is the 'compatibility' of several populations. Often we need to estimate parameters from large populations consisting of several sub-populations. For example, a number of machines might be producing identical components. Parameters for the whole component population could be best estimated by pooling the samples obtained from each machine output.

These types of situations are discussed more fully in the next two chapters, where we deal with special statistical tests.

8.3 DEGREES OF FREEDOM

We mentioned earlier in the text that the parameter involved with a χ^2 distribution is often referred to as 'degrees of freedom'. Degrees of freedom, in a broad sense, can be thought of as 'number of free choices'.

Consider the following illustration. We wish to choose three numbers a, b and c say, to add to 100. With a little thought, it is obvious that we do not have three free choices of the numbers. We can certainly give the first two numbers (a and b, say) any values we wish, but then c will be automatically determined. Algebraically, we have $a + b + c = 100$, giving $c = 100 - (a + b)$. That is, once a and b have been chosen, c is constrained to take the value $100 - (a + b)$; there is no choice in the matter. In this sense, we call the given relationship $a + b + c = 100$ a 'restriction' on the choice of a, b and c. In this particular case we have three variables (a, b and c) subject to 1 restriction ($a + b + c = 100$) giving just $3 - 1 = 2$ free choices (or degrees of freedom).

This is in fact generally true. That is:

no. of original variables $-$ no. of restrictions $=$ no. of degrees of freedom

An example of a type of situation involving degrees of freedom with which we will be particularly concerned later in the text is the following.

a	b	12
c	d	2
e	f	6
11	9	20

A 3×2 table (as shown) has to be completed by giving values to a, b, c, d, e and f, where the row and column totals together with the grand total are known. How many free choices have we of the 6 variables a to f? That is, how many degrees of freedom are available in filling the body of the table? First, we need to work out how many restrictions there are. There are three row totals, two column totals and a grand total given — seemingly 6 restrictions in all.

Namely:

(a) $a + b = 12$ (b) $c + d = 2$ (c) $e + f = 6$ (d) $a + c + e = 11$

(e) $b + d + f = 9$ and (f) $a + b + c + d + e + f = 20$

But, some of the above information is repeated.

For example, once the three row totals are known, the grand total is automatically determined. Also, once the two column totals are known, the grand total is again automatically determined.

In fact, we need only have been given either (a) the 3 row totals and any 1 column total (the grand total and missing column total then being automatically determined), (b) any 2 row totals and the 2 column totals (the grand total and missing row total then being automatically determined), or (c) any 2 row totals, any 1 column total and the grand total (the missing row and column totals then being automatically determined).

So that, although 6 totals are given for the table, only 4 of these are properly 'independent' restrictions on our choice of the 6 variables.

Hence, with 6 original variables and 4 (independent) restrictions we have just $6 - 4 = 2$ free choices (degrees of freedom) for the numbers a to f.

The reader might like to verify this by choosing values for just 2 of the 6 letters a to f (the 2 must be independent), and then using the surrounding information to determine the remaining 4.

For estimation purposes, degrees of freedom are associated with sample statistics. Formally:

DEFINITION 8.5

The *number* of degrees of freedom associated with a sample statistic T, derived from a random sample of size n, is defined as the number of random variables (n) minus the number of independent restrictions involved in the calculation of T.

Consider the sample statistic $S^2 = \dfrac{1}{n} \sum (X_i - \bar{X})^2$. Here the n random variables $X_1, \ldots, X_n$ together with the quantity $\bar{X}$ are involved. $\bar{X}$ is treated as a restriction simply because, by definition, $\bar{X} = \dfrac{1}{n}(X_1 + \ldots + X_n)$, constraining $X_1, \ldots, X_n$ to satisfy this relationship.

Thus with n random variables and 1 restriction ($\bar{X}$), we have $n - 1$ degrees of freedom in the calculation of S^2.

Note that the statistic that is an unbiased estimator for σ^2, namely:

$$\frac{n}{n - 1} S^2 = \frac{1}{n - 1} \sum (X_i - \bar{X})^2$$

can thus be described as 'the sum of squares of deviations from the mean *divided by the number of degrees of freedom involved*'.

Also, in the case of the combined two-sample unbiased estimator of a common population variance σ^2, namely:

$$\hat{\sigma}^2 = \frac{n_1 S_1^2 + n_2 S_2^2}{n_1 + n_2 - 2}$$

the denominator is again the number of degrees of freedom involved, since we have $n_1 + n_2$ original random variables together with the two restrictions $\bar{X}_1$ and $\bar{X}_2$, giving $n_1 + n_2 - 2$ degrees of freedom.

8.4 THE *T* DISTRIBUTION

As with the case of the χ^2 distribution, we look only briefly at the T distribution paying some attention to its definition and main characteristics and particularly at the interpretation and use of the special '*t*-tables'.

8.4.1 PROBABILITY DENSITY FUNCTION

The distribution has a p.d.f. of the form:

$$f(x) = T_\nu \left(1 + \frac{x^2}{\nu}\right)^{-\left(\frac{\nu+1}{2}\right)} \quad \text{for } -\infty < x < \infty$$

where T_ν is a special constant (depending on ν) and ν is an integral valued parameter.

The frequency curves for $f(x)$ for some values of ν are now given.

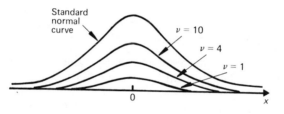

Fig. 8.2

The T curve can be seen to resemble a 'squashed' version of the standard Normal curve (which is included for comparison), the degree of 'squash' depending on how small ν is. In the limit, as $\nu \to \infty$, so the T distribution $\to$ standard Normal distribution.

The distribution is sometimes referred to as 'Student's t' distribution after the statistician W. S. Gossett who, while working for Guinness breweries, first published details of it under the pseudonym 'Student'.

8.4.2 SOME SPECIAL CHARACTERISTICS

Some important results and identities for this distribution are now given.

(a)   The distribution is *continuous*.

(b)   There is one parameter, ν.

(c)   The shorthand notation used for this distribution will be X is $T(\nu)$, translated as 'the random variable X has a T distribution with parameter ν.'

(d)   The distribution is *symmetric* about its mean.

STATEMENT 8.10

(a) If X is $T(\nu)$, then $E[X] = 0$.

(b) If X is $N(0, 1)$ and Y is $\chi^2(\nu)$, with X, Y independent, then the random variable $\dfrac{X}{\sqrt{Y/\nu}}$ is $T(\nu)$.

No Proof

Notice that the mean of the distribution is always 0, independent of the parameter ν. The parameter ν itself is, again, a number of degrees of freedom linked directly with the ν of the defining χ^2 distribution in (b) above.

An important statistic which is used in the next section and the next chapter has a T distribution and is now demonstrated.

EXAMPLE 8.9

A random sample $X_1, \ldots, X_n$ is drawn from a Normal population with mean μ and variance σ^2. Show that the statistic $\dfrac{\bar{X} - \mu}{S/\sqrt{n-1}}$ has a $T(n-1)$ distribution.

SOLUTION

We know that $\bar{X}$ is $N\left(\mu, \dfrac{\sigma^2}{n}\right)$.

Hence: $\dfrac{\bar{X} - \mu}{\sigma^2/n}$ is $N(0, 1)$ [1]

Also: $\dfrac{nS^2}{\sigma^2}$ is $\chi^2(n-1)$ [2]

(The above is true since the sample is drawn from a normal population.)

Now:

$$\frac{\bar{X} - \mu}{S/\sqrt{n-1}} = \frac{\dfrac{\bar{X} - \mu}{\sigma/\sqrt{n}}}{\dfrac{S}{\sqrt{n-1}}\dfrac{\sqrt{n}}{\sigma}} \quad \begin{pmatrix} \text{dividing numerator and} \\ \text{denominator by } \dfrac{\sigma}{\sqrt{n}} \end{pmatrix}$$

$$= \frac{\dfrac{\bar{X} - \mu}{\sigma/\sqrt{n}}}{\sqrt{\dfrac{nS^2}{\sigma^2(n-1)}}} = \frac{T_1}{\sqrt{\dfrac{T_2}{n-1}}} \quad \text{say}$$

But T_1 is $N(0, 1)$ and T_2 is $\chi^2(n-1)$ from [1] and [2].

Hence, by Statement 8.10(b):

$$\frac{T_1}{\sqrt{T_2/(n-1)}} = \frac{\bar{X} - \mu}{S/\sqrt{n-1}} \quad \text{is a} \; T(n-1) \; \text{variable}$$

8.4.3 USE OF TABLES

The values given in Appendix Table 5 are *percentage point* values for a $T(\nu)$ distributed variable for various values of parameter ν.

As an illustration of their use, consider column $P = 0.95$ for row $\nu = 10$. The value given for t is 1.81. This is interpreted as '95% of a $T(10)$ distribution lies to the left of value 1.81'. This is shown pictorially in Fig. 8.3.

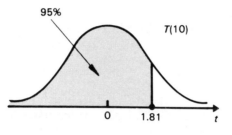

Fig. 8.3

Put another way, $T_{95\%}(10) = 1.81$. Similarly, the value given for t of 2.53 at the intersection of column $P = 0.99$ and row $\nu = 20$ can be interpreted at '99% of a $T(20)$ distribution lies to the left of $t = 2.53$'. This is demonstrated pictorially in Fig. 8.4.

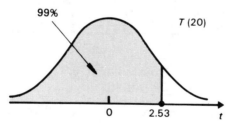

Fig. 8.4

It is sometimes necessary, with a T distribution (this will be seen later), to be able to identify the two values that have the central 95% (or 90% or 99%, for example) of the distribution lying between them. This is quite a straightforward procedure, once it is remembered that the T distribution (as with the Standard Normal) is symmetric about its mean of zero.

For instance, since 99% of a $T(20)$ distribution lies to the left of $t = 2.53$, we must have that 99% of a $T(20)$ distribution lies to the *right* of $t = -2.53$. In other words, the two values $t = -2.53$ and $t = 2.53$ enclose the central 98% of a $T(20)$ distribution.

EXAMPLE 8.10

Find the percentage of a $T(10)$ distribution that lies to the right of the value $t = 1.81$.

SOLUTION

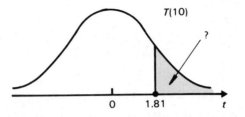

Looking along row $v = 10$ of Table 5, it can be seen that 95% of a $T(10)$ distribution lies to the left of 1.81.

Thus 5% of a $T(10)$ distribution must lie to the right of $t = 1.81$.

EXAMPLE 8.11

Find the two symmetrically placed values outside of which 5% of a $T(9)$ distribution lies.

SOLUTION

The two values required are $T_{0.025}(9)$, which has 2.5% of the distribution lying to the left, and $T_{0.975}(9)$, which has 2.5% of the distribution lying to the right.

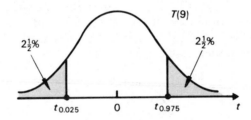

But $T_{0.975}(9) = 2.26$ from column 0.975, row 9 of Table 5. Thus, by symmetry, $T_{0.025}(9) = -2.26$.

The two values required are, therefore, -2.26 and $+2.26$.

EXAMPLE 8.12

If a random variable X is $T(7)$, find:

(a) $\Pr(X < -2.37)$ (b) $\Pr(X > 3.50)$ and hence

(c) $\Pr(-2.37 < X < 3.50)$

SOLUTION

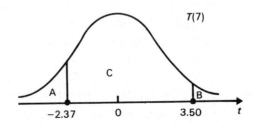

(a)   From Table 5, row $\nu = 7$ gives value $t = 2.37$ at column 0.975. That is, 97.5% of $T(7)$ lies to the left of 2.37, or, put another way, 2.5% of $T(7)$ lies to the left of -2.37.

Therefore, Area A $= \Pr(X < -2.37) = 0.025$.

(b)   From Table 5, row $\nu = 7$ gives value $t = 3.50$ at column 0.995. That is, 99.5% of $T(7)$ lies to the left of $3.50 =$ Area A + Area C. Thus, 0.5% (proportion 0.005) of $T(7)$ must lie to the right of 3.50.

Therefore, Area B $= \Pr(X > 3.50) = 0.005$.

(c)   Since, from (a), Area A $= 0.025$, and, from (b), Area A + Area C $= 0.995$, Area C $= 0.995 - 0.025 = 0.97$.

Therefore, $\Pr(-2.37 < X < 3.50) = 0.97$.

8.5 THE *F* DISTRIBUTION

8.5.1 PROBABILITY DENSITY FUNCTION

The distribution has a p.d.f. of the form:

$$f(x) = F_{\nu_1, \nu_2} \cdot \frac{x^{(\nu_1 - 2)/2}}{\left(1 + \dfrac{\nu_1}{\nu_2} \cdot x\right)^{(\nu_1 + \nu_2)/2}} \quad \text{where } 0 \leqslant x \leqslant \infty$$

F_{ν_1, ν_2} is a special constant depending on ν_1 and ν_2; ν_1 and ν_2 themselves are integral valued parameters.

The shape of the density curve for the distribution depends on the values of ν_1 and ν_2 and is, in general, skewed to the right.

8.5.2 SOME SPECIAL CHARACTERISTICS

Some important results and identities for the distribution are now given:

(a)   The distribution is *continuous*.

(b)   There are *two* parameters, ν_1 and ν_2.

(c)   The shorthand notation we will use for the distribution is X is $F(\nu_1, \nu_2)$, translated as 'the random variable X has an F distribution with parameters ν_1 and ν_2'.

(d)   The distribution *is not symmetric*, the amount of right skew depending on the values of ν_1 and ν_2. (The exact shape of the curve for particular values of ν_1 and ν_2 need not concern us.)

STATEMENT 8.11

(a)   If random variable X_1 is $\chi^2(\nu_1)$ and random variable X_2 is $\chi^2(\nu_2)$, where X_1 and X_2 are independent, then:

$$\frac{X_1/\nu_1}{X_2/\nu_2} \quad \text{is} \quad F(\nu_1, \nu_2)$$

(b)   If random variable Y is $F(\nu_1, \nu_2)$, then $\dfrac{1}{Y}$ is $F(\nu_2, \nu_1)$.

No Proof   Note that (b) is easily obtained using result (a). We can also think of (a) in terms of a definition of an F random variable. That is, the quotient of two χ^2 random variables, each divided by their respective degrees of freedom, is defined as an F variable. The two parameters involved (ν_1 and ν_2) are both a number of degrees of freedom.

8.5.3 USE OF TABLES

Since an F distribution has two associated parameters, an extensive coverage of percentage points for all parametric values is not possible. Also, there is no 'standard' form of the distribution that can be converted to (as in the case of the Normal distribution).

The tables given for the F distribution (Appendix Table 6) are for some selected percentage points of interest; namely 99.5, 99, 97.5 and 95. Each of these tables gives the points corresponding to a wide range of values for both ν_1 and ν_2.

As an example of these tables' use, consider table $P = 0.975$ (97.5% points). Column $\nu_1 = 10$ and row $\nu_2 = 9$ yields 3.964. This means that $F_{0.975}(10, 9) = 3.964$, i.e. 97.5% of an $F(10, 9)$ distribution lies to the left of 3.964. This is shown pictorially in Fig. 8.5(a).

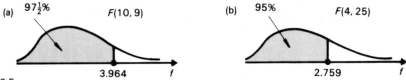

(a) $97\frac{1}{2}$% $F(10, 9)$ (b)   95% $F(4, 25)$

3.964 f 2.759 f

Fig. 8.5

Also, from table $P = 0.95$ (95% points), with $\nu_1 = 4$ and $\nu_2 = 25$, we obtain $f = 2.759$. This means that 95% of an $F(4, 25)$ distribution lies to the left of 2.759, i.e. $F_{0.95}(4, 25) = 2.759$. This is shown pictorially in Fig. 8.5(b).

Using the given four tables, together with Statement 8.11(b), enables us to obtain 5, 2.5, 1 and 0.5% points of an F distribution.

Suppose we required $F_{5\%}(\nu_1, \nu_2)$. Let X be $F(\nu_2, \nu_1)$. Then, from Statement 8.11(b), we have:

$$\frac{1}{X} \quad \text{as} \quad F(\nu_1, \nu_2) \qquad\qquad [1]$$

Now: $$\Pr(X < F_{0.95}(\nu_2, \nu_1)) = 95\% \quad \text{(by definition)}$$

Hence: $$\Pr\left(\frac{1}{X} > \frac{1}{F_{0.95}(\nu_1, \nu_2)}\right) = 95\%$$

Therefore $\dfrac{1}{F_{0.95}(\nu_1, \nu_2)}$ must represent the 5% point of the distribution of

of $\dfrac{1}{X}$ which, from [1], is $F(\nu_1, \nu_2)$,

i.e. $$F_{0.05}(\nu_1, \nu_2) = \frac{1}{F_{0.95}(\nu_2, \nu_1)}$$

Similarly: $$F_{0.025}(\nu_1, \nu_2) = \frac{1}{F_{0.975}(\nu_2, \nu_1)}$$

using the same argument as above.

In general, we have the following statement.

STATEMENT 8.12

$$F_{A\%}(\nu_1, \nu_2) = \frac{1}{F_{(100-A)\%}(\nu_2, \nu_1)}$$

Proof   Left as an exercise.

So that, if we have tables of the $(100 - A)\%$ points of an $F(\nu_2, \nu_1)$ distribution for various values of ν_2 and ν_1, we can obtain the $A\%$ points of an $F(\nu_1, \nu_2)$ distribution using the given relationship.

For example, we have $F_{0.975}(12, 5) = 6.525$ (from tables).

Therefore: $$F_{0.025}(5, 12) = \frac{1}{F_{0.975}(12, 5)} = \frac{1}{6.525} \quad \text{(from tables)}$$

$$= 0.153 \quad \text{(3D)}$$

EXAMPLE 8.13

Find the values of f_1 and f_2 such that the interval (f_1, f_2) encloses the central:

(a)   95%   (b)   99.8% of an $F(8, 4)$ distribution.

SOLUTION

(a) $f_2 = F_{0.975}(8, 4) = 8.98$

$$f_1 = F_{0.025}(8, 4) = \frac{1}{F_{0.975}(4, 8)} = \frac{1}{5.053}$$

$$= 0.198 \quad \text{(3D)}$$

(b) $f_2 = F_{0.995}(8, 4) = 21.35$

$$f_1 = F_{0.005}(8, 4) = \frac{1}{F_{0.995}(4, 8)} = \frac{1}{8.805}$$

$$= 0.114 \quad (3D)$$

EXAMPLE 8.14

If random variable X_1 is $\chi^2(6)$ and random variable X_2 is $\chi^2(10)$, where X_1 and X_2 are independent, find the probability that

$(0.6) \dfrac{X_2}{X_1} > 4.06$.

SOLUTION

From Statement 8.11(a), we have $\dfrac{X_2/10}{X_1/6}$ is $F(10, 6)$.

This gives $\left(\dfrac{6}{10}\right) \dfrac{X_2}{X_1} = (0.6) \dfrac{X_2}{X_1}$ is $F(10, 6)$.

Thus: $\Pr\left((0.6) \dfrac{X_2}{X_1} > 4.06\right) = \Pr(F(10, 6) \text{ random variable} > 4.06)$

$= 5\%$ (from Appendix Table 6)

$= 0.05$

8.6 INTERVAL ESTIMATION

The purpose of this section is to construct ranges of values within which population parameters are expected to lie with a given probability, based on the results of a random sample.

One of the important differences in the mechanics of setting up interval estimates (as opposed to point estimates) is the fact that we not only need the expectation and variance of the statistic involved, but its distribution also.

We usually refer to a given probability range for an unknown parameter as a 'confidence interval', the degree of confidence depending on the probability that the parameter lies in the particular given interval.

DEFINITION 8.6

A $B\%$ *confidence interval (C.I.)* for some unknown parameter θ is an interval constructed, based on the results of a random sample, so that the probability that θ lies in this interval is $\dfrac{B}{100}$.

The most commonly used is a 95% C.I., although any other degree of confidence can be used, such as 90 or 99%. So, if (a, b) constituted a 95% C.I. for some parameter θ, we have, in probability terms:

$$Pr(a \leqslant \theta \leqslant b) = 0.95$$

The construction of the intervals we consider will be based on the sample values of unbiased estimators for the particular parameters we are interested in.

As regards the interval estimation of a population mean, μ, there are three cases of particular interest. The first depends on the fact that the population being sampled has a known variance. The other two are concerned with samples from populations whose variances are unknown (a far more common case in practice) – one for small samples, the second as a large sample approximation.

Other interval estimates we consider are concerned with proportions, variances and correlation coefficients.

8.6.1 THE MEAN (KNOWN VARIANCE)

Consider a random sample of size n from a Normal population having *known* variance σ^2. We wish to construct a 95% C.I. for the (unknown) population mean, μ.

We know that the statistic $\bar{X} = \dfrac{1}{n} \sum X_i$ has a Normal distribution with mean μ and variance $\dfrac{\sigma^2}{n}$. Even if the parent population is not Normal, the statistic $\bar{X}$ has an approximately Normal distribution for a large enough sample size n (from the Central Limit Theorem).

So, $\bar{X}$ is $N\left(\mu, \dfrac{\sigma^2}{n}\right)$. Therefore $\dfrac{\bar{X} - \mu}{\sigma/\sqrt{n}}$ is $N(0, 1)$ by standardisation.

The technique is to obtain a 95% C.I. for the standardised statistic $Z = \dfrac{\bar{X} - \mu}{\sigma/\sqrt{n}}$, and then, using some algebraic manipulation, transform this to a 95% C.I. for μ.

Now, Z has a symmetric distribution about its mean of 0 and since we obviously require the C.I. to be centrally located we need to find the value of b such that 95% of a standard Normal distribution lies between $-b$ and $+b$. See Fig. 8.6. From Normal theory, $b = 1.96$. In probability terms, $Pr(-1.96 \leqslant Z \leqslant 1.96) = 0.95$.

That is:
$$Pr\left(-1.96 \leqslant \frac{\bar{X} - \mu}{\sigma/\sqrt{n}} \leqslant 1.96\right) = 0.95 \qquad [1]$$

Put another way, $(-1.96, 1.96)$ forms a 95% C.I. for $\dfrac{\bar{X} - \mu}{\sigma/\sqrt{n}}$.

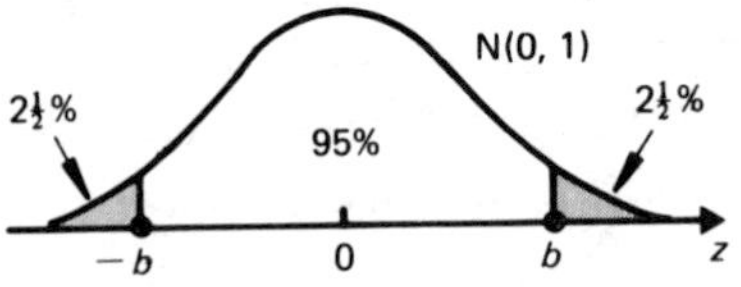

Fig. 8.6

As mentioned earlier, we need to manipulate the inequality in [1] for μ.

From [1]:

$$\Pr\left(-\frac{1.96\sigma}{\sqrt{n}} \leqslant \bar{X} - \mu \leqslant \frac{1.96\sigma}{\sqrt{n}}\right) = 0.95 \quad \left(\text{multiplying through by } \frac{\sigma}{\sqrt{n}}\right)$$

$$\therefore \Pr\left(-\frac{1.96\sigma}{\sqrt{n}} \leqslant \mu - \bar{X} \leqslant \frac{1.96\sigma}{\sqrt{n}}\right) = 0.95 \quad \begin{array}{l}\text{(multiplying through by } -1 \\ \text{and adjusting the inequality} \\ \text{accordingly)}\end{array}$$

By adding $\bar{X}$ to each part of this inequality, we obtain:

$$\Pr\left(\bar{X} - \frac{1.96\sigma}{\sqrt{n}} \leqslant \mu \leqslant \bar{X} + \frac{1.96\sigma}{\sqrt{n}}\right) = 0.95 \qquad [2]$$

which is the required form.

The interval is usually written as:

$$\bar{x} \pm \frac{1.96\sigma}{\sqrt{n}} \quad \text{or} \quad \left(\bar{x} - \frac{1.96\sigma}{\sqrt{n}}, \bar{x} + \frac{1.96\sigma}{\sqrt{n}}\right)$$

Note that we are using $\bar{X}$ in [2], since we can only associate probability with a random variable. However, we often use $\bar{x}$ when describing the limits of an interval since, in practice, these will be given numerically. (Remember, $\bar{X}$ is a random variable and $\bar{x}$ is a number.)

To summarise:

STATEMENT 8.13

If $\bar{x}$ is the mean of a random sample of size n from a Normal population with *known* variance σ^2, a central 95% confidence interval for μ, the population mean, is given by $\bar{x} \pm \dfrac{1.96\sigma}{\sqrt{n}}$.

That is: $$\Pr\left(\bar{X} - \frac{1.96\sigma}{\sqrt{n}} \leqslant \mu \leqslant \bar{X} + \frac{1.96\sigma}{\sqrt{n}}\right) = 0.95$$

Proof   In preceding text.

As an example of the use of the above interval estimate, suppose a random sample of size 12 is taken from a Normal population with variance 124, yielding a sample mean of 51.2. We wish to find a 95% confidence interval for the true population mean μ.

Here, $\bar{x} = 51.2$ and $\sigma^2 = 124$. $\dfrac{1.96\sigma}{\sqrt{n}} = \dfrac{(1.96)\sqrt{124}}{\sqrt{12}} = 6.301$ (3D).

Using the above statement, we have a 95% C.I. for μ as:

$$51.2 \pm 6.301 = (44.90, 57.50) \quad \text{(2D)}$$

Note that the value 1.96 is used in the general format for the interval by virtue of the fact that the range -1.96 to $+1.96$ encloses the central 95% of a $N(0, 1)$ distribution.

In the case of a 99% C.I. for instance, the interval is constructed in exactly the same way, except that the 1.96 in the previous statement needs to be changed to 2.58, since the central 99% of a $N(0, 1)$ distribution is enclosed by the interval -2.58 to $+2.58$. Hence a 99% C.I. would be given by $\bar{x} \pm \dfrac{2.58\sigma}{\sqrt{n}}$.

So, denoting a $N(0, 1)$ variable by Z,

(a) for a 95% C.I., $Z_{2\frac{1}{2}\%} = 1.96$ is multiplied by $\dfrac{\sigma}{\sqrt{n}}$, while

(b) for a 99% C.I., we use $Z_{\frac{1}{2}\%} = 2.58$ to multiply $\dfrac{\sigma}{\sqrt{n}}$. This is shown pictorially in Fig. 8.7.

$$\left(\text{Note} \quad 2\frac{1}{2} = \frac{100 - 95}{2} \quad \text{and} \quad \frac{1}{2} = \frac{100 - 99}{2} \right)$$

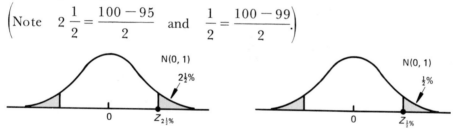

Fig. 8.7

In general, we have the following statement.

STATEMENT 8.14

If $\bar{x}$ is the mean of a random sample of size n from a Normal population with *known* variance σ^2, then *a central B% confidence interval for* μ, the population mean, is given by $\bar{x} \pm z\left(\dfrac{\sigma}{\sqrt{n}}\right)$, where z is the $\left(\dfrac{100 - B}{2}\right)$% point of a $N(0, 1)$ distribution.

That is: $\Pr\left(\bar{X} - z\left(\dfrac{\sigma}{\sqrt{n}}\right) \leqslant \mu \leqslant \bar{X} + z\left(\dfrac{\sigma}{\sqrt{n}}\right)\right) = \dfrac{B}{100}$

Proof   Left as an exercise.

Note that if the population being sampled from is not Normal, we would require n to be large for the above result to be used (via the Central Limit Theorem). It is not practical to give an exact value for n here, but usually $n \geqslant 30$ is considered adequate. However, for a symmetric, near-Normal population we would consider $n \geqslant 10$ as sufficient.

EXAMPLE 8.15

An experiment was carried out in which it was found that the lengths of span of ten particular animals' paws (in cm) were $12.3, 11.8, 11.6,$ $12.6, 13.4, 12.8, 11.1, 12.2, 14.8$ and 13.1 respectively.

Given that the spans are approximately Normally distributed with variance 1.44 cm, construct a 95% C.I. for μ, the mean of the population of spans.

SOLUTION

Here $\Sigma x = 125.7$. Therefore $\bar{x} = \dfrac{125.7}{10} = 12.57$.

Also, since $\sigma^2 = 1.44$, we have $\sigma = 1.2$.

So that a 95% C.I. for μ is:

$$\bar{x} \pm (1.96)\frac{\sigma}{\sqrt{n}} = 12.57 \pm (1.96)\frac{1.2}{\sqrt{10}}$$

$$= 12.57 \pm 0.74 \quad (2D)$$

$$= (11.83, 13.31) \quad cm$$

EXAMPLE 8.16

A random sample of size n is taken from a Normal distribution with unknown mean μ and variance $\sigma^2 = 10$. If $\bar{x}$ is the sample mean, find n (the sample size) so that $\bar{x} \pm 0.5$ is an approximate 95.4% C.I. for μ.

SOLUTION

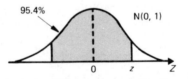

We have that X is $N(\mu, 10)$. Since there is a known variance of 10, using Statement 8.14, a 95.4% C.I. for μ is given by:

$$\bar{x} \pm z\left(\frac{\sigma}{\sqrt{n}}\right) = \bar{x} \pm z\left(\frac{\sqrt{10}}{\sqrt{n}}\right)$$

But z must be the $\left(\dfrac{100 - 95.4}{2}\right)\% = 2.3\%$ point of a $N(0, 1)$ distribution. Hence $z = Z_{2.3\%} = 1.9954$ (from Appendix Table 3).

So that the two intervals $\bar{x} \pm 0.5$ and $\bar{x} \pm (1.9954)\dfrac{\sqrt{10}}{\sqrt{n}}$ must be identical.

Therefore: $0.5 = (1.9954)\dfrac{\sqrt{10}}{\sqrt{n}}$ giving $n = 10\left(\dfrac{1.9954}{0.5}\right)^2$

Therefore $n = 159.26$.

We would take n as 159 here.

8.6.2 THE MEAN (UNKNOWN VARIANCE – SMALL SAMPLES)

We now formulate an interval estimate for the mean μ of a Normal population where the population variance, σ^2, is *not known*.

Given a random sample from a Normal population, we saw (in Example 8.9) that the statistic $\dfrac{\bar{X} - \mu}{S/\sqrt{n - 1}}$ has a $T(n - 1)$ distribution.

We will first construct a central $B\%$ C.I. for the statistic $\dfrac{\bar{X} - \mu}{S/\sqrt{n - 1}}$, then manipulate this for μ.

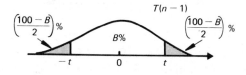

$T(n - 1)$

$\left(\dfrac{100 - B}{2}\right)\%$ $B\%$ $\left(\dfrac{100 - B}{2}\right)\%$

$-t$ 0 t

Fig. 8.8

Fig. 8.8 shows the situation pictorially, where we need to determine the value of t such that the interval $(-t, t)$ encloses the central $B\%$ of a $T(n - 1)$ distribution. Note that since any T distribution is symmetric about its mean of zero, we always get central intervals in the form $(-t, t)$. T tables are easily manipulated to give the value of t such that $(100 - B)\% = P\%$, say, lies outside the interval $(-t, t)$. So that, given B, we can determine t from tables.

Hence:
$$\Pr\left(-t \leqslant \frac{\bar{X} - \mu}{S/\sqrt{n - 1}} \leqslant t\right) = \frac{B}{100}$$

which (using the same algebraic technique as that of Section 8.6.1) easily transforms to:
$$\Pr\left(\bar{X} - t\frac{S}{\sqrt{n - 1}} \leqslant \mu \leqslant \bar{X} + t\frac{S}{\sqrt{n - 1}}\right) = \frac{B}{100}$$

To summarise:

STATEMENT 8.15

If $\bar{x}$ and s^2 are the mean and variance of a random sample of size n drawn from a Normal population with mean μ and variance σ^2 *both unknown*, a central $B\%$ confidence interval for μ is given by

$\bar{x} \pm t \dfrac{s}{\sqrt{n-1}}$, where t is such that the interval $(-t,\ t)$ encloses $B\%$ of a $T(n-1)$ distribution.

That is: $\Pr\left(\bar{X} - t \dfrac{s}{\sqrt{n-1}} \leqslant \mu \leqslant \bar{X} + t \dfrac{s}{\sqrt{n-1}}\right) = \dfrac{B}{100}$

Proof   In preceding text.

Note that the above form is also used for samples from populations that are approximately Normal.

This then is the form to be used to estimate μ when the population variance σ^2 is unknown. For the interval calculation we need to know the values of $\bar{x}$, s and n from the sample, together with t which is found from tables.

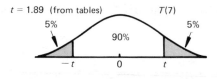

Fig. 8.9

For instance, suppose we require a 90% C.I. based on a random sample of size 8. Here $B = 90$ and $n = 8$. From t tables we require the value of t such that $(-t,\ t)$ encloses the central 90% of a $T(7)$ distribution. See Fig. 8.9. Now, $P = 100 - 90 = 10$. So we find the value of t corresponding to $P = 0.95(95\%)$ and $v = n - 1 = 7$, which from tables is $t = 1.89$.

Thus we have a 90% C.I. as $\bar{x} \pm (1.89)\dfrac{s}{\sqrt{7}}$ for a random sample of size 8 with $\bar{x}$ and s as the sample mean and standard deviation respectively.

EXAMPLE 8.17

Eight fish of a certain species are measured and their lengths are found to be 10.6, 11.2, 10.4, 12.2, 11.3, 10.2, 10.3 and 12.5 in respectively.

Find 95% confidence limits for the mean length of a fish of the species assuming these lengths form a Normal distribution.

Estimate the length which will be exceeded by only one fish in a thousand.

SOLUTION

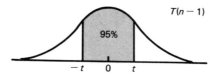

$T(n-1)$

95%

$-t$ 0 t

Here, we have both an unknown mean and variance, so the confidence limits take the form $\bar{x} \pm t \dfrac{s}{\sqrt{n-1}}$ where $(-t, t)$ is the interval of a $T(n-1)$ distribution enclosing the central 95% of the distribution. From t tables (with $v = n-1 = 7$) we have $t = 2.36$. We need to calculate $\bar{x}$ and s.

$$\sum x = 88.7 \qquad \therefore \quad \bar{x} = \frac{88.7}{8} = 11.09 \quad \text{(2D)}$$

$$\sum x^2 = 988.87 \qquad \therefore \quad s^2 = \frac{988.87}{8} - \left(\frac{88.7}{8}\right)^2$$

i.e.
$$s^2 = 0.68 \quad \text{giving} \quad s = 0.82 \quad \text{(2D)}$$

So, the 95% C.I. is:

$$11.09 \pm (2.36) \frac{(0.82)}{\sqrt{7}} = 11.09 \pm 0.73$$

$$= (10.36, 11.82)$$

We now require the length, y say, that is exceeded once in a thousand only. The distribution of lengths is Normal (we are told) but we will have to point estimate the distribution mean μ and variance σ^2.

Now $\hat{\mu} = \bar{x} = 11.09$ and

$$\hat{\sigma}^2 = \frac{ns^2}{n-1} = \frac{8}{7}(0.68) = 0.78 \quad \text{(2D)}$$

So, labelling the population random variable as X, we need to find y such that $\Pr(X > y) = 0.001$, the accompanying figure giving a picture of the situation.

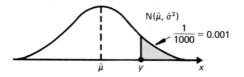

$N(\hat{\mu}, \hat{\sigma}^2)$

$\dfrac{1}{1000} = 0.001$

$\hat{\mu}$ y x

Since X is $N(11.09, 0.78)$, the above probability statement becomes:

$$\Pr\left(Z > \frac{y - 11.09}{\sqrt{0.78}}\right) = 0.001 \quad \text{after standardising.}$$

This gives:

$$1 - \Phi\left(\frac{y - 11.09}{\sqrt{0.78}}\right) = 0.001$$

and hence:

$$\Phi\left(\frac{y - 11.09}{\sqrt{0.78}}\right) = 0.999$$

From tables we obtain:

$$\frac{y - 11.09}{\sqrt{0.78}} = 3.09 \quad \text{(using linear interpolation)}$$

Rearranging, we have:

$$y = (3.09)\sqrt{(0.78)} + 11.09$$

$$= 13.82$$

Therefore, the length that will be exceeded only once in a thousand is 13.82 in.

Although the form of interval discussed in this sub-section can be used for a sample of any size (given the same situation), it is commonly used for small n only.

We deal next with a large sample approximation for just this same situation.

8.6.3 THE MEAN (UNKNOWN VARIANCE – LARGE SAMPLE APPROXIMATION)

It was noted in Section 8.4.1 that for large values of v, the T distribution closely resembles a N(0, 1) distribution. This is easily demonstrated using tables.

Since $Z_{2\frac{1}{2}\%} = 1.96$, we have that $(-1.96, 1.96)$ encloses the central 95% of a N(0, 1) distribution. Now, from column $P = 5\%$ and row $v = 120$ of t tables we obtain $t = 1.98$, That is, the central 95% of a $T(120)$ distribution lies within the interval $(-1.98, 1.98)$. The error in using 1.96 for 1.98 is only 1%.

Even for a T distribution with $v = 40$ having 95% limits of $(-2.02, 2.02)$, the N(0, 1) limits $(-1.96, 1.96)$ are still a good approximation giving an error of only about 2%. See Fig. 8.10.

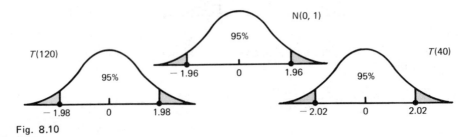

Fig. 8.10

In an estimating capacity, it would be in order to use N(0, 1) as an approximation to $T(\nu)$ for $\nu \geqslant 30$.

STATEMENT 8.16

If $\bar{x}$ and s^2 are the mean and variance of a random sample of size n (large) from a Normal population with mean μ and variance σ^2 *both unknown*, a central $B\%$ confidence interval for μ is given (approximately) by $\bar{x} \pm z\left(\dfrac{s}{\sqrt{n}}\right)$, where z is the $\left(\dfrac{100-B}{2}\right)\%$ point of a N(0, 1) distribution.

That is: $\Pr\left(\bar{X} - z\left(\dfrac{S}{\sqrt{n}}\right) \leqslant \mu \leqslant \bar{X} + z\left(\dfrac{S}{\sqrt{n}}\right)\right) = \dfrac{B}{100}$

Note $n \geqslant 30$ would be considered adequately large for this case.

No proof

As in previous cases, we would also use the above form for a sample from a population that is only approximately Normal.

The above form of confidence interval can be compared directly with that of Statement 8.14, where we had a known population variance σ^2. The only difference between the two forms is that σ (which, of course, is unknown in the above case) has been replaced by statistic S. In other words, we are using S as an estimator for σ.

But why haven't we considered the most efficient unbiased estimator for σ^2, namely $\dfrac{nS^2}{n-1}$, and used $\sqrt{\dfrac{nS^2}{n-1}} = \sqrt{\left(\dfrac{n}{n-1}\right)} S$ as an estimator for σ? This is easy to answer. The given form is meant only as a large sample approximation. But when n is large, $\sqrt{\left(\dfrac{n}{n-1}\right)} S$ approximates to S $\left(\text{since } \sqrt{\dfrac{n}{n-1}} \to 1 \text{ as } n \to \infty\right)$.

So, for a large sample, we can use statistic S as an estimator for σ.

Suppose that a random sample of size 80 yielded $\bar{x} = 124$ and $s = 14.6$.

Using the above statement, we form a 95% C.I. using $\bar{x} \pm z\left(\dfrac{s}{\sqrt{n}}\right)$.

Here, $z = Z_{2\frac{1}{2}\%} = 1.96$ $\left(\text{since } B = 95 \text{ and } \dfrac{100-B}{2} = 2\dfrac{1}{2}\right)$. Therefore a 95% confidence interval is given by:

$$124 \pm (1.96)\dfrac{(14.6)}{\sqrt{80}} = 124 \pm 3.2$$
$$= (120.8, 127.2)$$

EXAMPLE 8.18

One hundred men of the same race are taken at random, and their heights in inches are found to have a mean of 69 and variance 7.

Find 95% and 98% confidence limits for the mean height of the men of the race.

SOLUTION

From the sample, $n = 100$, $\bar{x} = 69$ and $s^2 = 7$.

Since $n = 100$ we can use the large sample approximation to obtain confidence limits for the unknown population mean. That is, we use

$$\bar{x} \pm z\left(\frac{s}{\sqrt{n}}\right).$$

For a 95% C.I., $z = 1.96$ giving the limits as:

$$69 \pm (1.96)\sqrt{\frac{7}{100}} = 69 \pm 0.52$$

For a 98% C.I., reference to Normal tables gives $z = Z_{1\%} = 2.33$ (approximately).

The limits in this case are:

$$69 \pm (2.33)\sqrt{\frac{7}{100}} = 69 \pm 0.62$$

8.6.4 A PROPORTION

We now construct a confidence interval for a population proportion, π, having a certain attribute, based on the result of a random sample of size n of which proportion p are found to have the given attribute.

From Statement 7.11 we know that statistic P is (approximately) $N\left(\pi, \dfrac{\pi(1-\pi)}{n}\right)$ for large n.

Now, it is not possible to manipulate this statement into an interval for π, while the variance involves π itself. To overcome this problem, we use $\dfrac{p(1-p)}{n}$ as an estimate of the variance $\dfrac{\pi(1-\pi)}{n}$. A discussion of the validity of this estimate is outside the scope of the text.

So, we consider P as $N\left(\pi, \dfrac{P(1-P)}{n}\right)$.

Standardising gives $\dfrac{P-\pi}{\sqrt{P(1-P)/n}}$ as $N(0, 1)$, or in terms of probability:

$$\Pr\left(-z \leqslant \frac{P-\pi}{\sqrt{P(1-P)/n}} \leqslant z\right) = \frac{B}{100}$$

which forms a $B\%$ C.I. for $\dfrac{P - \pi}{\sqrt{P(1 - P)/n}}$ where z is such that $(-z, z)$ contains the central $B\%$ of a $N(0, 1)$ distribution.

We then have: $\Pr\left(-z\sqrt{\dfrac{P(1 - P)}{n}} \leqslant P - \pi \leqslant z\sqrt{\dfrac{P(1 - P)}{n}}\right) = \dfrac{B}{100}$

Hence: $\Pr\left(-z\sqrt{\dfrac{P(1 - P)}{n}} \leqslant \pi - P \leqslant z\sqrt{\dfrac{P(1 - P)}{n}}\right) = \dfrac{B}{100}$

giving: $\Pr\left(P - z\sqrt{\dfrac{P(1 - P)}{n}} \leqslant \pi \leqslant P + z\sqrt{\dfrac{P(1 - P)}{n}}\right) = \dfrac{B}{100}$

which, of course, forms a $B\%$ C.I. for π.

To summarise:

STATEMENT 8.17

An approximate $B\%$ confidence interval for a population proportion π (having a certain attribute), based on a random sample of size n (large) yielding proportion p (with the given attribute), is given by:

$$p \pm z\sqrt{\dfrac{p(1 - p)}{n}}$$

where z is the $\left(\dfrac{100 - B}{2}\right)\%$ point of a $N(0, 1)$ distribution.

That is: $\Pr\left(P - z\sqrt{\dfrac{P(1 - P)}{n}} \leqslant \pi \leqslant P + z\sqrt{\dfrac{P(1 - P)}{n}}\right) = \dfrac{B}{100}$

Proof   In preceding text.

Note that the above interval is only approximate since:

(a)   the distribution of P as $N\left(\pi, \dfrac{\pi(1 - \pi)}{n}\right)$ is only a Normal approximation for large n, and

(b) $\dfrac{\pi(1 - \pi)}{n}$ is being estimated by $\dfrac{p(1 - p)}{n}$.

We would usually require $n \geqslant 30$ for a good approximation.

EXAMPLE 8.19

Before a by-election, for which there are two candidates A and B, inquiries are made of 400 voters, chosen at random, and it is found that 208 of them intend to vote for A. Give 95% confidence limits for the percentage of voters favourable to A at the time of the inquiry.

If in fact 55% of the voters were in favour of B, what is the probability that a random sample of 400 voters will contain at least as many in favour of A as there are in favour of B?   (AEB)'64

SOLUTION

We shall work with proportions to answer this question. Here $n = 400$ and the attribute we are measuring is 'an A voter' (we assume that the total population votes for either A or B — i.e., there are no 'don't knows'). Let the true population proportion who intend to vote for A be π.

The sample proportion p is $\dfrac{208}{400} = 0.52$.

A 95% C.I. for π is given by:

$$p \pm (1.96) \sqrt{\frac{p(1-p)}{n}} \quad \text{(using the previous statement)}$$

$$= 0.52 \pm (1.96) \sqrt{\frac{(0.52)(0.48)}{400}}$$

$$= 0.52 \pm 0.05 \quad \text{(2D)}$$

$$= (0.47, 0.57) \quad \text{or} \quad (47\%, 57\%)$$

We are now given π as $45\% = 0.45$ and we are interested in the distribution of P for a sample of 400.

We know that P is $N\left(\pi, \dfrac{\pi(1-\pi)}{n}\right)$ approximately

$$= N\left(0.45, \frac{(0.45)(0.55)}{400}\right) \quad \text{after substituting } \pi = 0.45$$

Notes

(a) We are using variance $\dfrac{\pi(1-\pi)}{n}$ since π is a known quantity here, and

(b) This approximation will be good since $n = 400$ is large.

Since p is the sample vote for A, $1-p$ will be the sample vote for B.

Hence, we require:

$$\Pr(P \geqslant 1-P) = \Pr(P \geqslant \tfrac{1}{2}) \quad \text{(rearranging for } P\text{)}$$

$$= \Pr\left(Z \geqslant \frac{(0.5 - \tfrac{1}{800}) - 0.45}{\sqrt{\dfrac{(0.45)(0.55)}{400}}}\right) \quad \begin{array}{l}\text{(Standardising,}\\ \text{using the}\\ \text{continuity correction)}\end{array}$$

$$= \Pr(Z \geqslant 1.96)$$

$$= 0.025$$

8.6.5 THE VARIANCE

The interval estimates we have considered up to now have involved statistics with either $N(0, 1)$ or $T(\nu)$ distributions, both of which are symmetric about a zero mean, facilitating a certain ease in fitting central confidence intervals.

Here, we have a slightly different case. We wish to find an interval estimate for an unknown population variance, σ^2 say, based on a random sample of size n from a Normal population.

Now, if we are sampling from a Normal distribution, the statistic $\dfrac{nS^2}{\sigma^2}$ has a $\chi^2(n-1)$ distribution.

First, a central $B\%$ C.I. will be found for $\dfrac{nS^2}{\sigma^2}$, then manipulated into a C.I. for σ^2.

Due to the fact that a χ^2 distribution is skewed and does not have a zero mean, each of the limits of an interval enclosing $\dfrac{nS^2}{\sigma^2}$ need to be considered separately. However, both of the limits, x_1 and x_2 say, are found easily from tables.

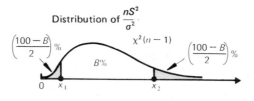

Fig. 8.11

Suppose then that we have the two values x_1 and x_2 that enclose the central $B\%$ of a $\chi^2(n-1)$ distribution. See Fig. 8.11.

We can write:
$$\Pr\left(x_1 \leqslant \frac{nS^2}{\sigma^2} \leqslant x_2\right) = \frac{B}{100}$$

Inverting all parts of the above inequality gives:
$$\Pr\left(\frac{1}{x_2} \leqslant \frac{\sigma^2}{nS^2} \leqslant \frac{1}{x_1}\right) = \frac{B}{100}$$

Finally we obtain:
$$\Pr\left(\frac{nS^2}{x_2} \leqslant \sigma^2 \leqslant \frac{nS^2}{x_1}\right) = \frac{B}{100} \quad \text{(multiplying through by } nS^2)$$

This then is the required $B\%$ C.I. for σ^2, where x_1 and x_2 are as detailed above.

For example, suppose we obtain $s^2 = 10.42$ from a random sample of size 25 from a Normal population, and we require a 95% C.I. for the population variance σ^2.

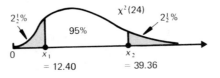

Fig. 8.12

First, we need to find x_1 and x_2, the two limits that enclose the central 95% of a $\chi^2(24)$ distribution. See Fig. 8.12.

Now: $x_1 = \chi^2_{0.025}(24) = 12.40$

and: $x_2 = \chi^2_{0.975}(24) = 39.36$ (both obtained from χ^2 tables)

Also: $ns^2 = 25(10.42) = 260.5$

So, using:

$$\Pr\left(\frac{ns^2}{x_2} \leqslant \sigma^2 \leqslant \frac{ns^2}{x_1}\right) = \frac{B}{100} \quad \text{(with } B = 95 \text{ here)}$$

we have:

$$\Pr\left(\frac{260.5}{39.36} \leqslant \sigma^2 \leqslant \frac{260.5}{12.40}\right) = 0.95$$

giving: $\Pr(6.62 \leqslant \sigma^2 \leqslant 21.00) = 0.95$ (2D)

That is, a 95% C.I. for σ^2 is (6.62, 21.00). To summarise:

STATEMENT 8.18

Let s^2 be the variance of a random sample of size n drawn from a Normal population with (unknown) variance σ^2.

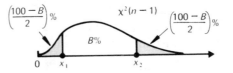

A B% confidence interval for σ^2 is given by $\left(\dfrac{ns^2}{x_2}, \dfrac{ns^2}{x_1}\right)$, where x_1 and x_2 $(x_1 < x_2)$ are the two values that enclose the central B% of a $\chi^2(n-1)$ distribution.

That is: $\Pr\left(\dfrac{ns^2}{x_2} \leqslant \sigma^2 \leqslant \dfrac{ns^2}{x_1}\right) = \dfrac{B}{100}$

Proof   In preceding text.

EXAMPLE 8.20

A random sample of 8 from an approximately Normal distribution gave the values 12, 11, 12, 8, 12, 10, 13, 10 (to the nearest whole number).

Calculate a   90% confidence interval   for the variance of the distribution.

SOLUTION

x	12	11	12	8	12	10	13	10	88
x^2	144	121	144	64	144	100	169	100	986

$$\bar{x} = \frac{88}{8} = 11.0; \ s^2 = \frac{986}{8} - (11.0)^2 = 123.25 - 121.0 = 2.25$$

The limits for the interval take the form:

$$\left(\frac{ns^2}{x_2}, \frac{ns^2}{x_1}\right) \quad \text{with } x_1 = \chi^2_{0.05}(7) = 2.17$$

and:

$$x_2 = \chi^2_{0.95}(7) = 14.07$$

Hence, a 90% C.I. is given by $\left(\dfrac{8(2.25)}{14.07}, \dfrac{8(2.25)}{2.17}\right) = (1.3, 8.3).$

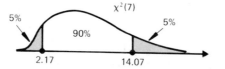

$\chi^2(7)$

5%   90%   5%

2.17   14.07

8.6.6 THE CORRELATION COEFFICIENT

We saw in Section 7.5.3 that if R is the correlation coefficient of a bivariate random sample of size N, drawn from a population with correlation coefficient ρ, then R^* is (approximately) $N\left(\rho^*, \dfrac{1}{n-3}\right)$, where $R^* = \text{Fi}(R); \ \rho^* = \text{Fi}(\rho).$

We would obviously use R as an estimator for parameter ρ, but note in particular that R^* is both an unbiased and consistent estimator for ρ^* since:

(a) $E[R^*] = \rho^*$ and   (b) $\text{Var}[R^*] \to 0$ as $n \to \infty$.

We are concerned here with obtaining an interval estimate of σ based on a sample coefficient R. Using the distribution of R^*, a $B\%$ confidence interval can be set up for ρ^* as follows.

Since R^* is $N\left(\rho^*, \dfrac{1}{n-3}\right)$, we have $\dfrac{R^* - \rho^*}{\sqrt{1/(n-3)}}$ as $N(0, 1)$.

Therefore: $\Pr\left(-z \leqslant \dfrac{R^* - \rho^*}{\sqrt{1/(n-3)}} \leqslant z\right) = \dfrac{B}{100}$

where z is the $\left(\dfrac{100 - B}{2}\right)\%$ point of a $N(0, 1)$ distribution.

From here, the inequality is easily arranged for ρ^* in the usual way to give:

$$\Pr\left(R^* - \frac{z}{\sqrt{n-3}} \leqslant \rho^* \leqslant R^* + \frac{z}{\sqrt{n-3}}\right) = \frac{B}{100}$$

That is, a $B\%$ C.I. for ρ^* is $r^* \pm \dfrac{z}{\sqrt{n-3}}$, where z is the $\left(\dfrac{100-B}{2}\right)\%$ point of a $N(0, 1)$ distribution.

Note that the limits given in the above interval *are for* ρ^*. To 're-transform' them to limits for ρ, we need to use inverse Fisher tables (Appendix Table 7). If, for example, we obtain limits for ρ^* of $(0.448, 1.528)$, using inverse Fisher tables in reverse gives the limits for ρ as $(0.42, 0.91)$.

As a more concrete example, suppose we wish to find 95% limits for ρ, given $r = 0.62$ from a sample of size 60.

We need to work out the two limits for ρ^* using $r^* \pm \dfrac{z}{\sqrt{n-3}}$.

Now: $r^* = \text{Fi}(r) = \text{Fi}(0.62)$ here
                      $= 0.725$ from tables

Also, $z = 1.96$ (the $2\frac{1}{2}\%$ point of a $N(0, 1)$ distribution) and, since $n = 12$, we have:

$$r^* \pm \frac{z}{\sqrt{n-3}} = 0.725 \pm \frac{1.96}{\sqrt{57}}$$

$$= 0.725 \pm 0.260 \quad (3\text{D})$$

$$= (0.465, 0.985) \quad (\text{which are limits for } \rho^*)$$

To obtain limits for ρ (using inverse Fisher tables), we find:

$$\rho^* = 0.465 \quad \text{corresponds to} \quad \rho = 0.434 \quad (2\text{D})$$

$$\rho^* = 0.985 \quad \text{corresponds to} \quad \rho = 0.755 \quad (2\text{D})$$

That is, $(0.43, 0.76)$ is a 95% confidence interval for ρ.

The form of this interval is now summarised.

STATEMENT 8.19

Let r be the correlation coefficient of a bivariate random sample of size n taken from a Normal population having (unknown) correlation coefficient ρ. Then, if $r^* = \text{Fi}(r)$ and $\rho^* = \text{Fi}(\rho)$ (Fisher's transformation), a *central $B\%$ confidence interval for ρ^** is given by $r^* \pm \dfrac{z}{\sqrt{n-3}}$, where z is the $\left(\dfrac{100-B}{2}\right)\%$ point of a $N(0, 1)$ distribution.

That is: $\Pr\left(R^* - \dfrac{z}{\sqrt{n-3}} \leqslant \rho^* \leqslant R^* + \dfrac{z}{\sqrt{n-3}}\right) = \dfrac{B}{100}$

(continued)

Notes (a) $Fi(-r) = -Fi(r)$.

(b) A set of tables for the Fisher transformation are given in Appendix Table 7.

(c) Inverse Fisher tables are given in Appendix Table 7.

Proof In preceding text.

EXAMPLE 8.21

The lengths (x) and breadths (y) of 243 cuckoo eggs were measured (in mm) with the results $\Sigma x = 5442.2$; $\Sigma x^2 = 122\,155.04$; $\Sigma y = 4019.6$; $\Sigma y^2 = 66\,588.92$; $\Sigma xy = 90\,113.83$.

Give a 95% confidence interval for the true correlation coefficient.

SOLUTION

First, we calculate r.

$$S_x^2 = \frac{122\,155.04}{243} - \left(\frac{5442.2}{243}\right)^2 = 1.120 \quad (3D)$$

$$S_y^2 = \frac{66\,588.92}{243} - \left(\frac{4019.6}{243}\right)^2 = 0.405 \quad (3D)$$

$$S_{xy} = \frac{90\,113.83}{243} - \left(\frac{5442.2}{243}\right)\left(\frac{4019.6}{243}\right) = 0.376 \quad (3D)$$

$$\therefore \quad r = \frac{S_{xy}}{\sqrt{S_x^2 S_y^2}} = \frac{0.376}{\sqrt{(1.120)(0.405)}} = 0.558 \quad (3D)$$

Hence: $r^* = Fi(0.558) = 0.630$ (interpolating from tables)

Statement 8.19 gives a 95% C.I. for ρ^* as

$$r^* \pm \frac{z}{\sqrt{n-3}} = 0.630 \pm \frac{1.96}{\sqrt{240}}$$

$$= 0.630 \pm 0.127 = (0.503, 0.757)$$

$$\therefore \quad \text{a 95\% C.I. for } \rho = (0.485, 0.640) \quad \text{(from tables)}$$

8.7 CONTROL CHARTS

8.7.1 INTRODUCTION

Quality Control can be described as the function or collection of duties which must be performed throughout an organisation in order to achieve its quality objectives. These objectives are to produce a quality of product which satisfies the user, customer or some general safety requirement, is as cheap as possible and can be achieved in time to meet delivery requirements.

One important function of Quality Control is the construction and use of *Quality Control charts*. These are based on the results of small samples (often no larger than size 5) which are taken periodically from the items being produced or constructed by some process.

These charts can be used for controlling *variables* (for example, weight of cattle feed pellets; diameters of heads of bolts) or *attributes*.

Charts for variables come in two forms. The first is used to record the means of samples, the second records sample ranges. The most common type of attribute chart records either number or proportion of the sample which is 'not acceptable' or 'defective'.

The basic principles and design of all control charts are similar. Particular statistics (or *control points*) obtained from each sample are plotted on the chart, using the vertical axis for the measurement and the horizontal axis for time. The control points will be means, ranges or numbers or proportion of defectives. Special *control lines* ('warning' and 'action') are also included on the chart. These are horizontal lines drawn such that only a certain percentage of control points will lie outside their limits when the process is under control. They include the following.

(a)   Warning lines. These are drawn on the chart so that, under normal circumstances, only 1 in 20 (5%) of control points will be expected to lie outside them. If breached, they act as a warning that the process may not be under control.

(b)   Action lines. These are drawn on the chart so that, under normal circumstances, only 1 in 500 (0.2%) of control points will be expected to lie outside them. If breached, they act as an indicator that action needs to be taken, since the process is out of control.

Note that both of the above assume a Normal distribution.

A typical control chart for sample means is shown in Fig. 8.12.

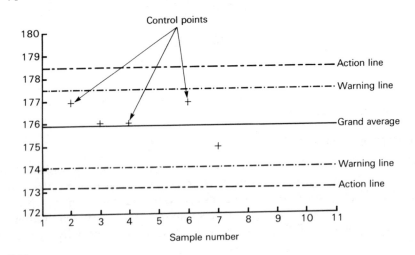

Fig. 8.12

The standard procedure for enabling a chart to become operational is:

(a)   draw an outline chart with horizontal and vertical axes labelled, but without control lines;

(b)   plot at least 10 (or more, up to 20) control points on the chart;

(c)   use the control points to calculate both warning and action control lines (the following subsections describe how this is done for the three main types of chart);

(d)   continue plotting control points.

8.7.2 CONTROL CHARTS FOR MEANS AND RANGES

Given a number of small samples of the same size, it is necessary to obtain both the mean and range for each sample. With regard to a chart for the means, the means themselves are plotted as the control points, the overall (or *grand*) mean being used as a central point on which to base the mean chart control lines. The ranges are used to calculate an average range, which then enables the process standard deviation to be estimated. From here, 95% and 99.8% confidence intervals are constructed, the control lines being drawn at the respective interval ends.

The theory on which range control lines are based is beyond the scope of this book. However, the technique for obtaining them is quite straightforward. The following details the procedure for obtaining control lines for each chart.

STATEMENT 8.20

The procedure for obtaining the control lines for sample *mean* and *range* control charts is given as follows:

(a)   For each sample, calculate:
(i)   total, and thus, mean;
(ii)   range.

(b)   Calculate the grand mean, μ.

(c)   Calculate the average range, $\bar{w}$ and use this to estimate the process standard deviation with $\sigma = k\bar{w}$. (k is found from Appendix Table 8.)

(d)   The mean chart control lines are drawn as:

(i)   Inner 95% (or warning) lines $= \mu \pm 1.96\dfrac{\sigma}{\sqrt{n}}$.

(ii)   Outer 99.8% (or action) lines $= \mu \pm 3.09\dfrac{\sigma}{\sqrt{n}}$.

(e)   The range chart control lines are drawn as:
(i)   Inner line $= r_i\bar{w}$.
(ii)   Outer line $= r_o\bar{w}$.

(Note: r_i and r_o are found from Appendix Table 8.)

Note 1   The mean control lines come in pairs, since it is important to check on both increases and decreases in the mean. They are constructed so that, under controlled circumstances, only 2.5% and 0.1% of the control points will lie outside *each* of the warning and action lines respectively. However, range control lines come singly, since only an increase in range needs to be checked on. The inner and outer lines are constructed so that, under controlled circumstances, only 2.5% and 0.1% respectively of the control points will lie outside them. Thus whereas the two mean warning lines together cut off 1 in 20 control points, the single range warning lines will only cut off 1 in 40. Similarly, the range action line will cut off only 1 in 1000 control points whereas the two mean action lines will together cut off 1 in 500.

Note 2   The information given in parts (c) and (d) above has been covered earlier in the chapter, namely, Section 8.2.3, Statement 8.5, and Section 8.6.1, Statements 8.13 and 8.14.

Note 3   In the above definition (and also when using Appendix Table 8), n is *the size of each sample* and *not* the number of samples taken.

Note 4   The above theory assumes that the population from which the samples have been drawn is (at least approximately) Normal.

EXAMPLE 8.22

A company produces small standard packs of panel pins. Random samples of size 4 are taken regularly from the packaging process line and the number of pins in each sample packet counted. The results for 9 consecutive samples are given below.

Sample no.	1	2	3	4	5	6	7	8	9
	368	386	375	373	378	369	372	369	376
	375	374	380	379	376	371	373	368	377
	368	376	381	369	379	371	372	371	368
	374	375	367	379	386	377	377	376	371

Use these results to draw standard control charts for the mean and range, which should include both control points and control lines.

SOLUTION

The totals, means and ranges for the samples are tabulated below.

Totals	1485	1511	1503	1500	1519
Means	371.25	377.75	375.75	375.00	379.75
Ranges	7	12	14	10	10

Totals	1488	1494	1484	1492
Means	372.00	373.50	371.00	373.00
Ranges	8	5	8	9

The grand mean is $(1485 + 1511 + \ldots + 1492)/36 = 374.3 = \mu$.

The average range is $(7 + 12 + 14 + \ldots + 9)/9 = 9.2$.

$\therefore \quad \sigma = (0.486)(9.2) = 4.47 \ (2D) \quad$ (using Appendix Table 8)

For mean chart:

Inner limits: $\qquad \mu \pm 1.96 \dfrac{\sigma}{\sqrt{n}}$

$$= 374.3 \pm (1.96)(4.47)/2$$

$$= 374.3 \pm 4.4$$

Thus, limits are drawn at 369.9 and 378.7.

Outer limits: $\qquad \mu \pm 3.09 \dfrac{\sigma}{\sqrt{n}}$

$$= 374.3 \pm (3.09)(4.47)/2$$

$$= 374.3 \pm 6.9$$

Thus, limits are drawn at 367.4 and 381.2.

For range chart:

Inner limit $= r_i \times$ mean range $= (1.93)(9.2) = 17.76$

Outer limit $= r_o \times$ mean range $= (2.57)(9.2) = 23.64$

The two charts are shown below (Figs. 8.13 and 8.14).

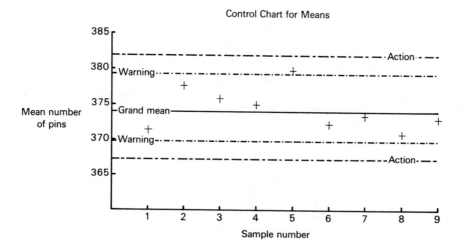

Fig. 8.13

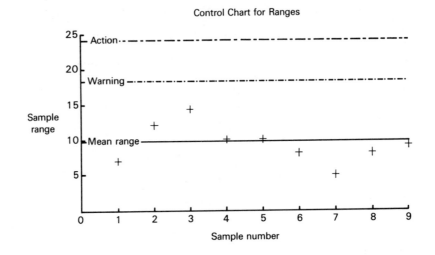

Fig. 8.14

Interpreting Mean and Range Charts In interpreting quality control charts, a mean chart that is 'out of limits' means something quite different to a range chart that is 'out of limits'. Note that the general term 'out of limits' can equally well apply to warning or action limits. Also it should be understood that interpretation of any one chart needs to be related to the particular application or type of production.

In the case of a chart for means, going out of limits will mean the machine or process in question has moved to a new average setting. This could occur as a *sudden change of average* which normally means a sudden cause. For example, unpredictable events such as a machine jamming, a process picking up some impurity, an unauthorised alteration of switches (taps or valves) or a sudden change in a temperature controller. More predictable sudden causes might include the start of a new batch of material having a quality different from usual or a new inexperienced operator. A *gradual drift of an average* indicates a progressive change which might be expected over time with tools wearing or chemicals losing their potency. More unexpected reasons might include loose nuts causing a little more movement in each operation or cycle of a machine, or impurities in batches where each piece (immersed in a chemical bath) decreases the potency of the solution in the bath.

In the case of a chart for ranges, going out of limits normally indicates a worsening of the overall condition of a machine or process. For example, play in bearings or tiredness of human operators, which would normally be a slow drift rather than a sudden change. It is not usual for a sample range to change suddenly, but if it did the change might be due to variable quality of new raw materials or a sudden worsening of the human environment (working-to-rule, bad weather, etc.).

Finally, as part of this particular subsection, it is worth mentioning the two main advantages that control charts for variable (i.e. means and ranges) have over control charts for attributes (i.e. OK v. defective, Good v. Bad taste, etc.). Firstly, measurements of a variable are far more precise and give more information about a process and secondly, variables require much smaller samples than attributes; for example 5 as opposed to 100.

8.7.3 CONTROL CHARTS FOR NUMBER OR PROPORTION DEFECTIVE

The construction of this type of (attribute) control chart follows the same lines as those for the previous mean and range charts. Significant factors involved are as follows.

(a) It has to be decided whether the plots are to be the actual number of defectives or the proportion of sample that is defective. Constructing a chart for the actual number of defectives involves fewer calculations but can only be used if all sample sizes are identical. If this is not so, then a proportion defective chart must be constructed.

(b) Both upper and lower warning and action lines need to be drawn on charts for defectives (unlike charts for ranges, which only have upper lines drawn). One reason that lower lines need to be drawn is that the number or proportion of defectives might suddenly drop owing to new operator inexperience in identifying flaws in products, and this would clearly need to be identified.

The procedure for constructing a chart for *proportion defectives* begins with calculating the overall proportion of defectives, p, say, and then using this to construct control lines based on 95% and 99.8% confidence limits.

That is: $\qquad p \pm (1.96) \sqrt{\dfrac{p(1-p)}{n}} \qquad$ for 95% (warning) lines

and $\qquad p \pm (3.09) \sqrt{\dfrac{p(1-p)}{n}} \qquad$ for 99.8% (action) lines

Note that, since it is unlikely that each sample size will be identical, it is necessary to identify n as the average sample size.

With regard to a chart for *number of defectives*, with d = average number of defectives and n = uniform sample size, multiplying through the latter 95% line by n gives:

$$np \pm (1.96)n \sqrt{\frac{p(1-p)}{n}}$$

$$= d \pm (1.96)\sqrt{d(1-p)} \quad (np = d)$$

$$= d \pm (1.96) \sqrt{d\left(1 - \frac{d}{n}\right)} \quad \text{for 95\% (warning) lines}$$

and correspondingly:

$$d \pm (3.09) \sqrt{d\left(1 - \frac{d}{n}\right)} \quad \text{for 99.8\% (action) lines}$$

To summarise:

STATEMENT 8.21

(a)   The procedure for obtaining the control lines for a *number of defectives* control chart is:

(i)   obtain d = average number of defectives per sample and identify n = uniform sample size;

(ii) calculate:

$$95\% \text{ (warning) lines as } d \pm (1.96) \sqrt{d\left(1 - \frac{d}{n}\right)}$$

$$99.8\% \text{ (action) lines as } d \pm (3.09) \sqrt{d\left(1 - \frac{d}{n}\right)}$$

(b) The procedure for obtaining the control lines for a *proportion of defectives* control chart is:

(i)   calculate p = overall proportion of defectives and

$\qquad\qquad n$ = average sample size;

(ii) calculate:

$$95\% \text{ (warning) lines as } p \pm (1.96) \sqrt{\frac{p(1-p)}{n}}$$

$$99.8\% \text{ (action) lines as } p \pm (3.09) \sqrt{\frac{p(1-p)}{n}}$$

Note 1   Proportion defective control charts are sometimes referred to as 'fraction defective' control charts.

Note 2   Proportions can be easily replaced by percentages in (b) above, the only change necessary is that $(1-p)$ needs to be replaced by $(100-p)$.

EXAMPLE 8.23

Ten successive samples of 50 items each gave the following number of defectives per sample:

$$3, 4, 5, 6, 5, 4, 5, 4, 6 \text{ and } 4$$

Draw up a control chart for number defective, showing 95% and 99.8% control lines, and plot the number of defectives per sample as the control points.

SOLUTION

Now, d = average number of defectives per sample = $46 \div 10 = 4.6$ and $n = 50$.

Thus, 95% (warning) lines are at:

$$4.6 \pm 1.96 \sqrt{4.6\left(1 - \frac{4.6}{50}\right)} = 4.6 \pm 4.0$$

$$= (0.6, 8.6)$$

and 99.8% (action) lines are at:

$$4.6 \pm 3.09 \sqrt{4.6\left(1 - \frac{4.6}{50}\right)} = 4.6 \pm 6.3$$

$$= 10.9 \text{ only}$$

Note that the lower action line is at $4.6 - 6.3 = -1.7$ which of course cannot be drawn on the chart. This discrepancy is due to the fact that a Normal approximation has been used in the formulation of the control limits (confidence intervals) where the distribution of number of defectives is essentially non-Normal. However, this is accepted, and in this type of case it is usual to draw only three control lines as shown in Fig. 8.15.

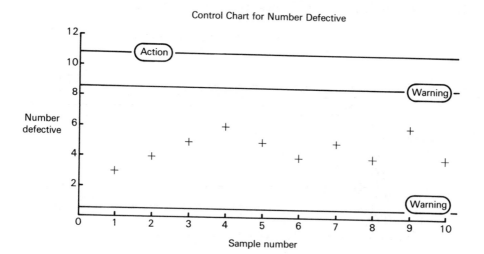

Control Chart for Number Defective

Fig. 8.15

EXAMPLE 8.24

The following information was obtained from seven successive samples taken from a production line.

Number of defectives	21	20	22	18	20	17	24
Size of sample	240	255	210	194	190	210	245
Proportion defective	0.09	0.08	0.10	0.09	0.11	0.08	0.10

Draw a control chart for proportion defective, showing 95% and 99.8% control lines.

SOLUTION

From the above table, total number of defectives = 142 and total number of items sampled = 1544.

Thus, p = proportion defective = 142 ÷ 1544 = 0.09 and n = 1544 ÷ 7 = 220.6 = 221.

Therefore, 95% warning lines are drawn at:

$$0.09 \pm 1.96 \sqrt{\frac{(0.09)(1-0.09)}{221}} = 0.09 \pm 0.04$$

$$= (0.05, 0.13)$$

Also, 99.8% action lines are drawn at:

$$0.09 \pm 3.09 \sqrt{\frac{(0.09)(1-0.09)}{221}} = 0.09 \pm 0.06$$

$$= (0.03, 0.15)$$

The control chart and control lines are shown in Fig. 8.16.

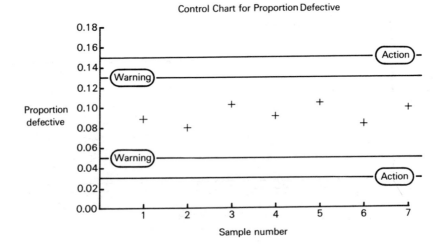

Control Chart for Proportion Defective

Fig. 8.16

The interpretation of control limits with respect to control points lying inside or outside them is exactly the same for attribute charts as it is for variable charts. However, it is worth realising some of the advantages that attribute charts have over variable charts, which include: certain qualities cannot easily be measured, such as taste or appearance; they are generally easier to perform, with less skill needed; they can be used to mix combinations of qualities.

8.8 EXERCISES

SECTION 8.2

1. Obtain the best unbiased estimates of the mean and variance of the population from which the following small sample was drawn: 47, 53, 72, 61, 46, 80, 59, 55, 64 and 43.

2. The marks of 150 students in a certain examination were distributed as follows:

Mark	1–10	11–20	21–30	31–40	41–50
No. of students	1	3	7	18	33

Mark	51–60	61–70	71–80	81–90	91–100
No. of students	46	20	12	8	2

If variable x denotes the mid-points of the classes, make the transformation $a = \dfrac{x - 55.5}{10}$ and calculate: (a) $\bar{a}$; (b) s_a^2. By decoding, calculate further: (c) $\bar{x}$; (d) s_x^2. Hence find the best unbiased estimates of the population mean and variance.

3. Find the best unbiased estimates of the mean and variance of the population from which the following sample is drawn.

x	0	1	2	3	4	5	6
f	2	8	24	83	32	4	1

4. A random sample of 25 children was taken from a large group and their heights measured (to the nearest inch) giving the distribution:

Height	58	60	62	63	64	65	66	67	68	69	70
Frequency	1	1	1	1	3	4	6	4	1	2	1

Estimate the group mean and standard deviation.

5. Obtain the most efficient (best) unbiased estimates of the mean and variance of the distribution from which the following random sample was drawn.

0–	5–	10–	15–	20–	25–	30–	35–40
2	6	12	27	21	14	8	3

6.   A distribution has a known mean μ with its variance, σ^2, unknown. A random sample $X_1, \ldots, X_n$ is drawn. Show that the statistic $\dfrac{1}{n} \sum X_i^2$ is an unbiased estimator for σ^2 only if $\mu = 0$, whereas the statistic $\dfrac{1}{n} \left(\sum X_i^2 - n\mu^2 \right)$ is an unbiased estimator for σ^2 for all values of μ.

     (*Hint:* Write $\mathrm{Var}[X_i] = \mathrm{E}[X_i^2] - \mathrm{E}^2[X_i]$ giving $\mathrm{E}[X_i^2] = \mathrm{Var}[X_i] + \mathrm{E}^2[X_i]$).

7.   Estimate the standard deviations of the populations from which the following samples are drawn: (a)   1, 2, 1, 0, 1, 1, 1, 2, 1, 0;   (b)   0.4, -0.1, 0.1, 0, 0.2, 0.3, 0.2, 0.3;   (c)   42, 84, 71, 83, 38, 70, 65, 51, 38, 80, 29.

8.   Estimate the variance of the population from which the following sample was drawn:

x	2 3 4 5 6 7
f	1 3 4 2 1 1

     by: (a)   using the range; and   (b)   using the best unbiased estimate.

9.   Calculate the percentage error if the variance of the population from which the following sample was drawn is estimated using the range rather than the best unbiased estimator. 8.1, 11.0, 9.4, 9.8, 10.6, 11.0, 9.2, 9.8, 8.9.

10.   Two samples are drawn from the same population as follows:

     Sample 1   0.4, 0.2, 0.2, 0.4, 0.3, 0.3

     Sample 2   0.2, 0.2, 0.1, 0.4, 0.2, 0.3, 0.1

     Find the best unbiased estimates of the population mean and variance.

11.   Estimate (unbiasedly, using best estimates) the mean and variance of the theoretical population from which the following two samples have been drawn:

Sample 1

x	0 1 2 3 4 5
f	4 10 12 8 5 1

Sample 2

x	0 1 2 3 4 5 6
f	5 14 17 16 12 7 1

12.   Calculate the percentage error if a sample variance, s^2, is taken as an approximation for the unbiased estimate of the parent population variance $\dfrac{ns^2}{n-1}$. Find the least sample size such that this error is less than 1%.

13.   Two samples are drawn from a population giving the following results: Sample 1: size 50; variance 18.6;   Sample 2: size 32; variance 20.1. Find estimates of the population variance by: (a)   taking the average of the two variances (ignoring sample sizes);   (b)   taking a weighted (by sample size) average of the two variances;   (c)   using the best estimate. Calculate the percentage errors in using (a) and (b) as approximations for (c).

14. If $\bar{x}$ is the arithmetic mean of the n numbers $x_1, \ldots, x_n$, prove that $\Sigma (x - \bar{x})^2 = \Sigma x^2 - n\bar{x}^2$. Independent random samples are taken as follows:

20 from a population having mean 2 and standard deviation 1

30 from a population having mean 5 and standard deviation 2

40 from a population having mean 4 and standard deviation 3

Find the expectations of the mean and variance of the whole 90, regarded as one sample.

15. Two random samples of sizes n_1 and n_2 respectively are taken from a Normal population with variance σ^2. Denoting the respective sample variances by S_1^2 and S_2^2: (a) show that $\dfrac{n_1 S_1^2 + n_2 S_2^2}{\sigma^2}$ has a χ^2 distribution and give its expectation. (b) Hence show that $\dfrac{n_1 S_1^2 + n_2 S_2^2}{n_1 + n_2 - 2}$ is an unbiased estimator for σ^2.

SECTION 8.4

16. Use t tables to find what percentage of a $T(12)$ distribution lies outside the interval -3.93 to $+3.93$.

17. Find the limits between which the central 90% of a $T(7)$ distribution lies, using t tables.

18. Find the value of P if: (a) $T_{P\%}(4) = 2.78$; (b) $T_{P\%}(40) = 2.42$; (c) $T_{P\%}(10) = -4.14$.

19. Find the value of x if: (a) $T_{5\%}(24) = x$; (b) $T_{95\%}(4) = x$; (c) $T_{87\frac{1}{2}\%}(15) = x$.

20. If a random variable Y is $T(10)$, find: (a) $\Pr(Y > 2.76)$; (b) $\Pr(Y > 3.17)$; (c) $\Pr(2.76 < Y < 3.17)$.

21. If the random variables X_1 and X_2 are independently distributed as $T(15)$ and $T(4)$ respectively, find: (a) $\Pr(X_1 < -1.20)$; (b) $\Pr(X_1 > 2.95)$; (c) $\Pr(-1.20 < X_1 < 2.95)$; (d) $\Pr(X_2 > 2.13)$; (e) $\Pr(X_2 > 3.75)$; (f) $\Pr(X_1 > -1.20$ and $X_2 < 2.13)$; (g) $\Pr(-1.20 < X_1 < 2.95$ and $2.13 < X_2 < 3.75)$.

SECTION 8.5

22. Find the values of the following using F tables:

(a) $F_{5\%}(3, 16)$ (b) $F_{5\%}(12, 25)$ (c) $F_{1\%}(4, 15)$ (d) $F_{0.1\%}(7, 4)$

23. Find the value of P (using F tables) that satisfies each of the following: (a) $F_{P\%}(6, 14) = 3.50$; (b) $F_{P\%}(10, 32) = 2.93$; (c) $F_{P\%}(24, 38) = 1.81$; (d) $F_{P\%}(2, 24) = 5.61$.

SECTION 8.6

24. A distribution is Normal with variance 0.042. A sample of 200 yields a mean of 0.824. Find: (a) 98%; (b) 90%; and (c) 99.73% confidence limits for the true mean of the distribution.

25.   A large school gave a certain examination to all of its students. It is
      known that the scores are distributed Normally with variance 250 and
      also that the boys and girls scores separately are distributed Normally
      with variances 120 and 350 respectively. A random sample of 7 boys
      and 5 girls scores was taken, with the results:

      Boys: 63, 81, 53, 68, 71, 49, 80.   Girls: 49, 73, 58, 32, 91.

      On the basis of these results, give 95% confidence intervals for the
      mean of: (a) the boys scores;   (b) the girls scores;   (c) the overall
      scores in the whole school.

26.   Show that, if $X_1, \ldots, X_n$ are n independent random variables with
      mean μ and variance σ^2, then $\Sigma X_i/n$ has mean μ and variance σ^2/n.

      The diameters of 25 steel rods are measured and found to have a mean
      of 0.980 cm and a standard deviation of 0.015 cm. Assuming this is a
      sample from a normal distribution with the same variance, find 99%
      confidence limits for the population mean.   (O & C)

27.   16 people measured the height of a large building using similar methods
      giving the results (in metres): 33.48, 33.46, 33.49, 33.50, 33.49, 33.51,
      33.48, 33.50, 33.47, 33.48, 33.49, 33.50, 33.47, 33.51, 33.50, 33.48.
      Give a 90% confidence interval for the real height of the building.

28.   A theodolite was set up at a certain distance from the outer wall of a
      factory to measure the angle subtended by the wall. Seven measurements
      were taken of this angle to give (in degrees): 58.6, 58.3, 58.3, 58.3,
      58.2, 58.5, 58.3. Find a central interval within which the measured
      angle is expected to lie with 99% confidence.

29.   The time between certain independent events occurring is approximately
      Normally distributed with mean μ. The exact times of occurrence of
      eight of these particular events (chosen randomly) were measured as:
      1.3, 2.2, 0.3, 1.8, 1.4, 1.2, 0.1, 2.0 seconds (the time origin being 0
      seconds). Obtain an unbiased estimate of μ and construct a 95% con-
      fidence interval for its true value.

30.   A sample of 400 students from a large college shows an average weight
      of 164 lb with a standard deviation of 6.2 lb. What can be said with
      95% confidence about the average weight of students in the college?

31.   A random sample of 100 people in a particular district indicated that
      55% of them were in favour of a particular idea. Find: (a) 95%;
      (b) 99%; and   (c) 99.73% confidence limits for the percentage of all
      of the people in that district who would be in favour of the idea.

32.   A random sample of size 15 from a Normal distribution gives $\bar{x} = 3.2$
      and $s^2 = 4.24$. Determine a 90% confidence interval for the distri-
      bution variance σ^2.

33.   In processing grain in the brewing industry, the percentage extract
      recovered is measured. A particular brewery introduces a new source
      of grain and the percentage extract on 11 separate days is as follows:

      95.2, 93.1, 93.5, 95.9, 94.0, 92.0, 94.4, 93.2, 95.5, 92.3, 95.4

      Regarding the data as a random sample from a normal population,
      calculate:

      (a) a 90% confidence interval for the population variance,
      (b) a 90% confidence interval for the population mean.

The previous source of grain gave daily percentage extract figures which were normally distributed with mean 94.2 and standard deviation 2.5. A high percentage extract is desirable but the brewery manager also requires as little day to day variation as possible. Without further calculation, compare the two sources of grain.   (AEB) '84

34.   A sweet shop sells chocolates which appear at first sight, to be identical. Of a random sample of 80 chocolates, 61 had hard centres and the rest soft centres. Calculate an approximate 99% confidence interval for the proportion of chocolates with hard centres.

The chocolates are in the shape of circular disks and the diameter (cm) of the 19 soft-centred chocolates were:

   2.79 2.63 2.84 2.77 2.81 2.69 2.66 2.71 2.62 2.75 2.77
   2.72 2.81 2.74 2.79 2.77 2.67 2.69 2.75

The mean diameter of the 61 hard-centred chocolates was 2.690 cm. If the diameters of both hard-centred and soft-centred chocolates are known to be normally distributed with standard deviation 0.042 cm, calculate a 95% confidence interval for the mean diameter of

(a) the soft-centred chocolates,
(b) the hard-centred chocolates.

Calculate also an interval within which approximately 95% of the diameters of hard-centred chocolates will lie.

Discuss, briefly, how useful knowledge of the diameter of a chocolate is in determining whether it is hard- or soft-centred.   (AEB) '86

35.   Find 90% confidence limits for the correlation coefficient of the population from which the following sample has been drawn:

x	25 30 35 40 45 50 55
y	82 73 81 85 79 87 95

36.   A bivariate random sample of size 20 yields a correlation coefficient of 0.92. Give a central 98% confidence interval for the correlation coefficient of the population from which the sample was drawn.

37.   Approximately, what size sample would be necessary in order to make the upper limit of a central 95% confidence interval (for a population correlation coefficient) 0.95, if the sample coefficient is 0.7? Given this sample size, what is the lower limit for the interval?

SECTION 8.7

38.   At regular intervals, an inspector took 10 samples of 5 items coming from a production line and weighed each item. The results (in grams) are given as follows.

Sample	1	2	3	4	5	6	7	8	9	10
	177	176	174	175	175	176	170	177	174	175
	176	178	177	178	175	177	175	177	181	175
	177	178	177	180	174	178	178	172	174	174
	178	180	176	172	173	178	177	176	176	175
	175	175	175	176	174	175	173	177	175	173

(a) Draw a control chart for means showing 95% and 99.8% control lines. Plot the control points (sample means) on the graph also.
(b) Draw a control chart for ranges showing standard warning (1 in 40) and action (1 in 1000) limits. Plot the control points on the graph also.

39.   The lengths (in inches) of pins produced by a machine were measured using 10 hourly samples of 5 pins. The results were as follows:

1	2	3	4	5	6	7	8	9	10
0.443	0.445	0.447	0.442	0.444	0.445	0.443	0.442	0.444	0.446
0.445	0.444	0.445	0.443	0.442	0.444	0.443	0.444	0.444	0.441
0.443	0.444	0.443	0.446	0.447	0.444	0.446	0.444	0.446	0.444
0.442	0.445	0.443	0.447	0.444	0.443	0.445	0.448	0.446	0.445
0.446	0.442	0.444	0.444	0.445	0.445	0.447	0.443	0.447	0.445

Draw control charts for both the mean and range, showing on each the standard warning and action lines. Also plot the relevant control points on the graphs and comment on the situation.

40.   12 samples are taken, each consisting of 100 items, and the number of defectives per sample recorded as follows.

Sample No.	1	2	3	4	5	6	7	8	9	10	11	12
Defectives	19	23	22	18	23	15	16	18	19	26	22	19

Draw a control chart for number defective, showing 95% and 99.8% control lines.

41.   A total of 1000 items are inspected using 10 separate samples of slightly differing sizes and exactly 100 defective items are found. Calculate the lower and upper, warning and action limits (95% and 99.8% respectively) for a proportion defective control chart.

42.   What is the main reason for setting up a control chart? In what particular type of situation would a control chart for fraction defective be used, and why?

Ballpoint pens are produced in large quantities by Quickwrite Industries. Samples of 150 pens are taken at regular intervals from the production line, and tested for correct operation. The table below gives the number of defective pens found in each of twelve consecutive samples.

Sample No.	1	2	3	4	5	6	7	8	9	10	11	12
Defectives	14	12	13	13	23	15	17	22	16	19	25	21

Use the first *six* samples to draw a control chart for the fraction defective showing 95% and 99.8% control lines. Plot the control points for all twelve samples. What is the state of the process, and what action, if any, would you recommend?   (AEB) '81

MISCELLANEOUS

43.   A Normal distribution has variance σ^2. A random sample yields the values 12.7, 13.3, 12.9, 13.0, 13.1. Find 90% confidence limits for the mean if: (a) $\sigma^2 = 0.06$;   (b) σ^2 is unknown.

44.   The length of life of a certain component is Normally distributed with unknown mean μ and unknown variance σ^2. Twenty components are tested and their times to failure are recorded. Given that $\Sigma x = 1050$; $\Sigma x^2 = 55\,296$, find a 90% confidence interval for mean μ.

45. As part of an energy conservation project, carried out in Sweden, the temperature of a bedroom and the kitchen in a random sample of 100 homes was automatically monitored by temperature sensors linked to a microcomputer. The following frequency distribution shows the average weekly temperature to the nearest °C.

Temperature of bedroom	Temperature of kitchen				
	19	20	21	22	23
18	4	6	0	0	0
19	0	14	10	1	0
20	0	2	22	12	3
21	0	0	8	9	0
22	0	0	2	4	3

(a) Calculate the product-moment correlation coefficient of these data.

(b) Calculate a 95% confidence interval for ρ, the population correlation coefficient, making any necessary assumptions.

(c) Comment briefly on the project director's suggestion that for future field work, it will only be necessary to monitor the temperature in the kitchen.

(AEB) '85

46. A sample of seven values $x_1, x_2, \ldots, x_7$ is taken from a population whose mean is μ and variance σ^2.

The estimators m_1, m_2 and m_3 of μ are defined by:

$$m_1 = (x_1 + x_2 + x_7)/4,$$
$$m_2 = (2x_1 - x_2 - x_6 + 3x_7)/3,$$
$$m_3 = (2x_1 + 3x_4 + x_7)/6.$$

Determine which of these estimators is unbiased.

The 'best' estimator is the one for which $E[(m_i - \mu)^2]$ is least. Discuss which of m_1, m_2 and m_3 is 'best'.

Author's note: $E[X_i X_j] = E[X_i] E[X_j]$ for X_i, X_j independent.

(Cambridge)

47. 24 out of a total of 72 experiments resulted in a successful outcome. What can be said with 95% confidence about the true success rate?

48. Eight trees of a rare species are found to have girths 2.51, 2.77, 2.27, 2.46, 2.04, 2.23, 2.31 and 2.29 ft. Estimate the mean and variance of the girth of this species of tree and give 95% confidence limits for the mean by using: (a) the unbiased estimate of the population variance; (b) the range of the sample to estimate the standard deviation of the population.

49. A Public Opinion survey has shown that 80 persons in a random sample of 400 persons answered 'yes' to a certain question. Within what limits can it be asserted, with 95% confidence, that the true population percentages of 'yeses' would lie?

50. Seven students were asked to measure the melting point of sodium chloride (common salt) under identical conditions. The results were as follows: 804, 801, 796, 796, 810, 804 and 802 (to the nearest degree Celsius). Using this data, estimate 95% and 99% confidence intervals for: (a) the mean; (b) the variance of the melting point of sodium chloride.

51.   Twenty measurements of the length of the wavelength of a spectral line
       gave a mean of 4122.7 Å with standard deviation 3.1 Å. By taking a
       sample of 20 as: (a) large;   (b) small; calculate a 99% confidence
       interval for the true mean wavelength comparing the two results.

52.   Explain what is meant by *the sampling distribution of an estimator*.

       Given that a random sample of n observations is drawn from a normal
       distribution having mean μ and variance σ^2, specify the sampling dis-
       tribution of the sample mean. State what can be said about this sampling
       distribution when n is large and the parent population is *not* normal.

       Explain what you understand by (a) an unbiased estimator, (b) a con-
       sistent estimator,   (c) the relative efficiency of two estimators of the
       same parameter.

       In order to estimate the mean μ of a population, random observations
       x_1, x_2, x_3 are taken of a random variable X which has variance σ^2.
       Find the relative efficiency of the two estimators $\hat{\mu}_1$ and $\hat{\mu}_2$, where

       $$\hat{\mu}_1 = \frac{x_1 + x_2 + x_3}{3}, \quad \hat{\mu}_2 = \frac{x_1 + 2x_2}{3} \qquad \text{(JMB)}$$

53.   Twelve cotton threads are taken at random from a large batch and the
       breaking strengths are found to be: 7.41, 7.01, 8.34, 8.29, 8.08, 6.60,
       6.59, 7.39, 4.72, 8.65, 8.51 and 8.02 ounces respectively. Assuming
       the breaking strengths of the threads form a Normal distribution, find
       95% confidence limits for the mean breaking strengths of threads in
       the batch, using the unbiased estimate of the population variance.

       Estimate the standard deviation of breaking strength from the range of
       the sample of twelve and comment on the agreement with the value of
       the standard deviation given by the unbiased estimate of the population
       variance.

54.   (a) In tossing a coin 100 times, 64 heads were obtained. Find 95%
       confidence limits for the proportion of heads which would be obtained
       in a large number of tosses of the coin.

       (b) In measuring reaction time, it is estimated that the standard devi-
       ation is 0.5 seconds. How many observations must be taken in order
       to be 99% confident that the error in the estimate of mean reaction
       time will not exceed 0.1 seconds.   (IOS)

55.   Give a 98% confidence interval for a population mean μ based on a
       random sample of n whose mean is $\bar{x}$ and whose standard deviation is
       s, when: (a) n is 100;   (b) n is 6. A paint manufacturer wishes
       to find the average drying time of a new wall paint. He arranges for 10
       test areas of equal size to be painted and records the following drying
       times, in minutes: 81.6, 97.2, 141.3, 112.9, 89.0, 124.5, 148.1, 128.1,
       71.6, 110.7. Construct a 98% confidence interval for the true average
       drying time of the paint. If he uses the mean value from the sample as
       an estimate of the true average drying time, what inference can be made
       about the proportion of drying times that should be within 15 min of
       the true average?

56.   Two paired columns, each consisting of n numbers are to be compiled
       subject to the constraints: (a) that the sum of the numbers in each
       row must add to a_1; and   (b) that the sum of the numbers in the first
       column must add to a_2. How many of the $2n$ numbers can be chosen
       independently? (i.e., how many degrees of freedom are available in
       choosing the $2n$ numbers?).

57. A random sample X_1, X_2, . . ., X_8 is to be taken from a population with mean μ and variance σ^2. The following statistics are to be taken for μ:

$$T_1 = \frac{X_1 + X_2 + X_3}{3}; \quad T_2 = \frac{2X_2 - X_4 + 4X_5 + X_8}{6}; \quad T_3 = \frac{X_8 - X_1}{8};$$

$T_4 = X_5$.

(a) Which of the above are unbiased? (b) Which of the above is most efficient?

58. Describe how the Normal distribution may be used as an approximation to the binomial distribution.

A random sample of 20 children in a large school were asked a question and twelve answered it correctly. Estimate the proportion of children in the school who would answer correctly and the standard error of this estimate. Calculate a 95% confidence interval for this proportion. (IOS)

59. A machine is making dimensioned parts, and a random sample of ten is found to have the following lengths (in inches):

1.504, 1.496, 1.492, 1.501, 1.503, 1.505, 1.495, 1.500, 1.493, 1.501

Estimate the variance of the population: (a) from the sum of squares; and (b) by using the range. By means of the t distribution, give 95% confidence limits for the mean of the population.

60. An athlete returns the following times, in seconds, for 220 yards in fifteen races: 23.0, 22.4, 23.5, 22.6, 22.0, 24.1, 23.7, 23.1, 22.2, 23.2, 22.4, 22.5, 24.2, 23.6, 23.3. Assuming his times form a Normal distribution and regarding the above as a random sample of 15 from this Normal distribution, estimate the mean and variance. Use these estimates to obtain the probability that, on a given occasion, the athlete will run 220 yards in less than 22.05 seconds. Comment on the validity of this procedure.

61. Two samples of size n_1 and n_2 respectively are drawn from Normal populations with known variances σ_1^2 and σ_2^2 respectively. Show that the statistic $\dfrac{\hat{\sigma}_1^2/\sigma_1^2}{\hat{\sigma}_2^2/\sigma_2^2}$ has an $F(n_1 - 1, n_2 - 1)$ distribution, where $\hat{\sigma}_1^2$ and $\hat{\sigma}_2^2$ are unbiased estimators for σ_1^2 and σ_2^2 respectively.

62. If X_1 is $N(\mu, \sigma^2)$ and X_2 is $\chi^2(\nu)$, show that:

(a) the random variable $\dfrac{\sqrt{\nu}}{\sigma} \dfrac{(X_1 - \mu)}{\sqrt{X_2}}$ has a $T(\nu)$ distribution;

(b) if $\nu = 9$ and $\sigma^2 = 100$, then $\Pr(X_1 - \mu > (9.4)\sqrt{X_2}) = 0.01$.

63. The two independent random variables X_1 and X_2 are distributed as $\chi^2(10)$ and $\chi^2(8)$ respectively. If $\Pr\left(\dfrac{X_1}{X_2} > \dfrac{5a}{4}\right) = 0.05$, find the value of a.

64. The independent random variables X and Y are distributed as $N(100, 36)$ and $N(150, 25)$ respectively. If the random variables U and V are

defined as $U = \dfrac{(X-100)^2}{36} + \dfrac{(Y-150)^2}{25}$ and $V = \dfrac{(X-100)^2}{36}$, find

$\Pr\left(\dfrac{U}{V} < 399\right)$.

65.   From each of two populations, the first having mean μ and variance σ^2, the second having mean $\mu/2$ and variance $b\sigma^2$ $(b>0)$, a random sample of size n is taken. Denoting the respective sample means by $\bar{X}_1$ and $\bar{X}_2$, two unbiased estimators for μ are defined by: $T_1 = 3\bar{X}_1 - a\bar{X}_2$ and $T_2 = \frac{1}{4}(3\bar{X}_1 + 2\bar{X}_2)$.

(a) Find the value of a.   (b) Show that $\mathrm{Var}[T_1] = \dfrac{\sigma^2}{n}(9 + 16b)$.

(c) Prove that T_2 is more efficient than T_1 for all valid values of b.

66.   Two samples of size n_1 and n_2 respectively are drawn from an infinite population having mean μ and variance σ^2. The two sample means are denoted by $\bar{X}_1$ and $\bar{X}_2$ respectively. Two estimators are defined as follows: $T_1 = \dfrac{n_1\bar{X}_1 + n_2\bar{X}_2}{n_1 + n_2}$ and $T_2 = \frac{1}{2}(\bar{X}_1 + \bar{X}_2)$.

(a) Show that, as n_1, $n_2 \to \infty$, $\mathrm{Var}[T_1] \to 0$ and hence T_1 can be thought of as a consistent pooled estimator for μ. (We know that $E[T_1] = \mu$.)

(b) (i) Show that T_2 is also an unbiased estimator for μ.   (ii) Prove that $\mathrm{Var}[T_2] = \dfrac{\sigma^2}{4}\left(\dfrac{1}{n_1} + \dfrac{1}{n_2}\right)$ and hence that T_2 is also a consistent estimator for μ.

(c) Show that $\dfrac{\mathrm{Var}[T_1]}{\mathrm{Var}[T_2]} = \dfrac{4n_1 n_2}{(n_1 + n_2)^2}$ and thus, in general, $\mathrm{Var}[T_1] < \mathrm{Var}[T_2]$.

(d) What condition on the sample sizes is necessary so that T_1 and T_2 have the same efficiency?

67.   Chocolate bars with a nominal net mass of 95.0 grams are produced on a continuous production line by a large company. Their quality control consists of taking a random sample of 5 bars every half hour and noting their masses. The masses of the bars in ten samples at the beginning of the production run are given below.

Sample	1	2	3	4	5	6	7	8	9	10
	97.4	91.6	93.3	94.6	91.9	96.3	96.0	95.5	95.4	93.7
	96.5	94.3	97.3	95.0	94.6	94.1	94.7	95.2	97.4	94.8
Masses	92.7	93.2	97.7	96.8	93.8	96.2	93.2	94.7	94.4	93.8
	94.3	95.9	97.6	95.9	94.7	96.5	94.2	96.3	96.6	96.4
	96.0	94.6	94.7	97.7	94.9	93.5	94.6	93.9	94.4	95.7

Compute the mean and range for each of these ten samples. Use these results to draw a control chart for the mean, showing 95% and 99.8% control lines and the control point corresponding to each sample.

More recently it has been found that 1 in 10 of the means fall outside the upper 95% control line. If the variance of the process is unchanged, find the shift in the mean.   (AEB) '80

68.   Reels of wire are wound automatically from a continuous source of
        wire. After each reel is wound the wire is cut and a new reel started.
        The aim is to wind 100 m on to each reel. The length of wire on twelve
        samples each of five reels was measured. The mean and range of each
        of the samples are given below.

Sample	1	2	3	4	5	6
Mean (m)	99.8	101.4	100.4	101.9	101.2	100.6
Range (m)	3.6	2.4	1.1	3.2	2.3	2.5

Sample	7	8	9	10	11	12
Mean (m)	101.7	100.9	100.2	101.4	101.5	101.0
Range (m)	1.7	0.9	1.9	2.2	2.3	1.6

Assuming a normal distribution, use the information to estimate the
standard deviation of the process and to calculate 95% and 99.8%
control lines for the sample means. Draw a control chart for means and
plot the twelve points.

Draw a control chart for ranges showing the upper 'action' and 'warning'
limits. Plot the twelve points and comment on the current state of the
process. If the next sample measured 97.3, 100.6, 103.3, 100.1, 100.9 m,
what action would you recommend and why?

Reels containing less than 95 m of wire are unacceptable to customers
and reels containing more than 105 m make the process uneconomic.
Comment on the ability of the process to produce reels which lie
consistently within these limits.   (AEB) '84

69.   A company manufactures a drug with a nominal potency of 5.0 mg cm^{-3}.
        For prescribing purposes, it is important that the mean potency should
        be accurate and the standard deviation low. To check this, ten samples
        each of size four were taken at regular intervals during a particular day.
        The potency was measured with the following results:

Sample	1	2	3	4	5	6	7	8	9	10
	4.97	4.98	5.13	5.03	5.19	5.13	5.16	5.11	5.07	5.11
	5.09	5.15	5.05	5.18	5.12	4.96	5.15	5.07	5.11	5.19
	5.08	5.08	5.12	5.06	5.10	5.02	4.97	5.09	5.01	5.13
	5.06	4.99	5.11	5.05	5.04	5.09	5.09	5.08	4.96	5.17

Estimate the standard deviation of the process, making clear the method
you have used.

Assuming a normal distribution, draw a control chart for means showing
95% and 99.8% control lines. Plot the sample means.

Draw a control chart for ranges showing upper warning and action
limits and plot the ranges. Comment on the current state of the process.

To satisfy the customer's requirements, the company must be able to
guarantee that at least 99.8% of potency measurements lie within
0.1 mg cm^{-3} of the nominal value. State, giving a reason, whether this
production process could meet this requirement.   (AEB) '86

70.   If r random samples of respective sizes $n_1, n_2, \ldots, n_r$ are taken from a population with mean μ and variance σ^2, where $\bar{X}_i$ and S_i^2 denote the mean and variance of sample i, prove that the following estimators are unbiased:

(a) $\hat{\mu} = \left(\sum_{i=1}^{r} n_i \bar{X}_i \right) \Big/ \left(\sum_{i=1}^{r} n_i \right)$ (b) $\hat{\sigma}^2 = \left(\sum_{i=1}^{r} n_i S_i^2 \right) \Big/ \left(\left(\sum_{i=1}^{r} n_i \right) - r \right)$.

71.   A random sample of size n is taken from a Normal population with known variance σ^2. If $\bar{X}$ represents the sample mean and z is the

$$\Pr\left(\bar{X} - z \frac{\sigma}{\sqrt{n}} \leqslant \mu \leqslant \bar{X} + z \frac{\sigma}{\sqrt{n}} \right) = \frac{P}{100}$$

where μ is the mean of the Normal population being sampled.

$\left(\dfrac{100 - P}{2} \right)$% point of a $N(0, 1)$ distribution, show that:

72.   Two random samples $X_1, X_2, \ldots, X_{n_1}$ and $Y_1, Y_2, \ldots, Y_{n_2}$ are drawn from a Normal population with mean μ and variance σ^2. Show

that the statistic $\dfrac{\bar{X} - \bar{Y}}{\sigma \sqrt{\dfrac{1}{n_1} + \dfrac{1}{n_2}}}$ has a standard Normal distribution,

where $\bar{X}$ and $\bar{Y}$ are the respective sample means. If S_1^2 and S_2^2 are

the respective sample variances, show that $\dfrac{n_1 S_1^2}{\sigma^2} + \dfrac{n_2 S_2^2}{\sigma^2}$ has a

$\chi^2(n_1 + n_2 - 2)$ distribution.   (*Hint:* Use Statements 7.8 and 7.9).
Hence, using Statement 8.10(2), show that the statistic

$$\frac{\bar{X} - \bar{Y}}{\sqrt{\left(\dfrac{n_1 S_1^2 + n^2 S_2^2}{n_1 + n_2 - 2} \right) \sqrt{\dfrac{1}{n_1} + \dfrac{1}{n_2}}}}$$

has a $T(n_1 + n_2 - 2)$ distribution.

Chapter 9

Statistical Tests 1

9.1 INTRODUCTION

This chapter is only part of a more general topic known as 'Statistical Inference'. It is concerned with making a certain assumption about a theoretical population and, using the results of one or more random samples drawn from the population, testing the truth or otherwise of the assumption.

The knowledge of statistical measures from samples and the distributions of certain important sample statistics is a necessary prerequisite for an understanding of the theory behind many of the concepts involved.

9.2 FORMULATION OF THE CONCEPTS

9.2.1 A STATISTICAL TEST

Suppose over a period of time a population was found to be Normally distributed with mean μ and variance σ^2 and now, for reasons of our own, we suspect that the mean has changed in some way. We would like to formulate a test in order to confirm or deny our suspicions.

To do this we draw a random sample $X_1, X_2, \ldots, X_n$ and calculate $\bar{x}$ for the realised values $x_1, x_2, \ldots, x_n$.

A very simple test would be as follows. Measure the difference between $\bar{x}$ and μ and

(a) if the difference is large, we will decide that the mean of the population is no longer μ,

(b) if the difference is small, we will accept that the population mean is still μ.

Clearly there are problems associated with this test. First, we have not taken account of the variance σ^2. If σ^2 were very small compared with the actual population mean (and it really *had* changed), the difference between μ and $\bar{x}$ would be expected to be small and (using (b) above) we could make the wrong decision very easily. Similarly, if σ^2 were large compared with the actual population mean (and the mean *had not* changed), the difference between μ and $\bar{x}$ would be expected to be large and (using (a) above) we could again very easily make the wrong decision.

Hence we need to take account of σ^2 in our test.

The second problem is that measuring differences as 'small' or 'large' is not necessarily easy. What one person might think of as large, another might consider small. We clearly need to define the idea of small and large more precisely.

To overcome both these difficulties we proceed as follows.

Let us assume that the population mean has *not* changed. That is, it *is* equal to μ. We will consider the distribution of the statistic $\bar{X}$ *under this assumption.*

Since the random sample $X_1, X_2, \ldots, X_n$ is assumed to be drawn from $N(\mu, \sigma^2)$, we have that $\bar{X}$ is $N\left(\mu, \dfrac{\sigma^2}{n}\right)$.

Hence:
$$\frac{\bar{X} - \mu}{\sigma/\sqrt{n}} = Z, \quad \text{say}, \quad \text{is } N(0, 1)$$

So that, given the realised values of the sample $x_1, x_2, \ldots, x_n$, we can calculate the quantity $z = \dfrac{\bar{x} - \mu}{\sigma/\sqrt{n}}$ (since we will know the numerical values of $\bar{x}, \mu, \sigma$ and n).

Now, a value of z which is 'close' to the standard Normal mean of zero would be compatible with the assumption that the true population mean was, in fact, μ. It is only when the value of z differs from 0 'significantly' that we would question (in fact reject) the above assumption.

We now need to select a set of values for z (far enough from 0) that tell us to reject the original assumption. These values are known as the *critical region* for the test. They are usually chosen so that the probability of Z falling within them is just $0.05 = 5\%$. See Fig. 9.1. Note that all other values of z will fall in what is known as the *acceptance region*, since these are the values that tell us to accept (or, alternatively, not to reject) the assumption that the true mean is in fact μ.

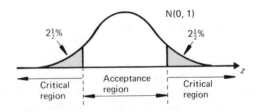

Fig. 9.1

From Normal tables, we find that the $2\frac{1}{2}\%$ point of a standard Normal distribution is 1.96. See Fig. 9.2. Note that we have split the critical region up into two equal portions, the probability that Z lies in either being just $2\frac{1}{2}\%$. Clearly this is necessary since a deviation of z from

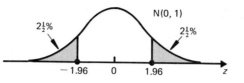

Fig. 9.2

zero can be in either direction. That is, the true mean of the population could be either less than or greater than μ on rejection of the original assumption that it *is* μ.

The procedure then to test whether the population mean is μ, consists of:

(a)   assuming that the mean is μ,

(b)   taking a random sample of size n from the population under consideration,

(c)   calculating the quantity $z = \dfrac{\bar{x} - \mu}{\sigma/\sqrt{n}}$ using the sample results, and

(d)   if $z < -1.96$ or $z > 1.96$ we reject the hypothesis in (a). Otherwise we accept it.

On rejection, we would say *there is evidence at the 5% level* that the population mean is not μ.

$Z = \dfrac{\bar{X} - \mu}{\sigma/\sqrt{n}}$ in this situation is called a *test statistic* and a realised value

lying in the critical region is said to be *significant at the 5% level*.

EXAMPLE 9.1

A machine produces components whose lengths should be distributed Normally with mean 0.16 cm and standard deviation 0.012 cm. A sample of 100 is drawn and the mean computed as 0.157 cm. Is there evidence at the 5% level that the mean of component lengths has changed?

SOLUTION

Here $\mu = 0.16$, $\sigma = 0.012$, $\bar{x} = 0.157$ and $n = 100$.

Our assumption in this case is that the component lengths *are* distributed as $N(0.16, (0.012)^2)$. We are looking for evidence that will cause us to reject this assumption (with respect to the mean), and if we don't find it we will accept the assumption as being correct. The

test statistic is $Z = \dfrac{\bar{X} - \mu}{\sigma/\sqrt{n}}$ and its realised value, here, is:

$$z = \frac{0.157 - 0.160}{0.012/\sqrt{100}} = -2.5$$

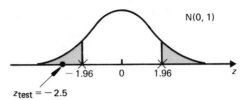

$z_{test} = -2.5$

The above test value lies outside the range $(-1.96, 1.96)$ and, hence, is significant at the 5% level (see the accompanying diagram).

Therefore the original assumption can be rejected and we have shown that there is evidence at the 5% level that the mean of the component lengths *has* changed.

Note that, in the above example, the difference between μ (the assumed population mean) and $\bar{x}$ (the sample mean) is very small, but due to the fact that σ is (comparatively) even smaller, this small difference has been shown to be significant.

EXAMPLE 9.2

Test at the 5% level whether a sample of size 25 with mean 42 could have been drawn from a Normal population with mean 40 and variance 20.

SOLUTION

We need to recognise here that the problem is essentially the same as the one in the previous example. That is, we are testing whether a sample mean is compatible with a given population mean.

As before, we assume that the sample *is* drawn from $N(40, 20)$.

The realised value of the test statistic is:

$$z = \frac{\bar{x} - \mu}{\sigma/\sqrt{n}} = \frac{42 - 40}{\sqrt{20/25}} = 2.24 \quad (2D)$$

This test value is significant, since $2.24 > 1.96$.

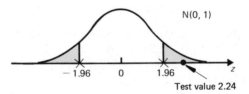

Test value 2.24

Hence we can reject the assumption $\mu = 40$ and state that there is evidence at the 5% level that the sample *does not* come from the given distribution.

9.2.2 SOME NOTATION AND DEFINITIONS

We now formally define some of the most important basic concepts involved in all types of statistical tests.

(a)   A *statistical hypothesis* is an assumption about the value of a parameter of some distribution under consideration. (In the case discussed in the previous sub-section the parameter involved was the mean.)

So far, we have considered two separate hypotheses concerning the mean of a Normal distribution. The first was that the mean *was* equal to a given value, μ, and the second was that the mean *was not* equal to μ.

(b)   A *null hypothesis* is a statistical hypothesis which can, in some way, be tested. Often it will be just that hypothesis that we are 'suspicious about' and wish to disprove. Of the two hypotheses mentioned in (a) above, the one that we chose as the null hypothesis (i.e., the one we assumed was true and wished to disprove) was the one that we could easily test. It would clearly be very difficult to assume that the mean was not μ and attempt to test this.

The symbol conveniently used for the null hypothesis is H_0, and we would write 'H_0: mean = μ' and read or say 'the null hypothesis is that the mean has value μ'.

(c)   An *alternative hypothesis* is that hypothesis that is automatically accepted if the null hypothesis is rejected. Of the two hypotheses mentioned in (a) above, the alternative hypothesis would be that the mean was not equal to μ.   The symbol used for this particular hypothesis is H_1.   Hence, in our test example so far considered, we have been testing H_0; mean = μ versus H_1; mean $\neq \mu$.

There are two possible types of alternative hypothesis, *one-tailed* and *two-tailed*. A two-tailed type is the alternative hypothesis that considers *any change* in a parameter (whether it be an increase or a decrease). The type of example we have been considering so far has been of this kind of alternative hypothesis. A one-tailed type looks *strictly for an increase* or, alternatively, *strictly for a decrease*.

For instance, suppose we took a random sample from a Normal distribution with a past mean of 15 and wished to test whether or not the mean had increased. Then the test might be $H_0 : \mu = 15$ versus $H_1 : \mu > 15$.

However, if we consider only values of μ as detailed in the above hypotheses, we are clearly ignoring the possibility $\mu < 15$ which, of course, can occur. (It can be argued that these values are irrelevant since we are really only concerned with an increase in μ over 15). But if the mean really was less than 15, we would obviously not wish to accept that $\mu > 15$. The only way this case can be incorporated into the above two hypotheses would be to have $H_0 : \mu \leqslant 15$ versus $H_1 : \mu > 15$ where, for the purposes of the test statistic, we would use the more specific assumption $\mu = 15$.

We will use this particular form of two opposing hypotheses for all the tests we consider in the text.

As another example, if we took a random sample from a supposed distribution $N(100, 20)$ and wished to test whether the mean, μ, had decreased, we would use $H_0 : \mu \geqslant 100$ versus $H_1 : \mu < 100$.

(d) A *test* of a null hypothesis is simply a rule, based on the results of a random sample, whereby we decide whether to accept or reject the particular null hypothesis under question. Tests are either *one- or two-tailed* depending on whether the alternative hypothesis is one- or two-tailed. So far we have conducted only two-tailed tests.

(e) Corresponding to every test, we need to set up a *critical region*. All the statistical tests that we will consider are based (under the assumption of some null hypothesis) on the distribution of a sample test statistic. In our examples so far this has been $Z = \dfrac{\bar{X} - \mu}{\sigma/\sqrt{n}}$, which has a $N(0, 1)$ distribution under H_0. The possible values that Z can take form a sample set. The critical region is just *the subset of this sample set that leads us to reject the null hypothesis.*

As noted in (c), in the case of a one-tailed test, H_0 considers more than one value of the parameter under test, i.e., $H_0 : \mu \leqslant \mu_0$ or $H_0 : \mu \geqslant \mu_0$ (where μ_0 is a specified value of μ).

Since we must have only a single value for μ (under H_0) in considering the distribution of a statistic, we always take that value of μ *closest to* the opposing H_1 values of μ (in the above two H_0 examples, this value is $\mu = \mu_0$). Thus, if we are testing $H_0 : \mu \leqslant 20$ versus $H_1 : \mu > 20$, we would use value $\mu = 20$ for the H_0 value of μ in the test statistic. If we are testing $H_0 : \mu \geqslant 12$ versus $H_1 : \mu < 12$, we would use value $\mu = 12$ for the H_0 value of μ in the test statistic.

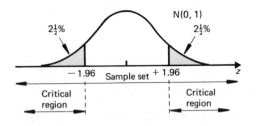

Fig. 9.3

In Fig. 9.3, we show the whole sample set for the test statistic (together with that subset of it that comprises the critical region) of the test detailed in Section 9.2.1, and standard (two-tailed) test for detecting any change in a Normal population mean.

A critical region will be *one- or two-tailed* depending on whether the test is one- or two-tailed. Fig. 9.3 shows a two-tailed critical region and since we were testing at the 5% level, $2\frac{1}{2}\%$ has been allocated to each tail. If we are dealing with a one-tailed test at the 5% level, we would choose the critical region at one particular end of the distribution; the end chosen depending on whether we were testing for an increase or a decrease in the respective population parameter. The whole 5% would then be allocated to this critical region.

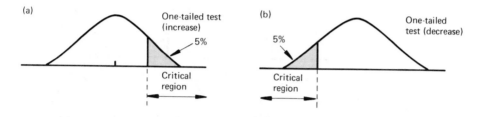

Fig. 9.4

Fig. 9.4(a) shows the case for a one-tailed increase and Fig. 9.4(b) for a one-tailed decrease at the 5% level.

(f)   The *critical value (or values)* of a test statistic is that value (or values) that lies on the boundary (or boundaries) of the critical region.

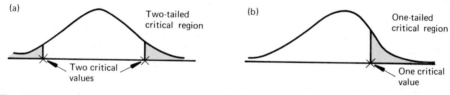

Fig. 9.5

Clearly, for a two-tailed critical region we will have two critical values as shown in Fig. 9.5(a) and for a one-tailed critical region only one critical value as shown in Fig. 9.5(b).

For information purposes, Fig. 9.6 gives the critical values for a standard Normally distributed test statistic at the most commonly used significance levels.

(g)   The *significance level of a test* is just the probability that the test statistic lies in the critical region under H_0. The most commonly used levels are 5% and 1%, although others (namely 10% and $\frac{1}{2}\%$) are sometimes used.

If the result of a test at the 5% level causes us to reject H_0, we say that the value of the test statistic is *significant*. If we reject H_0 at the 1% level however, we often use the term *highly significant* to describe the value of the test statistic.

(h)   A *decision rule* for a statistical test is a model giving values of a sample statistic (not necessarily the test statistic) that will lead to either

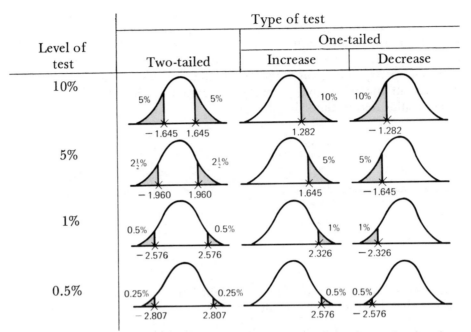

Fig. 9.6   Critical values for a standard Normal distribution. All of the above values have been obtained from standard Normal tables (Appendix Table 3) and the student should verify them.

Note   Percentage areas are not to scale.

acceptance or rejection of the stated null hypothesis. Here, we are not necessarily concerned with physically carrying out a test, rather with setting up a test in a situation that can be repeated.

For instance, in Example 9.2 we were testing at the 5% level whether a sample of size 25 could be considered as having been drawn from a Normal population with mean 40, variance 20.

The test statistic here is of course $\dfrac{\bar{X} - \mu}{\sigma/\sqrt{n}}$ with critical values -1.96 and $+1.96$ (the test is two-tailed). The values $\mu = 40$, $\sigma = \sqrt{20}$ and $n = 25$ are given.

We need to find the values of $\bar{X}$ that will lead to acceptance or rejection of $H_0: \mu = 40$. It is first necessary to find the two values of $\bar{X}$ that correspond to the critical values of -1.96 and $+1.96$.

We can find these from the two identities:

(a) $\dfrac{\bar{X} - 40}{\sqrt{20}/\sqrt{25}} = -1.96$ and   (b) $\dfrac{\bar{X} - 40}{\sqrt{20}/\sqrt{25}} = +1.96$

i.e.: $\bar{X} = \dfrac{(-1.96)\sqrt{20}}{\sqrt{25}} + 40$ and $\bar{X} = \dfrac{(1.96)\sqrt{20}}{\sqrt{25}} + 40$

giving $\bar{X} = 38.25$ and $\bar{X} = 41.75$.

Thus the decision rule is:

(a)   accept H_0 if $38.25 < \bar{X} < 41.75$,

(b)   reject H_0 otherwise.

EXAMPLE 9.3

It is known from experience that the weight of a certain species of fish (that is difficult to identify) is distributed approximately Normally with mean   9.72 oz and standard deviation   1.40 oz. Design a decision rule for testing at the   1% level   whether a sample of:

(a)   25,

(b)   36   alleged members of the species actually belong to it, as opposed to belonging to a less heavy species.

SOLUTION

This is a one-tailed test since we are looking for a significant decrease in order to reject a true population mean of 9.72 oz. That is, we test $H_0 : \mu \geqslant 9.72$ versus $H_1 : \mu < 9.72$. At a 1% level (one-tailed) the test statistic $\dfrac{\bar{X} - \mu}{\sigma/\sqrt{n}}$ has critical value -2.326.

For (a) we have $\mu = 9.72$, $\sigma = 1.40$ and $n = 25$ and we need to find $\bar{X}$ corresponding to the critical value -2.326.

Now:

$$\frac{\bar{X} - 9.72}{1.40/\sqrt{25}} = -2.326 \quad \text{gives} \quad \bar{X} = \frac{(-2.326)(1.40)}{\sqrt{25}} + 9.72$$

Hence $\bar{X} = 9.07$.

Thus the decision rule is:

(i)   reject H_0 if $\bar{X} < 9.07$,

(ii)   accept H_0 otherwise.

For (b), $\mu = 9.72$, $\sigma = 1.40$ and $n = 36$.

So that:

$$\frac{\bar{X} - 9.72}{1.40/\sqrt{36}} = -2.326 \quad \text{which gives} \quad \bar{X} = \frac{(-2.326)(1.40)}{\sqrt{36}} + 9.72$$

i.e. $\bar{X} = 9.18$.

The decision rule in this case is:

(i)   reject H_0 if $\bar{X} < 9.18$,

(ii)   accept H_0 otherwise.

9.2.3 FURTHER CONSIDERATIONS: ERRORS AND POWER

In all types of statistical tests it is possible to make errors. There are several reasons for this, not least the fact that we are forcing a decision to be made on the basis of a single sample from a population.

Other, more technical, reasons will become clear in this sub-section.

There are two important errors that we need to be aware of.

DEFINITION 9.1

(a)   A Type I error is made if we reject H_0 *when* H_0 *is true.*

(b)   A Type II error is made if we accept H_0 *when* H_0 *is false* (i.e., when H_1 is true.)

We usually write:

$$\Pr(\text{Type I error}) = \Pr(\text{rejecting } H_0/H_0 \text{ true}) = \alpha$$

$$\Pr(\text{Type II error}) = \Pr(\text{accepting } H_0/H_1 \text{ true}) = \beta$$

The value of α, that is the probability of making a Type I error, is just the significance level of the test.

As an illustration, consider the test statistic $Z = \dfrac{\bar{X} - \mu}{\sigma/\sqrt{n}}$ for testing whether a population mean has value μ.

Fig. 9.7 shows the set up for a one-tailed (increase) test at the 5% level. The realised value of the test statistic (z) will take on a value greater than the critical value 1.645 just 5% of the time *when* H_0 *is true.* Thus, the fact that we reject H_0 for $z > 1.645$ simply means that we will make a Type I error 5% of the time. That is, $\alpha = 0.05$. Note that if the test had been at the 1% level then we would have $\alpha = 0.01$.

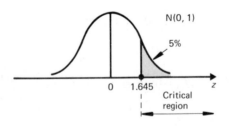

Fig. 9.7

When considering Type II errors, the fact that we are assuming H_0 is false, or alternatively H_1 is true, causes a particular problem for calculating a probability. This is just that the parameter involved has no fixed value. For instance, for a test of some parameter θ we might have

$H_0 : \theta = \theta_0$ versus $H_1 : \theta \neq \theta_0$ (two-tailed) or $H_0 : \theta \leqslant \theta_0$ versus $H_1 : \theta > \theta_0$ (one-tailed). In both of these cases, under the assumption that H_1 is true, θ has no definite value. So, in order to calculate the probability of a Type II error we need to 'fix' θ at a particular value (obviously compatible with H_1).

Fig. 9.8 shows some examples of different cases of parametric values under H_1, where θ ($= \mu$ here) is the mean of a Normal population.

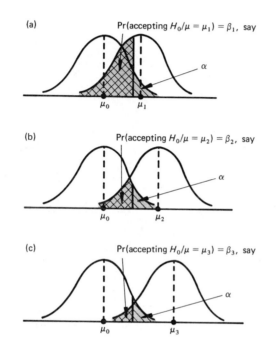

Fig. 9.8

In the three cases shown, we are considering $H_0 : \mu \leqslant \mu_0$ versus $H_1 : \mu > \mu_0$. The shaded (left hand side) area in each case is just $\Pr(\text{Type II error})$. Note that the further away from μ_0 the true mean (μ_1, μ_2 or μ_3) is, the smaller becomes β (the probability of a Type II error) whereas, in each case, α (the probability of a Type I error) is being held constant.

As a numerical example, consider $H_0 : \mu \leqslant 30$ versus $H_1 : \mu > 30$ (with a known population variance of $\sigma^2 = 15$).

We will calculate β for $\mu = 32, 34$ and 36 with the value of α set at 0.05 and consider a sample of size 4. Now, we know already that the critical value of test statistic Z for a one-tailed (increase) test at the 5% level is 1.645.

We first find the value of a $N(30, 3.75)$ distribution, $\bar{x}_0$ say, that corresponds to the standardised value 1.645. (Note that $\dfrac{15}{4} = 3.75$ is the variance for the distribution of the mean of a sample of size 4.)

We have that $\dfrac{\bar{x}_0 - 30}{\sqrt{3.75}} = 1.645$ giving $\bar{x}_0 = (1.645)\sqrt{3.75} + 30$. Hence
$\bar{x}_0 = 33.19$ (this technique can be thought of as 'de-standardising'). See
Fig. 9.9(a). Fig. 9.9(b) shows the pictorial situation that we have where
we need to find the three values of β (β_1, β_2 and β_3). For convenience,
we have identified three variables $\bar{X}_1, \bar{X}_2$ and $\bar{X}_3$ with each of the three
H_1 distributions so that $\bar{X}_1$ is $N(32, 3.75)$, $\bar{X}_2$ is $N(34, 3.75)$ and
$\bar{X}_3$ is $N(36, 3.75)$.

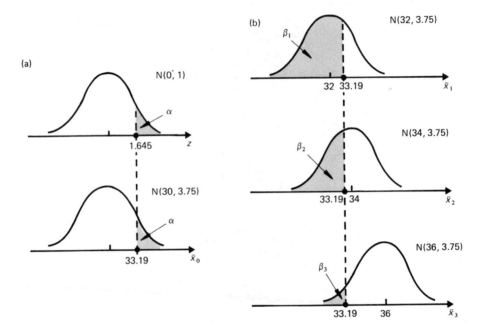

Fig. 9.9

The value of β when $\mu = 32$ (β_1, say) is just:

$$\Pr(\bar{X}_1 < 33.19) = \Pr\left(Z < \frac{33.19 - 32}{\sqrt{3.75}}\right)$$

$$= \Pr(Z < 0.61) = 0.729$$

Similarly:

$$\beta_2 = \Pr(\bar{X}_2 < 33.19) = \Pr\left(Z < \frac{33.19 - 34}{\sqrt{3.75}}\right)$$

$$= \Pr(Z < -0.42) = 1 - \Phi(0.42) = 0.337$$

and:

$$\beta_3 = \Pr(\bar{X}_3 < 33.19) = \Pr\left(Z < \frac{33.19 - 36}{\sqrt{3.75}}\right)$$

$$= \Pr(Z < -1.45) = 1 - \Phi(1.45) = 0.073$$

Note here that we have been holding α (the probability of a Type I error) constant while considering different values of μ under H_1, thereby varying the value of β.

We next consider a fixed value of μ under H_1 and note the effects of altering α.

It is clear (see Fig. 9.10) that, in this case, the smaller α is the larger β becomes (and vice versa).

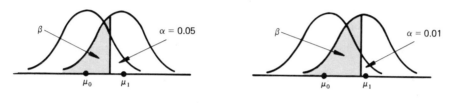

Fig. 9.10

Returning to the practical example on the previous page, consider increasing n (the sample size) to, say, 6 giving $\bar{X}$ as $N\left(30, \dfrac{15}{6}\right) = N(30, 2.5)$.

Thus $\bar{x}_0$ now satisfies:

$$\bar{x}_0 = (1.645)\sqrt{2.5} + 30 = 32.60$$

This gives:

$$\beta_1 = \Pr(\bar{X}_1 < 32.6) = \Pr\left(Z < \frac{32.6 - 32}{\sqrt{2.5}}\right)$$

$$= \Pr(Z < 0.38)$$

$$= 0.648$$

Similarly, $\beta_2 = 0.187$ and $\beta_3 = 0.016$.

Comparing the above values of β with the three obtained earlier shows a clear decrease.

That is for a fixed α, *increasing* the sample size n *decreases* the value of β (at a fixed parametric point).

EXAMPLE 9.4

A Normal distribution is known to have a variance of 2.8. A one-tailed (increase) test is proposed of the form $H_0 : \mu \leqslant 14$ versus $H_1 : \mu > 14$. Find the probability of making a Type II error with a sample of size 2 if the significance level of the test is:

(a)   0.05,

(b)   0.01 when the true population mean is 16.5

SOLUTION

Under H_0, the population we are considering is:

$$N\left(14, \frac{2.8}{2}\right) = N(14, 1.4).$$

(Remember, we take that distinct value of the parameter (under H_0) that is closest to the H_1 parametric values when we need to consider a distribution under H_0).

We have to find the value of β for the two cases:

(a) $\alpha = 0.05$, and

(b) $\alpha = 0.01$ when the population, under H_1, is $N(16.5, 1.4)$.

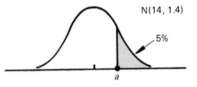

N(14, 1.4)

5%

a

For (a) we first require the value a of this population that corresponds to the standardised value 1.645 (the 5% point of a $N(0, 1)$ distribution). To find a, we have $\dfrac{a - 14}{\sqrt{1.4}} = 1.645$ giving:

$$a = (1.645)\sqrt{1.4} + 14$$

i.e.: $a = 15.95.$

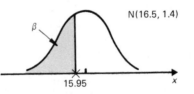

N(16.5, 1.4)

β

15.95 x

Associating $\bar{X}$ with the H_1 distribution $N(16.5, 1.4)$, we need to find:

$$\begin{aligned}
\beta &= \Pr(\text{Type II error}/\mu = 16.5) \\
&= \Pr(\bar{X} < 15.95) \\
&= \Pr\left(Z < \frac{15.95 - 16.5}{\sqrt{1.4}}\right) \\
&= \Phi(-0.46) \\
&= 1 - \Phi(0.46) \\
&= 0.323
\end{aligned}$$

For (b), using exactly the same technique:

$$\frac{a - 14}{\sqrt{1.4}} = 2.326 \quad \text{(the 1\% point of a N(0, 1) distribution)}$$

giving: $a = (2.326)\sqrt{1.4} + 14$

$= 16.75$

Hence: $\beta = \Pr(\bar{X} < 16.75)$

$= \Pr\left(Z < \frac{16.75 - 16.5}{\sqrt{1.4}}\right)$

$= \Phi(0.21)$

$= 0.583 \quad \text{(3D)}$

In the above example, notice that:

(a)   with the *decrease* in the probability of a Type I error from 5% to 1%, the probability of a Type II error *increases* from 32% to 58% (approximately) in this case, and also

(b)   the value of β in each case is unacceptably large. The reason for this of course is, first, that the mean μ of 16.5 is very close to the assumed mean, $\mu = 14$, under H_0 and secondly, the sample (size 2) is very small. Clearly, if the true mean were further away from 14, and/or the sample was larger, the corresponding values of β would be smaller.

In the formulation of standard tests (tests that we will be dealing with in this text) the theoretical aim is to make the risks of both types of errors as small as possible when selecting a critical region for the test. This is very difficult. Practically, since Type I errors are generally considered to be the most serious, the technique adopted is to fix the level of significance (the probability of a Type I error) at some acceptably low level. Subject to this constraint, a critical region is sought such that the probability of a Type II error is minimised. For the standard tests we will be dealing with, the test statistics, critical regions and significance levels all satisfy this criterion.

We look now at a useful concept, tied in with Type II errors.

DEFINITION 9.2

(a)   The *power function of a test* of some parameter θ, is the function $\Pr(\text{rejecting } H_0/H_1 \text{ true}) = \text{Pw}(\theta)$, which can be calculated and plotted on a graph for all values of θ *under H_1.*

(b)   The *power of a test* at some parametric point $\theta = \theta_1$ is just the value of the power function at $\theta = \theta_1$, i.e., $\text{Pw}(\theta_1)$.

Thus the power function, $Pw(\theta)$, measures the ability of a given test to reject H_0 for each value of θ in the critical region. The higher the power of the test (for particular values of θ), the *more powerful* the test is said to be. It is generally useful in judging the effectiveness of a proposed test.

Since:

$$Pr(\text{Type II error}/\theta = \theta_1) = Pr(\text{accepting } H_0/H_1 \text{ true with } \theta = \theta_1)$$

and:

$$Pw(\theta_1) = Pr(\text{rejecting } H_0/H_1 \text{ true with } \theta = \theta_1)$$

it should be clear that:

$$Pw(\theta_1) = 1 - Pr(\text{Type II error}/\theta = \theta_1)$$

That is, power (at $\theta = \theta_1$) $= 1 - \beta$.

Earlier in the text, we considered the test $H_0: \mu \leqslant 30$ versus $H_1: \mu > 30$ with the population variance, σ^2, fixed at 15.

β was calculated for the cases $\mu = 32, 34$ and 36 to obtain:

$$\beta_{\mu=32} = 0.729; \quad \beta_{\mu=34} = 0.337 \quad \text{and} \quad \beta_{\mu=36} = 0.073$$

Hence $Pw(32)$, the power of the test for $\mu = 32$, is $1 - 0.729 = 0.271$.

Also:

$$Pw(34) = 1 - 0.337 = 0.663 \quad \text{and} \quad Pw(36) = 1 - 0.073 = 0.927$$

Note that the test is more powerful the further away from 30 the true mean gets. That is, we are likely to reject $\mu \leqslant 30$ (when it is false) for values of μ greatly in excess of 30.

We shall calculate two further values of the power function, corresponding to $\mu = 33$ and 35 and so sketch the curve of the power function (often called the *power curve*).

We will associate variables $\bar{X}_1$ and $\bar{X}_2$ with $N(33, 3.75)$ and $N(35, 3.75)$ respectively (for convenience). Also, we know already that the critical value of $\bar{X}$ under $H_0: \mu \leqslant 30$ is 33.19. Fig. 9.11 shows the two cases pictorially.

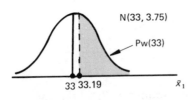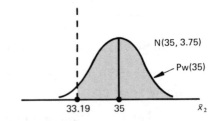

Fig. 9.11

So:

$$Pw(33) = Pr(\text{rejecting } H_0/\mu = 33)$$
$$= Pr(\bar{X} > 33.19)$$

$$= \Pr\left(Z > \frac{33.19 - 33}{\sqrt{3.75}}\right)$$

$$= \Pr(Z > 0.10)$$

$$= 1 - \Phi(0.10)$$

$$= 1 - 0.540$$

$$= 0.460$$

and:

$$Pw(35) = \Pr(\bar{X}_2 > 33.19)$$

$$= \Pr\left(Z > \frac{33.19 - 35}{\sqrt{3.75}}\right)$$

$$= \Pr(Z > -0.93)$$

$$= \Phi(0.93)$$

$$= 0.824$$

Fig. 9.12 shows a sketch of the power curve for the test based on the five values calculated. Also shown (as a dotted line) is the curve giving the probability of making a Type II error for different parametric values under H_1 which is called an *Operating Characteristic* (OC) curve (a mirror image of the power curve).

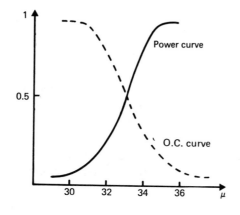

Fig. 9.12

We have only looked at one particular test so far (for a population mean to demonstrate the main principles of hypothesis testing. The test in question is one of a set of standard tests that are commonly used in applied statistics and, as mentioned earlier, we will be looking at the most important of the others in the following sections and chapters. However, to demonstrate the fact that the principles discussed can be used in any well-defined testing situation, we have the following example.

EXAMPLE 9.5

A test is constructed by tossing a coin independently four times with the test statistic defined as 'total number of heads' obtained. We wish to decide between the hypotheses $H_0: p = \dfrac{1}{2}$ versus $H_1: p = \dfrac{3}{4}$, where p is the probability of obtaining a 'head' with a single toss of the coin. Examine critical regions (one-tailed) w.r.t. Type I and Type II errors and discuss the suitability of each of them.

SOLUTION

Now, under H_0, $p = \dfrac{1}{2}$ and under H_1, $p = \dfrac{3}{4}$.

With four tosses of the coin, respective probabilities of obtaining r heads $(r = 0, 1, \ldots, 4)$ are given by the binomial probability function

$$\Pr(r \text{ heads}) = \Pr(r) = \binom{4}{r}\left(\frac{1}{2}\right)^r\left(\frac{1}{2}\right)^{4-r} \quad \text{under } H_0: p = \frac{1}{2}$$

$$= \binom{4}{r}\left(\frac{3}{4}\right)^r\left(\frac{1}{4}\right)^{4-r} \quad \text{under } H_1: p = \frac{3}{4}$$

Tabulating the probabilities for both cases gives the following:

	$H_0: p = \dfrac{1}{2}$	$H_1: p = \dfrac{3}{4}$
$\Pr(0)$	0.0625	0.0039
$\Pr(1)$	0.2500	0.0469
$\Pr(2)$	0.3750	0.2109
$\Pr(3)$	0.2500	0.4219
$\Pr(4)$	0.0625	0.3164

Bearing in mind that we are concerned with a one-tailed increase, we will examine the following four critical regions:

(a) $r = 1, 2, 3$ and 4

$$\alpha = \Pr\left(\text{rejecting } H_0/H_0: p = \frac{1}{2}\right)$$

$$= \Pr(1 \text{ or } 2 \text{ or } 3 \text{ or } 4/H_0)$$

$$= 0.2500 + 0.3750 + 0.2500 + 0.0625$$

$$= 0.9375$$

$$\beta = \Pr\left(\text{accepting } H_0/H_1: p = \frac{3}{4}\right)$$

$$= \Pr(0/H_1)$$

$$= 0.0039$$

Hence $\alpha + \beta = 0.9414$.

(b) $r = 2, 3$ and 4

$$\alpha = \Pr(2 \text{ or } 3 \text{ or } 4/H_0)$$

$$= 0.375 + 0.25 + 0.0625$$

$$= 0.6875$$

$$\beta = \Pr(0 \text{ or } 1/H_1)$$

$$= 0.0039 + 0.0469$$

$$= 0.0508$$

and $\alpha + \beta = 0.7383$.

(c) $r = 3$ and 4

$$\alpha = \Pr(3 \text{ or } 4/H_0)$$

$$= 0.25 + 0.0625$$

$$= 0.3125$$

$$\beta = \Pr(0 \text{ or } 1 \text{ or } 2/H_1)$$

$$= 0.0039 + 0.0469 + 0.2109$$

$$= 0.2617$$

and $\alpha + \beta = 0.5742$.

(d) $r = 4$

$$\alpha = \Pr(4/H_0)$$

$$= 0.0625$$

$$\beta = \Pr(0 \text{ or } 1 \text{ or } 2 \text{ or } 3/H_1)$$

$$= 0.0039 + 0.0469 + 0.2109 + 0.4219$$

$$= 0.6836$$

with $\alpha + \beta = 0.7461$.

In both (a) and (b), the level of α and the total $\alpha + \beta$ is unacceptably high. Although α is very small in (d), the total of $\alpha + \beta$ is
probably too high for practical considerations. Critical region (c)
then, would seem to be the best here.

9.3 TEST FOR A SAMPLE MEAN

In this section we are concerned with the problem of testing whether a given sample can be considered as coming from a given Normal (or near-Normal) distribution. To do this, we test the sample mean $\bar{x}$ against the distribution mean μ.

There are three separate cases to consider:

(a)   When the population variance, σ^2, is known — this is just the situation we investigated in Section 9.2.1.

(b)   When σ^2 is unknown and we are taking a small sample.

(c)   When σ^2 is unknown and we are taking a large sample.

We will be dealing with each of these cases in the following sub-sections. The null hypothesis in each of the three cases is the same. Namely, that the sample *is* drawn from a Normal population with mean μ. The alternative hypothesis (in each case) will depend on whether:

(a)   we are looking for any change in the population mean μ (i.e., a two-tailed H_1), or

(b)   we are looking for either a strict increase or, alternatively, a strict decrease in μ (i.e., a one-tailed H_1).

9.3.1 WITH KNOWN VARIANCE (Z TEST)

This is the test that we originally formulated in Section 9.2.1.

STATEMENT 9.1

To test whether a random sample of size n with mean $\bar{X}$ is drawn from a Normal population with known mean μ and known variance σ^2, we use the test statistic $Z = \dfrac{\bar{X} - \mu}{\sigma/\sqrt{n}}$ which is distributed as $N(0, 1)$ under the null hypothesis that the true population mean *is* μ.

This test is valid for all sample sizes.

Proof   In Section 9.2.1.

Note that the above test takes account of testing either:

(a)   whether a sample can be considered as coming from a particular given Normal distribution (as specifically stated above), or

(b)   whether the mean of the Normal distribution under consideration, formally having value μ, has now changed. For testing purposes, the two can be considered as equivalent cases.

EXAMPLE 9.6

A distribution is thought to have mean 9.72 and standard deviation 1.40 (take this as known). A sample of 36 gave a mean of 8.93. Is there evidence at:

(a)   the 5% level   (b)   the 1% level

that the distribution mean has decreased?

SOLUTION

Here we are testing $H_0: \mu \geqslant 9.72$ versus $H_1: \mu < 9.72$.

The test statistic has the form $Z = \dfrac{\bar{X} - \mu}{\sigma/\sqrt{n}}$ with a realised value of:

$$z_{test} = \frac{8.93 - 9.72}{1.40/6} = -3.386$$

But: critical z $(P = 5\%$, one-tailed$)$ $= -1.645$

and: critical z $(P = 1\%$, one-tailed$)$ $= -2.326$

Hence z_{test} is significant at both the 5% and 1% levels and we can reject H_0. So there is strong evidence that the mean *has* decreased.

EXAMPLE 9.7

A pair of dice are tossed 100 times and a total of seven is obtained 23 times. By using the Normal approximation to the binomial, test whether, on the basis of this sample result, the dice are not fair.

SOLUTION

There are 36 ways of throwing two dice. We can obtain a total of seven by:

(a)   a '1' and a '6' (2 ways, i.e., (1, 6) or (6, 1))
(b)   a '2' and a '5' (2 ways)
(c)   a '3' and a '4' (2 ways)

So there are six different ways of throwing a total of seven.

Thus:

$\text{Pr(throwing a total of seven)} = \dfrac{6}{36} = \dfrac{1}{6}$ (if the dice are 'fair')

We have, then, a binomial situation with $p = \dfrac{1}{6}$ and $n = 100$

giving a mean of:

$$np = \frac{100}{6} = 16.67 \quad (2D)$$

and standard deviation:

$$\sqrt{np(1-p)} = \sqrt{100\left(\frac{1}{6}\right)\left(\frac{5}{6}\right)} = 3.73 \quad (2D)$$

Using a Normal approximation, we have $\mu = 16.67$ and $\sigma = 3.73$.

Returning to the sample, we obtained a single binomial observation of 23 'sevens'. (Note that the sample size, n, is 1.)

So: $z_{test} = \dfrac{\bar{x} - \mu}{\sigma/\sqrt{n}} = \dfrac{23 - 16.67}{3.73} = 1.70$ (2D)

Now, the test must be two-tailed, since we are testing the hypothesis that the dice are unbiased against the alternative, that the dice are biased in some way, i.e., $H_0 : p = \dfrac{1}{6}$ versus $H_1 : p \neq \dfrac{1}{6}$ or, more specifically, $H_0 : \mu = 16.67$ versus $H_1 : \mu \neq 16.67$.

Critical z $(P = 5\%,\text{ two-tailed}) = \pm 1.96$.

So that $z_{test} = 1.697$ is not significant and thus we do not have sufficient evidence to conclude that the dice are biased.

9.3.2 WITH UNKNOWN VARIANCE: SMALL SAMPLES (*T* TEST)

In this situation, we are sampling from a population (assumed Normal) whose variance is unknown. The form of the test statistic is similar to that of the previous case (for known variance, σ^2) except for the fact that we need to estimate σ^2 with the quantity $\dfrac{ns^2}{n-1}$.

So, replacing σ by $\left(\dfrac{\sqrt{n}}{\sqrt{n-1}}\right)s$ in the previous test statistic, we can obtain the form given in the following statement. Note that the distribution of the test statistic under the null hypothesis has changed from a $N(0, 1)$ to a $T(n-1)$ distribution.

STATEMENT 9.2

To test whether a random sample of size n with mean $\bar{X}$ and sample variance S^2 is drawn from a Normal population with known mean μ and unknown variance, we use the test statistic

$T = \dfrac{\bar{X} - \mu}{S/\sqrt{n-1}}$ which is distributed as $T(n-1)$ under the null hypothesis that the true population mean *is* μ.

Proof   Essentially contained in Example 8.9.

The test procedure in this case is identical to the procedure in Section 9.2.1 except, of course, that we have a different form of distribution for the test statistic. To find the critical value(s), we need to refer to *T*-tables.

We know that the body of the table gives us t_P for different levels of P and different values of the parameter ν. Fig. 9.13 gives a particular

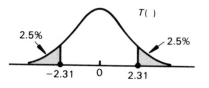

Fig. 9.13

situation for a two-tailed test. Section 8.4.3 should be referenced for a revision of the use of T-tables.

So, for instance, if we were interested in a two-tailed test at the 5% level for a sample of size $n = 9$, we would need $\nu = n - 1 = 8$ and $P = 0.975$ (97.5%) giving the value $t = 2.31$ (from tables). That is, -2.31 and $+2.31$ are the required critical values in this case. Fig. 9.13 shows this pictorially.

For a one-tailed test at the 5% level for $n = 9$, say, we would need to consider $P = 0.975$ and $\nu = 8$ in order to get a one-tailed critical region of size 5%. From tables we obtain $t = +1.86$ which can be used as $t = -1.86$, depending on whether we are interested in an increase or decrease of the respective Normal mean. See Fig. 9.14.

Fig. 9.14

EXAMPLE 9.8

It is claimed that the length of a certain species of animal is distributed Normally with mean 44 in. In order to test the truth of this, a sample of 21 such animals is taken and it is found that $\bar{x} = 42$ in and $s = 6$ in. Is there evidence at a 5% level to refute the claim?

SOLUTION

We are testing $H_0: \mu = 44$ against $H_1: \mu \neq 44$ (i.e., a two-tailed test) with $n = 21$.

The test statistic is $T = \dfrac{\bar{X} - \mu}{S/\sqrt{n-1}}$ and its realised value for this sample is:

$$t_{test} = \frac{42 - 44}{6/\sqrt{20}} = -1.49$$

The critical values for t with $\nu = 21 - 1 = 20$ (at the 5% level) are ± 2.09. Hence, we accept H_0. So there is no evidence on which to refute the claim.

EXAMPLE 9.9

A manufacturer claims that the ropes he produces have a mean breaking strength of 800 lb. A test on 6 ropes found a mean breaking strength of 775 lb with standard deviation 14.5 lb. Is there evidence that the manufacturer is over-estimating the strength of his ropes at the 1% level?

SOLUTION

We are obviously looking for a true breaking strength that is less than the manufacturer's claim. Hence we test $H_0 : \mu \geqslant 800$ versus $H_1 : \mu < 800$ (a one-tailed test).

Using the test statistic $T = \dfrac{\bar{X} - \mu}{S/\sqrt{n-1}}$, we have:

$$t_{\text{test}} = \frac{775 - 800}{14.5/\sqrt{5}} = -3.86$$

To find the critical value of t, we need to look in Appendix Table 5 at $p = 0.99$ (since we have a one-tailed test) and $\nu = 6 - 1 = 5$, which gives $t_{\text{critical}} = 3.365$. Since we are interested in a decrease, we need to use $t_{\text{critical}} = -3.365$.

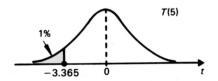

Hence, since $-3.86 < -3.365$, the test value of t is highly significant (at the 1% level), so we can reject H_0 confidently and conclude that the manufacturer is over-estimating the strength of his ropes, on the evidence given.

9.3.3 WITH UNKNOWN VARIANCE: LARGE SAMPLE APPROXIMATION (Z TEST)

STATEMENT 9.3

To test whether a random sample of size n (large), with mean $\bar{X}$ and sample variance S^2, is drawn from a Normal distribution with known mean μ and unknown variance, we use the test statistic

$Z = \dfrac{\bar{X} - \mu}{S/\sqrt{n}}$ which is approximately distributed as $N(0, 1)$ under the

null hypothesis that the true population mean *is* μ.

We would regard a large sample as one with $n > 30$.

No proof

Note that the above is a large sample approximation and is used under the conditions stated above. However, the previous test statistic (usually used with small samples) can in fact be used for all values of n.

The test procedure here is the same as with all previous tests, except for the form and distribution (under H_0) of the test statistic.

EXAMPLE 9.10

A manufacturer of matches claims that the average contents in a standard box is 45. A consumer organisation measures the number of matches in each of 100 randomly selected boxes and obtains a mean of 43.7 with standard deviation 0.53 matches. Is this enough evidence to complain (justifiably) that the manufacturer is over-estimating the contents of his standard boxes?

SOLUTION

We need to test whether a sample, size 100 with $\bar{x} = 43.7$ and $s = 0.53$, can be considered as coming from a distribution with mean 45 against the alternative that the sample comes from a distribution having a smaller mean. Symbolically, we test $H_0 : \mu \geqslant 45$ versus $H_1 : \mu < 45$ (strictly one-tailed).

Since the distribution variance, σ^2, is unknown, we use the test statistic $Z = \dfrac{\bar{X} - \mu}{S/\sqrt{n}}$ ($n = 100$ is large).

$$z_{test} = \frac{43.7 - 45}{0.53/\sqrt{100}} = -24.53$$

But critical z ($P = 1\%$, one-tailed) $= -2.326$.

Thus the value of $z_{test} = -24.53$ is highly significant and we can confidently reject the manufacturer's claim.

Note that with such a large sample, even a small difference between $\bar{x}$ and μ can give a highly significant result.

EXAMPLE 9.11

Metal rods are delivered in large batches to a customer. When production is under control, the rods have a mean length of 2.132 in with standard deviation 0.001 in. A random sample of 90 from a batch is found to have a mean length of 2.1316 in. Report on the likelihood that the production is still under control.

SOLUTION

We assume initially that the distribution of rod lengths will be at least approximately Normal.

Under H_0, the distribution is assumed to be $N(2.132, (0.001)^2)$, and since we are interested in general control of a mean, the test needs to be specifically of the form $H_0: \mu = 2.132$ versus $H_1: \mu \neq 2.132$ (two-tailed).

Since the population variance is known, the test statistic for a single mean takes the form $Z = \dfrac{\bar{X} - \mu}{\sigma/\sqrt{n}}$ with:

$$z_{test} = \frac{2.1316 - 2.1320}{0.001/\sqrt{90}} = -3.79$$

At the 1% level (two-tailed), the critical values of z are ± 2.576.

We therefore reject H_0 and conclude that (on the evidence) production is out of control.

EXAMPLE 9.12

Eight typists were given electric typewriters instead of their previous machines, and the increases in the number of finished quarto pages in a day were found to be as follows: $8, 12, -3, 4, 5, 13, 11, 0$. Do these figures indicate that the typists in general can work faster with electric typewriters? State the assumptions you make.

SOLUTION

The initial assumption we need to make is that the given figures are from an approximately Normal population.

Now, we are clearly looking for an increase in finished pages, so the test is one-tailed, and the null hypothesis must be that the electric typewriters make no difference (or worsen the situation!)

Since we expect the mean of the increases to be zero under H_0, symbolically we are testing $H_0: \mu \leqslant 0$ versus $H_1: \mu > 0$. We have as an H_0 model then, a Normal distribution with zero mean and unknown variance. Therefore we will need to use a T-test (small sample, since $n = 8$), and calculate $\bar{x}$ and s.

From the sample given $\Sigma x = 50$ and $\Sigma x^2 = 548$.

$$\bar{x} = \frac{50}{8} = 6.25 \quad \text{and} \quad s = \sqrt{\frac{548}{8} - (6.25)^2} = 5.426 \quad (3D)$$

$$t_{test} = \frac{\bar{x} - \mu}{s/\sqrt{n-1}} = \frac{6.25 - 0}{5.426/\sqrt{7}} = 3.05 \quad (2D)$$

Critical t (one-tailed, 5%) $= 1.895$

Hence t_{test} is highly significant and we reject H_0.

We conclude that there is evidence that the typists work faster with electric typewriters.

9.4 TEST FOR THE DIFFERENCE BETWEEN TWO MEANS

We now investigate the problem of whether two separate random samples can be considered as coming from the same population (assumed to be Normal). As we have done previously, we base the test on the means and variances of both the samples and the theoretical population under consideration.

Essentially, we test whether the means of the two samples are compatible while assuming that the samples come from Normal populations having a common variance. This assumption is both necessary and practical. It is necessary at this stage because to assume that the two population variances are different entails theoretical considerations beyond the scope of the text. In practice, all samples considered at this level will be of the afore-mentioned type.

The test procedure itself is the same as that used in previous tests except for the form and distribution (under H_0) of the test statistic.

Tests relating to the difference between two sample means fall readily into two categories — paired and unpaired. By paired samples, we are referring to two distinct samples, each member of one being paired in a natural (and obvious) way with a member of the other.

For example, if we were measuring temperatures in two towns at certain times of the day, we might obtain the following pair of samples:

Time	0900	1400	0910	1013	1600	1400	1100	1100
Town A	32	46	33	33	35	44	36	41
Town B	34	47	32	35	38	45	38	40

We might wish to test whether the temperatures in the two towns are the same.

The following times of reaction to a certain stimulus, obtained by timing eight people, each being treated with and without a certain drug, constitute paired samples:

Person		A	B	C	D	E	F	G	H
Reaction time (sec)	With drug	0.6	1.2	0.4	0.5	1.1	0.8	0.5	0.5
	Without drug	0.5	0.6	0.5	0.8	1.2	0.6	0.3	0.5

We might wish to test whether the drug really does affect reaction time.

In other cases, where we cannot pair observations in this way, we refer to the samples as unpaired. A different technique is used depending on whether the samples are paired or unpaired.

Although both the cases of known and unknown common population variance are considered, in most practical instances the latter is involved.

9.4.1 TEST FOR PAIRED SAMPLES

Suppose we are given two paired random samples from Normal populations $X_1, X_2, \ldots, X_n$ and $Y_1, Y_2, \ldots, Y_n$ where X_1 matches
Y_1, X_2 matches Y_2, etc., and we wish to test for any difference between the two.

Since we are interested in differences, it seems reasonable to consider
the variables:

$$D_1 = X_1 - Y_1, \ D_2 = X_2 - Y_2, \ldots, D_n = X_n - Y_n$$

and we end up with the single set of differences $D_1, D_2, \ldots, D_n$ (these
will also be distributed Normally from Normal theory).

For example, using the 'reaction time' data presented earlier, we obtain:

X	0.6	1.2	0.4	0.5	1.1	0.8	0.5	0.5
Y	0.5	0.6	0.5	0.8	1.2	0.6	0.3	0.5
$D = X - Y$	0.1	0.6	-0.1	-0.3	-0.1	0.2	0.2	0.0

Of course, once we transform the two samples to a single sample of
differences we have lost a certain amount of information. Namely, the
actual X and Y values. But this is not important in this case since the
whole test is concerned with differences.

We can consider the set of differences now as a sample from a theoretical Normal population of differences with mean and variance μ_D and
σ_D^2 respectively, say. We are now in exactly the same position as we
were in Section 9.3, since we wish to test whether a random sample
(albeit of differences) having a mean $\bar{D}$, say, can be considered as being
drawn from a Normal distribution (of differences) with mean μ_D and
variance σ_D^2.

The test statistic for this case, however, is always slightly simpler than
its analogue in Section 9.3. This is due to the following consideration.
The null hypothesis is that there is no difference between the means of
the distributions of X and Y, giving the theoretical mean of the differences (μ_D) as zero. Hence, in Section 9.3, where we had $\bar{X} - \mu$, we
now have $\bar{D} - \mu_D$, simplifying to just $\bar{D}$ under the above hypothesis.

We can now proceed to the following:

STATEMENT 9.4

To test whether two random samples (with paired values) are drawn
from the same theoretical Normal population, we use the following
test statistics:

(a) *For known variance of differences* (σ_D^2)

$$Z = \frac{\bar{D}}{\sigma_D / \sqrt{n}} \text{ distributed as } N(0, 1) \text{ under } H_0$$

(*continued*)

(b)   *For unknown variance of differences*

$$T = \frac{\bar{D}}{S/\sqrt{n-1}} \text{ distributed as } T(n-1) \text{ under } H_0$$

where, in both cases $\bar{D}$ is the mean of the paired sample differences, S_D is the standard deviation of the paired sample differences and n is the sample size. The null hypothesis is that both samples are drawn from the same Normal population or from Normal populations having the same mean and variance.

Proof   See preceding test and Statements 9.1, 9.2 and 9.3.

We can now return to the reaction time samples to test for any difference. Using $H_0: \mu_D = 0$ versus $H_1: \mu_D \neq 0$ (a two-tailed test), with the variance of differences unknown, we need to use the test statistic $T = \frac{\bar{D}}{S/\sqrt{n-1}}$.

d	0.1	0.6	-0.1	-0.3	-0.1	0.2	0.2	0	0.6
d^2	0.01	0.36	0.01	0.09	0.01	0.04	0.04	0	0.56

$$\bar{d} = \frac{0.6}{8} = 0.075$$

$$s = \sqrt{\frac{0.56}{8} - (0.075)^2} = 0.254$$

Hence:
$$t_{\text{test}} = \frac{0.075}{0.254/\sqrt{7}} = 0.78 \quad (2D)$$

But: $t_{\text{critical}}(7 \text{ d.f.}, 5\%) = \pm 2.36$ (the test is two-tailed)

The test result then is not significant at the 5% level, so we accept the null hypothesis that there is no essential difference between the two sets of reaction times. That is, there is no evidence that the drug affects the times.

EXAMPLE 9.13

The times taken by ten men to perform a task were measured. After training, designed to reduce the time, the men were tested again. The times in seconds were found to be as follows:

Man No.	1	2	3	4	5	6	7	8	9	10
Time before	27	45	51	33	49	22	23	35	41	25
Time after	22	37	45	26	44	20	16	35	42	23

Report on the effectiveness of the training as shown by this experiment.

SOLUTION

Since we are interested in the effectiveness of the training, it is a decrease in the 'time after' that we are looking for. These are clearly 'paired' samples, so we are investigating the mean of the theoretical population of differences, μ_D, of 'time before' minus 'time after'. The test involves the hypotheses $H_0: \mu_D \leqslant 0$ against $H_1: \mu_D > 0$ using the test statistic $T = \dfrac{\bar{D}}{S/\sqrt{n-1}}$ where $\nu = 10 - 1 = 9$.

At the 5% level, tables give us a critical value of 1.83.

d	5	8	6	7	5	2	7	0	-1	2	41
d^2	25	64	36	49	25	4	49	0	1	4	257

$$\bar{d} = \frac{41}{10} = 4.1$$

$$s = \sqrt{\frac{257}{10} - (4.1)^2} = 2.98 \quad \text{(2D)}$$

Hence: $t_{\text{test}} = \dfrac{4.1}{2.98/\sqrt{9}} = 4.13 \quad \text{(2D)}$

which is significant at the 5% level. We thus reject H_0. Therefore, on the evidence given, the training has been effective.

EXAMPLE 9.14

Groups of ten patients in a hospital are to be tested (on a voluntary basis) with a new 'sleep-inducing' drug over a period of time. For each patient, the average number of hours slept before (X) and after (Y) using the drug will be recorded. Design a decision rule, involving the sample difference mean and standard deviation, for testing the effectiveness of the drug at the 5% level.

The average increases $(y - x)$ in the number of hours slept for a particular ten patients were: 2.0, 0.2, $-$ 0.4, 0.3, 0.7, 1.2, 0.6, 1.8, $-$ 0.2 and 1.0. Use the above rule to test whether the drug is shown to be effective.

SOLUTION

Since we are interested in whether the drug is shown to be effective, we are clearly looking for an *increase* in sleeping hours. Putting $D = Y - X$, we test $H_0: \mu_D \leqslant 0$ versus $H_1: \mu_D > 0$, where μ_D is the theoretical mean of the differences in hours slept.

The test statistic is $T = \dfrac{\bar{D}}{S/\sqrt{n-1}}$ and (under H_0) is distributed as $T(10 - 1) = T(9)$. At a 5% level (one-tailed), the critical value for the test statistic is 1.83.

So putting:

$$\frac{\bar{D}}{S/\sqrt{n-1}} = 1.83 \quad \text{gives} \quad \frac{\bar{D}}{S} = \frac{1.83}{\sqrt{9}} = 0.61$$

The decision rule is, then:

(a)   reject H_0 if $\dfrac{\bar{D}}{S} > 0.61$,

(b)   accept H_0 otherwise.

From the given figures of $d = y - x$, we obtain $\Sigma d = 7.2$; $\Sigma d^2 = 10.86$.

Hence:

$$\bar{d} = 0.72 \quad \text{and} \quad s^2 = \frac{10.86}{10} - (0.72)^2 = 0.568, \text{ i.e., } s = 0.753$$

So that:

$$\frac{\bar{d}}{s} = \frac{0.72}{0.753} = 0.96 \quad \text{(2D)}$$

So, from (a) of the decision rule, we reject H_0, showing that there is evidence of an increase and hence the drug is shown to be effective in this case.

9.4.2 TEST FOR UNPAIRED SAMPLES

Here we are concerned with a test involving two unpaired, independent samples of sizes n_1 and n_2 (where n_1 and n_2 are not necessarily equal). The test is, specifically, whether the two samples can be considered as having been drawn from the same Normal population or, equivalently, from Normal populations with the same mean and variance. The latter is considered below, where we derive the form of the test statistic. We again consider both the cases of known and unknown common population variances although in practice, as in the case of paired samples, we are usually testing the compatibility of two means derived from *small samples* from populations where the common variance *is not known*.

Suppose we know the common population variance σ^2, say. We draw random samples $X_1, \ldots, X_n$ and $Y_1, \ldots, Y_n$ from Normal populations having respective means μ_X and μ_Y, say. We have then, for all i, X_i is $N(\mu_X, \sigma^2)$ and Y_i is $N(\mu_Y, \sigma^2)$.

Thus:

$$\bar{X} \text{ is } N\left(\mu_X, \frac{\sigma^2}{n_1}\right) \quad \text{and} \quad \bar{Y} \text{ is } N\left(\mu_Y, \frac{\sigma^2}{n_2}\right)$$

Therefore:

$$\bar{X} - \bar{Y} \text{ is } N\left(\mu_X - \mu_Y, \frac{\sigma^2}{n_1} + \frac{\sigma^2}{n_2}\right) = N\left(\mu_X - \mu_Y, \sigma^2\left(\frac{1}{n_1} + \frac{1}{n_2}\right)\right)$$

The null hypothesis for the test will be that there is no difference between the two population means, i.e., $H_0: \mu_X = \mu_Y$ or $H_0: \mu_X - \mu_Y = 0$.

So that, under H_0, $\bar{X} - \bar{Y}$ is $N\left(0, \sigma^2\left(\dfrac{1}{n_1} + \dfrac{1}{n_2}\right)\right)$.

Therefore:
$$Z = \frac{\bar{X} - \bar{Y}}{\sigma\sqrt{\dfrac{1}{n_1} + \dfrac{1}{n_2}}} \quad \text{is } N(0, 1)$$

and serves as the test statistic.

Note that if the samples are known to have been drawn from two populations having respective variances σ_1^2 and σ_2^2 the test statistic is adapted to:

$$Z = \frac{\bar{X} - \bar{Y}}{\sqrt{\dfrac{\hat{\sigma}_1^2}{n_1} + \dfrac{\hat{\sigma}_2^2}{n_2}}}$$

If σ^2 is unknown, we cannot use the above test statistic since it is given in terms of σ. What we do is simply replace σ in the above by an estimate of it, based on the sample results.

Using the form given in Section 8.2.4, with S_1^2 and S_2^2 as the respective sample variances, we have an unbiased estimate of the common population variance as:

$$\hat{\sigma}^2 = \frac{n_1 S_1^2 + n_2 S_2^2}{n_1 + n_2 - 2}$$

Hence, the test statistic in this case takes the form:

$$\frac{\bar{X} - \bar{Y}}{\hat{\sigma}\sqrt{\dfrac{1}{n_1} + \dfrac{1}{n_2}}} \quad \text{where} \quad \hat{\sigma} = \sqrt{\frac{n_1 S_1^2 + n_2 S_2^2}{n_1 + n_2 - 2}}$$

However, with this adjustment, the statistic now has a $T(n_1 + n_2 - 2)$ distribution (the proof of this is left as a directed exercise).

A test in this situation for the case of large samples (for unknown common population variance) is not generally used and we do not introduce it here.

To summarise:

STATEMENT 9.5

To test whether two unpaired random samples are drawn from Normal populations with the same mean, the following test statistics are used:

(a) *For known common population variance, σ^2*

$$Z = \frac{\bar{X} - \bar{Y}}{\sigma\sqrt{\dfrac{1}{n_1} + \dfrac{1}{n_2}}} \quad \text{distributed as } N(0, 1) \text{ under } H_0$$

(*continued*)

(b) *For unknown common population variance*

$$T = \frac{\bar{X} - \bar{Y}}{\hat{\sigma}\sqrt{\dfrac{1}{n_1} + \dfrac{1}{n_2}}} \quad \text{distributed as } T(n_1 + n_2 - 2) \text{ under } H_0$$

where:

$$\hat{\sigma} = \sqrt{\frac{n_1 S_1^2 + n_2 S_2^2}{n_1 + n_2 - 2}}$$

(c) *For known different population variances,* σ_1^2 *and* σ_2^2

$$Z = \frac{\bar{X} - \bar{Y}}{\sqrt{\dfrac{\hat{\sigma}_1^2}{n_1} + \dfrac{\hat{\sigma}_2^2}{n_2}}}$$

$\bar{X}$, $\bar{Y}$ are the respective sample means and n_1, n_2 the respective sample sizes. The null hypothesis to be tested is that the two theoretical population means are identical. That is, we test $H_0 : \mu_X = \mu_Y$ or $H_0 : \mu_X - \mu_Y = \mu_{X-Y} = \mu_D = 0$.

Proof (a)   In preceding text.

(b)   In preceding text and left as a directed exercise.

(c)   Similar technique to (a).

Note that in the above statement, if both theoretical populations considered above do not have the same variance, the results are invalidated.

EXAMPLE 9.15

It is thought that a certain species of small animal is lighter in weight in country B than in country A due to a different environment. To test this, two small samples of the species were taken, one from each of the two countries, as follows:

Sample from country A (x)	3.26 3.38 3.81 4.01 3.28 3.91 3.82	
Sample from country B (y)	3.51 3.42 2.93 3.41 3.28 3.37	

Perform a suitable test using the above results.

SOLUTION

We are obviously looking for a decrease in the y values over the x values. The test thus needs to be one-tailed using the competing hypotheses $H_0: \mu_D \leqslant 0$ versus $H_1: \mu_D > 0$, where μ_D is the difference between the means of the two theoretical (assumed Normal) populations of X and Y. We also assume that both populations have a common variance, and as this is unknown, the test statistic used is:

$$T = \frac{\bar{X} - \bar{Y}}{\hat{\sigma}\sqrt{\dfrac{1}{n_1} + \dfrac{1}{n_2}}}$$

where $\hat{\sigma}^2$ is the two-sample unbiased estimate of the common population variance. The critical value for t $(\nu = n_1 + n_2 - 2 = 11,$ $P = 5\%$, one-tailed) is 1.80 (approximately) from T tables.

The calculations for evaluating the test statistic are given as follows:

x	y	x^2	y^2
3.26	3.51	10.628	12.320
3.38	3.42	11.424	11.696
3.81	2.93	14.516	8.585
4.01	3.41	16.080	11.628
3.28	3.28	10.758	10.758
3.91	3.37	15.288	11.357
3.82		14.592	
25.47	19.92	93.287	66.345

$$\bar{x} = \frac{25.47}{7} = 3.639$$

$$\bar{y} = \frac{19.92}{6} = 3.32$$

$$s_1{}^2 = \frac{93.287}{7} - (3.639)^2 = 0.088$$

$$s_2{}^2 = \frac{66.345}{6} - (3.32)^2 = 0.035$$

$$\therefore \quad \hat{\sigma} = \sqrt{\frac{7(0.088) + 6(0.035)}{11}} = 0.274$$

Hence: $$t_{test} = \frac{3.639 - 3.32}{0.274\sqrt{\dfrac{1}{7} + \dfrac{1}{6}}} = 2.093$$

Since the test value of 2.093 is greater than the critical value of 1.80, it is significant at the 5% level and we reject the null hypothesis. Therefore, on the evidence given, the animals of country B are lighter in weight than those of country A.

EXAMPLE 9.16

Eight men of one tribe were found to weigh respectively 120, 174, 194, 122, 109, 144, 103, 161 lb (to the nearest lb). Ten men of another tribe weighed respectively 113, 115, 139, 145, 134, 110, 122, 134, 125, 113 lb. (a) Do these results indicate a difference in the mean weights of the men in the two tribes? (b) Given that the population variance is known to be 225, show that the decision made in (a) is reversed.

SOLUTION

(a) We test $H_0 : \mu_X = \mu_Y$ versus $H_1 : \mu_X \neq \mu_Y$ where we assume a common theoretical population variance, and variables X and Y are associated with the weights of the men in the first and second tribes respectively. From the given data, the following totals are easily obtained:

$$\sum x = 1127; \quad \sum y = 1250; \quad \sum x^2 = 166\,343; \quad \sum y^2 = 157\,630$$

So that:

$$\bar{x} = \frac{1127}{8} = 140.875 \quad \text{and} \quad \bar{y} = \frac{1250}{10} = 125$$

Also:

$$s_1{}^2 = \frac{166\,343}{8} - \left(\frac{1127}{8}\right)^2 = 947.109$$

$$s_2{}^2 = \frac{157\,630}{10} - (125)^2 = 138$$

giving:

$$\hat{\sigma} = \sqrt{\frac{8(947.109) + 10(138)}{8 + 10 - 2}} = 23.66 \quad \text{(2D)}$$

Hence:

$$t_{\text{test}} = \frac{140.875 - 125}{23.66\sqrt{\dfrac{1}{8} + \dfrac{1}{10}}} = 1.41 \quad \text{(2D)}$$

Here:

$$t_{\text{critical}} = 2.12 \quad \text{(approximately, with } \nu = 16 \text{ at the 5\% level)}$$

Hence, $t = 1.42$ is not significant at the 5% level, so there is no evidence that the mean weights of the men in the two tribes are different.

(b) Given that $\sigma^2 = 225$ (i.e. $\sigma = 15$), the test statistic is the one given in Statement 9.5(a). That is:

$$Z = \frac{\overline{X} - \overline{Y}}{\sigma\sqrt{\dfrac{1}{n_1} + \dfrac{1}{n_2}}} = \frac{140.875 - 125}{15\sqrt{\dfrac{1}{8} + \dfrac{1}{10}}} = 2.23 \text{ (2D)}$$

Here, $z_{\text{critical}} = 1.96$. Thus $z_{\text{test}} = 2.23$ *is significant* at the 5% level, giving evidence of a difference in weights. Thus the decision made in (a) above has been reversed.

9.5 TESTS INVOLVING SAMPLE PROPORTIONS

In Section 9.3.1 we tested an assertion about a Normal population mean, μ, using the sample statistic $\overline{X}$, an unbiased estimator. In just the same way we can use the sample statistic P (an unbiased estimator for population proportion π) to test assertions about π.

There are two common tests involving proportions which we shall deal with in turn.

9.5.1 TEST FOR A SAMPLE PROPORTION (Z TEST)

STATEMENT 9.6

To test whether a random sample of size n (large) with proportion P having a certain attribute is drawn from a population with proportion π having the given attribute, we use the test statistic:

$$Z = \frac{P - \pi}{\sqrt{\dfrac{\pi(1 - \pi)}{n}}}$$

which is distributed as $N(0, 1)$ under the null hypothesis that the sample *is* drawn from the given population.

Proof Trivial, since under the null hypothesis, we know that (for large n) P is (approximately): $N\left(\pi, \dfrac{\pi(1 - \pi)}{n}\right)$

giving $\dfrac{P - \pi}{\sqrt{\dfrac{\pi(1 - \pi)}{n}}}$ as $N(0, 1)$

Notice that we do not distinguish between a known and unknown variance as we do when a sample mean is tested, since once we assume the population proportion π, under H_0, the variance of P, $\pi(1 - \pi)$, is immediately known.

EXAMPLE 9.17

If a sample of 200 people, randomly chosen, contained 160 who found that a certain type of tablet was successful in curing their headaches, comment on the tablet manufacturer's claim that the tablet is 90% successful.

SOLUTION

Here, we have a sample size $n = 200$ with a proportion of $p = \dfrac{160}{200} = 0.8$ who find the tablet effective. We wish to test whether this sample could have come from a population with proportion 0.9 who find the tablet effective.

Let π be the true population proportion of the people who find the tablet effective. The test is $H_0: \pi \geqslant 0.9$ versus $H_1: \pi < 0.9$, and we are taking the implication from the manufacturer's claim that the success rate is 90% *or more*.

The test statistic here is $Z = \dfrac{P - \pi}{\sqrt{\dfrac{\pi(1 - \pi)}{n}}}$, and with $p = 0.8$, $\pi = 0.9$

and $n = 200$, we have:

$$z_{\text{test}} = \frac{0.8 - 0.9}{\sqrt{\dfrac{0.9(1 - 0.9)}{200}}} = -4.71$$

Now, $Z_{95\%} = -1.645$ and $Z_{99\%} = -2.326$ are the critical values of z at the 5% and 1% levels respectively. Comparing these with the test value of z, we clearly have a highly significant result.

Hence, on the basis of these results, we can confidently reject the manufacturer's claim.

EXAMPLE 9.18

A sample of 500 eggs was taken from a battery station to test whether (or not) the proportion of white eggs (as opposed to brown) is $\frac{1}{2}$. The result was just significant at the 5% level.

(a) How many white eggs were there in the sample?

(b) Find the probability of making a Type II error when the true proportion is $\pi = 0.44, 0.46, 0.48, 0.52, 0.54$ and 0.56, using the same critical region for the test.

Hence sketch the operating characteristic curve.

SOLUTION

This is a two-tailed test since we are interested only in whether the two proportions (of white and brown eggs) are different, and not whether one is larger than the other. The null hypothesis must be that both proportions are $\frac{1}{2}$ and the test statistic here is

$$Z = \frac{P - \pi}{\sqrt{\frac{\pi(1 - \pi)}{n}}} \quad \text{where} \quad \pi = \frac{1}{2} \quad \text{and} \quad n = 500. \text{ The value of } P \text{ is}$$

unknown of course since we do not know the number of white eggs in the sample. If we let the number of white eggs be W, then

$$p = \frac{W}{500}.$$

Now, for z_{test} to be just significant, we need it to take the value ± 1.96 (which one of the two will depend on whether W is above or below 250 — note that it can be either). Without loss of generality we will assume $W > 250$.

Hence, we have:
$$\frac{\frac{W}{500} - 0.5}{\sqrt{\frac{(0.5)(0.5)}{500}}} = 1.96$$

Therefore:

$$\frac{W}{500} = (1.96)\sqrt{\frac{(0.5)(0.5)}{500}} + 0.5 \quad \text{(by rearrangement)}$$

giving: $W = 500\left[(1.96)\sqrt{\frac{(0.5)(0.5)}{500}} + 0.5\right] = 271.9 \quad \text{(1D)}$

Thus: $W = 272$ (to the nearest whole number)

Alternatively, $W = 228$ (i.e., 22 *less* than 250).

The two values for P, (P_1 and P_2 say) that correspond to the two critical values for the test, can be found using the previous 'just significant' values of W.

That is:
$$P_1 = \frac{228}{500} = 0.456$$

and
$$P_2 = \frac{272}{500} = 0.544$$

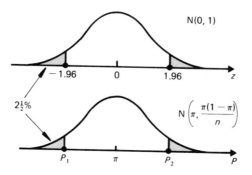

Hence:

$$\Pr(\text{Type II error}/\pi = 0.44) = \Pr(\text{accepting } H_0 : \pi = 0.5/\pi = 0.44)$$

$$= \Pr(0.456 < P < 0.544/\pi = 0.44)$$

$$= \Pr\left(\frac{0.456 - 0.440}{\sqrt{\dfrac{(0.44)(0.56)}{500}}} < Z < \frac{0.544 - 0.440}{\sqrt{\dfrac{(0.44)(0.56)}{500}}} \right) \quad \text{(standardising)}$$

$$= \Pr(0.72 < Z < 4.68)$$

$$= \Phi(4.68) - \Phi(0.72)$$

$$= 1 - 0.7642$$

$$= 0.236 \quad \text{(3D)}$$

$\Pr(\text{Type II error}/\pi = 0.46)$

$$= \Pr\left(\frac{0.456 - 0.460}{\sqrt{\dfrac{(0.46)(0.54)}{500}}} < Z < \frac{0.544 - 0.460}{\sqrt{\dfrac{(0.46)(0.54)}{500}}} \right)$$

$$= \Pr(-0.18 < Z < 3.77)$$

$$= \Phi(3.77) - \Phi(-0.18)$$

$$= 0.571$$

$\Pr(\text{Type II error}/\pi = 0.48)$

$$= \Pr\left(\frac{0.456 - 0.480}{\sqrt{\dfrac{(0.48)(0.52)}{500}}} < Z < \frac{0.544 - 0.480}{\sqrt{\dfrac{(0.48)(0.52)}{500}}} \right)$$

$$= \Pr(-1.07 < Z < 2.86)$$

$$= \Phi(2.86) - \Phi(-1.07)$$

$$= 0.856$$

Also, by symmetry:

$$\Pr(\text{Type II error}/\pi = 0.52) = \Pr(\text{Type II error}/\pi = 0.48)$$

$$\Pr(\text{Type II error}/\pi = 0.54) = \Pr(\text{Type II error}/\pi = 0.46)$$

and: $\Pr(\text{Type II error}/\pi = 0.56) = \Pr(\text{Type II error}/\pi = 0.44)$

Using the above probabilities, the operating characteristic curve is plotted in the following diagram.

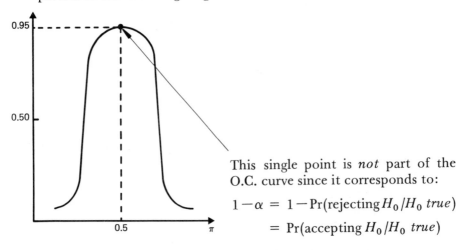

This single point is *not* part of the O.C. curve since it corresponds to:

$$1-\alpha = 1-\Pr(\text{rejecting } H_0/H_0 \ \textit{true})$$
$$= \Pr(\text{accepting } H_0/H_0 \ \textit{true})$$

9.5.2 TEST FOR THE DIFFERENCE BETWEEN TWO SAMPLE PROPORTIONS (Z TEST)

Given two samples of sizes n_1 and n_2 and respective proportions P_1 and P_2 (having a particular given attribute), a test statistic will be formulated to determine whether the two samples are drawn from the same population (or, equivalently, from two populations with the same proportion π, say).

The same sort of approach as that in Section 9.4 is used, where we consider the quantity $P_1 - P_2$.

Since P_1 and P_2 are both approximately Normally distributed, so is $P_1 - P_2$ (from Normal theory). Also, under a null hypothesis that the two samples *are* drawn from the same population, $P_1 - P_2$ must have a zero mean.

We need to estimate the variance of the distribution of $P_1 - P_2$. First, if π is the common proportion (w.r.t. the given attribute) of both populations, P_1 will have variance $\dfrac{\pi(1-\pi)}{n_1}$ and P_2 will have variance $\dfrac{\pi(1-\pi)}{n_2}$.

To estimate the value of π, the weighted mean of the realised values, p_1 and p_2 is used, i.e.:

$$\hat{\pi} = \frac{n_1 p_1 + n_2 p_2}{n_1 + n_2}$$

Hence: $\text{Var}[P_1 - P_2] = \text{Var}[P_1] + \text{Var}[P_2]$

$$= \frac{\hat{\pi}(1-\hat{\pi})}{n_1} + \frac{\hat{\pi}(1-\hat{\pi})}{n_2} \quad \text{(from above)}$$

$$= \hat{\pi}(1 - \hat{\pi}) \left(\frac{1}{n_1} + \frac{1}{n_2} \right)$$

So that $P_1 - P_2$ is $N\left(0, \hat{\pi}(1 - \hat{\pi}) \left(\frac{1}{n_1} + \frac{1}{n_2} \right) \right)$ approximately, where:

$$\hat{\pi} = \frac{n_1 p_1 + n_2 p_2}{n_1 + n_2}$$

Hence:

$$\frac{P_1 - P_2}{\sqrt{\hat{\pi}(1 - \hat{\pi}) \left(\frac{1}{n_1} + \frac{1}{n_2} \right)}}$$

is $N(0, 1)$ under the null hypothesis of a common population π or, alternatively, $\pi_1 = \pi_2$ if two populations are involved. This quantity serves as the test statistic.

EXAMPLE 9.19

A mail order firm sent out samples of two types of cloth, A and B, to two different randomly chosen groups of 1000 women. Of the 1000 to whom the firm sent cloth A, 100 orders were received. 120 orders were returned from the other group of 1000 who originally received cloth B. Test whether, in general, cloth B will prove to be more successful.

SOLUTION

Let p_A and p_B be the proportions of the groups that order cloths A and B respectively. Then clearly $p_A = 0.1$ and $p_B = 0.12$. The null hypothesis will be that there is no difference between the popularity of the two cloths, i.e., we test $H_0: \pi_A \geqslant \pi_B$ versus $H_1: \pi_A < \pi_B$. The alternative hypothesis here ties in with the fact that we are interested in whether a greater proportion of women prefer cloth B.

The test statistic is:

$$Z = \frac{P_A - P_B}{\sqrt{\hat{\pi}(1 - \hat{\pi}) \left(\frac{1}{n_A} + \frac{1}{n_B} \right)}}$$

which has a critical value of -1.645 $(P = 5\%$, one-tailed$)$.

Now, $n_A = n_B = 1000$, and:

$$\hat{\pi} = \frac{n_A p_A + n_B p_B}{n_A + n_B}$$

$$= \frac{1000(0.10) + 1000(0.12)}{1000 + 1000}$$

$$= \frac{220}{2000} = 0.11$$

Also:
$$\frac{1}{n_A} + \frac{1}{n_B} = \frac{2}{1000} = 0.002$$

Hence:
$$z_{\text{test}} = \frac{0.10 - 0.12}{\sqrt{(0.11)(1 - 0.11)(0.002)}} = -1.43$$

which, when compared with the critical value of -1.645, is not significant.

Hence, we accept H_0. In other words, there is no evidence that cloth B will be more successful than cloth A.

To summarise:

STATEMENT 9.7

To test whether two random samples with proportions P_1 and P_2 (having a certain common attribute) and sizes n_1 and n_2 are both drawn from a population having the attribute proportion π (unknown), we use the test statistic:

$$Z = \frac{P_1 - P_2}{\sqrt{\hat{\pi}(1 - \hat{\pi})\left(\frac{1}{n_1} + \frac{1}{n_2}\right)}} \quad \text{where} \quad \hat{\pi} = \frac{n_1 p_1 + n_2 p_2}{n_1 + n_2}$$

The above statistic has a $N(0, 1)$ distribution under the null hypothesis that both populations (from which the samples have been drawn) have the same theoretical proportion, π, with the given attribute.

Note $\hat{\pi}$ is an unbiased estimate of the unknown π.

Proof   In preceding text.

9.6 TESTS INVOLVING SAMPLE VARIANCES

There are two tests of interest involving sample variances.

The first is a straightforward test of the significance of a single sample variance. That is, having obtained a sample variance S^2, we ask whether the sample could have been drawn from a Normal population having variance σ^2.

The second test involves two samples; we ask whether they could have been drawn from Normal populations having the same variance (and base our decision on their sample variances). Recall that in order to carry out a two-sample (unpaired) T-test for the difference between two sample means, we had to assume that the two populations under consideration had a common variance. After studying the above latter test, we will be able to test the aforementioned assumption statistically.

9.6.1 TEST FOR A SAMPLE VARIANCE (χ^2 TEST)

If a random sample of size n, $X_1, X_2, \ldots, X_n$ is drawn from a Normal population having variance σ^2, we know from Section 7.5 that the statistic $\dfrac{nS^2}{\sigma^2}$ is distributed as a $\chi^2(n-1)$ variable. Thinking of this situation in terms of a statistical test, we have: given a random sample of size n and sample variance S^2, in order to see whether it could have been drawn from a Normal distribution, variance σ^2, we examine the statistic $\dfrac{nS^2}{\sigma^2}$ to see whether it behaves as a $\chi^2(n-1)$ variable. That is, $\dfrac{nS^2}{\sigma^2}$ is used as a test statistic, distributed as $\chi^2(n-1)$ under the null hypothesis that the sample *is* drawn from a Normal population with variance σ^2.

The actual test procedure is identical to that of previous tests. However, the determination of critical values is different here (as opposed to the Normal and T distributions) since we are dealing, under H_0, with a χ^2 distribution, which is strictly non-symmetric.

Recall the basic theory for the distribution and the use of χ^2 tables given in Section 7.4.

If we were conducting a one-tailed test (increase) at the 5% level with $n = 10$, say, the critical value for the test would be $\chi^2_{0.95}(9)$, which from tables is 16.92 (see Fig. 9.15).

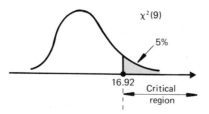

$\chi^2(9)$

5%

16.92

Critical region

Fig. 9.15

For a one-tailed (decrease) test at the 1% level with $n = 22$, say, the critical value would be $\chi^2_{0.01}(21)$ which, from tables, is 8.9 (see Fig. 9.16).

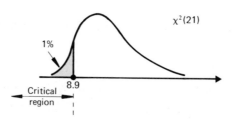

$\chi^2(21)$

1%

8.9

Critical region

Fig. 9.16

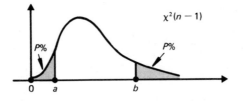

Fig. 9.17

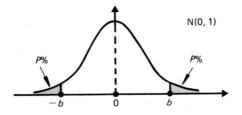

Fig. 9.18

To find the two critical values for use with a two-tailed test, note that since the χ^2 distribution is not symmetric, and for instance in Fig. 9.17, although the two tails of the distribution are equal in probability ($P\%$ each), there is no relationship between a and b. (Note that for a Normal distribution, a would equal $-b$, shown in Fig. 9.18.) So if the test was two-tailed at the 5% level with $n = 15$, the two critical values would be:

$$a = \chi^2_{0.025}(14) = 5.63 \quad \text{and} \quad b = \chi^2_{0.975}(14) = 26.12.$$

To summarise:

STATEMENT 9.8

To test whether a random sample of size n with sample variance S^2 is drawn from a Normal population with variance σ^2, we use

the test statistic $\chi^2 = \dfrac{nS^2}{\sigma^2}$ which is distributed as $\chi^2(n-1)$

under the null hypothesis that the sample *is* drawn from the given population.

Proof   See the discussion at the beginning of this sub-section.

EXAMPLE 9.20

A sample of size 9 has a variance of 8.01. Discuss whether it is likely to have been drawn from a Normal population with variance 9.0. Test at the 5% level.

SOLUTION

This is a two-tailed test, the two competing hypotheses being $H_0: \sigma^2 = 9.0$ versus $H_1: \sigma^2 \neq 9.0$.

$$\chi^2_{test} = \frac{nS^2}{\sigma^2} = \frac{9(8.01)}{9.0} = 8.01$$

Left critical $\chi^2 = \chi^2_{0.025}(8) = 2.18$

Right critical $\chi^2 = \chi^2_{0.975}(8) = 17.53$ (from tables)

Hence χ^2_{test} is not significant and we accept H_0. So that, on the evidence, it is likely that the sample has been drawn from a Normal population with variance 9.

EXAMPLE 9.21

From past experience it is known that a certain Normal population has variance 42. A sample of 16 yields a variance of 18.4. Is there any evidence to suggest that the variance of the population is now lower than 42?

SOLUTION

We are interested in detecting whether the population variance has decreased and so we need to conduct a one-tailed test, using the hypothesis $H_0: \sigma^2 \geq 42$ versus $H_1: \sigma^2 < 42$.

$$\chi^2_{test} = \frac{nS^2}{\sigma^2} = \frac{16(18.4)}{42} = 7.01 \quad (2D)$$

If we test at the 5% level, since we are concerned with a one-tailed decrease, the critical region is determined by the allocation of the 5% in the left-hand tail of a $\chi^2(15)$ distribution. The critical value itself, then, is just $\chi^2_{0.05}(15) = 7.26$, from tables.

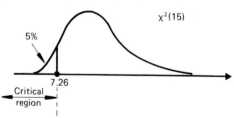

Thus the test value of 7.01 is just significant and so there is some evidence that the population variance has decreased.

9.6.2 TEST FOR THE RATIO OF TWO SAMPLE VARIANCES (F TEST)

We now consider a test for whether two samples could have been drawn from Normal populations having the same variance.

Let the two random samples be denoted by $X_1, \ldots, X_n$ and $Y_1, \ldots, Y_m$ respectively, where sample sizes n and m are not necessarily equal, with sample variances S_X^2 and S_Y^2.

Now, if the samples *are* drawn from Normal populations sharing a common variance, σ^2, say, we have that:

$$\frac{n \cdot S_X^2}{\sigma^2} \text{ is } \chi^2(n-1) \quad \text{and} \quad \frac{m \cdot S_Y^2}{\sigma^2} \text{ is } \chi^2(m-1)$$

using the results of Section 7.5.

Hence: $$\frac{n \cdot S_X^2}{(n-1)\sigma^2} \Big/ \frac{m \cdot S_Y^2}{(m-1)\sigma^2}$$

is an $F(n-1, m-1)$ variable, using the results of Section 8.5.2.

But: $$\frac{n \cdot S_X^2}{(n-1)\sigma^2} \Big/ \frac{m \cdot S_Y^2}{(m-1)\sigma^2} = \frac{n \cdot S_X^2}{n-1} \Big/ \frac{m \cdot S_Y^2}{m-1} = \frac{\hat{\sigma}_X^2}{\hat{\sigma}_Y^2}$$

So that, under the hypothesis that both samples come from Normal distributions having the same variance, the statistic $F = \dfrac{\hat{\sigma}_X^2}{\hat{\sigma}_Y^2}$ is $F(n-1, m-1)$. We can thus use F as a test statistic in the usual way.

This particular test for common variances is usually two-tailed. Symbolically the test takes the form $H_0: \sigma_X^2 = \sigma_Y^2$ versus $H_1: \sigma_X^2 \neq \sigma_Y^2$ usually.

As regards finding critical values for the test statistic, Section 8.5.3 discussed the method of obtaining percentage points using F tables (Appendix Table 6). For example, if we wanted the critical values for a 5% test where the test statistic was $F(10, 8)$, we would have (see Fig. 9.19):

$$F_2 = F_{0.975}(10, 8) = 4.30 \quad \text{(from } F \text{ tables)}$$

and: $$F_1 = F_{0.025}(10, 8)$$

$$= \frac{1}{F_{0.975}(8, 10)} \quad \text{(by Statement 8.12)}$$

$$= \frac{1}{3.85} \quad \text{(from } F \text{ tables)}$$

$$= 0.26$$

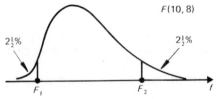

Fig. 9.19

However, the calculation of the second critical value (for the left hand tail) is obviated if the largest variance estimate is always placed in the numerator of the test statistic. This ensures that the test statistic value always lies on the right hand side of the curve, and hence we need only concern ourselves with a right critical value. Note, however, that the probability allocated to this critical region is still only $2\frac{1}{2}\%$ (for a 5% level test). It can be thought of as one half of the critical region superimposed over the other.

The justification for this approach is simply the reciprocal nature of the F distribution (see Statements 8.11(b) and 8.12). Referring to Fig. 9.19:

$$\Pr(F < F_1) = \Pr\left(\frac{1}{F} > \frac{1}{F_1}\right)$$

$$= \Pr\left(F' > \frac{1}{F_{0.025}(10, 8)}\right) \quad \text{where } F' = \frac{1}{F} = F(8, 10)$$

$$= \Pr(F' > F_{0.975}(8, 10))$$

and $F_{0.975}(8, 10)$ is just the corresponding F_2 point of a $\frac{1}{F}$ distribution.

In other words, inverting the test statistic is equivalent to 'switching the tails around'.

STATEMENT 9.9

To test whether two random samples of sizes n and m with respective sample variances S_X^2 and S_Y^2 are drawn from Normal populations having a common variance, we use the test statistic $F = \dfrac{\hat{\sigma}_X^2}{\hat{\sigma}_Y^2}$ which has an $F(n-1, m-1)$ distribution under the null hypothesis that the samples *are* drawn from the above mentioned populations.

$$\hat{\sigma}_X^2 = \frac{n \cdot S_X^2}{n-1} \quad \text{and} \quad \hat{\sigma}_Y^2 = \frac{m \cdot S_Y^2}{m-1}$$

are the respective unbiased estimates of the common population variance.

The test is usually two-tailed, of the form $H_0 : \sigma_X^2 = \sigma_Y^2$ versus $H_1 : \sigma_X^2 \neq \sigma_Y^2$.

It is usually found convenient to put the larger variance estimate in the numerator, but care must be taken in selecting the parameters for the hypothetical critical value for F. That is, the *first* parameter for F is taken from the number of degrees of freedom for the variance estimate in the *numerator*.

Proof   In preceding text.

Note   In the above test statistic, if the largest unbiased estimate of variance is put in the numerator then (for a $P\%$ level test) we need to use the $\dfrac{P}{2}$ % point   of the relevant F distribution   as the critical value.

EXAMPLE 9.22

Two random samples of sizes 8 and 11   yield variances of 12.4 and 19.3   respectively. Test at the   5% level   whether these two samples could have been drawn from Normal populations having the same variance.

SOLUTION

Putting $n = 8$; $m = 11$; $s_1^2 = 12.4$; $s_2^2 = 19.3$, we are testing the hypothesis $H_0: \sigma_1^2 = \sigma_2^2$ versus $H_1: \sigma_1^2 \neq \sigma_2^2$.

The two estimates of the common population variance are given by:

$$\hat{\sigma}_1^2 = \frac{ns_1^2}{n-1} = \frac{8(12.4)}{7} = 14.17 \quad (2D)$$

$$\hat{\sigma}_2^2 = \frac{ms_2^2}{m-1} = \frac{11(19.3)}{10} = 21.23$$

Hence: $F_{\text{test}} = \dfrac{\hat{\sigma}_2^2}{\hat{\sigma}_1^2} = \dfrac{21.23}{14.17} = 1.50 \quad (2D)$

But critical F is $F_{0.975}(10, 7) = 4.76$. (Note that parameter 10 comes first in the bracket, since this has to be linked to the variance estimate in the numerator of the test statistic.)

Clearly F_{test} is not significant here and thus we conclude that the two samples could have been drawn from Normal populations having the same variance.

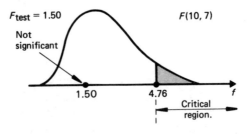

EXAMPLE 9.23

In an experiment on the reaction times in seconds of two individuals A and B, measured under identical conditions, the following results were obtained:

A	0.41	0.38	0.37	0.42	0.35	0.38
B	0.32	0.36	0.38	0.33	0.38	

Examine the hypothesis that there is no difference between the populations of reaction times for A and B.

SOLUTION

The hypothesis that there is no difference between the two populations is equivalent to the two separate hypotheses:

(a) No difference between the means.

(b) No difference between the variances.

A T-test is used for (a) and an F-test for (b). Since the truth of (b) has to be shown or assumed before (a) can be tested, we need to perform the F-test first.

x	y	x^2	y^2
0.41	0.32	0.1681	0.1024
0.38	0.36	0.1444	0.1296
0.37	0.38	0.1369	0.1444
0.42	0.33	0.1764	0.1089
0.35	0.38	0.1225	0.1444
0.38		0.1444	
2.31	1.77	0.8927	0.6297

$$\bar{x} = \frac{2.31}{6} = 0.385; \quad \bar{y} = \frac{1.77}{5} = 0.354;$$

$$s_x^2 = \frac{0.8927}{6} - (0.385)^2 = 0.000\,558 \quad (6D)$$

$$s_y^2 = \frac{0.6297}{5} - (0.354)^2 = 0.000\,624 \quad (6D)$$

For the F test, the competing hypotheses are $H_0: \sigma_x^2 = \sigma_y^2$ versus $H_1: \sigma_x^2 \neq \sigma_y^2$ with test statistic:

$$F = \frac{\text{larger unbiased variance estimate}}{\text{smaller unbiased variance estimate}}$$

$$\hat{\sigma}_x^2 = \frac{6(0.000\,558)}{5} = 0.000\,67$$

and:

$$\hat{\sigma}_y^2 = \frac{5(0.000\,624)}{4} = 0.000\,78$$

Thus:

$$F_{\text{test}} = \frac{0.000\,78}{0.000\,67} = 1.16 \quad (2D)$$

Since the critical value, $F_{0.975}(4, 5) = 7.39$, $F_{\text{test}} = 1.16$ is clearly not significant and we accept the null hypothesis that both samples are drawn from populations having the same variance.

Under this last assumption, we can use a T-test to look for a difference between the means, testing $H_0: \mu_x = \mu_y$ versus $H_1: \mu_x \neq \mu_y$.

The test statistic is:

$$T = \frac{\bar{X} - \bar{Y}}{\hat{\sigma}\sqrt{\dfrac{1}{n_x} + \dfrac{1}{n_y}}}$$

where, for this situation, we have:

$$\bar{x} = 0.385; \quad \bar{y} = 0.354; \quad n_x = 6; \quad n_y = 5$$

and:

$$\hat{\sigma} = \sqrt{\frac{n_x s_x^2 + n_y s_y^2}{n_x + n_y - 2}}$$

$$= \sqrt{\frac{6(0.000\,558) + 5(0.000\,624)}{9}}$$

$$= 0.0268 \quad (4D)$$

Hence: $$t_{\text{test}} = \frac{0.385 - 0.354}{0.0268\,\sqrt{\left(\dfrac{1}{6} + \dfrac{1}{5}\right)}} = 1.91 \quad (2D)$$

The test here is two-tailed, since we are looking for any difference between the means. Therefore critical t ($P = 0.975$, $\nu = 9$, two-tailed) is 2.26. The test value of 1.91 is thus not significant. Therefore there is no evidence to reject the null hypothesis and we conclude that the samples are drawn from populations having the same means.

Thus, we can finally assert that (on the evidence given) there is no difference between the populations of reaction times for A and B.

9.7 TESTS INVOLVING THE CORRELATION COEFFICIENT

In this section we look at two tests of importance.

The first is to see whether a bivariate random sample with correlation coefficient r could have been drawn from a bivariate Normal population having correlation coefficient ρ.

The second is to test whether two bivariate Normal random samples (with coefficients r_1 and r_2) come from the same Normal bivariate population or, equivalently, whether the two samples come from populations having the same correlation coefficient.

Both these tests are conducted in just the same way as other tests we have used so far and in the following sub-sections we simply derive the two respective test statistics (from previous theory) and give examples of their use.

9.7.1 TEST FOR A CORRELATION COEFFICIENT (Z AND T TESTS)

We know, from Statement 7.12, that if R is a correlation coefficient obtained from a sample of size n from a bivariate population with correlation coefficient ρ then, approximately, R^* is $N\left(\rho^*, \dfrac{1}{n-3}\right)$ where $*$ denotes Fisher's transformation.

Standardising the above gives: $\dfrac{R^* - \rho^*}{\sqrt{\dfrac{1}{n-3}}}$ is $N(0, 1)$

under the assumption that the sample is drawn from a population with coefficient $\rho(\neq 0)$. Note that for $H_0 : \rho = 0$ (or $\rho \leqslant 0$ or $\rho \geqslant 0$) a special t test statistic must be used, as given in Statement 9.10 below.

The above Z test statistic is used for ascertaining whether a sample yielding correlation coefficient r can be considered as having been drawn from a population having a correlation coefficient ρ.

For example, given a bivariate random sample of size 22 with a correlation coefficient of 0.78, could we say that the sample could have come from a population with coefficient 0.95?

Here, $r = 0.78$ giving $r^* = \mathrm{Fi}(0.78) = 1.045$.

Also, $\rho = 0.94$. Hence $\rho^* = \mathrm{Fi}(0.95) = 1.832$.

We are testing $H_0 : \rho = 0.95$ against $H_1 : \rho \neq 0.95$ (two-tailed).

The test statistic is:

$$Z = \frac{R^* - \rho^*}{\sqrt{\dfrac{1}{n-3}}} \quad \text{giving} \quad z_{\text{test}} = \frac{1.045 - 1.832}{\sqrt{1/19}} = -3.43$$

Critical z (5% level, two-tailed) $= \pm 1.96$.

Hence z_{test} is significant and, on the evidence, it is likely that the sample has not been drawn from the given population.

To summarise:

STATEMENT 9.10

Given a bivariate random sample of size n yielding a correlation coefficient r, we test whether the sample could have been drawn from a bivariate Normal population with correlation coefficient ρ:

(a) For $\rho \neq 0$, test statistic is $Z = \dfrac{R^* - \rho^*}{\sqrt{\dfrac{1}{n-3}}}$ (which is $N(0, 1)$ under H_0).

(b) For $\rho = 0$, test statistic is $T = R\sqrt{\dfrac{n-2}{1-R^2}}$ (which is $T(n-2)$ under H_0).

Notes   1. $R^* = \mathrm{Fi}(R)$, Fisher's transformation.

2. H_0 is that sample *is* drawn from the stated population.

Proof (a)   In preceding text.

(b)   None.

EXAMPLE 9.24

A bivariate random sample (involving variables X and Y) of size 10 yields a correlation coefficient of 0.32. Is there evidence of positive correlation between X and Y?

SOLUTION

Here we are looking for evidence of positive correlation between the two variables (X and Y). Thus the two competing hypotheses for the test are: $H_0 : \rho \leqslant 0$ against $H_1 : \rho > 0$. Notice that this is a one-tailed test of type (b) from Statement 9.10.

The test statistic is: $r \sqrt{\dfrac{n-2}{1-r^2}}$

But we are given that $r = 0.32$ and $n = 10$.

$$\therefore \quad t_{test} = r \sqrt{\frac{n-2}{1-r^2}} = (0.32) \sqrt{\frac{8}{1-(0.32)^2}} = 0.96 \quad (2D)$$

Critical t (5% level, 1-tailed, 8 d.f.) = 1.86, from tables.

Hence, $t = 0.96$ is *not significant* and thus there is no evidence of positive correlation between variables X and Y.

EXAMPLE 9.25

A bivariate random sample of size 14 yields a correlation coefficient of $r = 0.8$. (a) Is there any evidence that ρ (the true population correlation coefficient) is greater than 0.5? (b) Find the probability of making a type II error for this test if in fact $\rho = 0.7$.

SOLUTION

(a)   Since we are looking for evidence that $\rho > 0.5$, the competing hypotheses for the test can be set up as $H_0 : \rho \leqslant 0.5$ v. $H_1 : \rho > 0.5$ (a one-tailed test).

Notice that this is a test of type (a) from Statement 9.10.

That is, the test statistic is $z = \dfrac{r^* - \rho^*}{\sqrt{\dfrac{1}{n-3}}}$

But we are given that $n = 14$, $r = 0.8$ and ρ (under H_0) = 0.5.

Thus: $\quad r^* = \text{Fi}(0.8) = 1.0986 \quad$ and $\quad \rho^* = \text{Fi}(0.5) = 0.5493$

$$z_{\text{test}} = \frac{r^* - \rho^*}{\sqrt{\dfrac{1}{n-3}}} = \frac{1.0986 - 0.5493}{\sqrt{\dfrac{1}{11}}} = 1.82$$

But the critical value for Z (5%, 1-tailed) $= 1.645$. Therefore Z_{test} is *significant* and H_0 can thus be rejected. In other words, there *is* evidence that $\rho > 0.5$.

(b) $\text{Pr(Type II error)} = \text{Pr(accepting } H_0/\rho = 0.7)$.

Now, H_0 is accepted when $Z_{\text{test}} < 1.645$.

But $Z_{\text{test}} = 1.645$ when

$$\frac{R^* - \rho^*}{\sqrt{\dfrac{1}{n-3}}} = 1.645$$

i.e. when

$$\frac{R^* - 0.5493}{\sqrt{\dfrac{1}{11}}} = 1.645$$

This can be re-arranged to $R^* = (1.645)\sqrt{\dfrac{1}{11}} + 0.5493$

i.e. $R^* = 1.045$

That is, H_0 is accepted when $R^* < 1.045$.

Thus $\text{Pr(Type II error)} = \text{Pr(accepting } H_0/\rho = 0.7)$

$$= \text{Pr}(R^* < 1.045/\rho = 0.7)$$

(But R^* is $N(\rho^*, 1/(n-3))$ from Statement 7.12 and $\rho^* = \text{Fi}(0.7) = 0.8673$.)

$$= \text{Pr}\left(Z < \frac{1.045 - 0.8673}{\sqrt{\dfrac{1}{11}}}\right)$$

$$= \text{Pr}(Z < 0.59)$$

$$= \Phi(0.59)$$

$$= 0.72 \ (2\text{D})$$

EXAMPLE 9.26

Design a decision rule (involving the sample correlation coefficient R) to test, at the 5% level, whether the bivariate population from which a sample of size 20 is drawn has a correlation coefficient greater than 0.9

SOLUTION

This is a one-tailed test since we need to decide between the hypotheses $H_0 : \rho \leqslant 0.9$ versus $H_1 : \rho > 0.9$.

The test statistic is $Z = \dfrac{R^* - \rho^*}{\sqrt{\dfrac{1}{n-3}}}$.

With $\rho = 0.9$ we have $\rho^* = \text{Fi}(0.9) = 1.472$ and $n = 20$ gives:

$$\sqrt{\frac{1}{n-3}} = \sqrt{\frac{1}{17}} = 0.243$$

Hence: $$Z_{\text{test}} = \frac{R^* - 1.472}{0.243}$$

The critical value for Z at the 5% (one-tailed) level is 1.645.

To find the value of R^* corresponding to this critical value we have:

$$\frac{R^* - 1.472}{0.243} = 1.645$$

giving:

$$R^* = (1.645)(0.243) + 1.472 = 1.872 \quad (3D)$$

Hence: $R = 0.954$ (from Fisher transformation tables)

So the decision rule is:

(a)   reject H_0 if $R \geqslant 0.954$.

(b)   accept H_0 otherwise.

9.7.2 TEST FOR THE DIFFERENCE BETWEEN TWO CORRELATION COEFFICIENTS (Z TEST)

STATEMENT 9.11

Given two bivariate random samples of sizes n_1 and n_2 yielding respective correlation coefficients R_1 and R_2, to test whether they are drawn from the same bivariate population, the following test statistic is used:

$$Z = \frac{R_1^* - R_2^*}{\sqrt{\dfrac{1}{n_1 - 3} + \dfrac{1}{n_2 - 3}}}$$

(a) Z is distributed approximately as $N(0, 1)$ under the null hypothesis that both samples *are* drawn from the same bivariate population.

(b) $R^* = \text{Fi}(R)$ (Fisher's transformation).

Proof   Let the two samples be drawn from populations having respective correlation coefficients ρ_1 and ρ_2, where R_1 and R_2 are the respective sample correlation coefficients.

Hence: $R_1{}^*$ is $N\left(\rho_1{}^*, \dfrac{1}{n_1 - 3}\right)$ and $R_2{}^*$ is $N\left(\rho_2{}^*, \dfrac{1}{n_2 - 3}\right)$

approximately, using Statement 7.12.

So that $R_1{}^* - R_2{}^*$ is $N\left(\rho_1{}^* - \rho_2{}^*, \dfrac{1}{n_1 - 3} + \dfrac{1}{n_2 - 3}\right)$ using Normal theory.

Therefore: $\dfrac{R_1{}^* - R_2{}^* - (\rho_1{}^* - \rho_2{}^*)}{\sqrt{\dfrac{1}{n_1 - 3} + \dfrac{1}{n_2 - 3}}}$ is $N(0, 1)$.

But under the hypothesis that $\rho_1 = \rho_2$, we have $\rho_1{}^* = \rho_2{}^*$ giving $\rho_1{}^* - \rho_2{}^* = 0$.

Hence, from the above, $\dfrac{R_1{}^* - R_2{}^*}{\sqrt{\dfrac{1}{n_1 - 3} + \dfrac{1}{n_2 - 3}}}$ is $N(0, 1)$ under $H_0: \rho_1 = \rho_2$.

EXAMPLE 9.27

If two bivariate samples of 20 and 30 pairs have correlation coefficients 0.55 and 0.41 respectively, test, at the 5% level, the hypothesis that both samples are drawn from the same population.

SOLUTION

Assuming that the samples are drawn from populations with correlation coefficients ρ_1 and ρ_2, we test $H_0: \rho_1 = \rho_2$ versus $H_1: \rho_1 \neq \rho_2$ (a two-tailed test).

The test statistic is:

$$Z = \frac{R_1{}^* - R_2{}^*}{\sqrt{\dfrac{1}{n_1 - 3} + \dfrac{1}{n_2 - 3}}}$$

Here: $r_1 = 0.55$; $r_2 = 0.41$; $n_1 = 20$; $n_2 = 30$

$r_1{}^* = \text{Fi}(0.55) = 0.619$ and $r_2{}^* = \text{Fi}(0.41) = 0.436$

Hence: $z_{\text{test}} = \dfrac{0.619 - 0.436}{\sqrt{\dfrac{1}{17} + \dfrac{1}{27}}} = 0.59$ (2D)

Since the critical values of Z (5% level, two-tailed) are ± 1.96, the above test value is not significant. Thus we accept that both samples were drawn from the same population.

9.8 EXERCISES

SECTION 9.2

1. A distribution is Normal with mean 100 and variance 20. State whether the following samples can be considered as having been drawn from the above distribution. Use a two-tailed test at the 5% level. (a) Size 10, mean 102;   (b) Size 20, mean 98;   (c) Size 30, mean 104;   (d) Size 5, mean 97.

2. A company that packs salt prints the average contents in each packet as 453 gm and claims a variance of only 25 gm. A Food Ministry inspection consists of measuring the weights of a random sample of 100 packets of salt, finding that the sample mean of their weights is only 451.7 gm. Formulate a statistical test by: (a) determining H_0 (a null hypothesis);   (b) determining H_1 (an alternative hypothesis); (c) selecting a significance level;   (d) writing down the test statistic (in terms of μ and $\bar{X}$); and   (e) stating the critical value(s) for this statistic.

    Hence, decide whether the Ministry has any grounds for complaint.

3. The average age (in miles) of a particular make of motor car tyre is given as 20 000. A sample of tyres is taken and their age is scientifically measured. State four possible hypotheses that can be considered as suitable for use in a statistical test, stating which would be appropriate as null hypotheses.

4. With regard to the previous exercise, a rival manufacturer is considering constructing a test concerning the age of the above tyres. Which alternative hypothesis is he likely to formulate?

5. Two track athletes A and B have met 15 times in the same event, A having the faster time on 11 occasions, but B now claims that there is 'nothing between them'. At a forthcoming meeting they will be competing together in three races and it is decided to test whether indeed B is the same standard as A (H_0) or whether previous results are valid as a measure of their ability (H_1). H_0 will be accepted if B obtains a faster time than A *in any race*, otherwise H_1 will be assumed true. Find the probability of making both Type I and II errors for this test.

6. Dating of archaeological specimens is a difficult task. It is known that specimens emit a certain type of radioactive particle; the number of particles emitted in n minutes having a Poisson distribution with parameter $n\lambda$, where the value of λ depends upon the age of the specimen.

    Two hypotheses concerning the age of one particular specimen are put forward:

    H_A: specimen is 7000 years old (in which case $\lambda = 1.0$)
    H_B: specimen is 15 000 years old (in which case $\lambda = 4.0$)

    It is decided to count the number X of radioactive particles emitted in n minutes and

    accept H_A (and reject H_B) if $X \leqslant 1$

    and   accept H_B (and reject H_A) if $X \geqslant 2$

    If $n = 1$, what is

    (i) the probability of rejecting H_A when H_A is in fact true,
    (ii) the probability of rejecting H_B when H_B is in fact true?

If the probability of rejecting H_B when H_B is in fact true is to be less than 0.001, show that the minimum number of complete minutes for which counting should be recorded is three. What is the corresponding probability of rejecting H_A when H_A is in fact true?   (AEB) '80

7.   A sample of size 12 is drawn from a Normal population with known variance 2.2. A one-tailed test is proposed of the form H_0: $\mu \geqslant 24$ v H_1: $\mu < 24$ at the 5% level.   (a) Find the probability of a Type II error (β) for the cases: (i) $\mu = 23.75$; (ii) $\mu = 23.5$; (iii) $\mu = 23$; (iv) $\mu = 22$. Hence sketch the power curve for this test.   (b) Find the values of β corresponding to the above values of μ and superimpose a new power curve on the old for $\alpha = 1\%$.

SECTION 9.3

8.   The following sample was taken from a Normal distribution with variance 10:

Class	0 up to 10	10 up to 20	20 up to 30
Frequency	2	14	41
Class	30 up to 40	40 up to 50	50 up to 60
Frequency	28	12	3

Find the mean of the sample and test at the 5% level whether the mean of the parent distribution could have been 30.

9.   Examiners reported that students sitting a particular 'O' level G.C.E. examination had a mean mark of 74.5 with standard deviation 8.0. A particular school who sat 200 students for this examination boasted an average mark of 75.9. Is there any evidence (at the 5% level) that the students at this school scored better than the average?

10.   (a) A variable is expected to have a mean of 1.3 and standard deviation 0.04. A sample of 9 is taken and it is observed that the mean is 1.323. Is there significant evidence of a change and if so, at what level? (b) If the mean of the variable is now thought to have increased (while the standard deviation has remained the same), would a sample of 9 with a mean of 1.323 justify this hypothesis?

11.   Two dice were tossed 200 times and it was found that on 45 throws the two dice did not differ by more than 1. An observer deduced that the dice must be biased in some way. Find the probability of throwing two dice such that the number of pips shown on each differ by one at most, and by using a Normal approximation to the binomial, confirm or deny the observer's deduction.

12.   A random sample of n items is taken from a distribution whose mean, μ, is thought to be 48. If the sample mean and standard deviation are $\bar{x}$ and s respectively: (a) write down the test statistic, null hypothesis and alternative hypothesis for testing whether the distribution mean really is 48;   (b) calculate t_{test} and $t_{critical}$ for the following cases (where the level of significance is 5%):   (i) $n = 8, \bar{x} = 46, s = 6$; (ii) $n = 8$, $\bar{x} = 46$, $s = 2$;   (iii) $n = 5$, $\bar{x} = 46$, $s = 2$ and state your conclusions.

13. A certain population has a mean of 0.05. A sample of 10 gives $\bar{x} = 0.053$ and $s = 0.003$. Test the hypothesis that the sample does in fact come from the stated distribution at the: (a) 5% level;   (b) 1% level. (Note that this is a two-tailed test.)

14. It is known from experience that the marks obtained in a certain examination are approximately Normally distributed with mean 69. A sample of 10 students obtained an average mark of 64 with standard deviation 6.6. Does this provide evidence, at the 5% level, that the students have not done as well as might be expected?

15. A sample of 100 from an assumed Normal population gives a mean of 1570 and a standard deviation of 120. Test the hypothesis that the population mean is 1600 as opposed to the hypothesis that the mean is not 1600 at: (a) the 5% level;   (b) the 1% level.

16. A random sample of size 20 is drawn from a Normal population with known variance 5. Design a decision rule for rejecting at the 1% level a suspected population mean of 42 in favour of a larger one.

17. A process involving the manufacture of cylindrical steel canisters is thought to output the product with a mean diameter of 13.2 cm. A test is set up at the 5% level using $H_0: \mu = 13.2$ cm v $H_1: \mu \neq 13.2$ cm and a sample size of 10. Assuming the variance of the diameters of the canisters is constant at 0.8 cm: (a) design a decision rule (using the sample mean) for this test;   (b) find the power of the test when: (i) $\mu = 13.1$ cm;   (ii) $\mu = 13.4$ cm.

SECTION 9.4

18. The differences in the weights of ten similar dogs, after being fed a particular type of food over a certain period, were measured and gave an average of -1.2 lb with a standard deviation of 2.6 lb. Is there any evidence that this type of food significantly alters the dogs' weights?

19. The average fuel consumption of ten small cars, before and after a certain additive is introduced into their fuel, is as follows:

Car	1	2	3	4	5	6	7	8	9	10
With additive	47	38	44	48	52	55	44	52	60	44
Without additive	40	39	32	33	40	27	36	56	50	40

Test whether the additive really does increase fuel consumption.

20. Nine students were tested before and after having private tuition in a certain subject and the differences in their marks for the tests (mark after minus mark before) were: $7, -2, 5, 4, 22, 15, -5, 1$ and 12. Is there evidence that the tuition is effective?

21. The 'slump' of a wet concrete mix is measured by the traditional method using a conical mould to produce a pillar of wet concrete. The slump is the distance (in mm) between the top of the pillar and the top of the pile of concrete into which it subsides when the mould is removed. The

abilities of an experienced operator A and a novice B to measure slump are to be compared. Describe how an experiment requiring a paired sample t-test for its analysis could be set up. The results of such an experiment are

Sample	1	2	3	4	5	6	7
A	12	23	43	81	104	135	150
B	10	18	38	84	105	139	147

Perform the t-test.   (Cambridge)

22. The differences (d) between six pairs in a paired sample T-test yielded $\bar{d} = 2.1$ and the result was just significant at the 1% level. Find s (the standard deviation of the differences) if the test was: (a) two-tailed; (b) one-tailed.

23. Two small random samples, from each of two factories, are taken to measure the time (in minutes) to complete a particular job. The results were as follows: Factory A; sample size 6; mean 60; variance 292. Factory B; sample size 4; mean 72; variance 74. Test whether there is a difference in the times to complete the jobs between the two factories.

24. The barometric pressures in millibars at two observatories A and B, taken at random times, are as follows:

A: 1002, 996, 998, 1027, 1031, 1015, 991

B: 972, 984, 987, 999, 979, 965

Examine the hypothesis that the mean barometric pressure is the same at A as at B. Suggest a better way of making observations to test this hypothesis.

25. Prove that

$$\sum_{i=1}^{n}(x_i - \bar{x})^2 = \sum_{i=1}^{n} x_i^2 - \frac{\left(\sum_{i=1}^{n} x_i\right)^2}{n} \quad \text{where} \quad \bar{x} = \frac{\sum_{i=1}^{n} x_i}{n}$$

The lengths of the femur in samples of *mus homunculus* from two sources (Britain and North Africa) are given in the table.

Source	Lengths in mm
Britain	12.3 12.7 13.1 10.8 11.3 11.8 12.4 13.2
N. Africa	10.6 9.8 11.5 10.0 11.1

The mean length of the femur is known to be characteristic of each breed of *mus homunculus*. Test whether the data are consistent with the assumption that *mus homunculus* in Britain and North Africa are of the same breed.   (Cambridge)

26. Two samples of 10 and 12 observations respectively taken from two Normal populations gave the following sums and sums of squares:

Sample 1 $\sum_{i=1}^{10} x_i = 17.5 \quad \sum_{i=1}^{10} x_i^2 = 100.625$

Sample 2 $\sum_{i=1}^{12} y_i = 30.6 \quad \sum_{i=1}^{12} y_i^2 = 188.03$

Assuming that the samples are independent and that the population variances are unknown but equal, test the hypothesis that the two population means are equal.

27.   A firm is to buy a fleet of cars for use by its salesmen and wishes to choose between two alternative models, A and B. It places an advertisement in a local paper offering four free gallons of petrol to anyone who has bought a new car of either model in the last year. The offer is conditional on being willing to answer a questionnaire and to note how far a car goes, under typical driving conditions, on the free petrol supplied.

The following data were obtained:

	Miles driven on four gallons of petrol								
Model A	117	136	108	147					
Model B	98	124	96	117	115	126	109	91	108

(a) Test, at the 5% significance level, whether there is a difference between the mean petrol consumption of the two models. Treat the data as random samples from two normal distributions with equal variances. ·

(b) List good and bad features of the experimental method and suggest how it could be improved.   (AEB) '85

28.   In an investigation into the effectiveness of a particular course in speed reading a group of 500 students was split into two groups, A and B, of sizes 300 and 200 respectively, thought to have been chosen at random. Those in group A were given no special instruction; those in group B were given a course in speed reading. Each student was asked to read the same passage and the time taken was measured. The results were

Group A: mean time 78.4 sec, variance $14 \sec^2$

Group B: mean time 77.4 sec, variance $15 \sec^2$

Carry out a significance test to see if there is evidence that the course has improved reading speed. State carefully your null hypothesis, alternative hypothesis and final conclusion.

You learn later that, of the original 500 students, 200 students had decided for themselves that they wanted to take the course in speed reading and that these students became group B. Discuss briefly how this might affect your previous conclusion.   (JMB)

SECTION 9.5

29.   A sample of 300 from a large population produced a proportion 0.67 with a particular attribute. Test whether this result is compatible with the claim that the true proportion of the population with the attribute is 0.75, at the 5% level.

30.   The percentage of grade 'A's given in a biology examination at a certain college over a number of years was 9%. During one year there were 39 grade 'A's out of 290 students who sat the examination. Test this result at the: (a) 5%;   (b) 1% level and state any conclusions reached.

31.   At an election, party A obtained 48% of the electoral vote and some weeks later a random sample of 260 voters gave 112 in favour of party A. Does this provide evidence of a decrease in the popularity of A?

32.   The numbers of male and female in a particular city are supposed to be equally divided. A random sample of 1000 people gave 521 female. (a) Is there any evidence at the 5% level that the population is not equally divided?   (b) How many *more* females would be required for the sample (assuming the same proportion as previously) to make us reverse the decision in (a)?

33.   As a statistician, you are asked to act as an expert witness for the defence in a prosecution under the Gaming Acts. The facts are that during observation by a police witness of 900 consecutive spins of a roulette wheel the number zero won on 34 occasions. The prosecution allege that this constitutes an unfair bias since the wheel should give equal chances to any of the 37 numbers winning. Write a short statement explaining how you think the casino could be defended against the allegation of bias.   (IOS)

34.   A group of 243 people have a minor disease and 108 are given a special serum with the effect that 77 recover. If 88 of the group who have not been given the serum recover, is there evidence that the serum is effective?

35.   To test whether one bus route was utilised more than another the following observations were made. Over a period of time there were 180 buses that were over one-half full out of 500 random checks for the first route. The second route had 210 more than half-full out of 700 random checks. Does this provide evidence that the first route is utilised more fully than the second? (Note that this is a one-tailed test.)

36.   In the manufacture of a certain kind of tile, it is found that 8% of the first 1000 produced are defective. Of the next 3000, 6.8% are found to be defective. Calculate 95% confidence limits for the fraction defective on the basis of the combined set of 4000 tiles. Is there any reason to suppose the process improved after the first 1000?   (AEB)'75

37.   Out of a random sample of 250 men and 500 women, 100 men and 160 women were found to be holders of current driving licences. Carry out a test of significance to test the hypothesis that there is no difference between the proportions of men and women who hold licences.

38.   (a) Design a decision rule (w.r.t. a sample proportion) for rejecting the hypothesis that a random sample of size 50 is drawn from a population having proportion 0.3 of its members with a certain attribute in favour of a larger theoretical proportion, at the 5% level.

        (b) For the above test, calculate the probability of a Type II error when $\pi = 0.35, 0.4, 0.45, 0.5, 0.55$. Hence sketch an operating characteristic curve for the test.

SECTION 9.6

39.   A test is suggested at a 5% level to see whether a sample of 9 items is drawn from a Normal population with a variance of 11 (or higher). If the sample variance is calculated as 12.1, determine the result of the test.

40.   A sample of 15 from a certain population gives the results 9.1, 14.3, 11.2, 8.4, 8.5, 14.0, 9.9, 8.9, 11.0, 10.2, 10.8, 11.4, 13.0, 9.9 and 10.4. (a) Find the mean and variance of the sample and test at the 5% level whether the sample could have been drawn from a parent Normal population whose variance is 1.9.   (b) If the test had been to see whether the parent population variance was greater than 1.9, would a different conclusion have been reached?

41.   A test was made at the 5% level to see whether a sample of 7 items could have been drawn from a Normal population with variance 12.3, the test statistic taking the value 12.18.   (a) Find the variance of the sample.   (b) If the sample variance remains constant, find the minimum size of the sample necessary to obtain a significant value for the test statistic.

42.   The variance of a sample of 10 items is 8.4 while the variance of another sample of 12 items is 4.2. Could these two samples have been drawn from Normal populations with a common variance?

43.   Two samples were taken of sizes 14 and 19. Their respective variances were 12.8 and 20.4.   (a) Could these results be used to estimate a common parent population variance?   (b) If so, use the two sample variances to calculate this estimate.

44.   Random samples were taken of the output of two machines producing thin metal bolts and the diameter of each bolt measured in millimetres. The results were:

Machine A: 3.8, 4.2, 3.9, 3.3, 3.9, 5.1

Machine B: 2.7, 3.8, 2.9, 2.8, 3.2

Is there any evidence to show that the variation in the bolt diameters differs for each machine?

45.   A new method of treating steel sheet to delay rusting was to be compared with the process of galvanising. A laboratory experiment was set up using the new treatment on 12 pieces of steel sheet and galvanising on 10 pieces. The times (x days) to the appearance of the first signs of rust were x_1 for the new treatment and x_2 for galvanising. The following values of y_1 ($= x_1 - 480$) and y_2 ($= x_2 - 480$) were obtained:

y_1 $-12, 36, -3, -8, 30, 21, 10, -1, 8, 23, 18, 22$

y_2 $-19, 26, 5, -16, 20, -11, -3, 4, 12, 14$

and from these: $\Sigma y_1 = 144$, $\Sigma y_2 = 32$, $\Sigma y_1^2 = 4356$ and $\Sigma y_2^2 = 2204$. Use an appropriate test to determine whether the new treatment is an improvement on galvanising.   (Cambridge)

SECTION 9.7

46.   A random sample of size 25 is to be taken from a bivariate population with correlation coefficient ρ to test $H_0: \rho \leqslant 0.7$ v $H_1: \rho > 0.7$. If the probability of a Type II error when $\rho = 0.83$ is 0.672;   (a) design a decision rule to reject H_0, based on the sample correlation coefficient, R; and   (b) find the significance level of the test.

47.   In a research project, data are collected on the intelligence quotient, X, of a child and the age of its mother when the child was born, Y. From 100 observations the correlation coefficient between X and Y is calculated as 0.35. Test this for significance, and explain exactly what you are testing. A further examination of the data reveals that better educated mothers tend to have married later. To what extent would this modify your interpretation of the correlation coefficient between X and Y?

    (IOS)

48.   In a survey involving 25 salesmen, the correlation coefficient between number of successful sales and mileage expenses claimed was calculated as 0.66.

(a) Is there any evidence that salesmen of this type have a sales and expenses correlation coefficient of less than 0.85?

(b) One year later, another 38 salesmen were surveyed giving a corresponding correlation coefficient of 0.81. Is there any evidence of an increase in the population correlation coefficient?

49.   80 housewives in Durham were sampled and the correlation coefficient between age at marriage and age of mother at marriage was found to be 0.68. A similar sample of 65 Kent housewives yielded a coefficient of 0.61. Is there any evidence of a difference in the corresponding population correlation coefficients at the 5% level?

50.   A physics teacher takes two different groups of students for an Advanced Level course. One group spends only a small amount of time on practical work and receives a lot of classroom teaching, while the other group does a lot of practical work and receives relatively little classroom teaching. At the beginning of the course, both groups took the same physics examination and just before the end they took the same mock Advanced Level examination. For the first group, which consisted of 28 students, the product-moment correlation coefficient between their marks in the two examinations was 0.81, while for the second group, which consisted of 15 students, it was 0.47. The teacher asks you the following question 'Is there a significant difference between the correlation coefficients and, if so, which teaching method is better?'

(a) Making any necessary assumptions, test at the 5% significance level, whether there is a difference between the correlation coefficients for the two methods.

(b) Explain why the correlation coefficient is of no direct use in deciding which method is better, and suggest what calculations could be applied to the examination marks in an attempt to decide between the methods. (*No detailed formulae are required.*)

(c) Point out some of the features of the experiment which might affect the validity of any conclusions and suggest how these problems could be reduced or overcome in future trials.

    (AEB) '86

MISCELLANEOUS

51.   Describe what is meant by a *random sample* of observations of a random variable.

A battery manufacturer claims that his batteries have a mean life of 8 hours when used in a particular model of calculating machine. Describe how you would use a significance test to examine this claim on the basis of a large random sample of observed lifetimes. You should explain clearly the hypotheses you would consider, the choice of significance level, and the details of the test. State what is meant by concluding that the null hypothesis is rejected at the 5% significance level.

A random sample of 121 batteries has a mean life of 7.56 hours and, for this sample, $s^2 = 5.30$ hours^2. Test whether these data would lead to rejection of the manufacturer's claim, at the 5% significance level.

(JMB)

52.   A two-tailed two-sample T-test is taken to test for a difference between the means of two populations where: (a) the size, mean and variance of the first sample are 9, 9.4 and 10.1 respectively; and   (b) the size and variance of the second sample are 7 and 11.4 respectively. Find the mean of the second sample if the test statistic is just significant at the 5% level.

53.   The voltages of ten dry cells of Type A are found to be 1.50, 1.46, 1.52, 1.52, 1.47, 1.52, 1.49, 1.51, 1.49, 1.45 and those of eight dry cells of Type B are 1.48, 1.46, 1.57, 1.54, 1.59, 1.46, 1.56, 1.58. Is there reason to believe that the distributions of voltages in the two types of batteries are different and, if so, in what way?

54.   In a psychological experiment a memory test was taken by a large number of students. The mean score was 100 with a standard deviation of 12. An analysis of a sample of 30 of these students gave the following results:

Score	101	102	103	104	105	106	107
Frequency	2	4	6	8	5	3	2

On this evidence determine whether the mean score of this group was significantly different from 100, using a 5% significance level. If the mean scores for two groups of 30 students were 95 and 105, comment on the significance of the difference between these two mean scores.

(Cambridge)

55.   State what you understand by the terms: (a) null hypothesis; (b) alternative hypothesis;   (c) significance level.

A group of 10 tomato plants was grown with a certain fertilizer and the mean yield of tomatoes was found to be 1.200 kg. Records show that the same variety of plant grown under similar conditions but without the fertilizer shows a mean yield of 1.160 kg with standard deviation 0.070 kg. Stating your null and alternative hypotheses clearly, test at the 5% level of significance whether it can be claimed that the fertilizer improves the yield.

A further sample of 6 plants treated with the fertilizer gave a mean yield of 1.168 kg. Pool the results of the two samples to see whether or not the extra sample alters the previous results.

(You may assume that the yields are distributed normally.)   (SUJB)

56.   Two machines are making components to a particular dimension. Assuming that the average dimension is a function only of the setting of the machine, do the following observed values indicate that the machines are set differently?

Dimensions in inches

Machine A: 2.145 2.142 2.147 2.143 2.148 2.146 2.143 2.144

Machine B: 2.144 2.149 2.147 2.144 2.145 2.143 2.148 2.146

If the two machines should have been working to a nominal dimension of 2.146 inches, is there any evidence that either machine is offset? In what circumstances would the significance test not be strictly valid in comparing two such machines with each other?   (IOS)

57.   (a) A manufacturer based a money-back clause in his guarantee on a report which stated that at least 90% of the items he supplied were satisfactory. A random sample of 400 items contained 60 defectives. Does this suggest that there has been a change in the proportion of defectives being produced?   (b) A random sample of 300 items of another product is taken and found to contain 50 defectives. Test whether the proportions of defectives for the two products differ.   (IOS)

58.   The masses of loaves from a certain bakery are normally distributed with mean 500 g and standard deviation 20 g. (a) Determine what percentage of the output would fall below 475 g and what percentage would be above 530 g.   (b) The bakery produces 1000 loaves daily at a cost of 8p per loaf and can sell all those above 475 g for 20p each but is not allowed to sell the rest. Calculate the expected daily profit.   (c) A sample of 25 loaves yielded a mean mass of 490 g. Does this provide evidence of a reduced population mean? Use the 5% level of significance and state whether the test is one-tailed or two.   (IOS)

59.   A motoring organisation announces, after an investigation, that 1 in every 85 cars is in a dangerous condition. A road safety committee however claim that the probability of finding a car in a dangerous condition is as high as 1/50. A random inspection of 30 cars is conducted in order to decide whether to reject the former claim in favour of the latter. The former hypothesis is accepted only if none of the cars is in a dangerous condition. Find the size of the Type I and II errors for this test.

60.   Show that the estimate of $\hat{\sigma}^2 = \dfrac{1}{n-1} \displaystyle\sum_{i=1}^{n} (x_i - \bar{x})^2$ of the population variance σ^2, obtained from a random sample of size n whose mean is $\bar{x}$, is an unbiased estimate.

A random sample of size n, drawn from a Normal population whose mean and standard deviation are unknown has a variance of s^2, while a second random sample, size $(n + 4)$, drawn from the same population has a variance of $(s^2 + 100)$. The means of the two samples differ by 10. A t-test carried out on the difference between means has a t value of 2.06 and is just significant at the 5% level. Find: (a) the value of n;   (b) the unbiased estimate of the population variance based on the combined samples;   (c) the value of s.

N.B. For (a) make use of the table of percentage points of the t distribution for a two-tailed test.

61.   A sample of ten values of a continuous variable x yields the following data: $\Sigma x = 52$, $\Sigma (x - \bar{x})^2 = 33.3$. A second sample of 16 values has $\Sigma x = 72$, $\Sigma (x - \bar{x})^2 = 26.7$. The samples are assumed to come from normal populations. Show that it is a reasonable assumption that they come from the same population and find 95% confidence limits for the estimate of the population variance found from the two samples combined.   (Cambridge)

62.   The manufacturer of brand A of washing powder claims that 70% of households use brand A. A quick check showed that only 2 out of 5 of my neighbours used it. Determine whether this provides significant evidence that the manufacturer's claim is false.

A sample survey of 100 households found that 60 used brand A. Determine whether this provides significant evidence that the manufacturer's claim is false.

If, in a sample of n households, only 60% use brand A what is the smallest value of n for which this provides evidence significant at the 5% level that the manufacturer's claim is false?   (Cambridge)

63.   (a) Given that sample A (size N_1, mean $\bar{x}_1$) and sample B (size N_2, mean $\bar{x}_2$) are drawn at random, supposedly from the same Normal population (mean μ, s.d. σ).   (i) Write down the separate estimates of σ^2 which may be obtained from samples A and B respectively and use these to derive an estimate of σ^2 for the pooled sample $(A + B)$. (ii) Derive the standard error of the difference between the means of two samples of size N_1 and N_2 respectively drawn from a Normal population, using the known result for the variance of the difference of two independent variables.

(b) A group of 120 first-year students at a certain university take a test, obtaining a mean score of 67 with standard deviation 12. A group of 80 first-year students from another university take the same test and obtain a mean score of 62 with standard deviation 14. Test the hypothesis that the two samples come from a homogeneous Normal population. Use a 5% level of significance.

64.   A random sample $x_1, x_2, \ldots, x_{25}$ taken from a population having a Normal distribution with unknown mean μ and known standard deviation 1.5 gives $\sum\limits_{i=1}^{25} x_i = 62.5$.   (a) Test the null hypothesis $H_0: \mu = 2.0$ against the alternative hypothesis $H_1: \mu > 2.0$ at the 5% level of significance   (b) Calculate the size of sample necessary to ensure that the sample mean lies within 0.5 units of μ with a probability of at least 0.95.   Comment on the validity of your calculations if the population is not known to be Normal.   (IOS)

65.   A sample of 124 from a certain population found that 27 favoured a particular party for a forthcoming election. A second sample, one half the size of the first, found that the proportion in favour of the same party was greater than that of the first sample. If the difference in the two proportions was just significant at the 5% level, show that the proportion who favoured the party in the second sample satisfies the equation: $(13.52 - p) = 0.01(27 + p)(159 - p)$, where the constants are correct to 2D.

66.   In a manufacturing process, on the average one article in 20 is defective. Use the Poisson distribution to estimate the probability that a box of 100 articles will contain: (a) no defective article;   (b) fewer than three defectives.   A random sample of 20 of the articles is found to contain three defectives. Is it likely that on the average only one article in 20 is defective?

67.   With regard to a set of games of chess that they might play together, A and B make the following statements. A claims that he can win twice as many games as B can win, while conceding that half of the games should be drawn. B insists that they will draw three-quarters of all games and that he can win as many as A. (A chess game can end in a win for either player or a draw.)   (a) Find the probability that A will win any single game according to: (i) A's claim;   (ii) B's claim.

To decide between the two hypotheses (H_0: A's claim v H_1: B's claim), they decide to play six games together. Three competing tests are set up based on the decision rule: (1) accept H_0 if $X \geqslant r$; (2) reject H_0 otherwise. (X is the sample statistic 'number of wins for A' and $r = 1, 2$ and 3.)

(b) Calculate α and β (the probabilities of Type I and II errors) for each of the cases $r = 1, 2$ and 3 and hence choose that test as best for which $\alpha + \beta$ is a minimum.

68.   Two bivariate random samples of sizes 15 and 20 are to be taken from populations having correlation coefficients ρ_1 and ρ_2 respectively.

Three tests are proposed at the 5% level.

Test 1: H_0: $\rho_1 \leqslant 0.5$ v H_1: $\rho_1 > 0.5$

Test 2: H_0: $\rho_2 \leqslant 0.4$ v H_1: $\rho_2 > 0.4$

Test 3: H_0: $\rho_1 = \rho_2$ v H_1: $\rho_1 \neq \rho_2$

Find the power of each test if $\rho_1 = 0.6$ and $\rho_2 = 0.5$.

69.   In a recent promotion of a new cat food it was announced that 10 cats were offered the choice of the new cat food and an existing brand. Of the 10 cats 7 chose the new food. Set up appropriate hypotheses and determine what conclusions can be drawn from the results. If 25 cats were used find the smallest number that would have to prefer the new food in order to achieve results significant at the 5% level. (Use the normal approximation.)
                                                                                          (SUJB)

70.   What is meant by the level of significance of a test? A sample of 16 observations from a Normal population with mean μ and variance σ^2 gave the following sum and sum of squares of the observations: $\Sigma x = 64$, $\Sigma x^2 = 344$. Test the null hypothesis that $\mu = 4.7$ against the alternative hypothesis that $\mu < 4.7$ when: (a) σ is unknown; (b) $\sigma = 2$.
                                                                                          (IOS)

71.   Two brands of motor car tyres were tested for length of life on 20 cars. Do the results of these tests show a significant difference between these two brands of tyres?

1000's of kilometres driven before tyre worn out.

Brand X	38 45 37 44 39 41 39 32 51 34
Brand Y	41 37 43 33 40 38 31 39 34 44

State your assumptions and comment on the design of this experiment.
                                                                                          (IOS)

72.   A sample is drawn from a normal population with unknown mean μ and known variance 20 000. An investigator wishes to test the hypothesis, H_0, that $\mu = 0$ against the alternative hypothesis, H_1, that $\mu \neq 0$. He wishes to have a probability of 0.05 that he will make a Type I error (i.e., reject H_0 when in fact it is true). He takes a sample of 50 values. Find the range R of values of the sample mean for which he will accept H_0.

If in fact the true value of μ is 30, what is the probability that the investigator will make a Type II error (i.e, accept H_0 when it is false)?

Find: (a) the greatest probability of a Type II error if it is known that the true mean is outside R; (b) the positive value of the true mean that will bring the probability of a Type II error down to 0.05.

Show, with the aid of diagrams if you wish, that decreasing the probability of the Type I error will increase the probability of the Type II error. Discuss the significance of this effect for the practical statistician.

(Cambridge)

73.   The metallic content of six truck loads of ore was determined by the supplier before despatch and again by an industrial consumer on receipt. The results, in percentages, are shown in the table below:

Truck	A	B	C	D	E	F
Supplier	10.8	11.1	11.4	12.2	12.1	11.8
Consumer	10.1	10.3	11.1	12.4	11.2	11.4

You should assume that the supplier and the consumer used different methods of testing (so that one may be more variable than the other) but you may also assume that each tested all six consignments by the same method and that the errors of determination followed a normal distribution in both places. (a) Are the supplier's and the consumer's findings mutually consistent? State a null hypothesis and state briefly how you intend to test it. (b) Make the test you have proposed, showing your working in full, and state your conclusion. (O & C)

74.   A textbook has two authors. Author A wrote some of the chapters and author B the rest. It is known from previous work that the number of words on a page written by A is approximately normally distributed with mean 790 and standard deviation 50. Author B tends to use slightly longer words and the number of words on a page is approximately normally distributed with mean 720 and standard deviation 45.

A reviewer wishes to decide which author has written a particular chapter and decides to test

H_0: chapter written by A, against
H_1: chapter written by B

by counting the number of words on a page and rejecting H_0 if the result is less than a constant, k. Find k to give a risk of type I error of approximately 0.05. What is the probability of type II error?

X_A and X_B are the total number of words on a random sample of ten pages written by authors A and B respectively. What is the distribution of

(a) X_A
(b) $X_A - X_B$?

Calculate an interval within which $X_A - X_B$ will lie with a probability of 0.99.

The reviewer knows that two chapters have been written by different authors but she is not sure which. She counts the total number of words on a random sample of 10 pages from each chapter and assigns the chapter with the larger total to author A. Explain why the interval calculated shows that this procedure will almost certainly give her the correct answer. (AEB) '86

Chapter 10

Statistical Tests 2

10.1 INTRODUCTION

This chapter deals with some other important statistical tests. They have no particular relationship to each other or anything in common, other than they did not fit conveniently into the development of the last chapter. This was of course concerned with tests involving population parameters from assumed Normal populations derived from either one or two single samples.

The first type of test dealt with here is a non-parametric χ^2 test. It is described as non-parametric simply because it does not involve any parameters of an associated population. The test statistic uses the frequencies with which particular items occur. The test itself splits naturally into two main types: a 'goodness of fit' test, concerned with measuring how like a particular null distribution an observed frequency distribution is and a 'contingency table' test, concerned with testing whether or not two particular attributes (or factors) of a population are associated. The form of the test statistic, under a given null hypothesis, has a χ^2 distribution.

The one-way Analysis of Variance (test) can be considered as a natural extension (w.r.t. three or more samples) of a two-sample T-test. Here we test whether or not respective samples can be considered as coming from the same population, using an F-distributed test statistic. A two-way analysis is a further development where two sets of samples are taken, each set pertaining to a particular factor, and two separate one-way analyses are performed.

Finally, we look at a particular alternative to a T test for single-sample and two-sample (paired) situations. It is the distribution-free sign test. 'Distribution-free' is the phrase used when the population that we are sampling from is not constrained to be of a particular type. The parameter involved in this test is the median.

10.2 χ^2 TEST: GOODNESS OF FIT

10.2.1 FOR ANY DISTRIBUTION

In Section 9.6.1 the test statistic $\dfrac{nS^2}{\sigma^2}$, which, under a defined H_0 is a $\chi^2(n-1)$ variable, was used in testing for the significance of a sample variance. In this section we will be formulating and using another χ^2 test statistic, used far more widely than the aforementioned.

Its formulation and use is illustrated with an elementary example. Suppose we throw a die 120 times, to obtain the following frequency distribution.

Number shown on face	1	2	3	4	5	6	Total
Frequency	20	10	10	20	20	40	120

We wish to test whether or not the die is biased.

Taking the obvious null hypothesis that the die is fair, we can immediately set up a theoretical distribution under this assumption for comparison with the observed frequency distribution.

Clearly, if an unbiased die is thrown 120 times, we would have an expected distribution as follows:

Number shown on face	1	2	3	4	5	6	Total
Frequency	20	20	20	20	20	20	120

since, theoretically, the die should show each of the faces an equal number of times.

The test statistic itself is formed by pairing the observed (O) and expected (E) frequencies, and

(a) squaring the differences,

(b) dividing by E, and

(c) adding these quantities for all pairs.

Combining the previous two tables and performing the necessary calculations, we obtain:

Number on die	1	2	3	4	5	6	Total
Observed frequency (O)	20	10	10	20	20	40	120
Expected frequency (E)	20	20	20	20	20	20	120
$O-E$	0	-10	-10	0	0	20	
$\dfrac{(O-E)^2}{E}$	0	5	5	0	0	20	30

So that $\sum \frac{(O-E)^2}{E} = 30$ here.

It can be shown that the statistic $\sum \frac{(O-E)^2}{E}$ has a χ^2 distribution under H_0.

However, we now need a parameter. Recall from the theory that the in-volved parameter, ν, represents a number of degrees of freedom. To calculate a test value in this case, we used the information from 6 pairs of values. These 6 were not independent, however, since their total had to be 120 or, alternatively, the differences of each pair had to add to zero. Hence, we really only have $6-1 = 5$ independent paired observations (degrees of freedom) with which to calculate the χ^2 test value. Put another way, we have 6 cells to fill subject to 1 restriction — namely, that the total of the six must add to 120. Hence 6 choices -1 restriction $= 5$ degrees of freedom.

So, under the null hypothesis, the test statistic $\sum \frac{(O-E)^2}{E}$ should be dis-tributed as a $\chi^2(5)$ variable and we can now proceed to test in the usual way.

In this particular case, for a 5% test, we have the decision rule:

(a)   reject H_0 if $\sum \frac{(O-E)^2}{E} > \chi^2_{0.95}(5)$

(b)   otherwise accept H_0.
Note that the whole critical region is put in the right hand tail (see Fig. 10.1).

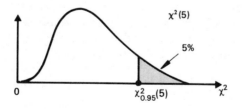

Fig. 10.1

Now, $\chi^2_{0.95}(5) = 11.07$ but, also, $\chi^2_{0.99}(5) = 15.09$. Hence the test value of 30 is highly significant and there is strong evidence for reject-ing the assumption of unbiasedness. We thus conclude that the die is biased in some way. To say in *what* particular way the die is biased is not the purpose of this test.

The test is used quite generally as follows.

STATEMENT 10.1

To test whether a set of n values having 'observed' frequencies (O) varies significantly from the same set of n values having 'expected' frequencies (E), the latter being calculated according to some null hypothesis, we use the test statistic $\chi^2 = \sum \dfrac{(O-E)^2}{E}$, which is distributed as a $\chi^2 (n-r)$ variable under the given null hypothesis. The value r is the number of restrictions that are placed on the E's according to the situation.

Notes (a) Any 'expected' frequencies that are less than 5 need to be combined with neighbouring class frequencies.

(b) The whole of the critical region is put in the right tail of the relevant null distribution.

No proof

Note that in the previous working example, the restriction placed on the E's was that they should sum to 120, as the O values did. Here, then, $r = 1$ and the number of pairs involved was $n = 6$. Hence $\nu = n - r = 6 - 1 = 5$.

For this test statistic, the critical region is always the right tail of the relevant null distribution since it is only unreasonably large values of χ^2 that make us suspect that the expected frequencies do not reasonably agree with the observed frequencies. If the χ^2 test value is small, this implies that the 'fit' of the expected distribution to the observed distribution is good.

However, the χ^2 test described is unreliable when any 'expected' frequency is less than 5. When this occurs, the frequency in question is combined with a neighbouring one, the same happening to the two corresponding 'observed' frequencies. This of course will cause the number of degrees of freedom for the test to be decreased.

EXAMPLE 10.1

A coin was tossed 200 times and it was observed that it fell down heads 116 times. Test at the 5% level whether the coin is unbiased.

SOLUTION

Under the null hypothesis that the coin is unbiased, we have $\Pr(\text{head}) = \Pr(\text{tail}) = \frac{1}{2}$ and from 100 throws we would clearly expect 100 heads and 100 tails.

The given figures and the calculations for the test statistic are shown in the following table.

	Heads	Tails	Total
O	116	84	200
E	100	100	200
$O - E$	16	-16	
$\dfrac{(O-E)^2}{E}$	2.56	2.56	5.12

$$\chi^2_{test} = \Sigma \frac{(O-E)^2}{E} = 5.12$$

Now, there are two pairs and the calculation of the E values was subject to the single restriction that they both add to 200.

Therefore there is just one degree of freedom.

So: $$\chi^2_{critical} = \chi^2_{0.95}(1) = 3.84$$

Therefore the test result of 5.12 is significant at the 5% level and we can reject H_0. Thus there is evidence that the coin is not unbiased.

EXAMPLE 10.2

Observations were made of the number of minor faults in each of 82 pieces of steel plate produced by a machine, with the following result.

No. of faults	0	1	2	3	4	5	6	7
No. of pieces	28	25	12	8	6	2	1	0

Under ideal conditions, the following 'expected' distribution (calculated based on 2 restrictions) should hold:

No. of faults	0	1	2	3	4	5	6	7
No. of pieces	26	22	15	8	5	3	2	1

Test whether the two distributions are essentially the same, using an appropriate test.

SOLUTION

In the expected table, each of the last three classes have frequencies which are less than 5, so we combine them (together with the corresponding observed classes) to obtain:

No. of faults	0	1	2	3	4	5 or more
No. of pieces (O)	28	25	12	8	6	3
No. of pieces (E)	26	22	15	8	5	6

The test is based on this table.

$$\chi^2_{\text{test}} = \frac{(28-26)^2}{26} + \frac{(25-22)^2}{22} + \frac{(12-15)^2}{15}$$

$$+ \frac{(8-8)^2}{8} + \frac{(6-5)^2}{5} + \frac{(3-6)^2}{6}$$

$$= 0.15 + 0.41 + 0.60 + 0 + 0.20 + 1.5$$

i.e. $\chi^2_{\text{test}} = 2.86$

There are 6 cells with 2 restrictions, giving the number of degrees of freedom as $6 - 2 = 4$.

$$\chi^2_{\text{critical}} = \chi^2_{0.95}(4) = 9.49$$

Hence the test result is not significant.

We therefore conclude that both distributions are essentially the same.

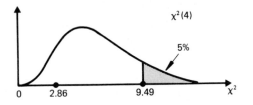

10.2.2 FOR A BINOMIAL

This sub-section and the following two are concerned with testing whether a given frequency distribution can be considered as being either Normal, binomial or Poisson, based on the observed frequencies from the distribution given. The test statistic used is just that given in the previous sub-section, namely $\chi^2 = \sum \dfrac{(O-E)^2}{E}$.

Now, in order to use a χ^2 test, we need of course to calculate an 'expected' distribution for comparison purposes. The way we generate one in general is to use one or more sample statistics from the given distribution.

For instance, consider the following sampling distribution:

x	0	1	2	3	4	5 or 6	Total
O	12	16	8	3	1	0	40

We want to see whether the frequencies are distributed in such a way that we might consider the distribution as binomial.

The technique here is, by using sample statistics from this (null hypothetical binomial) distribution, to generate a theoretical binomial distribution. The two sets of frequencies are then compared using a χ^2 test. A non-significant result means that the differences between the two sets of frequencies are not large enough to suspect any real disagreement. Hence we accept the original binomial null hypothesis. A significant result for χ^2_{test} clearly means that the differences between frequency sets are too large to be attributed to random variation. Hence we reject H_0.

So, we have now to generate a theoretical or 'expected' binomial distribution based on the given sample values. But in order to generate a binomial distribution, we need to know:

(a) n, the number of trials (for the hypothetical binomial situation), and

(b) p, the probability of a 'success'.

Since the largest value that x can take for a binomial variable is n, using the given distribution we have immediately $n = 6$. However, p is unknown here and will consequently need to be estimated.

Now, the mean of a theoretical binomial distribution is $n \cdot p$, and we can calculate the mean $(\bar{x})$ of the given frequency distribution.

Hence, by putting $n \cdot p = \bar{x}$ we have $p = \dfrac{\bar{x}}{n}$.

From the given table we easily calculate $\Sigma f = 40$ and $\Sigma fx = 45$.

Hence: $\bar{x} = \dfrac{45}{40} = 1.125$,   giving $p = \dfrac{1.125}{6} = 0.1875$

Armed with values $n = 6$ and $p = 0.1875$, we can now generate a theoretical binomial distribution (as previously demonstrated in Section 6.3.5). We thus obtain:

x	0	1	2	3	4	5 or 6
E	11.51	15.93	9.19	2.83	0.49	0.05

Note that in the above distribution there are three frequencies that have a value less than 5, and it is not until the last four classes are combined that the resulting total is greater than 5 (12.56). So, combining classes and comparing corresponding frequencies directly, we have:

x	0	1	2 or more	Total
O	12	16	12	40
E	11.51	15.93	12.56	40

and: $\chi^2_{test} = \dfrac{(12 - 11.51)^2}{11.51} + \dfrac{(16 - 15.93)^2}{15.93} + \dfrac{(12 - 12.56)^2}{12.56}$

$= 0.05$

For calculating the number of degrees of freedom (ν), we must use the above distributions and not the original ones given.

In the above table there are 3 frequency cells. However, there are two restrictions that have been placed on the E values:

(a)   that their total is 40, and

(b)   they are chosen so that $\bar{x} = 1.125$. We end up then with only one degree of freedom.

The critical value for the test is $\chi^2_{0.95}(1) = 3.84$.

Hence, the test result of $\chi^2 = 0.05$ is clearly not significant and we thus accept H_0. That is, there is no reason to suppose that the given distribution is not binomial.

There are, basically, two important cases of the χ^2 test for a binomial distribution, these differing in their number of degrees of freedom and the method for calculating p.

The first is where *n is known and p is unknown*, as in the above example, and we need to make use of the sample mean, $\bar{x}$, to find p, and the total frequency, ΣO, to multiply calculated probabilities in order to find 'expected' frequencies. Here, *there are two restrictions* (mean and total).

The second is the case when *both n and p are known*. Here, binomial probabilities are generated without needing to use $\bar{x}$ from the given distribution. Only ΣO is necessary, in order to convert probabilities into expected frequencies. Thus there is *only one restriction* here (total). See Example 10.4.

EXAMPLE 10.3

Test whether the sampling distribution:

x	0	1	2	3	4	5
O	1	6	14	33	31	15

can be considered to be binomial.

SOLUTION

Taking n as 5 (the maximum x value), p has to be determined. Under the null hypothesis that the above distribution *is* binomial, $\bar{x}$ needs to be determined in order to estimate p.

From the above distribution, we easily find $\Sigma f = 100$; $\Sigma fx = 332$.

Therefore $\bar{x} = \dfrac{332}{100} = 3.32$.

Thus, putting $np = 5p = 3.32$, we have $p = \dfrac{3.32}{5} = 0.664$.

We have then $n = 5$ and $p = 0.664$.

For the calculations we use the recurrence formula:

$$\Pr(r + 1) = \frac{n - r}{r + 1} \left(\frac{p}{1 - p} \right) \Pr(r)$$

or, more conveniently:

$$f(r + 1) = \frac{n - r}{r + 1} \frac{p}{1 - p} f(r)$$

where $f(r)$ means 'the frequency of $x = r$'.

Now:

$$\Pr(0) = (1 - 0.664)^5 = 0.0043$$

Therefore:

$$f(0) = 100(0.0043) = 0.43$$

With $\frac{p}{1 - p} = \frac{0.664}{0.336} = 1.976$, we have:

$$f(1) = \frac{5}{1}(1.976)(0.43) = 4.2484$$

$$f(2) = \frac{4}{2}(1.976)(4.2484) = 16.790$$

$$f(3) = \frac{3}{3}(1.976)(16.790) = 33.176$$

$$f(4) = \frac{2}{4}(1.976)(33.176) = 32.778$$

and: $$f(5) = \frac{1}{5}(1.976)(32.778) = 12.954$$

Rounding the above frequencies to 1D and adjusting, we have the following table with the two sets of frequencies for comparison:

x	0	1	2	3	4	5
O	1	6	14	33	31	15
E	0.4	4.2	16.7	33.1	32.7	12.9

The first two cells have expected frequencies that are less than 5, and even when the two are added the total is still less than 5. Thus we need to add the first three cells (for the two sets of frequencies) before the E total is first greater than 5. Hence:

x	2 or less	3	4	5
O	21	33	31	15
E	21.3	33.1	32.7	12.9

This gives: $$\chi^2_{\text{test}} = \sum \frac{(O - E)^2}{E} = 0.43.$$

Since we had to estimate p using $\bar{x}$, there are two restrictions. Thus we have $4 - 2 = 2$ degrees of freedom.

The test critcal value $= \chi^2_{0.95}(2) = 5.99$. Hence the test result of $\chi^2 = 0.43$ is not significant and we accept H_0, that the original distribution is binomial.

EXAMPLE 10.4

4 coins were tossed 200 times with the following results:

No. of heads	0	1	2	3	4	Total
No. of times	9	42	73	61	15	200

Decide whether the coins are biased, using a χ^2 goodness of fit test.

SOLUTION

Under a null hypothesis that the 4 coins are 'normal' (i.e., unbiased), we can generate binomial distribution probabilities corresponding to each 'number of heads' value. We know that there are 4 coins ($n = 4$) and the probability of a head at a single toss, under H_0, is $\frac{1}{2}$ ($p = \frac{1}{2}$). Thus we need to use only the total frequency (200 in this case) from the given information in order to generate expected frequencies, i.e., we have only one restriction.

Calculating the expected frequencies:

$$f(0) = 200(\tfrac{1}{2})^4 = 12.5; \qquad f(1) = 200(4)(\tfrac{1}{2})^4 = 50.0;$$
$$f(2) = 200(6)(\tfrac{1}{2})^4 = 75.0; \qquad f(3) = 200(4)(\tfrac{1}{2})^4 = 50.0;$$

and:
$$f(4) = 200(\tfrac{1}{2})^4 = 12.5$$

We thus have the table:

No. of heads	0	1	2	3	4	Total
O	9	42	73	61	15	200
E	12.5	50.0	75.0	50.0	12.5	200

$$\chi^2_{test} = \frac{(9-12.5)^2}{12.5} + \frac{(42-50)^2}{50} + \dots$$

$$= 0.98 + 1.28 + \dots$$

$$= 5.23$$

Since there are 5 cells and 1 restriction, we have 4 degrees of freedom for the test. So that the test critical value $= \chi^2_{0.95}(4) = 9.49$. Hence, $\chi^2 = 5.23$ is not significant and we conclude that the binomial gives a good fit to the sampling distribution given. This means that the coins have behaved in line with H_0 and so they are (probably) unbiased.

10.2.3 FOR A POISSON

The situation and procedure for this test is virtually the same as those for the binomial test. We are given a frequency distribution to test for its likeness to a Poisson distribution having the same mean. The theoretical

Poisson is generated using sampling information from the given data. We then compare the two distributions by means of the usual χ^2 test statistic, $\Sigma \frac{(O-E)^2}{E}$. However, whereas the binomial test can take two forms ((a) p unknown (b) p known) the Poisson test has generally only one.

As in the previous case, to obtain expected frequencies under H_0, we need to generate a theoretical distribution. But to generate a Poisson we need:

(a)   parameter μ, the mean, and

(b)   a total frequency.

μ is obtained (or rather, estimated) by the mean of the given distribution. It would rarely be known in advance. We have then, 2 restrictions on the choice of the E values, mean and total, in this situation. The technique is illustrated by the following example.

EXAMPLE 10.5

Test whether a good fit is given by a Poisson distribution to the following frequency distribution:

x	0	1	2	3	4	5	6 or more
f	19	26	27	13	11	2	0

SOLUTION

Under a null hypothesis that the given distribution is Poisson, we first calculate $\bar{x}$.

$$\Sigma f = 98 \quad \text{and} \quad \Sigma fx = 173, \quad \text{hence} \quad \bar{x} = \frac{173}{98} = 1.765$$

Taking this value as the mean of a theoretical Poisson distribution, we have:

$$f(0) = 98 \cdot e^{-1.765} = 16.78 \quad \text{(2D)}$$

The recurrence formula $f(r + 1) = \frac{\mu}{r + 1} \cdot f(r)$ will be used to generate the other expected frequencies.

$$f(1) = (1.765)(16.776) = 29.61$$

$$f(2) = \frac{1.765}{2}(29.61) = 26.13$$

$$f(3) = \frac{1.765}{3}(26.13) = 15.37$$

$$f(4) = \frac{1.765}{4}(15.37) = 6.78$$

and:

$$f(5 \text{ or more}) = 98 - (6.78 + 15.37 + 26.13 + 29.61 + 16.78)$$
$$= 3.33$$

Notice that this last expected frequency is less than 5 and so must be combined with the previous one. We now have the following table for comparing the frequencies.

x	0	1	2	3	4 or more	Total
O	19	26	27	13	13	98
E	16.8	29.6	26.1	15.4	10.1	98

Hence:

$$\chi^2_{test} = \frac{(19 - 16.8)^2}{16.8} + \ldots$$

$$= 0.288 + \ldots$$

$$= 1.964$$

5 frequency pairs and 2 restrictions gives just 3 degrees of freedom.

The critical value for the test is then $\chi^2_{0.95}(3) = 7.81$, which means that the test value of 1.964 is not significant. Thus we accept H_0 and conclude that the Poisson distribution gives a good fit.

10.2.4 FOR A NORMAL

The goodness of a fit test for a Normal follows along the same lines as those for a binomial or Poisson. However, in general, given a frequency distribution to test for Normality, both the sample values $\bar{x}$ and s, together with the total frequency, will need to be used. That is, we will generally have *three restrictions* on the choice of the E values.

The procedure is:

(a) Calculate $\bar{x}$ and s for the given frequency distribution.

(b) Using $\bar{x}$ and s as 'estimates' of the population μ and σ, together with the given total frequency, set up a theoretical Normal distribution.

(c) Compare observed and expected frequencies in the usual way, using the χ^2 test statistic with three restrictions.

If it *does* happen that both μ and σ are known (or separately given) for the theoretical distribution, then it will not of course be necessary to use the given frequency distribution to estimate them.

For example, for the following distribution:

x	Up to 15	15 to 20	20 to 25	25 to 30	30 to 35	35 and over
O	3	7	15	20	9	4

We have already generated a theoretical Normal distribution to fit in Section 6.5.3, which is given as follows:

x	Up to 15	15 to 20	20 to 25	25 to 30	30 to 35	35 and over
E	2.4	7.9	16.2	17.5	10.2	3.8

We wish to test the fit.

Now for the χ^2 test procedure, it is necessary to combine the first two and last two classes in order to obtain an expected frequency of greater than 5. We thus obtain the following table for comparison of frequencies:

O	10	15	20	13	58
E	10.3	16.2	17.5	14.0	58

$$\chi^2_{test} = \sum \frac{(O-E)^2}{E} = \frac{(10-10.3)^2}{10.3} + \dots$$

$$= 0.53 \quad (2D)$$

The critical value for the test is $\chi^2_{0.95}(1) = 3.84$ and we see that 0.53 is not significant. Thus, it is accepted that the given distribution is Normal.

EXAMPLE 10.6

The following information relates to the height (measured to the nearest cm) of 694 nine-year old girls.

Height	117–120	121–124	125–128	129–132	133–136
Frequency	8	28	82	140	188

Height	137–140	141–144	145–148	149–152
Frequency	148	69	15	16

Test this sample for Normality.

SOLUTION

We easily calculate $\bar{x} = \mu = 134.356$ and $s = \sigma = 6.195$.

To calculate the expected frequencies under the assumption of Normality using the above μ and σ, we have the following table:

u.c.b. (x)	$z = \dfrac{x - 134.356}{6.195}$	$Pr(Z < z)$	p	E	O (given)
120.5	− 2.24	0.013	0.013	9.0	8
124.5	− 1.59	0.056	0.043	29.9	28
128.5	− 0.95	0.171	0.115	79.8	82
132.5	− 0.30	0.382	0.211	146.4	140
136.5	0.35	0.637	0.255	177.0	188
140.5	0.99	0.839	0.202	140.2	148
144.5	1.64	0.950	0.111	77.0	69
148.5	2.28	0.989	0.039	27.1	15
−	−	1.000	0.011	7.6	16

The test statistic is:

$$\chi^2 = \Sigma \frac{(O-E)^2}{E}$$

$$= \frac{(8-9)^2}{9} + \frac{(28-29.9)^2}{29.9} + \ldots$$

$$= 17.21$$

$\chi^2_{0.95}(6) = 12.59$. (There are 9 original cells with 3 restrictions — mean, s.d. and total — giving $\nu = 9 - 3 = 6$).

Therefore result of 17.21 is significant and we reject the assumption of Normality.

10.3 χ^2 TEST: CONTINGENCY TABLES

Sometimes we need to consider data from populations that are classified with respect to two different attributes. For example, in a hospital we might, for some reason, classify a certain group of patients as to their susceptibility to a new drug and at the same time record their sex in order to determine whether or not sex affects susceptibility.

We might be interested in determining if the weather has any effect on a cricket team's performance by measuring games as 'win', 'lose' or 'draw' against 'good' or 'bad' weather.

The test considered here has just the same test statistic as the previous 'goodness of fit' tests and concerns itself with whether or not the two involved factors are independent of each other. The observed information comes in the form of a *contingency table* as follows:

Factor A

	a_1	a_2	$\ldots$	a_n
b_1	N_{11}	N_{12}	$\ldots$	N_{1n}
b_2	N_{21}	N_{22}	$\ldots$	N_{2n}
$\vdots$	$\vdots$			
b_m	N_{m1}	N_{m2}	$\ldots$	N_{mn}

Factor B

$a_1, a_2, \ldots, a_n$ are often called the *levels* of factor A and $b_1, b_2, \ldots, b_m$, the *levels* of factor B and are simply subdivisions of interest. For example, the factor 'sex' will usually have the two levels 'male' and 'female', and the factor 'weather' might be split into the levels 'stormy', 'rainy', 'overcast' and 'dry'. Deciding on the levels to be used will clearly depend on the situation, in particular on what the experimenter considers important.

The body of the table gives the number of observations pertaining to each combination of the levels for the two factors. For instance, N_{12} shows the number of observations of level a_2 against b_1, and so on.

We look at the test for a 2 × 2 table first, then generalise the technique to an $m \times n$ table.

10.3.1 A 2 × 2 TABLE

Suppose a hospital researcher randomly selects 30 male and 30 female patients for susceptibility to a particular drug and obtains the following results:

Observed table		Susceptibility		Totals
		Good	Bad	
Sex	Male	17	13	30
	Female	20	10	30
	Totals	37	23	60

Here, the two factors being compared are 'susceptibility' and 'sex', each split into two levels, giving a 2 × 2 contingency table. The test is whether the sex of the patient affects susceptibility to the drug. As in all these types of test, the null hypothesis is that there is no effect; that is, that both factors (sex and susceptibility) are independent.

If, under this H_0, we compile an 'expected' table using the same row and column totals, we can use the χ^2 statistic $\sum \dfrac{(O-E)^2}{E}$ to compare these four expected frequencies (E) against the four given frequencies (O).

As regards the calculation of the E values, consider first the expected frequency corresponding to the observed value 17. The row totals contain the information that $\dfrac{30}{60}$ of all the patients are men, i.e., $\Pr(\text{man}) = \dfrac{30}{60}$.

Similarly, from the column totals, we have that $\dfrac{37}{60}$ of all the patients have good susceptibility, i.e., $\Pr(\text{good susceptibility}) = \dfrac{37}{60}$.

Hence:

$\Pr(\text{man } and \text{ good susceptibility}) = \Pr(\text{man}) \Pr(\text{good susceptibility})$

since, under H_0, sex is independent of susceptibility

$$= \frac{30}{60}\left(\frac{37}{60}\right)$$

Thus, since there are 60 patients in all, we have the expected frequency:

$$E(\text{man } and \text{ good susceptibility}) = 60\left(\frac{30}{60}\right)\left(\frac{37}{60}\right)$$

$$= \frac{30 \times 37}{60} = 18.5$$

That is, under the hypothesis that sex is independent of susceptibility, we would expect 18.5 of the 60 patients to be men of good susceptibility to the drug.

In a similar way we can calculate the expected number of men out of the 60 who have a bad susceptibility to the drug as:

$$E(\text{man } and \text{ bad susceptibility}) = 60\left(\frac{30}{60}\right)\left(\frac{23}{60}\right)$$

$$= \frac{30 \times 23}{60} = 11.5$$

Notice the form that the calculation reduces to, namely, the product of the respective row and column totals divided by the grand total. The other two expected values are easily calculated to give the following expected table:

Expected table		Susceptibility		Totals
		Good	Bad	
Sex	Male	18.5	11.5	30
	Female	18.5	11.5	30
	Totals	37	23	60

We now need to determine the number of degrees of freedom there are for the χ^2 test. Recall, from Section 8.3, the example concerning calculating the number of degrees of freedom for a table with 3 rows and 2 columns (3 x 2), where we had 6 cells to fill with 4 restrictions giving $6 - 4 = 2$ degrees of freedom. In the above case, there are $2 \times 2 = 4$ cells to fill and clearly only 3 of the row and column totals are necessary. That is, once any one E is calculated, all others are then determined.

Hence there is $4 - 3 = 1$ degree of freedom for the test. This is always the case for a 2 x 2 table. We will thus be referring to a $\chi^2(1)$ null distribution.

The test statistic is:

$$\chi^2 = \sum \frac{(O - E)^2}{E}$$

$$= \frac{(17 - 18.5)^2}{18.5} + \frac{(13 - 11.5)^2}{11.5} + \frac{(20 - 18.5)^2}{18.5} + \frac{(10 - 11.5)^2}{11.5}$$

$$= 0.63 \quad (2D)$$

The critical value is $\chi^2_{0.95}(1) = 3.84$.

Thus $\chi^2 = 0.63$ is not significant and we accept H_0. That is, there is not enough evidence of association between sex and susceptibility to the drug.

To summarise the discussion:

STATEMENT 10.2

Given the following observed 2×2 contingency table:

		1st factor		Totals
		(1)	(2)	
2nd	(A)	$O(A1)$	$O(A2)$	n_A
factor	(B)	$O(B1)$	$O(B2)$	n_B
Totals		n_1	n_2	n

a test for whether the two factors are independent or not can be carried out, using:

(a)　H_0: the two factors *are* independent, and

(b)　calculating corresponding expected cell frequencies using:

$$E(A1) = \frac{n_A n_1}{n}; \quad E(A2) = \frac{n_A n_2}{n}; \ldots$$

Finally:

(c)　perform test using the statistic $\chi^2 = \sum \dfrac{(O-E)^2}{E}$, which is distributed as $\chi^2(1)$ under H_0.

Proof for (b)　Given the row and column totals n_A, n_B, n_1 and n_2, together with grand total n, we have:

$$\Pr(A) = \frac{n_A}{n}; \quad \Pr(B) = \frac{n_B}{n}; \quad \Pr(1) = \frac{n_1}{n}; \quad \Pr(2) = \frac{n_2}{n}$$

Assuming the two factors are independent (under H_0), we have further:

$$\Pr(A1) = \Pr(A)\Pr(1) = \left(\frac{n_A}{n}\right)\left(\frac{n_1}{n}\right)$$

$$\therefore \quad E(A1) = n \Pr(A1) = \frac{n_A n_1}{n}$$

as required. The three other values of E are calculated in a similar way.

When using a continuous distribution (in this case χ^2) to describe data that is discrete (in this case the cell frequencies), an error can creep into the calculations. For instance, when considering the class '8 or more' in the discrete sense, it would need to be adjusted to '7.5 or more' in the continuous sense (assuming integral numbers). 'More than 8' (discrete) would be adjusted to '8.5 or more' (continuous). In general when dealing with large classes or groups of data, the error is negligible, but in the case of a 2×2 table we are dealing with only one independent quantity (the single degree of freedom) and often the error cannot be ignored.

We often use a corrected χ^2 test statistic incorporating what is known as *Yates' continuity correction* as follows:

$$\chi^2 \text{ (corrected)} = \Sigma \frac{(|O - E| - 0.5)^2}{E}$$

The above corrected statistic can not only be used for a 2×2 table but for any data undergoing a test (using the χ^2 test statistic) having only one degree of freedom. Sometimes it occurs that this value is so different from the ordinary χ^2 value that while one causes us to reject H_0 the other leads us to accept H_0. In this sort of case, we might increase the sample size and retest or use another method.

Note that the continuity correction will always *decrease* the value of χ^2, a fact that can be verified by examining carefully the above form of χ^2 (corrected). Hence, if a χ^2 test value is obtained that caused us to accept H_0, we would not need to consider calculating a corrected value since the test result could not be affected.

EXAMPLE 10.7

Two forms, A and B, in a school put their students in for an exam, the results of which were as follows:

	Form A	Form B
Passed	72	64
Failed	17	23

Test the hypothesis that there is no difference between the results of the two forms using a χ^2 test (with and without a continuity correction).

SOLUTION

The observed table (dropping headings) can be written:

72	64	136
17	23	40
89	87	176

and under a null hypothesis of 'no difference', we obtain the expected table (for comparison with the above) as follows:

$$
\begin{array}{cc|c}
68.8 & 67.2 & 136 \\
20.2 & 19.8 & 40 \\
\hline
89 & 87 & 176
\end{array}
\qquad \left(68.8 = \frac{(136)(89)}{176}; \dots \text{etc.}\right)
$$

Hence:
$$
\chi^2_{test} = \frac{(72 - 68.8)^2}{68.8} + \dots
$$

$$
= 0.15 + \dots
$$

$$
= 1.33 \quad (2D)
$$

Also:
$$
\chi^2 (\text{corrected})_{test} = \frac{(|72 - 68.8| - 0.5)^2}{68.8} + \dots
$$

$$
= 0.11 + \dots
$$

$$
= 0.94 \quad (2D)
$$

(Notice the difference in the above two values of χ^2.)

$$
\chi^2_{critical} = \chi^2_{0.95}(1) = 3.84
$$

Thus neither of the χ^2 test values are significant and we therefore accept H_0. That is, there is no evidence of association between 'form' and 'examination result'.

10.3.2 AN $m \times n$ TABLE

We now generalise the test technique to take account of a general $m \times n$ table. Note that the data is still classified in two ways (or into two factors) but instead of each classification being divided into two, one is divided into m parts the other into n.

For example, consider the members of the cricket team who are interested in whether the weather has an effect on their results. They play 65 matches, broken down as follows:

		Weather Good	Weather Bad	Total
Result	Win	17	5	22
	Draw	6	10	16
	Lose	4	23	27
Total		27	38	65

This is a 3×2 *contingency table*. The expected table, under the H_0 of 'match result is independent of weather', is as follows:

		Weather		Total
		Good	Bad	
Result	Win	9.14	12.86	22
	Draw	6.65	9.35	16
	Lose	11.21	15.79	27
Total		27	38	65

The calculations for the above table are exactly the same as for the 2×2 table. That is:

$$9.14 = \frac{(22)(27)}{65}; \quad 6.65 = \frac{(16)(27)}{65}; \quad \ldots \quad \text{etc.}$$

The number of degrees of freedom for this table is 2. This can be worked out from the fact that there are 6 observations and just 4 (independent) restrictions. Statement 10.3 which follows gives a general expression which can always be used to obtain the correct number of degrees of freedom.

So here:
$$\chi^2_{test} = \frac{(17 - 9.14)^2}{9.14} + \ldots$$

$$= 19.60$$

The critical value for the test is $\chi^2_{0.99}(2) = 9.21$.

Hence the result is highly significant and we can confidently reject H_0.

We thus conclude that the weather does have a significant effect on the team's results!

The technique is generalised by the following:

STATEMENT 10.3

Given the following observed $m \times n$ contingency table:

		1st factor					Totals
		1	2	3	$\ldots$	n	
2nd factor	1	$O(1, 1)$	$O(1, 2)$	$O(1, 3)$	$\ldots$	$O(1, n)$	T_1
	2	$O(2, 1)$	$O(2, 2)$	$O(2, 3)$	$\ldots$	$O(2, n)$	T_2
	3	$O(3, 1)$	$O(3, 2)$	$O(3, 3)$	$\ldots$	$O(3, n)$	T_3
	.	.	.	.	.	.	.
	m	$O(m, 1)$	$O(m, 2)$	$O(m, 3)$	$\ldots$	$O(m, n)$	T_m
Totals		S_1	S_2	S_3	$\ldots$	S_n	N

(continued)

a test for whether the two factors are independent or not can be carried out, using:

(a) H_0: The two factors *are* independent, and

(b)   calculating corresponding expected cell frequencies using

$$E(1, 1) = \frac{T_1 \cdot S_1}{N}; \quad E(1, 2) = \frac{T_1 \cdot S_2}{N}; \quad \ldots \quad \text{etc.}$$

Finally

(c)   perform test using the test statistic $\chi^2 = \sum \frac{(O - E)^2}{E}$, which is distributed as $\chi^2 ((m - 1)(n - 1))$ under H_0.

Note that every E used must be greater than 5 for the test to be valid.

Proof   We only prove here that the number of degrees of freedom for the test is $(m - 1)(n - 1)$ as stated. (The rest follows naturally as an extension of the previous proof.) Now, there are $m \times n = mn$ cells to fill in the table. We need to determine how many (independent) restrictions there are on the choice of values. We have given m row totals and n column totals together with the grand total. That is, $m + n + 1$ totals altogether. But, once we know all the row totals, say, the grand total is automatically determined. Thus there are m independent values here. Also, given the grand total, we can only choose $n - 1$ of the column totals independently. Altogether then, there are $m + n - 1$ independent restrictions (totals). Therefore, the number of degrees of freedom is:

$$mn - (m + n - 1) = mn - m - n + 1 = (m - 1)(n - 1)$$

as required, *completing the proof.*

EXAMPLE 10.8

Students at a college were given two tests; one in English, another in French, each subject result being split into the three grades A, B and C as follows:

		English			Total
		A	B	C	
French	A	55	72	12	139
	B	48	162	38	248
	C	14	42	85	141
Total		117	276	135	528

Use a χ^2 test to decide whether the performance in English is independent of the performance in French.

SOLUTION

Under the null hypothesis of independence of the two subjects, we can draw up an expected table as follows for comparison:

30.8	72.7	35.5	139
55.0	129.6	63.4	248
31.2	73.7	36.1	141
117	276	135	528

The test statistic is calculated as:

$$\chi^2 = \Sigma \frac{(O-E)^2}{E}$$

$$= \frac{(55-30.8)^2}{30.8} + \frac{(72-72.7)^2}{72.7} + \ldots$$

$$= 19.0 + 0 + 15.6 + 0.9 + 8.1 + 10.2 + 9.5 + 13.6 + 66.2$$

$$= 143.1$$

The number of degrees of freedom for the test (with $m = n = 3$) is $(3-1)(3-1) = 2 \times 2 = 4$.

Hence critical $\chi^2 = \chi^2_{0.95}(4) = 9.49$ and $\chi^2_{0.99}(4) = 13.28$.

The result therefore is very highly significant and we can reject the hypothesis of independence.

10.4 EXPERIMENTAL DESIGN AND ANALYSIS OF VARIANCE

10.4.1 INTRODUCTION TO EXPERIMENTAL DESIGN

Experimental Design is an area of statistical study which is concerned with investigating the effects of different factors which might (either artificially or naturally) influence the values of a variable. The involved factors are technically known as *treatments, observations* being made on *experimental units*.

For example, uniform iron bars might be painted with various types of protective substances, immersed in sea water for a period of time and then measured for degree of corrosion. Here, the units are iron bars, the treatments are the various types of protective substances and the observations are the measurements of corrosion.

The label 'treatment' can have a wide variety of uses and applications.

For example: types of feed given to animals; varieties of agricultural crop; various processes or machines used to manufacture identical goods; the testing of different drugs; and so on.

There are a number of ways to analyse data in Experimental Design. The one which is of interest to us for the purpose of this text is the investigation of differences (as measured by variation) in the values of the data caused by the various treatments. The group of statistical tests which is used to investigate differences is known as *Analysis of Variance* (ANOVA). In order to carry out this type of test, it is necessary to consider the nature and cause of variation.

As a simple example of the type of problem that can be encountered when considering variation, consider the following. A farmer plants crop variety A in one field and variety B in another and measures the total yield of both (for comparison) at the end of one growing season. The results of this experiment however will give little information in respect of which variety of crop is best! This is because, irrespective of whether or not he finds a difference in yields, there may be other significant factors present which will confound the comparison. For example:

(a)   one field might have much better natural growing conditions in respect of soil texture or drainage;

(b)   the fields might be located at a distance from each other with a possible difference in environmental factors such as the effect of wind, traffic or people;

(c)   the crops may have been planted at two different times, using two different methods or applying different types of fertiliser.

Thus, variation in the values of a variable can be due to a variety of causes.

Some structures, within which variation can be described and explained, are now given.

If a random sample is taken from $N(\mu, \sigma^2)$, the variable involved, X, say, can be described using the following *model*:

$$X_i = \mu + E_i \quad \text{where variable } E \text{ is } N(0, \sigma^2)$$

E is sometimes called *experimental error* and can be thought of in practical terms as a measure of variation of each item from the mean, μ.

For example, if a sample of 3 from $N(20, \sigma^2)$ yields values 20.2, 23.1 and 18.7, then we have $e_1 = 0.2$, $e_2 = 3.1$ and $e_3 = -1.3$.

In this type of situation, the only source of variation between the observed x values is from *within the sample*.

However, if we now consider a number of sub-samples (considered as treatments here) which might be assumed to come from some homogeneous Normal population, there will be *two* possible sources of variation that any x value will be subject to; firstly, experimental error and secondly, variation *between samples (or treatment groups)*. The model (sometimes known as the *one-way model*) to explain the variation in this situation is:

$$X_{ij} = \mu + G_i + E_{ij} \qquad \text{one-way model}$$

where:

X_{ij} = the jth observation in the ith sub-sample;

E_{ij} = experimental error, distributed as $N(0, \sigma^2)$;

G_i = variation factor solely attributable to treatment (or sub-sample or Group) i and expressed as a deviation from μ. (i.e. $\Sigma G_i = 0$.)

For example, measurements of diameters (in mm) of components which have been produced by a similar process might be taken from 3 different machines, A, B and C say. Suppose that the overall effect of production from A causes diameters on average to be higher by 0.2 mm; from machine B, higher by 0.1 mm and from machine C, lower by 0.3 mm. Thus the three different samples taken (1, 2 and 3, say) will have respectively: $G_1 = 0.2$, $G_2 = 0.1$ and $G_3 = -0.3$.

Differences between treatments can often be seen by simple inspection of the data.

For example, the following data describe the weight of crops (in lb/plot) resulting from various plots of the same plant being treated with three different types of fertiliser, A, B and C.

A: 15, 17, 22, 29, 18, 18, 24, 17

B: 30, 45, 50, 37, 39, 46

C: 27, 28, 36, 41, 26, 30

Fertiliser B seems to give the best results and A the worst and it seems fairly clear that there are differences between the fertilisers.

The particular test that is used to determine (statistically) whether or not there are differences between various treatments is the *one-way* ANOVA which is described in Subsection 10.4.2. This test can be considered as an extension of the two-sample (unpaired) T test. That is, given sets of observations from more than two samples, a one-way ANOVA will test whether or not the samples can be considered as being drawn from the same population *or not*.

Consider now that it is known (or suspected) that another source of systematic variation exists which is quite independent of the variation between treatments.

For example, in an investigation of possible differences between machine operators working on some industrial process, it might be decided that the machines themselves will also contribute to the variation in times taken to complete a particular job. In this type of case, it is usual for a single observation to be made (a job completion time here) for each combination of the two variation factors. The observed data might be recorded as follows, where values relate to times in hours:

		Machine			
		A	B	C	D
	1	2.2	2.4	2.6	2.3
	2	2.1	2.5	2.8	1.9
Operator	3	2.3	1.9	2.1	2.1
	4	1.8	2.0	2.0	1.8
	5	2.3	2.3	2.2	2.0

Sometimes the term 'blocks' is used to describe the non-treatment factor. Thus in the example above, the operators are the treatments and the machines are the blocks.

In order to take account of the variation involved when considering two independent factors (one described by the rows, the other by the columns), the one-way model is adapted as follows:

$$X_{ij} = \mu + R_i + C_j + E_{ij} \qquad \text{two-way model}$$

where:

X_{ij} = the observation in the ith row and jth column;

E_{ij} = experimental error, distributed as $N(0, \sigma^2)$;

R_i = variation factor solely attributable to row factor i;

C_j = variation factor solely attributable to column factor j;

Notes (a) Both R and C are expressed as deviations from μ. (i.e. $\Sigma R_i = \Sigma C_j = 0$);

(b) It is irrelevant whether rows or columns are used to describe the treatments;

(c) Sometimes the symbols T (for treatments) and B (for blocks) are used instead of R and C.

The particular test that is used to determine (statistically) whether or not there are differences between the various levels of *either* of the two factors is the *two-way* ANOVA, which is described in Subsection 10.4.3. In certain respects, this test can be considered as an extension of the two-sample (paired) T test.

There are certain principles involved in Experimental Design (particularly for comparing treatments) which are described as follows:

A. Replication   More than one observation should be made on a single treatment. This is known as *replication*. This of course is so that the treatment mean and variation within each treatment can be estimated with a reasonable measure of confidence.

B. Randomisation   Since it will be virtually impossible that all the experimental units will be identical, it follows that it is quite likely that there will be numbers of units (either in groups or scattered) which will be better, or perhaps worse, than the majority in their reception of any treatment. Thus it is necessary to allocate randomly which treatment is given to a particular unit. This is known as *randomisation*. It is accomplished quite easily using random number tables.

For example, suppose that 4 treatments (A, B, C and D, say) were to be randomly allocated to 15 units. For each unit in turn, a random digit from 1 (representing treatment A) through to 4 (representing treatment D) is chosen. If the 15 digits selected turned out as:

$$4 \ 3 \ 2 \ 4 \ 3 \ 4 \ 2 \ 2 \ 1 \ 2 \ 3 \ 2 \ 1 \ 4 \ 4$$

this would be translated as follows:

Unit	1	2	3	4	5	6	7	8	9	10	11	12	13	14	15
Treatment	D	C	B	D	C	D	B	B	A	B	C	B	A	D	D

Experimental designs which incorporate both replication and randomisation are known as *completely randomised* layouts. A one-way ANOVA is used to test for differences between treatments with this sort of design.

If, however, another source of systematic variation is known to be acting (other than experimental error or variation between treatments), the completely randomised layout is not suitable since a random allocation of treatments would pay no particular regard to the new variation.

C. Blocking   To take account of another source of variation, the technique of *blocking* should be used. This involves separating the units into (homogeneous) blocks; that is, each block will have the same characteristics with respect to the new source of variation.

As an example, an agricultural experiment might consist of plots (units) which are being used to test various types of fertiliser (treatments). If the land on which the plots are located has a slope, this is probably enough to count as a variation factor, since plots on higher ground will have different characteristics to their counterparts on lower ground. Thus a number of plots situated in a row at right angles to the slope would constitute a block. Various plots (blocks) could be located at intervals along the field.

Each block consists of exactly one complete set of treatments; that is, there are as many units in the block as there are treatments. If the treatments are allocated to the units within the block randomly, this is known as a *randomised block* layout. For the same reasons as in 1 above, blocks should be replicated.

For example, if five types of fertiliser, A, B, C, D and E, are being tested and seven blocks are selected for comparison, a randomised block layout might look something like the following:

		\multicolumn{5}{c}{Units}				
		1	2	3	4	5
	1	B	C	D	A	E
	2	C	A	D	E	B
	3	E	A	D	B	C
Blocks	4	D	C	A	B	E
	5	E	A	C	B	D
	6	B	C	A	D	E
	7	B	C	E	D	A

The treatments in each block have been allocated randomly.

10.4.2 ONE-WAY ANALYSIS

The general situation is that we are testing for any differences between the means of k samples, having sizes $n_1, n_2, \ldots, n_k$ respectively, with $\Sigma n_i = n$. The data can be shown in tabular form as follows:

\multicolumn{6}{c}{Sample}					
1	2	3	$\ldots$	j $\ldots$	k
x_{11}	x_{12}	x_{13}	$\ldots$	x_{1j} $\ldots$	x_{1k}
x_{21}	x_{22}	x_{23}	$\ldots$	x_{2j} $\ldots$	x_{2k}
x_{31}	x_{32}	x_{33}	$\ldots$	x_{3j} $\ldots$	x_{3k}
$\ldots$	$\ldots$	$\ldots$	$\ldots$	$\ldots$ $\ldots$	$\ldots$
$\ldots$	$\ldots$	$\ldots$	$\ldots$	$\ldots$ $\ldots$	$\ldots$
$\ldots$	$\ldots$	$\ldots$	$\ldots$	x_{ij} $\ldots$	$\ldots$
$\ldots$	$\ldots$	$\ldots$	$\ldots$	$\ldots$ $\ldots$	$\ldots$
T_1	T_2	T_3	$\ldots$	T_j $\ldots$	T_k

where: x_{ij} represents the ith item in the jth sample;

T_i is the total of the n_i items in the ith sample;

and: $T = \Sigma T_i = T_1 + T_2 + \ldots + T_k$

The null hypothesis for the test is the obvious one – that each sample comes from the same (assumed Normal) population; the alternative hypothesis being that *one or more* samples come from different populations. Note that if the test statistic, which we will be dealing with later, proves to be significant, we would thus reject H_0 and accept H_1. That is, all we could say in this particular case is that one or more samples come from different populations.

Under H_0, we can regard the complete set of n items as a single sample from some (Normal) population with a variance of σ^2, say. A major part of the analysis is concerned with estimating this unknown variance.

There are two ways of estimating σ^2: (a) using all the given values, and (b) using the sample means.

(a)   Using the assumption that $X_{ij} = X$ is $N(\mu, \sigma^2)$:

$$\text{then} \quad \hat{\sigma}^2 = \frac{nS^2}{n-1} = \frac{\Sigma(x-\bar{x})^2}{n-1} \quad \begin{array}{l} \rightarrow \text{'sum of squares'} \\ \rightarrow \text{degrees of freedom} \end{array}$$

The expression marked 'sum of squares' is referred to in this context as the 'total sum of squares' (SS_T) and can be calculated from the given table of values using:

$$SS_T = \Sigma(x-\bar{x})^2 = \Sigma x^2 - n\bar{x}^2 = \Sigma x^2 - n\left(\frac{T}{n}\right)^2$$

$$= \underset{i}{\Sigma}\underset{j}{\Sigma} x_{ij}{}^2 - \frac{T^2}{n}$$

(b)   Since T_i is the total of the items in sample i, the mean of the items in sample i is given by T_i/n_i.

So that $\dfrac{T_i}{n_i}$ is $N\left(\mu, \dfrac{\sigma^2}{n_i}\right)$ and thus $\text{Var}\left[\dfrac{T_i}{\sqrt{n_i}}\right] = \sigma^2$

$$\text{Hence, } \hat{\sigma}^2 = \frac{kS^2}{k-1} = \frac{\Sigma\left(\dfrac{T}{\sqrt{n}} - \bar{x}\right)^2}{k-1} \quad \begin{array}{l} \rightarrow \text{'sum of squares'} \\ \rightarrow \text{degrees of freedom} \end{array}$$

where S^2 is the variance of the k *sample means*. This particular 'sum of squares' is known as the 'between sample sum of squares' (SS_B) and it can be shown that:

$$SS_B = \Sigma\left(\frac{T}{\sqrt{n}} - \bar{x}\right)^2$$

$$= \Sigma\frac{T_i^2}{n_i} - \frac{T^2}{n}$$

We now have two estimates of the hypothetical Normal population variance, σ^2.

In general any observations of a random variable that are made are going to vary. For the observations in the table, there are many reasons for variation. For instance, inaccuracies in measuring instruments (not to mention human error in reading them). But, ignoring these types of error (which, in a controlled experiment, will be kept to the very minimum anyway), a large contribution to variation will come from between the samples themselves, or more specifically, between the means of the samples. The other contributor will be the intrinsic variability within each sample itself. The former is called 'between sample' variation, the latter 'within sample' variation.

So, we effectively consider that the total variation of all the given items (as measured by SS_T) consists of the two components (a) SS_B (the *between* sample variation) and (b) SS_W (the *within* sample variation).

A simple way of calculating SS_W is given in the following statement, which also shows the relationship between the three associated numbers of degrees of freedom.

STATEMENT 10.4

(a) (i) The total sum of squares of the observations involved in a one-factor analysis of variance table is the sum of the be-tween sample sum of squares and the within sample sum of squares, i.e., $SS_T = SS_B + SS_W$. So that, in particular, $SS_W = SS_T - SS_B$.

  (ii) The respective numbers of degrees of freedom associated with the above sums of squares are also additive, i.e., $df_T = df_B + df_W$.

  So that, in particular:

$$df_W = df_T - df_B$$
$$= n - 1 - (k - 1)$$
$$= n - k$$

(b) With $\hat{\sigma}_T^2 = \dfrac{SS_T}{df_T}$; $\hat{\sigma}_B^2 = \dfrac{SS_B}{df_B}$ and $\hat{\sigma}_W^2 = \dfrac{SS_W}{df_W}$:

  (i) $\hat{\sigma}_T^2$ and $\hat{\sigma}_B^2$ are both unbiased estimates of σ^2 under H_0,

  (ii) $\hat{\sigma}_W^2$ is an unbiased estimate of σ^2 *independent of H_0*,

where σ^2 is the unknown variance of the population from which, under H_0, all the samples have been drawn.

Using (b) above, the ratio of the two estimates $\hat{\sigma}_B^2$ and $\hat{\sigma}_W^2$ is distri-buted as an F variable (see Section 8.5) and in fact forms the test statistic in this situation. A significant result would cause us to reject the hypothesis that the samples came from the same Normal population; otherwise we accept this H_0.

The choice of $\hat{\sigma}_B^2$ as one member of the test ratio should be clear, since it is based on the means of the samples; if the means vary considerably, so $\hat{\sigma}_B^2$ will be larger. It might seem at first obvious (it is certainly con-venient) to use $\hat{\sigma}_T^2$ as the second of the two parts of the ratio for F, since this would clearly obviate the calculation of $\hat{\sigma}_W^2$ (a quantity that we will have gone to some lengths to obtain). However we use $\hat{\sigma}_W^2$ instead because of its independence w.r.t. H_0 (see the previous state-ment). Note that since $\hat{\sigma}_T^2$ and $\hat{\sigma}_B^2$ depend on H_0, their values *are affected by differences in the sample means*.

Also, since the significance of an F test statistic depends on the difference between its two variance estimates, it is quite likely that even though the sample means might be very different the quantities $\hat{\sigma}_T^2$ and $\hat{\sigma}_B^2$ could be similar, causing a possible acceptance of H_0 when it should be clearly rejected. The fact that $\hat{\sigma}_W^2$ is independent of H_0 ensures that in the case of the sample means varying considerably it will remain small compared with $\hat{\sigma}_B^2$ causing a correct rejection of H_0. If the means do not vary significantly (i.e., what small variation there is, is acceptable as 'chance') then the two estimates will be similar and H_0 will be correctly accepted.

We summarise the discussion as follows:

STATEMENT 10.5(a)

To test whether k samples, of sizes $n_1, n_2, \ldots n_k$ (where $\Sigma n_i = n$) are drawn from the same Normal population, we use the test statistic

$F = \dfrac{\sigma_B^2}{\sigma_W^2}$ which has an $F(k-1, n-k)$ distribution under the null

hypothesis that the samples *are* drawn from the same population.

(a) $\hat{\sigma}_B^2$ is an unbiased estimate of the hypothetical population variance σ^2, based on the means of the samples.

(b) $\hat{\sigma}_W^2$ is an unbiased estimate of σ^2, independent of H_0, based on the observations of all the items in the samples after the 'between samples' variation has been eliminated.

Note   The critical value for the test is always one-tailed (increase). That is, the *whole* of the significance level is placed in the right-hand tail.

Proof (in outline)   In previous discussion.

For computational purposes however, the above is not very useful. The following text explains and states the practical techniques employed in the analysis, working through a numerical example. It is rounded off finally with a practical analogue of the previous statement. (See Statement 10.5(b).)

Consider the following three samples of sizes 4, 5 and 3 respectively. We wish to test whether there is any difference between the (mean) values of the samples. Various statistics have been calculated and show below, together with their standard symbols.

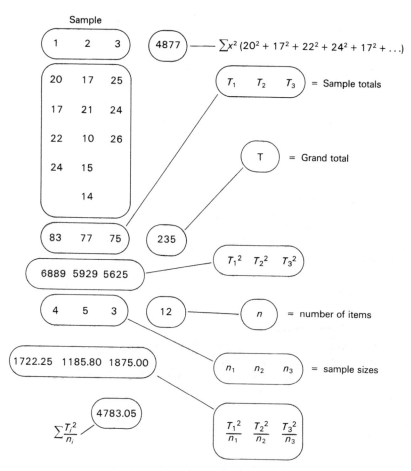

(a)   Given the sample data, immediately calculate all the special statistics shown (and highlighted) above.

(b)
$$SS_T = \Sigma x^2 - \frac{T^2}{n} = 4877 - \frac{(235)^2}{12} = 274.92$$

$$df_T = n - 1 = 12 - 1 = 11$$

(c)
$$SS_B = \Sigma_i \frac{T_i^2}{n_i} - \frac{T^2}{n} = 4783.05 - \frac{(235)^2}{12} = 180.97$$

$$df_B = k - 1 = 3 - 1 = 2$$

(d)
$$SS_W = SS_T - SS_B = 274.92 - 180.97 = 93.9$$

$$df_W = df_T - df_B = 11 - 2 = 9$$

(e)
$$\hat{\sigma}_B^2 = \frac{SS_B}{df_B} = \frac{180.97}{2} = 90.48$$

and
$$\hat{\sigma}_W^2 = \frac{SS_W}{df_W} = \frac{93.95}{9} = 10.44$$

(f) Thus $F = \dfrac{\hat{\sigma}_B{}^2}{\hat{\sigma}_W{}^2} = \dfrac{90.48}{10.44} = 8.67$

From Statement 10.5(a), the critical value of F is $F_{95\%}(2,9) = 4.256$ from tables. Thus $F_{\text{test}} = 8.67$ is *significant* at the 5% level, giving evidence of a difference between samples.

Statement 10.5(b) summarises the procedure for performing a one-way analysis of variance test.

STATEMENT 10.5(b)

Given k samples with sizes $n_1, n_2, \ldots, n_k$ respectively, the following steps are carried out in order to perform a one-way Analysis of Variance test.

(a)   Evaluate:
   (i)   Sample totals, T_i;
   (ii)   Grand total, T;
   (iii) Squares of sample totals, T_i^2.
   (iv)   Sum of values T_i^2/n_i, $\Sigma(T_i^2/n_i)$
   (v)   Sum of squares of original x values, Σx^2.

(b)   Calculate $SS_T = \Sigma x^2 - \dfrac{T^2}{n}$ and $\mathrm{df}_T = n - 1$

(c)   Calculate $SS_B = \Sigma_i \dfrac{T_i^2}{n_i} - \dfrac{T^2}{n}$ and $\mathrm{df}_B = k - 1$

(d)   Hence find $SS_W = SS_T - SS_B$ and $\mathrm{df}_W = \mathrm{df}_T - \mathrm{df}_B$

(e)   Evaluate the two variance estimates:

$$\hat{\sigma}_B{}^2 = \frac{SS_B}{\mathrm{df}_B} \qquad \text{and} \qquad \hat{\sigma}_W{}^2 = \frac{SS_W}{\mathrm{df}_W}$$

(f)   Thus calculate the value of test statistic F as: $F = \dfrac{\hat{\sigma}_B{}^2}{\hat{\sigma}_W{}^2}$

which, under the H_0 of 'no difference between sample means', has an $F(k-1, n-k)$ distribution.

Note   The critical value for the test is always one-tailed (increase). That is, the *whole* of the significance level is placed in the right-hand tail.

It is usual to accompany calculations for a one-way analysis of variance with a table as follows:

Source of variation	Sums of squares	Degrees of freedom	Variance estimates
Between samples	SS_B	$k - 1$	$SS_B/k - 1$
Residual	SS_W	$(n - k)$	$SS_W/n - k$
Total	SS_T	$n - 1$	

The values in the body of the table are entered as they are calculated.

If the magnitude of the original table items is large or they consist of too many decimal places, a coding method can be used. The technique is simple and straightforward. A linear coding of the form $X = \dfrac{x - a}{b}$ is chosen (for convenience) transforming the set of x values into X values. The value of F_{test} is then calculated in the usual way *without any decoding*. This latter point is easy to prove, as follows.

Let $F_X = \dfrac{\hat{\sigma}_B^2(X)}{\hat{\sigma}_W^2(X)}$ say, where $\hat{\sigma}_B^2(X)$ means that $\hat{\sigma}_B^2$ has been calculated using the coded X values, etc. Now, from Section 2.6.4, we know that:

$$\hat{\sigma}_B^2(x) = b^2 \cdot \hat{\sigma}_B^2(X) \text{ and } \hat{\sigma}_W^2(x) = b^2 \cdot \hat{\sigma}_W^2(X)$$

Hence:
$$F_X = \frac{\hat{\sigma}_B^2(X)}{\hat{\sigma}_W^2(X)} = \frac{(1/b^2)\hat{\sigma}_B^2(x)}{(1/b^2)\hat{\sigma}_W^2(x)}$$

$$= \frac{\hat{\sigma}_B^2(x)}{\hat{\sigma}_W^2(x)}$$

$$= F_x$$

That is, the value of F_{test} is not affected by a linear coding and a coding method should be used wherever it will cut down the arithmetic.

EXAMPLE 10.9

Six specimens of fuel gas had a number of check tests performed on them to measure their thermal unit values with the following results:

Specimen					
1	2	3	4	5	6
533	541	526	541	549	520
537	545	536	519	537	520
535	549	534	528	538	518
546	532	533	536		530
	540				522

Is there any evidence of a difference between the specimens?

SOLUTION

The competing hypotheses for the test are:

H_0: No difference between (the means of) the samples;
H_1: One or more (of the means) of the samples are different.

Since, for calculation purposes, the given values are large, we shall subtract 530 from each of them. That is, we will make the transformation
$X = x - 530$. Remember that no decoding is necessary.

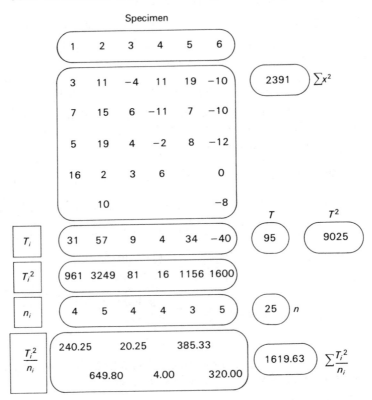

Specimen

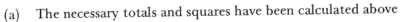

(a)   The necessary totals and squares have been calculated above

(b) $SS_T = \Sigma x^2 - \dfrac{T^2}{n} = 2391 - \dfrac{9025}{25} = 2030; \quad df_T = 24$

(c) $SS_B = \displaystyle\sum_i \dfrac{T_i^2}{n_i} - \dfrac{T^2}{n} = 1619.63 - \dfrac{9025}{25} = 1258.63; \quad df_B = 5$

(d) $SS_W = SS_T - SS_B = 2030 - 1258.63 = 771.37;$
       $df_W = df_T - df_B = 24 - 5 = 19$

(e) $\hat{\sigma}_B^2 = \dfrac{SS_B}{df_B} = \dfrac{1258.63}{5} = 251.73; \quad \hat{\sigma}_W^2 = \dfrac{SS_W}{df_W} = \dfrac{771.37}{19} = 40.60$

(f)   Thus $F = \dfrac{\hat{\sigma}_B^2}{\hat{\sigma}_W^2} = \dfrac{251.73}{40.60} = 6.20$

Critical F is $F_{0.95}(5.19) = 2.74$. Thus $F_{test} = 6.20$ is significant at the 5% level and H_0 is rejected. That is, there is evidence of a difference between specimens.

Looking back at the original data, it is fairly clear that specimen 6 is significantly different from the others. However, to show this statistically it would be necessary to perform an analysis of variance test on the first 5 specimens and obtain a non-significant result. This is left as an exercise (Exercise 32).

10.4.3 TWO-WAY ANALYSIS

Basically a two-way analysis of variance can be thought of as two separate one-way analyses. The theory behind this particular analysis will not be covered in as much detail as for the 'one-way' case since it is basically an extension.

We begin with a theoretical model as follows. The test is for the differences between the means of n samples (each of m items) with respect to factor A *and* testing for differences between the means of m samples (each of n items) with respect to factor B.

		Columns (factor A samples)			
		1	2	...	n
Rows (factor B samples)	1 2 . . .	x_{11} x_{21}	x_{12} x_{22}	... ...	x_{1n} x_{2n}
	m	x_{m1}	x_{m2}	...	x_{mn}

The null hypothesis for both tests is that all the items come from the same (assumed Normal) population.

Alternatively, we can consider the two separate null hypotheses in the form:

H_0^A: No difference between the means of the column samples

H_0^B: No difference between the means of the row samples

This latter approach is preferable since we will be conducting two separate tests (as mentioned earlier) and hence need two separate critical regions, corresponding to two separate alternative hypotheses.

As in the one factor case, we consider a 'total variation' and split it into several components:

(a)   a variation between column (factor A) samples;
(b)   a variation between row (factor B) samples; and
(c)   a residual (or *error*) variation.

The residual variation can be thought of as what is left over after the 'between column' and 'between row' variations have been eliminated.

σ^2, the theoretical population variance envisaged under the null hypothesis, can be estimated with the three measures:

$\hat{\sigma}_R^2$ — obtained from the means of the row samples

$\hat{\sigma}_C^2$ — obtained from the means of the column samples

$\hat{\sigma}_T^2$ — obtained by taking all items into consideration

These variance estimates can be expressed in the form of a sum of squares divided by a number of degrees of freedom as follows:

$$\hat{\sigma}_R^2 = \frac{SS_R}{df_R}; \qquad \hat{\sigma}_C^2 = \frac{SS_C}{df_C}; \qquad \hat{\sigma}_T^2 = \frac{SS_T}{df_T}$$

Statement 10.6 gives formulae for calculating these estimates.

Another measure used as an unbiased estimator of σ^2 is $\hat{\sigma}_E^2$.

It can be expressed in the form $\quad \hat{\sigma}_E^2 = \dfrac{SS_E}{df_E}.$

The residual or error sum of squares, SS_E and its degrees of freedom, df_E, are calculated by subtraction.

In a two-way analysis of variance, there are two F_{test} statistics. One tests for differences between column samples and *independently* the other tests for differences between row samples.

The following statement lays out the order of the calculations and statistics necessary to perform both tests.

STATEMENT 10.6

Given the following tabular data structure (which has had column and row totals and other statistics added):

		Columns						
		1	2	...	...	n	T_i	T_i^2
	1	x_{11}	x_{12}	...	...	x_{1n}	T_1	T_1^2
	2	x_{21}	x_{22}	...	...	x_{2n}	T_2	T_2^2
Rows	...	...	...	...		...	...	...
	...	...	...	...		...	...	...
	m	x_{m1}	x_{m2}	...	...	x_{mn}	T_m	T_m^2
	T_j	T_1	T_2	...	...	T_n	T	ΣT_i^2
	T_j^2	T_1^2	T_2^2	...	...	T_n^2	ΣT_j^2	Σx^2

the following steps are carried out in calculating the statistics necessary to perform a two-way analysis of variance.   *(continued)*

(a)   As shown in the table above, construct row and column totals (T_i and T_j respectively), their squares and grand totals. Also calculate the sum of squares of the individual values (Σx^2).

(b)   Calculate $SS_T = \Sigma x^2 - \dfrac{T^2}{mn}$ and $df_T = mn - 1$

(c)   Calculate $SS_R = \dfrac{\Sigma T_i^2}{n} - \dfrac{T^2}{mn}$ and $df_R = m - 1$

(d)   Calculate $SS_C = \dfrac{\Sigma T_j^2}{m} - \dfrac{T^2}{mn}$ and $df_C = n - 1$

(e)   Hence find: $SS_E = SS_T - SS_R - SS_C$ and $df_E = df_T - df_R - df_C$

(f)   Evaluate the three variance estimates:

$$\hat{\sigma}_R^2 = \frac{SS_R}{df_R}; \qquad \hat{\sigma}_C^2 = \frac{SS_C}{df_C}; \qquad \hat{\sigma}_E^2 = \frac{SS_E}{df_E}$$

(g)   Thus evaluate the two F_{test} statistics as:

$$F_{test}\,(\text{rows}) = \frac{\hat{\sigma}_R^2}{\hat{\sigma}_E^2}; \qquad F_{test}(\text{columns}) = \frac{\hat{\sigma}_C^2}{\hat{\sigma}_E^2}$$

F_{test} (rows) tests for differences between rows (with $H_0 = no\ differ$-$ences$ between the row means) and the critical value has an $F(df_R,\ df_E)$ distribution under H_0.

F_{test} (columns) tests for differences between columns (with $H_0 = no\ differences$ between the column means) and the critical value has an $F(df_C,\ df_E)$ distribution under H_0.

Note that the two separate tests are independent of one another and are always one-tailed (increase).

Notice the similarities between this analysis and that of the last subsection. Also note the use of σ_E^2 in both test statistic ratios above, since, as in the one way case, $\hat{\sigma}_E^2$ is an unbiased estimate of the hypothetical population σ^2.

A method of coding (linear) can be conveniently employed in the two-way case when the data from the samples is large and/or unwieldy. This of course is because no 'decoding' procedure is involved when using F test statistics.

EXAMPLE 10.10

Samples were taken involving two interactive factors A and B in a two-way analysis of variance experiment, giving the following results:

	Factor A			
	9	6	8	8
	0	2	1	3
Factor B	0	4	9	0
	8	0	5	3
	1	1	4	2

Is there any evidence at differences between the samples with respect to either factor?

SOLUTION

Setting up a table for the calculations, we have:

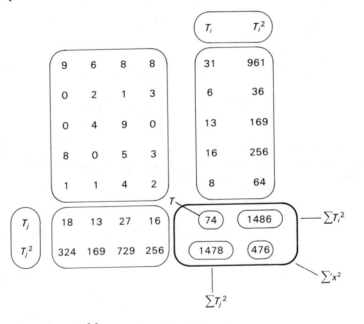

So, from the above table:

$T = 74$; $T^2 = 5476$; $\Sigma T_i^2 = 1486$; $\Sigma T_j^2 = 1478$; $\Sigma x^2 = 476$; $m = 5$; $n = 4$ and thus $mn = 20$.

$$SS_T = \Sigma x^2 - \frac{T^2}{mn} = 476 - \frac{5476}{20} = 202.2$$

$$SS_R = \frac{\Sigma T_i^2}{n} - \frac{T^2}{mn} = \frac{1486}{4} - \frac{5476}{20} = 97.7$$

$$SS_C = \frac{\Sigma T_j^2}{m} - \frac{T^2}{mn} = \frac{1478}{5} - \frac{5476}{20} = 21.8$$

Also: $df_T = mn - 1 = 19$; $df_R = m - 1 = 4$;
           $df_C = n - 1 = 3$

Thus: $SS_E = SS_T - SS_R - SS_C = 202.2 - 97.7 - 21.8 = 82.7$

and: $df_E = df_T - df_R - df_C = 19 - 4 - 3 = 12$

Hence: $\hat{\sigma}_R^2 = \dfrac{SS_R}{df_R} = \dfrac{97.7}{4} = 24.43$; $\hat{\sigma}_C^2 = \dfrac{SS_C}{df_C} = \dfrac{21.8}{3} = 7.27$

and $\hat{\sigma}_E^2 = \dfrac{SS_E}{df_E} = \dfrac{82.7}{12} = 6.89$

Thus: $$F_{\text{test}}(\text{rows}) = \frac{\hat{\sigma}_R^2}{\hat{\sigma}_E^2} = \frac{24.43}{6.89} = 3.54$$

The critical value for F here is $F_{0.95}(4, 12) = 3.26$ from tables. There-fore F_{test} is significant at the 5% level and thus H_0 can be *rejected*, i.e. there is evidence of differences between the row (factor B) samples.

Also: $$F_{\text{test}}(\text{columns}) = \frac{\hat{\sigma}_C^2}{\hat{\sigma}_E^2} = \frac{7.27}{6.89} = 1.05$$

The critical value for F here is $F_{0.95}(3, 12) = 3.49$ from tables. There-fore F_{test} is not significant at the 5% level and thus H_0 is *accepted*, i.e. there is no evidence of differences between the column (factor A) samples.

10.5 THE SIGN TEST

For statistical tests related to continuous populations that are not concerned with involving the mean, the median is a convenient and practical alternative. For any distribution (not necessarily Normal), the median has the useful property that the probability that a random value is less than it is just $\frac{1}{2}$. Using this fact, there are certain tests that can be employed. This section is concerned with just one of these, commonly called the 'sign test', where we look at its use in two specific situations.

First, given a random sample of size n, can it be considered as having been drawn from a population having median $m = m_0$, say? Second, given two paired samples, are they drawn from the same population or equivalently, are they drawn from populations having the same median?

10.5.1 FOR A SINGLE SAMPLE

We consider the former situation first. Given a random sample X_1, $X_2, \ldots, X_n$, we are interested in whether it can be considered as coming from a population having median $m = m_0$. Assuming that it does, and taking this as the null hypothesis, the probability of any sample member exceeding m_0 is just $\frac{1}{2}$. This is of course by virtue of the fact that the median lies exactly half way along an arrayed population.

So that:
$$\Pr(X_i > m_0) = \frac{1}{2} = \Pr(X_i < m_0)$$

or:
$$\Pr(X_i - m_0 > 0) = \frac{1}{2} = \Pr(X_i - m_0 < 0)$$

We now introduce what is known as an *indicator variable* I defined as follows:
$$I_i = \begin{cases} 1; & \text{if } X_i - m_0 > 0 \\ 0; & \text{if } X_i - m_0 < 0 \end{cases}$$

with:
$$\Pr(I_i = 0) = \frac{1}{2} = \Pr(I_i = 1)$$

Using the definition in Section 6.6.1, I_i is clearly a Bernoulli random variable with parameter $p = \frac{1}{2}$, for each i. Hence, putting $U = \sum_{i=1}^{n} I_i$, we have U as a binomial variable with parameters n and $\frac{1}{2}$, from Statement 6.15.

In other words, given a set of sample values from a distribution with suspected median m_0, we count up how many are greater than m_0, giving the value of U. In practice, it is convenient to discard sample values that are equal to m_0 and adjust the value of n accordingly.

Given that U is binomial with parameters n and $\frac{1}{2}$, the Normal approximation can be used (Statement 6.14) to give U as $N\left(\dfrac{n}{2}, \dfrac{n}{4}\right)$ approximately, under H_0. Note that the approximation is very good here even for small n, since $p = \frac{1}{2}$.

Standardising gives:
$$\frac{U - \dfrac{n}{2}}{\sqrt{\dfrac{n}{4}}} = Z \quad \text{is} \quad N(0, 1)$$

and this is used as the test statistic.

There is a slight adjustment necessary to the above statistic however, namely a continuity correction of $\frac{1}{2}$. This is because U is strictly discrete, while the transformed Z (as a $N(0, 1)$ variable) is, of course, continuous. The correction of $\frac{1}{2}$ is either added or subtracted depending on the situation. For a one-tailed test, the $\frac{1}{2}$ is added for a decrease (left critical region) and subtracted for an increase (right critical region). For a two-tailed test, the $\frac{1}{2}$ is used so that the value of Z_{test} is pushed towards zero *away from the nearest critical region*.

The test statistic can thus be written in the form:

$$Z = \frac{U - \dfrac{n}{2} \pm \dfrac{1}{2}}{\sqrt{\dfrac{n}{4}}}$$

In practice, X values that are less than hypothetical median m_0 are replaced by 'minus' signs and those that are greater are replaced by 'plus' signs. U is then obtained by finding the total number of plus signs.

To summarise:

STATEMENT 10.7

To test whether a random sample of size n $X_1, X_2, \ldots, X_n$ can be considered as being drawn from any population (not necessarily Normal) having median m_0, the test statistic used is:

$$Z = \frac{U - \dfrac{n}{2} \pm \dfrac{1}{2}}{\sqrt{\dfrac{n}{4}}}$$

which has a $N(0, 1)$ distribution under the null hypothesis that the true population median *is* m_0.

Here:

(a) U is the number of X_i that are greater than m_0.

(b)   The continuity correction of $\frac{1}{2}$ in the test statistic numerator is either added or subtracted so that Z_{test} is moved away from the nearest critical region.

(c)   Any sample values equal to m_0 are discarded and n adjusted accordingly.

Proof   In preceding text.

Consider the following data obtained from testing the strength of a certain type of concrete developed by a manufacturer:

$$42, \quad 38, \quad 38, \quad 41, \quad 41, \quad 39, \quad 43, \quad 42, \quad 42, \quad 38$$

Is there evidence of an increase over the strength of the previous concrete used, which was known to have a median of 39?

Subtracting 39 from each of the given values and recording only the signs of the numbers obtained gives the set $+ - - + + 0 + + + -$. Excluding the zero in the count gives $u = 6$. Note now $n = 9$. Since we are looking for an increase in strength, we test $H_0 : m_0 \leqslant 39$ versus

H_1: $m_0 > 39$, a one-tailed (increase) test. Thus, since the critical region will be in the right tail of the null $N(0, 1)$ distribution, we need to *subtract* the correction of $\frac{1}{2}$ in the test statistic numerator.

Hence:
$$Z_{test} = \frac{6 - 4.5 - \frac{1}{2}}{\sqrt{9/4}} = 0.67 \quad (2D)$$

The critical value for the test is $Z_{5\%} = 1.645$. Hence the above test result is not significant and we accept H_0. That is, there is not enough evidence of an increase in the population median.

EXAMPLE 10.11

Cylinders of median diameter 4.04 in are being produced by a manufacturing company. A sample of ten cylinders was taken from a particular machine and their diameters measured giving 3.88, 4.09, 3.92, 3.97, 4.02, 3.95, 4.03, 3.92, 3.98 and 4.06 in. Is there any reason to believe that this particular machine is producing cylinders whose diameters are smaller than the company average?

SOLUTION

We use a sign test to decide between the hypotheses H_0: $m_0 \geqslant 4.04$ and H_1: $m_0 < 4.04$ (one-tailed) using the test statistic:

$$Z = \frac{U - \frac{n}{2} \pm \frac{1}{2}}{\sqrt{\frac{n}{4}}} \quad \text{which is } N(0, 1) \text{ under } H_0$$

Subtracting 4.04 from each of the given values and taking the signs of the results gives $- + - - - - - - - +$. Therefore $u = 2$, $n = 10$ and since we have a one-tailed test, the critical value for Z_{test} is $Z_{95\%} = -Z_{5\%} = -1.645$. Note that since there is a left critical region, the correction of $\frac{1}{2}$ needs to be added to the test statistic numerator.

Thus $Z_{test} = \dfrac{2 - 5 + \frac{1}{2}}{\sqrt{10/4}} = -1.58$ which is not quite significant.

Hence we accept H_0 and conclude that there is not enough evidence to suppose that the diameters are smaller in this case.

10.5.2 FOR THE DIFFERENCE BETWEEN TWO SAMPLES

In the situation where two (paired) samples have been drawn, the test is concerned with whether the two medians (of the theoretical populations from which the samples are drawn) are compatible.

There is a direct analogy here between this situation and that necessary for the application of a paired T test. In the latter case, of course, we need both the assumption of Normality and identical population variances. See Fig. 10.2. This latter test is sometimes referred to as a Normal distribution *slippage test*, i.e., we can 'slide' one distribution on to the other.

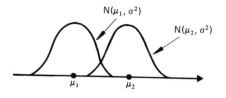

$N(\mu_1, \sigma^2)$

$N(\mu_2, \sigma^2)$

μ_1 μ_2

Fig. 10.2

The paired sign test is directly analogous to this in the sense that the two population density functions are identical except for a translation of one along the horizontal axis (the x-axis, say).

If we take the density function of the two populations as $f_1(x)$ and $f_2(x)$ respectively, then in testing the equality (or otherwise) of the two medians m_1 and m_2, we are really testing $H_0 : f_1(x) = f_2(x)$. The test can be either one or two tailed, depending on the situation.

For a two-tailed test, we have $H_1 : f_1(x) \neq f_2(x)$. For a one-tailed (decrease) test we have $H_1 : f_1(x) = f_2(x - a)$, where the second distribution is shifted a units to the left. For a one-tailed (increase) test, we have $H_1 : f_1(x) = f_2(x + a)$, where the second distribution is shifted a units to the right (see Fig. 10.3).

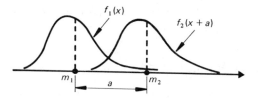

$f_1(x)$

$f_2(x + a)$

m_1 a m_2

Fig. 10.3

The form and derivation of the test statistic in this case is similar to the previous one.

STATEMENT 10.8

To test whether two paired random samples $X_1, X_2, \ldots, X_n$ and $Y_1, Y_2, \ldots, Y_n$ of size n can be considered as having been drawn from the same population, the following test statistic is used:

$$Z = \frac{U - \dfrac{n}{2} \pm \dfrac{1}{2}}{\sqrt{\dfrac{n}{4}}}$$

(continued)

which has a   N(0, 1) distribution   under the null hypothesis that both samples *are* drawn from the same population.

Here, U is the number of X_i that are greater than their corresponding pair-partners, Y_i.

If, for any j, $X_j = Y_j$, these two values are discarded and n adjusted accordingly.

Note that the continuity correction of $\frac{1}{2}$ is again applied here (see Statement 10.7).

Proof   Under H_0,   both samples come from the same distribution. In particular, X_i and Y_i constitute a random sample of   2   from the same (hypothetical) population.

Hence:
$$\Pr(X_i > Y_i) \;=\; \frac{1}{2} \;=\; \Pr(X_i < Y_i)$$

Thus:
$$\Pr(X_i - Y_i > 0) \;=\; \frac{1}{2} \;=\; \Pr(X_i - Y_i < 0)$$

Now introduce indicator variable I,   by putting:
$$I_i \;=\; \begin{cases} 1; & \text{if } X_i - Y_i > 0 \\ 0; & \text{if } X_i - Y_i < 0 \end{cases} \qquad \text{for all } i$$

This gives $\Pr(I_i = 0) = \frac{1}{2} = \Pr(I_i = 1)$ (for all i)   and I_i is clearly a Bernoulli random variable.

Thus $U = \sum_{i=1}^{n} I_i$ is $\text{Bin}(n, \frac{1}{2})$ under H_0.

Using a Normal approximation, we have U as $N\left(\dfrac{n}{2}, \dfrac{n}{4}\right)$ giving the statistic:
$$\frac{U - \dfrac{n}{2}}{\sqrt{\dfrac{n}{4}}} \quad \text{as } N(0, 1)$$

Introducing a continuity correction, since we are approximating a discrete variable with a continuous one, we have the test statistic as:
$$Z \;=\; \frac{U - \dfrac{n}{2} \pm \dfrac{1}{2}}{\sqrt{\dfrac{n}{4}}}$$

which has a   N(0, 1) distribution   under H_0.

This completes the proof.

EXAMPLE 10.12

The following scores relate to a set of 22 students who sat an examination at half term (x) and at the end of term (y) in the same subject.

x	67 71 83 70 69 68 36 52 72 72 80
y	58 62 80 70 67 72 28 63 70 81 62

x	63 73 70 82 51 82 90 83 62 47 81
y	62 82 51 80 37 72 69 81 64 46 80

Is there any evidence that the students perform better in the half term examination?

SOLUTION

If m_x and m_y are the two respective population medians then we are testing $H_0: m_x \leqslant m_y$ against $H_1: m_x > m_y$.

The test statistic is:

$$Z = \frac{U - \frac{n}{2} \pm \frac{1}{2}}{\sqrt{\frac{n}{4}}}$$

which is $N(0, 1)$ under $m_x = m_y$.

The critical value for Z at the 5% level (one-tailed, increase) is 1.645.

The signs of $x - y$ for the given set are:

$$+ + + 0 + - + - + - +$$
$$+ - + + + + + + - + +$$

Excluding the single '0', $u = 16$.

Thus, with $n = 21$, discarding 0 we have:

$$Z_{test} = \frac{16 - \frac{21}{2} - \frac{1}{2}}{\sqrt{\frac{21}{4}}} = 2.18$$

The test result is therefore significant and we accordingly reject H_0. We conclude that there is evidence, at the 5% level, that higher scores are obtained in the half term examinations.

10.6 EXERCISES

SECTION 10.2

1.   50 observations were made of a random variable and on the basis of these an 'expected' distribution was fitted (subject to 2 restrictions) as follows:

Value of Variable	0	5	10	15	20
Observed frequency	7	6	12	19	6
Expected frequency	5	5	15	15	10

Comment on the fit with the aid of a χ^2 test.

2.   Two fair dice are thrown 432 times. Find the expected frequencies of scores of 2, 3, 4, ..., 12. Two players A and B are each given two dice and told to throw them 432 times, recording the results. The frequencies reported are:

Scores	2	3	4	5	6	7	8	9	10	11	12
A's frequency	18	33	28	54	62	65	66	42	30	27	7
B's frequency	14	22	34	51	58	73	63	45	38	25	9

Is there any evidence that either pair of dice is biased? What can be said about B's alleged results?   (AEB)'76

3.   A cubical die was thrown 300 times, and the results were as follows:

Score	1	2	3	4	5	6
Frequency	48	51	51	58	54	38

Do these results indicate that the die is biased?

4.   3000 observations were made of a random variable and a theoretical 'expected' distribution was calculated using the total frequency, to obtain the following table for comparison:

	x_1	x_2	x_3	x_4	x_5	x_6
O	620	530	490	410	370	580
E	500	500	500	500	500	500

Comment on the fit, using a χ^2 test.

5.   At a public library, during a given week, the following numbers of books were borrowed

Day	Mon.	Tues.	Wed.	Thurs.	Fri.	Sat.
Number issued	204	292	242	283	252	275

Is there reason to believe that more books are generally borrowed on one weekday than on another?

6. A bag contains 9 red balls, 7 white balls, 5 yellow balls and 4 blue balls. A ball is drawn at random from the bag and its colour noted. The ball is then returned to the bag and this process is repeated 175 times. The observed frequencies are given below.

Colour	Frequency
Red	60
White	51
Yellow	43
Blue	21

Use the χ^2 test at the 5% significance level to determine whether there is close agreement between the observed results and the theoretical results you would expect.

If the process had been repeated 350 times instead of 175 times with observed frequencies 120, 102, 86 and 42, would your conclusion be the same?

[Table of 5% points of the χ^2 distribution:

Degrees of freedom (f)	1	2	3	4	5
Critical value of χ^2	3.84	5.99	7.81	9.49	11.1]

(Cambridge)

7. Two dice were tossed 216 times and the results were reported to be distributed as follows:

Total score	2	3	4	5	6	7	8	9	10	11	12
Frequency	8	20	18	20	32	30	30	27	17	11	3

Determine whether these results are likely to have come from fair dice and fair throws. How many of the 30 throws of eight would you expect to have arisen from double fours?

8. In a seed viability test, 600 seeds were planted in 100 rows of 6. The number of seeds that germinated in each row was counted and the results are shown in the table below:

No. of seeds germinating per row	0	1	2	3	4	5	6
Observed no. of rows	1	4	7	29	33	18	8

Calculate: (a) the mean number of seeds germinating per row; and (b) the expected frequencies corresponding to these observed values for a binomial distribution with the same mean as that found in (a).

Carry out the appropriate χ^2 test, at the 5% level of significance, to determine whether the observed results are sufficiently close to the theoretical results to support the assumption of a binomial distribution with mean as calculated above.

9. If $\Pr(x)$ denotes the probability of x successes in a binomial distribution for n trials for which the probability of a success in each trial is p, show that $\Pr(x + 1)/\Pr(x) = p(n - x)/(1 - p)(x + 1)$.

A bag contains a very large number of black marbles and white marbles. 8192 random samples of 6 marbles are drawn from the bag. The frequencies of the numbers of black marbles in these samples are tabulated below:

Number of black marbles per sample	0	1	2	3	4	5	6	Total
Frequencies		3	42	255	1115	2505	2863	1409	8192

Test the hypothesis that the ratio of the numbers of black to white marbles in the bag is $3:1$.
(AEB)'72

10. Four coins were tossed 4096 times, and the number of heads obtained was noted each time, with the following results:

Number of heads	0	1	2	3	4
Observed frequency	249	1000	1552	1050	245

Examine the hypothesis that the coins are unbiased.

11. The owner of a small country inn observes that during the holiday season the demand for rooms is as follows:

Rooms required	0	1	2	3	4	5	6	7	8
Number of nights	2	9	16	26	33	25	20	11	5

Calculate the mean demand for rooms per night. Use a χ^2 test with a 5% significance level to determine whether the Poisson distribution is an adequate model for the data.

The inn has only four rooms available to let. Assuming demand for rooms follows a Poisson distribution with mean 4.17, calculate the mean and variance of the number of rooms occupied per night.
(AEB)'86

12. Four players meet weekly and play eight hands of cards. In a year one of the players finds that he has won x of the eight hands with a frequency given in the following table.

x	0	1	2	3	4	5	6	7	8
f	4	13	12	12	6	3	2	0	0

Find the frequencies of the number of hands he would expect to win if the probability of winning any hand were $\frac{1}{4}$. Use the χ^2 distribution to test this hypothesis.
(AEB)'75

13. In a period of 100 days, the frequencies with which varying numbers of calls arrived at a telephone switchboard in the period 8 a.m.—9 a.m. were as follows:

Number of calls	0	1	2	3	4	5 or more
Number of days	34	37	25	3	1	0

Use a suitable test to examine whether these frequencies differ significantly from those expected on the hypothesis that calls arrive independently and at random.

14. Describe briefly an experiment (if possible one you have carried out) which would lead to the use of a test based on the chi-squared distribution. Describe how you would use the test, making particular reference to the proper choice of the number of degrees of freedom and to any constraints that arise from the sizes of the expected frequencies. The number of defective items found in a box of 144 items is denoted by X. The following table gives the distribution of defective items found in a random sample of 100 boxes of manufactured items.

X	0	1	2	3	4	5
Frequency	22	37	20	13	6	2

(a) Test whether the data may reasonably be regarded as conforming to a Poisson distribution with mean 1.5, (b) The binomial distribution with $n = 5$ and $p = 0.3$ fits the above data. Without further calculation state why, nevertheless, this binomial model is not appropriate.

(JMB)

15. The results of twenty football matches, played on the same occasion, revealed the following frequency distribution for the numbers of goals scored by the forty teams:

Goals scored	0 1 2 3 4 5 6 7 8
Teams scoring	3 5 14 9 3 4 1 1 0

Use the χ^2 test to compare these observed frequencies with the frequencies which would be expected from the Poisson distribution having the same mean and total frequency. Comment on the agreement.

16. (a) Over a period of 500 working days the numbers of telephone calls received by an office telephone exchange during the lunch hour were recorded. The results were as follows:

No. of calls	0	1	2	3 or more
Frequency	287	149	46	18

The expected Poisson distribution with the same mean and total frequency is:

No. of calls	0	1	2	3 or more
Frequency	275	164	49	12

Using the χ^2 test at the 5% significance level, test the likelihood of the recorded distribution being Poisson.

(b) A coin was tossed 200 times and 120 heads appeared. Is this evidence that the coin was biased? (Use the 1% significance level.)

[Table of the 5% and 1% points of the χ^2 distribution:

Degrees of freedom (f)		1	2	3	4	5
Critical value of χ^2	5%	3.84	5.99	7.81	9.49	11.1
	1%	6.64	9.21	11.34	13.28	15.09]

(Cambridge)

17.   The table below gives the distribution for the number of heavy rain-storms reported by 330 weather stations in the United States of America over a one year period.

Number of rainstorms (x)	0	1	2	3	4	5	more than 5
Number of stations (f) reporting x rainstorms	102	114	74	28	10	2	0

(a) Find the expected frequencies of rainstorms given by the Poisson distribution having the same mean and total as the observed distribution.

(b) Use the χ^2 distribution to test the adequacy of the Poisson distribution as a model for these data.   (AEB)'77

18.   It is observed that the frequencies with which the number (x) of fertile seeds per pod of a certain legume occur are:

x	0	1	2	3	4	5	6	7	8 or more
f	24	60	52	32	22	5	3	2	0

It may be calculated from these that $\Sigma f = 200$, $\Sigma fx = 405$ and $\Sigma fx^2 = 1239$. Estimate the mean and variance of x and say why it is reasonable to consider the Poisson distribution with mean 2 as a suitable model for x. Perform a suitable test for the goodness of fit of the data to this Poisson distribution.   (Cambridge)

19.   Use a χ^2 goodness of fit test to say whether the following distribution can be considered as Normal:

Weight (gm)	50–59.9	60–69.9	70–79.9	80–89.9
No. of specimens	8	10	16	14

Weight (gm)	90–99.9	100–109.9	110–119.9
No. of specimens	10	5	2

20.   An experiment to estimate the nature of the distribution of gamma emission rates, from a radioactive source, produced the following frequency distribution:

No. of photons registered by counter in 15-second interval	No. of intervals
1800–1819	1
1820–1839	4
1840–1859	8
1860–1879	9
1880–1899	13
1900–1919	13
1920–1939	7
1940–1959	17
1960–1979	10
1980–1999	2
2000–2019	4
2020–2039	3

Compare, by the χ^2 test, these observed frequencies with those frequencies expected from a normal distribution based on the sample mean and variance. State your conclusions.

21.   A sample of 230 observations of the variable x has the frequency
distribution

x	below 5.999	6.000 to 7.999	8.000 to 9.999	10.000 to 11.999	12.000 to 13.999
f	3	16	21	38	50

x	14.000 to 15.999	16.000 to 17.999	18.000 to 19.999	above 19.999
f	48	36	10	8

The variable x is believed to be distributed as $N(13, 3.4^2)$ which, for
the same intervals and total frequency, gives the frequencies

f' 5 12 27 45 52 45 27 12 5

Draw a suitable diagram to illustrate all this information. Test whether
the data justify the belief.   (Cambridge)

SECTION 10.3

22.   A firm uses three similar machines to produce a large number of components. On a particular day a random sample of 99 from the defective
components produced on the early shift were traced back to the
machines which produced them. The same was done with a random
sample of 65 from the defectives produced on the late shift. The table
shows the number of defectives found to be from each machine on each
shift.

		Machine		
		A	B	C
Shift	Early	37	29	33
	Late	13	16	36

(a) Use the χ^2 distribution to test, at the 5% significance level, the
hypothesis that the probability of a defective coming from a particular
machine is independent of the shift on which it was produced.

(b) Using the early shift figures *only*, test, at the 5% significance level,
the hypothesis that a defective is equally likely to have come from any
machine.

(c) Comment briefly on your results' given the following additional
information.
(i) All machines produce components at a similar rate.
(ii) During the early shift, machines B and C were overhauled and so
were in production for a much shorter time than A.
(iii) All machines were in production for the same time on the late
shift.
                                                              (AEB)'84

23.   Show that, for the contingency table as given below, a coefficient of association between attributes A and B is given by

$$Q = \frac{ps - qr}{ps + qr}$$

	B	not-B	Total
A	p	q	m_1
not-A	r	s	m_2
Total	n_1	n_2	N

In a survey covering 5000 individuals, the possible association between colour blindness and sex was investigated. The following results were obtained:

	Normal Colour Vision	Colour Blind
Male	2210	190
Female	2540	60

Use the χ^2-test at the 5% level of significance to establish whether or not there is evidence of association between colour blindness and sex. If so, find the coefficient of association Q, as defined above, between these two attributes.

24.   The following fictitious data refer to hair and eye colours for 200 persons:

Eye Colour	Hair colour			Total
	Dark	Medium	Fair	
Blue	3	42	30	75
Grey	2	18	5	25
Brown	45	40	15	100
Total	50	100	50	200

Use the χ^2 test to examine the hypothesis that there is no association between hair and eye colours.

25.   Oral tests are conducted by three examiners separately. The numbers of candidates in the categories credit, pass, fail are as shown in the table

	Examiner			Total
	A	B	C	
Credit	10	5	13	28
Pass	31	38	28	97
Fail	29	20	26	75
Total	70	63	67	200

Use a χ^2 test to examine the hypothesis that the examiners do not differ in their standards of awards. State the assumptions made.

26.   Four treatments for a common skin disease were tested on a number of
        patients. The results were as follows:

                              Treatment
                         A   B   C   D
        Patients cured   36   53   11   29
        Total treated   60   80   20   40

        Is there any evidence of a difference between the effectiveness of the
        treatments?

27.   A thousand households are taken at random and divided into three
        groups A, B and C, according to the total weekly income. The follow-
        ing table shows the numbers in each group having a colour television
        receiver, a black and white receiver, or no television at all.

                              A   B   C
        Colour television   56   51   93
        Black and white   118   207   375
        None   26   42   32

        Calculate the expected frequencies if there is no association between
        total income and television ownership. Apply a test to find whether the
        observed frequencies suggest that there is such an association. (AEB)'74

28.   A company which manufactures Spiv washing machines conducted a
        survey of 500 women who bought their machines last year. Only 150
        of these 500 women replied to the postal questionnaire, so the
        company decided to send reminders to the remaining 350 women. This
        produced a further 200 replies and a second reminder produced another
        50 replies. The remaining 100 women were visited by an interviewer
        and they all gave the required information at this stage. As a result of
        this work, the company suspects that there is some association between
        a woman's degree of satisfaction with her Spiv washing machine and her
        willingness to complete the questionnaire. Do the figures given below
        support this hypothesis?

	Replied immediately	Replied to first reminder	Replied to second reminder	Visited by interviewer	Total
Satisfied	100	134	21	45	300
Indifferent	40	48	20	42	150
Not Satisfied	10	18	9	13	50
Total	150	200	50	100	500

                                                                    (IOS)

29.   Fergatron Electrics manufacture video cassette recorders at four different
        factories and the reliability of these machines is of great interest to the
        company. The table below relates to the performance of a number of
        similar machines during their first year of operation, together with the
        factory that had manufactured them.

		Factory			
		A	B	C	D
Reliability (first year)	Repair needed	4	15	9	12
	Repair not needed	8	10	6	6

Use the χ^2 distribution and a 5% level of significance to test the hypothesis that there is no association between the factory of manufacture and the reliability of the product in the first year.

Subsequently Sam Pling, the company statistician, discovered that the above table was not actually the raw data, but that a junior assistant had divided all the original entries by 10 to reduce the arithmetic, and that this table was the result. How will the calculated value of your test statistic have to change to reflect this information, and what effect will this have on the result of the above test of association? Comment briefly on these results. (AEB)'84

SECTION 10.4

30.　Explain briefly what is meant by replication, randomisation and experimental error in the context of experimental design.

Two neighbouring farmers wish to compare the yield of two different varieties of seed potato. They agree that for the coming season one of them will plant the first variety and the other will plant the second variety. At the end of the season they will compare the yield per acre obtained.

Explain what is wrong with this experiment, and how it could be improved. Suggest factors which might affect the yield, and responses other than the yield which might be relevant. (AEB)'84

31.　A large chain of department stores plan to replace their existing mechanical cash tills with modern electronic devices. The planning section has compiled a short list of five models which seem to be the most suitable and has arranged for a number of each to be obtained on approval for a limited period. Design an experiment, the results of which will enable you to recommend the appropriate course of action to the management. Your answer should include details of:

(a)　all factors which may affect the experimental outcome,
(b)　the treatments to be used,
(c)　the experimental layout chosen,
(d)　the response measures to be recorded,
(e)　an outline analysis. (AEB)'82

32.　The following five specimens of fuel gas had a number of check tests performed on them to measure their thermal unit values with the following results:

Specimen	1	2	3	4	5
	533	541	526	541	549
	537	545	536	519	537
	535	549	534	528	538
	546	532	533	536	

Is there any evidence of a difference between the specimens?

33. Samples from 6 different batches of a particular acid were taken and from each, a number of preparations were made and measured in terms of individual yield with respect to standard colour determined by dye-trial. Is there any evidence of a difference between the batches with respect to yield?

Sample	1	2	3	4	5	6
	1545	1540	1595	1445	1595	1520
	1440	1555	1605	1440	1630	1455
	1440	1490	1510	1595	1515	1450
	1520	1560	1560	1465	1635	1445
	1580	1495		1545	1625	

34. In an examination in which the marks awarded formed approximately a normal distribution, there were some candidates from each of three schools. The marks obtained were as follows:

School A   51, 24, 45, 30, 13, 15, 9, 21, 91.
School B   94, 43, 80, 22, 99, 84, 97, 60.
School C   65, 0, 6, 67, 2, 55, 18, 49, 32.

Carry out an analysis of variance, and examine the hypothesis that there is no essential difference between the schools.

35. Four plots of land were each divided into five sections. For each plot, five different treatments were allocated randomly to the sections on which potatoes were grown. The yields were as follows:

		Treatment				
		A	B	C	D	E
	1	310	353	366	299	367
Plot	2	284	293	335	264	314
	3	307	306	339	311	377
	4	267	312	312	266	342

36. Four diets were compared on premature babies with three types of respiratory disease. The table below gives the increases in mass, in kilograms, for these babies.

		Diets			
		1	2	3	4
Respiratory	A	3.2	3.9	2.7	2.0
Disease	B	2.3	3.0	3.9	4.5
	C	2.7	3.4	5.7	6.3

Carry out an analysis of variance and test for differences between diets and between diseases.

(AEB)'77

37. In a workshop there are four workers and five machines, producing similar articles. The production per shift was observed, every worker using each of the five machines for a whole shift. The results were as follows:

		Machine			
Worker	I	II	III	IV	V
A	27	41	22	62	32
B	34	31	26	64	35
C	34	50	26	66	39
D	22	47	13	57	32

Carry out an analysis of variance and test for differences between workers and between machines. (There is no need to test differences between individual pairs.) (AEB)'76

38. Five varieties of corn are planted in six randomized blocks and the yields, in kg per plot, are as follows

		Variety				
		A	B	C	D	E
	1	45	30	36	36	39
	2	44	35	38	37	41
	3	43	36	35	38	40
Block	4	44	29	36	35	41
	5	40	31	34	34	35
	6	38	27	33	36	36

Carry out an analysis of variance and test for differences between blocks and between varieties. (You are not required to test for individual differences between two blocks or between two varieties.) (AEB)'75

39. A textile factory produces a silicone proofed nylon cloth for making into rainwear. The chief chemist thought that a silicone solution of about 12% strength would yield a cloth with a maximum waterproofing index. He also suspected there may be some batch to batch variation because of slight differences in the cloth.

To test this, five different strengths of solution were used on each of three different batches of cloth. The following values of the waterproofing index were obtained.

		Strength of Silicone Solution (%)				
		6	9	12	15	18
	1	20.8	20.6	22.0	22.6	20.9
Cloth	2	19.4	21.2	21.8	23.9	22.4
	3	19.9	21.1	22.7	22.7	22.1

You may assume the sum of squares of the observations $(\Sigma x^2) = 7022.79$.

Carry out an analysis of variance to test at the 5% significance level for differences between strengths of silicone solution and between cloths.

Comment on the chief chemist's original beliefs in the light of these results and suggest what actions he might take. (AEB)'86

40. Four kinds of metal finishes a, b, c, d, for outside use, are tested in three different towns A, B, C, by leaving a panel of each kind exposed to the weather for one year. The panels are examined at the end of the year and given a merit score out of 100 for their condition. The results are given in the table.

Metal	Town		
finish	A	B	C
a	70	86	66
b	61	72	59
c	29	48	22
d	48	68	46

Carry out an analysis of variance and test for differences between towns and between finishes. State your conclusions.

41.　Four investigators A, B, C, D buy test portions of chips at three fish-and-chip shops and the following table shows the weight of a shilling portion to the nearest ounce:

Investigator	Smoky Bill's	Snooty's Grill	Grease & Batter Ltd.
A	7	5	9
B	8	5	10
C	5	6	8
D	7	4	8

Carry out an analysis of variance and test for differences between shops and between investigators.

42.　The following table gives the marks of candidates, one from each of four Colleges, who sat five examinations in Chemistry.

College	Inorganic		Organic		
	Theory	Practical	Theory	Practical	Physical
1	42	32	40	73	45
2	41	18	79	64	57
3	40	55	49	36	40
4	42	29	69	50	46

Source: *Chemistry in Britain*, August 1968

It is desired to ascertain whether these results provide any evidence of differences in Examiners' marks in the five subjects and of differences between the four students' performances, which may be considered significant. Use the analysis of variance to examine these results.

SECTION 10.5

43.　13 athletes compete in a long distance race, their times (to the nearest minute) being recorded as: 32, 30, 26, 32, 29, 38, 29, 25, 33, 34, 36, 31, 44. In the past, approximately one half of all competitors have completed the course in 35 minutes or less. Are these athletes above-average on the given times (assume the times given do not conform to a Normal distribution).

44.　18 children from a special school were given an IQ test and they obtained the following scores: 103, 121, 98, 104, 116, 83, 88, 96, 104, 101, 98, 108, 102, 102, 127, 103, 100, 111. Use the sign test to examine the hypothesis that the above scores are not standard. (*Note:* Standard score = 100).

45.　During negotiations between the union and the management of a large firm two alternative offers are put to the union side:　(a) old rates plus 20% increase across the board,　(b) new rates based on a production bonus scheme.

The union statistician, Percy Glum, considers these offers and calculates the pay that 25 typical employees would receive from each offer. These are summarised overleaf.

Employee	1	2	3	4	5	6	7	8	9	10	11	12	13
Old rates plus 20% (£)	56	43	59	62	38	49	53	37	71	53	47	39	37
New rates (£)	67	58	58	75	47	51	52	49	75	59	56	41	42

Employee	14	15	16	17	18	19	20	21	22	23	24	25
Old rates plus 20% (£)	68	27	68	75	42	53	61	56	58	35	46	37
New rates (£)	65	31	72	84	45	54	65	61	57	39	49	39

By use of the sign test, or otherwise, determine whether the new rates will lead, on average, to an increase of more than the 20% on the old rates.   (AEB)'77

46.   In a particular institution, two different methods were used on 9 mice to measure their resistance to a particular drug. Is there any evidence of a difference, using the sign test at a 5% level?

Method 1	68	60	72	80	67	74	82	64	64
Method 2	66	62	73	78	63	73	79	63	66

47.   Test whether there is any difference between the sets of samples from the following machines:

Sample	1	2	3	4	5	6	7	8	9	10	11
Machine A	1.3	1.2	1.4	1.6	1.3	1.4	1.1	1.2	1.2	1.7	1.3
Machine B	1.1	1.1	1.2	1.8	1.3	1.2	1.2	1.1	1.3	1.5	1.4

assuming that the samples are drawn from: (a) a Normal distribution; (b) a distribution that is not Normal.

MISCELLANEOUS

48.   A loom is used to produce cloth. Occasionally, at random, the machine falters and a flaw appears across the cloth. In the past it has been found that the distance between flaws in the cloth follows an exponential distribution with mean 1 metre. State briefly why an exponential distribution would be expected to serve as a reasonable model in this situation.

The 100 distances between successive flaws on a given day are shown in the following table.

Distance between flaws	Frequency
0 to 0.5 m	44
0.5 to 1.0 m	20
1.0 to 2.0 m	22
2.0 to 3.0 m	10
more than 3 m	4

(a) Plot these data as a histogram. On the same diagram draw an appropriately scaled graph of the density function of the exponential distribution with mean 1.

(b) Calculate the corresponding expected frequencies for an exponential distribution with mean 1 and test whether this model gives a reasonable fit to the data. (Two significant figure accuracy for expected frequencies is accepted.)   (JMB)

49. A cubical metal die has one face painted white. It is believed that this will bias the die so that when it is thrown the painted face is the least likely to land uppermost. In 360 throws the frequencies with which the faces land uppermost are

White   Opposite white   Others
 43   86   231

The theory is put forward that the probabilities of landing uppermost are: white, $\frac{1}{12}$; opposite white, $\frac{1}{4}$; others, $\frac{1}{6}$ each. Use a χ^2-test at the 5% significance level to show that this theory is not acceptable.

A new theory retains the assumption that the painting only affects the probabilities for the white face and the one opposite and estimates these from the observed frequencies. Test the new theory.   (Cambridge)

50. According to genetic theory the number of the colour-strains red, yellow, blue and white in a certain flower should appear in the proportions $4:12:5:4$. Observed frequencies of red, yellow, blue and white strains amongst 800 plants were 110, 410, 150, 130 respectively. Are these differences from the expected frequencies significant at the 5% level?

If the number of plants had been 1600 and the observed frequencies 220, 820, 300, 260, would the differences have been significant at the 5% level?

(Table of 5% points of the χ^2 distribution

Degrees of freedom (f)	1	2	3	4	5
Critical value of χ^2	3.84	5.99	7.81	9.49	11.1)

(Cambridge)

51. A table of random digits gives, for the first 200, the ten digits with the following frequencies:

0	1	2	3	4	5	6	7	8	9
19	18	16	26	21	24	19	22	16	19

Test the hypothesis that the digits appear with equal probability.

Successive sets of 25 digits contain the following numbers of eights: 2, 0, 5, 2, 0, 6, 2, 5, 2, 3, 1, 1, 2, 2, 3, 7, 3, 2, 3, 3, 4, 2, 1, 1, 2, 3, 4, 2, 2, 3, 3, 0, 2, 1, 5, 5, 4, 2, 2, 3. Use the Poisson distribution to find the probabilities of 0, 1, 2, . . . eights in a random set of 25 and compare observed with expected frequencies.

52. In dyeing cloth the temperature of the dyeing vat may affect the amount by which the cloth shrinks during the process. The table shows the observations of the percentage shrinkage (x) from 20 pieces of cloth dyed at five different temperatures.

	x	Σx	$(\Sigma x)^2$	Σx^2
98 °C	5.3 5.5 5.9 6.5	23.2	538.24	135.40
94 °C	5.0 5.5 6.0 6.3	22.8	519.84	130.94
90 °C	4.2 4.8 5.0 5.6	19.6	384.16	97.04
86 °C	5.5 4.1 4.6 4.2	18.4	338.56	85.86
82 °C	3.9 5.0 5.3 4.8	19.0	361.00	91.34

State the assumptions about x which should be satisfied if an F-test is to be used.

(a) Perform an F-test to determine whether the mean shrinkage for different temperatures differs significantly.

(b) It has been suggested that there is a difference in the shrinkage of cloth according to whether the temperature of the vat is above or below $92\,^{\circ}$C. Investigate whether the data support this theory.   (Cambridge)

53. Four different breeds of cow are fed with 3 different types of ration over a fixed period and the (coded) weight gain is given below. Is there evidence of a difference in either breeds or rations with respect to the weight gain?

		Ration		
		A	B	C
	1	7	14	8.5
Breed	2	16	15.5	16.5
	3	10.5	15	9.5
	4	13.5	21	13.5

54. The table below summarises the incidence of cerebral tumours in 141 neurosurgical patients.

		Type of tumour		
		Benign	Malignant	Others
Site of	Frontal lobes	23	9	6
tumour	Temporal lobes	21	4	3
	Elsewhere	34	24	17

Find the expected frequencies on the hypothesis that there is no association between the type and site of a tumour. Use the χ^2 distribution to test this hypothesis.   (AEB)'77

55. At St. Trinian's College for Young Ladies there are 1000 pupils. Of these 75 have represented the College at both hockey and netball, 10 have represented the College at hockey but do not play netball, 35 have represented the College at netball but do not play hockey, and 100 do not play games at all. In all 100 girls have represented the College at hockey, and 150 at netball. The number who do not play hockey is 200 and the number who do not play netball is 125.

Arrange the above data in the form of a 3×3 contingency table, and state how many pupils play both hockey and netball but have not represented the College in either.

Apply the χ^2 test to your 3×3 table, and state the hypothesis which it tests.   (AEB)'76

56. Four fields are used to test 3 new fertiliser mixtures and the average time (in weeks) taken for the crop to reach a particular stage is noted. Analyse the given results using an analysis of variance test.

	Mixture		
Field	A	B	C
1	13	16	21
2	17	13	20
3	6	18	21
4	11	19	6

57.   A large food processing firm is considering introducing a new recipe for its ice cream. In a preliminary trial, a panel of 11 tasters were asked to score ice cream made from both the existing and the new recipe for sweetness. The results, on a scale from 0 to 100 with the sweeter ice cream being given the higher score, were as follows:

Taster	A	B	C	D	E	F	G	H	I	J	K
Existing recipe	88	35	67	17	24	32	8	44	73	47	25
New recipe	94	49	66	82	25	96	14	56	27	44	79

Use the sign test, at the 5% significance level, to test whether the new recipe is sweeter than the existing one.

Because of the erratic nature of the scores obtained, it was decided to repeat the trial with a new panel of 10 tasters, this time giving some guidance as to the scores to allocate. Two other ice creams were tasted first. One was very sweet and the tasters were told that it had a score of 90. The other was not sweet and had a score of 10.

The new trial gave the following results:

Taster	L	M	N	O	P	Q	R	S	T	U
Existing recipe	52	44	57	49	61	55	49	69	64	46
New recipe	74	65	66	47	71	55	62	66	73	59

Use a paired t-test, making any necessary assumptions, to test the hypothesis that there is no difference in sweetness between the two recipes at the 1% significance level.

Discuss briefly the suitability of the choice of the sign test for the first set of data and the paired t-test for the second.   (AEB)'84

58.   For the 2 X 2 contingency table shown overleaf prove that

$$a - \frac{n_A n_1}{n} = \frac{ad - bc}{n}$$

and hence that χ^2_{test} can be written in the form:

$$\frac{(ad - bc)^2}{n} \left[\frac{1}{n_A n_1} + \frac{1}{n_A n_2} + \frac{1}{n_B n_1} + \frac{1}{n_B n_2} \right]$$

Thus show that $\chi^2_{test} = \dfrac{n(ad - bc)^2}{n_1 n_2 n_A n_B}$.

a	b	n_A
c	d	n_B
n_1	n_2	n

59.   For the contingency table shown in the above exercise, show (using Yates' correction) that:

$$\chi^2(\text{corrected})_{test} = \frac{n\left(|ad - bc| - \dfrac{n}{2}\right)^2}{n_1 n_2 n_A n_B}$$

60.   In a one-way Analysis of Variance, n random samples are taken, each of m items (where x_{ij} is the ith item in the jth sample). If S^2 is the sum of squares of all items and T is their grand total, by using

$$SS_T = \sum_{i=1}^{m} \sum_{j=1}^{n} (x_{ij} - \bar{x})^2, \text{ show that:} \qquad \text{(a)} \ SS_T = \sum_i \sum_j x_{ij}^2 - mn\bar{x}^2;$$

and hence   (b) $SS_T = S^2 - \dfrac{T^2}{mn}$.

61.   For the above case, using the definition $SS_B = m \sum_{j=1}^{n} (\bar{x}_j - \bar{x})^2$, show that: (a) $SS_B = m[\sum \bar{x}_j^2 - 2\bar{x} \sum \bar{x}_j + n\bar{x}^2]$; and hence   (b) using identity $\bar{x}_j = \dfrac{T_j}{m}$, show that $SS_B = \dfrac{\sum T_j^2}{m} - \dfrac{T^2}{mn}$.

$$\left[Note: \ T_j = \sum_{i=1}^{m} x_{ij}. \right]$$

Answers

Chapter 1 Answers

1 (a) Continuous (b) Continuous
(c) Discrete

2 2, 14, 23, 7, 2

6 (a) Limits: (lower) 26, 31, 36, 41, 46, 56,
66, 76, 86 (upper) 31, 36, 41, 46, 56, 66,
76, 86, 106 Boundaries: 26, 31, 36, 41, 56,
66, 76, 86, 106 Widths: 5, 5, 5, 5, 10, 10,
10, 10, 20 (b) Limits: (lower) 100.0, 110.0,
120.0, 130.0, 140.0 (upper) 109.5, 119.5,
129.5, 139.5, 149.5 Boundaries: 99.75,
109.75, 119.75, 129.75, 139.75, 149.75
Widths: 10, 10, 10, 10, 10.

7 (a) Upper: 0.381, 0.388, 0.395, 0.402,
0.409 Lower: 0.375, 0.382, 0.389, 0.396,
0.403 (b) 0.3745, 0.3815, 0.3885, 0.3955,
0.4025, 0.4095 (c) 0.007 for *all* classes

8 (a) 3.0; 1.5, 4.5, 7.5, 10.5, 13.5, 16.5
(b) 0.5; 109.45, 110.45, 110.95, 111.45,
111.95 (c) 0.1; 0.4445, 0.5445, 0.6445,
0.7445, 0.8445, 0.9445

9 Male: 8.21, 7.15, 6.97, 7.32, 7.68, 7.62,
8.05, 8.17, 7.75, 6.61, 5.69, 5.20, 4.50,
3.75, 2.74, 2.59 Female: 7.30, 6.44, 6.32,
6.75, 7.33, 7.16, 7.64, 7.83, 7.50, 6.88,
6.35, 5.80, 5.22, 4.42, 3.29, 3.76

10 0.05, 0.05, 0.11, 0.23, 0.19, 0.14, 0.11,
0.08, 0.04

11 0.040, 0.203, 0.348, 0.308, 0.101

12 (a) 0.05, 0.20, 0.40, 0.30, 0.05
(b) 5, 20, 40, 30, 5.

17 108.0, 180.0, 8.0, 0.4, 0.20

19 Plotted points: 144, 117, 221, 92, 38, 11,
1.5, 0.2

20 Plotted points: 0.5, 5, 9, 17, 12, 3, 1.25

21 Angles: 156, 26, 70, 35, 10, 63°

22 Angles: 128.5, 12.0, 197.1, 5.2, 17.2°

23 Circle 1: 115, 109, 90, 46°
Circle 2: 113, 85, 85, 77°. $r_2 = (0.87)r_1$.

25 (a)

u.c.b.	9.5	10.5	11.5	12.5	13.5	14.5
Cum f	0	2	10	36	77	88

(b)

u.c.b.	11.5	14.5	17.5	20.5	23.5	26.5	29.5
Cum f	0	1	4	11	23	41	47

(c)

u.c.b.	−0.05	9.95	19.95	39.95	59.95
Cum f	0	21	59	70	74

(d)

u.c.b.	−0.505	−0.405	−0.305	−0.205
Cum f	0	13	39	50

u.c.b.	−0.105	−0.005	0.095	0.195	0.295
Cum f	60	72	80	83	84

26 563, 887, 1343, 1614, 1739, 1798

Chapter 2 Answers

1 (a) 14 (b) 88 (c) 38

2 (a) 21 (b) 49/6 (c) −120 (d) 0.95 (2D)

3 (a) 22 (b) 33 (c) 53 (d) −54

4 (a) $\sum_{i=1}^{25} x_i^2$ (b) $3\sum_{i=1}^{n} y_i^2$ (c) $2\sum_{i=1}^{4} x_i y_i^2$

(d) $\sum_{i=1}^{n} f_i x_i \Big/ \sum_{i=1}^{n} f_i$

(e) $\dfrac{1}{n}\sum_{i=1}^{n} x_i y_i - \dfrac{1}{n^2}\left(\sum_{i=1}^{n} x_i\right)\left(\sum_{i=1}^{n} y_i\right)$

5 (a) 80.375 (b) 0.85 (c) 4.8 (d) 0.50(2S)
(e) 3.32

6 10.5

7 (a) 10 (b) 11

8 (a) 1.43 (2D) (b) 2.43 (2D)

9 4.55

10 4.2

11 (a) 12.14 (2D) (b) 19.91 (2D)
(c) 4.05 (2D) (d) 5.20 (2D)

12 Weights are 3, 2 and 1 respectively. 80.5

13 56.94

14 14.6 (1D)

15 117.48

16 13.85 (taking the last class as 30−45)

17 (a) 25.69 (2D) (b) 19.34 (2D)

18 33.6 (1D)

19 0; 67.5

20 500.6 (1D)

21 67.45

22 28.0

23 1.675 mm

24 0.9764 inches (4D)

25 (a) 0.05 (b) 4.22, 4.27, ..., 4.57
   (d) 4.379 (3D)

26 (a) 81 (b) 0 (c) 8.01 (d) 3.75

27 (a) 22 (b) 12.1

28 15.0 (1D)

29 1.69 (2D)

30 5.98 (2D)

31 1.25; 1

32 146.8

33 67.5 (1D)

34 1.4

35 -2.0

36 (a) 5 (b) 4.83 (2D)

37 2.7

38 (a) 45 to 49 (b) 45.1

39 1175; 1250; 1331.8 (1D)

40 (a) 3.73 (2D) (b) 4.67 (2D) (c) 4 (d) 27

41 (a) 0.413 (3D) (b) 0.411 (3D)

42 (a) 1.33 (2D) (b) 1.64 (2D) (c) 3.89 (2D)
   (d) 0.164

43 35.32 (2D) (Harmonic mean)

44 1; 3; 5

45 (a) 4.25 (b) 4.25 (c) 2.25

46 (a) 1.84 (b) 1.80

47 $\bar{x} = 26.2$ (1D); median $= 25.42$ (2D)
   (a) 5.97 (2D) (b) 6.00 (2D)

48 $\bar{x} = 4.2$; 0.91 (2D)

49 0.75; 0.75; 0.75

50 (a) $\bar{x} = 2.26$ (2D); median $= 2.30$
   (i) 0.93 (2D) (ii) 0.93 (2D)
   (b) $\bar{x} = 2.22$ (2D); median $= 2.10$
   (i) 0.36 (2D) (ii) 0.35

51 (a) 68.43; 70.75 (b) 67.99; 71.02
   (c) 66.25; 72.99 (d) 1.52 (e) 6.75

52 14.2; 47.6

53 0.51 (2D)

54 10.3; 9.74 (2D)

55 (a) $Q_1 = 65.66$; $Q_3 = 69.69$; 0.03
   (b) $Q_1 = 200.72$; $Q_3 = 302.28$; 0.20

56 1.7

57 $Q_1 = 14.35$; $Q_2 = 14.69$; $Q_3 = 14.95$;
   SIR $= 0.3$

58 9.5; 4.87 (2D)

59 (a) 2.16 (b) 0.68 (2D) (c) 32.78 (2D)

60 4.5; 2.87 (2D)

61 36.32 (2D); 9.95 (2D)

62 31.4; 12.3 (1D)

63 (a) $\dfrac{n+1}{2}$; $\sqrt{\dfrac{n^2-1}{12}}$

   (b) $\bar{x}$, s are *independent* of a.

64 $s_X = 0.973$; $s_x = 2.920$ (3D)

65 3.34 (2D)

66 147.0 (1D); 13.7 (1D)

67 3.13 (2D); 1.95 (2D)

68 1.81 (2D); 2.91 (2D)

69 (a) 0.167 12; 0.002 21
   (b) 0.165 71; 0.002 74

70 0; 2.67; 0.37; 13.48 (all to 2D)

71 (a) 25.69; 697.63; 19828.66
   (b) 0; 37.67; -28.70

72 (a) 2.39; 6.22; 17.26; 50.57
   (b) 1.06%, 1.86%, 2.35%, 2.51%

73 (a) 5.2 (b) 4.577 (c) 3.972

74 (b) 46 600 (c) 46 731 (nwn); 20 990

75 153.75; 58.08 (2D)

76 (a) 25.87 (b) 26.45 (c) 27.25

77 Median $= 10.90$; $D_7 = 11.98$. $\bar{x} = 10.99$;
   Mode $= 10.72$

78 (b) 164.83; 7.65

79 (a) 63.89; 1.6 (b) 63.66

80 m.d. $= 4.47$; s.d. $= 5.53$; $\dfrac{\text{md}}{\text{sd}} = 0.81$

81 (a) (i) 2 (ii) 8 (iii) 2 (b) (i) $w = 4x$
   (ii) $v = x + 10$ (iii) $w = 4v - 40$

82 (a) $\dfrac{2}{5}n$; $\dfrac{3}{5}n$ (b) 5.24; 6.38

83 (a) (i) 14.38 (ii) 19.84 (iii) 16.25
   (iv) 27.01 (all to 2D) (c) 0.18 (2D)

84 (a) 0.0445 (b) 0.0336

85 (b) 46.17; 5.23 (c) 48.66; 13.54

87 $\sqrt[N]{(x_1{}^{f_1})(x_2{}^{f_2}) \ldots (x_n{}^{f_n})}$

89 (a) $a + \dfrac{(n-1)d}{2}$ (b) $\dfrac{d^2}{6}n(n-1)(2n-1)$

   (c) $\dfrac{d^2}{12}(n^2 - 1)$

90 (a) 4 (b) 4.17 (2D)

91 35.70; 23.09

95 (a) 3 (b) 2 (c) 2.18 (2D) (d) 2.59 (2D); 6

96 91.40; 35.81; 149.99 (all to 2D)

97 86.57; 44.07; 188.2

98 (a) (i) £55.67 (ii) £54.63 (b) (i) 4
   (ii) 4.48 (2D)

99 (a) *1961*: 4.69, 1.59, 33.9%;
   1971: 4.92, 1.64, 33.3%
   (8 or more is taken as '8–12')

Chapter 3 Answers

1 High, positive correlation

2 Positive correlation

4 Negative correlation

5 (a) 2, $-\frac{3}{2}$ (b) $\frac{1}{4}$, 14 (c) $\frac{1}{4}$, $\frac{1}{2}$ (d) $\frac{1}{16}$, $-\frac{1}{8}$

7 (a) 1 (b) -1 (c) $-\frac{1}{3}$ (d) 4

8 $y = \frac{1}{3}(2x - 6)$

9 $y = \frac{1}{6}(3x - 7)$

11 $x = 1, y = 2$ (b) $x = -3, y = -5$
(c) $x = 16, y = -24$

12 (a) $y = 3x$ (b) $y = 7 - 2x$ (c) $y = 2x - 5$
(d) $y = \frac{3}{2}x + 6$

13 (a) $x = 2, y = 2$ (b) $x = \frac{9}{4}, y = \frac{3}{4}$
(c) $x = 9, y = 1$ (d) $x = 8, y = 6$
(e) $x = \frac{1}{2}, y = -\frac{1}{2}$

14 $x = \frac{3}{2}, y = -\frac{5}{2}$

15 (a) $6x$ (b) 5 (c) $-6x^2$ (d) $6x - 2$
(e) $-8 + 3x^2$

16 (a) $27x^2$ (b) $-16x^3 + 8x^7$ (c) $-4x^{-2} - 3$

17 (a) $9x^2 - 20x$ (b) $4x + 9x^{-4}$

18 (a) $4x - 4y$ (b) $3x^2 + 4xy$
(c) $\frac{1}{y} - \frac{y}{x^2} + 2\left(\frac{y^2}{x^3}\right)$

19 $\frac{10}{3}$

20 min $\frac{17}{27}$; max 3

21 (a) $9y^2 + 2yz - z^2$ (b) $y^2 - 2yz + 3z^2$

23 $y = 0.72x + 14.76; x = 0.67y + 17.75$

24 $y = 0.61x + 2.41; \ x = 0.50y + 4.00$

25 $y = 0.48x + 35.82; x = 1.04y - 3.38$

26 $y = 2.79x - 5451$

27 (a) 0.3; 1.65 (c) $y = 1.257 + 1.31x$
(e) 3.42

28 (b) $y = 0.353x - 0.88$

29 $Y = -2.312X + 1.432;$
$X = -0.325Y + 0.610;$
$y = -23.12x + 484.50$
$x = -0.0325y + 18.25$

30 $y = 114.38 - 1.446x$; (a) No
(b) 60.89 (2D)

31 $Y = 10.81X + 3.82; y = 10.81x - 20\,880;$
275.5, 286.3

32 2.75

33 -4.67

35 $r = -0.654$

36 0.72

37 $r = 0.48$

38 0.92

39 (a) (i) $y = -0.047x + 54.90$
(ii) $x = -0.033y + 52.87$ (b) $r = -0.04$

40 0.46 (2D)

41 0.05

42 (a) 0.60 (2D) (b) 0.7

43 0.95

44 (b) $A = 1.1543; B = 0.4960; a = 14.3$ (3S);
$b = 3.13$ (3S) (c) 79.2 (3S)

45 (a) $b = 1.405; a = 4.019 \times 10^6$ (b) 245

46 (a) $a = 1.24; b = 6.18$ (b) 7.66°C
(c) 2.03 hrs
(d) Assumes that the given law will hold for
values of x greater than 1.84

47 $y = 0.56x + 2.90; 0.79$

48 (a) -0.9997 (4D); $y = -1.17x + 19.38$;
$x = -0.85y + 16.51$ (b) 7.68 millions

49 $y = 18.77 - 0.80x; x = 18.68 - 0.82y$;
negative correlation

50 0.997 (3D)

51 $y = 0.74x - 0.36; x$ is a controlled variable.

52 (a) -0.93 (2D) (b) 0.14 (2D)

53 (a) $y = 0.738x + 6.92$ (b) 0.58 (2D)

54 (a) 31.4 (b) $y = 31.4 - 3.1t$
(c) $40.7 - 6.2x$ (d) 6.56 (2D)

55 (b) (i) 0.81 (2D) (ii) 0.81 (2D)

56 $y \equiv$ yield; $x \equiv$ fertiliser; $y = 3.05x + 166.8$

57 0.60 (2D); $w = 0.888h - 75.9$

58 (d) 0.60

59 (a) 0.78 (b) -0.05 (2D)

60 (a) 0.40 (2D) (b) 0.43 (2D)

61 (c) $\log_e y = 0.16 + 2.02 \log_e x$

64 (a) $y = -1.31x + 231.1;$
$x = -0.56y + 157.4$ (b) -0.86 (2D)

Chapter 4 Answers

1 (a) $S \subset T$ (b) $2 \in S$ (c) & (d) $3 \notin \emptyset$
(e) $R \supset \emptyset$ or $\emptyset \subset R$

2 $A \cup B = A; A \cap B = B;$
$A \cup C = \{1, 2, 3, 4, 5, 6\}; A \cap C = \{3, 4, 5\};$
$B \cup C = \{1, 2, 3, 4, 5, 6\}; B \cap C = \{3, 4\}$

4 $\emptyset, (3), (4), (6), (7), (3, 4), (3, 6), (3, 7),$
$(4, 6), (4, 7), (6, 7), (3, 4, 6), (3, 4, 7),$
$(3, 6, 7), (4, 6, 7), (3, 4, 6, 7)$

5 $A' = \{1, 3, 5, 7, 9\}; B' = \{2, 4, 6, 7, 8\};$
$A \cup B = \{1, 2, 3, 4, 5, 6, 8, 9\};$
$(A \cup B)' = \{7\}; A \cap B = \emptyset$

6 *One* possible answer:
$Q = \{3, 4, 5, 6, 7, 8, 9\};$
$S = \{6\}; T = \{2, 3\}$

7 $a \in B$ (b) $C \not\subset B$ (c) $2 \notin B \cup A$
(d) $\{2, 4\} \subset \{2, 4, 5, 8, 10\}$ (e) $B \subset \{1, 2, 7\}$
(f) $3 \notin \emptyset$

8 (a) None (b) None (c) B or D
(d) A, B, C, D, E or $\emptyset$ (e) B (f) A or C

10 (a) 47 (b) $\frac{7}{2}$ (c) $-115, -2075$ (d) 5500

11 (a) $3, 729$ (b) $3, 243$ (c) $\frac{1}{2}, \frac{3}{32}$, (d) $\frac{1}{4}, \frac{1}{1024}$

12 (a) 1023 (b) $1\frac{127}{128}$ (c) $3\frac{255}{256}$ (d) $\frac{171}{256}$

13 (a) 15 (b) 325 (c) 5050 (d) $\frac{n}{2}(n + 1)$

14 (a) 1661 (b) 288 (c) 11 204
(d) $\frac{5}{3}(2^{10} - 1) = 1705$

15 (a) $a < b$ (b) $4 > 3.836$ (c) $a \not> c$ (d) $6 \not< 2$
(e) $4 \leqslant 4$

16 (a) $x \leqslant 3$ (b) $x < -2$ (c) $x \geqslant 2$ (d) $x > 3$
(e) $x \geqslant \frac{6}{7}$

17 (a) $-4 \leqslant x < -1$ (b) $-2 \leqslant x < 3$
(c) $-3 \leqslant x \leqslant 4$

18 (a) $1 < x < 4$ (b) $-1 \leqslant x \leqslant 4$
(c) $-5 \leqslant x \leqslant -2$ (d) $-4 \leqslant x \leqslant 4$
or $|x| \leqslant 4$ (d) $-3 < y < 7$

20 The outcome set S will consist of 36 *ordered pairs* where the first of the pair will represent the no. of spots shown on the 1st die, the second of the pair, the spots shown on the 2nd die.

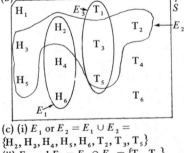

E_1 and E_2 *are* m.e.

21 (a) the 26 letters of the alphabet
(b) E_1 and E_2 only (c) $\bar{E}_1 = E_2 ; \bar{E}_2 = E_1$;
$\bar{E}_4$ is the set 'last 16 letters of alphabet'
(d) E_2 and E_4

22 (a) Experiment is 'a football game between A and B and the observation of the result'.
The outcome set could be:
$S = \{A \text{ wins, a draw}, A \text{ loses}\}$
(b) E is $\{A \text{ wins, a draw}\}$ (c) A loses.

23 (a) {HHH, HHT, HTH, THH, HTT, THT, TTH, TTT} (c) E_1 and E_2; E_2 and E_3.

25 (a) An infinite number (b) E_1 and E_2 only (c) One element, namely HHH (d) None

26 (a) Sum of points is odd or at least one 3 is shown (b) Sum of points is odd and at least one 3 is shown (c) Sum of points is even or no 3's are shown

27 (a) $\frac{1}{36}$ (b) $\frac{1}{9}$ (c) $\frac{1}{2}$

28 (a) $\frac{1}{4}$ (b) $\frac{1}{4}$ (c) $\frac{1}{13}$ (d) $\frac{1}{52}$ (e) $\frac{12}{13}$ (f) $\frac{9}{13}$

29 (a) $\frac{3}{25}$ (b) $\frac{18}{25}$ (c) $\frac{4}{25}$ (d) $\frac{18}{25}$ (e) $\frac{23}{50}$

30 (a) $\frac{2}{3}$ (b) $\frac{1}{3}$

32

33 (a) $S = \{H_1, H_2, H_3, H_4, H_5, H_6, T_1, T_2, T_3, T_4, T_5, T_6\}$
(b)

(c) (i) E_1 or $E_2 = E_1 \cup E_2 =$
$\{H_2, H_3, H_4, H_5, H_6, T_2, T_3, T_5\}$
(ii) E_2 and $E_3 = E_2 \cap E_3 = \{T_3, T_5\}$
(d) E_1 and E_3

34 (a) $\frac{3}{16}$ (b) $\frac{1}{8}$

35 (a) $\frac{3}{5}$ (b) $\frac{3}{5}$

36 (a) (i) $\frac{1}{6}$ (ii) $\frac{5}{9}$ (iii) $\frac{5}{18}$ (iv) $\frac{1}{18}$ (v) $\frac{1}{18}$;
(b) (i) $\frac{2}{3}$ (ii) $\frac{5}{6}$ (iii) $\frac{7}{18}$

37 $\frac{1}{8}$

38 (a) $\frac{1}{2}$ (b) $\frac{3}{4}$ (c) $\frac{1}{2}$ (d) $\frac{2}{5}$ (e) $\frac{3}{7}$ (f) 0

39 (a) $\frac{1}{6}$ (b) $\frac{1}{4}$ (c) $\frac{3}{8}$ (d) $\frac{1}{2}$

40 (a) $\frac{1}{12}$ (b) $\frac{1}{4}$ (c) $\frac{5}{12}$

41 (a) $\frac{1}{18}$ (b) $\frac{1}{18}$ (c) $\frac{5}{9}$ (d) $\frac{1}{3}$ (e) $\frac{5}{36}$

42 $\frac{23}{45}$

43 $\frac{1}{2}$

44 $\Pr(E_1 \cap E_2) = \Pr(\emptyset) = 0$
$\therefore \Pr(E_1/E_2) = 0; E_1 \subset E_2$

45 (a) 1 (b) $\frac{2}{5}$

46 $\frac{15}{56}$

47 0.7762 (4D)

48 $\frac{45}{173}$

49 0.714 (3D)

50 $\frac{3}{5}$

51 Bag 2

52 (a) $\frac{1}{3}$ (b) $\frac{1}{4}$ (c) $\frac{1}{4}$ (d) $\frac{1}{3}$

53 (a) $\frac{1}{12}$ (b) $\frac{1}{9}$ (c) $\frac{1}{3}$

54 (a) $\frac{5}{8}$ (b) $\frac{3}{32}$

55 (a) 0.80 (b) 0.35 (c) 0.85 (d) 0.60
(e) 0.57 (2D) (f) 0.25

56 $\frac{2}{3}$

57 (a) 362 880 (b) 3 628 800 (c) 4 (d) 3.75
(e) n

58 (a) $n + 1$ (b) $(n + 2)(n + 1)$
(c) $(n + r)(n + r - 1)(n + r - 2) \cdots$
$(n + 2)(n + 1)$

59 $6^4 = 1296$

60 (a) $\frac{16}{81}$ (b) $\frac{1}{4}$ (c) $\frac{1}{2}$

61 $^8P_3 \times {}^5P_2 = 6720$

62 $^{52}C_8 ; {}^{51}C_7$

63 (a) 210 (b) 56 (c) 1365 (d) 1365 (e) 336

64 (a) $(n - r + 1)(n - r)$ (b) $(n - 1)(n - 2)$

65 (a) 720 (b) 72 (c) 216

66 241 920; 282 240

67 $10^7 - 10^4 - 10^3 = 9\,989\,000$

68 1134 (if 0 is *not allowed* as 1st digit) or 2520 (*if it is*)

69 0.61 (2D)

70 19 448

71 0.121 (3D)

72 $\left(\frac{1}{13} \cdot \frac{16}{51}\right)^2 = 0.000\,58$

73 (a) $\left(\dfrac{26}{51}\right)\left(\dfrac{25}{50}\right) = 0.25$ (2D)
(b) $\left(\dfrac{25}{51}\right)\left(\dfrac{24}{50}\right) = 0.24$ (2D)
(c) $1 - \left(\dfrac{38}{51}\right)\left(\dfrac{37}{50}\right) = 0.45$ (2D)
(d) $\dfrac{(13)(12) + (13)(13)}{(51)(50)} = 0.13$ (2D)

74 $(^{10}P_6)/10^6 = 0.1512$

75 $\frac{7}{33}$

76 $\frac{25}{41}$

77 $\frac{1}{3}$

78 $\frac{3}{4}$; (a) 0.313 (b) 0.687 (c) 0.212

79 (i)

THTHH	
THHTH	E_2
E_1 (HTHTH	THTHTH)
TTHHH	TTTHHH
HTTHH	HTTTHH
HHTTH	HHTTTH

E_3

(ii) E_2 and E_3

81 $\frac{1}{7}$, $\frac{5}{56}$

82 9

84 5

85 $\frac{1}{20}$

86 (a) $A \cap B = \emptyset$ (b) $A \subset B$
(Note: this includes the case $A = B$)

88 $\frac{35}{128}$

89 $\frac{5}{9}$

90 (a) $\left(\frac{k}{N}\right)^n$ (b) $\dfrac{k^n - (k-1)^n}{N^n}$

91 $\frac{3}{5}$; $\frac{2}{5}$

93 (a) (i) $\dbinom{w}{r}$ (ii) $\dbinom{b}{r}$ (iii) $\dbinom{w+b}{r}$

94 (a) $\frac{1}{18}$ (b) $\frac{1}{12}$ (c) $\frac{1}{18}$ *No pairs are independent. The two pairs B, C and A, C are mutually exclusive.*

95 (a) 0.19 (b) 0.238 (c) 0.798 (d) 0.080 (3D)

96 (a) (i) $\frac{2}{5}$ (ii) $\frac{7}{15}$ (b) $\frac{4}{9}$

97 (a) $1 - \left(\frac{35}{36}\right)^3 = 0.08$ (2D); $1 - \left(\frac{35}{36}\right)^n$; with 25 throws, the probability of a double six is slightly better than $\frac{1}{2}$ (b) $x + y - xy$; $x + y^2 - xy^2$

98 (a) 0.029 (b) 5 cm

99 Loss of 0.49 p

100 (a) $\frac{1}{36}$ (b) $\frac{5}{12}$; 0.113 (3D), 0.309 (3D)

101 (a) $q^5(6 - 5q)$ (b) $6q^5(5 - 4q)$ (c) $q^9(30 - 54q + 30q^2 - 5q^3)$
Pr(0.9) = 0.796 (3D);
Pr(0.8) = 0.462 (3D)

102 (a) $\frac{9}{20}$ (b) (i) $\frac{1}{100}$ (ii) $\frac{7}{400}$

103 0.8125; 0.25

104 No advantage (a) $\frac{15}{36}$ (b) $\left(\frac{1}{6}\right)^{r-1}\left(\frac{15}{36}\right)$ (c) $\frac{1}{2}$

Chapter 5 Answers

1 (a) $2x + 3$ (b) $-3x^2 + 16x - 21$ (c) $\dfrac{1}{(x+1)^2}$ (d) $-\dfrac{69x^2}{(4x^3 - 3)^2}$ (e) $-\dfrac{2x}{(x-1)^3}$

2 $9(3t - 2)^2$

3 (a) $4(2x + 4)$ (b) $16(4x - 3)^3$ (c) $18x(3x^2 - 4)^2$ (d) $\dfrac{-9}{(3x - 7)^4}$

4 (a) $6(2x - 4)^2$ (b) $4(2x - 4)^2(10x^2 - 8x - 3)$

5 $\dfrac{8}{(2 - 2t)^2}$

6 $2(7x^2 + 12x + 3)(3 + x^2)^2$

7 (a) $\frac{3}{4}x^4 + c$ (b) $-x^5 + c$ (c) $-\dfrac{3}{4x^4} + c$ (d) $-\dfrac{4}{3}x^{-3} + c$ (e) $-x^{-3.3} + c$

8 (a) $\frac{2}{3}x^3 - \frac{3}{2}x^2 + c$ (b) $-\dfrac{1}{x^2} + \frac{1}{2}x^2 - \frac{4}{3}x^3 + c$

9 $x^3 - \frac{1}{2}x^4 + c$

10 (a) $\frac{2}{3}$ (b) -6 (c) $\frac{13}{4}$ (d) 0

11 $\frac{1}{12}$

12 (a) $-20\frac{5}{6}$ (b) $20\frac{5}{6}$ (c) $\frac{1}{4}$

13 (a) $2e^{2x}$ (b) $5e^{5x+2}$ (c) $12(x-1)e^{2x^2-4x}$ (d) $-b\,e^{a-bx}$ (e) $-\dfrac{4}{3e^{4x}}$

14 (a) $e^x(x+1)$ (b) $x\,e^{-x}(2-x)$ (c) $x^2 e^{-2x}(3 - 2x)$ (d) $2x\,e^{2x^2}(2x^2 + 5)$ (e) $\dfrac{(2x-1)e^{2x}}{x^2}$ (f) $12x(x^2 - 2)e^{3(x^2-2)^2}$

17 $a = \frac{3}{2}$

18 (a) $A = \frac{20}{19}$ (b) (i) $\frac{5}{19}$ (ii) $\frac{4}{19}$ (iii) $\frac{25}{57}$ (iv) $\frac{14}{19}$

19 (a) Yes (b) No (negative probability) (c) No (Sum of p's $\neq 1$) (d) Yes

20 $\frac{3}{8}$

21 (a) $A = \frac{3}{13}$ (b) 0.53 (2D)

22 (a) $\frac{1}{12}$ (b) $\frac{1}{3}$

23 $\frac{1}{4}$

25 $(x; \Pr(X=x))$: $(1; \frac{11}{36})$, $(2; \frac{9}{36})$, $(3; \frac{7}{36})$, $(4; \frac{5}{36})$, $(5; \frac{3}{36})$, $(6; \frac{1}{36})$

26 $\sqrt{\dfrac{2}{k} + 4}$

27 (a) $\frac{7}{32}$ (b) $\frac{7}{32}$ (c) $\frac{49}{1024}$ (d) $\frac{39}{64}$

28 (a) $\Pr(X=x)$ for $x = 2, 3, \ldots, 12$ is: $\frac{1}{36}$, $\frac{2}{36}$, $\frac{3}{36}$, $\frac{4}{36}$, $\frac{5}{36}$, $\frac{6}{36}$, $\frac{5}{36}$, $\frac{4}{36}$, $\frac{3}{36}$, $\frac{2}{36}$, $\frac{1}{36}$ (b) $\frac{1}{36}$, $\frac{3}{36}$, $\frac{6}{36}$, $\frac{10}{36}$, $\frac{15}{36}$, $\frac{21}{36}$, $\frac{26}{36}$, $\frac{30}{36}$, $\frac{33}{36}$, $\frac{35}{36}$, $\frac{36}{36}$ (c) $F(9) = \frac{30}{36}$, the probability that two dice (when rolled) will show a total of 9 or less.

29 3.01

30 (a) 4 (b) 2.5 (c) 8.5

31 $\frac{2}{3}$

32 $\Pr(X=2) = \frac{1}{4}$; $\Pr(X=3) = \frac{3}{4}$

34 (a) $\dfrac{1}{b-a}$ (b) $\dfrac{a+b}{2}$

35 $a = 20$

36 $\frac{13}{3}$; $\frac{20}{9}$; $\frac{26}{3}$; $\frac{80}{9}$

37 (a) 4.45 (b) 3.45 (c) 10.35

38 (a) $\frac{17}{8}$ (b) $\frac{9}{8}$ (c) $\frac{45}{8}$

39 (a) 1 (b) 2 (c) 6

41 $E[X] = 0.6$; $E[X^2] = 9.6$; $E[f(X)] = 12.6$

42 $E[X] = 4$; $E[X^2] = 18.1$; $E[2X^2 - 3X + 4] = 28.2$

43 $E[X] = 1.47$; $\mathrm{Var}[X] = 0.05$ $\Pr(X > 1.8) = 0.078$

44 $M = 3$; $Q_1 = 2$; $Q_3 = 4$

45 (a) 2.5 (b) 5.43

46 £12.50

47 1.9 (1D)

48 1.9; 2.1

49 (a) £2.25 (b) £0.86

50 $E[X] = \frac{11}{3}$; $Var[X] = \frac{80}{9}$; $E[Y] = 6$; $Var[Y] = 80$

51 (a) 7 (b) 5.83 (2D)

52 (a) -0.02; 2.92 (b) 2.94

53 (a) $Q_1 = 1.73$; $Q_3 = 2.65$ (b) 1.52 (2D)

54 $G(t) = 1 - p + pt$; $E[X] = p$

55 (a) 1.7 (b) 1.0 (1D)

56 (a) $K = \frac{4}{27}$ (b) 1.8; 0.36 (c) 0.0272

57 (b) $E[X] = \frac{1}{2}$; $Var[X] = \frac{1}{20}$ (c) $a^2(3 - 2a) = \frac{1}{2}$

58 (a) 2 (b) 2.44 (2D) (c) 2.22 (2D); X, Y *not* independent

59 £3

60 (c) (i) $\frac{5}{8}$ (ii) $\frac{7}{8}$ (iii) $\frac{31}{64}$

61 (a) 5; $\frac{25}{3}$ (b) 10; $\frac{50}{3}$

63 (a) $a = \frac{3}{8}$; $E[X] = \frac{13}{15}$ (b) 0.241 (3D)

69 $\mu_s = \mu'_5 - 5\mu'_4\mu'_1 + 10\mu'_3(\mu'_1)^2 - 10\mu'_2(\mu'_1)^3 + 4(\mu'_1)^5$

71 (a) $\frac{60}{77}$ (b) $\frac{27}{77}$ (c) $\frac{86}{77}$

72 (a) $c = \frac{1}{4}$ (b) $Pr(E_1) = \frac{3}{4}$; $Pr(E_2) = \frac{63}{256}$

73 (a) 2 (b) 2.29 (2D) (c) 0.25 (d) 0.183

74 (c) $\frac{4}{5}$

75 $k = 4$; $E[X] = 8/15$ (2D); $Var[X] = \dfrac{11}{225}$ (2D); median = 0.54 (2D)

76 (a) 0.216 (b) 5; $0 \leqslant x \leqslant 10$

77 (b) $\frac{2}{15}$ (c) $\frac{11}{15}$ (d) 24

78 $\frac{12}{5}$

79 0.894 (2D)

Chapter 6 Answers

1 (a) $x^6 + 12x^5y + 60x^4y^2 + 160x^3y^3 + 240x^2y^4 + 192xy^5 + 64y^6$

(b) $2^{11} + 11(2^{10})(3x) + \dfrac{(11)(10)}{2!}$
$\times (2^9)(3x)^2 + \dfrac{(11)(10)(9)}{3!}(2^8)(3x)^3 +$
$\ldots + (3x)^{11}$

(c) $(3v)^{12} + 12(3v)^{11}(4) + 66(3v)^{10}(4)^2 + \ldots + (4)^{12}$

2 (a) b^6 (b) $\binom{10}{4}3^6y^4$ (c) $\binom{9}{3}y^6$

3 $\dfrac{3}{5x}$

4 (a) 9 (b) $\frac{495}{16}$ (c) $10(\frac{7}{8})^9(\frac{1}{8})$ (d) $\binom{24}{12}(\frac{1}{2})^{24}$

5 (a) $- e^{-ax}\left[\dfrac{ax + 1}{a^2}\right] + c$

(b) $e^x[x^2 - 2x + 2] + c$

6 (a) $\frac{1}{2}(x^2 - 1)e^{x^2} + c$ (b) $c - \frac{1}{2}(x^2 + 1)e^{-x^2}$

7 2

8 (a) $\frac{80}{243}$ (b) $\frac{5}{16}$ (c) $\frac{3}{64}$

9 $n = 5$; $p = \frac{1}{4}$ (a) 0.237 (b) 0.016 (c) 0.896

10 (a) $\binom{10}{x}\left(\dfrac{1}{2}\right)^{10}$ (b) $\binom{n}{x}\left(\dfrac{1}{n}\right)^x\left(\dfrac{n-1}{n}\right)^{n-x}$

(c) $\binom{\frac{m+1}{2}}{x}\left(\dfrac{3}{1+y}\right)^x\left(\dfrac{y-2}{y+1}\right)^{\frac{m+1}{2}-x}$

11 (a) $\frac{1}{8}$ (b) $\frac{3}{8}$ (c) $\frac{3}{8}$ (d) $\frac{1}{8}$

12 $\frac{1}{16}, \frac{4}{16}, \frac{6}{16}, \frac{4}{16}, \frac{1}{16}$

13 (a) 20, 0.01 (b) 0.8179, 0.1652, 0.0159, 0.0010 (c) 0.0001

14 (a) 0.017 (b) 0.299 (c) 0.020

15 4

16 6

17 (a) 0.7220 (b) 0.9392

18 0.888

19 (a) $\frac{1}{8}$ (b) $\frac{5}{32}$

20 (a) 19/144 (b) 5; 2.5

21 0.1719 (4D)

22 (a) 0.11 (b) 0.27 (c) 0.30; 0.68

23 100, 99

24 270; 15

25 (a) $a = 2.5$; $b = 0.5$ (b) $n = 12$; $p = 0.3$ (c) n must be integral

26 Average $= 1$; $n = 4$; $p = \frac{1}{4}$; 63, 84, 42, 9, 1.

27 $\bar{x} = 1.24$ (2D); $p = 0.21$ (2D); frequencies: 2348, 3744, 2488, 882, 176 19, 1

28 (a) 0.082 (b) 0.205 (c) 0.257 (d) 0.214 (e) 0.891 (f) 0.014

29 2.5

30 (a) 0.0183; 0.0733; 0.1465; 0.1954; 0.1954; 0.1563
(b) 0.1225; 0.2572; 0.2700; 0.1890; 0.0992; 0.0417

31 0.368; 0.368; 0.184; 0.061; 0.015; 0.003

32 (a) 0.3233 (4D) (b) 0.3712 (4D)

33 (a) 0.0025 (b) 0.2851 (c) 0.5543

34 (a) 0.607 (b) 0.303 (c) 0.076; 0.287

35 0.125

36 (a) 0.0067 (b) 42 seconds (n.w.n)

37 0.61; 0.09

38 0.195; fatal accidents occur at random.

39 0.8781 (4D)

40 (a) 40 mins (b) 2.4 (d) 12 (e) Morning and afternoon traffic will generally display a different pattern

41 (a) (i) 1.04 (ii) 53, 55, 29, 10, 3 (b) 0.121 (3D)

42 (a) 0.779 (3D) (b) 0.002 (3D); 0.827

43 Poisson frequencies: 339, 504, 375, 186, 64, 27; accidents are assumed to occur independently

44 (a) 0.5 (b) Poisson frequencies: 303, 152, 38, 6, 1 (c) 0.53 (Obs.); 0.5 (Poisson)

45 Frequencies: 327, 393, 236, 94, 28, 7, 1, 0; relative deviations appear to be random — fit is fairly good.

46 (a) 548, 330, 99, 20, 3
   (b) 521, 369, 98, 11, 1
47 (a) 0.5832 (b) 0.9842 (c) 0.1314
   (d) 0.1112 (e) 0.7910 (f) 0.7574
   (g) 0.1664
49 (a) 0.83 (b) 1.16
50 (a) 0.0014 (b) 0.4986
51 0.2266; 5 (nwn)
52 (a) (i) 0.0913 (ii) 0.0913 (b) 0.8174
53 24
54 (a) 1(0.6) (b) 13 (c) 88 (d) 2(2.4)
55 $\mu = 46.12$; $\sigma = 11.68$ (a) 0.37 (b) 0.216
   (c) 0.253
56 0.048
57 $\sigma = 1.56$; 0.1
58 (a) 95; 9 (b) 1% (c) 0.21%

59

x	0	1	3
$\Pr(X = x)$	0.1975	0.3951	0.2963

x	3	4
$\Pr(X = x)$	0.0988	0.0123

; 0.728 (3D)

60 (a) 9.1% (b) 99.69 cm (c) 0.036 cm
61 (a) 33.7 (1D) (b) $\sigma_A = 31.2$; $\sigma_B = 19.4$
62 (a) 0.0333 (b) 0.9088
63 (a) 34 (b) 3 (c) (21.08, 28.92) yr
64 (a) ± 2.58 (b) ± 3.09
65 (a) 0.048 (b) 0.217 (3D)
66 1.56; 0.0273
67 (0.3)1.0, 3.0, 7.4, 15.1, 25.0, 33.9, 37.6,
   33.9, 25.0, 15.1, 7.4, 3.0, 1.0(0.3)
68 (a) 0.227 (b) 0.302 (c) 0.683
69 (a) 29 (b) 18 (c) 11
70 1, 8, 49, 145, 205, 139, 45, 7, 1
71 (4), 7, 16, 31, 48, 66, 80, 78, 66, 49, 29, 15,
   7, 3, (1)
72 2, 4, 13, 21, 25, 19, 11, 4, 1
74 $\frac{7}{2}$; 3; $\frac{2}{3}$
76 (a) $\frac{1}{5}$ (b) $\frac{1}{2}$
77 6
78 0.05
79 $\frac{1}{8}$
80 (a) 0.22 (b) $2\frac{1}{2}$ months (approx)
81 300
82 0.0803
83 (a) 0.977 (b) 0.914
84 (a) 0.25 (b) 0.199 (3D)
85 0.125; 0.133 (3D); Poisson is likely
86 (a) 0.175 (b) 0.082
87 (b) $E[X] = n/2$
88 (a) 0.80; 0.80 (b) Poisson (c) 44.93,
   35.95, 14.38, 3.83, 0.77 (2D); 0.1438 (4D)
   (d) 0.011 (3D)
89 (a) $\frac{1}{64}$ (b) $\frac{1}{8}$
90 (a) 0.037 (b) 0.222 (c) 0.368 (d) 0.368
   (e) 0.109 (f) 0.891
91 0.9982; £1.40

93 Y is bin$(4, \frac{5}{8})$; 0.481 (3D)
94 (a) 0.18 (2D) (b) $\dfrac{e^{-n/2}\left(\dfrac{n}{2}\right)^x}{x!}$
95 2.58, 0.99 (2D);

   (a) $\Pr(X_2 = r + 1) = 1.0833\left(\dfrac{5-r}{r+1}\right)$
   $\times \Pr(X_2 = r)$ (b) 0.0255, 0.1380, 0.2990,
   0.3240, 0.1755, 0.0380 ($p = 0.52$)
   (c) 63%, 26%, 17%, 1%, 2% and 143%; largest
   errors at extremes
96 0.185 (3D); 0.202 (3D) using a normal
   approximation
97 $Q_1 = -0.675$; $Q_3 = 0.675$ (approx.)
98 2.7
99 (a) −1.96 and 1.96 (b) 808.1 and
   815.9 gm (c) 4.2%
100 (a) 14.9 (b) 88
101 (a) 0.019 (3D) (b) 0.798 (3D)
102 (a) 0.0062 (4D) (b) 0.0994 (4D); £2534.20
103 (a) 0.2436 (4D) (b) $l = 148$; 0.6065 (4D)
104 95%; 253.5
105 (a) 5.75×10^{-19} (b) 0.9997 (c) 10
106 (a) 0.058 (b) $10\{1 + 5(1 - q)q^9\}$
107 $f(x) = \frac{1}{4}$; $2 \leqslant x \leqslant k$, $f(x) = 0$ elsewhere
   (b) 0.3 (c) 4
108 11; 0.349 (3D)
109 (a) 0.040 (3D) (b) 11
110 0.9996; mean = 40.1, sd = 0.8;
   max sample mean = 39.92
111 (b) 0.001
112 (a) $Y = \dfrac{b_2}{b_1}(X - a_1) + a_2$ (b) (i) 96
   (ii) 43.9

Chapter 7 Answers

3 In general, no, except in a trivial case
4 (a) N(220, 50) (b) N(−20, 50)
   (c) N(20, 50)
5 (a) N(100, 15) (b) N(0, 15) (c) N(0, 15)
   (d) 0.3982
6 (a) N(3.9, 0.004) (b) 37.6%
7 (a) 0.673 (3D) (b) 0.138 (3D)
8 0.0826; 0.0002
9 Normal; mean 118.79 cm, s.d. 0.74 cm
10 (a) N(91, 15) (b) N(148, 35) (c) N(5, 15)
   (d) N(191, 45) (e) N(263, 50)
   (f) N(105, 45) (g) N(−25, 55)
11 (a) 249.2, 11.37 (b) 20.92% (c) 0.95
12 N(6.4, 1.55)
13 (a) N(140, 27) (b) N(25, 20)
   (c) N(185, 35) (d) N(240, 108)
   (e) N(220, 335)
14 (a) N(a + 100, b + 10)
   (b) N(100 − a, b + 10) (c) $a = 115$, $b = 8$
15 N(6π, $0.01\pi^2$)

16 (a) 50; 2.2 (b) 10; 2.2 (c) 20; 28.0

17 (a) $N(100, 2)$ (b) $N(50, \frac{1}{2})$ (c) $N(a^2, 1)$

18 $N(880, 58)$

19 (a) 0.6687 (b) 0.0038

20 105.6 (106)

21 (a) 0.9750 (b) 0.0071

22 0.0254

23 (a) 0.90; 0.079 (b) 0.016 (3D)

24 None (0.1 (1D))

25 (a) 0.057 (b) 0.013 (c) 0.930

26 4

27 (a) $\chi^2(6)$ (b) $\chi^2(11)$

28 1, 2

29 (a) (i) 2.5 (ii) 97.5 (b) (i) 52.34 (ii) 0.297

30 (a) 0.1 (b) 0.01 (c) 0.005 (d) 0.05

32 (a) 0.95 (b) 0.1 (c) 0.995

33 0.9

34 (a) 0.05 (b) 0.05 (c) 0.9

35 (a) 8.69, 0.0734 (b) 0.05

36 0.493 (3D)

37 (a) 0.113 (b) 0.005

38 0.028 (3D)

39 7818, 3443, 9611, 0649, 2153, 9629,
3008, 2521, 0797, 8337, 1973, 9136,
8470, 2092, 3316

40 3.3592, 4.5845, 5.2015, 7.2030,
2.3071, 9.2933, 8.4219, 3.9455,
5.7792, 1.4216

41 Theoretical mean and variance is 4.5 and 8.25

42 Propn: 0.0453, 0.0736, 0.2679, 0.2151,
0.1528, 0.1170, 0.1075, 0.0208
Freq: 1, 0, 10, 11, 3, 4, 1, 0

45 Sample of 30: mean = 1.73 (2D),
variance = 1.80 (2D)
Theoretical distbn: mean = 2.0
variance = 2.0
Sample of 60: mean = 1.87 (2D),
variance = 1.75 (2D)

46 (a) 0.994 (b) 0.967

47 $(1-p)^{13}[1 + 13p + 91p^2]$

48 $(1-p)^4[1 + 4p(1-p)^3]$

49 $(1-p)^7[1 + 7p(1 + 6p)(1-p)^5]$
(i) A: 0.668; B: 0.993
(ii) A: 0.122; B: 0.677

51 (a) Plan 1: 0.797; Plan 2: 0.745
(b) Plan 1: 0.050; Plan 2: 0.053. Plan 1 is
better since there is a greater chance of
accepting a batch with a defective rate as
low as 0.2 and a lesser chance of accepting
a batch with a defective rate as high as 0.6.

52 6.18, 6.59, 7.94, 7.41, 6.83

53

p	0.05	0.1	0.2	0.3
Pr(accept)	0.977	0.919	0.737	0.528

p	0.4	0.5	0.6
Pr(accept)	0.337	0.188	0.087

Estimate of AQL = 0.075

54 Single, since it has a higher probability of
accepting batches with a relatively low
defective rate and a higher probability of
rejecting batches with a relatively high
defective rate.

55 AQL = 0.08; LTPD = 0.58 (both to 2D)

56 $k = 6.41$

57 (a) $N(93, 1)$ (b) $N(11, 1)$

58 0.159

59 (a) Increased, by a factor of 4
(b) Divided, by 3

60 Pr(late/old scheme) = 0.3085;
Pr(late/new scheme) = 0.0122

61 $a = 0.085$

62 4.283 m; 4.317 m

63 (a) 0.599 (b) 11.72 (c) 0.131
(d) $N(241, 0.89^2)$; Pr = 0.242

64 17

65 $\bar{x} = 71.8, s = 1.71; \mu = 70.62, \sigma = 1.93$

66 $N(600, 9.5^2)$; 0.319; 0.348; 0.038

67 (a) (i) $N(170, 41)$ (ii) $N(10, 41)$
(iii) $N(240, 225)$ (iv) $N(120, 481)$
(b) (i) $a = 50, b = 11$ (ii) 0.921 (3D)

68 (a) 0.95 (b) 0.95

69 (a) 11 (b) 30

70 (b) 0.332; 0.059; 0.01

71 mean = $p\mu_1 + (1-p)\mu_2$,
variance = $p^2\sigma_1^2 + (1-p)^2\sigma_2^2$;
mean = $\mu_1 - \mu_2$, variance = $\sigma_1^2 + \sigma_2^2$

72 $\bar{x}_2$:

x	1	2	3	4	5	6	7
$Pr(X = x)$	$\frac{1}{16}$	$\frac{2}{16}$	$\frac{3}{16}$	$\frac{4}{16}$	$\frac{3}{16}$	$\frac{2}{16}$	$\frac{1}{16}$

$\bar{x}_3$:

x	1	$\frac{5}{3}$	$\frac{7}{3}$	3	$\frac{11}{3}$	$\frac{13}{3}$	5	$\frac{17}{3}$	$\frac{19}{3}$	7
$Pr(X = x)$	$\frac{1}{64}$	$\frac{3}{64}$	$\frac{6}{64}$	$\frac{10}{64}$	$\frac{12}{64}$	$\frac{12}{64}$	$\frac{10}{64}$	$\frac{6}{64}$	$\frac{3}{64}$	$\frac{1}{64}$

73 (a) 37.8% (b) (125.5, 194.5)
(c) $(0.7977)^4 = 0.405$ (3D)

74 (a) 0.013 (3D) (b) 0.037 (3D); £3647

75 (a) 323.4 (b) 2.04 (c) 323.7 (1D)
(d) 104 589.6 (e) 649.8 (f) 8.16 (g) 26.52

76 $x = 108.4$ (a) 105.6 (b) 0.640 (3D)
(c) 0.994

77 (b)

p	0.05	0.1	0.2	0.3
$A(p)$	0.99	0.93	0.68	0.38

p	0.4	0.5	0.6
$A(p)$	0.17	0.06	0.01

(c) 23.3, 71.2, 127.2

78 (a) A: (i) 0.545 (ii) 0.082
B: (i) 0.879 (ii) 0.296
C: (i) 0.978 (ii) 0.565
(c) (i) C (ii) A (iii) B

79 (i) $(1-p)^5[1 + 5p(1 + 4p)(1-p)^3]$
(ii) A: 0.678; 0.055 B: 0.630; 0.061
(iii) A is better.

81 (a) $N(0, 1)$ (b) $N(0, 1)$ (c) $\chi^2(1)$ (d) $\chi^2(1)$

Chapter 8 Answers

1 $\hat{\mu} = \bar{x} = 58; \hat{\sigma}_2 = \dfrac{ns^2}{n-1} = 138.9$

2 (a) -0.21 (2D) (b) 2.71 (2D) (c) 53.4
(d) 271.1; 53.4, 272.9

3 2.98 (2D); 0.86 (2D)

4 65.4; 2.69

5 20.46; 59.37

6 $E\left[\dfrac{1}{n}\sum X_i^2\right] = \sigma^2 + \mu^2;$

$E\left[\dfrac{1}{n}\left\{\sum X_i^2 - n\mu^2\right\}\right] = \sigma^2$

7 (a) 0.67 (b) 0.17 (c) 20.19 (all 2D)

8 (a) 2.35 (b) 1.97 (2D)

9 Using range: 0.976 (3D).
Best estimate: 0.965 (3D). Error = 1.14%

10 0.25 (2D); 0.01 (2D)

11 $\hat{\mu} = 2.393$ (3D); $\hat{\sigma}^2 = 1.968$ (3D)

12 $\dfrac{100}{n}$; 101

13 (a) 19.35 (b) 19.185 (3D) (c) 19.665.
1.6; 2.4

14 3.89; 6.77

15 (a) $n_1 + n_2 - 2$

16 0.2%

17 -1.89 and 1.89

18 (a) $2\frac{1}{2}$ (b) 1 (c) 99.9

19 (a) 1.71 (b) -2.13 (c) -1.20

20 (a) 0.01 (b) 0.005 (c) 0.005

21 (a) 0.125 (b) 0.005 (c) 0.87 (d) 0.05
(e) 0.01 (f) 0.831 (g) 0.035

22 (a) 3.24 (b) 2.16 (c) 4.89 (d) 49.66

23 (a) $2\frac{1}{2}$ (b) 1 (c) 5 (d) 1

24 (a) 0.824 ± 0.034 (b) 0.824 ± 0.024
(c) 0.824 ± 0.044

25 (a) 66.4 ± 8.12 (b) 60.6 ± 16.40
(c) 64.0 ± 8.95

26 (0.972, 0.988)

27 (33.482, 33.495)

28 (58.16, 58.55)

29 1.2875; (0.65, 1.92)

30 The mean lies in the interval
(163.39, 164.61)

31 (a) 0.55 ± 0.10 (b) 0.55 ± 0.13
(c) 0.55 ± 0.15

32 (2.69, 9.68)

33 (a) (0.982, 4.56) (b) (90.88, 97.21)

34 99% CI = (0.640, 0.885) (a) (2.717, 2.755)
(b) (2.679, 2.701); 95% CI = (2.608, 2.772)

35 $r = 0.73$; (0.11, 0.94)

36 (0.771, 0.973)

37 $n = 7$; -0.10 (2D)

38 Grand mean = 175.8; mean range = 4.6
Means chart: warning lines = 175.8 ± 1.75;
action lines = 175.8 ± 2.72
Ranges chart: warning line = 8.3;
action line = 10.8

39 Grand mean = 0.4444;
mean range = 0.0041
Means chart: warning lines = 0.4444
± 0.0015; action lines = 0.4444 ± 0.0024
Ranges chart: warning = 0.0074;
action = 0.0096
Variability of process is under control.
However process setting needs to be
investigated urgently since many control
points break the action limits.

40 Average defective, $d = 20.0$ with $n = 100$.
Warning = (12.2, 27.8); action = (7.6, 32.4)

41 Warning = (0.041, 0.159);
action = (0.007, 0.193)

42 Average defectives = 15.0 (1D).
Warning = (8, 22); action = (4, 26).
Process seems to be slowly drifting
towards producing more defectives. Three
samples have exceeded the warning line,
thus system should be monitored
closely.

43 (a) 13.0 ± 0.18 (b) 13.0 ± 0.21

44 (51.34, 53.66)

45 (a) 0.700 (3D) (b) (0.586, 0.787) (3D)

46 $E[m_1] = \frac{3}{4}\mu$ ∴ biased
$E[m_2] = E[m_3] = \mu$ ∴ unbiased; m_1 is best

47 (0.22, 0.44)

48 $\hat{\mu} = 2.36$; $\hat{\sigma}^2 = 0.048$; (a) $\hat{\sigma} = 0.219$;
2.36 ± 0.18 (b) $\hat{\sigma} = 0.256$; 2.36 ± 0.21

49 (0.1608, 0.2392)

50 (a) 95% CI: 801.86 ± 4.54
99% CI: 801.86 ± 6.88
(b) 95% CI: (10.03, 117.07)
99% CI: (7.81, 214.37)

51 (a) (4120.9, 4124.5) (b) (4120.7, 4124.7)

52 $E[\hat{\mu}_1] = E[\hat{\mu}_2] = \mu$ ∴ unbiased.
relative efficiency is $\frac{1}{3} : \frac{5}{9}$ ∴ $\hat{\mu}_1$ is
"best".

53 7.47 ± 0.71; 1.21

54 (a) (0.546, 0.734) (b) $n = 166$

55 (a) $\bar{x} \pm (2.33)\dfrac{s}{10}$ (large sample

approximation) (b) $\bar{x} \pm (3.36)\dfrac{s}{\sqrt{5}}$;

110.50 ± 27.12; 0.465

56 $n - 1$

57 (a) T_1, T_2 and T_4 (b) T_1

58 0.6; 0.110; 0.6 ± 0.215 (3D)

59 (a) 0.000 021 8 (b) 0.000 017 8;
1.499 ± 0.003

60 23.05 (2D); 0.44 (2D); 0.065 (3D)

63 $a = 3.35$

64 0.95

65 (a) $a = 4$; $\mathrm{Var}[T_2] = \dfrac{\sigma^2}{16n}(9 + 4b)$

66 (a) $\mathrm{Var}[T_1] = \dfrac{\sigma^2}{n_1 + n_2}$ (d) $n_1 = n_2$

67 Grand mean $= 95.09$; mean range $= 3.34$
Warning $= 95.09 \pm 1.26$;
action $= 95.09 \pm 1.98$.
Shift in mean $= +0.43$.

68 $\sigma(\text{est}) = 0.92$ (2D);
grand mean $= 101.00$ (2D);
mean range $= 2.142$ (4S)
Means chart: warning $= 101.00 \pm 0.81$;
action $= 101.00 \pm 1.27$
Ranges chart: warning $= 3.88$;
action $= 5.01$

69 $\sigma(\text{est}) = 0.063$; grand mean $= 5.08$;
mean range $= 0.13$
Means chart: warning $= (5.02, 5.14)$;
action $= (4.98, 5.18)$
Ranges chart: warning $= 0.25$;
action $= 0.34$
Process OK for both setting and variability.
Process could NOT meet requirement,
since probability of lying within stated
range is only 0.89.

Chapter 9 Answers

1 (a) $z = 1.414$ 3D; N.Sig; Yes (b) $z = -2$;
Sig; No (c) $z = 4.90$ (2D); Sig; No
(d) $z = -1.5$; N.Sig; Yes

2 (a) $H_0: \mu \geqslant 453$ (b) $H_1: \mu < 453$ (c) 5%
(d) $\dfrac{\bar{x} - \mu}{\sigma/\sqrt{n}}$ (e) -1.645. $z = -2.6$ and Sig at
5% level. $\therefore$ ministry has grounds for
complaint.

3 $\mu < 20\,000$; $\mu > 20\,000$; $\mu \geqslant 20\,000$;
$\mu \neq 20\,000$; H_0 would be $\geqslant 20\,000$

4 $\mu < 20\,000$

5 $\alpha = 0.125$; $\beta = 0.606$

6 (i) 0.264 (ii) 0.092; 0.801

7 (a) Critical $\bar{x} = 23.296$; (i) 0.855 (ii) 0.683
(iii) 0.245 (iv) 0.001 (b) Critical $\bar{x} = 23.004$;
(i) 0.959 (ii) 0.877 (iii) 0.496 (iv) 0.009

8 $\bar{x} = 29.3$; $z_{\text{test}} = -2.21$;
critical $z = \pm 1.96$ $\therefore$ Sig at 5% level

9 $z_{\text{test}} = 2.47$;
critical $z = 1.645$ $\therefore$ Sig at 5% level

10 $z_{\text{test}} = 1.725$
(a) critical $z = \pm 1.96$ $\therefore$ N. Sig at 5% level
$\therefore$ No evidence of change
(b) critical $z = 1.64$ $\therefore$ Sig at 5% level
$\therefore$ H_0 justified

11 $\frac{4}{9}$; $z_{\text{test}} = -6.25$
critical $z = \pm 1.96$ $\therefore$ Sig and deduction is
confirmed.

12 (a) $t = \dfrac{\bar{x} - 48}{s/\sqrt{n-1}}$; $H_0: \mu = 48$; $H_1: \mu \neq 48$
(two-tailed test) (b) (i) $t = -0.88$;
critical $t = -2.36$ (ii) $t = -2.65$
critical $t = -2.36$ (iii) $t = -2$;

critical $t = -2.78$; Accept (i) $\mu = 48$
(ii) $\mu \neq 48$ (iii) $\mu = 48$

13 $H_0: \mu = 0.05$ vs $H_1: \mu \neq 0.05$; $t = 3.00$
(a) critical t is 2.26 $\therefore$ reject H_0
(b) critical t is 3.25 $\therefore$ accept H_0

14 $t_{\text{test}} = -2.27$;
critical $t = -1.83$ $\therefore$ evidence given

15 $H_0: \mu = 1600$ vs $H_1: \mu \neq 1600$;
$z_{\text{test}} = -2.49$
(a) Critical $z = \pm 1.96$ $\therefore$ Sig and reject H_0
(b) Critical $z = \pm 2.58$ $\therefore$ N. Sig. and accept
H_0

16 $H_0: \mu \leqslant 42$ vs $H_1: \mu > 42$ (a) reject H_0 if
$\bar{X} > 43.16$ (b) otherwise accept H_0

17 (a) Accept H_0 if $12.65 < \bar{X} < 13.75$, reject
H_0 otherwise (b) (i) 0.066 (ii) 0.112

18 $t_{\text{test}} = -1.385$; critical $t = \pm 2.26$ $\therefore$ No
evidence

19 $\bar{d} = 9.1$; $s = 8.455$; $t_{\text{test}} = 3.23$ $\therefore$ Sig

20 $\bar{d} = 6.56$ (2D); $s = 8.07$ (2D);
$t_{\text{test}} = 2.30$ (2D)
Critical $t = 1.86$ $\therefore$ Sig at 5% level $\therefore$ evidence

21 $\bar{d} = 1$; $s = 3.423$; $t_{\text{test}} = 0.72$;
critical $t = \pm 2.45$ $\therefore$ N. Sig.

22 (a) 1.165 (b) 1.398

23 $\hat{\sigma} = 16.0$; $t_{\text{test}} = 1.16$;
critical $t = 2.31$ $\therefore$ differences not significant

24 $\bar{x} = 1008.57$; $\bar{y} = 981.0$; $s_1^2 = 215.10$;
$s_2^2 = 118.33$; $\hat{\sigma} = 14.19$; $t_{\text{test}} = 3.49$
Critical $t = 2.20$ (5% level) $= 3.11$ (1% level)
$\therefore$ Sig at 1% level. Paired samples would be
better.

25 $\bar{x}_1 = 12.2$; $\bar{x}_2 = 10.6$; $s_1^2 = 0.63$;
$s_2^2 = 0.412$; $t_{\text{test}} = 3.49$;
critical $t = 2.2$ $\therefore$ Sig

26 $\bar{x}_1 = 1.75$; $\bar{x}_2 = 2.55$; $s_1^2 = 7$; $s_2^2 = 9.167$;
$t_{\text{test}} = -0.62$; critical $t = \pm 2.09$; $\therefore$ N. Sig.

27 $t_{\text{test}} = 2.09$; $t(5\%, 2$ tail, 11 df$) = 2.20$
$\therefore$ N. Sig.

28 $H_0: \mu_A \leqslant \mu_B$ vs $H_1: \mu_A > \mu_B$. $t_{\text{test}} = 2.88$;
critical $t = 1.65$ $\therefore$ Sig.

29 $z_{\text{test}} = -3.2$; critical $z = \pm 1.96$ $\therefore$ Sig
$\therefore$ Not compatible

30 $z_{\text{test}} = 2.750$, critical $z = 1.645$ (5%) or
2.33 (1%) $\therefore$ result Sig.

31 $z_{\text{test}} = -1.59$ (2D); critical $z = -1.645$;
No evidence of decrease at 5% level

32 (a) $z_{\text{test}} = 1.33$; critical $z = 1.96$ $\therefore$ No
evidence (b) at least 10

33 Probability of a proportion as high as this
(or higher) is 0.0233, which is not
significant at a 1% level.

34 $p_1 =$ proportion who recover. $p_1 = 0.713$;
$p_2 = 0.652$ $\hat{\pi} = 0.679$; $z_{\text{test}} = 1.01$;
critical $z = 1.64$ $\therefore$ No evidence

35 $z_{\text{test}} = 2.19$; critical $z = 1.645$ $\therefore$ result Sig

36 0.071 ± 0.008; $z_{\text{test}} = 1.28$.
Critical $z = 1.645$ $\therefore$ N. Sig.

37 $z_{test} = 2.17$; critical $z = \pm 1.96$ ∴ Sig

38 (a) Reject H_0 if $P \geqslant 0.407$, otherwise accept H_0 (b) 0.801, 0.540, 0.271, 0.094, 0.021

39 $\chi^2_{test} = 9.9$; $\chi^2_{5\%}(8) = 15.51$ ∴ Result N. Sig.

40 (a) $\bar{x} = 10.73$ $s = 3.17$; $\chi^2_{test} = 25.03$; $\chi^2_{2\frac{1}{2}\%}(14) = 26.12$ ∴ Result N. Sig. (b) Yes. $\chi^2_{5\%}(14) = 23.68$

41 (a) $s^2 = 21.402$ (b) 9

42 $\hat{\sigma}_1^2 = 9.333$; $\hat{\sigma}_2^2 = 4.582$; $F = 2.04$ (2D); critical $F = 3.59$ ∴ Result N. Sig

43 (a) Yes; $F_{test} = 1.56$; $F_{2\frac{1}{2}\%}(18, 13) = 3.0$; differences N. Sig. (b) 18.28

44 No; $\hat{\sigma}_A^2 = 0.359$; $\hat{\sigma}_B^2 = 0.197$; $F_{test} = 1.82$, $F_{2\frac{1}{2}\%}(5, 4) = 9.36$

45 $H_0: \mu_1 \leqslant \mu_2$ vs $H_1: \mu_1 > \mu_2$; $t_{test} = 1.34$, $t(5\%, \nu = 20, 1\text{ tail}) = 1.72$ ∴ N. Sig.

46 (a) Reject H_0 if $R > 0.857$, otherwise accept H_0 (b) 2.6%

47 $H_0: \rho = 0$ vs $H_1: \rho \neq 0$. $t_{test} = 3.7$, $t_{critical} = 2.63$ ∴ Sig. at 1% level

48 (a) $z_{test} = -2.17$; $z_{crit} = -1.645$ ∴ Sig; thus evidence (b) $z_{test} = -1.23$; $z_{crit} = -1.645$ ∴ N. Sig; thus no evidence

49 $z_{test} = 0.70$; $z_{crit} = \pm 1.96$ ∴ N. Sig; thus no evidence

50 (a) $z_{test} = 1.76$; $z(5\%, 2\text{-tail}) = 1.96$ ∴ N. Sig.

51 $H_0: \mu \geqslant 8\text{ hr}$ vs $H_1: \mu < 8\text{ hr}$. $t_{test} = -2.09$, $t(5\%, \nu = 120, 1\text{ tail}) = -1.66$ ∴ Sig.

52 5.62 or 13.18

53 $\bar{x} = 1.493$; $\bar{y} = 1.530$; $s_x^2 = 0.000601$; $s_y^2 = 0.002625$; $\hat{\sigma}_x^2 = 0.0006677$; $\hat{\sigma}_y^2 = 0.003$; $F_{test} = 4.493$ $F_{2\frac{1}{2}\%}(7, 9) = 4.20$ ∴ Sig at 5% level. Thus, dispersion is different. (Note: T test is nullified.)

54 (a) $z_{test} = 1.78$, critical $z = \pm 1.96$ ∴ N. Sig. (b) $z_{test} = 3.23$, critical $z = \pm 1.96$ ∴ Sig.

55 $H_0: \mu \leqslant 1.16\text{ kg}$ vs $H_1: \mu > 1.16\text{ kg}$ (a) $z_{test} = 1.81$ $z_{5\%} = 1.64$ ∴ Sig. (b) $z_{test} = 1.60$ $z_{5\%} = 1.64$ ∴ N. Sig.

56 $H_0: \mu_A = \mu_B$ vs $H_1: \mu_A \neq \mu_B$; $t_{test} = -0.94$; $t(5\%, \nu = 14, 2\text{ tail}) = -2.14$ ∴ N. Sig.
$H_0: \mu_A = 2.146$ vs $H_1: \mu_A \neq 2.146$; $t_{test} = -1.66$; $t(5\%, \nu = 7, 2\text{ tail}) = -2.36$ ∴ N. Sig.
$H_0: \mu_B = 2.146$ vs $H_1: \mu_B \neq 2.146$; $t_{test} = -0.33$; $t(5\%, \nu = 7, 2\text{ tail}) = -2.36$ ∴ N. Sig.

57 (a) $z_{test} = +3.33$, Sig at 1% level (b) $z_{test} = 0.60$ ∴ N. Sig.

58 (a) 10.56%; 6.68% (b) £98.88 (c) 1 tail test $z_{test} = -2.5$ $z_{95\%} = -1.64$ ∴ Sig.

59 0.299; 0.545

60 (a) 11 (b) 149.5 (c) 9.0 (1D)

61 $F_{test} = 2.08$; critical $F = 3.12$ ∴ N. Sig: $t_{test} = 1.10$; critical $t = 2.06$ ∴ N. Sig 95% limits for σ^2 are given by (1.52, 4.84)

62 $H_0: \pi \geqslant 0.7$ vs $H_1: \pi < 0.7$. Critical $z = -1.64$; $z_{test} = -1.46$ (N. Sig); $z_{test} = -2.18$ (Sig); $n = 57$

63 (a) (i) $\hat{\sigma}_A^2 = \dfrac{N_1 S_A^2}{N_1 - 1}$; $\hat{\sigma}_B^2 = \dfrac{N_2 S_B^2}{N_2 - 1}$; $\hat{\sigma}^2 = \dfrac{N_1 S_A^2 + N_2 S_B^2}{N_1 + N_2 - 2}$
(ii) S. Error $= \hat{\sigma}\sqrt{\dfrac{1}{N_1} + \dfrac{1}{N_2}}$
(b) $F_{test} = 1.37$; critical $F = 1.60$ ∴ N. Sig. For T test: $\hat{\sigma} = 12.9$, $t_{test} = 2.68$, $t_{critical} = \pm 1.96$ ∴ Sig

64 (a) $z_{test} = 1.67$; $z_{5\%} = 1.645$ ∴ Sig ∴ reject H_0 (b) $n = 35$

66 (a) 0.00674 (b) 0.125 (3D); Sample too small for Normal approximation. Exact Poisson gives $Pr(X \geqslant 3/m = 1) = 0.08$ ∴ N. Sig. ∴ it *is* likely that $p = \frac{1}{20}$.

67 (a) $\frac{1}{3}, \frac{1}{8}$ (b) $\alpha = (0.088, 0.351, 0.680)$; $\beta = (0.551, 0.166, 0.029)$. Test 2 is best.

68 0.127; 0.129; 0.067

69 (a) $H_0: \pi \leqslant 0.5$ vs $H_1: \pi > 0.5$; $z_{test} = 1.26$, critical $z = 1.645$ ∴ N. Sig. ∴ Not enough evidence that cats prefer new food. (b) 17

70 (a) $t_{test} = -1.16$; critical $t = -1.75$ ∴ N. Sig. (b) $z_{test} = -1.4$; critical $z = -1.645$ ∴ N. sig.

71 $t_{test} = 0.90$; critical $= 2.10$ ∴ N. Sig. Paired sample test would be better.

72 $R = (-39.2, 39.2)$; 0.677; (a) 0.5 (b) 72.1

73 $H_0: \mu_x = \mu_y$ vs $\mu_x \neq \mu_y$. $F_{test} = 2.18$. $F_{2\frac{1}{2}\%}(5, 5) = 7.15$ ∴ N. Sig. (Common variance) $t_{test} = 1.18$, $t(5\%, \nu = 10, 2\text{ tail}) = 2.23$ ∴ N. Sig. (Consistent findings)

74 $k = 708$; $Pr(\text{type 2 error}) = 0.394$ (a) N(7900, 25 000) (b) N(700, 45 250); 700 ± 548

Chapter 10 Answers

1 $\chi^2_{test} = 4.27$; $\chi^2_{0.95}(3) = 7.81$ ∴ N. Sig.

2 Expected: 12, 24, 36, 48, 60, 72, 60, 48, 36, 24, 12
$A: \chi^2_{test} = 14.46$; $\chi^2_{0.95}(10) = 18.31$ ∴ N. Sig.
$B: \chi^2_{test} = 2.12$, $\chi^2_{0.95}(10) = 18.31$ ∴ N. Sig.

3 $\chi^2_{test} = 4.6$; $\chi^2_{0.95}(5) = 11.07$ ∴ N. Sig.

4 $\chi^2_{test} = 93.6$; $\chi^2_{0.99}(5) = 15.09$ ∴ Highly Sig.

5 $\chi^2_{test} = 20.46$; $\chi^2_{0.99}(5) = 15.09$ ∴ Highly Sig.

6 $\chi^2\text{test} = 3.80$, N. Sig;
   $\chi^2\text{test} = 7.61$, N. Sig. also
7 $\chi^2\text{test} = 9.81$; $\chi^2_{0.95}(10) = 18.31$ ∴ N. Sig.
   ∴ Fair dice and throws; 6
8 (i) 3.75 (ii) E: 0.3, 2.8, 11.6, 25.7, 32.2, 21.5
   5.9. $\chi^2_{0.95}(3) = 7.81$; $\chi^2\text{test} = 2.26$ ∴ N. Sig.
9 E: 2, 36, 270, 1080, 2430, 2916, 1458.
   $\chi^2\text{test} = 8.18$; $\chi^2_{0.95}(5) = 11.07$ ∴ N. Sig.
10 $\chi^2\text{test} = 2.05$; $\chi^2_{0.95}(4) = 9.49$ ∴ N. Sig.
11 Mean $= 4.17$; $\chi^2\text{test} = 3.85$,
    $\chi^2(5\%, 6\,\text{df}) = 12.59$ ∴ N. Sig. ∴ Poisson is
    adequate. Mean $= 3.37$, variance $= 0.98$.
12 $\chi^2\text{test} = 5.97$; $\chi^2_{0.95}(4) = 9.49$ ∴ N. Sig.
13 $\chi^2\text{test} = 4.58$; $\chi^2_{0.95}(2) = 5.99$ ∴ N. Sig.
14 (a) $\chi^2\text{test} = 1.76$; $\chi^2_{0.95}(4) = 9.49$ ∴ N. Sig.
    (b) The binomial situation here gives
    $n = 144$. ∴ Given model is not appropriate.
15 $\chi^2\text{test} = 3.64$; $\chi^2_{0.95}(3) = 7.81$ ∴ N. Sig.
16 (a) $\chi^2\text{test} = 5.08$; $\chi^2_{0.95}(2) = 5.99$ ∴ N. Sig.
    (b) $\chi^2\text{test} = 8.00$; $\chi^2_{0.99}(1) = 6.64$
    ∴ Highly Sig.
17 (a) 99.4, 119.3, 71.6, 28.6, 8.6, 2.1, 0.4
    (b) $\chi^2\text{test} = 0.47$; $\chi^2_{0.95}(3) = 7.81$ ∴ N. Sig.
18 Mean $= 2.0$ (1D); Variance $= 2.1$ (1D);
    Poisson, since mean $\triangleq$ variance;
    $\chi^2\text{test} = 2.47$, $\chi^2_{0.95}(5) = 11.07$ ∴ N. Sig.
19 E: 6.7, 10.6, 15.6, 15.5, 10.3, 4.6, 1.7,
    $\chi^2\text{test} = 0.53$, $\chi^2_{0.95}(3) = 7.81$ ∴ N. Sig.
20 E: 2.4, 3.2, 5.7, 9.0, 12.0, 13.9, 13.8, 11.8,
    8.6, 5.5, 3.0, 2.1; $\chi^2\text{test} = 7.16$
    $\chi^2_{0.95}(6) = 12.59$ ∴ N. Sig.
21 $\chi^2\text{test} = 9.96$; $\chi^2_{0.95}(8) = 15.51$ ∴ N. Sig.
22 (a) $\chi^2\text{test} = 8.73$; $\chi^2(5\%, 2\,\text{df}) = 5.79$
    ∴ Sig. ∴ NOT independent
    (b) $\chi^2\text{test} = 0.97$ ∴ N. Sig. ∴ Equally likely
23 $\chi^2\text{test} = 82.7$; $\chi^2_{0.99}(1) = 6.63$ ∴ Highly Sig.;
    $Q = -0.57$
24 $\chi^2\text{test} = 48.1$; $\chi^2_{0.99}(4) = 13.28$ ∴ Sig.
25 $\chi^2\text{test} = 6.62$; $\chi^2_{0.95}(4) = 9.49$ ∴ N. Sig.
26 $\chi^2\text{test} = 2.54$; $\chi^2_{0.95}(3) = 7.81$ ∴ N. Sig.
27 $\chi^2\text{test} = 26.61$; $\chi^2_{0.99}(4) = 13.28$
    ∴ Highly Sig.
28 $\chi^2\text{test} = 24.58$; $\chi^2_{0.99}(6) = 16.81$
    ∴ Highly Sig.
29 (i) $\chi^2\text{test} = 3.58$; $\chi^2(5\%, 3\,\text{df}) = 7.82$
    ∴ N. Sig. ∴ No association
    (ii) $\chi^2\text{test} = 35.8$ ∴ Sig. ∴ Strong evidence
    of association
32 $F\text{test} = 2.13$; $F\text{crit} = F_{0.95}(4, 15) = 3.06$
    ∴ N. Sig.
33 $F\text{test} = 4.06$; $F\text{crit} = F_{0.95}(5, 22) = 2.7$
    ∴ Sig.
34 $F\text{test} = 5.89$; $F\text{crit} = F_{0.95}(2, 23) = 3.4$
    ∴ Sig.
35 $F(\text{treatments}) = 15.39$; $F_{0.99}(4, 12) = 5.41$
    ∴ Highly Sig.
    $F(\text{plots}) = 10.17$; $F_{0.99}(3, 12) = 5.95$
    ∴ Highly Sig.

36 $F(\text{diets}) = 0.93$; $F_{0.95}(3, 6) = 4.76$
    ∴ N. Sig.;
    $F(\text{diseases}) = 1.65$, $F_{0.95}(2, 6) = 5.14$
    ∴ N. Sig.
37 $F(\text{machines}) = 38.4$; $F_{0.99}(4, 12) = 5.41$
    ∴ Highly Sig.
    $F(\text{workers}) = 2.73$; $F_{0.99}(3, 12) = 3.49$
    ∴ N. Sig.
38 Variety $F = 33.23$; $F_{0.99}(4, 20) = 4.43$
    ∴ Highly Sig.
    Block $F = 6.44$; $F_{0.99}(5, 20) = 4.10$
    ∴ Highly Sig.
39 $F\text{test}(\text{cloths}) = 16.77$; $F(5\%, 3, 12) = 4.47$
    ∴ Sig.
    $F\text{test}(\text{strength}) = 1.42$; $F(5\%, 4, 12) = 4.12$
    ∴ N. Sig.
40 $F(\text{towns}) = 54.09$; $F_{0.99}(2, 8) = 8.65$
    ∴ Highly Sig;
    $F(\text{finishes}) = 107.27$; $F_{0.99}(3, 8) = 7.59$
    ∴ Highly Sig.
41 $F(\text{investigators}) = 1.26$; $F_{0.95}(3, 6) = 4.76$
    ∴ N. Sig.
    $F(\text{shops}) = 14.49$; $F_{0.99}(2, 6) = 10.92$
    ∴ Highly Sig.
42 Subjects $F = 2.14$; $F_{0.95}(4, 2) = 3.26$
    ∴ N. Sig.
    Colleges $F = 0.26$; $F_{0.95}(3, 12) = 3.49$
    ∴ N. Sig.
43 One tailed test. $z\text{test} = -1.664$;
    critical $z = -1.645$ ∴ (Just) Sig at 5% level.
44 $z\text{test} = 1.455$; $z_{2\frac{1}{2}\%} = 1.96$ ∴ N. Sig.
45 One tailed test. $z\text{test} = 3.2$;
    $z_{0.99} = 2.33$ ∴ Highly Sig.
46 $z\text{test} = 0.67$; $z_{2\frac{1}{2}\%} = 1.96$ ∴ N. Sig.
47 (a) $t\text{test} = 1.00$; critical $t = 2.23$ ∴ N. Sig.
    (b) $z\text{test} = 0.32$; critical $z = 1.96$ ∴ N. Sig.
48 Theoretically, the exponential distribution
    describes the time/distance between
    'Poisson' events.
    (b) E: 39, 24, 23, 8.6, 5.0 $\chi^2\text{test} = 1.66$;
    $\chi^2_{0.95}(4) = 9.49$ ∴ N. Sig.
49 (a) E: 30, 90, 240. $\chi^2\text{test} = 6.15$;
    $\chi^2_{0.95}(2) = 5.99$ ∴ Sig. (b) H_0 is now that
    die is unbiased. e: 60, 60, 240.
    $\chi^2\text{test} = 16.42$; $\chi^2_{0.99}(2) = 9.21$ ∴ Highly Sig.
    ∴ New theory accepted.
50 (a) $\chi^2\text{test} = 4.95$; $\chi^2_{0.95}(3) = 7.81$ ∴ N. Sig.
    (b) $\chi^2\text{test} = 9.90$ ∴ Sig.
51 (a) $\chi^2\text{test} = 4.8$; $\chi^2_{0.95}(9) = 16.92$ ∴ N. Sig.
    (b) $\chi^2\text{test} = 2.45$; $\chi^2_{0.95}(2) = 5.99$ ∴ N. Sig.
52 A of V test: $F = 3.65$; $F_{0.95}(4, 15) = 3.06$
    ∴ Sig. T test: $t = 3.75$; $t_{0.99}(18) = 2.90$
    ∴ Highly Sig.
53 $F(\text{rations}) = 5.76$; $F_{0.95}(2, 6) = 5.14$ ∴ Sig.
    $F(\text{breeds}) = 6.22$; $F_{0.95}(3, 6) = 4.76$ ∴ Sig.
54 $\chi^2\text{test} = 7.84$; $\chi^2_{0.95}(4) = 9.49$ ∴ N. Sig.
55 645; $\chi^2\text{test} = 693.64$ ∴ Highly Sig.
    ($\chi^2_{0.99}(4) = 18.47$)

H_0: Hockey and netball are not associated (with respect to standard of play).

56 F(mixtures) = 1.00; $F_{0.95}(2, 6) = 5.14$ ∴ N. Sig. F(fields) = 0.43 ∴ N. Sig.

57 (i) $z_{test} = 1.21$; $z(5\%, 1\text{-tail}) = 1.645$ ∴ N. Sig.

(ii) $t_{test} = -3.31$; $t(1\%, 2\text{-tail}, 9 \text{ df})$ = -3.25 ∴ Highly Sig.

Appendix

TABLE 1   THE BINOMIAL DISTRIBUTION FUNCTION

$p =$		0.05	0.1	0.15	0.2	0.25	0.3	0.35	0.4	0.45	0.5
$n = 2$	0	.9025	.8100	.7225	.6400	.5625	.4900	.4225	.3600	.3025	.2500
	1	.9975	.9900	.9775	.9600	.9375	.9100	.8775	.8400	.7975	.7500
$n = 3$	0	.8574	.7290	.6141	.5120	.4219	.3430	.2746	.2160	.1664	.1250
	1	.9928	.9720	.9393	.8960	.8438	.7840	.7183	.6480	.5747	.5000
	2	.9999	.9990	.9966	.9920	.9844	.9730	.9571	.9360	.9089	.8750
$n = 4$	0	.8145	.6561	.5220	.4096	.3164	.2401	.1785	.1296	.0915	.0625
	1	.9860	.9477	.8905	.8192	.7383	.6517	.5630	.4752	.3910	.3125
	2	.9995	.9963	.9880	.9728	.9492	.9163	.8735	.8208	.7585	.6875
	3	1.000	1.000	.9995	.9984	.9961	.9919	.9850	.9744	.9590	.9375
$n = 5$	0	.7738	.5905	.4437	.3277	.2373	.1681	.1160	.0778	.0503	.0313
	1	.9774	.9185	.8352	.7373	.6328	.5282	.4284	.3370	.2562	.1875
	2	.9988	.9914	.9734	.9421	.8965	.8369	.7648	.6826	.5931	.5000
	3	1.000	.9995	.9978	.9933	.9844	.9692	.9460	.9130	.8688	.8125
	4	1.000	1.000	1.000	.9997	.9990	.9976	.9947	.9898	.9815	.9688
$n = 6$	0	.7351	.5314	.3771	.2621	.1780	.1176	.0754	.0467	.0277	.0156
	1	.9672	.8857	.7765	.6554	.5339	.4202	.3191	.2333	.1636	.1094
	2	.9978	.9841	.9527	.9011	.8306	.7443	.6471	.5443	.4415	.3438
	3	1.000	.9987	.9941	.9830	.9624	.9295	.8826	.8208	.7447	.6563
	4	1.000	1.000	.9996	.9984	.9954	.9891	.9777	.9590	.9308	.8906
	5	1.000	1.000	1.000	1.000	.9998	.9993	.9982	.9959	.9917	.9844
$n = 7$	0	.6983	.4783	.3206	.2097	.1335	.0824	.0490	.0280	.0152	.0078
	1	.9556	.8503	.7166	.5767	.4449	.3294	.2338	.1586	.1024	.0625
	2	.9962	.9743	.9262	.8520	.7564	.6471	.5323	.4199	.3164	.2266
	3	.9998	.9973	.9879	.9667	.9294	.8740	.8002	.7102	.6083	.5000
	4	1.000	.9998	.9988	.9953	.9871	.9712	.9444	.9037	.8471	.7734
	5	1.000	1.000	1.000	.9996	.9987	.9962	.9910	.9812	.9643	.9375
	6	1.000	1.000	1.000	1.000	1.000	.9998	.9994	.9984	.9963	.9922
$n = 8$	0	.6634	.4305	.2725	.1678	.1001	.0576	.0319	.0168	.0084	.0039
	1	.9428	.8131	.6572	.5033	.3671	.2553	.1691	.1064	.0632	.0352
	2	.9942	.9619	.8948	.7969	.6785	.5518	.4278	.3154	.2201	.1445
	3	.9996	.9950	.9786	.9437	.8862	.8059	.7064	.5941	.4770	.3633
	4	1.000	.9996	.9971	.9896	.9727	.9420	.8939	.8263	.7396	.6367
	5	1.000	1.000	.9998	.9988	.9958	.9887	.9747	.9502	.9115	.8555
	6	1.000	1.000	1.000	1.000	.9996	.9987	.9964	.9915	.9819	.9648
	7	1.000	1.000	1.000	1.000	1.000	1.000	.9998	.9993	.9983	.9961

$p =$	0.05	0.1	0.15	0.2	0.25	0.3	0.35	0.4	0.45	0.5
$n = 9$ 0	.6302	.3874	.2316	.1342	.0751	.0404	.0207	.0101	.0046	.0020
1	.9288	.7748	.5995	.4362	.3003	.1960	.1211	.0705	.0385	.0195
2	.9916	.9470	.8591	.7382	.6007	.4628	.3373	.2318	.1495	.0898
3	.9994	.9917	.9661	.9144	.8343	.7297	.6089	.4826	.3614	.2539
4	1.000	.9991	.9944	.9804	.9511	.9012	.8283	.7334	.6214	.5000
5	1.000	1.000	.9994	.9969	.9900	.9747	.9464	.9006	.8342	.7461
6	1.000	1.000	1.000	.9997	.9987	.9957	.9888	.9750	.9502	.9102
7	1.000	1.000	1.000	1.000	.9999	.9996	.9986	.9962	.9909	.9805
8	1.000	1.000	1.000	1.000	1.000	1.000	1.000	.9997	.9992	.9980
$n = 10$ 0	.5987	.3487	.1969	.1074	.0563	.0282	.0135	.0060	.0025	.0010
1	.9139	.7361	.5443	.3758	.2440	.1493	.0860	.0464	.0233	.0107
2	.9885	.9298	.8202	.6778	.5256	.3828	.2616	.1673	.0996	.0547
3	.9990	.9872	.9500	.8791	.7759	.6496	.5138	.3823	.2660	.1719
4	1.000	.9984	.9901	.9672	.9219	.8497	.7515	.6331	.5044	.3770
5	1.000	.9999	.9986	.9936	.9803	.9527	.9051	.8338	.7384	.6230
6	1.000	1.000	.9999	.9991	.9965	.9894	.9740	.9452	.8980	.8281
7	1.000	1.000	1.000	1.000	.9996	.9984	.9952	.9877	.9726	.9453
8	1.000	1.000	1.000	1.000	1.000	.9999	.9995	.9983	.9955	.9893
9	1.000	1.000	1.000	1.000	1.000	1.000	1.000	.9999	.9997	.9990
$n = 11$ 0	.5688	.3138	.1673	.0859	.0422	.0198	.0088	.0036	.0014	.0005
1	.8981	.6974	.4922	.3221	.1971	.1130	.0606	.0302	.0139	.0059
2	.9848	.9104	.7788	.6174	.4552	.3127	.2001	.1189	.0652	.0327
3	.9984	.9815	.9306	.8389	.7133	.5696	.4256	.2963	.1911	.1133
4	.9999	.9972	.9841	.9496	.8854	.7897	.6683	.5328	.3971	.2744
5	1.000	.9997	.9973	.9883	.9657	.9218	.8513	.7535	.6331	.5000
6	1.000	1.000	.9997	.9980	.9924	.9784	.9499	.9006	.8262	.7256
7	1.000	1.000	1.000	.9998	.9988	.9957	.9878	.9707	.9390	.8867
8	1.000	1.000	1.000	1.000	.9999	.9994	.9980	.9941	.9852	.9673
9	1.000	1.000	1.000	1.000	1.000	1.000	.9998	.9993	.9978	.9941
10	1.000	1.000	1.000	1.000	1.000	1.000	1.000	1.000	.9998	.9995
$n = 12$ 0	.5404	.2824	.1422	.0687	.0317	.0138	.0057	.0022	.0008	.0002
1	.8816	.6590	.4435	.2749	.1584	.0850	.0424	.0196	.0083	.0032
2	.9804	.8891	.7358	.5583	.3907	.2528	.1513	.0834	.0421	.0193
3	.9978	.9744	.9078	.7946	.6488	.4925	.3467	.2253	.1345	.0730
4	.9998	.9957	.9761	.9274	.8424	.7237	.5833	.4382	.3044	.1938
5	1.000	.9995	.9954	.9806	.9456	.8822	.7873	.6652	.5269	.3872
6	1.000	1.000	.9993	.9961	.9857	.9614	.9154	.8418	.7393	.6128
7	1.000	1.000	1.000	.9994	.9972	.9905	.9745	.9427	.8883	.8062
8	1.000	1.000	1.000	1.000	.9996	.9983	.9944	.9847	.9644	.9270
9	1.000	1.000	1.000	1.000	1.000	.9998	.9992	.9972	.9921	.9807
10	1.000	1.000	1.000	1.000	1.000	1.000	1.000	.9997	.9989	.9968
11	1.000	1.000	1.000	1.000	1.000	1.000	1.000	1.000	1.000	.9998

$p =$		0.05	0.1	0.15	0.2	0.25	0.3	0.35	0.4	0.45	0.5
$n = 13$	0	.5133	.2542	.1209	.0550	.0238	.0097	.0037	.0013	.0004	.0001
	1	.8646	.6213	.3983	.2336	.1267	.0637	.0296	.0126	.0049	.0017
	2	.9755	.8661	.6920	.5017	.3326	.2025	.1132	.0579	.0269	.0112
	3	.9969	.9658	.8820	.7473	.5843	.4206	.2783	.1686	.0929	.0461
	4	.9997	.9935	.9658	.9009	.7940	.6543	.5005	.3530	.2279	.1334
	5	1.000	.9991	.9925	.9700	.9198	.8346	.7159	.5744	.4268	.2905
	6	1.000	1.000	.9987	.9930	.9757	.9376	.8705	.7712	.6437	.5000
	7	1.000	1.000	.9998	.9988	.9944	.9818	.9538	.9023	.8212	.7095
	8	1.000	1.000	1.000	.9998	.9990	.9960	.9874	.9679	.9302	.8666
	9	1.000	1.000	1.000	1.000	.9999	.9993	.9975	.9922	.9797	.9539
	10	1.000	1.000	1.000	1.000	1.000	1.000	.9997	.9987	.9959	.9888
	11	1.000	1.000	1.000	1.000	1.000	1.000	1.000	.9999	.9995	.9983
	12	1.000	1.000	1.000	1.000	1.000	1.000	1.000	1.000	1.000	.9999

TABLE 2 THE POISSON DISTRIBUTION FUNCTION

$m =$	.1	.2	.3	.4	.5	.6	.7	.8	.9	1.0
0	.9048	.8187	.7408	.6703	.6065	.5488	.4966	.4493	.4066	.3679
1	.9953	.9825	.9631	.9384	.9098	.8781	.8442	.8088	.7725	.7358
2	.9998	.9989	.9964	.9921	.9856	.9769	.9659	.9526	.9371	.9197
3	1.000	1.000	.9997	.9992	.9982	.9966	.9942	.9909	.9865	.9810
4	1.000	1.000	1.000	1.000	.9998	.9996	.9992	.9986	.9977	.9963
5	1.000	1.000	1.000	1.000	1.000	1.000	1.000	.9998	.9997	.9994
6	1.000	1.000	1.000	1.000	1.000	1.000	1.000	1.000	1.000	1.000

$m =$	1.1	1.2	1.3	1.4	1.5	1.6	1.7	1.8	1.9	2.0
0	.3329	.3012	.2725	.2466	.2231	.2019	.1827	.1653	.1496	.1353
1	.6990	.6626	.6268	.5918	.5578	.5249	.4932	.4628	.4337	.4060
2	.9004	.8795	.8571	.8335	.8088	.7834	.7572	.7306	.7037	.6767
3	.9743	.9662	.9569	.9463	.9344	.9212	.9068	.8913	.8747	.8571
4	.9946	.9923	.9893	.9857	.9814	.9763	.9704	.9636	.9559	.9473
5	.9990	.9985	.9978	.9968	.9955	.9940	.9920	.9896	.9868	.9834
6	.9999	.9997	.9996	.9994	.9991	.9987	.9981	.9974	.9966	.9955
7	1.000	1.000	1.000	.9999	.9998	.9997	.9996	.9994	.9992	.9989
8	1.000	1.000	1.000	1.000	1.000	1.000	1.000	.9999	.9998	.9998
9	1.000	1.000	1.000	1.000	1.000	1.000	1.000	1.000	1.000	1.000

$m =$	2.1	2.2	2.3	2.4	2.5	2.6	2.7	2.8	2.9	3.0
0	.1225	.1108	.1003	.0907	.0821	.0743	.0672	.0608	.0550	.0498
1	.3796	.3546	.3309	.3084	.2873	.2674	.2487	.2311	.2146	.1991
2	.6496	.6227	.5960	.5697	.5438	.5184	.4936	.4695	.4460	.4232
3	.8386	.8194	.7993	.7787	.7576	.7360	.7141	.6919	.6696	.6472
4	.9379	.9275	.9162	.9041	.8912	.8774	.8629	.8477	.8318	.8153
5	.9796	.9751	.9700	.9643	.9580	.9510	.9433	.9349	.9258	.9161
6	.9941	.9925	.9906	.9884	.9858	.9828	.9794	.9756	.9713	.9665
7	.9985	.9980	.9974	.9967	.9958	.9947	.9934	.9919	.9901	.9881
8	.9997	.9995	.9994	.9991	.9989	.9985	.9981	.9976	.9969	.9962
9	1.000	.9999	.9999	.9998	.9997	.9996	.9995	.9993	.9991	.9989
10	1.000	1.000	1.000	1.000	1.000	1.000	.9999	.9998	.9998	.9997
11	1.000	1.000	1.000	1.000	1.000	1.000	1.000	1.000	1.000	1.000
12	1.000	1.000	1.000	1.000	1.000	1.000	1.000	1.000	1.000	1.000

$m =$	3.2	3.4	3.6	3.8	4.0	4.2	4.4	4.6	4.8	5.0
0	.0408	.0334	.0273	.0224	.0183	.0150	.0123	.0101	.0082	.0067
1	.1712	.1468	.1257	.1074	.0916	.0780	.0663	.0563	.0477	.0404
2	.3799	.3397	.3027	.2689	.2381	.2102	.1851	.1626	.1425	.1247
3	.6025	.5584	.5152	.4735	.4335	.3954	.3594	.3257	.2942	.2650
4	.7806	.7442	.7064	.6678	.6288	.5898	.5512	.5132	.4763	.4405
5	.8946	.8705	.8441	.8156	.7851	.7531	.7199	.6858	.6510	.6160
6	.9554	.9421	.9267	.9091	.8893	.8675	.8436	.8180	.7908	.7622
7	.9832	.9769	.9692	.9599	.9489	.9361	.9214	.9049	.8867	.8666
8	.9943	.9917	.9883	.9840	.9786	.9721	.9642	.9549	.9442	.9319

$m =$	3.2	3.4	3.6	3.8	4.0	4.2	4.4	4.6	4.8	5.0
9	.9982	.9973	.9960	.9942	.9919	.9889	.9851	.9805	.9749	.9682
10	.9995	.9992	.9987	.9981	.9972	.9959	.9943	.9922	.9896	.9863
11	.9999	.9998	.9996	.9994	.9991	.9986	.9980	.9971	.9960	.9945
12	1.000	1.000	.9999	.9998	.9997	.9996	.9993	.9990	.9986	.9980
13	1.000	1.000	1.000	1.000	1.000	.9999	.9998	.9997	.9995	.9993
14	1.000	1.000	1.000	1.000	1.000	1.000	1.000	1.000	.9999	.9998
15	1.000	1.000	1.000	1.000	1.000	1.000	1.000	1.000	1.000	1.000

$m =$	5.5	6.0	6.5	7.0	7.5	8.0	8.5	9.0	9.5	10.0
0	.0041	.0025	.0015	.0009	.0006	.0003	.0002	.0001	.0001	.0000
1	.0266	.0174	.0113	.0073	.0047	.0030	.0019	.0012	.0008	.0005
2	.0884	.0620	.0430	.0296	.0203	.0138	.0093	.0062	.0042	.0028
3	.2017	.1512	.1118	.0818	.0591	.0424	.0301	.0212	.0149	.0103
4	.3575	.2851	.2237	.1730	.1321	.0996	.0744	.0550	.0403	.0293
5	.5289	.4457	.3690	.3007	.2414	.1912	.1496	.1157	.0885	.0671
6	.6860	.6063	.5265	.4497	.3782	.3134	.2562	.2068	.1649	.1301
7	.8095	.7440	.6728	.5987	.5246	.4530	.3856	.3239	.2687	.2202
8	.8944	.8472	.7916	.7291	.6620	.5925	.5231	.4557	.3918	.3328
9	.9462	.9161	.8774	.8305	.7764	.7166	.6530	.5874	.5218	.4579
10	.9747	.9574	.9332	.9015	.8622	.8159	.7634	.7060	.6453	.5830
11	.9890	.9799	.9661	.9467	.9208	.8881	.8487	.8030	.7520	.6968
12	.9955	.9912	.9840	.9730	.9573	.9362	.9091	.8758	.8364	.7916
13	.9983	.9964	.9929	.9872	.9784	.9658	.9486	.9261	.8981	.8645
14	.9994	.9986	.9970	.9943	.9897	.9827	.9726	.9585	.9400	.9165
15	.9998	.9995	.9988	.9976	.9954	.9918	.9862	.9780	.9665	.9513
16	1.000	.9998	.9996	.9990	.9980	.9963	.9934	.9889	.9823	.9730
17	1.000	1.000	.9998	.9996	.9992	.9984	.9970	.9947	.9911	.9857
18	1.000	1.000	1.000	.9999	.9997	.9993	.9987	.9976	.9957	.9928
19	1.000	1.000	1.000	1.000	.9999	.9997	.9995	.9989	.9980	.9965
20	1.000	1.000	1.000	1.000	1.000	1.000	.9998	.9996	.9991	.9984
21	1.000	1.000	1.000	1.000	1.000	1.000	1.000	.9998	.9996	.9993
22	1.000	1.000	1.000	1.000	1.000	1.000	1.000	1.000	.9999	.9997
23	1.000	1.000	1.000	1.000	1.000	1.000	1.000	1.000	1.000	.9999
24	1.000	1.000	1.000	1.000	1.000	1.000	1.000	1.000	1.000	1.000

TABLE 3 (a) THE NORMAL DISTRIBUTION FUNCTION – $\Phi(x)$

x	.00	.01	.02	.03	.04	.05	.06	.07	.08	.09
0.0	.5000	.5040	.5080	.5120	.5160	.5199	.5239	.5279	.5319	.5359
0.1	.5398	.5438	.5478	.5517	.5557	.5596	.5636	.5675	.5714	.5753
0.2	.5793	.5832	.5871	.5910	.5948	.5987	.6026	.6064	.6103	.6141
0.3	.6179	.6217	.6255	.6293	.6331	.6368	.6406	.6443	.6480	.6517
0.4	.6554	.6591	.6628	.6664	.6700	.6736	.6772	.6808	.6844	.6879
0.5	.6915	.6950	.6985	.7019	.7054	.7088	.7123	.7157	.7190	.7224
0.6	.7257	.7291	.7324	.7357	.7389	.7422	.7454	.7486	.7517	.7549
0.7	.7580	.7611	.7642	.7673	.7704	.7734	.7764	.7794	.7823	.7852
0.8	.7881	.7910	.7939	.7967	.7995	.8023	.8051	.8078	.8106	.8133
0.9	.8159	.8186	.8212	.8238	.8264	.8289	.8315	.8340	.8365	.8389
1.0	.8413	.8438	.8461	.8485	.8508	.8531	.8554	.8577	.8599	.8621
1.1	.8643	.8665	.8686	.8708	.8729	.8749	.8770	.8790	.8810	.8830
1.2	.8849	.8869	.8888	.8907	.8925	.8944	.8962	.8980	.8997	.9015
1.3	.9032	.9049	.9066	.9082	.9099	.9115	.9131	.9147	.9162	.9177
1.4	.9192	.9207	.9222	.9236	.9251	.9265	.9279	.9292	.9306	.9319
1.5	.9332	.9345	.9357	.9370	.9382	.9394	.9406	.9418	.9429	.9441
1.6	.9452	.9463	.9474	.9484	.9495	.9505	.9515	.9525	.9535	.9545
1.7	.9554	.9564	.9573	.9582	.9591	.9599	.9608	.9616	.9625	.9633
1.8	.9641	.9649	.9656	.9664	.9671	.9678	.9686	.9693	.9699	.9706
1.9	.9713	.9719	.9726	.9732	.9738	.9744	.9750	.9756	.9761	.9767
2.0	.9772	.9778	.9783	.9788	.9793	.9798	.9803	.9808	.9812	.9817
2.1	.9821	.9826	.9830	.9834	.9838	.9842	.9846	.9850	.9854	.9857
2.2	.9861	.9864	.9868	.9871	.9875	.9878	.9881	.9884	.9887	.9890
2.3	.9893	.9896	.9898	.9901	.9904	.9906	.9909	.9911	.9913	.9916
2.4	.9918	.9920	.9922	.9925	.9927	.9929	.9931	.9932	.9934	.9936
2.5	.9938	.9940	.9941	.9943	.9945	.9946	.9948	.9949	.9951	.9952
2.6	.9953	.9955	.9956	.9957	.9959	.9960	.9961	.9962	.9963	.9964
2.7	.9965	.9966	.9967	.9968	.9969	.9970	.9971	.9972	.9973	.9974
2.8	.9974	.9975	.9976	.9977	.9978	.9978	.9979	.9979	.9980	.9981
2.9	.9981	.9982	.9982	.9983	.9984	.9984	.9985	.9985	.9986	.9986

x	3.0	3.1	3.2	3.3	3.4	3.5	3.6	3.7	3.8	3.9
$\Phi(x)$	.9987	.9990	.9993	.9995	.9997	.9998	.9998	.9999	.9999	1.000

(b) PERCENTAGE POINTS OF THE NORMAL DISTRIBUTION

p	.5	.75	.875	.9	.95	.975	.99	.999
x	0.00	0.67	1.15	1.28	1.645	1.96	2.33	3.09

TABLE 4   PERCENTAGE POINTS OF THE χ^2 DISTRIBUTION

$p =$	0.005	0.01	0.025	0.05	0.1	0.9	0.95	0.975	0.99	0.995
$\nu = 1$	0.044	0.032	0.021	0.024	0.02	2.71	3.84	5.02	6.64	7.88
2	0.01	0.02	0.05	0.10	0.21	4.61	5.99	7.38	9.21	10.60
3	0.07	0.12	0.22	0.35	0.58	6.25	7.82	9.35	11.35	12.84
4	0.21	0.30	0.48	0.71	1.06	7.78	9.49	11.14	13.28	14.86
5	0.41	0.55	0.83	1.15	1.61	9.24	11.07	12.83	15.09	16.75
6	0.68	0.87	1.24	1.64	2.20	10.65	12.59	14.45	16.81	18.55
7	0.99	1.24	1.69	2.17	2.83	12.02	14.07	16.01	18.48	20.28
8	1.34	1.65	2.18	2.73	3.49	13.36	15.51	17.54	20.09	21.96
9	1.74	2.09	2.70	3.33	4.17	14.68	16.92	19.02	21.67	23.59
10	2.16	2.56	3.25	3.94	4.87	15.99	18.31	20.48	23.21	25.19
11	2.60	3.05	3.82	4.58	5.58	17.28	19.68	21.92	24.73	26.76
12	3.07	3.57	4.40	5.23	6.30	18.55	21.03	23.34	26.22	28.30
13	3.57	4.11	5.01	5.89	7.04	19.81	22.36	24.74	27.69	29.82
14	4.08	4.66	5.63	6.57	7.79	21.06	23.69	26.12	29.14	31.32
15	4.60	5.23	6.26	7.26	8.55	22.31	25.00	27.49	30.58	32.80
16	5.14	5.81	6.91	7.96	9.31	23.54	26.30	28.85	32.00	34.27
17	5.70	6.41	7.56	8.67	10.09	24.77	27.59	30.19	33.41	35.72
18	6.27	7.02	8.23	9.39	10.87	25.99	28.87	31.53	34.81	37.16
19	6.84	7.63	8.91	10.12	11.65	27.20	30.14	32.85	36.19	38.58
20	7.43	8.26	9.59	10.85	12.44	28.41	31.41	34.17	37.57	40.00
25	10.52	11.52	13.12	14.61	16.47	34.38	37.65	40.65	44.31	46.93
30	13.79	14.95	16.79	18.89	20.60	40.26	43.77	46.98	50.89	53.67
50	27.99	29.71	32.36	34.76	37.69	63.17	67.51	71.42	76.15	79.49
100	67.33	70.07	74.22	77.93	82.36	118.5	124.3	129.6	135.8	140.2

TABLE 5 PERCENTAGE POINTS OF THE *T* DISTRIBUTION

$p =$	0.9	0.95	0.975	0.99	0.995	$p =$	0.9	0.95	0.975	0.99	0.995
$v =$						$v =$					
1	3.08	6.31	12.71	31.82	63.66	11	1.36	1.80	2.20	2.72	3.11
2	1.89	2.92	4.30	6.97	9.93	12	1.36	1.78	2.18	2.68	3.06
3	1.64	2.35	3.18	4.54	5.84	13	1.35	1.77	2.16	2.65	3.01
4	1.53	2.13	2.77	3.75	4.60	14	1.35	1.76	2.15	2.62	2.98
5	1.48	2.02	2.57	3.37	4.03	15	1.34	1.75	2.13	2.60	2.95
6	1.44	1.94	2.45	3.14	3.71	20	1.33	1.73	2.09	2.53	2.85
7	1.42	1.90	2.37	3.00	3.50	30	1.31	1.70	2.04	2.46	2.75
8	1.40	1.86	2.31	2.90	3.36	50	1.30	1.68	2.01	2.40	2.68
9	1.38	1.83	2.26	2.82	3.25	100	1.29	1.66	1.98	2.36	2.63
10	1.37	1.81	2.23	2.76	3.17	∞	1.28	1.65	1.96	2.33	2.58

TABLE 6   (a) 95% POINTS OF THE *F* DISTRIBUTION

v_1 v_2	1	2	3	4	5	6	7	8	9	10	12	15	20
1	161.4	199.5	215.7	224.6	230.2	234.0	236.8	238.9	240.5	241.9	243.9	246.0	248.0
2	18.51	19.00	19.16	19.25	19.30	19.33	19.35	19.37	19.39	19.40	19.41	19.43	19.45
3	10.13	9.552	9.277	9.117	9.013	8.941	8.887	8.845	8.812	8.786	8.745	8.703	8.660
4	7.709	6.944	6.591	6.388	6.256	6.163	6.094	6.041	5.999	5.964	5.912	5.858	5.803
5	6.608	5.786	5.409	5.192	5.050	4.950	4.876	4.818	4.772	4.735	4.678	4.619	4.558
6	5.987	5.143	4.757	4.534	4.387	4.284	4.207	4.147	4.099	4.060	4.000	3.938	3.874
7	5.591	4.737	4.347	4.120	3.972	3.866	3.787	3.726	3.677	3.637	3.575	3.511	3.445
8	5.318	4.459	4.066	3.838	3.687	3.581	3.500	3.438	3.388	3.347	3.284	3.218	3.150
9	5.117	4.256	3.863	3.633	3.482	3.374	3.293	3.230	3.179	3.137	3.073	3.006	3.936
10	4.965	4.103	3.708	3.478	3.326	3.217	3.135	3.072	3.020	2.978	2.913	2.845	2.774
11	4.844	3.982	3.587	3.357	3.204	3.095	3.012	2.948	2.896	2.854	2.788	2.719	2.646
12	4.747	3.885	3.490	3.259	3.106	2.996	2.913	2.849	2.796	2.753	2.687	2.617	2.544
15	4.543	3.682	3.287	3.056	2.901	2.790	2.707	2.641	2.588	2.544	2.475	2.403	2.328
20	4.351	3.493	3.098	2.866	2.711	2.599	2.514	2.447	2.393	2.348	2.278	2.203	2.124

(b) 97.5% POINTS OF THE *F* DISTRIBUTION

v_1 v_2	1	2	3	4	5	6	7	8	9	10	12	15	20
1	647.8	799.5	864.2	899.6	921.8	937.1	948.2	956.7	963.3	968.6	976.7	984.9	993.1
2	38.51	39.00	39.17	39.25	39.30	39.33	39.36	39.37	39.39	39.40	39.42	39.43	39.45
3	17.44	16.04	15.44	15.10	14.89	14.74	14.62	14.54	14.47	14.42	14.34	14.25	14.17
4	12.22	10.65	9.979	9.605	9.364	9.197	9.074	8.980	8.905	8.844	8.751	8.657	8.560
5	10.01	8.434	7.764	7.388	7.146	6.978	6.853	6.757	6.681	6.619	6.525	6.428	6.329
6	8.813	7.260	6.599	6.227	5.988	5.820	5.695	5.600	5.523	5.461	5.366	5.269	5.168
7	8.073	6.542	5.890	5.523	5.285	5.119	4.995	4.899	4.823	4.761	4.666	4.568	4.467
8	7.571	6.059	5.416	5.053	4.817	4.652	4.529	4.433	4.357	4.295	4.200	4.101	3.999
9	7.209	5.715	5.078	4.718	4.484	4.320	4.197	4.102	4.026	3.964	3.868	3.769	3.667
10	6.937	5.456	4.826	4.468	4.236	4.072	3.950	3.855	3.779	3.717	3.621	3.522	3.419
11	6.724	5.256	4.630	4.275	4.044	3.881	3.759	3.664	3.588	3.526	3.430	3.330	3.226
12	6.554	5.096	4.474	4.121	3.891	3.728	3.607	3.512	3.436	3.374	3.277	3.177	3.073
15	6.200	4.765	4.153	3.804	3.576	3.415	3.293	3.199	3.123	3.060	2.963	2.862	2.756
20	5.871	4.461	3.859	3.515	3.289	3.128	3.007	2.913	2.837	2.774	2.676	2.573	2.464

(c) 99% POINTS OF THE *F* DISTRIBUTION

v_1 v_2	1	2	3	4	5	6	7	8	9	10	12	15	20
1	4052	5000	5403	5625	5764	5859	5928	5981	6022	6056	6106	6158	6209
2	98.50	99.00	99.17	99.25	99.30	99.33	99.36	99.37	99.39	99.40	99.42	99.43	99.45
3	34.12	30.82	29.46	28.71	28.24	27.91	27.67	27.49	27.35	27.23	27.05	26.87	26.69
4	21.20	18.00	16.69	15.98	15.52	15.21	14.98	14.80	14.66	14.55	14.37	14.20	14.02
5	16.26	13.27	12.06	11.39	10.97	10.67	10.46	10.29	10.16	10.05	9.888	9.722	9.553
6	13.75	10.93	9.780	9.148	8.746	8.466	8.260	8.102	7.976	7.874	7.718	7.559	7.396
7	12.25	9.547	8.451	7.847	7.460	7.191	6.993	6.840	6.719	6.620	6.469	6.314	6.155
8	11.26	8.649	7.591	7.006	6.632	6.371	6.178	6.029	5.911	5.814	5.667	5.515	5.359
9	10.56	8.022	6.992	6.422	6.057	5.802	5.613	5.467	5.351	5.257	5.111	4.962	4.808
10	10.04	7.559	6.552	5.994	5.636	5.386	5.200	5.057	4.942	4.849	4.706	4.558	4.405
11	9.646	7.206	6.217	5.668	5.316	5.069	4.886	4.744	4.632	4.539	4.397	4.251	4.099
12	9.330	6.927	5.953	5.412	5.064	4.821	4.640	4.499	4.388	4.296	4.155	4.010	3.858
15	8.683	6.359	5.417	4.893	4.556	4.318	4.142	4.004	3.895	3.805	3.666	3.522	3.372
20	8.096	5.849	4.938	4.431	4.103	3.871	3.699	3.564	3.457	3.368	3.231	3.088	2.938

(d) 99.5% POINTS OF THE *F* DISTRIBUTION

ν_1 ν_2	1	2	3	4	5	6	7	8	9	10	12	15	20
1	16211	20000	21615	22500	23056	23437	23715	23925	24091	24224	24426	24630	24836
2	196.5	199.0	199.2	199.2	199.3	199.3	199.4	199.4	199.4	199.4	199.4	199.4	199.4
3	55.55	49.80	47.47	46.19	45.39	44.84	44.43	44.13	43.88	43.69	43.39	43.09	42.78
4	31.33	26.28	24.26	23.15	22.46	21.97	21.62	21.35	21.14	20.97	20.71	20.44	20.17
5	22.78	18.31	16.53	15.56	14.94	14.51	14.20	13.96	13.77	13.62	13.38	13.15	12.90
6	18.64	14.54	12.92	12.03	11.46	11.07	10.79	10.57	10.39	10.25	10.03	9.814	9.589
7	16.24	12.40	10.88	10.05	9.522	9.155	8.885	8.678	8.514	8.380	8.176	7.968	7.754
8	14.69	11.04	9.596	8.805	8.302	7.952	7.694	7.496	7.339	7.211	7.015	6.814	6.608
9	13.61	10.11	8.717	7.956	7.471	7.134	6.885	6.693	6.541	6.417	6.227	6.032	5.832
10	12.83	9.427	8.081	7.343	6.872	6.545	6.302	6.116	5.968	5.847	5.661	5.471	5.274
11	12.23	8.912	7.600	6.881	6.422	6.102	5.865	5.682	5.537	5.418	5.236	5.049	4.855
12	11.75	8.510	7.226	6.521	6.071	5.757	5.525	5.345	5.202	5.085	4.906	4.721	4.530
15	10.80	7.701	6.476	5.803	5.372	5.071	4.847	4.674	4.536	4.424	4.250	4.070	3.883
20	9.944	6.986	5.818	5.174	4.762	4.472	4.257	4.090	3.956	3.847	3.678	3.502	3.318

TABLE 7   (a) THE FISHER *Z*-TRANSFORMATION

r	.00	.01	.02	.03	.04	.05	.06	.07	.08	.09
0.00	0.000	0.010	0.020	0.030	0.040	0.050	0.060	0.070	0.080	0.090
0.10	0.100	0.110	0.121	0.131	0.141	0.151	0.161	0.172	0.182	0.192
0.20	0.203	0.213	0.224	0.234	0.245	0.255	0.266	0.277	0.288	0.299
0.30	0.310	0.321	0.332	0.343	0.354	0.365	0.377	0.388	0.400	0.412
0.40	0.424	0.436	0.448	0.460	0.472	0.485	0.497	0.510	0.523	0.536
0.50	0.549	0.563	0.576	0.590	0.604	0.618	0.633	0.648	0.662	0.678
0.60	0.693	0.709	0.725	0.741	0.758	0.775	0.793	0.811	0.829	0.848
0.70	0.867	0.887	0.908	0.929	0.950	0.973	0.996	1.020	1.045	1.071
0.80	1.099	1.127	1.157	1.188	1.221	1.256	1.293	1.333	1.376	1.422

r	.00	.01	.02	.03	.04	.05	.06	.07	.08	.09
0.900	1.472	1.478	1.483	1.488	1.494	1.499	1.505	1.510	1.516	1.522
0.910	1.528	1.533	1.539	1.545	1.551	1.557	1.564	1.570	1.576	1.583
0.920	1.589	1.596	1.602	1.609	1.616	1.623	1.630	1.637	1.644	1.651
0.930	1.658	1.666	1.673	1.681	1.689	1.697	1.705	1.713	1.721	1.730
0.940	1.738	1.747	1.756	1.764	1.774	1.783	1.792	1.802	1.812	1.822
0.950	1.832	1.842	1.853	1.863	1.874	1.886	1.897	1.909	1.921	1.933
0.960	1.946	1.959	1.972	1.986	2.000	2.014	2.029	2.044	2.060	2.076
0.970	2.092	2.110	2.127	2.146	2.165	2.185	2.205	2.227	2.249	2.273
0.980	2.298	2.323	2.351	2.380	2.410	2.443	2.477	2.515	2.555	2.599
0.990	2.647	2.700	2.759	2.826	2.903	2.994	3.106	3.250	3.453	3.800

(b) THE INVERSE FISHER *Z*-TRANSFORMATION

Z	.00	.01	.02	.03	.04	.05	.06	.07	.08	.09
0.00	0.000	0.010	0.020	0.030	0.040	0.050	0.060	0.070	0.080	0.090
0.10	0.100	0.110	0.119	0.129	0.139	0.149	0.159	0.168	0.178	0.188
0.20	0.197	0.207	0.217	0.226	0.235	0.245	0.254	0.264	0.273	0.282
0.30	0.291	0.300	0.310	0.319	0.327	0.336	0.345	0.354	0.363	0.371
0.40	0.380	0.388	0.397	0.405	0.414	0.422	0.430	0.438	0.446	0.454
0.50	0.462	0.470	0.478	0.485	0.493	0.501	0.508	0.515	0.523	0.530
0.60	0.537	0.544	0.551	0.558	0.565	0.572	0.578	0.585	0.592	0.598
0.70	0.604	0.611	0.617	0.623	0.629	0.635	0.641	0.647	0.653	0.658
0.80	0.664	0.670	0.675	0.680	0.686	0.691	0.696	0.701	0.706	0.711
0.90	0.716	0.721	0.726	0.731	0.735	0.740	0.744	0.749	0.753	0.757
1.00	0.762	0.766	0.770	0.774	0.778	0.782	0.786	0.789	0.793	0.797
1.10	0.800	0.804	0.808	0.811	0.814	0.818	0.821	0.824	0.827	0.831
1.20	0.834	0.837	0.840	0.843	0.845	0.848	0.851	0.854	0.856	0.859
1.30	0.862	0.864	0.867	0.869	0.872	0.874	0.876	0.879	0.881	0.883
1.40	0.885	0.887	0.890	0.892	0.894	0.896	0.898	0.900	0.901	0.903

Z	.00	.01	.02	.03	.04	.05	.06	.07	.08	.09
1.50	0.905	0.907	0.909	0.910	0.912	0.914	0.915	0.917	0.919	0.920
1.60	0.922	0.923	0.925	0.926	0.927	0.929	0.930	0.932	0.933	0.934
1.70	0.935	0.937	0.938	0.939	0.940	0.941	0.943	0.944	0.945	0.946
1.80	0.947	0.948	0.949	0.950	0.951	0.952	0.953	0.954	0.954	0.955
1.90	0.956	0.957	0.958	0.959	0.960	0.960	0.961	0.962	0.963	0.963
2.00	0.964	0.965	0.965	0.966	0.967	0.967	0.968	0.969	0.969	0.970
2.10	0.970	0.971	0.972	0.972	0.973	0.973	0.974	0.974	0.975	0.975
2.20	0.976	0.976	0.977	0.977	0.978	0.978	0.978	0.979	0.979	0.980
2.30	0.980	0.980	0.981	0.981	0.982	0.982	0.982	0.983	0.983	0.983
2.40	0.984	0.984	0.984	0.985	0.985	0.985	0.986	0.986	0.986	0.986
2.50	0.987	0.987	0.987	0.987	0.988	0.988	0.988	0.988	0.989	0.989
2.60	0.989	0.989	0.989	0.990	0.990	0.990	0.990	0.990	0.991	0.991
2.70	0.991	0.991	0.991	0.992	0.992	0.992	0.992	0.992	0.992	0.992
2.80	0.993	0.993	0.993	0.993	0.993	0.993	0.993	0.994	0.994	0.994
2.90	0.994	0.994	0.994	0.994	0.994	0.995	0.995	0.995	0.995	0.995

	3.0	3.1	3.2	3.3	3.4	3.5	3.6	3.7	3.8	3.9
	0.995	0.996	0.997	0.997	0.998	0.998	0.999	0.999	0.999	0.999

TABLE 8   SAMPLE RANGE AND CONTROL CHART VALUES

n	2	3	4	5	6	7	8	9	10
k	0.886	0.591	0.486	0.430	0.395	0.370	0.351	0.337	0.325
r_1	2.809	2.176	1.935	1.804	1.721	1.662	1.617	1.583	1.555
r_0	4.124	2.992	2.579	2.358	2.217	2.119	2.045	1.988	1.941

TABLE 9   THE NEGATIVE EXPONENTIAL FUNCTION (e^{-x})

x	.00	.01	.02	.03	.04	.05	.06	.07	.08	.09
0.0	1.0000	.99005	.98020	.97045	.96079	.95123	.94176	.93239	.92312	.91393
0.1	.90484	.89583	.88692	.87810	.86936	.86071	.85214	.84366	.83527	.82696
0.2	.81873	.81058	.80252	.79453	.78663	.77880	.77105	.76338	.75578	.74826
0.3	.74082	.73345	.72615	.71892	.71177	.70469	.69768	.69073	.68386	.67706
0.4	.67032	.66365	.65705	.65051	.64404	.63763	.63128	.62500	.61878	.61263
0.5	.60653	.60050	.59452	.58860	.58275	.57695	.57121	.56553	.55990	.55433
0.6	.54881	.54335	.53794	.53259	.52729	.52205	.51685	.51171	.50662	.50158
0.7	.49659	.49164	.48675	.48191	.47711	.47237	.46767	.46301	.45841	.45384
0.8	.44933	.44486	.44043	.43605	.43171	.42741	.42316	.41895	.41478	.41066
0.9	.40657	.40252	.39852	.39455	.39063	.38674	.38289	.37908	.37531	.37158
1.0	.36788	.36422	.36059	.35701	.35345	.34994	.34646	.34301	.33960	.33622
1.1	.33287	.32956	.32628	.32303	.31982	.31664	.31349	.31037	.30728	.30422
1.2	.30119	.29820	.29523	.29229	.28938	.28650	.28365	.28083	.27804	.27527
1.3	.27253	.26982	.26714	.26448	.26185	.25924	.25666	.25411	.25158	.24908
1.4	.24660	.24414	.24171	.23931	.23693	.23457	.23224	.22993	.22764	.22537
1.5	.22313	.22091	.21871	.21654	.21438	.21225	.21014	.20805	.20598	.20393
1.6	.20190	.19989	.19790	.19593	.19398	.19205	.19014	.18825	.18637	.18452
1.7	.18268	.18087	.17907	.17728	.17552	.17377	.17204	.17033	.16864	.16696
1.8	.16530	.16365	.16203	.16041	.15882	.15724	.15567	.15412	.15259	.15107
1.9	.14957	.14808	.14661	.14515	.14370	.14227	.14086	.13946	.13807	.13670
2.0	.13534	.13399	.13266	.13134	.13003	.12873	.12745	.12619	.12493	.12369
2.1	.12246	.12124	.12003	.11884	.11765	.11648	.11533	.11418	.11304	.11192
2.2	.11080	.10970	.10861	.10753	.10646	.10540	.10435	.10331	.10228	.10127
2.3	.10026	.09926	.09827	.09730	.09633	.09537	.09442	.09348	.09255	.09163
2.4	.09072	.08982	.08892	.08804	.08716	.08629	.08543	.08458	.08374	.08291
2.5	.08208	.08127	.08046	.07966	.07887	.07808	.07730	.07654	.07577	.07502
2.6	.07427	.07353	.07280	.07208	.07136	.07065	.06995	.06925	.06856	.06788
2.7	.06721	.06654	.06587	.06522	.06457	.06393	.06329	.06266	.06204	.06142
2.8	.06081	.06020	.05961	.05901	.05843	.05784	.05727	.05670	.05613	.05558
2.9	.05502	.05448	.05393	.05340	.05287	.05234	.05182	.05130	.05079	.05029
3.0	.04979	.04929	.04880	.04832	.04783	.04736	.04689	.04642	.04596	.04550

TABLE 10   RANDOM DIGITS

01745 20669 35800 47030 46598 50647 97207 36793 29776 06824
74223 14890 95539 90090 53295 49077 81068 67798 23355 84824
33592 45845 52015 72030 23071 92933 84219 39455 57792 14216
49113 61668 24320 37194 18010 32174 34896 26928 45687 49648
67998 25514 58478 66770 85407 46619 88514 92435 84483 29166

78183 44396 11064 99153 96293 00825 21079 78337 19739 13684
70209 23316 32828 00927 61841 53554 23344 32032 66577 94934
87024 74221 69721 07928 64801 77457 96271 04019 77405 50532
36675 28363 01572 64172 04454 12865 63095 08034 57650 19136
61593 89224 88276 20612 15015 40638 19004 79237 47667 02256

63226 17106 19469 24740 45704 94347 21399 20151 51834 37587
88504 21620 07292 71021 80929 45042 08703 45894 24521 49942
33186 49273 87542 41086 29615 81101 43707 87031 36101 15137
40068 35043 05280 62921 92351 06770 59945 89675 41952 56605
66455 41570 05302 19556 97546 64447 52627 59245 79685 21826

09282 95147 41704 42733 05421 98438 11014 38446 30889 97150
85828 53859 45783 50780 78301 40658 87688 41336 10147 23966
93874 04444 79311 72429 81113 54543 36416 96488 73046 66733
33612 68316 09353 01816 04317 60752 43578 24300 25848 21395
34694 04578 11685 92091 95275 88676 35638 49291 48935 58962

95128 11866 86386 07664 50383 65072 41089 50645 84624 29294
19820 70965 11856 91196 23382 17628 01137 76531 21264 85346
87822 15097 11472 71912 47222 32976 53911 76071 86332 30973
78813 93305 78486 42329 67172 68361 31212 70492 16081 18900
81823 56098 40198 02275 22044 90538 76366 62092 65314 15548

03128 98987 84668 47494 35289 14624 70318 27477 56706 34453
81338 04729 99737 83522 55769 75115 25360 88317 64554 69620
63016 45733 30541 90585 90419 56140 49489 51816 69411 29580
99424 48137 86763 58910 11166 53430 84169 86210 11359 20088
95148 95547 31538 34348 82784 74550 47231 76631 14519 40159

13331 33660 30542 13630 91393 19109 72124 22588 65798 72923
27974 64118 59826 56801 97805 39125 85077 70809 24679 69182
49731 92532 82635 01608 35977 74427 26423 36979 09041 70691
50534 34596 50237 44211 38214 65043 04124 39834 89355 15470
21332 18287 92257 17384 09959 45840 51596 25823 99535 08652

21217 87012 75383 77998 76867 35484 68525 48424 06056 67692
70009 27227 24998 57644 25889 99123 77744 66782 71179 04869
49732 80208 09885 69424 13023 88984 87445 66970 61863 24621
75608 22779 77120 97242 68066 41704 51640 23635 70701 48338
72878 84155 37921 86792 98424 66842 49129 98939 34173 49883

Index